ADVANCED CALCULUS

ADVANCED CALCULUS

FOURTH EDITION

WILFRED KAPLAN

University of Michigan

ADDISON-WESLEY PUBLISHING COMPANY

Reading, Massachusetts • Menlo Park, California • New York
Don Mills, Ontario • Wokingham, England • Amsterdam • Bonn
Sydney • Singapore • Tokyo • Madrid • San Juan • Milan • Paris

Reprinted with corrections, October 1993.

Publisher: *Allan M. Wylde*
Mathematics Editor: *Barbara Holland*
Promotions Manager: *Laura Likely*
Production Manager: *Pam Suwinsky*

Library of Congress Cataloging-in-Publication Data

Kaplan, Wilfred, 1915-
      Advanced calculus / Wilfred Kaplan. -- 4th ed.
            p.   cm.
      Includes index.
      ISBN 0-201-57888-3
QA303,K33   1991
515--dc20

Production services by Black Hole Publishing

6 7 8 9 10-MA-00999897

Contents

2
Differential Calculus of Functions of Several Variables

3
Vector Differential Calculus

4

Integral Calculus of Functions of Several Variables

5

Vector Integral Calculus

Two-Dimensional Theory

Three-Dimensional Theory and Applications

6
Infinite Series

7
Fourier Series
and Orthogonal
Functions

8

Functions of A Complex Variable

9
Conformal
Mapping

10
Partial
Differential
Equations

Answers
to Problems

Index

Preface
to the Fourth Edition

This edition differs from the previous one in two major aspects and in a number of minor ones:

The chapter on ordinary differential equations (chapter 8) has been omitted and a chapter on conformal mapping (chapter 9) has been added, using mainly material from the first edition (the topic was not included in the second and third editions).

Omission of the chapter on ordinary differential equations required adding to chapter 10, on partial differential equations, a brief discussion of ordinary linear differential equations.

Chapter 1, on linear algebra, has been shortened, especially in the treatment of Gaussian elimination. In chapter 2, the treatment of the implicit and inverse function theorems has been made more complete and the discussion of functional dependence has been improved. In chapter 3, sections have been added providing a fuller treatment of tensors. In chapters 2 and 3, some classical geometry of curves and surfaces is now provided, mainly in problems. In chapter 5, there is now a concise introduction to differential forms.

Throughout, problems have been added; in a number of cases, these significantly extend the coverage of the theory. For example, problems 11–15 following section 7–13 provide many additional complete orthogonal systems.

As with the previous editions, the purpose of this book is to provide sufficient material for a course in advanced calculus up to one year in length. It is hoped that the great variety of topics covered will also make this work useful as a reference book.

The background assumed is that usually obtained in the freshman–sophomore calculus sequence. Linear algebra is not assumed known but is developed in the first chapter.

Subjects discussed include all the topics usually found in texts on advanced calculus. However, there is more than the usual emphasis on applications and on physical motivation. Vectors are introduced at the outset and serve at many points to indicate geometrical and physical significance of mathematical relations.

Numerical methods are touched upon at various points, both because of their practical value and because of the insights they give into the theory.

A sound level of rigor is maintained throughout. Definitions are clearly labeled as such and all important results are formulated as theorems. A few of the finer points of real variable theory are treated at the ends of chapters 2, 4, and 6.

A large number of problems (with answers) are distributed throughout the text. These include simple exercises as well as complex ones planned to stimulate critical reading. Some points of the theory are relegated to the problems, with hints given where appropriate.

Generous references to the literature are given, and each chapter concludes with a list of books for supplementary reading.

Topical Summary

Chapter 1 opens with a review of vectors in space, determinants, and linear equations, and then develops matrix algebra, including Gaussian elimination, and n-dimensional geometry, with stress on linear mappings. The second chapter takes up partial derivatives and develops them with the aid of vectors (gradient, for example) and matrices; partial derivatives are applied to geometry and to maximum–minimum problems. The third chapter introduces the divergence and curl and the basic identities; orthogonal coordinates are treated concisely; final sections provide an introduction to tensors in n-dimensional space.

The fourth chapter, on integration, reviews definite and indefinite integrals, using numerical methods to show how the latter can be constructed; multiple integrals are treated carefully, with emphasis on the rule for change of variables; Leibnitz's Rule for differentiating under the integral sign is proved. Improper integrals are also covered; the discussion of these is completed at the end of chapter 6, where they are related to infinite series. Chapter 5 is devoted to line and surface integrals. Although the notions are first presented without vectors, it very soon becomes clear how natural the vector approach is for this subject. Line integrals are used to provide an exceptionally complete treatment of transformation of variables in a double integral. Many physical applications, including potential theory, are given.

Chapter 6 studies infinite series without assumption of previous knowledge. The notions of upper and lower limits are introduced and used sparingly as a simplifying device; with their aid, the theory is given in almost complete form. The usual tests are given: in particular, the root test. With its aid, the treatment of power series is greatly simplified. Uniform convergence is presented with great care and applied to power series. Final sections point out the parallel with improper integrals; in particular, power series are shown to correspond to the Laplace transform.

Chapter 7 is a complete treatment of Fourier series at an elementary level. The first sections give a simple introduction with many examples; the approach is gradually deepened and a convergence theorem is proved. Orthogonal functions are then studied, with the aid of inner product, norm, and vector procedures. A general theorem on complete systems enables one to deduce completeness of the trigonometric system and Legendre polynomials as a corollary. Closing sections cover Bessel functions, Fourier integrals, and generalized functions.

Chapter 8 develops the theory of analytic functions with emphasis on power series, Laurent series and residues, and their applications.

Chapter 9 provides a full treatment of conformal mapping, with many examples and physical applications and extensive discussion of the Dirichlet problem.

The final chapter, on partial differential equations, lays great stress on the relationship between the problem of forced vibrations of a spring (or a system of springs) and the partial differential equation $\rho u_{tt} + h u_t - k^2 \nabla^2 u = F(x,y,z,t)$. By pursuing this idea vigorously the discussion uncovers the physical meaning of the partial differential equation and makes the mathematical tools used become natural. Numerical methods are also motivated on a physical basis.

Throughout, a number of references are made to the text *Calculus and Linear Algebra* by Wilfred Kaplan and Donald J. Lewis (2 vols., New York, John Wiley & Sons, 1970–1971), cited simply as CLA.

Suggestions on the Use of this Book as the Text for a Course

The chapters are independent of each other in the sense that each can be started with a knowledge of only the simplest notions of the previous ones. The later portions of the chapter may depend on some of the later portions of earlier ones. It is thus possible to construct a course using just the earlier portions of several chapters. The following is an illustration of a plan for a one-semester course, meeting four hours a week: 1.1 to 1.8, 1.13, 1.15, 2.1 to 2.10, 2.12 to 2.18, 3.1 to 3.6, 4.1 to 4.9, 5.1 to 5.13, 6.1 to 6.7, 6.11 to 6.19.

If it is desired that one topic be stressed, then the corresponding chapters can be taken up in full detail. For example. chapters 1, 3, and 5 together provide a very substantial training in vector analysis; chapters 7 and 10 together contain sufficient material for a one-semester course in partial differential equations; chapters 8 and 9 provide sufficient text for a one-semester course in complex variables.

The author expresses his appreciation to the many colleagues who gave advice and encouragement in the preparation of this book. Professors R. C. F. Bartels, F. E. Hohn, and J. Lehner deserve special thanks and recognition for their thorough criticisms of the manuscript; a number of improvements were made on the basis of their suggestions. Others whose counsel has been of value are Professors R. V. Churchill, C. L. Dolph, G. E. Hay, M. Morkovin, G. Piranian, G. Y. Rainich, L. L. Rauch, M. O. Reade, E. Rothe, H. Samelson, R. Buchi, A. J. Lohwater, and W. Johnson, and Dr. G. Beguin. To his wife the author expresses his deeply felt appreciation for her aid and counsel in every phase of the arduous task.

For the preparation of the third edition, valuable advice was provided by Professors James F. Arnold, Jr., Douglas Cameron, Ronald Guenther, Joseph Horowitz, and David O. Lomen.

For the fourth edition similar help was given by Professors William M. Boothby, Harold Parks, B. K. Sachveva, and Nashed.

To the Addison-Wesley Publishing Company the author expresses his appreciation for their unfailing cooperation and for the high standards of publishing that they have set and maintained.

Wilfred Kaplan
Ann Arbor, Michigan

April 1991

1
Vectors and Matrices

INTRODUCTION

Our main goal in this book is to develop higher-level aspects of the calculus. The calculus deals with functions of one or more variables. The simplest such functions are the linear ones: for example, $y = 2x + 5$ and $z = 4x + 7y + 1$. Normally, one is forced to deal with functions that are not linear. A central idea of the differential calculus is the approximation of a nonlinear function by a linear one. Geometrically, one is approximating a curve or surface or similar object by a tangent line or plane or similar linear object built of straight lines. Through this approximation, questions of the calculus are reduced to ones of the algebra associated with lines and planes—linear algebra.

This first chapter develops linear algebra with these goals in mind. The first four sections of the chapter review vectors in space, determinants, and simultaneous linear equations. The following sections then develop the theory of matrices and some related geometry. A final section shows how the concept of vector can be generalized to the objects of an arbitrary "vector space."

1.1 VECTORS IN SPACE

We assume that mutually perpendicular x, y, and z axes are chosen as in Fig. 1.1, and we assume a common unit of distance along these axes. Then every point P in space has coordinates (x, y, z) with respect to these axes, as in Fig. 1.1. The origin O has coordinates $(0, 0, 0)$.

Figure 1.1　Coordinates in space.

A vector **v** in space has a magnitude (length) and direction but no fixed location. We can thus represent **v** by any one of many directed line segments in space, all having the same length and direction (Fig. 1.1). In particular, we can represent **v** by the directed line segment from O to a point P, provided that the direction from O to P is that of **v** and that the distance from O to P equals the length of **v**, as suggested in Fig. 1.1. We write simply

$$\mathbf{v} = \overrightarrow{OP}. \tag{1.1}$$

The figure also shows the components v_x, v_y, v_z of **v** along the axes. When (1.1) holds, we have

$$v_x = x, \qquad v_y = y, \qquad v_z = z. \tag{1.2}$$

We assume the reader's familiarity with addition of vectors and multiplication of vectors by numbers (scalars). With the aid of these operations a general vector **v** can be represented as follows:

$$\mathbf{v} = v_x \mathbf{i} + v_y \mathbf{j} + v_z \mathbf{k}. \tag{1.3}$$

Here **i**, **j**, **k** are *unit vectors* (vectors of length 1) having the directions of the coordinate axes, as in Fig. 1.2. By the Pythagorean theorem, **v** then has magnitude, denoted by $|\mathbf{v}|$, given by the equation

$$|\mathbf{v}| = \sqrt{v_x^2 + v_y^2 + v_z^2}. \tag{1.4}$$

In particular, for $\mathbf{v} = \overrightarrow{OP}$ the distance of P: (x, y, z) from O is

$$|\overrightarrow{OP}| = \sqrt{x^2 + y^2 + z^2}. \tag{1.5}$$

More generally, for $\mathbf{v} = \overrightarrow{P_1 P_2}$, where P_1 is (x_1, y_1, z_1) and P_2 is (x_2, y_2, z_2), one has

Figure 1.2 Vector **v** in terms of **i, j, k**.

$\mathbf{v} = \overrightarrow{OP_2} - \overrightarrow{OP_1} = (x_2 - x_1)\,\mathbf{i} + \cdots$ and the distance between P_1 and P_2 is

$$d = |\mathbf{v}| = |\overrightarrow{P_1 P_2}| = \sqrt{(x_2 - x_1)^2 + (y_2 - y_1)^2 + (z_2 - z_1)^2}. \qquad (1.6)$$

The vector **v** can have 0 length, in which case $\mathbf{v} = \overrightarrow{OP}$ only when P coincides with O. We then write

$$\mathbf{v} = \mathbf{0} \qquad (1.7)$$

and call **v** the zero vector.

The vector **v** is completely specified by its components v_x, v_y, v_z. It is often convenient to write

$$\mathbf{v} = (v_x, v_y, v_z) \qquad (1.8)$$

instead of Eq. (1.3). Thus we think of a vector in space as an *ordered triple of numbers*. Later we shall consider such triples as matrices (row vectors or column vectors).

The *dot product* (or *inner product*) of two vectors **v, w** in space is the number

$$\mathbf{v} \cdot \mathbf{w} = |\mathbf{v}|\,|\mathbf{w}| \cos\theta, \qquad (1.9)$$

where $\theta = \sphericalangle(\mathbf{v}, \mathbf{w})$, chosen between 0 and π inclusive (see Fig. 1.3). When **v** or **w** is **0**, the angle θ is indeterminate, and $\mathbf{v} \cdot \mathbf{w}$ is taken to be 0. We also have $\mathbf{v} \cdot \mathbf{w} = 0$ when **v, w** are *orthogonal* (perpendicular) vectors, $\mathbf{v} \perp \mathbf{w}$. We agree to say that the **0** vector is orthogonal to all vectors (and parallel to all vectors). With this convention we can state:

$$\mathbf{v} \cdot \mathbf{w} = 0 \quad \text{precisely when} \quad \mathbf{v} \perp \mathbf{w}. \qquad (1.10)$$

Figure 1.3 Definition of dot product.

The dot product satisfies some algebraic rules:

$$\mathbf{u} \cdot \mathbf{v} = \mathbf{v} \cdot \mathbf{u}, \qquad \mathbf{u} \cdot (\mathbf{v} + \mathbf{w}) = \mathbf{u} \cdot \mathbf{v} + \mathbf{u} \cdot \mathbf{w},$$

$$\mathbf{u} \cdot (c\mathbf{v}) = (c\mathbf{u}) \cdot \mathbf{v} = c(\mathbf{u} \cdot \mathbf{v}), \qquad \mathbf{u} \cdot \mathbf{u} = |\mathbf{u}|^2. \tag{1.11}$$

Here c is a scalar.

In Eq. (1.9) the quantity $|\mathbf{v}| \cos \theta$ is interpreted as the component of $\mathbf{v}$ in the direction of $\mathbf{w}$ (see Fig. 1.4):

$$\operatorname{comp}_w \mathbf{v} = |\mathbf{v}| \cos \theta. \tag{1.12}$$

This can be positive, negative, or 0.

The angles α, β, γ between $\mathbf{v}$ (assumed to be nonzero) and the vectors $\mathbf{i}, \mathbf{j}, \mathbf{k}$ are called *direction angles* of $\mathbf{v}$; the corresponding cosines $\cos \alpha, \cos \beta, \cos \gamma$ are the *direction cosines* of $\mathbf{v}$. By Eqs. (1.9) and (1.12) and Fig. 1.2,

$$\mathbf{v} \cdot \mathbf{i} = |\mathbf{v}| \cos \alpha = \operatorname{comp}_i \mathbf{v} = v_x,$$

$$\mathbf{v} \cdot \mathbf{j} = |\mathbf{v}| \cos \beta = \operatorname{comp}_j \mathbf{v} = v_y,$$

$$\mathbf{v} \cdot \mathbf{k} = |\mathbf{v}| \cos \gamma = \operatorname{comp}_k \mathbf{v} = v_z. \tag{1.13}$$

Accordingly,

$$\cos \alpha = \frac{v_x}{|\mathbf{v}|}, \qquad \cos \beta = \frac{v_y}{|\mathbf{v}|}, \qquad \cos \gamma = \frac{v_z}{|\mathbf{v}|}. \tag{1.14}$$

Thus the vector $(1/|\mathbf{v}|)\mathbf{v}$ has components $\cos \alpha, \cos \beta, \cos \gamma$; we observe that this vector is a unit vector, since its length is $(1/|\mathbf{v}|)|\mathbf{v}| = 1$.

Since $\mathbf{i} \cdot \mathbf{i} = 1, \mathbf{i} \cdot \mathbf{j} = 0$, etc., we can compute the dot product of

$$\mathbf{u} = u_x \mathbf{i} + u_y \mathbf{j} + u_z \mathbf{k} \quad \text{and} \quad \mathbf{v} = v_x \mathbf{i} + v_y \mathbf{j} + v_z \mathbf{k}$$

as follows:

$$\mathbf{u} \cdot \mathbf{v} = (u_x \mathbf{i} + u_y \mathbf{j} + u_z \mathbf{k}) \cdot (v_x \mathbf{i} + v_y \mathbf{j} + v_z \mathbf{k})$$

$$= u_x v_x \mathbf{i} \cdot \mathbf{i} + u_x v_y \mathbf{i} \cdot \mathbf{j} + \cdots .$$

Here we use the rules (1.11). We conclude:

$$\mathbf{u} \cdot \mathbf{v} = u_x v_x + u_y v_y + u_z v_z. \tag{1.15}$$

The *vector product* or *cross product* $\mathbf{u} \times \mathbf{v}$ of two vectors $\mathbf{u}, \mathbf{v}$ is defined with reference to a chosen *orientation of space*. This is usually specified by a right-handed *xyz*-coordinate system in space. An ordered triple of vectors is then called a *positive* triple if the vectors can be moved continuously to attain the respective directions of $\mathbf{i}, \mathbf{j}, \mathbf{k}$ eventually without making one of the vectors lie in a plane parallel to the other two; a practical test for this is by aligning the vectors with the thumb, index finger, and middle finger of the right hand. The triple is called *negative* if the test can be satisfied by using $\mathbf{j}, \mathbf{i}, \mathbf{k}$ instead of $\mathbf{i}, \mathbf{j}, \mathbf{k}$. If one of the vectors is $\mathbf{0}$ or all three vectors are coplanar (can be represented in one plane), the definition is not applicable.

Figure 1.4   Component.

Figure 1.5   Vector product.

Now we define $\mathbf{u} \times \mathbf{v} = \mathbf{w}$, where

$$\mathbf{u} \cdot \mathbf{w} = 0, \qquad \mathbf{v} \cdot \mathbf{w} = 0,$$

$$|\mathbf{w}| = |\mathbf{u}|\,|\mathbf{v}|\sin\theta, \qquad \theta = \sphericalangle(\mathbf{u}, \mathbf{v}),$$

$\mathbf{u}, \mathbf{v}, \mathbf{w}$ form a positive triple. (1.16)

This is illustrated in Fig. 1.5.

The definition breaks down when $\mathbf{u}$ or $\mathbf{v}$ is $\mathbf{0}$ or when $\theta = 0$ or π ($\mathbf{u}, \mathbf{v}$ *collinear*). In these cases we write $\mathbf{u} \times \mathbf{v} = \mathbf{0}$. We can say simply:

$$\mathbf{u} \times \mathbf{v} = \mathbf{0} \quad \text{precisely when} \quad \mathbf{u} \,\|\, \mathbf{v}. \tag{1.17}$$

From Eq. (1.16) we observe that

$$|\mathbf{w}| = |\mathbf{u} \times \mathbf{v}| = |\mathbf{u}|\,|\mathbf{v}|\sin\theta$$

$$= \text{area of parallelogram of sides } \mathbf{u}, \mathbf{v}, \tag{1.18}$$

as illustrated in Fig. 1.5.

The vector product satisfies algebraic rules:

$$\mathbf{u} \times \mathbf{v} = -\mathbf{v} \times \mathbf{u}, \; \mathbf{u} \times (\mathbf{v} + \mathbf{w}) = \mathbf{u} \times \mathbf{v} + \mathbf{u} \times \mathbf{w},$$

$$\mathbf{u} \times (c\mathbf{v}) = (c\mathbf{u}) \times \mathbf{v} = c(\mathbf{u} \times \mathbf{v}),$$

$$\mathbf{u} \times \mathbf{u} = \mathbf{0},$$

$$\mathbf{u} \times (\mathbf{v} \times \mathbf{w}) = (\mathbf{u} \cdot \mathbf{w})\mathbf{v} - (\mathbf{u} \cdot \mathbf{v})\mathbf{w},$$

$$(\mathbf{u} \times \mathbf{v}) \times \mathbf{w} = (\mathbf{u} \cdot \mathbf{w})\mathbf{v} - (\mathbf{v} \cdot \mathbf{w})\mathbf{u}. \tag{1.19}$$

The last two rules are described as the identities for *vector triple products*.

Since $\mathbf{i} \times \mathbf{i} = \mathbf{0}$, $\mathbf{i} \times \mathbf{j} = \mathbf{k}$, $\mathbf{i} \times \mathbf{k} = -\mathbf{j}$, and so on, we can calculate $\mathbf{u} \times \mathbf{v}$ as

$$(u_x\mathbf{i} + u_y\mathbf{j} + u_z\mathbf{k}) \times (v_x\mathbf{i} + v_y\mathbf{j} + v_z\mathbf{k}) = u_x v_x \mathbf{i} \times \mathbf{i} + u_x v_y \mathbf{i} \times \mathbf{j} + \cdots$$

and conclude:

$$\mathbf{u} \times \mathbf{v} = (u_y v_z - u_z v_y)\mathbf{i} + (u_z v_x - u_x v_z)\mathbf{j} + (u_x v_y - u_y v_x)\mathbf{k}. \tag{1.20}$$

This can also be written as a determinant (Section 1.3):

$$\mathbf{u} \times \mathbf{v} = \begin{vmatrix} \mathbf{i} & \mathbf{j} & \mathbf{k} \\ u_x & u_y & u_z \\ v_x & v_y & v_z \end{vmatrix}. \tag{1.21}$$

Here we expand by minors of the first row.

From the rules (1.19) we see that, in general, $\mathbf{u} \times \mathbf{v} \neq \mathbf{v} \times \mathbf{u}$ and $\mathbf{u} \times (\mathbf{v} \times \mathbf{w})$ $\neq (\mathbf{u} \times \mathbf{v}) \times \mathbf{w}$.

For further discussion of vectors, see Chapter 11 of CLA.*

1.2   LINEAR INDEPENDENCE ▪ LINES AND PLANES

Two vectors $\mathbf{u}, \mathbf{v}$ in space are said to be *linearly independent* if they cannot be represented by directed line segments on the same line. Otherwise, they are said to be *linearly dependent* or *collinear* (see Fig. 1.6). When $\mathbf{u}$ or $\mathbf{v}$ is $\mathbf{0}$, the vectors are considered to be linearly dependent. We thus see that $\mathbf{u}, \mathbf{v}$ are linearly dependent precisely when $\mathbf{u} \times \mathbf{v} = \mathbf{0}$.

(a)   (b)

Figure 1.6   (a) Linearly independent vectors $\mathbf{u}, \mathbf{v}$.   (b) Linearly dependent vectors $\mathbf{u}, \mathbf{v}$.

Three vectors $\mathbf{u}, \mathbf{v}, \mathbf{w}$ in space are said to be linearly independent when they cannot be represented by directed line segments in the same plane. Otherwise, they are said to be linearly dependent or *coplanar* (see Fig. 1.7).

(a)   (b)

Figure 1.7   (a) Linearly independent vectors $\mathbf{u}, \mathbf{v}, \mathbf{w}$.   (b) Linearly dependent vectors $\mathbf{u}, \mathbf{v}, \mathbf{w}$.

We can include both these cases in a general definition: Vectors $\mathbf{u}_1, \ldots, \mathbf{u}_k$ in space are linearly independent if the only scalars $c_1, \ldots, c_k$ such that

$$c_1\mathbf{u}_1 + \cdots + c_k\mathbf{u}_k = \mathbf{0} \tag{1.22}$$

are $c_1 = 0, c_2 = 0, \ldots, c_k = 0$.

---

* The work *Calculus and Linear Algebra* by the author and Donald J. Lewis, 2 vols. (New York: John Wiley and Sons, 1970–1971), will be referred to throughout as CLA.

For $k = 2$, $\mathbf{u}_1$ and $\mathbf{u}_2$ are thus linearly *dependent* if $c_1\mathbf{u}_1 + c_2\mathbf{u}_2 = \mathbf{0}$ for some scalars c_1, c_2 that are not both 0. If, say, $c_2 \neq 0$, then

$$\mathbf{u}_2 = -\frac{c_1}{c_2}\mathbf{u}_1.$$

Thus $\mathbf{u}_2$ is a scalar times $\mathbf{u}_1$ and is collinear with $\mathbf{u}_1$ (if $c_1 = 0$ or $\mathbf{u}_1 = \mathbf{0}$, then $\mathbf{u}_2$ would be $\mathbf{0}$). Conversely, if $\mathbf{u}_1, \mathbf{u}_2$ are collinear, then $\mathbf{u}_2 - k\mathbf{u}_1 = \mathbf{0}$ or $\mathbf{u}_1 - k\mathbf{u}_2 = \mathbf{0}$ for some scalar k. Thus the new definition agrees with the old one.

Similarly, for $k = 3$, $\mathbf{u}_1, \mathbf{u}_2, \mathbf{u}_3$ are linearly dependent if $c_1\mathbf{u}_1 + c_2\mathbf{u}_2 + c_3\mathbf{u}_3 = \mathbf{0}$ for some scalars c_1, c_2, c_3 that are not all 0. If, for example, $c_3 \neq 0$, then

$$\mathbf{u}_3 = -\frac{c_1}{c_3}\mathbf{u}_1 - \frac{c_2}{c_3}\mathbf{u}_2.$$

Thus $\mathbf{u}_3$ is a *linear combination* of $\mathbf{u}_1$ and $\mathbf{u}_2$, so the three must be coplanar (Fig. 1.8).

Conversely, if the three vectors are coplanar, then it can be verified that one must equal a linear combination of the other two, say, $\mathbf{u}_3 = k_1\mathbf{u}_1 + k_2\mathbf{u}_2$, and then $k_1\mathbf{u}_1 + k_2\mathbf{u}_2 - 1\mathbf{u}_3 = \mathbf{0}$, so the vectors are linearly dependent by the new definition. Again the two definitions agree.

What about four vectors in space? Here the answer is simple: They must be linearly dependent. Let the vectors be $\mathbf{u}_1, \mathbf{u}_2, \mathbf{u}_3, \mathbf{u}_4$. There are then two possibilities: (a) $\mathbf{u}_1, \mathbf{u}_2, \mathbf{u}_3$ are linearly dependent, and (b) $\mathbf{u}_1, \mathbf{u}_2, \mathbf{u}_3$ are linearly independent. In case (a),

$$c_1\mathbf{u}_1 + c_2\mathbf{u}_2 + c_3\mathbf{u}_3 = \mathbf{0}$$

for some scalars c_1, c_2, c_3 not all 0. But then

$$c_1\mathbf{u}_1 + c_2\mathbf{u}_2 + c_3\mathbf{u}_3 + 0\mathbf{u}_4 = \mathbf{0},$$

with not all of c_1, c_2, c_3 equal to 0. Thus $\mathbf{u}_1, \mathbf{u}_2, \mathbf{u}_3, \mathbf{u}_4$ are linearly dependent. In case (b), $\mathbf{u}_1, \mathbf{u}_2, \mathbf{u}_3$ are not coplanar and hence can be represented by the directed edges of a parallelepiped in space, as in Fig. 1.9. From this it follows that $\mathbf{u}_4$ can be represented as $c_1\mathbf{u}_1 + c_2\mathbf{u}_2 + c_3\mathbf{u}_3$ for appropriate c_1, c_2, c_3, as in the figure; this is analogous to the representation of $\mathbf{v}$ in terms of $\mathbf{i}, \mathbf{j}, \mathbf{k}$ in Eq. (1.3) and Fig.

Figure 1.8   The vector $\mathbf{u}_3$ as a linear combination of $\mathbf{u}_1, \mathbf{u}_2$.

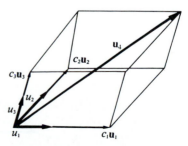

Figure 1.9   Expression of $\mathbf{u}_4$ as a linear combination of $\mathbf{u}_1, \mathbf{u}_2, \mathbf{u}_3$.

1.2. Now

$$c_1 \mathbf{u}_1 + c_2 \mathbf{u}_2 + c_3 \mathbf{u}_3 - \mathbf{u}_4 = \mathbf{0},$$

so that again $\mathbf{u}_1, \ldots, \mathbf{u}_4$ are linearly dependent.

Accordingly, there cannot be four linearly independent vectors in space. By similar reasoning we see that for every k greater than 3 there is no set of k linearly independent vectors in space.

However, for $k \le 3$ there are k linearly independent vectors in space. For example, $\mathbf{i}, \mathbf{j}$ is such a set of two vectors, and $\mathbf{i}, \mathbf{j}, \mathbf{k}$ is such a set of three vectors. (We can also consider $\mathbf{i}$ by itself—or any nonzero vector—as a set of one linearly independent vector.)

Every triple $\mathbf{u}_1, \mathbf{u}_2, \mathbf{u}_3$ of linearly independent vectors in space serves as a *basis* for vectors in space; that is, every vector in space can be expressed uniquely as a linear combination $c_1 \mathbf{u}_1 + c_2 \mathbf{u}_2 + c_3 \mathbf{u}_3$, as in Fig. 1.9.

We call $\mathbf{i}, \mathbf{j}, \mathbf{k}$ the standard basis. The equation $\mathbf{v} = v_x \mathbf{i} + v_y \mathbf{j} + v_z \mathbf{k}$ is the representation of $\mathbf{v}$ in terms of the standard basis.

We observe that one could specialize the discussion of linear independence to *two*-dimensional space—that is, the xy-plane. Here there are pairs of linearly independent vectors, and each such pair forms a basis; $\mathbf{i}, \mathbf{j}$ is the standard basis. Every set of more than two vectors in the plane is linearly dependent.

Planes in space.   If $P_1: (x_1, y_1, z_1)$ is a point of a plane and $\mathbf{n} = A\mathbf{i} + B\mathbf{j} + C\mathbf{k}$ is a nonzero normal vector (perpendicular to the plane), then $P: (x, y, z)$ is in the plane precisely when

$$\mathbf{n} \cdot \overrightarrow{P_1 P} = 0 \qquad (1.23)$$

or

$$A(x - x_1) + B(y - y_1) + C(z - z_1) = 0 \qquad (1.24)$$

(see Fig. 1.10). Equation (1.24) can be written as a linear equation

$$Ax + By + Cz + D = 0 \qquad (1.25)$$

and every linear equation (1.25) (A, B, C not all 0) represents a plane, with $\mathbf{n} = A\mathbf{i} + B\mathbf{j} + C\mathbf{k}$ as normal vector.

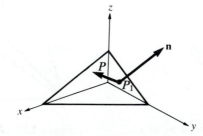

Figure 1.10   Plane.

Lines in space. If $P_1: (x_1, y_1, z_1)$ is a point of a line and $\mathbf{v} = a\mathbf{i} + b\mathbf{j} + c\mathbf{k}$ is a nonzero vector along the line (that is, representable by a directed line segment joining two points of the line), then $P: (x, y, z)$ is on the line precisely when

$$\mathbf{v} \times \overline{P_1 P} = \mathbf{0}, \tag{1.26}$$

that is, when $\mathbf{v}$ and $\overline{P_1 P}$ are linearly dependent. Since $\mathbf{v} \neq \mathbf{0}$, $\overline{P_1 P}$ must be a scalar times $\mathbf{v}$:

$$\overline{P_1 P} = t\mathbf{v}, \tag{1.27}$$

where t can be any number. From Eq. (1.27) we obtain *parametric equations* of the line:

$$x = x_1 + at, \qquad y = y_1 + bt, \qquad z = z_1 + ct, \qquad -\infty < t < \infty. \tag{1.28}$$

If $\mathbf{v}$ happens to be a unit vector, then $|\overline{P_1 P}| = |t|$, so that t can be regarded as a distance coordinate along the line. In this case we usually replace t by s, so that

$$x = x_1 + as, \qquad y = y_1 + bs, \qquad z = z_1 + cs, \qquad -\infty < s < \infty, \tag{1.29}$$

as in Fig. 1.11.

Figure 1.11 Line, distance s as parameter.

1.3 DETERMINANTS

For second-order determinants, one has the formula

$$\begin{vmatrix} a & b \\ c & d \end{vmatrix} = ad - bc. \tag{1.30}$$

Then higher-order determinants are reduced to those of lower order. For example,

$$\begin{vmatrix} a_1 & b_1 & c_1 \\ a_2 & b_2 & c_2 \\ a_3 & b_3 & c_3 \end{vmatrix} = a_1 \begin{vmatrix} b_2 & c_2 \\ b_3 & c_3 \end{vmatrix} - b_1 \begin{vmatrix} a_2 & c_2 \\ a_3 & c_3 \end{vmatrix} + c_1 \begin{vmatrix} a_2 & b_2 \\ a_3 & b_3 \end{vmatrix}; \tag{1.31}$$

$$\begin{vmatrix} a_1 & b_1 & c_1 & d_1 \\ a_2 & b_2 & c_2 & d_2 \\ a_3 & b_3 & c_3 & d_3 \\ a_4 & b_4 & c_4 & d_4 \end{vmatrix} = a_1 \begin{vmatrix} b_2 & c_2 & d_2 \\ b_3 & c_3 & d_3 \\ b_4 & c_4 & d_4 \end{vmatrix} - b_1 \begin{vmatrix} a_2 & c_2 & d_2 \\ a_3 & c_3 & d_3 \\ a_4 & c_4 & d_4 \end{vmatrix} + \cdots . \tag{1.32}$$

From these formulas, one sees that a determinant of order n is a sum of terms, each of which is ± 1 times a product of n factors, one each from the n columns of the array and one each from the n rows of the array. Thus from (1.31) and (1.30), one obtains the six terms

$$a_1 b_2 c_3 - a_1 b_3 c_2 - b_1 a_2 c_3 + b_1 a_3 c_2 + c_1 a_2 b_3 - c_1 a_3 b_2 .$$

We now state six rules for determinants:

I. Rows and columns can be interchanged. For example,

$$\begin{vmatrix} a_1 & b_1 & c_1 \\ a_2 & b_2 & c_2 \\ a_3 & b_3 & c_3 \end{vmatrix} = \begin{vmatrix} a_1 & a_2 & a_3 \\ b_1 & b_2 & b_3 \\ c_1 & c_2 & c_3 \end{vmatrix}.$$

Hence in every rule the words *row* and *column* can be interchanged.

II. Interchanging two rows (or columns) multiplies the determinant by -1. For example,

$$\begin{vmatrix} a_1 & b_1 & c_1 \\ a_2 & b_2 & c_2 \\ a_3 & b_3 & c_3 \end{vmatrix} = - \begin{vmatrix} a_2 & b_2 & c_2 \\ a_1 & b_1 & c_1 \\ a_3 & b_3 & c_3 \end{vmatrix}.$$

III. A factor of any row (or column) can be placed before the determinant. For example,

$$\begin{vmatrix} ka_1 & kb_1 & kc_1 \\ a_2 & b_2 & c_2 \\ a_3 & b_3 & c_3 \end{vmatrix} = k \begin{vmatrix} a_1 & b_1 & c_1 \\ a_2 & b_2 & c_2 \\ a_3 & b_3 & c_3 \end{vmatrix}.$$

IV. If two rows (or columns) are proportional, the determinant equals 0. For example,

$$\begin{vmatrix} ka_1 & kb_1 & kc_1 \\ a_1 & b_1 & c_1 \\ a_2 & b_2 & c_2 \end{vmatrix} = 0.$$

V. Determinants differing in only one row (or column) can be added by adding corresponding elements in that row and leaving the other elements unchanged. For example,

$$\begin{vmatrix} a_1 & b_1 & c_1 \\ a_2 & b_2 & c_2 \\ a_3 & b_3 & c_3 \end{vmatrix} + \begin{vmatrix} A_1 & b_1 & c_1 \\ A_2 & b_2 & c_2 \\ A_3 & b_3 & c_3 \end{vmatrix} = \begin{vmatrix} a_1 + A_1 & b_1 & c_1 \\ a_2 + A_2 & b_2 & c_2 \\ a_3 + A_3 & b_3 & c_3 \end{vmatrix}.$$

VI. The value of a determinant is unchanged if the elements of one row are multiplied by the same quantity k and added to the corresponding elements of another row. For example,

$$\begin{vmatrix} a_1 & b_1 & c_1 \\ a_2 & b_2 & c_2 \\ a_3 & b_3 & c_3 \end{vmatrix} = \begin{vmatrix} a_1 + ka_2 & b_1 + kb_2 & c_1 + kc_2 \\ a_2 & b_2 & c_2 \\ a_3 & b_3 & c_3 \end{vmatrix}.$$

By a suitable choice of k, one can use this rule to introduce zeros; by repetition of the process, one can reduce all elements but one in a chosen row to 0. This procedure is basic for numerical evaluation of determinants. (See Section 1.9.)

From Rule II. one deduces that a succession of an even number of interchanges of rows (or of columns) leaves the determinant unchanged, whereas an odd number of interchanges reverses the sign. In each case we end up with a *permutation* of the rows (or columns) which we term even or odd according to the number of interchanges.

From an arbitrary determinant, one obtains others, called *minors* of the given one, by deleting k rows and k columns. Equations (1.31) and (1.32) indicate how a given determinant can be *expanded by minors* of the first row. There is a similar expansion by minors of the first column or by minors of any chosen row or column. In the expansion, each element of the row or column is multiplied by its minor (obtained by deleting the row and column of the element) and by ± 1. The $\pm$ signs follow a checkerboard pattern, starting with $+$ in the top left corner.

From three vectors $\mathbf{u}, \mathbf{v}, \mathbf{w}$ in space, one obtains a determinant

$$D = \begin{vmatrix} u_x & u_y & u_z \\ v_x & v_y & v_z \\ w_x & w_y & w_z \end{vmatrix}. \tag{1.33}$$

One has the identities

$$D = \mathbf{u} \cdot \mathbf{v} \times \mathbf{w} = \mathbf{w} \cdot \mathbf{u} \times \mathbf{v} = \mathbf{v} \cdot \mathbf{w} \times \mathbf{u}. \tag{1.34}$$

The vector expressions here are called *scalar triple products*. The equality $D = \mathbf{u} \cdot \mathbf{v} \times \mathbf{w}$ follows from expansion of D by minors of the first row and the formula (1.20) applied to $\mathbf{v} \times \mathbf{w}$. The other equalities are consequences of Rule II for interchanging rows.

In (1.34), one can also interchange $\cdot$ and $\times$. For example,

$$\mathbf{u} \cdot \mathbf{v} \times \mathbf{w} = \mathbf{u} \times \mathbf{v} \cdot \mathbf{w},$$

since the right-hand side equals $\mathbf{w} \cdot \mathbf{u} \times \mathbf{v}$. Also, interchanging two vectors in one of the scalar triple products changes the sign:

$$\mathbf{u} \cdot \mathbf{w} \times \mathbf{v} = \mathbf{u} \cdot (-1)(\mathbf{v} \times \mathbf{w}) = -\mathbf{u} \cdot \mathbf{v} \times \mathbf{w}.$$

The number D in (1.34) can be interpreted as plus or minus the *volume* of a parallelepiped whose edges, properly directed, represent $\mathbf{u}, \mathbf{v}, \mathbf{w}$ as in Fig. 1.12. For

$$D = |\mathbf{w}| \, |\mathbf{u} \times \mathbf{v}| \cos \phi,$$

where $|\mathbf{w}| \cos \phi$ is the altitude h of the parallelepiped (or the negative of h if

Figure 1.12 Scalar triple product as volume.

$\phi > \pi/2$), as in Fig. 1.12. Also, $|\mathbf{u} \times \mathbf{v}|$ is the area of the base, so that D is indeed $\pm$ the volume. One sees that the $+$ holds when $\mathbf{u}, \mathbf{v}, \mathbf{w}$ form a positive triple and that the $-$ holds when they form a negative triple. When the vectors are linearly independent, one of these two cases must hold. When they are linearly dependent, the parallelepiped collapses, and $D = 0$; in the case of linear dependence, either $\mathbf{u}$ or $\mathbf{v} \times \mathbf{w}$ is $\mathbf{0}$, or else the angle ϕ is $\pi/2$.

Thus we have a useful test for linear independence of three vectors $\mathbf{u}, \mathbf{v}, \mathbf{w}$ in space: They are linearly independent precisely when $D \neq 0$.

This discussion can be specialized to two dimensions. For two vectors $\mathbf{u}, \mathbf{v}$ in the xy-plane, one can form

$$D = \begin{vmatrix} u_x & u_y \\ v_x & v_y \end{vmatrix}.$$

Now $\mathbf{u} \times \mathbf{v} = (u_x v_y - u_y v_x)\mathbf{k} = D\mathbf{k}$. Thus

$$D = \pm|\mathbf{u} \times \mathbf{v}| = \pm(\text{area of parallelogram}), \qquad (1.35)$$

where the parallelogram has edges $\mathbf{u}, \mathbf{v}$ as in Fig. 1.13. Again $D = 0$ precisely when $\mathbf{u}, \mathbf{v}$ are linearly dependent. We observe that

$$D = \mathbf{u} \times \mathbf{v} \cdot \mathbf{k}$$

and hence D is positive or negative according to whether $\mathbf{u}, \mathbf{v}, \mathbf{k}$ form a positive triple. We verify that if ϕ is the angle from $\mathbf{u}$ to $\mathbf{v}$ (measured in the usual counterclockwise sense for angles in the plane), then the triple is positive for $0 < \phi < \pi$ and negative for $\pi < \phi < 2\pi$. In Fig. 1.13, $\mathbf{u} = 3\mathbf{i} + \mathbf{j}$, $\mathbf{v} = \mathbf{i} - \mathbf{j}$, clearly $\pi < \phi < 2\pi$, and $D = \begin{vmatrix} 3 & 1 \\ 1 & -1 \end{vmatrix} = -4$; the triple is negative.

Figure 1.13   Parallelogram formed by $\mathbf{u}, \mathbf{v}$.

For proofs of rules for determinants, see Chapter 10 of CLA and the book by Cullen listed at the end of the chapter.

1.4   SIMULTANEOUS LINEAR EQUATIONS

We consider a system of three equations in three unknowns:

$$a_{11}x + a_{12}y + a_{13}z = k_1,$$
$$a_{21}x + a_{22}y + a_{23}z = k_2,$$
$$a_{31}x + a_{32}y + a_{33}z = k_3. \qquad (1.36)$$

With this system we associate the determinants

$$D = \begin{vmatrix} a_{11} & a_{12} & a_{13} \\ a_{21} & a_{22} & a_{23} \\ a_{31} & a_{32} & a_{33} \end{vmatrix}, \qquad D_1 = \begin{vmatrix} k_1 & a_{12} & a_{13} \\ k_2 & a_{22} & a_{23} \\ k_3 & a_{32} & a_{33} \end{vmatrix},$$

$$D_2 = \begin{vmatrix} a_{11} & k_1 & a_{13} \\ a_{21} & k_2 & a_{23} \\ a_{31} & k_3 & a_{33} \end{vmatrix}, \qquad D_3 = \begin{vmatrix} a_{11} & a_{12} & k_1 \\ a_{21} & a_{22} & k_2 \\ a_{31} & a_{32} & k_3 \end{vmatrix}. \tag{1.37}$$

Cramer's Rule asserts that the unique solution of (1.36) is given by

$$x = \frac{D_1}{D}, \qquad y = \frac{D_2}{D}, \qquad z = \frac{D_3}{D}, \tag{1.38}$$

provided that $D \neq 0$.

We can derive the rule by multiplying the first equation of (1.36) by $\begin{vmatrix} a_{22} & a_{23} \\ a_{32} & a_{33} \end{vmatrix}$ (that is, by the minor of a_{11} in D), the second equation by minus the minor of a_{21}, and the third by the minor of a_{31}. If we then add the equations, we obtain

$$x \left(a_{11} \begin{vmatrix} a_{22} & a_{23} \\ a_{32} & a_{33} \end{vmatrix} - a_{21} \begin{vmatrix} a_{12} & a_{13} \\ a_{32} & a_{33} \end{vmatrix} + a_{31} \begin{vmatrix} a_{12} & a_{13} \\ a_{22} & a_{23} \end{vmatrix} \right)$$

$$+ y \left(a_{12} \begin{vmatrix} a_{22} & a_{23} \\ a_{32} & a_{33} \end{vmatrix} - \cdots \right) + z \left(a_{13} \begin{vmatrix} a_{22} & a_{23} \\ a_{32} & a_{33} \end{vmatrix} - \cdots \right)$$

$$= k_1 \begin{vmatrix} a_{22} & a_{23} \\ a_{32} & a_{33} \end{vmatrix} - \cdots$$

The coefficient of x is the expansion of D by minors of the first column. The coefficient of y is the expansion of

$$\begin{vmatrix} a_{12} & a_{12} & a_{13} \\ a_{22} & a_{22} & a_{23} \\ a_{32} & a_{32} & a_{33} \end{vmatrix} = 0$$

and similarly the coefficient of z is 0. The right-hand side is the expansion of D_1 by minors of the first column. Hence

$$Dx = D_1 \tag{1.39}$$

and similarly

$$Dy = D_2, \qquad Dz = D_3. \tag{1.40}$$

Thus each solution x, y, z of (1.36) must satisfy (1.39) and (1.40). If $D \neq 0$, these are the same as (1.38); we can verify that, in this case, (1.38) does provide a solution of (1.36) (Problem 15). Thus the rule is proved.

If $D = 0$, then (1.39) and (1.40) show that, if there is a solution, then $D_1 = 0$, $D_2 = 0$, $D_3 = 0$. We reserve discussion of the general case $D = 0$ to Section 1.9

and here consider only the *homogeneous* system

$$a_{11}x + a_{12}y + a_{13}z = 0,$$
$$a_{21}x + a_{22}y + a_{23}z = 0,$$
$$a_{31}x + a_{32}y + a_{33}z = 0. \tag{1.41}$$

Here $D_1 = 0$, $D_2 = 0$, $D_3 = 0$. Thus if $D \neq 0$, Eqs. (1.41) have the unique solution

$$x = 0, \qquad y = 0, \qquad z = 0 \tag{1.42}$$

called the *trivial solution*.

On the other hand, if $D = 0$, then Eqs. (1.41) have infinitely many solutions. To show this, we introduce the vectors

$$\mathbf{v}_1 = a_{11}\mathbf{i} + a_{21}\mathbf{j} + a_{31}\mathbf{k},$$
$$\mathbf{v}_2 = a_{12}\mathbf{i} + a_{22}\mathbf{j} + a_{32}\mathbf{k},$$
$$\mathbf{v}_3 = a_{13}\mathbf{i} + a_{23}\mathbf{j} + a_{33}\mathbf{k}.$$

Then $x\mathbf{v}_1 + y\mathbf{v}_2 + z\mathbf{v}_3$ has components $a_{11}x + a_{12}y + a_{13}z$, $a_{21}x + a_{22}y + a_{23}z$, $a_{31}x + a_{32}y + a_{33}z$. Hence Eqs. (1.41) are equivalent to the vector equation

$$x\mathbf{v}_1 + y\mathbf{v}_2 + z\mathbf{v}_3 = \mathbf{0}. \tag{1.43}$$

Now we have assumed that $D = 0$. It follows that $\mathbf{v}_1, \mathbf{v}_2, \mathbf{v}_3$ are linearly dependent. For the corresponding determinant,

$$\mathbf{v}_1 \cdot \mathbf{v}_2 \times \mathbf{v}_3$$

equals D with rows and columns interchanged; by Rule I of Section 1.3, this determinant equals D and thus is 0. Therefore $\mathbf{v}_1, \mathbf{v}_2, \mathbf{v}_3$ are linearly dependent, and numbers c_1, c_2, c_3 that are not all 0 can be found such that

$$c_1\mathbf{v}_1 + c_2\mathbf{v}_2 + c_3\mathbf{v}_3 = \mathbf{0}.$$

Thus $x = c_1t$, $y = c_2t$, $z = c_3t$, where t is arbitrary, provides infinitely many solutions of (1.43) and hence of (1.41).

The results established here extend to the general case of n equations in n unknowns:

$$a_{11}x_1 + a_{12}x_2 + \cdots + a_{1n}x_n = k_1,$$
$$\vdots \tag{1.44}$$
$$a_{n1}x_1 + a_{n2}x_2 + \cdots + a_{nn}x_n = k_n.$$

Here

$$D = \begin{vmatrix} a_{11} & a_{12} & \cdots & a_{1n} \\ \vdots & \vdots & & \vdots \\ a_{n1} & a_{n2} & \cdots & a_{nn} \end{vmatrix} \tag{1.45}$$

and $D_1, \ldots, D_n$ are obtained from D by replacing the entries of the first, second, $\ldots, n$th columns of D by $k_1, \ldots, k_n$ as in (1.37). Cramer's Rule holds: If

$D \neq 0$, then Eqs. (1.44) have the unique solution

$$x_1 = \frac{D_1}{D}, \ldots, x_n = \frac{D_n}{D}. \tag{1.46}$$

If $k_1 = 0, \ldots, k_n = 0$, then Eqs. (1.44) are homogenous and always have the *trivial solution*

$$x_1 = 0, x_2 = 0, \ldots, x_n = 0. \tag{1.47}$$

If $D \neq 0$, this is the only solution. If $D = 0$, then there are infinitely many solutions of the homogeneous equations.

For further discussion of this topic, see Section 1.9.

PROBLEMS

1. Let points P_1: $(1, 0, 2)$, P_2: $(2, 1, 3)$, P_3: $(1, 5, 4)$ be given in space.

   a) From a rough graph, verify that the points are vertices of a triangle.

   b) Find the lengths of the sides of the triangle.

   c) Find the angles of the triangle.   d) Find the area of the triangle.

   e) Find the length of the altitude on side $P_1 P_2$.

   f) Find the midpoint of side $P_1 P_2$.   g) Find the point where the medians meet.

2. Let vectors $\mathbf{u} = \mathbf{i} - \mathbf{j} + 2\mathbf{k}$, $\mathbf{v} = 3\mathbf{i} + \mathbf{j} - \mathbf{k}$, $\mathbf{w} = \mathbf{i} + 5\mathbf{j} + 2\mathbf{k}$ be given.

   a) Find $\mathbf{u} \cdot \mathbf{v}$.   b) Find $\mathbf{u} \times \mathbf{v}$.   c) Find $|\mathbf{u}|$.   d) Find $\sphericalangle(\mathbf{u}, \mathbf{v})$.

   e) Find $\mathbf{u} \times (2\mathbf{v} + 3\mathbf{w})$.   f) Find $\mathbf{u} \cdot \mathbf{v} \times \mathbf{w}$ and $\mathbf{v} \times \mathbf{u} \cdot \mathbf{w}$.   g) Find $\mathbf{u} \times (\mathbf{v} \times \mathbf{w})$.

   h) Find the direction cosines of $\mathbf{u}$.   i) Find $\text{comp}_w \mathbf{v}$.

3. a) Show that an equation of the plane through (x_1, y_1, z_1), (x_2, y_2, z_2), (x_3, y_3, z_3) is given by

$$\begin{vmatrix} x - x_1 & y - y_1 & z - z_1 \\ x_2 - x_1 & y_2 - y_1 & z_2 - z_1 \\ x_3 - x_1 & y_3 - y_1 & z_3 - z_1 \end{vmatrix} = 0.$$

   Are there any exceptions?

   b) Find an equation of the plane through the points of Problem 1.

4. Let points P_1: $(1, 3, -1)$, P_2: $(2, 1, 4)$, P_3: $(1, 3, 7)$, P_4: $(5, 0, 2)$ be given.

   a) Show that the points do not lie in a plane and hence form the vertices of a tetrahedron.

   b) Find the volume of the tetrahedron.

5. Test for linear independence:

   a) $\mathbf{u} = 15\mathbf{i} - 21\mathbf{k}$, $\mathbf{v} = 20\mathbf{i} - 28\mathbf{k}$

   b) $\mathbf{u} = \mathbf{i} + \mathbf{j} - \mathbf{k}$, $\mathbf{v} = 2\mathbf{i} + \mathbf{j} + \mathbf{k}$, $\mathbf{w} = 7\mathbf{i} + 5\mathbf{j} - \mathbf{k}$

   c) $\mathbf{u} = 2\mathbf{i} + \mathbf{j}$, $\mathbf{v} = 2\mathbf{j} + \mathbf{k}$, $\mathbf{w} = 2\mathbf{i} + \mathbf{k}$, $\mathbf{p} = \mathbf{i} + \mathbf{j} + \mathbf{k}$

6. Find parametric equations for the line satisfying the given conditions:

   a) passes through $(2, 1, 0)$ and $(3, 2, 5)$.

   b) passes through $(1, 1, 2)$ and is parallel to the line $x = 2 - 5t, y = 1 + 2t, z = 3t$.

c) passes through $(0,0,0)$ and is perpendicular to the plane $5x - y + z = 2$.

d) passes through $(1,2,2)$ and is perpendicular to the line $x = 1 + t$, $y = 2 - t$, $z = 3 + t$ and to the line $x = 2 + t$, $y = 5 + 2t$, $z = 7 + 4t$.

7. Find an equation for the plane satisfying the given conditions:

a) passes through the z-axis and the point $(1,2,5)$.

b) passes through $(1,2,2)$ and is perpendicular to the line $x = 2 - t$, $y = 3 + 5t$, $z = 2 - 4t$.

c) passes through the line $x = 1 + t$, $y = 2 + 3t$, $z = 1 + 5t$ and is perpendicular to the plane $2x + y - z = 0$.

8. Let line L_1 pass through P_1 and have nonzero vector $\mathbf{v}_1$ along the line; let line L_2 pass through P_2 and have nonzero vector $\mathbf{v}_2$ along the line. Let $\mathbf{u} = \overrightarrow{P_1 P_2}$. Use geometric reasoning for the following:

a) Show that L_1 and L_2 coincide precisely when $\mathbf{u} \times \mathbf{v}_1 = \mathbf{0}$ and $\mathbf{v}_1 \times \mathbf{v}_2 = \mathbf{0}$.

b) Show that L_1 and L_2 are parallel and noncoincident precisely when $\mathbf{u} \times \mathbf{v}_1 \neq \mathbf{0}$ and $\mathbf{v}_1 \times \mathbf{v}_2 = \mathbf{0}$.

c) Show that L_1 and L_2 intersect at one point precisely when $\mathbf{u} \cdot \mathbf{v}_1 \times \mathbf{v}_2 = 0$ and $\mathbf{v}_1 \times \mathbf{v}_2 \neq \mathbf{0}$.

d) Show that L_1 and L_2 are skew precisely when $\mathbf{u} \cdot \mathbf{v}_1 \times \mathbf{v}_2 \neq \mathbf{0}$. Show that in this case the shortest distance between two points on L_1, L_2 is

$$d = \frac{|\mathbf{u} \cdot \mathbf{v}_1 \times \mathbf{v}_2|}{|\mathbf{v}_1 \times \mathbf{v}_2|}.$$

9. Let L be the line through $P_1: (1,2,3)$ and $P_2: (10,11,15)$.

a) Find the trisection points of the segment $P_1 P_2$.

b) Find a point P on L that is not on $P_1 P_2$ and is two units from P_2.

10. Evaluate the determinants:

a) $\begin{vmatrix} 3 & 5 \\ 1 & -4 \end{vmatrix}$ b) $\begin{vmatrix} 2 & 1 \\ 7 & 0 \end{vmatrix}$ c) $\begin{vmatrix} 1 & 0 & 1 \\ 0 & 1 & 2 \\ 1 & 0 & 3 \end{vmatrix}$

d) $\begin{vmatrix} 2 & 2 & 1 \\ 1 & 1 & 1 \\ 3 & 4 & 2 \end{vmatrix}$ e) $\begin{vmatrix} 5 & 1 & 7 & 2 \\ 3 & 1 & 4 & 1 \\ 2 & 1 & -2 & 3 \\ 0 & 1 & 4 & 1 \end{vmatrix}$ f) $\begin{vmatrix} a & b & c \\ a^2 & b^2 & c^2 \\ a^3 & b^3 & c^3 \end{vmatrix}$

11. Determine whether the ordered triple $\mathbf{u}, \mathbf{v}, \mathbf{w}$ is positive or negative:

a) $\mathbf{u} = \mathbf{i} - 2\mathbf{j} - 5\mathbf{k}$, $\mathbf{v} = 2\mathbf{i} + \mathbf{j} - \mathbf{k}$, $\mathbf{w} = 3\mathbf{i} + 4\mathbf{j} + 2\mathbf{k}$

b) $\mathbf{u} = 2\mathbf{i} + 3\mathbf{j} + 4\mathbf{k}$, $\mathbf{v} = 4\mathbf{i} + 3\mathbf{j} + 2\mathbf{k}$, $\mathbf{w} = \mathbf{i} + \mathbf{j} + \mathbf{k}$

12. Prove the identities:

a) $|\mathbf{u} \times \mathbf{v}|^2 = \begin{vmatrix} \mathbf{u} \cdot \mathbf{u} & \mathbf{u} \cdot \mathbf{v} \\ \mathbf{v} \cdot \mathbf{u} & \mathbf{v} \cdot \mathbf{v} \end{vmatrix}$ b) $(\mathbf{u} \times \mathbf{v}) \cdot (\mathbf{w} \times \mathbf{z}) = (\mathbf{u} \cdot \mathbf{w})(\mathbf{v} \cdot \mathbf{z}) - (\mathbf{u} \cdot \mathbf{z})(\mathbf{v} \cdot \mathbf{w})$

c) $(\mathbf{u} \times \mathbf{v}) \times (\mathbf{u} \times \mathbf{w}) = (\mathbf{u} \cdot \mathbf{v} \times \mathbf{w})\mathbf{u}$

d) $\mathbf{u} \times (\mathbf{v} \times \mathbf{w}) + \mathbf{v} \times (\mathbf{w} \times \mathbf{u}) + \mathbf{w} \times (\mathbf{u} \times \mathbf{v}) = \mathbf{0}$

13. Solve the simultaneous linear equations:

a) $3x - 2y = 4$, $x + 2y = 4$

b) $5x - y = 4$, $x + 2y = 3$

c) $x - y + z = 1$, $x + y - z = 2$, $3x + y + z = 0$

d) $x - y + z = 0$, $3x - y + 2z = 0$, $6x - 4y + 5z = 0$

14. Consider the simultaneous equations

$$a_{11}x + a_{12}y = k_1, a_{21}x + a_{22}y = k_2.$$

a) Show that if $D \neq 0$, Cramer's Rule provides the unique solution. Interpret geometrically.

b) Let $D = 0$, $D_1 \neq 0$. Show geometrically why there is no solution.

c) Let $D = 0$, $D_1 = 0$, $D_2 = 0$. Show geometrically various cases that can arise, some yielding solutions and others yielding no solutions.

15. Show that for the system (1.36) with $D \neq 0$, Cramer's Rule (1.38) does provide a solution. [Hint: It suffices to check the first equation of (1.36), since the others are similar. Show that after substitution from (1.38) it can be written $a_{11}D_1 + a_{12}D_2 + a_{13}D_3 = k_1D$. Show that the left-hand side can be written as

$$k_1(a_{11}\Delta_{11} - a_{12}\Delta_{12} + a_{13}\Delta_{13}) + k_2(-a_{11}\Delta_{21} + a_{12}\Delta_{22} - a_{13}\Delta_{13})$$

$$+ k_3(a_{11}\Delta_{31} - a_{12}\Delta_{32} + a_{13}\Delta_{33}).$$

Now interpret the coefficients of k_1, k_2, k_3 as determinants expanded by minors.]

16. Let D be as in (1.45).

a) Let $a_{ij} = \delta_{ij}$, where $\delta_{ij} = 1$ for $i = j$ and $\delta_{ij} = 0$ for $i \neq j$. Show that $D = 1$. [This is the rule $\det I = 1$, in the notation of Section 1.8 below.]

b) Let $a_{ij} = 1$ if $i + j = n + 1$ and $a_{ij} = 0$ otherwise. Show that $D = -1$.

c) Let $a_{ij} = 0$ for $i > j$, so that the array has "upper triangular form." Show that $D = a_{11}a_{22} \cdots a_{nn}$.

d) It can be shown that

$$D = \sum \varepsilon_{j_1 j_2 \cdots j_n} a_{1j_1} a_{2j_2} \cdots a_{nj_n},$$

where the sum is over all permutations $(j_1, \ldots, j_n)$ of $(1, 2, \ldots, n)$ and $\varepsilon_{j_1 j_2 \cdots j_n}$ is 1 for a permutation which is even (obtainable from $(1, 2, \ldots, n)$ by an *even* number of interchanges of two integers) and is -1 for an *odd* permutation (odd number of interchanges). (See Chapter 4 of the book by Perlis listed at the end of chapter.) Verify this rule for $n = 2$ and for $n = 3$.

1.5 MATRICES

By a *matrix* we mean a rectangular array of m rows and n columns:

$$\begin{bmatrix} a_{11} & \cdots & a_{1n} \\ \vdots & & \vdots \\ a_{m1} & \cdots & a_{mn} \end{bmatrix}. \tag{1.48}$$

For this chapter (with a very few exceptions) the objects $a_{11}, a_{12}, \ldots, a_{mn}$ will be real numbers. In some applications they are complex numbers, and in some they are functions of one or more variables. We call each a_{ij} an *entry* of the matrix; more specifically, a_{ij} is the *ij-entry*.

We can denote a matrix by a single letter such as $A, B, C, X, Y, \ldots$. If A denotes the matrix (1.48), then we write also, concisely, $A = (a_{ij})$.

Let A be the matrix (a_{ij}) of (1.48). We say that A is an $m \times n$ matrix. When $m = n$, we say that A is a *square matrix of order n*. The following are examples of matrices:

$$A = \begin{bmatrix} 2 & 3 & 5 \\ 1 & 2 & 3 \end{bmatrix}, \qquad B = \begin{bmatrix} 1 & 2 \\ 4 & 3 \end{bmatrix}, \qquad C = \begin{bmatrix} 1 & 4 \\ 2 & -3 \end{bmatrix}. \tag{1.49}$$

Here A is 2×3, and B and C are 2×2; B and C are *square matrices of order 2*.

An important square matrix is the *identity matrix* of order n, denoted by I:

$$I = \begin{bmatrix} 1 & 0 & \cdots & 0 \\ 0 & 1 & \cdots & 0 \\ \vdots & & & \vdots \\ 0 & 0 & \cdots & 1 \end{bmatrix} = (\delta_{ij}), \qquad \delta_{ij} = \begin{cases} 1 & \text{for } i = j, \\ 0 & \text{for } i \neq j. \end{cases} \tag{1.50}$$

We call δ_{ij} the *Kronecker delta symbol*. One sometimes writes I_n to indicate the order of I, but normally the context makes this unnecessary.

The *principal diagonal* of a square matrix is formed of the entries $a_{11}, a_{22}, \ldots, a_{nn}$. For I_n the principal diagonal is $1, 1, \ldots, 1$ (n times).

For each m and n we define the $m \times n$ *zero matrix*

$$O = \begin{bmatrix} 0 & \cdots & 0 \\ \vdots & & \vdots \\ 0 & \cdots & 0 \end{bmatrix}. \tag{1.51}$$

One sometimes denotes this matrix by O_{mn} to indicate the *size*—that is, the number of rows and columns.

In general, two matrices $A = (a_{ij})$ and $B = (b_{ij})$ are said to be equal, $A = B$, when A and B have the same size and $a_{ij} = b_{ij}$ for all i and j.

A $1 \times n$ matrix A is formed of one row: $A = (a_{11}, \ldots, a_{1n})$. We call such a matrix a *row vector*. In a general $m \times n$ matrix (1.48), each of the successive rows forms a row vector. We often denote a row vector by a boldface symbol: $\mathbf{u}, \mathbf{v}, \ldots$ (or, in handwriting, by an arrow). Thus the matrix A in (1.49) has the row vectors $\mathbf{u}_1 = (2, 3, 5)$ and $\mathbf{u}_2 = (1, 2, 3)$.

Similarly, an $m \times 1$ matrix A is formed of one column:

$$A = \begin{bmatrix} a_{11} \\ \vdots \\ a_{m1} \end{bmatrix}. \tag{1.52}$$

We call such a matrix a *column vector*. For typographical reasons we sometimes denote this matrix by $\mathrm{col}(a_{11},\ldots,a_{m1})$ or even by $(a_{11},\ldots,a_{m1})$, if the context makes clear that a column vector is intended. We also denote column vectors by boldface letters: $\mathbf{u},\mathbf{v},\ldots$. The matrix B in (1.49) has the column vectors $\mathbf{v}_1 = \mathrm{col}(1,4)$ and $\mathbf{v}_2 = \mathrm{col}(2,3)$.

We denote by $\mathbf{0}$ the row vector or column vector $(0,\ldots,0)$. The context will make clear whether $\mathbf{0}$ is a row vector or a column vector and the number of entries.

The vectors occurring here can be interpreted geometrically as vectors in k-dimensional space, for appropriate k. For example, the row vectors or column vectors with three entries are simply ordered triples of numbers and, as in Section 1.1, they can be represented as vectors $a\mathbf{i} + b\mathbf{j} + c\mathbf{k}$ in 3-dimensional space. This interpretation is discussed in Section 1.13.

Matrices often arise in connection with simultaneous linear equations. Let such a set of equations be given:

$$a_{11}x_1 + \quad \cdots \quad a_{1n}x_n = y_1,$$
$$\vdots \tag{1.53}$$
$$a_{m1}x_1 + \quad \cdots \quad a_{mn}x_n = y_m.$$

Here we may think of $y_1,\ldots,y_m$ as given numbers and $x_1,\ldots,x_n$ as unknown numbers, to be found; however, we may also think of $x_1,\ldots,x_n$ as variable numbers and $y_1,\ldots,y_m$ as "dependent variables" whose values are determined by the values chosen for the "independent variables" $x_1,\ldots,x_n$. Both points of view will be important in this chapter. In either case we call $A = (a_{ij})$ the *coefficient matrix* of the set of equations. The numbers $y_1,\ldots,y_m$ can be considered as the entries in a column vector $\mathbf{y} = \mathrm{col}(y_1,\ldots,y_m)$. The numbers $x_1,\ldots,x_n$ can be thought of as the entries in a row vector or column vector $\mathbf{x}$; in this chapter, we usually write $\mathbf{x} = \mathrm{col}(x_1,\ldots,x_n)$.

1.6 ADDITION OF MATRICES ▪ SCALAR TIMES MATRIX

Let $A = (a_{ij})$ and $B = (b_{ij})$ be matrices of the *same size*, both $m \times n$. Then one defines the sum $A + B$ to be the $m \times n$ matrix $C = (c_{ij})$ such that $c_{ij} = a_{ij} + b_{ij}$ for all i and j; that is, *one adds two matrices by adding corresponding entries*. For example,

$$\begin{bmatrix} 3 & 5 & 1 \\ 1 & 0 & -2 \end{bmatrix} + \begin{bmatrix} 0 & 1 & 0 \\ 2 & 3 & 5 \end{bmatrix} = \begin{bmatrix} 3 & 6 & 1 \\ 3 & 3 & 3 \end{bmatrix}.$$

Let c be a number (scalar); let $A = (a_{ij})$ be an $m \times n$ matrix. Then one defines cA to be the $m \times n$ matrix $B = (b_{ij})$ such that $b_{ij} = ca_{ij}$ for all i and j; that is, *cA is obtained from A by multiplying each entry of A by c*. For example,

$$5\begin{bmatrix} 2 & 0 \\ 1 & -2 \end{bmatrix} = \begin{bmatrix} 10 & 0 \\ 5 & -10 \end{bmatrix}.$$

We denote $(-1)A$ by $-A$ and $B + (-A)$ by $B - A$.

From these definitions we can deduce the following rules governing the two operations:

1. $A + B = B + A$.
2. $A + (B + C) = (A + B) + C$.
3. $c(A + B) = cA + cB$.
4. $(a + b)C = aC + bC$.
5. $a(bC) = (ab)C$.
6. $1A = A$.
7. $0A = O$.
8. $A + O = A$.
9. $A + C = B$ if and only if $C = B - A$.   (1.54)

Throughout, we assume that the sizes of the matrices are such that the operations have meaning. The proofs are obtained by simply applying the definitions. For example, for Rule 1 we write

$$A + B = (a_{ij}) + (b_{ij}) = (a_{ij} + b_{ij})$$
$$= (b_{ij} + a_{ij}) = (b_{ij}) + (a_{ij}) = B + A.$$

PROBLEMS

In these problems the following matrices are given:

$$A = \begin{bmatrix} 1 \\ 3 \end{bmatrix}, \quad B = \begin{bmatrix} 2 \\ 0 \end{bmatrix}, \quad C = \begin{bmatrix} 2 & 3 \\ 4 & 1 \end{bmatrix}, \quad D = \begin{bmatrix} 1 & -1 \\ 2 & 0 \end{bmatrix}, \quad E = \begin{bmatrix} 1 & 2 \\ 2 & 4 \end{bmatrix},$$

$$F = \begin{bmatrix} 1 & 4 & 5 \\ 2 & 0 & 7 \end{bmatrix}, G = \begin{bmatrix} 3 & 1 & 4 \\ -1 & 0 & -1 \end{bmatrix}, H = (1,0,1), J = (3,5,2), K = (3,5),$$

$$L = \begin{bmatrix} 3 & 1 & 0 \\ 2 & 5 & 6 \\ 1 & 4 & 3 \end{bmatrix}, M = \begin{bmatrix} 2 & -1 & 0 \\ 1 & 2 & 1 \\ 3 & 2 & -1 \end{bmatrix}, N = \begin{bmatrix} 1 & 4 \\ 0 & 3 \\ 7 & 1 \end{bmatrix}, P = \begin{bmatrix} 2 & 2 \\ -1 & -1 \\ 3 & 3 \end{bmatrix}.$$

1. a) Give the number of rows and columns for each of the matrices A, F, H, L, and P.
   b) Writing $A = (a_{ij})$, $B = (b_{ij})$, and so on, give the values of the following entries: $a_{11}, a_{21}, c_{21}, c_{22}, d_{12}, e_{21}, f_{11}, g_{23}, g_{21}, h_{12}, m_{23}$.
   c) Give the row vectors of C, G, L, and P.
   d) Give the column vectors of D, F, L, and N.
2. Evaluate each expression that is meaningful:
   a) $A + B$.   b) $C + D$.   c) $E + F$.   d) $L + M$.   e) $N - P$.   f) $G - F$.
   g) $5C$.   h) $2E$.   i) $3E + 4D$.   j) $2C + D - E$.   k) $3L - N$.
3. Solve for X:
   a) $C + X = D$,   b) $F - 5X = G$.
4. Solve for X and Y:
   a) $X + Y = N, X - Y = P$.   b) $2X - 3Y = L, X - 2Y = M$.
5. Prove each of the following rules of (1.54):
   a) Rule 2.   b) Rule 3.   c) Rule 4.   d) Rule 5.
   e) Rule 6.   f) Rule 7.   g) Rule 8.   h) Rule 9.

1.7   MULTIPLICATION OF MATRICES

In order to motivate the definition of the product AB of two matrices A and B, we consider two systems of simultaneous equations:

$$
\left.
\begin{aligned}
a_{11}u_1 + \quad \cdots \quad + a_{1p}u_p &= y_1, \\
&\vdots \\
a_{m1}u_1 + \quad \cdots \quad + a_{mp}u_p &= y_m.
\end{aligned}
\right\} \tag{1.55}
$$

$$
\left.
\begin{aligned}
b_{11}x_1 + \quad \cdots \quad + b_{1n}x_n &= u_1, \\
&\vdots \\
b_{p1}x_1 + \quad \cdots \quad + b_{pn}x_n &= u_p.
\end{aligned}
\right\} \tag{1.56}
$$

Such pairs of systems arise in many practical problems. A typical situation is that in which $x_1,\ldots,x_n$ are known numbers and $y_1,\ldots,y_m$ are sought, all coefficients b_{ij} and a_{ij} being known. The second set of equations allows us to compute $u_1,\ldots,u_p$; if we substitute the values found in the first set of equations, we can then find $y_1,\ldots,y_m$. We carry this out for the general case:

$$
\begin{aligned}
y_1 &= a_{11}u_1 + \cdots + a_{1p}u_p \\
&= a_{11}(b_{11}x_1 + \cdots + b_{1n}x_n) + \cdots + a_{1p}(b_{p1}x_1 + \cdots + b_{pn}x_n) \\
&= (a_{11}b_{11} + \cdots + a_{1p}b_{p1})x_1 + \cdots + (a_{11}b_{1n} + \cdots + a_{1p}b_{pn})x_n;
\end{aligned}
$$

and, in general, for $i = 1,\ldots,m$,

$$
y_i = (a_{i1}b_{11} + \cdots + a_{ip}b_{p1})x_1 + \cdots + (a_{i1}b_{1n} + \cdots + a_{ip}b_{pn})x_n
$$

or

$$
y_i = c_{i1}x_1 + \cdots + c_{in}x_n, \quad i = 1,\ldots,m,
$$

where

$$
c_{ij} = a_{i1}b_{1j} + \cdots + a_{ip}b_{pj} \quad \text{for } i = 1,\ldots,m, \quad j = 1,\ldots,n. \tag{1.57}
$$

Thus from the coefficient matrix $A = (a_{ij})$ and the coefficient matrix $B = (b_{ij})$ we obtain the coefficient matrix $C = (c_{ij})$ by the rule (1.57). We write $C = AB$ and have thereby defined the *product of the matrices A and B*.

We observe that $(a_{i1},\ldots,a_{ip})$ is the ith *row* vector of A and that $\text{col}(b_{1j},\ldots,b_{pj})$ is the jth *column* vector of B. Hence to form the product $AB = C = (c_{ij})$, we obtain each c_{ij} by multiplying corresponding entries of the ith row of A and the jth column of B and adding. The process is suggested in Fig. 1.14.

We remark that the product AB is defined only when the number of *columns* of A equals the number of *rows* of B; that is, when A is $m \times p$ and B is $p \times n$, AB is defined and is $m \times n$. Also, when AB is defined, BA need not be defined, and even when it is, AB is generally *not equal* to BA; that is, there is *no commutative law* for multiplication.

Figure 1.14 Product of two matrices.

EXAMPLE 1

$$\begin{bmatrix} a & b \\ c & d \end{bmatrix}\begin{bmatrix} e & f & g \\ h & i & j \end{bmatrix} = \begin{bmatrix} ae + bh & af + bi & ag + bj \\ ce + dh & cf + di & cg + dj \end{bmatrix}.$$

Here $m = 2$, $p = 2$, $n = 3$. ■

EXAMPLE 2

$$\begin{bmatrix} 1 & 2 & 1 \\ 3 & 0 & 5 \\ 2 & 1 & -7 \end{bmatrix}\begin{bmatrix} 1 \\ 4 \\ 2 \end{bmatrix} = \begin{bmatrix} 1 + 8 + 2 \\ 3 + 0 + 10 \\ 2 + 4 - 14 \end{bmatrix} = \begin{bmatrix} 11 \\ 13 \\ -8 \end{bmatrix}.$$

Here $m = 3$, $p = 3$, $n = 1$. ■

The second example illustrates the important case of the product $A\mathbf{v}$, where A is an $m \times n$ matrix and $\mathbf{v}$ is an $n \times 1$ column vector. The product $A\mathbf{v}$ is again a column vector $\mathbf{u}$, $m \times 1$.

In the general product $AB = C$, as defined above, we note that the jth column vector of C is formed from A and the jth column vector of B, for the jth column vector of C is

$$\begin{bmatrix} a_{11}b_{1j} + & \cdots & + a_{1p}b_{pj} \\ a_{21}b_{1j} + & \cdots & + a_{2p}b_{pj} \\ & \vdots & \\ a_{m1}b_{1j} + & \cdots & + a_{mp}b_{pj} \end{bmatrix} = \begin{bmatrix} a_{11} & \cdots & a_{1p} \\ a_{21} & \cdots & a_{2p} \\ \vdots & & \vdots \\ a_{m1} & \cdots & a_{mp} \end{bmatrix} \cdot \begin{bmatrix} b_{1j} \\ \vdots \\ b_{pj} \end{bmatrix}.$$

Hence if we denote the successive column vectors of B by $\mathbf{u}_1, \ldots, \mathbf{u}_n$, then the column vectors of $C = AB$ are $A\mathbf{u}_1, \ldots, A\mathbf{u}_n$. Symbolically,

$$\begin{bmatrix} A \end{bmatrix}\begin{bmatrix} \mathbf{u}_1 & \mathbf{u}_2 & \cdots & \mathbf{u}_n \end{bmatrix} = \begin{bmatrix} A\mathbf{u}_1 & A\mathbf{u}_2 & \cdots & A\mathbf{u}_n \end{bmatrix}.$$

EXAMPLE 3 To calculate AB, where

$$A = \begin{bmatrix} 3 & 1 & 2 \\ 1 & 0 & 5 \end{bmatrix}, \qquad B = \begin{bmatrix} 1 & 0 & 3 & 1 & 3 \\ 5 & 2 & 1 & 5 & 1 \\ -1 & 2 & 4 & -1 & 4 \\ \mathbf{u}_1 & \mathbf{u}_2 & \mathbf{u}_3 & \mathbf{u}_4 & \mathbf{u}_5 \end{bmatrix},$$

we calculate

$$Au_1 = \begin{bmatrix} 6 \\ -4 \end{bmatrix}, \qquad Au_2 = \begin{bmatrix} 6 \\ 10 \end{bmatrix}, \qquad Au_3 = \begin{bmatrix} 18 \\ 23 \end{bmatrix},$$

and then note that $u_4 = u_1$ and $u_5 = u_3$, so that $Au_4 = Au_1$ and $Au_5 = Au_3$. Therefore

$$AB = \begin{bmatrix} 6 & 6 & 18 & 6 & 18 \\ -4 & 10 & 23 & -4 & 23 \end{bmatrix}. \quad \blacksquare$$

EXAMPLE 4   The simultaneous equations

$$3x + 2y + 5z = u,$$
$$4x - 5y - 8z = v,$$
$$7x + 2y + 9z = w$$

are equivalent to the matrix equation

$$\begin{bmatrix} 3 & 2 & 5 \\ 4 & -5 & -8 \\ 7 & 2 & 9 \end{bmatrix} \begin{bmatrix} x \\ y \\ z \end{bmatrix} = \begin{bmatrix} u \\ v \\ w \end{bmatrix},$$

for the product on the left-hand side equals the column vector

$$\begin{bmatrix} 3x + 2y + 5z \\ 4x - 5y - 8z \\ 7x + 2y + 9z \end{bmatrix}$$

and this equals col (u, v, w) precisely when the given simultaneous equations hold.

In the same way the two sets of simultaneous equations (1.55) and (1.56) can be replaced by the equations

$$Au = y \quad \text{and} \quad Bx = u.$$

The elimination process at the beginning of this section is equivalent to replacing **u** by $B\mathbf{x}$ in the first equation to obtain $\mathbf{y} = A(B\mathbf{x})$. *Our definition of the product AB is then such that* $\mathbf{y} = A(B\mathbf{x}) = (AB)\mathbf{x}$. Therefore *for every column vector* **x**,

$$A(B\mathbf{x}) = (AB)\mathbf{x}. \quad \blacksquare$$

Powers of a square matrix.   If A is a square matrix of order n, then the product AA has meaning and is again $n \times n$; we write A^2 for this product. Similarly, $A^3 = A^2A$, $A^4 = A^3A, \ldots, A^{s+1} = A^sA, \ldots$. We also define A^0 to be the $n \times n$ identity matrix I. Negative powers can also be defined for certain square matrices A; see Section 1.8 below.

Rules for multiplication.   Multiplication of matrices obeys a set of rules, which we adjoin to those of the preceding section:

10. $A(BC) = (AB)C$.

11. $AI = A$.

12. $IA = A$.

13. $A(B + C) = AB + AC$.

14. $c(AB) = A(cB)$.

15. $AO = O$.

16. $OA = O$.

17. $A^0 = I$.

18. $A^k A^l = A^{k+l}$.

19. $(A^k)^l = A^{kl}$.

20. $A\mathbf{x} = B\mathbf{x}$ for all $\mathbf{x}$ if and only if $A = B$. $\hfill(1.58)$

Here the sizes of the matrices must again be such that the operations are defined. For example, in Rule 13, if A is $m \times p$, then B and C must be $p \times n$.

To prove Rule 10 (associative law), we let C have the column vectors $\mathbf{u}_1,\ldots,\mathbf{u}_k$. Then BC has the column vectors $B\mathbf{u}_1,\ldots,B\mathbf{u}_k$, and hence $A(BC)$ has the column vectors $A(B\mathbf{u}_1),\ldots,A(B\mathbf{u}_k)$. But as was remarked above, $A(B\mathbf{x}) = (AB)\mathbf{x}$ for every $\mathbf{x}$. Therefore $A(BC)$ has the column vectors $(AB)\mathbf{u}_1,\ldots,(AB)\mathbf{u}_k$. But these are the column vectors of $(AB)C$. Hence $A(BC) = (AB)C$.

For Rule 11, A is, say, $m \times p$, and I is $p \times p$, so that

$$
AI = \begin{bmatrix} a_{11} & \cdots & a_{1p} \\ \vdots & & \vdots \\ a_{m1} & \cdots & a_{mp} \end{bmatrix} \begin{bmatrix} 1 & 0 & \cdots & 0 \\ 0 & 1 & \cdots & 0 \\ \vdots & \vdots & & \vdots \\ 0 & 0 & \cdots & 1 \end{bmatrix} = \begin{bmatrix} a_{11} & \cdots & a_{1p} \\ \vdots & & \vdots \\ a_{m1} & \cdots & a_{mp} \end{bmatrix} = A.
$$

We can also write $AI = C = (c_{ij})$, where

$$
c_{ij} = a_{i1}\delta_{1j} + \cdots + a_{ip}\delta_{pj} = a_{ij},
$$

since $\delta_{ii} = 1$ but $\delta_{ij} = 0$ for $i \neq j$. Rule 12 is proved similarly. For Rule 13 we have $A(B + C) = D$, where

$$
d_{ij} = a_{i1}(b_{1j} + c_{1j}) + \cdots + a_{ip}(b_{pj} + c_{pj})
$$
$$
= (a_{i1}b_{1j} + \cdots + a_{ip}b_{pj}) + (a_{i1}c_{1j} + \cdots + a_{ip}c_{pj}),
$$

and hence $D = AB + AC$. Rule 14 is proved similarly. Rules 15 and 16 follow from the fact that all entries of O are 0; here again the size of O must be such that the products have meaning.

In Rules 17, 18, and 19, A is a square matrix, and k and l are nonnegative integers. Rule 17 is true by definition of A^0; and Rules 18 and 19 are true for $l = 0$ and $l = 1$ by definition. They can be proved for general l by induction (see Problem 4 below).

For Rule 20, let A and B be $m \times n$, and let $\mathbf{e}_1,\ldots,\mathbf{e}_n$ be the column vectors of the identity matrix I of order n. Then AI is a matrix whose columns are $A\mathbf{e}_1,\ldots,A\mathbf{e}_n$, and BI is a matrix whose column vectors are $B\mathbf{e}_1,\ldots,B\mathbf{e}_n$. If $A\mathbf{x} = B\mathbf{x}$ for all $\mathbf{x}$, then we have

$$
A\mathbf{e}_1 = B\mathbf{e}_1,\ldots,A\mathbf{e}_n = B\mathbf{e}_n.
$$

Therefore $AI = BI$ or, by Rule 11, $A = B$. Conversely, if $A = B$, then $Ax = Bx$ for all $\mathbf{x}$, by the definition of equality of matrices.

Remark.   Because of the associative law, Rule 10, we can generally drop parentheses in multiple products of matrices. For example, we replace $[A(BC)]D$ by $ABCD$. No matter how we group the factors, the same result is obtained.

PROBLEMS

Let the matrices $A,\dots,P$ be given as at the beginning of the set of problems following Section 1.6.

1. Evaluate each expression that is meaningful:

   a) AB.

   b) CA.

   c) AC.

   d) CD and DC.

   e) CE and EC.

   f) AI.

   g) IL.

   h) $2I + 3GL$.

   i) $C^2 - 3C - 10C^0$.

   j) $E(E - 5I)$.

   k) LNI.

   l) MPG.

   m) $HL + J$.

   n) KA.

   o) $OC + N$.

   p) E^2.

   q) E^3.

   r) E^4.

   s) $HELP$.

2. Calculate RS for each of the following choices of R and S:

   a) $R = \begin{bmatrix} 3 & 1 \\ 5 & 2 \end{bmatrix}$, $S = \begin{bmatrix} 1 & 0 & 2 & 0 & 2 \\ 0 & 1 & 3 & 1 & 3 \end{bmatrix}$.

   b) $R = \begin{bmatrix} 1 & 4 \\ 2 & 1 \\ 5 & 0 \end{bmatrix}$, $S = \begin{bmatrix} 2 & 1 & 1 & 2 & 1 \\ 3 & 2 & 2 & 3 & 2 \end{bmatrix}$.

3. Consider each of the following pairs of simultaneous equations as cases of (1.55) and (1.56) and express $y_1,\dots$ in terms of $x_1,\dots$ (i) by eliminating $u_1,\dots$ and (ii) by multiplying the coefficient matrices:

   a) $\begin{cases} 3u_1 + 2u_2 = y_1 \\ 5u_1 + 6u_2 = y_2 \end{cases}$ and $\begin{cases} 5x_1 - x_2 = u_1 \\ x_1 + 2x_2 = u_2 \end{cases}$.

   b) $\begin{cases} 2u_1 - u_2 = y_1 \\ 5u_1 + u_2 = y_2 \end{cases}$ and $\begin{cases} x_1 + 2x_2 - x_3 = u_1 \\ 2x_1 + 3x_2 + x_3 = u_2 \end{cases}$.

4. Prove each of the following rules of Section 1.7:

   a) Rule 12.

   b) Rule 14.

   c) Rule 18, by induction with respect to l.

   d) Rule 19, by induction with respect to l and Rule 18.

5. Let A be a square matrix. Prove:

   a) $A^2 - I = (A + I)(A - I)$.

   b) $A^3 - I = (A - I)(A^2 + A + I)$.

   c) $A^2 - 2A - 3I = (A - 3I)(A + I)$.

   d) $6A^2 - A - 2I = (2A + I)(3A - 2I)$.

6. If A and B are $n \times n$ matrices, is $A^2 - B^2$ necessarily equal to $(A - B)(A + B)$? When must this be true?

7. Find nonzero 2×2 matrices A and B such that $A^2 + B^2 = O$.

8. Prove: If A is a 2×2 matrix such that $AB = BA$ for all 2×2 matrices B, then $A = cI$ for some scalar c.

1.8   INVERSE OF A SQUARE MATRIX

Let A be an $n \times n$ matrix. If an $n \times n$ matrix B exists such that $AB = I$, then we call B an *inverse* of A. We shall see below that A can have at most one inverse. Hence if $AB = I$, we call B *the* inverse of A and write $B = A^{-1}$.

For a general $n \times n$ matrix A we denote by det A the determinant formed from A; that is,

$$\det A = \begin{vmatrix} a_{11} & \cdots & a_{1n} \\ \vdots & & \vdots \\ a_{n1} & \cdots & a_{nn} \end{vmatrix}. \tag{1.59}$$

We stress that det A is a number, whereas A itself is a square array—that is, a matrix. The principal properties of determinants are summarized in Section 1.3.

If A and B are $n \times n$ matrices, then one has the rule

$$\det A \det B = \det(AB). \tag{1.60}$$

For a proof, see CLA, Sections 10–13 and 10–14 or page 80 of the book by Perlis listed at the end of the chapter; see also Problem 9 below. From this rule it follows that if A has an inverse, then det $A \neq 0$. For $AB = I$ implies

$$\det A \det B = \det I = 1,$$

so that det $A \neq 0$; also det $B = \det A^{-1} \neq 0$ and, in fact,

$$\det B = \det A^{-1} = \frac{1}{\det A}. \tag{1.61}$$

Conversely, *if* det $A \neq 0$, *then* A *has an inverse.* For if det $A \neq 0$, then the simultaneous linear equations

$$a_{11}x_1 + \quad \cdots \quad + a_{1n}x_n = y_1$$
$$\vdots \tag{1.62}$$
$$a_{n1}x_1 + \quad \cdots \quad + a_{nn}x_n = y_n$$

can be solved for $x_1, \ldots, x_n$ by Cramer's Rule (Section 1.4). For example,

$$x_1 = \begin{vmatrix} y_1 & a_{12} & \cdots & a_{1n} \\ \vdots & \vdots & & \vdots \\ y_n & a_{n2} & \cdots & a_{nn} \end{vmatrix} \div D,$$

where $D = \det A$. Upon expanding the first determinant on the right, we obtain an expression of the form

$$x_1 = b_{11} y_1 + \cdots + b_{1n} y_n,$$

with appropriate constants $b_{11}, \ldots, b_{1n}$. In general,

$$x_i = b_{i1} y_1 + \cdots + b_{in} y_n, \qquad i = 1, \ldots, n. \tag{1.63}$$

Now our given equations (1.62) are equivalent to the matrix equation

$$A\mathbf{x} = \mathbf{y},$$

and the solution (1.63) is given by

$$\mathbf{x} = B\mathbf{y}.$$

The fact that this is a solution is expressed by the relation

$$AB\mathbf{y} = \mathbf{y} \quad \text{or} \quad AB\mathbf{y} = I\mathbf{y}.$$

This relation holds for all $\mathbf{y}$. Hence by Rule 20 of Section 1.7 we must have $AB = I$, so that B is an inverse of A.

The reasoning just given also provides a constructive way of finding A^{-1}. One simply forms the equations (1.62) and solves for $x_1, \ldots, x_n$. The solution can be written as $\mathbf{x} = B\mathbf{y}$, where $B = A^{-1}$.

EXAMPLE 1 $A = \begin{bmatrix} 2 & 5 \\ 1 & 3 \end{bmatrix}$. The simultaneous equations are

$$2x_1 + 5x_2 = y_1, \qquad x_1 + 3x_2 = y_2.$$

We solve by elimination, and find

$$x_1 = 3y_1 - 5y_2, \qquad x_2 = -y_1 + 2y_2.$$

Therefore $A^{-1} = \begin{bmatrix} 3 & -5 \\ -1 & 2 \end{bmatrix}$. We check by verifying that $AA^{-1} = I$. ∎

Nonsingular matrices.   A matrix A having an inverse is said to be *nonsingular*. Hence we have shown that A is nonsingular precisely when $\det A \neq 0$. A square matrix having *no* inverse is said to be *singular*.

Now let A have an inverse B, so that $AB = I$. Then as remarked above, also $\det B \neq 0$, so that B also has an inverse B^{-1}, and $BB^{-1} = I$. We can now write

$$BA = BAI = BABB^{-1} = B(AB)B^{-1} = BIB^{-1} = BB^{-1} = I.$$

Therefore, also, $BA = I$. Furthermore, if $AC = I$, then

$$C = IC = BAC = BI = B.$$

This shows that the inverse of A is unique. Furthermore, if $CA = I$, then

$$C = CI = CAB = IB = B.$$

These results can be summarized as follows: *The inverse $B = A^{-1}$ of A is unique and B satisfies the two equations*

$$AB = I \quad \text{and} \quad BA = I;$$

furthermore, if a matrix B satisfies either one of these two equations, then B must satisfy the other equation, and $B = A^{-1}$.

The inverse satisfies several additional rules:

21. $(AD)^{-1} = D^{-1}A^{-1}$.
22. $(cA)^{-1} = c^{-1}A^{-1} \ (c \neq 0)$.
23. $(A^{-1})^{-1} = A$.

Here A and D are assumed to be nonsingular $n \times n$ matrices. To prove Rule 21, we write

$$(AD)(D^{-1}A^{-1}) = A(DD^{-1})A^{-1} = AIA^{-1} = AA^{-1} = I.$$

Therefore $D^{-1}A^{-1}$ must be the inverse of AD. The proof of Rule 22 is left as an exercise (Problem 5 below). For Rule 23 we reason that A^{-1} is nonsingular and hence A^{-1} has an inverse. But $A^{-1}A = I$, so that A is the inverse of A^{-1}; that is, $A = (A^{-1})^{-1}$.

Rule 21 extends to more than two factors, for example,

$$(ABCD)^{-1} = D^{-1}C^{-1}B^{-1}A^{-1}.$$

The proof is as above. In this way we see that the product of two or more nonsingular matrices is nonsingular.

Negative powers of a square matrix.   Let A be nonsingular, so that A^{-1} exists. For each positive integer p we now define A^{-p} to mean $(A^{-1})^p$. Since A is nonsingular, A^p is also nonsingular; in fact, A^p has the inverse

$$(AA \cdots A)^{-1} = A^{-1}A^{-1} \cdots A^{-1} = (A^{-1})^p.$$

Therefore

$$A^{-p} = (A^p)^{-1} \quad \text{or} \quad A^pA^{-p} = I. \tag{1.64}$$

Rules 18 and 19,

$$A^pA^q = A^{p+q}, \tag{1.65}$$

$$(A^p)^q = A^{pq}, \tag{1.66}$$

are now satisfied, for A nonsingular, for arbitrary integers p and q, positive, negative, or zero. These rules are proved for p and q nonnegative in Section 1.7.

To prove (1.65) for general p and q, we first reason that, since

$$(A^{p+q})^{-1} = A^{-p-q},$$

(1.65) is equivalent to $A^pA^qA^{-p-q} = I$. This in turn is equivalent to the statement

(*) $\qquad\qquad A^pA^qA^r = I \quad$ whenever $\quad p + q + r = 0$.

Now for s and t nonnegative we know that $A^sA^t = A^{s+t}$. This implies

$$A^sA^tA^{-s-t} = I \quad \text{and} \quad A^{-s}A^{-t}A^{s+t} = I;$$

and hence

$$A^{-s-t}A^sA^t = I, \qquad A^tA^{-s-t}A^s = I,$$
$$A^{s+t}A^{-s}A^{-t} = I, \qquad A^{-t}A^{s+t}A^{-s} = I.$$

The last six equations state that (*) holds in each of the following cases: $p \geq 0$, $q \geq 0$, $r \leq 0$; $p \leq 0$, $q \leq 0$, $r \geq 0$; $p \leq 0$, $q \geq 0$, $r \geq 0$; $p \geq 0$, $q \leq 0$, $r \geq 0$; $p \geq 0$, $q \leq 0$, $r \leq 0$; $p \leq 0$, $q \geq 0$, $r \leq 0$. These are *all possible cases*. Hence (*) is proved, and (1.65) follows.

We note that, by (1.65), $A^pA^q = A^qA^p$. Hence A^p and A^q commute (that is, obey the commutative law) under multiplication.

The proof of (1.66) is left as an exercise (Problem 6 below).

The procedure of Example 1 for finding the inverse of a matrix can be much improved. The calculation of inverses of matrices is very important in numerical work, and various procedures have been devised for this purpose, especially adapted to digital computers (see Section 1.9).

Inverses can be used to solve matrix equations—for example, equations of the form $AX = B$ or $XA = B$, where A and B are known and X is sought; if A is $n \times n$ and nonsingular, then we find, respectively,

$$X = A^{-1}B \quad \text{and} \quad X = BA^{-1},$$

and verify in each case that the equation is satisfied. As a special case, we solve the equation $Ax = y$ (A being $n \times n$) to obtain $\mathbf{x} = A^{-1}\mathbf{y}$.

Remark. If $k < n$, A is an $n \times k$ matrix, and B is a $k \times n$ matrix, then AB is an $n \times n$ matrix and AB is *singular*. To see this, we complete A and B to $n \times n$ matrices by adding extra columns and rows of zeros. Let A_1 and B_1 be the expanded matrices obtained. Then $AB = A_1B_1$, since the added zeros contribute nothing to the product. But A_1 and B_1 are singular, so that $A_1B_1 = AB$ is singular, since det A_1B_1 is 0 by (1.60).

EXAMPLE 2

$$\begin{bmatrix} 2 & 1 \\ 4 & 1 \\ 5 & 3 \end{bmatrix} \begin{bmatrix} 1 & 2 & 4 \\ 2 & 1 & 5 \end{bmatrix} = \begin{bmatrix} 2 & 1 & 0 \\ 4 & 1 & 0 \\ 5 & 3 & 0 \end{bmatrix} \begin{bmatrix} 1 & 2 & 4 \\ 2 & 1 & 5 \\ 0 & 0 & 0 \end{bmatrix} = \begin{bmatrix} 4 & 5 & 13 \\ 6 & 9 & 21 \\ 11 & 13 & 35 \end{bmatrix}. \quad \blacksquare$$

PROBLEMS

1. Verify that each of the following matrices is nonsingular and find the inverse of each:

a) $A = \begin{bmatrix} 3 & 5 \\ 2 & 4 \end{bmatrix}$.

b) $B = \begin{bmatrix} 4 & 7 \\ 1 & 6 \end{bmatrix}$.

c) $C = \begin{bmatrix} 1 & 0 & 1 \\ 2 & 2 & 1 \\ 0 & 1 & -1 \end{bmatrix}$.

d) $D = \begin{bmatrix} 2 & 0 & 1 \\ 3 & 1 & 2 \\ 4 & 0 & 3 \end{bmatrix}$.

2. Let $A, \ldots, D$ be as in Problem 1. With the aid of the answers to Problem 1, solve for X or x:

a) $AX = \begin{bmatrix} 0 & 1 \\ 1 & 0 \end{bmatrix}$.

b) $AX = \begin{bmatrix} 1 & 5 \\ 6 & 2 \end{bmatrix}$.

c) $BXA = \begin{bmatrix} 7 & 2 \\ 0 & 5 \end{bmatrix}$.

d) $B^2 XA^2 = \begin{bmatrix} 3 & 4 \\ 2 & 1 \end{bmatrix}$.

e) $Ax = \text{col}(4, 3)$.

f) $Bx = \text{col}(1, 7)$.

g) $BX = \begin{bmatrix} 1 & 6 & 2 \\ 5 & 0 & -3 \end{bmatrix}$.

h) $XC = \begin{bmatrix} 0 & 5 & 4 \\ 5 & 0 & 1 \end{bmatrix}$.

i) $xC = (1, 2, 3)$.

j) $xD = (2, 1, 0)$.

3. Simplify:

a) $[(AB)^{-1} A^{-1}]^{-1}$.

b) $(ABC)^{-1} (C^{-1} B^{-1} A^{-1})^{-1}$.

c) $\{[(A^{-1})^2 B]^{-1} A^{-2} B^{-1}\}^{-2}$.

4. Prove:

a) If A and B are nonsingular and $AB = BA$, then $A^{-1} B^{-1} = B^{-1} A^{-1}$.

b) If $ABC = I$, then $BCA = I$ and $CAB = I$.

5. Prove Rule 22.

6. Prove the rule (1.66) for arbitrary integers p, q. [Hint: For $s \geq 0, t \geq 0$, show by (1.64) that $(A^s)^{-t} = (A^{-s})^t = A^{-st}$ and that $(A^{-s})^{-t} = A^{st}$.]

7. Let $A = (a_{ij})$ be a nonsingular square matrix of order n. Prove:

a) If $n = 2$, then $A^{-1} = \dfrac{1}{\det A} \begin{bmatrix} a_{22} & -a_{12} \\ -a_{21} & a_{11} \end{bmatrix}$.

b) If $n = 3$, then A^{-1} is the matrix

$$\frac{1}{\det A} \begin{bmatrix} \begin{vmatrix} a_{22} & a_{23} \\ a_{32} & a_{33} \end{vmatrix} & \begin{vmatrix} a_{13} & a_{12} \\ a_{33} & a_{32} \end{vmatrix} & \begin{vmatrix} a_{12} & a_{13} \\ a_{22} & a_{23} \end{vmatrix} \\ \begin{vmatrix} a_{23} & a_{21} \\ a_{33} & a_{31} \end{vmatrix} & \begin{vmatrix} a_{11} & a_{13} \\ a_{31} & a_{33} \end{vmatrix} & \begin{vmatrix} a_{13} & a_{11} \\ a_{23} & a_{21} \end{vmatrix} \\ \begin{vmatrix} a_{21} & a_{22} \\ a_{31} & a_{32} \end{vmatrix} & \begin{vmatrix} a_{12} & a_{11} \\ a_{32} & a_{31} \end{vmatrix} & \begin{vmatrix} a_{11} & a_{12} \\ a_{21} & a_{22} \end{vmatrix} \end{bmatrix}.$$

c) Let A_{ij} denote the minor determinant of A obtained by deleting the ith row and jth column from A. Let $b_{ij} = (-1)^{i+j} A_{ji}$. The matrix $B = (b_{ij})$ is called the *adjoint* of A and is denoted by adj A. Show that

$$A^{-1} = \frac{1}{\det A} \text{ adj } A.$$

8. Let all matrices occurring in the following equations be *square* of order n. Solve for X and Y, stating which matrices are assumed to be nonsingular:

a) $X + Y = A, X - Y = B$.

b) $X + Y = A, X + BY = C$.

c) $X + AY = B, X + CY = D$.

d) $AX + BY = C, DX + EY = F$.

e) $XA + YB = C, XD + YE = F$.

9. a) Prove the rule (1.60) for 2×2 matrices; that is, prove:

$$\begin{vmatrix} a_{11} & a_{12} \\ a_{21} & a_{22} \end{vmatrix} \begin{vmatrix} b_{11} & b_{12} \\ b_{21} & b_{22} \end{vmatrix} = \begin{vmatrix} a_{11}b_{11} + a_{12}b_{21} & a_{11}b_{12} + a_{12}b_{22} \\ a_{21}b_{11} + a_{22}b_{21} & a_{21}b_{12} + a_{22}b_{22} \end{vmatrix}.$$

[Hint: Use Rule V of Section 1.3 for adding determinants to write the right-hand side as a sum of four determinants. Then factor $b_{11},\ldots$ from these and show that two terms are 0 and the other two have sum

$$\begin{vmatrix} a_{11} & a_{12} \\ a_{21} & a_{22} \end{vmatrix} (b_{11}b_{22} - b_{21}b_{12}).]$$

b) Prove the rule (1.60) for 3×3 matrices.

c) Prove the rule (1.60) for $n \times n$ matrices.

10. Let $A = \text{col}(u_1, \ldots, u_n)$ (column vector) and $B = (v_1, \ldots, v_n)$ (row vector); let $n \geq 2$.

a) Show directly that the $n \times n$ matrix AB is singular (cf. the Remark at the end of Section 1.8).

b) Show that BA is 1×1—that is, a scalar.

11. (*Permutation matrices*) Let P_{ij} be the $n \times n$ matrix obtained from I by interchanging the ith and jth rows ($i < j$). Thus for $n = 3$,

$$P_{13} = \begin{bmatrix} 0 & 0 & 1 \\ 0 & 1 & 0 \\ 1 & 0 & 0 \end{bmatrix}.$$

a) Show that $P_{ij}^2 = I$.

b) Show that P_{ij} can be obtained from I by interchanging the ith and jth columns of I.

c) Show that, if A is an $n \times n$ matrix, then $P_{ij}A$ is obtained from A by interchanging the ith and jth rows of A, and AP_{ij} is obtained from A by interchanging the ith and jth columns of A.

d) Show that P_{ij} is nonsingular.

1.9 GAUSSIAN ELIMINATION

For solving simultaneous linear equation, finding inverses of matrices and evaluating determinants, a number of practical techniques have been developed and implemented in computer software. We here summarize the standard Gauss and Gauss-Jordan procedures and refer for details to the book of Gerald and Wheatley and that of Perlis, listed at the end of the chapter.

Gauss procedure. The system (1.44) of Section 1.4 is replaced by the matrix equation

$$Ax = k, \tag{1.67}$$

where $A = (a_{ij})$ and $\mathbf{k} = \text{col}(k_1, \ldots, k_n)$. One forms the "augmented matrix"

$$B = [A \quad \mathbf{k}] \tag{1.68}$$

formed from A by adjoining $\mathbf{k}$ as $(n+1)$-st column. One operates on the matrix B by

I. adding multiples of one row to another row,

II. interchanging two rows.

When A is nonsingular, suitable applications of these steps replace B by a new matrix

$$C = [D \quad \mathbf{h}], \tag{1.69}$$

obtained by augmentation from an *upper triangular* matrix D—that is, a matrix with 0's below the principal diagonal. The steps described correspond to replacement of the given simultaneous equations (1.44) by new ones with *the same solutions*. Since D is upper triangular, the new equations have the form

$$d_{11}x_1 + \cdots + d_{1n}x_n = h_1, \quad d_{22}x_2 + \cdots + d_{2n}x_n = h_2, \ldots, \quad d_{nn}x_n = h_n \tag{1.70}$$

and are easily solved for $x_n, x_{n-1}, \ldots, x_1$ in turn. Thus we have found the unique solution.

We observe that by Rules II. and VI. of Section 1.3,

$$\det D = (-1)^\ell \det A, \tag{1.71}$$

where ℓ is the number of row interchanges. Thus, for nonsingular A, $\det A$ and $\det D$ are both nonzero. By Problem 16 following Section 1.3, $\det D = d_{11}d_{22}\ldots d_{nn} \neq 0, \ldots$, as is needed to solve (1.70), and by (1.71) one can at once obtain $\det A$.

Gauss-Jordan procedure. In addition to steps I. and II., we use a step:

III. multiplying a row by a nonzero scalar.

By applications of these three steps we can obtain from the matrix B a new matrix

$$E = [I \quad \mathbf{p}], \tag{1.72}$$

and the new equations (still having the same solutions) have the form

$$x_1 = p_1, \quad x_2 = p_2, \ldots, \quad x_n = p_n; \tag{1.73}$$

we have again obtained the unique solution.

The inverse of A can be obtained by the same procedure, applied to the matrix

$$F = [A \quad I], \tag{1.74}$$

obtained from A by adjoining to it the successive columns of the identity matrix of order n. The steps described can be applied to F to reduce it to the form

$$G = [I \quad H], \tag{1.75}$$

and one verifies that $H = A^{-1}$ (Problem 9 below).

Singular case.   When A is singular, one cannot proceed from (1.68) to (1.72). However, one can always use steps I. and II. to reduce B to the form

$$J = [K \ \mathbf{c}]. \tag{1.76}$$

where K is an $n \times n$ matrix in "row echelon form." By this we mean that in each row the first nonzero entry (the "leading entry") has column index which increases as the row index increases. An example is

$$\begin{bmatrix} 0 & 3 & 5 & 1 \\ 0 & 0 & 2 & 5 \\ 0 & 0 & 0 & 4 \\ 0 & 0 & 0 & 0 \end{bmatrix} \tag{1.77}$$

In the singular case, the first r rows ($0 \le r \le n-1$) contain nonzero entries, whereas the last $n - r$ rows are all 0; here r is the *rank* of the matrix. In the nonsingular case $r = n$.

By steps I., II. and III. one can in fact achieve a form (1.76) in which K has row echelon form and each leading entry is 1 and is the only nonzero entry in its column. The first r of the corresponding linear equations can now be solved for r of the unknowns in terms of the remaining $n - r$ unknowns, which can be chosen arbitrarily. These yield infinitely many solutions, provided that the remaining $n - r$ equations contain no contradictions. Since the last $n - r$ rows of K are zero, one must have $c_{r+1} = c_{r+2} = \cdots = c_n = 0$. Otherwise there is no solution.

If the given simultaneous equations are homogenous, then the steps described preserve that property, so that $\mathbf{c}$ must be $\mathbf{0}$ and no contradiction can arise. Thus *if* $\det A = 0$, *so that A is singular, the simultaneous equations $A\mathbf{x} = \mathbf{0}$ have infinitely many solutions.*

In the extreme case $r = 0$, A must be the zero matrix and there is no solution if $\mathbf{k} \ne \mathbf{0}$, all $\mathbf{x}$ are solutions if $\mathbf{k} = \mathbf{0}$.

PROBLEMS

1. Solve by Gaussian elimination:

a) $x + y - z = 1, \quad 2x + y + z = 5, \quad 2x + y = 0$

b) $x + 2y + z = 4, \quad 2x - y + z = 4, \quad x - y + 2z = 0$

c) $y + 2z = 2, \quad x + y - z = 3, \quad 2x + z = 4$

d) $2x + y + 3z = 6, \quad 2x + y + 5z = 8, \quad -2x + 3y - 4z = -3$

2. Evaluate the determinant as in Gaussian elimination:

a) $\begin{vmatrix} 1 & 2 & 1 \\ 2 & 1 & 5 \\ 3 & 0 & 2 \end{vmatrix}$

b) $\begin{vmatrix} 1 & 2 & 1 & 2 \\ 3 & 1 & 3 & 1 \\ 1 & 2 & 2 & 1 \\ 2 & 1 & 1 & 2 \end{vmatrix}$

3. Solve by Gauss-Jordan elimination:

   a) $2x + y - z = 3$, $x + y - z = 1$, $x - 2y + 2z = 4$

   b) $x + y - z + w = 5$, $x - y - z + 2w = 4$,
   $2x - y + z - w = 7$, $x + 3y - z - w = 5$

4. Find the inverse matrix:

   a) $\begin{bmatrix} 1 & 2 & 3 \\ 2 & 0 & -1 \\ 4 & 2 & 2 \end{bmatrix}$
   b) $\begin{bmatrix} 1 & -1 & 2 \\ 2 & 0 & 2 \\ 3 & 1 & 3 \end{bmatrix}$
   c) $\begin{bmatrix} 1 & 2 & 1 & 2 \\ 1 & 2 & 2 & 1 \\ 2 & 1 & 1 & 0 \\ 1 & 0 & 1 & 1 \end{bmatrix}$

5. In the following systems the elimination process shows that the determinant of coefficients is 0. Carry out the process and find all solutions:

   a) $2x - y + z = 3$, $x + 2y - z = 1$, $5x - 5y + 4z = 8$

   b) $x + y - z = 1$, $2x - y - z = 2$, $x + 4y - 2z = 2$

   c) $x - y + z + w = 0$, $x + 2y - z - w = 0$,
   $3x - y - z + 2w = 0$, $x + 3y + z - 2w = 0$

   d) $x + y - 2z + 3w = 0$, $2x - y + z - w = 0$,
   $3x - z + 2w = 0$, $5x + 2y - 5z + 8w = 0$

6. Consider two equations in three unknowns:

   $$a_1 x + b_1 y + c_1 z = k_1, \qquad a_2 x + b_2 y + c_2 z = k_2.$$

   a) Show that if Gaussian elimination can be carried out to solve for x and y, then by writing $z = t$ the solutions became parametric equations for a line in space (Section 1.2).

   b) Assume that the two equations represent two planes in space. Interpret geometrically the case in which the equations have no solution and the case in which elimination leads to a second equation $0 = 0$.

7. Consider four equations in three unknowns:

   $$a_i x + b_i y + c_i z = k_i \qquad (i = 1, \ldots, 4).$$

   Assume that the pair of equations for $i = 1, 2$ and the pair for $i = 3, 4$ each represents a line in space as in Problem 6(a). Thus the solutions of the four equations represent the points common to two lines in space. Discuss the geometrical alternatives that can occur and relate them to the set of solutions of the given equations.

8. Let A be an $m \times n$ matrix. Let $C = (c_{ij})$ be the $m \times m$ matrix which differs from the identity only in that $c_{hh} = \lambda \neq 0$. Let B be the $m \times m$ matrix which differs from the identity only in the k-th column, in which $b_{kk} = 1$.

   a) Show that CA is obtained from A by multiplying the h-th row by λ.

   b) Show that BA is obtained from A by adding b_{jk} times the k-th row to the j-th row for all j except k.

   c) Show that B and C are nonsingular.

Remark. These results show that steps I. and III. of Section 1.9 are obtained by multiplying A on the left by a nonsingular matrix. The same assertion applies to step II. (Problem 11 following Section 1.8)

9. Let $AH = I$ and let H have column vectors $\mathbf{k}_1, \ldots, \mathbf{k}_n$, so that $A\mathbf{k}_j$ equals $\mathbf{q}_j$, the j-th column of I. Show that the Gauss-Jordan process of going from (1.74) to (1.75) is equivalent to solving the n systems $A\mathbf{k}_j = \mathbf{q}_j$ to obtain H.

*1.10   EIGENVALUES OF A SQUARE MATRIX

Let A be an $n \times n$ matrix. For some *nonzero* column vector $\mathbf{v} = \text{col}(v_1, \ldots, v_n)$ it may happen that, for some scalar λ,

$$A\mathbf{v} = \lambda\mathbf{v}. \tag{1.78}$$

If this occurs, we say that λ is an *eigenvalue* of A and that $\mathbf{v}$ is an *eigenvector* of A, associated with the eigenvalue λ. The concept of eigenvalue has important applications in many branches of physics. An important example is the *spectrum* —of light, of an atom, of a nucleus. The frequencies occurring in the spectrum correspond to the eigenvalues of a matrix (or of a suitable generalization of a matrix). Eigenvalues are also of importance in the solution of linear differential equations (see Chapter 10).

We can write Eq. (1.78) in the form $A\mathbf{v} = \lambda I\mathbf{v}$ or in the form

$$(A - \lambda I)\mathbf{v} = \mathbf{0}.$$

Thus $\mathbf{v} = \text{col}(v_1, \ldots, v_n)$ is an eigenvector of A associated with the eigenvalue λ precisely when $v_1, \ldots, v_n$ form a nontrivial solution of the set of homogeneous linear equations

$$
\begin{aligned}
(a_{11} - \lambda)v_1 + a_{12}v_2 + &\quad \cdots \quad + &a_{1n}v_n &= 0, \\
a_{21}v_1 + (a_{22} - \lambda)v_2 + &\quad \cdots \quad + &a_{2n}v_n &= 0, \\
&\;\;\vdots & &\;\;\vdots \\
a_{n1}v_1 + a_{n2}v_2 + &\quad \cdots \quad +(a_{nn} - \lambda)v_n &= 0.
\end{aligned}
\tag{1.79}
$$

Now we know (Section 1.9) that Eqs. (1.79) have a nontrivial solution precisely when the determinant of the coefficients is 0, that is, when $\det(A - \lambda I) = 0$ or

$$
\begin{vmatrix}
a_{11} - \lambda & a_{12} & \cdots & a_{1n} \\
a_{21} & a_{22} - \lambda & \cdots & a_{2n} \\
\vdots & \vdots & & \vdots \\
a_{n1} & a_{n2} & \cdots & a_{nn} - \lambda
\end{vmatrix} = 0.
\tag{1.80}
$$

When expanded, (1.80) becomes an algebraic equation of degree n for λ, called the *characteristic equation* of the matrix A. The eigenvalues of A are simply the real roots of the characteristic equation. (The complex roots of the characteristic equation can also be interpreted as eigenvalues of A; see the end of this section).

EXAMPLE 1   Let $A = \begin{bmatrix} 1 & 2 \\ 3 & 2 \end{bmatrix}$. Then the characteristic equation is

$$\begin{vmatrix} 1 - \lambda & 2 \\ 3 & 2 - \lambda \end{vmatrix} = 0 \quad \text{or} \quad \lambda^2 - 3\lambda - 4 = 0 \quad \text{or} \quad (\lambda - 4)(\lambda + 1) = 0.$$

The roots are $\lambda_1 = 4$ and $\lambda_2 = -1$. To find $\mathbf{v}$ for $\lambda = \lambda_1 = 4$, we form the equations

$$(1 - 4)v_1 + 2v_2 = 0, \quad 3v_1 + (2 - 4)v_2 = 0,$$

and find that the solutions are all vectors $k(2, 3)$. Hence the eigenvectors associated with the eigenvalue $\lambda_1 = 4$ are all vectors $k(2, 3)$ for nonzero k. In the same way we find that all eigenvectors associated with the eigenvalue $\lambda_2 = -1$ are all vectors $k(1, -1)$ for nonzero k.   ∎

EXAMPLE 2   Let

$$B = (\lambda_i \delta_{ij}) = \begin{bmatrix} \lambda_1 & 0 & \cdots & 0 \\ \vdots & \vdots & & \vdots \\ 0 & 0 & \cdots & \lambda_n \end{bmatrix}.$$

We call B a *diagonal* matrix and write $B = \text{diag}(\lambda_1, \dots, \lambda_n)$. Then B has the characteristic equation

$$\begin{vmatrix} \lambda_1 - \lambda & \cdots & 0 \\ & \vdots & \\ 0 & \cdots & \lambda_n - \lambda \end{vmatrix} = 0 \quad \text{or} \quad (\lambda_1 - \lambda)(\lambda_2 - \lambda) \cdots (\lambda_n - \lambda) = 0.$$

Hence B has the eigenvalues $\lambda_1, \dots, \lambda_n$. We leave to Problem 5 below the discussion of the eigenvectors of B. We remark here that when $\lambda_1, \dots, \lambda_n$ are n different numbers, the eigenvectors associated with the eigenvalue λ_k are the vectors $c(0, \dots, 0, 1, 0, \dots, 0)$, with 1 as kth entry and c nonzero.   ∎

Similar matrices.   Let A and B be $n \times n$ matrices. We say that B is *similar* to A if

$$B = C^{-1}AC$$

for some nonsingular $n \times n$ matrix C. If this holds, then

$$A = CBC^{-1} = (C^{-1})^{-1}BC^{-1},$$

so that A is also similar to B. Hence we speak of similar matrices A, B.
   If A and B are similar, then A and B have the same characteristic equation. For

$$\det(B - \lambda I) = \det(C^{-1}AC - \lambda I) = \det(C^{-1}AC - \lambda C^{-1}IC)$$

$$= \det C^{-1}(A - \lambda I)C$$

$$= \det C^{-1} \det(A - \lambda I) \det C = \det(A - \lambda I),$$

since $\det C^{-1} \det C = 1$. It follows also that A and B have the same eigenvalues.

Matrices with n distinct real eigenvalues.   Let the $n \times n$ matrix A have n distinct (real) eigenvalues $\lambda_1, \ldots, \lambda_n$ and let $\mathbf{v}_1, \ldots, \mathbf{v}_n$ be eigenvectors associated with $\lambda_1, \ldots, \lambda_n$, respectively: $A\mathbf{v}_i = \lambda_i \mathbf{v}_i,\ i = 1, \ldots, n$. Now let C be the matrix whose column vectors are $\mathbf{v}_1, \ldots, \mathbf{v}_n$, respectively. Write $\mathbf{v}_j = \mathrm{col}\,(v_{1j}, \ldots, v_{nj})$ for $j = 1, \ldots, n$. Then

$$
AC = A \begin{bmatrix} v_{11} & \cdots & v_{1n} \\ \vdots & & \vdots \\ v_{n1} & \cdots & v_{nn} \end{bmatrix} = \begin{bmatrix} \lambda_1 v_{11} & \cdots & \lambda_n v_{1n} \\ \vdots & & \vdots \\ \lambda_1 v_{n1} & \cdots & \lambda_n v_{nn} \end{bmatrix}
$$

$$
= \begin{bmatrix} v_{11} & \cdots & v_{1n} \\ \vdots & & \vdots \\ v_{n1} & \cdots & v_{nn} \end{bmatrix} \begin{bmatrix} \lambda_1 & \cdots & 0 \\ \vdots & & \vdots \\ 0 & \cdots & \lambda_n \end{bmatrix} = CB,
$$

where B is a diagonal matrix, $B = \mathrm{diag}\,(\lambda_1, \ldots, \lambda_n)$. It can be shown that C must be nonsingular (see Problem 12 following Section 1.15 below). Hence $B = C^{-1}AC$, and A *is similar to the diagonal matrix* $\mathrm{diag}\,(\lambda_1, \ldots, \lambda_n)$.

EXAMPLE 3

$$
A = \begin{bmatrix} 1 & 2 & 2 \\ 2 & 3 & -2 \\ -5 & 3 & 8 \end{bmatrix}.
$$

Here A has the characteristic equation

$$
\begin{vmatrix} 1-\lambda & 2 & 2 \\ 2 & 3-\lambda & -2 \\ -5 & 3 & 8-\lambda \end{vmatrix} = 0 \quad \text{or} \quad (\lambda - 3)(\lambda - 4)(\lambda - 5) = 0.
$$

For $\lambda = 3$ the eigenvector $\mathbf{v} = (v_1, v_2, v_3)$ must satisfy the equations

$$
-2v_1 + 2v_2 + 2v_3 = 0, \qquad 2v_1 + 0v_2 - 2v_3 = 0, \qquad -5v_1 + 3v_2 + 5v_3 = 0.
$$

Hence $\mathbf{v}_1 = (1, 0, 1)$ is an associated eigenvector. Similarly, for $\lambda_2 = 4$ an associated eigenvector is $(2, 2, 1)$, and for $\lambda_3 = 5$ an associated eigenvector is $(0, 1, -1)$. Hence $C^{-1}AC = B$, with

$$
C = \begin{bmatrix} 1 & 2 & 0 \\ 0 & 2 & 1 \\ 1 & 1 & -1 \end{bmatrix}, \qquad B = \begin{bmatrix} 3 & 0 & 0 \\ 0 & 4 & 0 \\ 0 & 0 & 5 \end{bmatrix} = \mathrm{diag}\,(3, 4, 5).
$$

We check by verifying that $AC = CB$. Thus A is similar to the diagonal matrix B.   ■

In general, some of the eigenvalues may coincide (multiple roots of the characteristic equation), and there may be less than n real roots of the characteristic equation (1.80). To handle the general case, it is best to extend the theory of matrices to the case of *complex matrices*, whose elements are complex numbers. All concepts generalize easily to this case, and, in particular, one can define complex eigenvalues and complex eigenvectors as above. For example, the matrix

$$\begin{bmatrix} 1 & -2 \\ 1 & -1 \end{bmatrix}$$

has the characteristic equation $\lambda^2 + 1 = 0$. The eigenvalues are $\pm i$, and associated eigenvectors are $(2, 1 \mp i)$. One proves, exactly as above, that when the matrix A has n distinct complex eigenvalues, A is similar to a diagonal matrix.

PROBLEMS

1. Find all eigenvalues and associated eigenvectors:

   a) $\begin{bmatrix} 3 & 1 \\ 4 & 3 \end{bmatrix}$,   b) $\begin{bmatrix} 1 & 3 \\ 2 & 6 \end{bmatrix}$,   c) $\begin{bmatrix} 0 & 1 & -2 \\ 2 & 1 & 0 \\ 4 & -2 & 5 \end{bmatrix}$,   d) $\begin{bmatrix} 5 & -2 & 8 \\ -4 & 0 & -5 \\ -4 & 2 & -7 \end{bmatrix}$.

2. (a), (b), (c), (d) For each of the matrices of Problem 1, call the matrix A and find a nonsingular matrix C and a diagonal matrix B such that $B = C^{-1}AC$.

3. (*Complex case*)   For each of the following choices of matrix A, find all eigenvalues and associated eigenvectors, and find a diagonal matrix B such that $B = C^{-1}AC$:

   a) $\begin{bmatrix} 1 & -1 \\ 4 & 1 \end{bmatrix}$,   b) $\begin{bmatrix} 3 & -2 \\ 1 & 5 \end{bmatrix}$,   c) $\begin{bmatrix} 0 & 0 & 0 \\ 0 & 1 & -2 \\ 0 & 1 & -1 \end{bmatrix}$.

4. (*Repeated roots*)   Find all eigenvalues and associated eigenvectors:

   a) I_3,   b) O_{44},   c) $\begin{bmatrix} 3 & -4 \\ 4 & -5 \end{bmatrix}$,   d) $\begin{bmatrix} 0 & 1 & -2 \\ -6 & 5 & -4 \\ 0 & 0 & 3 \end{bmatrix}$.

5. (*Diagonal matrices*)

   a) Let $B = \text{diag}(\lambda, \mu)$. Show that, if $\lambda \neq \mu$, then the eigenvectors associated with λ are all nonzero vectors $c(1, 0)$ and those associated with μ are all nonzero vectors $c(0, 1)$; show that, if $\lambda = \mu$, then the eigenvectors associated with λ are all nonzero vectors (v_1, v_2).

   b) Let $B = \text{diag}(\lambda_1, \lambda_2, \lambda_3)$ and let $e_1 = (1, 0, 0)$, $e_2 = (0, 1, 0)$, $e_3 = (0, 0, 1)$ (column vectors). Show that if $\lambda_1, \lambda_2, \lambda_3$ are distinct, then for each λ_k the associated eigenvectors are the vectors ce_k for $c \neq 0$; show that if $\lambda_1 = \lambda_2 \neq \lambda_3$, then the eigenvectors associated with λ_1 are all nonzero vectors $c_1 e_1 + c_2 e_2$ and those associated with λ_3 are all nonzero vectors ce_3; show that if $\lambda_1 = \lambda_2 = \lambda_3$, then the eigenvectors associated with λ_1 are all nonzero vectors $v = (v_1, v_2, v_3)$.

   c) Let $B = \text{diag}(\lambda_1, \ldots, \lambda_n)$. Show that the eigenvectors associated with the eigenvalue λ_k are all nonzero vectors $v = (v_1, \ldots, v_n)$ such that $v_i = 0$ for all i such that $\lambda_i \neq \lambda_k$.

6. Let A and B be similar matrices. Let $\lambda_1, \ldots, \lambda_n$ be the (not necessarily distinct) eigenvalues of A.

   a) Prove that $\lambda_1 \lambda_2 \cdots \lambda_n = \det A$.

b) Conclude from the result of (a) that $\det A = \det B$.

c) Prove that $\lambda_1 + \cdots + \lambda_n = a_{11} + \cdots + a_{nn}$. The number $a_{11} + \cdots + a_{nn}$ is called the *trace* of A.

d) Prove from the result of (c) that A and B have equal traces.

7. Prove that the matrix $A = \begin{bmatrix} 1 & 1 \\ 0 & 1 \end{bmatrix}$ is not similar to a diagonal matrix. [Hint: Show by the results of Problem 6 that the diagonal matrix would have to be I.]

8. Prove the following:

a) Every square matrix is similar to itself.

b) If A is similar to B and B is similar to C, then A is similar to C.

*1.11   THE TRANSPOSE

Let $\mathbf{A} = (a_{ij})$ be an $m \times n$ matrix. We denote by A' the $n \times m$ matrix $B = (b_{ij})$ such that $b_{ij} = a_{ji}$ for $i = 1, \ldots, n, j = 1, \ldots, m$. Thus $B = A'$ is obtained from A by interchanging rows and columns. The following pair is an illustration:

$$A = \begin{bmatrix} 3 & 1 & 2 \\ 5 & 0 & 7 \end{bmatrix}, \qquad A' = \begin{bmatrix} 3 & 5 \\ 1 & 0 \\ 2 & 7 \end{bmatrix}.$$

The first row of A becomes the first column of A'; the second row of A becomes the second column of A'. In general, we call A' the *transpose* of A. We observe that $I' = I$. The transpose of a matrix obeys several rules, which we adjoin to our list:

24. $(A + B)' = A' + B'$.

25. $(cA)' = cA'$.

26. $(A')' = A$.

27. $(AB)' = B'A'$.

28. If A is nonsingular, then $(A^{-1})' = (A')^{-1}$.   (1.81)

To prove Rule 24, we write $D = (A + B) = (d_{ij})$, so that

$$d_{ij} = a_{ij} + b_{ij}$$

for all i and j. Then $D' = E = (e_{ij})$, where $e_{ij} = d_{ji}$ for all i and j, or

$$e_{ij} = a_{ji} + b_{ji}.$$

Thus $E = A' + B'$ or $D' = A' + B'$. The proofs of Rules 25 and 26 are left as exercises (Problem 4 below).

To prove Rule 27, we let A be $m \times p$, B be $p \times n$. We then write $C = AB = (c_{ij})$, $D = A' = (d_{ij})$, $E = B' = (e_{ij})$. Then

$$B'A' = ED = F = (f_{ij}),$$

where

$$f_{ij} = e_{i1}d_{1j} + \cdots + e_{ip}d_{pj} = b_{1i}a_{j1} + \cdots + b_{pi}a_{jp}$$

$$= a_{j1}b_{1i} + \cdots + a_{jp}b_{pj} = c_{ji} \quad (i = 1,\ldots,n,\; j = 1,\ldots,m).$$

Hence $F = C'$ or $B'A' = (AB)'$.

To prove Rule 28, we write $AA^{-1} = I$. Then by Rule 27,

$$(A^{-1})'A' = I' = I.$$

From this equation it follows, as in Section 1.8, that $(A')^{-1} = (A^{-1})'$.

A matrix A such that $A = A'$ is called a *symmetric* matrix. Here A must be a square matrix. The matrix I is symmetric, as are the following matrices:

$$\begin{bmatrix} 1 & 2 \\ 2 & 3 \end{bmatrix}, \qquad \begin{bmatrix} 3 & -1 & 0 \\ -1 & 7 & 2 \\ 0 & 2 & 4 \end{bmatrix}.$$

Also, every diagonal matrix is symmetric.

Symmetric matrices are useful in discussing quadratic forms, that is, algebraic expressions of the form

$$\sum_{i=1}^{n} \sum_{j=1}^{n} a_{ij}x_i x_j. \tag{1.82}$$

For $n = 2$ the expression is

$$a_{11}x_1^2 + a_{12}x_1x_2 + a_{21}x_2x_1 + a_{22}x_2^2.$$

Here x_1x_2 is the same as x_2x_1, so we could combine the second and third terms. However, it is preferable to split the combined term into two equal terms, each having as coefficient the average of a_{12} and a_{21}. For example, $3x_1^2 + 5x_1x_2 + 7x_2x_1 + 4x_2^2$ is replaced by

$$3x_1^2 + 6x_1x_2 + 6x_2x_1 + 4x_2^2. \tag{1.83}$$

By proceeding similarly for the general quadratic form (1.82) we can always assume that the *coefficient matrix* (a_{ij}) is symmetric, and it is standard practice to write quadratic forms in this way. For the example just considered, this matrix is $\begin{bmatrix} 3 & 6 \\ 6 & 4 \end{bmatrix}$.

Now let the $n \times n$ symmetric matrix $A = (a_{ij})$ be given, and consider the quadratic form (1.82). Here we can consider $x_1,\ldots,x_n$ as variables. For each assignment of numerical values to $x_1,\ldots,x_n$ the form (1.82) has a numerical value Q. Hence Q is a function of the n variables $x_1,\ldots,x_n$. However, it is simpler to think of Q as a function of the vector $\mathbf{x} = \text{col}(x_1,\ldots,x_n)$:

$$Q(\mathbf{x}) = \sum_{i=1}^{n} \sum_{j=1}^{n} a_{ij}x_i x_j. \tag{1.84}$$

Furthermore, for each $\mathbf{x}$ we can compute the number $Q(\mathbf{x})$ by matrix multiplications:

$$Q(\mathbf{x}) = \mathbf{x}'A\mathbf{x}. \tag{1.85}$$

Here $\mathbf{x}'$ is the transpose of the column vector $\mathbf{x}$ and is therefore the row vector $(x_1,\ldots,x_n)$. Accordingly, (1.85) is the same as

$$Q(\mathbf{x}) = (x_1,\ldots,x_n) \begin{bmatrix} a_{11} & \cdots & a_{1n} \\ \vdots & & \vdots \\ a_{n1} & \cdots & a_{nn} \end{bmatrix} \begin{bmatrix} x_1 \\ \vdots \\ x_n \end{bmatrix}.$$

The product of the last two factors is an $n \times 1$ column vector whose ith entry is $a_{i1}x_1 + \cdots + a_{in}x_n$. The product of the $1 \times n$ row vector $(x_1,\ldots,x_n)$ and this $n \times 1$ column vector is a 1×1 matrix—that is, a number; in fact, it is precisely the number on the right-hand side of (1.84). Therefore (1.85) is indeed another way of writing (1.84).

As an illustration, we write the quadratic form $Q(\mathbf{x})$ of (1.83) as follows:

$$Q(\mathbf{x}) = \mathbf{x}' \begin{bmatrix} 3 & 6 \\ 6 & 4 \end{bmatrix} \mathbf{x} \qquad (\mathbf{x} = \mathrm{col}\,(x_1,\, x_2)).$$

When expanded, this becomes

$$Q(\mathbf{x}) = (x_1,\, x_2) \begin{bmatrix} 3 & 6 \\ 6 & 4 \end{bmatrix} \begin{bmatrix} x_1 \\ x_2 \end{bmatrix} = (x_1,\, x_2) \begin{bmatrix} 3x_1 + 6x_2 \\ 6x_1 + 4x_2 \end{bmatrix}$$

$$= x_1(3x_1 + 6x_2) + x_2(6x_1 + 4x_2) = 3x_1^2 + 6x_1x_2 + 6x_2x_1 + 4x_2^2,$$

as expected.

*1.12   ORTHOGONAL MATRICES

Let A be a real $n \times n$ matrix. Then A is said to be *orthogonal* if

$$AA' = I. \tag{1.86}$$

Hence, A is orthogonal if and only if $A^{-1} = A'$, that is, if and only if the inverse of A equals the transpose of A. Thus every orthogonal matrix is nonsingular. The following are examples of orthogonal matrices:

$$A = \begin{bmatrix} \frac{3}{5} & \frac{4}{5} \\ -\frac{4}{5} & \frac{3}{5} \end{bmatrix}, \qquad B = \begin{bmatrix} \frac{2}{3} & \frac{2}{3} & \frac{1}{3} \\ \frac{2}{3} & -\frac{1}{3} & -\frac{2}{3} \\ \frac{1}{3} & -\frac{2}{3} & \frac{2}{3} \end{bmatrix}.$$

Let us consider the row vectors $\mathbf{u}_1, \mathbf{u}_2$ of A as vectors in the xy-plane:

$$\mathbf{u}_1 = \tfrac{3}{5}\mathbf{i} + \tfrac{4}{5}\mathbf{j}, \qquad \mathbf{u}_2 = -\tfrac{4}{5}\mathbf{i} + \tfrac{3}{5}\mathbf{j}.$$

Then we observe that $\mathbf{u}_1$ and $\mathbf{u}_2$ are both unit vectors and that $\mathbf{u}_1 \cdot \mathbf{u}_2 = 0$, so that $\mathbf{u}_1, \mathbf{u}_2$ are perpendicular. A similar statement applies to the column vectors of A:

$$\mathbf{v}_1 = \tfrac{3}{5}\mathbf{i} - \tfrac{4}{5}\mathbf{j}, \qquad \mathbf{v}_2 = \tfrac{4}{5}\mathbf{i} + \tfrac{3}{5}\mathbf{j}.$$

We can proceed similarly with the row vectors or column vectors of B, regarding them as vectors in space: $\mathbf{u}_1 = \frac{2}{3}\mathbf{i} + \frac{2}{3}\mathbf{j} + \frac{1}{3}\mathbf{k}, \ldots$. Again we verify that the row vectors, or the column vectors, are mutually perpendicular unit vectors.

The geometrical concepts used here can be generalized to n dimensions (Section 1.13). Here we phrase them algebraically. For an $n \times n$ matrix $A = (a_{ij})$ the crucial conditions are as follows:

$$a_{i1}^2 + \cdots + a_{in}^2 = 1, \qquad i = 1, \ldots, n, \tag{1.87}$$

$$a_{i1}a_{j1} + \cdots + a_{in}a_{jn} = 0, \qquad i \neq j, i = 1, \ldots, n, \, j = 1, \ldots, n, \tag{1.88}$$

$$a_{1j}^2 + \cdots + a_{nj}^2 = 1, \qquad j = 1, \ldots, n, \tag{1.89}$$

$$a_{1j}a_{1k} + \cdots + a_{nj}a_{nk} = 0, \qquad j \neq k, \, j = 1, \ldots, n, \, k = 1, \ldots, n. \tag{1.90}$$

(Here (1.87) states that the row vectors are unit vectors, (1.88) states that different row vectors are orthogonal, and (1.89) and (1.90) express the analogous conditions on the column vectors.)

THEOREM

a) Let A be an $n \times n$ orthogonal matrix. Then conditions (1.87) through (1.90) all hold.

b) If A is an $n \times n$ matrix such that (1.87) and (1.88) hold or such that (1.89) and (1.90) hold, then A is orthogonal.

Proof

a) Let A be orthogonal, so that (1.86) holds. By the definition of matrix multiplication, (1.86) asserts that

$$a_{i1}a_{j1} + \cdots + a_{in}a_{jn} = \delta_{ij} = \begin{cases} 1, & i = j, \\ 0, & i \neq j. \end{cases} \tag{1.91}$$

Hence (1.87) and (1.88) follow. From (1.86) we have also, by the properties of inverses (Section 1.8),

$$A'A = I, \tag{1.92}$$

and hence

$$a_{1i}a_{1j} + \cdots + a_{ni}a_{nj} = \delta_{ij}; \tag{1.93}$$

and this implies (1.89) and (1.90).

b) If (1.87) and (1.88) hold, then (1.91) holds, so that $AA' = I$; if (1.89) and (1.90) hold, then (1.93) holds, so that $A'A = I$. In either case, A is orthogonal. $\square$

Orthogonal matrices are important in studying changes of coordinates (see Problem 13 following Section 1.14).

If two $n \times n$ matrices B, C are such that

$$B = A^{-1}CA \tag{1.94}$$

for some orthogonal matrix A, then B is said to be *orthogonally congruent* to C. It then follows that C is also orthogonally congruent to B, so that we refer to B, C as *orthogonally congruent matrices* (Problem 9 following this section). We observe that orthogonally congruent matrices are similar.

If C is symmetric, then C is orthogonally congruent to a diagonal matrix B, $B = diag(\lambda_1, \lambda_2, \ldots, \lambda_n)$, *where* $\lambda_1, \ldots, \lambda_n$ *are the (necessarily real) eigenvalues of C (with multiple eigenvalues repeated in accordance with their multiplicities).* For a proof of this theorem, see Section 5–3 of the book by Perlis listed at the end of the chapter (see also Problems 12 and 16 following this section).

This result is very important for the study of quadratic forms. Let C be a symmetric matrix and let A be an orthogonal matrix such that $A^{-1}CA = B = diag(\lambda_1, \ldots, \lambda_n)$. Then let us consider the quadratic form

$$Q(\mathbf{x}) = \sum_{i=1}^{n} \sum_{j=1}^{n} c_{ij} x_i x_j = \mathbf{x}'C\mathbf{x}, \qquad \mathbf{x} = col(x_1, \ldots, x_n).$$

We express $x_1, \ldots, x_n$ in terms of new variables $y_1, \ldots, y_n$ by writing

$$\mathbf{x} = A\mathbf{y}, \qquad \mathbf{y} = col(y_1, \ldots, y_n).$$

Then $\mathbf{x}' = \mathbf{y}'A' = \mathbf{y}'A^{-1}$, and hence $Q(\mathbf{x})$ becomes a quadratic form $Q_1(\mathbf{y})$:

$$Q(\mathbf{x}) = Q_1(\mathbf{y}) = \mathbf{y}'A^{-1}CA\mathbf{y} = \mathbf{y}'B\mathbf{y} = \sum_{i=1}^{n}\sum_{j=1}^{n} b_{ij} y_i y_j$$

$$= \lambda_1 y_1^2 + \cdots + \lambda_n y_n^2.$$

Hence $Q(\mathbf{x})$ *can be written in terms of* $y_1, \ldots, y_n$ *as a quadratic form containing only the squares of the unknowns.* The coefficients $\lambda_1, \ldots, \lambda_n$ are the eigenvalues of B. But B and C have the same eigenvalues. Hence once we know the eigenvalues of C (with their multiplicities), we can write Q in terms of $y_1, \ldots, y_n$. It is not necessary to find the matrix A.

We can interpret $x_1, \ldots, x_n$ as Cartesian coordinates in n-dimensional space (Section 1.13). Then $y_1, \ldots, y_n$ are simply new Cartesian coordinates in n-dimensional space with the same origin (see Problem 14 following Section 1.14). For $n = 2$ the equation $Q(\mathbf{x}) = 1$—that is, the equation

$$c_{11} x_1^2 + c_{12} x_1 x_2 + c_{21} x_2 x_1 + c_{22} x_2^2 = 1,$$

—represents a conic section in the $x_1 x_2$-plane. In terms of the new coordinates y_1, y_2 this equation becomes $Q_1(\mathbf{y}) = 1$, or

$$\lambda_1 y_1^2 + \lambda_2 y_2^2 = 1,$$

and hence represents an ellipse or hyperbola. In fact, in the language of plane analytic geometry the theorem on symmetric matrices for the case $n = 2$ is equivalent to the familiar statement that every second-degree equation $Ax^2 + Bxy + Cy^2 = 1$ can be reduced to the standard form $Ax^2 + By^2 = 1$ by an appropriate rotation of axes in the xy-plane. (If $\lambda_1 \lambda_2 = 0$, the curve is degenerate.)

PROBLEMS

1. Find the transpose of each of the matrices:

a) $\begin{bmatrix} 1 & 2 & 3 \\ 3 & 0 & 5 \end{bmatrix}$ b) $\begin{bmatrix} 3 & 1 \\ 0 & 2 \\ 1 & 0 \end{bmatrix}$ c) $(1,5,0,4)$ d) $\begin{bmatrix} 1 \\ 0 \\ 7 \end{bmatrix}$

2. Choose a and b so that each of the following matrices becomes symmetric:

a) $\begin{bmatrix} 1 & 3a-1 \\ 2a & 3 \end{bmatrix}$ b) $\begin{bmatrix} 2 & a & 3 \\ b-a & 0 & 4+a \\ 3 & b & 5 \end{bmatrix}$

3. For each of the following quadratic forms, obtain the coefficient matrix when the form is written in such a way that the coefficient matrix is symmetric:

a) $5x_1^2 + 4x_1x_2 + 3x_2^2$.

b) $7x_1^2 + 2x_1x_2 - x_2^2$.

c) $x_1^2 + 3x_2^2 - x_3^2 + 4x_1x_2 + 6x_1x_3 + 2x_2x_3$,

d) $2x_1^2 + x_2^2 + x_3^2 + 2x_1x_3 - 4x_2x_3$.

4. Prove each of the following parts of (1.81).

a) Rule 25,   b) Rule 26.

5. Show that each of the following matrices is orthogonal:

a) $\dfrac{1}{13}\begin{bmatrix} 5 & 12 \\ -12 & 5 \end{bmatrix}$,   b) $\begin{bmatrix} \cos\omega & \sin\omega \\ -\sin\omega & \cos\omega \end{bmatrix}$,

c) $\dfrac{1}{7}\begin{bmatrix} 2 & 3 & 6 \\ 6 & 2 & -3 \\ 3 & -6 & 2 \end{bmatrix}$,   d) $\dfrac{1}{2}\begin{bmatrix} 1 & 1 & 1 & 1 \\ 1 & -1 & -1 & 1 \\ 1 & 1 & -1 & -1 \\ 1 & -1 & 1 & -1 \end{bmatrix}$.

6. (a), (b), (c). Represent the row vectors of each of the matrices in Problem 5 (a), (b), (c) as vectors in the plane or in space, graph them, and verify that they are mutually perpendicular unit vectors.

7. (a), (b), (c). Proceed as in Problem 6 with column vectors instead of row vectors.

8. Let A and B be $n \times n$ orthogonal matrices. Prove:

a) $\det A = \pm 1$.

b) AB is also orthogonal.

c) A' and A^{-1} are orthogonal.

9. Prove:

a) Every square matrix is orthogonally congruent to itself.

b) If B is orthogonally congruent to C, then C is orthogonally congruent to B.

c) If B is orthogonally congruent to C and C is orthogonally congruent to D, then B is orthogonally congruent to D.

10. We consider *complex matrices*, that is, matrices with complex entries. If $A = (a_{ij})$ is such a matrix, we denote by $\bar{A}$ the *conjugate* of A, that is, the matrix $(\bar{a}_{ij})$, where $\bar{z}$ is the conjugate of the complex number z (for example, if $z = 3 + 5i$, then $\bar{z} = 3 - 5i$). Prove the following:

a) If z_1 and z_2 are complex numbers, then $\overline{(z_1 + z_2)} = \bar{z}_1 + \bar{z}_2$ and $\overline{z_1 z_2} = \bar{z}_1 \bar{z}_2$.

b) If A and B are matrices and c is a complex scalar, then $\overline{(A + B)} = \bar{A} + \bar{B}$, $\overline{AB} = \bar{A}\,\bar{B}$, $\overline{(cA)} = \bar{c}\bar{A}$.

c) Every complex matrix A can be written uniquely as $A_1 + iA_2$, where A_1 and A_2 are real matrices, and $A_1 = \frac{1}{2}(A + \bar{A})$, $A_2 = (2i)^{-1}(A - \bar{A})$. We call A_1 the real part of A, A_2 the imaginary part of A.

d) If A is a square matrix, then $(\bar{A})' = \overline{(A')}$ and, if A is nonsingular, then $\overline{(A^{-1})} = (\bar{A})^{-1}$.

e) A is a real matrix if and only if $A = \bar{A}$.

11. Let A be a real square matrix, let λ be real, and let $Av = \lambda v$ for a nonzero *complex* column vector v. Show that $Au = \lambda u$ for a real nonzero vector u, so that λ is an eigenvalue of A considered as a real matrix. [Hint: Let $v = p + iq$, where p and q are real and not both zero. Show, with the aid of the results of Problem 10, that $Ap = \lambda p$ and $Aq = \lambda q$ and hence that u can be chosen as one of p, q.]

12. Let A be a real symmetric matrix. Show, with the aid of the results of Problem 10, that all eigenvalues of A are real. [Hint: Let $Av = \lambda v$ for some complex λ and some nonzero complex vector $v = \text{col}(v_1, \ldots, v_n)$. Consider the product $Q = v'A\bar{v}$. Show that Q is real and that

$$Q = \lambda v'\bar{v} = \lambda \left(|v_1|^2 + \cdots + |v_n|^2\right).$$

Conclude that λ is real.]

Remarks. By Problem 11, λ is actually an eigenvalue of A as a real matrix. The expression $v'A\bar{v}$ is a special case of a Hermitian quadratic form $Q(v)$. See Section 5–7 of the book by Perlis listed at the end of the chapter.

13. Let A be as in Problem 12 and let $Q(x) = x'Ax$ be the corresponding quadratic form. We call this form *positive definite* if $Q(x) > 0$ for $x \neq 0$. Show that, if $Q(x)$ is positive definite, then $\det A > 0$. [Hint: Use Problem 6 following Section 1.10.]

Remark. It can be shown that Q is positive definite if and only if $\det A_k > 0$ for $k = 1, \ldots, n$, where $A_k = (a_{ij})$, $i = 1, \ldots, k, j = 1, \ldots, k$. See Chapter 6 of G. Strang, *Linear Algebra and its Applications*, 3rd ed., Academic Press, New York, 1987.

14. Let A and Q be as in Problem 13. Show that $Q(x)$ can be written as $u_1^2 + \ldots + u_n^2$ in terms of appropriate coordinates $(u_1, \ldots, u_n)$, where $x = Du$ and D is nonsingular.

15. The *inner product* of $u = \text{col}(u_1, \ldots, u_n)$ and $v = \text{col}(v_1, \ldots, v_n)$ is defined in Section 1.13 below as $u_1v_1 + \cdots + u_nv_n$ and hence equals $u'v = v'u$. Also $|u|$ is defined as $(u'u)^{1/2}$; u and v are termed orthogonal if $u'v = 0$. Now let A be a real symmetric $n \times n$ matrix, so that $A = A'$. Let λ_1, λ_2 be eigenvalues of A which are unequal and let v_1, v_2 be corresponding eigenvectors. Show that v_1 and v_2 are orthogonal. [Hint: First show that $x'Ay = y'Ax$ for all x, y. Then from $Av_1 = \lambda_1 v_1$, $Av_2 = \lambda_2 v_2$ deduce that $\lambda_1 v_2' v_1 = \lambda_2 v_1' v_2$ and hence that $v_1' v_2 = 0$.]

16. For A as in Problem 15, let A have n distinct eigenvalues $\lambda_1, \ldots, \lambda_n$. Show that A is orthogonally congruent to $B = \text{diag}(\lambda_1, \ldots, \lambda_n)$. [Hint: As in Section 1.10, $B = C^{-1}AC$, where the columns of C are eigenvectors $v_1, \ldots, v_n$ of A. From the results of Problem 15 show that the v_j are pairwise orthogonal and can be chosen so that $|v_j| = 1$ for $j = 1, \ldots, n$.]

1.13   ANALYTIC GEOMETRY AND VECTORS IN n-DIMENSIONAL SPACE

The formal operations on vectors and coordinates in Section 1.1 suggest that the restriction to 3-dimensional space, and hence to three coordinates or components, is not necessary.

We define *Euclidean n-dimensional space* E^n to be a space having n coordinates $x_1, \ldots, x_n$. Throughout this section, n will be a fixed positive integer. A *point* P of the space is by definition an ordered n-tuple $(x_1, \ldots, x_n)$; all points of the space are obtained by allowing $x_1, \ldots, x_n$ to take on all real values. The point $(0, \ldots, 0)$ is the *origin*, O. The *distance* between two points A: $(a_1, \ldots, a_n)$ and B: $(b_1, \ldots, b_n)$ is defined to be the number

$$d = \sqrt{(a_1 - b_1)^2 + \cdots + (a_n - b_n)^2}. \tag{1.95}$$

A vector $\mathbf{v}$ in n-dimensional space is defined to be an ordered n-tuple $(v_1, \ldots, v_n)$ of real numbers; $v_1, \ldots, v_n$ are the *components* of $\mathbf{v}$ (with respect to the given coordinates). In particular, we define a zero vector

$$\mathbf{0} = (0, \ldots, 0). \tag{1.96}$$

To the pair of points A, B in that order corresponds the vector

$$\overrightarrow{AB} = (b_1 - a_1, \ldots, b_n - a_n). \tag{1.97}$$

Remarks. Both points and vectors are represented by n-tuples, but this will be seen to cause no confusion. In fact, there are advantages in being able to go back and forth freely between the two points of view, that of points and that of vectors. With each point P: $(x_1, \ldots, x_n)$ we can associate the vector $\overrightarrow{OP} = (x_1, \ldots, x_n) = \mathbf{x}$. Conversely, to each vector $\mathbf{x} = (x_1, \ldots, x_n)$ we can assign the point P whose coordinates are $(x_1, \ldots, x_n)$, and then $\mathbf{x} = \overrightarrow{OP}$. Later we shall also interpret our vectors as matrices, either as row vectors or as column vectors.

The *sum* of two vectors, *multiplication* of a vector by a scalar, and the *scalar product* (or *inner product* or *dot product*) of two vectors are defined by the equations:

$$\mathbf{u} + \mathbf{v} = (u_1 + v_1, \ldots, u_n + v_n), \tag{1.98}$$

$$h\mathbf{u} = (hu_1, \ldots, hu_n), \tag{1.99}$$

$$\mathbf{u} \cdot \mathbf{v} = u_1 v_1 + \cdots + u_n v_n. \tag{1.100}$$

The *length* or *norm* of $\mathbf{v} = (v_1, \ldots, v_n)$ is defined as the scalar

$$|\mathbf{v}| = \sqrt{\mathbf{v} \cdot \mathbf{v}} = \sqrt{v_1^2 + \cdots + v_n^2}. \tag{1.101}$$

If $|\mathbf{v}| = 1$, $\mathbf{v}$ is called a *unit vector*.

The vector product of two vectors can be generalized to n dimensions only with the aid of *tensors* (see, however, Section 11–9 of CLA). The set of all vectors

$(v_1, \ldots, v_n)$ with the operations (1.98) through (1.101) will be denoted by V^n.
    In V^n the following properties can then be verified (Problem 6 below):

I. $\mathbf{u} + \mathbf{v} = \mathbf{v} + \mathbf{u}$.   II. $(\mathbf{u} + \mathbf{v}) + \mathbf{w} = \mathbf{u} + (\mathbf{v} + \mathbf{w})$.
III. $h(\mathbf{u} + \mathbf{v}) = h\mathbf{u} + h\mathbf{v}$.   IV. $(a + b)\mathbf{u} = a\mathbf{u} + b\mathbf{u}$.
V. $(ab)\mathbf{u} = a(b\mathbf{u})$.   VI. $1\mathbf{u} = \mathbf{u}$.
VII. $0\mathbf{u} = \mathbf{0}$.   VIII. $\mathbf{u} \cdot \mathbf{v} = \mathbf{v} \cdot \mathbf{u}$.
IX. $(\mathbf{u} + \mathbf{v}) \cdot \mathbf{w} = \mathbf{u} \cdot \mathbf{w} + \mathbf{v} \cdot \mathbf{w}$.   X. $(a\mathbf{u}) \cdot \mathbf{v} = a(\mathbf{u} \cdot \mathbf{v})$.
XI. $\mathbf{u} \cdot \mathbf{u} \geq 0$.   XII. $\mathbf{u} \cdot \mathbf{u} = 0$ if and only if $\mathbf{u} = \mathbf{0}$.   (1.102)

A set of k vectors $\mathbf{v}_1, \ldots, \mathbf{v}_k$ of V^n is said to be *linearly independent* if an equation

$$c_1\mathbf{v}_1 + \cdots + c_k\mathbf{v}_k = \mathbf{0} \tag{1.103}$$

can hold only if $c_1 = \cdots = c_k = 0$. If a relation (1.103) does hold with not all c's equal to 0, then the vectors are said to be *linearly dependent*.

Remarks.   We observe that for every $n \times k$ matrix A,

$$A \operatorname{col}(c_1, \ldots, c_k) = c_1\mathbf{v}_1 + \cdots + c_k\mathbf{v}_k, \tag{1.104}$$

where $\mathbf{v}_1, \ldots, \mathbf{v}_k$ are the column vectors of A. For

$$\begin{bmatrix} a_{11} & \cdots & a_{1k} \\ \vdots & & \vdots \\ a_{n1} & \cdots & a_{nk} \end{bmatrix} \begin{bmatrix} c_1 \\ \vdots \\ c_k \end{bmatrix} = \begin{bmatrix} c_1 a_{11} + & \cdots & + c_k a_{1k} \\ & \vdots & \\ c_1 a_{n1} + & \cdots & + c_k a_{nk} \end{bmatrix} = c_1\mathbf{v}_1 + \cdots + c_k\mathbf{v}_k.$$

Thus the linear combinations of $\mathbf{v}_1, \ldots, \mathbf{v}_k$ can be expressed as $A\mathbf{c}$. Accordingly, to state that $\mathbf{v}_1, \ldots, \mathbf{v}_k$ are linearly independent is the same as to state that $A\mathbf{c} = \mathbf{0}$ is satisfied only for $\mathbf{c} = \mathbf{0}$. In particular, for $k = n$, A is a square matrix, and $A\mathbf{c} = \mathbf{0}$ is equivalent to n homogeneous linear equations in n unknowns; thus the column vectors $\mathbf{v}_1, \ldots, \mathbf{v}_n$ of A are linearly independent precisely when these equations have only the trivial solution $c_1 = 0, \ldots, c_n = 0$—that is, when $\det A \neq 0$, or A is nonsingular. Accordingly, n *vectors* $\mathbf{v}_1, \ldots, \mathbf{v}_n$ *of* V^n *are linearly independent if and only if* A *is nonsingular, where* A *is the matrix whose column vectors are* $\mathbf{v}_1, \ldots, \mathbf{v}_n$.
    A set of k vectors $\mathbf{v}_1, \ldots, \mathbf{v}_k$ of V^n is said to be a *basis* for V^n if every vector $\mathbf{v}$ of V^n can be expressed in unique fashion as a linear combination of $\mathbf{v}_1, \ldots, \mathbf{v}_k$, that is, if

$$\mathbf{v} = c_1\mathbf{v}_1 + \cdots + c_k\mathbf{v}_k$$

for unique choices of the scalars $c_1, \ldots, c_k$. We call these scalars the *components* of $\mathbf{v}$ with respect to the basis $\mathbf{v}_1, \ldots, \mathbf{v}_k$.
    We can now state a number of rules concerning linear independence and basis:

a) The vectors $v_1, \ldots, v_k$ are linearly dependent if and only if one of these vectors is expressible as a linear combination of the others.

b) If one of the vectors $v_1, \ldots, v_k$ is 0, then $v_1, \ldots, v_k$ are linearly dependent.

c) If $v_1, \ldots, v_k$ are linearly independent, but $v_1, \ldots, v_k, v_{k+1}$ are linearly dependent, then v_{k+1} is expressible as a linear combination of $v_1, \ldots, v_k$.

d) If $v_1, \ldots, v_k$ are linearly independent and $h < k$, then $v_1, \ldots, v_h$ are linearly independent.

e) (Rule for comparing coefficients). If $v_1, \ldots, v_k$ are linearly independent and

$$a_1 v_1 + \cdots + a_k v_k = b_1 v_1 + \cdots + b_k v_k,$$

then $a_1 = b_1, a_2 = b_2, \ldots, a_k = b_k$.

f) There exist n linearly independent vectors in V^n: for example, the vectors

$$e_1 = (1, 0, \ldots, 0), \qquad e_2 = (0, 1, 0, \ldots, 0), \ldots,$$

$$e_n = (0, \ldots, 0, 1). \tag{1.105}$$

g) There do not exist $n + 1$ linearly independent vectors in V^n.

h) If $v_1, \ldots, v_n$ are linearly independent vectors in V^n, then $v_1, \ldots, v_n$ form a basis for V^n; in particular, $e_1, \ldots, e_n$ form a basis for V^n.

i) Every basis for V^n consists of n linearly independent vectors.

j) If $k < n$ and $v_1, \ldots, v_k$ are linearly independent, then there exist $v_{k+1}, \ldots, v_n$ such that $v_1, \ldots, v_n$ form a basis for V^n.

k) If $v_1, \ldots, v_k$ are linearly independent vectors in V^n, and $u_1, \ldots, u_{k+1}$ are all expressible as linear combinations of $v_1, \ldots, v_k$, then $u_1, \ldots, u_{k+1}$ are linearly dependent.

We prove Rules (g) and (i) here and leave the remaining proofs to Problem 10 below.

Proof of (g).   Let $v_1, \ldots, v_{n+1}$ be linearly independent vectors in V^n. Then by rule (d), $v_1, \ldots, v_n$ are also linearly independent. Hence the matrix A, whose columns are $v_1, \ldots, v_n$, is nonsingular. Therefore the equation $Ac = v_{n+1}$ has a unique solution for c; that is,

$$v_{n+1} = c_1 v_1 + \cdots + c_n v_n.$$

By Rule (a), $v_1, \ldots, v_{n+1}$ would then be linearly dependent, contrary to assumption. Therefore there cannot be $n + 1$ linearly independent vectors in V^n. $\square$

Proof of (i).   Let $v_1, \ldots, v_k$ be a basis for V^n. Then $v_1, \ldots, v_k$ are linearly independent. For if $c_1 v_1 + \cdots + c_k v_k = 0$, then

$$c_1 = 0, \ldots, c_k = 0,$$

since, by definition of a basis, 0 can be represented in only one way as a linear combination of $v_1, \ldots, v_k$. Hence by rules (d) and (g) we must have $k \leq n$. Let us suppose $k < n$. Then we can express $e_1, \ldots, e_n$ as linear combinations of $v_1, \ldots, v_k$ —say, $A b_1 = e_1, \ldots, A b_n = e_n$, where A is the $n \times k$ matrix whose column vectors are $v_1, \ldots, v_k$. Accordingly, $AB = I$, where B is the $k \times n$ matrix whose column vectors are $b_1, \ldots, b_n$. But for $k < n$ a product of an $n \times k$ matrix and a $k \times n$

matrix is always singular (see the Remark at the close of Section 1.8). Hence we have a contradiction, and we must have $k = n$. Thus Rule (i) is proved. $\quad\square$

The basis $\mathbf{e}_1, \ldots, \mathbf{e}_n$ defined in (1.105) is called the *standard basis* of V^n. For $n = 3$ this is the familiar basis $\mathbf{i}, \mathbf{j}, \mathbf{k}$. In general,

$$\mathbf{v} = (v_1, \ldots, v_n) = v_1\mathbf{e}_1 + \cdots + v_n\mathbf{e}_n. \tag{1.106}$$

We now prove a theorem of much importance for the geometry of E^n:

THEOREM   If $\mathbf{u}$ and $\mathbf{v}$ are vectors of V^n, then

$$|\mathbf{u} \cdot \mathbf{v}| \le |\mathbf{u}|\,|\mathbf{v}| \quad \text{(Cauchy-Schwarz inequality)}. \tag{1.107}$$

The equality holds precisely when $\mathbf{u}$ and $\mathbf{v}$ are linearly dependent. Furthermore,

$$|\mathbf{u} + \mathbf{v}| \le |\mathbf{u}| + |\mathbf{v}| \quad \text{(triangle inequality)}. \tag{1.108}$$

The equals sign holds precisely when $\mathbf{u} = h\mathbf{v}$ or $\mathbf{v} = h\mathbf{u}$ with $h \ge 0$.

Proof.   If $\mathbf{v} \ne \mathbf{0}$ and $\mathbf{u}$ is not a scalar multiple of $\mathbf{v}$, then $\mathbf{u} + t\mathbf{v} \ne \mathbf{0}$ for every scalar t, and hence $|\mathbf{u} + t\mathbf{v}|^2 > 0$ for all t, so that

$$(\mathbf{u} + t\mathbf{v}) \cdot (\mathbf{u} + t\mathbf{v}) > 0 \quad \text{or} \quad P(t) = |\mathbf{u}|^2 + 2t(\mathbf{u} \cdot \mathbf{v}) + t^2|\mathbf{v}|^2 > 0$$

for all t. Hence the discriminant of the quadratic function $P(t)$ must be negative; that is,

$$4(\mathbf{u} \cdot \mathbf{v})^2 - 4|\mathbf{u}|^2|\mathbf{v}|^2 < 0,$$

so that $|\mathbf{u} \cdot \mathbf{v}| < |\mathbf{u}|\,|\mathbf{v}|$. Similarly, $|\mathbf{u} \cdot \mathbf{v}| < |\mathbf{u}|\,|\mathbf{v}|$ if $\mathbf{u} \ne \mathbf{0}$ and $\mathbf{v}$ is not a scalar multiple of $\mathbf{u}$. If $\mathbf{u} = c\mathbf{v}$ (in particular, if $\mathbf{u} = \mathbf{0}$), then

$$|\mathbf{u} \cdot \mathbf{v}| = |c\mathbf{v} \cdot \mathbf{v}| = |c(\mathbf{v} \cdot \mathbf{v})|$$

$$= |c|\,|\mathbf{v}|^2 = |c\mathbf{v}|\,|\mathbf{v}| = |\mathbf{u}|\,|\mathbf{v}|,$$

by Rule X in (1.102) and the definition (1.101). Similarly, equality holds if $\mathbf{v} = c\mathbf{u}$ (in particular, if $\mathbf{v} = \mathbf{0}$. Thus (1.107) is proved, along with the conditions for equality.

To prove (1.108), we write:

$$|\mathbf{u} + \mathbf{v}|^2 = (\mathbf{u} + \mathbf{v}) \cdot (\mathbf{u} + \mathbf{v}) = |\mathbf{u}|^2 + |\mathbf{v}|^2 + 2\mathbf{u} \cdot \mathbf{v}$$

$$= (|\mathbf{u}| + |\mathbf{v}|)^2 + 2\mathbf{u} \cdot \mathbf{v} - 2|\mathbf{u}|\,|\mathbf{v}| \le (|\mathbf{u}| + |\mathbf{v}|)^2. \tag{1.109}$$

The last inequality follows from (1.107). By taking square roots we obtain (1.108). The investigation of the case of equality is left to Problem 8 below. $\quad\square$

Now let P_1, P_2, P_3 be three points of E^n. By (1.95), (1.97) and (1.101), $|\overrightarrow{P_1P_2}|$ equals the distance d from P_1 to P_2. We write $d(P_1, P_2) = |\overrightarrow{P_1P_2}|$. It follows that $d(P_1, P_2) \ge 0$ and that $d(P_1, P_2) = 0$ only for $P_1 = P_2$. Also, $d(P_1, P_2) = d(P_2, P_1)$, since $|\overrightarrow{P_2P_1}| = |-\overrightarrow{P_1P_2}| = |\overrightarrow{P_1P_2}|$. Finally, from (1.97), $\overrightarrow{P_1P_3} = \overrightarrow{P_1P_2} + \overrightarrow{P_2P_3}$, so that, by (1.108),

$$|\overrightarrow{P_1P_3}| \leq |\overrightarrow{P_1P_2}| + |\overrightarrow{P_2P_3}| \quad \text{or} \quad d(P_1, P_3) \leq d(P_1, P_2) + d(P_2, P_3);$$

that is, the length of one side of a triangle is less than or equal to the sum of the lengths of the other two sides. This explains the term *triangle inequality* in (1.108).

Because of the theorem, we can define the *angle* θ between two nonzero vectors $\mathbf{u}, \mathbf{v}$ by the conditions

$$\cos \theta = \frac{\mathbf{u} \cdot \mathbf{v}}{|\mathbf{u}||\mathbf{v}|}, \qquad 0 \leq \theta \leq \pi, \qquad (1.110)$$

for by (1.107), $\mathbf{u} \cdot \mathbf{v}/(|\mathbf{u}||\mathbf{v}|)$ lies between -1 and 1 (inclusive). With the aid of distance and angle, one can now develop geometry in n-dimensional space in familiar fashion.

We define two vectors $\mathbf{u}$ and $\mathbf{v}$ to be *orthogonal* (or *perpendicular*) if $\mathbf{u} \cdot \mathbf{v} = 0$, that is, by (1.110) if they form an angle of $\pi/2$ (or if one of the vectors is $\mathbf{0}$). We define an *orthogonal system* of vectors as a system of k nonzero vectors $\mathbf{v}_1, \ldots, \mathbf{v}_k$ such that each two vectors of the system are orthogonal; the set of vectors is necessarily linearly independent (Problem 9 below), so that by Rule (g) above, $k \leq n$. An orthogonal system of *unit* vectors is called an *orthonormal system*. There exists an orthonormal system of n vectors, namely, $\mathbf{e}_1, \ldots, \mathbf{e}_n$. Every orthogonal system of n vectors forms a basis for V^n, by the preceding Rule (h); we call such a system an *orthogonal basis* (or, for a system of unit vectors, an *orthonormal basis*).

Given k linearly independent vectors $\mathbf{v}_1, \ldots, \mathbf{v}_k$, it is always possible to construct an orthogonal system of k vectors $\mathbf{u}_1, \ldots, \mathbf{u}_k$, each of which is a linear combination of $\mathbf{v}_1, \ldots, \mathbf{v}_k$ (Gram-Schmidt orthogonalization process). We illustrate this for $k = 3$; the construction is then easily pictured in 3-dimensional space:

$$\mathbf{u}_1 = \mathbf{v}_1$$

$$\mathbf{u}_2 = \mathbf{v}_2 - (\mathbf{v}_2 \cdot \mathbf{u}_1)\frac{\mathbf{u}_1}{|\mathbf{u}_1|^2}, \qquad (1.111)$$

$$\mathbf{u}_3 = \mathbf{v}_3 - (\mathbf{v}_3 \cdot \mathbf{u}_1)\frac{\mathbf{u}_1}{|\mathbf{u}_1|^2} - (\mathbf{v}_3 \cdot \mathbf{u}_2)\frac{\mathbf{u}_2}{|\mathbf{u}_2|^2}.$$

To obtain $\mathbf{u}_2$, we subtract from $\mathbf{v}_2$ the vector obtained by "projecting" $\mathbf{v}_2$ on $\mathbf{v}_1$; to obtain $\mathbf{u}_3$, we subtract from $\mathbf{v}_3$ its projections on $\mathbf{u}_1$ and $\mathbf{u}_2$. By (1.111), $\mathbf{u}_1 = \mathbf{v}_1$, $\mathbf{u}_2$ is a linear combination of $\mathbf{v}_1$ and $\mathbf{v}_2$, and $\mathbf{u}_3$ is a linear combination of $\mathbf{v}_1, \mathbf{v}_2, \mathbf{v}_3$; also, $\mathbf{v}_1 = \mathbf{u}_1$, $\mathbf{v}_2$ is a linear combination of $\mathbf{u}_1$ and $\mathbf{u}_2$, and $\mathbf{v}_3$ is a linear combination of $\mathbf{u}_1$, $\mathbf{u}_2$, and $\mathbf{u}_3$. The proof of these assertions and of the orthogonality is left to Problem 11 below. The process clearly extends by induction to any finite number of linearly independent vectors. Furthermore, in general the process is such that, for each $h, \mathbf{u}_h$ is a linear combination of $\mathbf{v}_1, \ldots, \mathbf{v}_h$, and $\mathbf{v}_h$ is a linear combination of $\mathbf{u}_1, \ldots, \mathbf{u}_h$.

Remark. With the aid of the Gram-Schmidt process we can extend Rule (j) as follows:

l) If $k < n$ and $\mathbf{u}_1, \ldots, \mathbf{u}_k$ form an orthogonal system in V^n, then there exist vectors $\mathbf{u}_{k+1}, \ldots, \mathbf{u}_n$ such that $\mathbf{u}_1, \ldots, \mathbf{u}_n$ form an orthogonal basis.

The proof is left as an exercise (Problem 10(j)).

Volume in E^n.   In 3-dimensional space, we saw that $\pm D = \pm \mathbf{u} \cdot \mathbf{v} \times \mathbf{w}$ gives the volume of a parallelepiped whose edges, properly directed, represent $\mathbf{u}, \mathbf{v}, \mathbf{w}$ (Section 1.3). Furthermore, D equals a third-order determinant, as in Eq. (1.33). These results generalize to E^n as follows: If $\mathbf{v}_1, \ldots, \mathbf{v}_n$ are vectors of V^n, then

$$D = \begin{vmatrix} v_{11} & \cdots & v_{1n} \\ v_{21} & & v_{2n} \\ \vdots & & \vdots \\ v_{n1} & \cdots & v_{nn} \end{vmatrix} \tag{1.112}$$

equals plus or minus the volume of an n-dimensional parallelotope whose edges, properly directed, represent $\mathbf{v}_1, \ldots, \mathbf{v}_n$. In Eq. (1.112), $\mathbf{v}_1 = (v_{11}, \ldots, v_{1n}), \ldots, \mathbf{v}_n = (v_{n1}, \ldots, v_{nn})$. The parallelotope can be chosen to have vertices $O, P_1, \ldots, P_n$, where $\mathbf{v}_1 = \overrightarrow{OP}_1, \ldots, \mathbf{v}_n = \overrightarrow{OP}_n$ and the n-dimensional solid consists of all points P of E^n such that

$$\overrightarrow{OP} = t_1\mathbf{v}_1 + \cdots + t_n\mathbf{v}_n, \tag{1.113}$$

where $0 \le t_1 \le 1, \ldots, 0 \le t_n \le 1$ (see Sections 11–15 and 11–16 of CLA).

*1.14   AXIOMS FOR V^n

From Rule (h) of Section 1.13 we can state:

XIII. V^n has a basis consisting of n vectors.

It is of considerable interest that Rules I through XII of (1.102) plus Rule XIII can be regarded as a set of axioms, from which one can deduce all other properties of the vectors in n-dimensional space, without reference to the sets of components. For example, $\mathbf{u} + \mathbf{0} = \mathbf{u}$, since

$$\mathbf{u} + \mathbf{0} = 1\mathbf{u} + 0\mathbf{u} = (1 + 0)\mathbf{u} = \mathbf{u}$$

by Rules VI, VII, and IV. Also, the equation $\mathbf{v} + \mathbf{u} = \mathbf{w}$ has a solution for $\mathbf{u}$—namely, the vector $\mathbf{w} + (-1)\mathbf{v}$, which we also denote by $\mathbf{w} - \mathbf{v}$; for

$$\mathbf{v} + \left[\mathbf{w} + (-1)\mathbf{v}\right] = \mathbf{v} + \left[(-1)\mathbf{v} + \mathbf{w}\right] = \left[1\mathbf{v} + (-1)\mathbf{v}\right] + \mathbf{w}$$
$$= 0\mathbf{v} + \mathbf{w} = \mathbf{0} + \mathbf{w} = \mathbf{w}$$

by Rules II, VI, IV, VII, I, and the rule $\mathbf{u} + \mathbf{0} = \mathbf{u}$ just proved. The solution is unique, for $\mathbf{v} + \mathbf{u} = \mathbf{w}$ implies $-\mathbf{v} + (\mathbf{v} + \mathbf{u}) = -\mathbf{v} + \mathbf{w} = \mathbf{w} - \mathbf{v}$ or $(-\mathbf{v} + \mathbf{v}) + \mathbf{u} = \mathbf{w} - \mathbf{v}$ or $\mathbf{u} = \mathbf{w} - \mathbf{v}$; here we wrote $-\mathbf{v}$ for $(-1)\mathbf{v}$, as is usual. Other algebraic properties can be deduced in similar fashion. Also, the norm $|\mathbf{u}|$ is defined in terms of the scalar product by Eq. (1.101); its properties derive from those of the scalar product. In particular, the Cauchy-Schwarz inequality and triangle inequality follow as before, without use of components.

One can introduce coordinates or components on the basis of the axioms themselves, for Rule XIII guarantees existence of a basis $\mathbf{v}_1, \ldots, \mathbf{v}_n$. The Gram-

Schmidt process can then be used to construct an orthogonal basis $\mathbf{u}_1,\ldots,\mathbf{u}_n$. We "normalize" these vectors by dividing by their norms to obtain an orthonormal basis:

$$\mathbf{e}_1^* = \frac{\mathbf{u}_1}{|\mathbf{u}_1|},\ldots,\ \mathbf{e}_n^* = \frac{\mathbf{u}_n}{|\mathbf{u}_n|}.$$

For an arbitrary vector $\mathbf{v}$ we can now write, in unique fashion,

$$\mathbf{v} = v_1^*\mathbf{e}_1^* + \cdots + v_n^*\mathbf{e}_n^*, \tag{1.114}$$

and call $(v_1^*,\ldots,v_n^*)$ the components of $\mathbf{v}$ with respect to the basis $\mathbf{e}_1^*,\ldots,\mathbf{e}_n^*$. We note that

$$\mathbf{v} \cdot \mathbf{e}_1^* = v_1^*(\mathbf{e}_1^* \cdot \mathbf{e}_1^*) + v_2^*(\mathbf{e}_2^* \cdot \mathbf{e}_1^*) + \cdots + v_n^*(\mathbf{e}_n^* \cdot \mathbf{e}_1^*) = v_1^*,$$

because the basis is orthonormal. In general,

$$v_1^* = \mathbf{v} \cdot \mathbf{e}_1^*,\ldots,v_n^* = \mathbf{v} \cdot \mathbf{e}_n^*. \tag{1.115}$$

On the basis of Rule I through XII we now verify (see Problem 12 below) that

$$\mathbf{u} + \mathbf{v} = (u_1^* + v_1^*)\mathbf{e}_1^* + \cdots + (u_n^* + v_n^*)\mathbf{e}_n^*,$$
$$h\mathbf{v} = (hv_1^*)\mathbf{e}_1^* + \cdots + (hv_n^*)\mathbf{e}_n^*, \tag{1.116}$$
$$\mathbf{u} \cdot \mathbf{v} = u_1^*v_1^* + \cdots + u_n^*v_n^*.$$

This simply means that the vector operations can be defined in terms of components just as in (1.98), (1.99), and (1.100). Accordingly, while the new basis may be different from the original one (this is simply a choice of new axes), all properties described with the old components can be found just as well with the new ones. This means that *all* properties of the vectors can be found from (1.102) and Rules XIII alone. This leads to the following definition:

DEFINITION   A Euclidean n-dimensional vector space is a collection of objects $\mathbf{u},\mathbf{v},\ldots$ called vectors, including a zero vector $\mathbf{0}$, for which the operations of addition, multiplication by scalars, and scalar product are defined and obey Rules I through XIII.

By virtue of our discussion there is, except for notation, really only one Euclidean n-dimensional vector space, namely, V^n.

We have emphasized the axiomatic approach to vectors. There is a similar discussion for points. In fact, given a Euclidean n-dimensional vector space, we can assign components $(v_1^*,\ldots,v_n^*)$ to each vector as above. To the zero vector $(0,\ldots,0)$, we then assign an origin O; to each vector $(v_1^*,\ldots,v_n^*)$ we assign a point P and interpret $v_1^*,\ldots,v_n^*$ as the coordinates of P. All the concepts of geometry can now be introduced as above, and we recover our n-dimensional Euclidean space, the same (except for notation) as E^n.

One can ask whether there is any value in the generalization of vectors to n-dimensional space, since we live in a 3-dimensional world. The answer is that the generalization has proved to be exceedingly valuable. The equations of a mechanical system having "N degrees of freedom" are easier to describe and understand in terms of a vector space of dimension $2N$. In the kinetic theory of gases an n-dimensional space (the "phase space") is of fundamental importance; here n may be as large as 10^{23}. In relativity a 4-dimensional space is needed. In quantum mechanics, one is in fact forced to consider the limiting case $n = \infty$; surprisingly enough, this theory of a vector space of infinite dimension turns out to be closely related to the theory of Fourier series (Chapter 7).

PROBLEMS

1. In V^4, let $u = (3,2,1,0)$, $v = (1,0,1,2)$, $w = (5,4,1,-2)$.
   a) Find $u + v$, $u + w$, $2u$, $-3v$, $0w$.
   b) Find $3u - 2v$, $2v + 3w$, $u - w$, $u + v - 2w$.
   c) Find $u \cdot v$, $u \cdot w$, $|u|$, $|v|$.
   d) Show that w can be expressed as a linear combination of u and v and hence that u, v, w are linearly dependent.

2. In E^n the line segment $P_1 P_2$ joining points P_1, P_2 is formed of all points P such that $\overrightarrow{P_1 P} = t\overrightarrow{P_1 P_2}$, where $0 \le t \le 1$.
   a) In E^4, show that $(5, 10, -1, 8)$ is on the line segment from $(1, 2, 3, 4)$ to $(7, 14, -3, 10)$.
   b) In E^4, is $(2, 8, 1, 6)$ on the line segment joining $(1, 7, 2, 5)$ to $(9, 2, 0, 7)$?
   c) In E^5, find the *midpoint* of the line segment from P_1: $(2, 0, 1, 3, 7)$ to P_2: $(10, 4, -3, 3, 5)$; that is, find the point P on the segment such that $|\overrightarrow{P_1 P}| = |\overrightarrow{P_2 P}|$.
   d) In E^5, trisect the line segment joining $(7, 1, 7, 9, 6)$ to $(16, -2, 10, 3, 0)$.
   e) In E^n, show that if P is on the line segment $P_1 P_2$, then $|\overrightarrow{P_1 P}| + |\overrightarrow{P_2 P}| = |\overrightarrow{P_1 P_2}|$.

3. Prove the Pythagorean theorem in E^n: If $\overrightarrow{P_1 P_2}$ is orthogonal to $\overrightarrow{P_2 P_3}$, then $|\overrightarrow{P_1 P_3}|^2 = |\overrightarrow{P_1 P_2}|^2 + |\overrightarrow{P_2 P_3}|^2$. [Hint: Write the left side as
$$\left(\overrightarrow{P_1 P_2} + \overrightarrow{P_2 P_3}\right) \cdot \left(\overrightarrow{P_1 P_2} + \overrightarrow{P_2 P_3}\right).]$$

4. Prove the Law of Cosines in E^n: If $\overrightarrow{P_1 P_2}$ and $\overrightarrow{P_1 P_3}$ are nonzero vectors, then
$$|\overrightarrow{P_2 P_3}|^2 = |\overrightarrow{P_1 P_2}|^2 + |\overrightarrow{P_1 P_3}|^2 - 2|\overrightarrow{P_1 P_2}| \, |\overrightarrow{P_1 P_3}| \cos \theta,$$
where θ is the angle between $\overrightarrow{P_1 P_2}$ and $\overrightarrow{P_1 P_3}$.

5. a) In E^5, find the sides and angles of the triangle with vertices
$$(1, 2, 3, 4, 5), \quad (5, 4, 2, 3, 1), \quad \text{and} \quad (2, 2, 2, 2, 2).$$
   b) Prove: In E^n, the sum of the angles of a triangle is π. [Hint: For a triangle with sides a, b, c in E^n, construct the triangle with the same sides in the plane E^2; why is this possible? Then show, with the aid of the Law of Cosines, that corresponding angles of the two triangles are equal.]

6. Prove the rules (1.102), with the aid of (1.96) through (1.101).

7. Prove Cauchy's inequality:
$$\left[\sum_{i=1}^{n} u_i v_i \right]^2 \le \sum_{i=1}^{n} u_i^2 \sum_{i=1}^{n} v_i^2.$$

[Hint: Use (1.107).]

8. Prove the rule for equality in (1.108) as stated in the theorem. [Hint: Use (1.109) to reduce the problem to that for $|u \cdot v| = |u| \, |v|$. Use the first part of the theorem and the meaning of linear dependence to prove the desired result.]

9. Prove: The vectors of an orthogonal system are linearly independent.
10. Prove the following rules for vectors in V^n, with the aid of the hints given.
    a) Rule (a) [Hint: If $c_1 v_1 + \cdots + c_k v_k = 0$ and, for example, $c_k \neq 0$, then express v_k as a linear combination of $v_1, \ldots, v_{k-1}$. For the converse, suppose $v_k = c_1 v_1 + \cdots + c_{k-1} v_{k-1}$ and write this equation in the form $c_1 v_1 + \cdots + c_k v_k = 0$.]
    b) Rule (b) [Hint: Suppose $v_k = 0$ and choose $c_1, \ldots, c_k$ not all 0 so that $c_1 v_1 + \cdots + c_k v_k = 0$.]
    c) Rule (c) [Hint: One has $c_1 v_1 + \cdots + c_{k+1} v_{k+1} = 0$ with not all c's 0. Show that c_{k+1} cannot be 0.]
    d) Rule (d)
    e) Rule (e)
    f) Rule (f)
    g) Rule (h) [Hint: Apply Rules (g) and (e).]
    h) Rule (j) [Hint: By Rule (i), $v_1, \ldots, v_k$ do not form a basis. Hence one can choose v_{k+1} so that $v_1, \ldots, v_{k+1}$ are linearly independent. If $k + 1 < n$, repeat the process.]
    i) Rule (k) [Hint: Apply Rule (j) to obtain a basis $v_1, \ldots, v_n$. By Rule (g), $u_1, \ldots, u_{k+1}$, $v_{k+1}, \ldots, v_n$ are linearly dependent, so that

$$c_1 u_1 + \cdots + c_{k+1} u_{k+1} + c_{k+2} v_{k+1} + \cdots + c_{n+1} v_n = 0.$$

Show from the hypotheses that $c_{k+2} = 0, \ldots, c_{n+1} = 0$.]
    j) Rule (l) [Hint: Let $v_1 = u_1, \ldots, v_k = u_k$. Then choose $v_{k+1}, \ldots, v_n$ as in Rule (j) so that $v_1, \ldots, v_n$ form a basis. Now apply the Gram-Schmidt process to $v_1, \ldots, v_n$.]
11. a) Let v_1, v_2, v_3 be linearly independent. Prove that the vectors (1.111) are nonzero and form an orthogonal system. Prove also that, for $h = 1, 2, 3$, u_h is a linear combination of $v_1, \ldots, v_h$, and v_h is a linear combination of $u_1, \ldots, u_h$.
    b) Carry out the process (1.111) in V^3 with $v_1 = i$, $v_2 = i + j + k$, $v_3 = 2j + k$, and graph.
12. Prove that if $u_1^*, \ldots, u_n^*$ and $v_1^*, \ldots, v_n^*$ are the components of u and v with respect to the orthonormal basis $e_1^*, \ldots, e_n^*$, then (1.116) holds.
13. (*Change of basis*)   Let $e_1^*, \ldots, e_n^*$ be an orthonormal basis of V^n, so that (1.114) and (1.115) hold for an arbitrary vector v. In particular, we can write

$$e_j = a_{1j} e_1^* + \cdots + a_{nj} e_n^*,$$

$$a_{ij} = e_j \cdot e_i^*, \qquad i = 1, \ldots, n, \quad j = 1, \ldots, n.$$

    a) Show that, with $A = (a_{ij})$,

$$\mathrm{col}\,(v_1^*, \ldots, v_n^*) = A\,\mathrm{col}\,(v_1, \ldots, v_n).$$

    Thus the matrix A provides the link between the components of v with respect to the two orthonormal bases. [Hint: Write $v = v_1 e_1 + \cdots + v_n e_n = v_1^* e_1^* + \cdots + v_n^* e_n^*$ and dot both sides with e_i^* for $i = 1, \ldots, n$.]
    b) Show that the jth column of A gives the components of e_j with respect to the basis $e_1^*, \ldots, e_n^*$ and the ith row of A gives the components of e_i^* with respect to the basis $e_1, \ldots, e_n$.
    c) Show that A is an orthogonal matrix (Section 1.12).

14. (*Change of coordinates in E^n*)   We choose a new origin O^* and a new orthonormal basis $\mathbf{e}_1^*,\ldots,\mathbf{e}_n^*$ for V^n as in Problem 13. Then each point P obtains new coordinates by the equation

$$\overrightarrow{O^*P} = x_1^* \mathbf{e}_1^* + \cdots + x_n^* \mathbf{e}_n^*.$$

a) Let O^* have old coordinates $(h_1,\ldots,h_n)$. Show with the aid of the results of Problem 13 that

$$\operatorname{col}\left(x_1^*,\ldots,x_n^*\right) = A\operatorname{col}\left(x_1 - h_1,\ldots,x_n - h_n\right),$$

where A is an orthogonal matrix.

b) Show that if the origin is changed but there is no change of basis (so that one has a translation of axes), then

$$x_i^* = x_i - h_i, \qquad i = 1,\ldots,n.$$

1.15   LINEAR MAPPINGS

A basic tool of the calculus is the concept of *function*. We can describe this in general terms as the assignment to each object in one set (the *domain* of the function) of an object in a second set (the *target set* of the function). The *range* of the function is the set of objects in the target set that are actually used (see Fig. 1.15). A function is also called a *mapping*. We speak of the function as a mapping of the domain *into* the target set. When the range is the whole target set, we replace "into" by "onto."

We now consider the simultaneous linear equations

$$
\begin{aligned}
a_{11}x_1 + &\quad \cdots \quad + a_{1n}x_n = y_1, \\
&\ \vdots \qquad\qquad\qquad \vdots \\
a_{m1}x_1 + &\quad \cdots \quad + a_{mn}x_n = y_m
\end{aligned}
\tag{1.117}
$$

or, in matrix form,

$$A\mathbf{x} = \mathbf{y}, \tag{1.117'}$$

where $\mathbf{x} = \operatorname{col}(x_1,\ldots,x_n)$, $\mathbf{y} = \operatorname{col}(y_1,\ldots,y_m)$. (In this section, all vectors will be written as column vectors.) Through (1.117) or (117') a vector $\mathbf{y}$ of V^m is assigned to each vector $\mathbf{x}$ of V^n. Thus we have a function or mapping whose domain is V^n and whose range is contained in V^m. Furthermore, if $\mathbf{x}_1$ and $\mathbf{x}_2$ are in V^n and c_1, c_2 are scalars, then

$$A(c_1\mathbf{x}_1 + c_2\mathbf{x}_2) = c_1(A\mathbf{x}_1) + c_2(A\mathbf{x}_2); \tag{1.118}$$

Figure 1.15   Function or mapping.

that is, *the mapping* (1.117′) *assigns to each linear combination* $c_1x_1 + c_2x_2$ *the corresponding linear combination fo the values assigned to* x_1 *and* x_2. We call such a mapping linear:

DEFINITION   Let T be a mapping of V^n into V^m. Then T is *linear* if for every choice of x_1 and x_2 in V^n and every pair of scalars c_1, c_2, one has

$$T(c_1x_1 + c_2x_2) = c_1T(x_1) + c_2T(x_2). \tag{1.119}$$

Thus for every $m \times n$ matrix A, the equation $y = Ax$ defines a linear mapping T of V^n into V^m with $T(x) = Ax$.

Conversely, *if T is a linear mapping of V^n into V^m, then there is an $m \times n$ matrix A such that $T(x) = Ax$ for all x in V^n.* To prove this assertion, we first note that (1.119) implies the general rule

$$T(c_1x_1 + \cdots + c_kx_k) = c_1T(x_1) + \cdots + c_kT(x_k). \tag{1.120}$$

Now let $e_1, \ldots, e_n$ be the standard basis in V^n and let

$$T(e_j) = u_j, \qquad j = 1, \ldots, n. \tag{1.121}$$

For each vector $x = (x_1, \ldots, x_n) = x_1e_1 + \cdots + x_ne_n$ in V^n it then follows from (1.120) that

$$T(x) = x_1T(e_1) + \cdots + x_nT(e_n) = x_1u_1 + \cdots + x_nu_n$$

$$= A\,\mathrm{col}\,(x_1, \ldots, x_n) = Ax,$$

where A is the $m \times n$ matrix whose column vectors are $u_1, \ldots, u_n$.

It follows that the study of linear mappings of V^n into V^m is equivalent to the study of mappings of the form $y = Ax$, where A is an $m \times n$ matrix. For each linear mapping T we call A *the matrix of T*, or *the matrix representing T*. For each matrix A we call the corresponding linear mapping T the *linear mapping T determined by A*; we also write, more concisely, the linear mapping $y = Ax$, or even, the linear mapping A.

From the proof just given we see that for each linear mapping T the column vectors $u_1, \ldots, u_n$ of the matrix A representing T are the vectors $T(e_1), \ldots, T(e_n)$ of V^m. Thus if $A = (a_{ij})$, then

$$Ae_j = \mathrm{col}\,(a_{1j}, \ldots, a_{mj}) = u_j. \tag{1.122}$$

Figure 1.16   Linear mapping of V^2 into V^3.

Figure 1.17   Linear mapping of V^2 into V^2.

Furthermore, as in the proof, we can write

$$y = Ax = A \, col \, (x_1, \ldots, x_n) = x_1 u_1 + \cdots + x_n u_n.$$

Thus the linear mapping A assigns to each vector $x = col(x_1, \ldots, x_n)$ the linear combination $x_1 u_1 + \cdots + x_n u_n$ of the column vectors

$$u_1, \ldots, u_n \quad (\text{vectors of } V^m).$$

It is very helpful to think of a linear mapping as suggested in Fig. 1.16, in which the vectors are represented by directed segments from the origin; here $n = 2, m = 3$.

EXAMPLE 1   Let the linear mapping T have the matrix $A = \begin{bmatrix} 2 & 3 \\ 1 & 2 \end{bmatrix}$. Evaluate $T(x)$ for $x = (1, 0) = x_1, x = (0, 1) = x_2, x = (2, -1) = x_3, x = (1, -1) = x_4$ and graph.   ∎

Solution. $T(x_1) = A \quad col \, (1, 0) = col \, (2, 1), \quad T(x_2) = A \quad col \, (0, 1) = col \, (3, 2),$
$T(x_3) = A \, col \, (2, -1) = col \, (1, 0), T(x_4) = A \, col \, (1, -1) = col \, (-1, -1)$ (see Fig. 1.17).

We note that $x_1 = e_1$ and $x_2 = e_2$, so that

$$T(x_1) = T(e_1) = u_1 = col \, (2, 1),$$

the first column vector of A. Similarly, $T(x_2) = u_2 = col \, (3, 2)$, the second column vector of A. Also $x_3 = 2e_1 - e_2$, so that

$$T(x_3) = 2u_1 - u_2 = 2(2, 1) - (3, 2) = (1, 0).$$

Similarly, $T(x_4) = T(e_1 - e_2) = u_1 - u_2 = (-1, -1)$.

EXAMPLE 2   Find a linear mapping T of V^2 into V^3 such that

$$T((1, 0)) = (2, 1, 2) \quad \text{and} \quad T((0, 1)) = (5, 3, 7). \quad ∎$$

Solution. T has the matrix $A = \begin{bmatrix} 2 & 5 \\ 1 & 3 \\ 2 & 7 \end{bmatrix}$.

Notation. We shall write $T(x_1, \ldots, x_n)$ for $T((x_1, \ldots, x_n))$, since the extra parentheses are not necessary.

For each particular linear mapping T of V^n into V^m, one is concerned with such properties as the following:

a) the *range* of T—the set of all $\mathbf{y}$ for which $T(\mathbf{x}) = \mathbf{y}$ for at least one $\mathbf{x}$;

b) whether T maps V^n *onto* V^m—that is, whether the range of T is all of V^m;

c) whether T is one-to-one—that is, whether, for every $\mathbf{y}$, $T(\mathbf{x}) = \mathbf{y}$ has at most one solution $\mathbf{x}$;

d) the set of all $\mathbf{x}$ for which $T(\mathbf{x}) = \mathbf{0}$; this set is called the *kernel* of T;

e) the set of all $\mathbf{x}$ for which $T(\mathbf{x}) = \mathbf{y}_0$, where $\mathbf{y}_0$ is a given element of V^m.

We remark that $T(\mathbf{0})$ must be $\mathbf{0}$. For $T(\mathbf{0}) = T(0\mathbf{0}) = 0T(\mathbf{0}) = \mathbf{0}$. Hence *the kernel of T always contains the zero vector of V^n*. With regard to (c) and (e) we have a useful rule:

THEOREM   Let T be a linear mapping of V^n into V^m and let $T(\mathbf{x}_0) = \mathbf{y}_0$ for a particular $\mathbf{x}_0$ and $\mathbf{y}_0$. Then all solutions $\mathbf{x}$ of the equation $T(\mathbf{x}) = \mathbf{y}_0$ are given by

$$\mathbf{x} = \mathbf{x}_0 + \mathbf{z},$$

where $\mathbf{z}$ is an arbitrary vector in the kernel of T. Hence T is one-to-one precisely when the kernel of T consists of $\mathbf{0}$ alone.

Proof. We are given that $T(\mathbf{x}_0) = \mathbf{y}_0$. If also $T(\mathbf{x}) = \mathbf{y}_0$, then

$$T(\mathbf{x} - \mathbf{x}_0) = T(\mathbf{x}) - T(\mathbf{x}_0) = \mathbf{y}_0 - \mathbf{y}_0 = \mathbf{0}.$$

Hence $\mathbf{x} - \mathbf{x}_0$ is an element $\mathbf{z}$ of the kernel of T:

$$\mathbf{x} - \mathbf{x}_0 = \mathbf{z} \quad \text{or} \quad \mathbf{x} = \mathbf{x}_0 + \mathbf{z}.$$

Conversely, if $\mathbf{x} = \mathbf{x}_0 + \mathbf{z}$, where $\mathbf{z}$ is in the kernel of T, then

$$T(\mathbf{x}) = T(\mathbf{x}_0) + T(\mathbf{z}) = T(\mathbf{x}_0) + \mathbf{0} = \mathbf{y}_0.$$

The last sentence of the theorem follows from the previous result, since T is one-to-one precisely when the equation $T(\mathbf{x}) = \mathbf{y}_0$ has exactly one solution, for every $\mathbf{y}_0$ for which there is a solution. $\square$

EXAMPLE 3   Let T map V^2 into V^3 and have the matrix

$$A = \begin{bmatrix} 2 & 4 \\ 3 & 6 \\ 1 & 2 \end{bmatrix}.$$

Then $A\mathbf{x} = \mathbf{0}$ is equivalent to

$$2x_1 + 4x_2 = 0, \qquad 3x_1 + 6x_2 = 0, \qquad x_1 + 2x_2 = 0.$$

This is satisfied for $x_1 = -2x_2$, that is, by all vectors $t(2, -1)$ of V^2. We note that $A \operatorname{col}(2, 3) = (16, 24, 8)$; that is, $T(\mathbf{x}) = (16, 24, 8)$ is satisfied for $\mathbf{x} = \mathbf{x}_0 = (2, 3)$. Hence all vectors $\mathbf{x}$ such that $T(\mathbf{x}) = (16, 24, 8)$ are given by

$$\mathbf{x} = (2, 3) + t(2, -1),$$

where t is arbitrary.   ■

EXAMPLE 4 T maps V^3 into V^2 and

$$T(x_1, x_2, x_3) = (y_1, y_2) = (x_1 - x_2, x_2 - x_3).$$

To find the kernel, we must solve the equations $x_1 - x_2 = 0$, $x_2 - x_3 = 0$. The solutions are given by all (x_1, x_2, x_3) such that $x_1 = x_2 = x_3$, hence by all vectors $t(1, 1, 1)$, where t is an arbitrary scalar. Here T is not one-to-one. However, the range of T is all of V^2, for the equations

$$x_1 - x_2 = y_1, \qquad x_2 - x_3 = y_2$$

can always be solved for x_1, x_2, x_3—for example, by taking $x_3 = 0$, $x_2 = y_2$, and $x_1 = y_1 + y_2$. Therefore T maps V^3 *onto* V^2 but is not one-to-one.   ∎

EXAMPLE 5 T maps V^n into V^m and $T(x) = 0$ for every x. Thus T has the matrix $O = O_{mn}$. We call T the *zero mapping* and sometimes also denote this mapping by O. T is clearly linear and is neither one-to-one nor onto.   ∎

EXAMPLE 6 T maps V^n into V^n and $T(x) = x$ for every x. Thus T has the matrix $I = I_n$. We call T the *identity mapping* and sometimes also denote this mapping by I. T is clearly linear and is one-to-one and onto.   ∎

EXAMPLE 7   Let T map V^3 into V^3 and have the matrix

$$A = \begin{bmatrix} 2 & 3 & 1 \\ 1 & 0 & 2 \\ 1 & 2 & 0 \end{bmatrix}.$$

Find the range of T.   ∎

Solution.   The range of T consists of all $y = x_1 u_1 + x_2 u_2 + x_3 u_3$, where u_1, u_2, u_3 are the column vectors of A:

$$u_1 = \text{col}\,(2, 1, 1), \qquad u_2 = \text{col}\,(3, 0, 2), \qquad u_3 = \text{col}\,(1, 2, 0).$$

Thus the range consists of all linear combinations of u_1, u_2, u_3. If u_1, u_2, u_3 are linearly independent, then they form a basis for V^3, and the range is V^3. However, we verify that $2u_1 - u_2 - u_3 = 0$, so that these vectors are linearly dependent. Furthermore, we see that u_1, u_2 are linearly independent and that u_3 is expressible as a linear combination of u_1, u_2. Hence the range is given by all linear combinations of u_1 and u_2, and this set is not all of V^3. Thus T does not map V^3 onto V^3, and the equation $T(x) = y_0$ has no solution for x, for some choices of y_0; for example, $T(x) = (1, 0, 0)$ has no solution, as one can verify (the vector $(1, 0, 0)$ is not a linear combination of u_1 and u_2).

PROBLEMS

In these problems, all vectors are written as column vectors. Also, the following matrices are referred to:

$$A = \begin{bmatrix} 2 & 1 \\ 3 & 5 \end{bmatrix}, \quad B = \begin{bmatrix} 2 & 3 \\ 4 & 6 \end{bmatrix}, \quad C = \begin{bmatrix} 1/\sqrt{2} & -1/\sqrt{2} \\ 1/\sqrt{2} & 1/\sqrt{2} \end{bmatrix}, \quad D = \begin{bmatrix} 1 & 0 \\ 0 & -1 \end{bmatrix},$$

$$E = \begin{bmatrix} -1 & 0 \\ 0 & -1 \end{bmatrix}, \quad F = \begin{bmatrix} 3 & 1 & 2 \\ 0 & 2 & 4 \end{bmatrix}, \quad G = \begin{bmatrix} 1 & 4 & 3 \\ 2 & 8 & 6 \end{bmatrix}.$$

$$H = \begin{bmatrix} 2 & 1 \\ 4 & 2 \\ -6 & -3 \end{bmatrix}, \quad J = \begin{bmatrix} 2 & 1 \\ 1 & 2 \\ 1 & 2 \end{bmatrix}, \quad K = \begin{bmatrix} 3 & 1 & 2 \\ 1 & 0 & 1 \\ 5 & 2 & 3 \end{bmatrix}, \quad L = \begin{bmatrix} 1 & 0 & 1 \\ 0 & 1 & 0 \\ 0 & 1 & 1 \end{bmatrix},$$

$$M = \begin{bmatrix} 2 & 0 & 0 \\ 0 & 2 & 0 \\ 0 & 0 & 2 \end{bmatrix} = 2I, \quad N = \begin{bmatrix} 1 & 0 & 0 \\ 0 & 2 & 0 \\ 0 & 0 & 3 \end{bmatrix}.$$

1. Let the linear mapping T have the matrix A.

   a) Evaluate $T(1,0)$, $T(0,1)$, $T(2, -1)$, $T(-1,1)$ and graph.

   b) Find the kernel of T, determine whether T is one-to-one, and find all $\mathbf{x}$ such that $T(\mathbf{x}) = (2, 3)$.

   c) Find the range of T and determine whether T maps V^2 onto V^2.

2. Let the linear mapping T have the matrix B.

   a) Evaluate $T(1,0)$, $T(0,1)$, $T(1, -1)$, $T(-1, -1)$ and graph.

   b) Find the kernel of T, determine whether T is one-to-one, and find all $\mathbf{x}$ such that $T(\mathbf{x}) = (2, 4)$.

   c) Find the range of T and determine whether T maps V^2 onto V^2.

3. Let T map V^n into V^m and have the matrix F.

   a) Find n and m.

   b) Find the kernel of T, and determine whether T is one-to-one.

   c) Find the range of T and determine whether T maps V^n onto V^m.

4. (a), (b), (c) Proceed as in Problem 3 with matrix G.

5. (a), (b), (c) Proceed as in Problem 3 with matrix H.

6. (a), (b), (c) Proceed as in Problem 3 with matrix J.

7. (a), (b), (c) Proceed as in Problem 3 with matrix K.

8. (a), (b), (c) Proceed as in Problem 3 with matrix L.

9. Let the linear mapping T have the equation $\mathbf{y} = C\mathbf{x}$. For general $\mathbf{x}$, find the angle between $\mathbf{x}$ and $T(\mathbf{x}) = \mathbf{y}$, as vectors in V^2, and also compare $|\mathbf{x}|$ and $|T(\mathbf{x})|$. From these results, interpret T geometrically. [Hint: Consider $\mathbf{x}$ as $\overrightarrow{OP}$ and $\mathbf{y}$ as $\overrightarrow{OQ}$, where O is the origin of E^2 and P and Q are points of E^2.]

10. Let the linear mapping T have the equation $\mathbf{y} = D\mathbf{x}$. Regard $\mathbf{x}$ as $\overrightarrow{OP}$, $\mathbf{y}$ as $\overrightarrow{OQ}$, as in Problem 9, and describe geometrically the relation between $\mathbf{x}$ and $\mathbf{y} = T(\mathbf{x})$.

11. Interpret each of the following linear mappings geometrically, as in Problems 9 and 10:

    a) $\mathbf{y} = E\mathbf{x}$,   b) $\mathbf{y} = M\mathbf{x}$.

12. If T maps V^n into V^n, then $\mathbf{x}$ and $\mathbf{y} = T(\mathbf{x})$ are vectors in the same n-dimensional vector space and can be compared (see Problems 9, 10, and 11). If $T(\mathbf{x})$ is a scalar multiple of $\mathbf{x}$, for some nonzero $\mathbf{x}$, say $T(\mathbf{x}) = \lambda\mathbf{x}$, then we say that $\mathbf{x}$ is an *eigenvector* of T, associated with the *eigenvalue* λ. If T has the matrix A, then $\mathbf{x}$ is also an eigenvector of A, associated with the eigenvalue λ, as in Section 1.10.

    Prove: If $\mathbf{v}_1, \ldots, \mathbf{v}_k$ are eigenvectors of the linear mapping T, associated with distinct eigenvalues $\lambda_1, \ldots, \lambda_k$, then $\mathbf{v}_1, \ldots, \mathbf{v}_k$ are linearly independent. [Hint: Let T have the matrix A. Use induction. Verify that the assertion is true for $k = 1$. Assume that it is true for $k = m < n$ and prove it for $k = m + 1$ by assuming $c_1\mathbf{v}_1 + \cdots +$

$c_{m+1}\mathbf{v}_{m+1} = \mathbf{0}$, multiplying by A on the left, and then eliminating $\mathbf{v}_{m+1}$ to obtain

$$c_1(\lambda_{m+1} - \lambda_1)\mathbf{v}_1 + \cdots + c_m(\lambda_{m+1} - \lambda_m)\mathbf{v}_m = \mathbf{0}.$$

Now use the induction hypothesis to conclude that $c_1 = 0, \ldots, c_m = 0$ and hence also $c_{m+1} = 0$.]

13. **a)** Let T be a linear mapping of V^n into V^m. Prove: If $\mathbf{v}_1, \ldots, \mathbf{v}_k$ are linearly dependent vectors of V^n, then $T(\mathbf{v}_1), \ldots, T(\mathbf{v}_k)$ are linearly dependent vectors of V^m.

    Is the converse true? Explain.

    b) Let A be a nonsingular $n \times n$ matrix. Show that if $\mathbf{v}_1, \ldots, \mathbf{v}_k$ are vectors of V^n, then they are linearly independent if and only if $A\mathbf{v}_1, \ldots, A\mathbf{v}_k$ are linearly independent.

14. **a)** Let $\mathbf{u}_1, \ldots, \mathbf{u}_n$ be a basis for V^n and let $\mathbf{v}_1, \ldots, \mathbf{v}_m$ be a basis for V^m. Let T be a linear mapping of V^n into V^m, so that T assigns to each vector $\mathbf{u} = p_1\mathbf{u}_1 + \cdots + p_n\mathbf{u}_n$ of V^n a vector $\mathbf{v} = q_1\mathbf{v}_1 + \cdots + q_m\mathbf{v}_m$ of V^m. Show that

$$\text{col}(q_1, \ldots, q_m) = B \, \text{col}(p_1, \ldots, p_n), \qquad B = (b_{ij}),$$

    where B is the $m \times n$ matrix such that

$$T(\mathbf{u}_j) = b_{1j}\mathbf{v}_1 + \cdots + b_{mj}\mathbf{v}_m, \qquad j = 1, \ldots, n.$$

    Thus for each choice of bases in V^n and V^m there is a matrix B representing the linear mapping T.

    b) With reference to part (a), let

$$\mathbf{u}_j = u_{1j}\mathbf{e}_1 + \cdots + u_{nj}\mathbf{e}_n \quad \text{for} \quad j = 1, \ldots, n,$$

$$\mathbf{v}_i = v_{i1}\mathbf{e}_1 + \cdots + v_{im}\mathbf{e}_m \quad \text{for} \quad i = 1, \ldots, m,$$

    in terms of the standard bases in V^n and V^m, respectively. Show that

$$B = V^{-1}AU,$$

    where V is the $m \times m$ matrix (v_{ij}), U is the $n \times n$ matrix (u_{ij}), and $\mathbf{y} = A\mathbf{x}$ is the representation of T in terms of standard bases in V^n and V^m as in Section 1.15. (Why is V nonsingular?)

    c) With reference to parts (a) and (b), let T be a linear mapping from V^n to V^n and let only one basis be used, so that $\mathbf{v}_1 = \mathbf{u}_1, \ldots, \mathbf{v}_n = \mathbf{u}_n$. Show that in part (b), B is similar to A. Show further that if the basis $\mathbf{u}_1, \ldots, \mathbf{u}_n$ is orthonormal, then B is orthogonally congruent to A.

*1.16   SUBSPACES · RANK OF A MATRIX

By a *subspace* of V^n we mean a collection W of vectors of V^n such that

(i) W contains $\mathbf{0}$, and

(ii) if $\mathbf{u}, \mathbf{v}$ are in W, then so is each linear combination $a\mathbf{u} + b\mathbf{v}$.

Let W be a subspace and let W contain vectors besides $\mathbf{0}$. We then seek sets of linearly independent vectors in W. Each such set contains at most n vectors but may in fact always contain fewer than n. We let k be the largest integer such that

W contains k linearly independent vectors, and call k the *dimension* of W. When W contains $\mathbf{0}$ alone, we assign the dimension 0 to W.

When $k = n$, W necessarily coincides with V^n. For then W contains linearly independent vectors $\mathbf{u}_1, \ldots, \mathbf{u}_n$, and by repeatedly applying (ii) we conclude that W contains each linear combination of $\mathbf{u}_1, \ldots, \mathbf{u}_n$. Since $\mathbf{u}_1, \ldots, \mathbf{u}_n$ form a basis for V^n, $W = V^n$.

For $k = 1$, W consists of all scalar multiples of a single vector $\mathbf{u}_1$, that is, all $\mathbf{w} = w_1 \mathbf{u}_1$. If, say, $n = 3$, then the corresponding points P such that $\overrightarrow{OP} = w_1 \mathbf{u}_1$ trace out a line L through O in 3-dimensional space (Fig. 1.18a).

If $k = 2$, then W contains two linearly independent vectors $\mathbf{u}_1, \mathbf{u}_2$ and, by (ii), contains all the linear combinations

$$\mathbf{w} = w_1 \mathbf{u}_1 + w_2 \mathbf{u}_2.$$

These linear combinations exhaust W, since any vector not expressible in this form would have to be linearly independent of $\mathbf{u}_1, \mathbf{u}_2$ (Rule (c) of Section 1.13). If again $n = 3$, then the corresponding points P such that $\overrightarrow{OP} = \mathbf{w}$ fill a plane in space, as in Fig. 1.18b.

In general, we see that if W has dimension k, then W consists of all linear combinations

$$\mathbf{w} = w_1 \mathbf{u}_1 + \cdots + w_k \mathbf{u}_k, \tag{1.123}$$

where $\mathbf{u}_1, \ldots, \mathbf{u}_k$ are linearly independent vectors of W. We term $\mathbf{u}_1, \ldots, \mathbf{u}_k$ a *basis* of W. There are many choices of basis for a given W, but all have the same number of vectors (Problem 4 which follows below).

Geometrically, the subspaces of V^n correspond to the point O ($k = 0$), and lines, planes, and "hyperplanes" through O, in E^n, and to E^n itself ($k = n$).

Now let $\mathbf{v}_1, \ldots, \mathbf{v}_k$ be k vectors of V^n, not necessarily linearly independent, $k \geq 1$. We then form the set W of all linear combinations $w_1 \mathbf{v}_1 + \cdots + w_k \mathbf{v}_k$ of these vectors. We see at once that conditions (i) and (ii) are satisfied. Hence W is a subspace. If the vectors are in fact linearly independent, then, as before, $\mathbf{v}_1, \ldots, \mathbf{v}_k$ must be a basis of W, and k is the dimension of W.

If, however, they are linearly dependent, then some of the vectors $\mathbf{v}_1, \ldots, \mathbf{v}_k$ can be omitted, and we see that W has dimension $k' < k$. We can choose k' as the

(a)   (b)

Figure 1.18   (a) One-dimensional subspace of V^3.   (b) Two-dimensional subspace of V^3.

largest integer such that k' of the vectors $v_1, \ldots, v_k$ are linearly independent. For example, we can suppose by renumbering that $v_1, \ldots, v_{k'}$ are such a linearly independent set (as large as possible) in W. Then $v_1, \ldots, v_{k'}$ form a basis for W.

The practical determination of a basis for W can be carried out by Gaussian elimination as in Section 1.9. We form the $k \times n$ matrix A whose row vectors are the given vectors $v_1, v_2, \ldots, v_k$. Thus W consists of all linear combinations of the rows of A; we call W the *row space* of A. We then use the steps I. (adding multiples of one row to the other rows) and II. (interchanging two rows) of Section 1.9. Neither of these changes the row space of A. For II. no proof is needed. For I., $v_1, \ldots, v_k$ are replaced by $v_1 + c_1 v_h, \ldots, v_k + c_k v_h$, where $c_h = 0$. Every linear combination of these vectors is a linear combination of $v_1, \ldots, v_k$ and is hence in W. Also, every vector in W can be written as a linear combination of the new vectors, since v_h is one of the new vectors and $v_i = (v_i + c_i v_h) - c_i v_h$ for $i \neq h$. Hence the row space is unchanged.

As in Section 1.9, we can apply the two steps to reduce A to row echelon form with nonzero row vectors $u_1, \ldots, u_r$ and (for $r < k$), the following rows are 0. The vectors $u_1, \ldots, u_r$ are linearly independent as one sees at once from the row echelon form. Thus $u_1, \ldots, u_r$ are a basis for the subspace W and W has dimension r, equal to the rank of A, as in Section 1.9. We conclude that *the rank of a matrix A is the maximal number of linearly independent row vectors of A.* From the above analysis it follows that if A is $k \times n$, then the rank r cannot exceed n or k.

EXAMPLE 1   Let $v_1 = (2, 4, -1, 1)$, $v_2 = (4, -2, 1, 1)$, $v_3 = (2, 14, -4, 2)$, and let W be the set of all linear combinations of v_1, v_2, v_3, so that W is a subspace of V^4. We seek a basis for W and form

$$A = \begin{bmatrix} 2 & 4 & -1 & 1 \\ 4 & -2 & 1 & 1 \\ 2 & 14 & -4 & 2 \end{bmatrix}.$$

Gaussian elimination gives

$$\begin{bmatrix} 2 & 4 & -1 & 1 \\ 0 & -10 & 3 & -1 \\ 0 & 10 & -3 & 1 \end{bmatrix} \begin{bmatrix} 2 & 4 & -1 & 1 \\ 0 & -10 & 3 & -1 \\ 0 & 0 & 0 & 0 \end{bmatrix}.$$

Hence $u_1 = (2, 4, -1, 1)$, $u_2 = (0, -10, 3, -1)$ form a basis, and W has dimension $2 = r =$ rank of A.   ■

One can also show that *the rank of A equals the maximum number of linearly independent column vectors of A.* If A is $k \times n$, the column vectors are n vectors of V^k, say, $v_1, \ldots, v_n$, and the corresponding linear combinations form a subspace, which is the range of the mapping $y = Ax$ from V^n to V^k (Section 1.15). Thus our assertion is equivalent to the statement that *the rank of A equals the dimension of the range of A.* For a proof see Problem 7 following this section.

We can summarize our results as follows:

column rank of A = row rank of A = dimension of range of A.

EXAMPLE 2   We again consider the matrix A of Example 1. The range of A consists of all linear combinations of

$$\mathbf{w}_1 = \text{col}\,(2, 4, 2), \qquad \mathbf{w}_2 = \text{col}\,(4, -2, 14),$$

$$\mathbf{w}_3 = \text{col}\,(-1, 1, -4), \qquad \mathbf{w}_4 = \text{col}\,(1, 1, 2).$$

We find a basis for the range by Gaussian elimination applied to the matrix (transpose of A):

$$A' = \begin{bmatrix} 2 & 4 & 2 \\ 4 & -2 & 14 \\ -1 & 1 & -4 \\ 1 & 1 & 2 \end{bmatrix}.$$

We obtain

$$\begin{bmatrix} 2 & 4 & 2 \\ 0 & -10 & 10 \\ 0 & 3 & -3 \\ 0 & -1 & 1 \end{bmatrix} \quad \begin{bmatrix} 2 & 4 & 2 \\ 0 & -10 & 10 \\ 0 & 0 & 0 \\ 0 & 0 & 0 \end{bmatrix}.$$

Therefore $\mathbf{z}_1 = \text{col}(2, 4, 2)$, $\mathbf{z}_2 = \text{col}(0, -10, 10)$ are a basis for the range; the dimension of the range is $2 = r$, as expected.   ■

The kernel.   The kernel of A consists of all $\mathbf{x}$ in V^n such that $A\mathbf{x} = \mathbf{0}$. The kernel satisfies (i) and (ii) and is hence a subspace of V^n. For $A\mathbf{0} = \mathbf{0}$ and if $A\mathbf{u} = \mathbf{0}$, $A\mathbf{v} = \mathbf{0}$, then $A(a\mathbf{u} + b\mathbf{v}) = aA\mathbf{u} + bA\mathbf{v} = \mathbf{0}$.

We let h be the dimension of the kernel of A; h is called the *nullity* of A.

Relation between rank and nullity.   Let A be a $k \times n$ matrix of rank r and nullity h. Then we have

$$r + h = n. \tag{1.124}$$

To show this, we consider the corresponding homogeneous simultaneous linear equations

$$a_{11}x_1 + \quad \cdots \quad + a_{1n}x_n = 0,$$

$$\vdots \tag{1.125}$$

$$a_{k1}x_1 + \quad \cdots \quad + a_{kn}x_n = 0,$$

or

$$A\mathbf{x} = \mathbf{0}. \tag{1.125'}$$

As we did in Section 1.9, we can solve these by Gaussian elimination, reducing A to the row echelon form. The solutions are given by linear expressions for r of the unknowns in terms of the remaining unknowns. By renumbering we can assume that $x_1, \ldots, x_r$ are expressed in terms of $x_{r+1}, \ldots, x_n$:

$$x_1 = d_{1, r+1} x_{r+1} + \quad \cdots \quad + d_{1, n} x_n$$

$$\vdots$$

$$x_r = d_{r, r+1} x_{r+1} + \quad \cdots \quad + d_{r, n} x_n.$$

Here we write

$$x_{r+1} = w_{r+1},$$

$$\vdots$$

$$x_n = w_n.$$

Then the solutions are given by

$$\mathbf{x} = w_{r+1} \operatorname{col}(d_{1, r+1}, \ldots, d_{r, r+1}, 1, 0, \ldots, 0)$$

$$+ \cdots + w_n \operatorname{col}(d_{1, n}, \ldots, d_{r, n}, 0, \ldots, 0, 1)$$

$$= w_{r+1} \mathbf{u}_{r+1} + \cdots + w_n \mathbf{u}_n, \tag{1.126}$$

where $w_{r+1}, \ldots, w_n$ are arbitrary scalars. Thus the kernel of A is represented as the set of linear combinations of $n - r$ vectors $\mathbf{u}_{r+1}, \ldots, \mathbf{u}_n$. We see that these vectors are linearly independent (see Problem 5 below). Therefore the kernel has dimension $n - r$ or $h = n - r$. Equation (1.124) is proved. [Note: As in Section 1.9, if $r = 0$, $A\mathbf{x} = \mathbf{0}$ is satisfied by all $\mathbf{x}$, so that $h = n$.] □

Maximum rank. Determinant definition of rank. For a $k \times n$ matrix A we have seen that the rank r of A cannot exceed k or n. When r is the largest integer satisfying this condition, A is said to have *maximum rank*. If $k \leq n$, this means that $r = k$; if $k > n$, it means that $r = n$.

By a *minor* of A we mean a determinant formed from the array A by striking out certain rows and columns to obtain a square array. If, for example, A is 3×3, then A has one 3×3 minor det A, nine 2×2 minors, and nine 1×1 minors (the nine entries of A).

One can show that r, the rank of A, is the largest integer such that some $r \times r$ minor of A is nonzero (see Problem 6 which follows).

EXAMPLE 3   We again consider the matrix A of Example 1. The kernel of A is found from the Gaussian elimination applied to A, as was done in Example 1. Hence it consists of all $\mathbf{x}$ such that

$$2x_1 + 4x_2 - x_3 + x_4 = 0,$$

$$- 10x_2 + 3x_3 - x_4 = 0.$$

We solve for x_1, x_2, letting $x_3 = w_3$, $x_4 = w_4$. We find

$$x_1 = -\tfrac{1}{10} w_3 - \tfrac{3}{10} w_4, \qquad x_2 = \tfrac{3}{10} w_3 - \tfrac{1}{10} w_4.$$

Thus

$$\mathbf{x} = (x_1, x_2, x_3, x_4) = \left(-\tfrac{1}{10} w_3 - \tfrac{3}{10} w_4, \tfrac{3}{10} w_3 - \tfrac{1}{10} w_4, w_3, w_4\right)$$

$$= w_3\left(-\tfrac{1}{10}, \tfrac{3}{10}, 1, 0\right) + w_4\left(-\tfrac{3}{10}, -\tfrac{1}{10}, 0, 1\right),$$

where $-\infty < w_3 < \infty$, $-\infty < w_4 < \infty$. Accordingly,

$$\mathbf{p_1} = \left(-\tfrac{1}{10}, \tfrac{3}{10}, 1, 0\right), \qquad \mathbf{p_2} = \left(-\tfrac{3}{10}, -\tfrac{1}{10}, 0, 1\right)$$

are a basis for the kernel. Thus $h = 2 = 4 - r$ as in Eq. (1.124).
We can also check that the 3×3 minors of A are all 0:

$$\begin{vmatrix} 2 & 4 & -1 \\ 4 & -2 & 1 \\ 2 & 14 & -4 \end{vmatrix} = 0, \qquad \begin{vmatrix} 4 & -1 & 1 \\ -2 & 1 & 1 \\ 14 & -4 & 2 \end{vmatrix} = 0, \dots$$

but there are nonzero 2×2 minors:

$$\begin{vmatrix} 2 & 4 \\ 4 & -2 \end{vmatrix} = -20.$$

Hence again we find $r = 2$. ∎

PROBLEMS

1. Find the rank of the matrix in two different ways and verify equality:

a) $\begin{bmatrix} 3 & 1 & 2 \\ 2 & 5 & 0 \\ 5 & -7 & 6 \end{bmatrix}$

c) $\begin{bmatrix} 0 & 1 & 3 & 4 \\ 1 & 0 & 0 & 5 \\ 3 & 0 & 2 & 3 \end{bmatrix}$

b) $\begin{bmatrix} 1 & 2 & 2 \\ 3 & 5 & 4 \\ -1 & 4 & 2 \end{bmatrix}$

d) $\begin{bmatrix} 2 & 1 & 0 & 1 & 2 \\ 1 & 0 & 1 & 1 & 3 \\ 5 & 3 & -1 & 2 & 3 \\ 0 & -1 & 2 & 1 & 4 \end{bmatrix}$

2. (a), (b), (c), (d) Find the nullity h of each matrix of Problem 1 and check that $h + r = n$; also find a basis for the kernel if $h > 0$.

3. Let W be the subspace formed of all linear combinations of the vectors given. Find a basis for W:

a) $(1,3,2)$, $(2,-1,2)$, $(1,10,4)$, $(0,1,1)$,

b) $(3,6,2)$, $(1,3,1)$, $(5,1,7)$,

c) $(1,2,2,3)$, $(3,2,2,1)$, $(1,1,1,1)$, $(2,1,1,0)$,

d) $(3,1,5,0,1)$, $(2,0,4,2,0)$, $(1,-1,3,4,-1)$, $(1,1,0,0,0)$.

4. Let W be a subspace of V^n of dimension k. Prove: Every basis of W has k vectors. [Hint: Show first that no basis can have more than k vectors. Next suppose that a basis has fewer than k vectors. By completing the given basis of k vectors to a basis for V^n, show that one could then obtain a basis for V^n of less than n vectors.]

5. Show that the vectors $\mathbf{u}_{r+1}, \dots, \mathbf{u}_n$ in Eq. (1.126) are linearly independent.

6. Prove that the evaluation of rank by nonzero minors is correct. [Hint: Show that this rank is unaffected by Gaussian elimination and then find its value for a matrix in row echelon form.]

7. Prove: the row rank of A equals the column rank. [Hint: Use the result of Problem 6 and Rule I. of Section 1.3 to show that A and A' have the same row rank.]

8. For a given λ and given $n \times n$ matrix A, let W be the set of vectors $\mathbf{x}$ such that $A\mathbf{x} = \lambda\mathbf{x}$. Show that W is a subspace of V^n. (If λ is an eigenvalue of A, W consists of $\mathbf{0}$ plus all eigenvectors for this λ.)

9. Prove: For a given **y**, the equation $A\mathbf{x} = \mathbf{y}$ has a solution if and only if the matrices A and $[A \ \mathbf{y}]$ have the same rank. Here $[A \ \mathbf{y}]$ is obtained from A by adjoining the column vector **y** at the right. [Hint: Consider the relationship between **y** and the range of A as a subspace.]

*1.17 OTHER VECTOR SPACES

Thus far in this chapter we have considered only Euclidean n-dimensional vector spaces. One is often led to consider sets of objects that can be added and multiplied by scalars, in accordance with the familiar rules, but for which there may be no scalar product or norm. We call such a set, with the given operations, a *vector space*. The term "Euclidean" is reserved for a vector space having a scalar product for which *all* the rules (1.102) of Section 1.13 are satisfied. In each vector space, one can define linear independence and dependence in the usual way. One then calls the vector space *n-dimensional* if it contains n but no more than n linearly independent vectors. For some vector spaces, one can find n linearly independent vectors for *every* positive integer n; one then speaks of an *infinite-dimensional* vector space. For technical reasons it is also useful to introduce a zero-dimensional vector space V^0, consisting of **0** alone; this vector space has no sets of linearly independent vectors.

For convenience we define a vector space formally:

Definition. A vector space V is a collection of objects $\mathbf{u}, \mathbf{v}, \ldots$ called vectors, including a zero vector **0**, for which addition and multiplication by real scalars are defined in accordance with the following rules:

$$
\begin{aligned}
&\text{I. } \mathbf{u} + \mathbf{v} = \mathbf{v} + \mathbf{u}. &&\text{II. } (\mathbf{u} + \mathbf{v}) + \mathbf{w} = \mathbf{u} + (\mathbf{v} + \mathbf{w}).\\
&\text{III. } h(\mathbf{u} + \mathbf{v}) = h\mathbf{u} + h\mathbf{v}. &&\text{IV. } (a + b)\mathbf{u} = a\mathbf{u} + b\mathbf{u}.\\
&\text{V. } (ab)\mathbf{u} = a(b\mathbf{u}). &&\text{VI. } 1\mathbf{u} = \mathbf{u}. &&&(1.128)\\
&\text{VII. } 0\mathbf{u} = \mathbf{0}.
\end{aligned}
$$

By a *subspace* of V we mean a subset W of V satisfying the two conditions (i) and (ii) of Section 1.16. We observe that the rules (1.128) must also hold in such a subspace W. Thus W is also a vector space.

For many vector spaces the objects called vectors do not arise in a geometrical context, and our standard boldface or arrow notations are not appropriate; see, for instance, Examples 1 and 2 below.

As was pointed out in Section 1.14, other rules can be deduced from Rules I through VII—for example, the rule stating that subtraction is always possible and is unique and the rule $\mathbf{u} + \mathbf{0} = \mathbf{u}$. Furthermore, the discussion of basis and linear independence carries over to an arbitrary vector space. In particular, Rules (a), (b), (c), (d), and (e) of Section 1.13 remain valid. Rules (f) and (g) are essentially the definition of an n-dimensional vector space. Rules (h), (i), (j), and (k) remain valid in the following forms:

h') If $\mathbf{v}_1, \ldots, \mathbf{v}_n$ are linearly independent vectors in an n-dimensional vector space V, then $\mathbf{v}_1, \ldots, \mathbf{v}_n$ form a basis for V.

i') Every basis for an n-dimensional vector space consists of n linearly independent vectors.

j') If $k < n$ and $v_1, \ldots, v_k$ are linearly independent vectors in an n-dimensional vector space V, then there exist $v_{k+1}, \ldots, v_n$ so that $v_1, \ldots, v_n$ form a basis for V.

k') If $v_1, \ldots, v_k$ are linearly independent vectors in a vector space, V and $u_1, \ldots, u_{k+1}$ are all expressible as linear combinations of $v_1, \ldots, v_k$, then $u_1, \ldots, u_{k+1}$ are linearly dependent.

The discussion of a general n-dimensional vector space V can in fact be referred back to the case of V^n by the following procedure. Let $v_1, \ldots, v_n$ be n linearly independent vectors in V. Since V does not contain $n + 1$ linearly independent vectors, it follows that every vector of V is expressible uniquely as a linear combination of $v_1, \ldots, v_n$, so that $v_1, \ldots, v_n$ form a basis. Hence for every vector u in V we can write

$$u = u_1 v_1 + \cdots + u_n v_n$$

and, as in Section 1.13, we call $u_1, \ldots, u_n$ the components of u with respect to the basis $v_1, \ldots, v_n$. We can identify the vectors of V with the corresponding n-tuples, writing $u = (u_1, \ldots, u_n)$. Then

$$cu = (cu_1, \ldots, cu_n), \qquad u + w = (u_1 + w_1, \ldots, u_n + w_n)$$

exactly as in V^n. Hence (apart from considerations related to the scalar product or norm), V is the same as V^n.

For a general, not necessarily finite-dimensional, vector space V, one can take advantage of the theory of V^n in the following way. Let $v_1, \ldots, v_n$ be linearly independent vectors in V. Then all the linear combinations of $v_1, \ldots, v_n$ themselves form a vector space V', as in the preceding paragraph, and we can identify V' with V^n by replacing each vector of V' by the corresponding n-tuple of components with respect to $v_1, \ldots, v_n$. Hence V' forms an n-dimensional vector space. We call V' an n-dimensional *subspace* of V.

As a special case of the preceding paragraph, we conclude that if all vectors in a vector space V are expressible as linear combinations of n linearly independent vectors $v_1, \ldots, v_n$, then V is n-dimensional and $v_1, \ldots, v_n$ form a basis for V.

We proceed to give some examples of vector spaces:

EXAMPLE 1　Let V consist of *all* polynomials in x: $3 - x$, $5 + x + 2x^2$, $7 - 3x^2 + 9x^3, \ldots$. The sum of two polynomials is again a polynomial; a scalar times a polynomial is again a polynomial:

$$3(5 + x + 2x^2) = 15 + 3x + 6x^2.$$

There is a zero polynomial: 0. Rules I through VII are satisfied, as one sees at once. For example, if $p(x)$ and $q(x)$ are polynomials, then $p(x) + q(x) = q(x) + p(x)$ (equality meaning identity). Thus Rule I holds. Hence V is a *vector space*. We observe that the n polynomials $1, x, x^2, \ldots, x^{n-1}$ are linearly independent. For suppose that

$$c_1 1 + c_2 x + \cdots + c_n x^{n-1} = 0;$$

that is, let $c_1 + c_2 x + \cdots + c_n x^{n-1}$ coincide with the 0 polynomial and hence have the value 0 for all x. Now a polynomial of degree k has at most k roots, and hence $c_1, \ldots, c_n$ must all be 0. Therefore 1, $x, \ldots, x^{n-1}$ are linearly independent. Accordingly, V is an *infinite-dimensional* vector space. ■

EXAMPLE 2   Let V consist of all polynomials of degree at most 3, that is, of all polynomials of the form $a_0 + a_1 x + a_2 x^2 + a_3 x^3$. The sum of two such polynomials is again a polynomial of degree at most 3; there is a similar statement for a scalar times such a polynomial. Rules I through VII all hold. Therefore V is a vector space. The four polynomials 1, x, x^2, x^3 are in V and are linearly independent. Thus we can consider V as the space V', as before, formed of all linear combinations of four linearly independent vectors in the vector space of Example 1. Accordingly, as before, V is a 4-dimensional subspace of the vector space of all polynomials. ■

EXAMPLE 3   Let V consist of all 2×2 matrices. Then again addition and multiplication by scalars yield matrices of the same size and, by the rules of Section 1.6, all of Rules I through VII are satisfied. The "zero vector" is $O = O_{22}$. In V the four matrices

$$E_{11} = \begin{bmatrix} 1 & 0 \\ 0 & 0 \end{bmatrix}, \qquad E_{12} = \begin{bmatrix} 0 & 1 \\ 0 & 0 \end{bmatrix}, \qquad E_{21} = \begin{bmatrix} 0 & 0 \\ 1 & 0 \end{bmatrix}, \qquad E_{22} = \begin{bmatrix} 0 & 0 \\ 0 & 1 \end{bmatrix}$$

are linearly independent. For the equation

$$c_1 E_{11} + c_2 E_{12} + c_3 E_{21} + c_4 E_{22} = O$$

is equivalent to

$$\begin{bmatrix} c_1 & c_2 \\ c_3 & c_4 \end{bmatrix} = O = \begin{bmatrix} 0 & 0 \\ 0 & 0 \end{bmatrix}$$

and hence to $c_1 = 0, \ldots, c_4 = 0$. Furthermore, every 2×2 matrix A is expressible uniquely as a linear combination of $E_{11}, \ldots, E_{22}$: $A = (a_{ij}) = a_{11} E_{11} + \cdots + a_{22} E_{22}$. It follows, as before, that V is 4-dimensional and that $E_{11}, \ldots, E_{22}$ form a basis for V.

By similar reasoning we show that for each fixed choice of m and n the set V of all $m \times n$ matrices forms a vector space of dimension mn. The mn matrices E_{ij}, having 1 as the ij-entry and all other entries 0, form a basis for V. ■

Complex vector spaces.   One can extend the concept of vector space by allowing the scalars to be complex numbers. One then obtains a *complex vector* space. The previous theory carries over with the obvious changes.

EXAMPLE 4   The set of all complex numbers forms a complex vector space, whose dimension is 1, with a basis consisting of any one nonzero complex number. ■

EXAMPLE 5 The set of all polynomials with complex coefficients forms an infinite-dimensional complex vector space. ∎

EXAMPLE 6 The set of all functions of form $ae^{2ix} + be^{-2ix}$, where a and b are complex constants, forms a complex vector space of dimension 2, with basis e^{2ix}, e^{-2ix}. ∎

PROBLEMS

1. Show that each of the following sets of objects, with the usual operations of addition and multiplication by scalars, forms a vector space. Give the dimension in each case and, if the dimension is finite, give a basis.

   a) All polynomials of degree at most 2.

   b) All polynomials containing no term of odd degree: $3 + 5x^2 + x^4$, $x^2 - x^{10}, \ldots$.

   c) all *trigonometric polynomials*:

   $$a_0 + a_1 \cos x + b_1 \sin x + \cdots + a_n \cos nx + b_n \sin nx.$$

   d) All functions of the form $ae^x + be^{-x}$.

   e) All 3×3 diagonal matrices.

   f) All 4×4 symmetric matrices A; that is, all matrices A such that $A = A'$.

   g) All functions $y = f(x)$, $-\infty < x < \infty$, such that $y'' + y = 0$.

   h) All functions $y = f(x)$, $-\infty < x < \infty$, such that $y''' - y' = 0$.

   i) All functions $f(x)$ which are defined and continuous for $0 \leq x \leq 1$.

   j) All functions $f(x)$ which are defined and have a continuous derivative for $0 \leq x \leq 1$.

   k) All infinite sequences: $x_1, x_2, \ldots, x_n, \ldots$.

   l) All convergent sequences.

2. Show that the four polynomials 1, $1 + x$, $1 + x + x^2$, $1 + x + x^3$ form a basis for the vector space of Example 2.

3. Show that the following matrices form a basis for the vector space of Example 3:

   $$\begin{bmatrix} 1 & 2 \\ 1 & -3 \end{bmatrix}, \begin{bmatrix} 0 & 1 \\ 2 & -1 \end{bmatrix}, \begin{bmatrix} 5 & 2 \\ 7 & 1 \end{bmatrix}, \begin{bmatrix} 0 & 1 \\ 1 & 0 \end{bmatrix}.$$

4. Show that the functions $\cos 2x, \sin 2x$ form a basis for the complex vector space of Example 6.

Suggested References

Birkhoff, G., and S. MacLane, Survey of Modern Algebra, 4th ed. New York: Macmillan, 1977.

Cullen, Charles G., Matrices and Linear Transformation, 2nd ed. Reading, Mass.: Addison-Wesley, 1972.

Curtis, Charles W., Linear Algebra: An Introductory Approach. New York: Springer-Verlag, 1984.

Gantmacher, F. R., The Theory of Matrices, transl. by K. A. Hirsch, 2 vols. New York: Chelsea Publishing Co., 1960.

Gerald, Curtis F., and Wheatley, Patrick O., Applied Numerical Analysis, 4th ed. Reading, Mass.: Addison-Wesley, 1989.

Kaplan, Wilfred, and Donald J. Lewis, Calculus and Linear Algebra, 2 vols. New York: John Wiley and Sons, Inc., 1970–1971.

Kuiper, Nicolaas H., Linear Algebra and Geometry, transl. by A. van der Sluis. New York: Interscience, 1962.

Perlis, Sam, Theory of Matrices. Reading, Mass.: Addison-Wesley, 1952.

2

Differential Calculus of Functions of Several Variables

2.1 FUNCTIONS OF SEVERAL VARIABLES

If to each point (x, y) of a certain part of the xy-plane is assigned a real number z, then z is said to be given as a function of the two real variables x and y. Thus

$$z = x^2 - y^2, \qquad z = x \sin xy \qquad [\text{all } (x, y)]$$

are such functions. Many such functions are considered, without special mention, in the theory of functions of one variable. For example, the function $y = a^x$ is a function of both a and x, as is the function $y = \log_a x$. The basic theorems on differentiation relate to the functions $y = u + v$, $y = u \cdot v$, $y = u/v$, that is, to certain functions of u and v, where u and v are functions of x. In many cases a function of one variable can be considered as a function of two variables with one of the two variables held fixed; for example, $y = x^3$ is obtained from $y = x^n$ (a function of x and n) by assigning to n the value 3.

Similar remarks apply to functions of three or more variables. Thus

$$u = xyz, \qquad u = x^2 + y^2 + z^2 - t^2$$

give u as a function of three and four variables, respectively.

It will be seen throughout the following discussion that the theory of functions of three or more variables differs only slightly from that of functions of two variables. For this reason, most of the emphasis will be on functions of two variables. On the other hand, there are basic differences between the calculus of functions of one variable and that of functions of two variables.

Functions of two, three, four, and even millions of variables occur in physics. The following are simple examples:

$$p = \frac{RT}{V} \quad \text{(ideal gas law)},$$

$$L = \frac{\pi r^4 \theta n}{2} \quad \text{(torque on a wire)},$$

$$E = \frac{m}{2} \sum_{i=1}^{N} \left(u_i^2 + v_i^2 + w_i^2 \right) \quad \left(\begin{array}{l} \text{energy of an ideal gas in terms of velocity} \\ \text{components of the } N \text{ molecules} \end{array} \right)$$

2.2 DOMAINS AND REGIONS

Most of the theory of functions of one variable is given in terms of a function defined over an interval: $a \le x \le b$. For functions of x and y a corresponding concept is needed. A natural one would be a rectangle: $a \le x \le b, c \le y \le d$. But many problems require more complicated sets, such as circles, ellipses, and so on. In order to have sufficient generality to cover all practical cases, it is necessary to formulate the concept of a domain.

The general term *set of points* in the xy-plane means any sort of collection of points, finite or infinite in number: the points $(0,0)$ and $(1,0)$; the points on the line $y = x$; the points inside the circle $x^2 + y^2 = 1$.

A *neighborhood* of a point (x_1, y_1) will mean the set of points inside a circle having center (x_1, y_1) and radius δ; we can thus speak of a *neighborhood of radius* δ. Each point (x, y) of the neighborhood satisfies the inequality:

$$(x - x_1)^2 + (y - y_1)^2 < \delta^2. \tag{2.1}$$

A set of points is called *open* if every point (x_1, y_1) of the set has a neighborhood lying wholly within the set. The interior of a circle is open, as is the interior of an ellipse or a square; these open sets are defined by inequalities such as the following:

$$x^2 + y^2 < 1; \qquad \frac{x^2}{2} + \frac{y^2}{3} < 1; \qquad |x| < 1 \quad \text{and} \quad |y| < 1.$$

The entire xy-plane is open, as is a half-plane such as the "right half-plane": $x > 0$. However, the interior of a circle plus the circumference is not open, for no neighborhood of a point on the circumference lies entirely in the set.

A set E is called *closed* if the points of the plane that are not in E form an open set. Thus the points on and outside the circle $x^2 + y^2 = 1$ form a closed set. The points of the circumference themselves form a closed set, as do the points on and interior to the circumference.

A set is called *bounded* if the whole set can be enclosed in a circle of sufficiently large radius. Thus the points of the square, $|x| \le 1, |y| \le 1$, form a bounded set; this set is also closed. The points interior to an ellipse, $x^2 + 2y^2 < 1$, form a bounded open set.

A nonempty open set is called a *connected open set* or a *domain* if, besides being open, it has the property that any two points P, Q of the set can be joined by a broken line lying wholly within the set. Thus the interior of a circle is a domain. (As in Section 1.15, the word *domain* is also used for the set on which a given function is defined. The context will make clear in which sense the word is to be understood. Furthermore, in many cases the domain of definition of a function is also a domain as defined here.)

We remark that a *domain D cannot be formed of two nonoverlapping open sets.* For example, the points for which $|x| > 0$ form an open set E composed of two parts: the set of points for which $x > 0$ and the set for which $x < 0$. The set E is not a domain, since the points $(-1, 0)$ and $(1, 0)$ lie in E but cannot be joined by a broken line in E (see Problem 8 after Section 2.4).

A *boundary point* of a set is a point every neighborhood of which contains at least one point in the set and at least one point not in the set. Thus the boundary points of the circular domain $x^2 + y^2 < 1$ are the points of the circumference: $x^2 + y^2 = 1$. No boundary point of an open set can belong to the set; however, every boundary point of a closed set belongs to the set.

An *interior point* of a set is a point having a neighborhood that is contained in the set. Thus every point of an open set E is an interior point of E. A boundary point of a set cannot be an interior point.

The term *region* will be used to describe a set consisting of a domain plus, perhaps, some or all of its boundary points. Thus a region may be a domain (if no boundary points are included). If all boundary points are included, the region is called a *closed region*; it then necessarily forms a *closed* set. (See Problem 13 following Section 2.4). Thus a circle plus interior, $x^2 + y^2 \leq 1$, is a closed region. A domain is sometimes called an *open region.*

It will be found that for most practical problems a domain is defined by one or more inequalities, and the boundary of a domain is defined by one or more equations, whereas a closed region is given by a combination of the two; for example,

$$xy < 1 \text{ is a domain,}$$

$$xy = 1 \text{ is its boundary,}$$

$$xy \leq 1 \text{ is a closed region.}$$

These concepts are illustrated in Fig. 2.1.

The extension of these ideas to three or more dimensions is not difficult; for four or more dimensions, graphical representation is essentially hopeless. Thus a *neighborhood* of a point (x_1, y_1, z_1) in space is the set of points (x, y, z) inside a sphere,

$$(x - x_1)^2 + (y - y_1)^2 + (z - z_1)^2 < \delta^2,$$

and the other definitions can be repeated without change. In general, in n-dimensional space E^n (Section 1.13) a neighborhood of radius δ of a point A: $(a_1, \ldots, a_n)$ consists of all points P: $(x_1, \ldots, x_n)$ such that $d(A, P) < \delta$, that is,

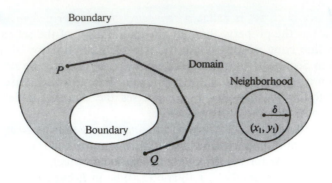

Figure 2.1   Set concepts.

such that

$$|\overrightarrow{AP}|^2 = (x_1 - a_1)^2 + \cdots + (x_n - a_n)^2 < \delta^2.$$

The other concepts, such as open set and closed set, are defined in terms of neighborhoods as before.

The definitions can also be adapted to the case of one dimension. A *neighborhood* of a point x_1 of the x axis is an interval: $x_1 - \delta < x < x_1 + \delta$. A *domain* on the x axis is one of the following four types of sets: (1) an *open interval*: $a < x < b$; (2) an *infinite open interval*: $a < x$; (3) an *infinite open interval*: $x < b$; (4) the entire x axis. A *bounded closed region* on the x axis is a *closed interval*: $a \leq x \leq b$.

Occasionally it is convenient to use the term *neighborhood of a point* to mean an open set containing the point.

2.3   FUNCTIONAL NOTATION · LEVEL CURVES AND LEVEL SURFACES

Most of the functions to be considered will be defined in a domain or, occasionally, in a closed region. The notation "$z = f(x, y)$ in domain D" will mean that z is given as a function of x and y for all points in a domain D of the xy-plane. The variables x and y are called *independent variables*, while z is *dependent*. Similarly, one writes: "$u = f(x, y, z)$ in domain D" or "$w = f(x, y, z, u)$ in domain D" for functions of more than two variables. As with functions of one variable, the functional notation also serves to indicate corresponding values for a given function. Thus if $z = f(x, y)$ is defined by the equation $z = \sqrt{1 - x^2 - y^2}$ in the domain $x^2 + y^2 < 1$, then $f(0,0) = 1$, $f(\frac{1}{2}, \frac{1}{2}) = \sqrt{\frac{1}{2}}$, and so on.

The functional relationship $z = f(x, y)$ is sometimes written thus: $z = z(x, y)$. For functions of three or more variables one writes, similarly, $u = u(x, y, z)$, $w = w(x, y, z, u)$, $y = y(x_1, \ldots, x_n)$.

A function of two variables can be represented graphically by a surface in 3-dimensional space, as shown in Fig. 2.2. For functions of three or more variables this representation is not available.

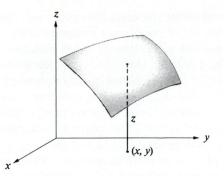

Figure 2.2   Function of two variables.

Another method for representing functions of two variables is that of *level curves* or *contour lines*. This is the method used in making a contour map or topographical map. One plots the loci,

$$f(x, y) = c_1, \qquad f(x, y) = c_2, \ldots$$

for various choices of the constants $c_1, c_2, \ldots$; each locus $f(x, y) = c$ is called a *level curve* of $f(x, y)$; it may actually be formed of several distinct curves. This is illustrated in Fig. 2.3, in which $f(x, y) = x^2y + x^2 + 2y^2$; the value of c is shown on each curve. The level curves often provide a better understanding of the function than a sketch of the surface $z = f(x, y)$.

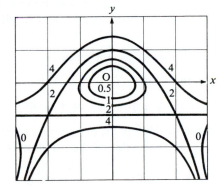

Figure 2.3   Level curves.

The method just described is available in principle for functions of three variables; here one draws the *level surfaces*: $f(x, y, z) = c_1, f(x, y, z) = c_2, \ldots$ for appropriate choices of $c_1, c_2, \ldots$. The surfaces of constant gravitational potential (approximately spheres) about the earth illustrate this. The surfaces of constant temperature or pressure are of importance in meteorology.

One can also represent a function of three variables with the aid of level *curves*; if, for example, z is fixed, then $f(x, y, z)$ becomes a function of x and y and can be represented by its level curves in an xy-plane as earlier. If this is done for several values of z, one obtains a corresponding number of diagrams that together represent the function. This is common practice in meteorology, in which lines of constant pressure ("isobars") are plotted for various altitudes.

For functions of four or more variables the loci, $f = $ const, are "hyper-surfaces" in a space of four or more dimensions. These level hypersurfaces are mainly of theoretical interest. To represent the function graphically, one is forced to fix one or more of the variables and thereby obtain level surfaces in 3-dimensional space or level curves in a plane.

A function $f(x, y)$ is said to be *bounded* when (x, y) is restricted to a set E, if there is a number M such that $|f(x, y)| < M$ when (x, y) is in E. For example, $z = x^2 + y^2$ is bounded, with $M = 2$, if $|x| < 1$ and $|y| < 1$. The function $z = \tan(x + y)$ is not bounded for $|x + y| < \frac{1}{2}\pi$.

2.4   LIMITS AND CONTINUITY

Let $z = f(x, y)$ be given in a domain D, and let (x_1, y_1) be a point of D or a boundary point of D. Then the equation

$$\lim_{\substack{x \to x_1 \\ y \to y_1}} f(x, y) = c \tag{2.2}$$

means the following: given any $\epsilon > 0$, a $\delta > 0$ can be found such that for every (x, y) in D and within the neighborhood of (x_1, y_1) of radius δ, except possibly for (x_1, y_1) itself, one has

$$|f(x, y) - c| < \epsilon. \tag{2.3}$$

In other terms, if (x, y) is in D and

$$0 < (x - x_1)^2 + (y - y_1)^2 < \delta^2, \tag{2.4}$$

then (2.3) holds. Thus if the variable point (x, y) is sufficiently close to (but not at) its limiting position (x_1, y_1), the value of the function is as close as desired to its limiting value c.

If the point (x_1, y_1) is in D and

$$\lim_{\substack{x \to x_1 \\ y \to y_1}} f(x, y) = f(x_1, y_1), \tag{2.5}$$

then $f(x, y)$ is said to be *continuous* at (x_1, y_1). If this holds for every point (x_1, y_1) of D, then $f(x, y)$ is said to be *continuous in* D.

The notions of limits and continuity can be extended to more complicated sets—for example, to closed regions. The preceding definitions can be repeated essentially without change. Thus if $f(x, y)$ is defined in a closed region R and (x_1, y_1) is in R, then (2.2) is said to hold if, for given $\epsilon > 0$, a $\delta > 0$ can be found such that (2.3) holds whenever (x, y) *is in R* and is within distance δ of (x_1, y_1), but not at (x_1, y_1). If (2.5) holds, then $f(x, y)$ is continuous at (x_1, y_1). Similar definitions hold if $f(x, y)$ is defined only on a curve in the xy-plane. The notions of limits and continuity must always be considered *in relation to the set in which the function is defined.*

Continuity for functions of two variables is a more subtle requirement than for functions of one variable. Such a simple function as

$$z = \frac{x^2 - y^2}{x^2 + y^2}$$

is badly discontinuous at the origin, without becoming infinite there. For z has limit 0 if (x, y) approaches the origin on the line $x = y$, has limit 1 if (x, y) approaches the origin on the x axis, and has limit -1 if (x, y) approaches the origin on the y axis. Thus no limiting value can be assigned at $(0, 0)$. It should be noted that the level curves of this function are straight lines, all passing through $(0, 0)$; this alone shows that there is a discontinuity at the origin.

However, the fundamental theorem on limits and continuity holds without change:

THEOREM Let $u = f(x, y)$ and $v = g(x, y)$ both be defined in the domain D of the xy-plane. Let

$$\lim_{\substack{x \to x_1 \\ y \to y_1}} f(x, y) = u_1, \qquad \lim_{\substack{x \to x_1 \\ y \to y_1}} g(x, y) = v_1. \tag{2.6}$$

Then

$$\lim_{\substack{x \to x_1 \\ y \to y_1}} [f(x, y) + g(x, y)] = u_1 + v_1, \tag{2.7}$$

$$\lim_{\substack{x \to x_1 \\ y \to y_1}} [f(x, y) \cdot g(x, y)] = u_1 \cdot v_1, \tag{2.8}$$

$$\lim_{\substack{x \to x_1 \\ y \to y_1}} \frac{f(x, y)}{g(x, y)} = \frac{u_1}{v_1} \qquad (v_1 \neq 0). \tag{2.9}$$

If $f(x, y)$ and $g(x, y)$ are continuous at (x_1, y_1), then so also are the functions

$$f(x, y) + g(x, y), \quad f(x, y) \cdot g(x, y), \quad \frac{f(x, y)}{g(x, y)},$$

provided, in the last case, $g(x_1, y_1) \neq 0$.

Let $F(u, v)$ be defined and continuous in a domain D_0 of the uv-plane and let $F[f(x, y), g(x, y)]$ be defined for (x, y) in D. Then, if (u_1, v_1) is in

D_0,

$$\lim_{\substack{x \to x_1 \\ y \to y_1}} F[f(x, y), g(x, y)] = F(u_1, v_1). \tag{2.10}$$

If $f(x, y)$ and $g(x, y)$ are continuous at (x_1, y_1), then so also is $F[f(x, y), g(x, y)]$.

Proof.   We first consider the composite function $F[f(x, y), g(x, y)]$, which is the most fundamental notion of the whole theorem. Since $F[u, v]$ is assumed to be continuous in D_0, one has

$$\lim_{\substack{u \to u_1 \\ v \to v_1}} F[u, v] = F[u_1, v_1]; \tag{2.11}$$

by (2.6), as (x, y) approaches (x_1, y_1), (u, v) approaches (u_1, v_1), so that by (2.11),

$$\lim_{\substack{x \to x_1 \\ y \to y_1}} F[f(x, y), \quad g(x, y)] = F[\lim_{\substack{x \to x_1 \\ y \to y_1}} f(x, y), \quad \lim_{\substack{x \to x_1 \\ y \to y_1}} g(x, y)] = F[u_1, v_1].$$

Thus (2.10) is established. If f and g are continuous at (x_1, y_1), then $f(x_1, y_1) = u_1$ and $g(x_1, y_1) = v_1$, so that by (2.10),

$$\lim_{\substack{x \to x_1 \\ y \to y_1}} F[f(x, y), g(x, y)] = F[f(x_1, y_1), g(x_1, y_1)];$$

that is, $F[f(x, y), g(x, y)]$ is continuous at (x_1, y_1).

Now one verifies easily that the particular $F[u, v] \equiv u + v$ is continuous for all values of u and v. If (2.10) is applied to this choice of F, one finds

$$\lim_{\substack{x \to x_1 \\ y \to y_1}} [f(x, y) + g(x, y)] = u_1 + v_1,$$

which is (2.7); by the same reasoning, one concludes that if $f(x, y)$ and $g(x, y)$ are continuous at (x_1, y_1), then so is $f(x, y) + g(x, y)$.

The statements about products and quotients follow in the same way by consideration of the special functions $F \equiv u \cdot v$ and $F \equiv u/v$. One need only show that these functions are continuous (for $v \neq 0$ in the second case). This can be done directly by applying the preceding ϵ, δ definition or as follows. One shows that the functions $u + v$ and $u - v$ are continuous functions of u and v and also that $\frac{1}{4}w$ and w^2 are continuous functions of w (theorems on functions of one variable). It follows from the function-of-function rule just proved that $(u + v)^2$ and $(u - v)^2$ are continuous and hence that

$$u \cdot v \equiv \tfrac{1}{4}\big[(u + v)^2 - (u - v)^2\big]$$

is continuous for all (u, v) and we obtain the conclusions about the product $f \cdot g$.

Next one shows that $1/v$ is continuous function of v for $v \neq 0$ (function of one variable) and hence that $u/v \equiv u \cdot (1/v)$ is a continuous function of u and v for $v \neq 0$. If, for example, $v_1 > 0$, then we take D_0 to be the half-plane $v > 0$ in the uv-plane. From (2.6) we conclude that $g(x, y) > v_1/2 > 0$ in some neighborhood D_1 of radius δ of (x_1, y_1) (except possibly at (x_1, y_1) itself). In the discussion of limits and continuity, (x, y) can be restricted to D_1, so that the values $(u, v) = (f(x,y), g(x,y))$ lie in D_0, where $F = u/v$ is continuous. Hence the previous reasoning applies to u/v.   □

The theorem above can be restated in analogous form for functions of three or more variables and, in the case of the composite function, for combinations of functions of one and two variables, one and three variables, and so on:

$$F[f(x, y)], \quad F[f(t), g(t)], \quad F[f(x, y, z)], \quad F[f(t), g(t), h(t)], \quad \dots .$$

By virtue of this theorem, one can conclude that *polynomial* functions such as

$$w = x^3 y + 3xz^2 - xyz$$

are continuous for all values of the variables, whereas *rational* functions such as

$$w = \frac{x^2 y - x}{1 - x^2 - y}$$

are continuous except where the denominator is 0.

Mappings from V^n to V^m. At times, one considers sets of functions of n variables:

$$y_1 = f_1(x_1, \dots, x_n),$$

$$\vdots$$

$$y_m = f_m(x_1, \dots, x_n). \tag{2.12}$$

We assume that all functions are defined in a domain D of n-dimensional space. Equations (2.12) then describe a *mapping* from D to m-dimensional space. A special case of such a mapping is a linear mapping, as discussed in Section 1.15, with D all of n-dimensional space.

EXAMPLE

$$y_1 = \frac{x_1}{x_1^2 + x_2^2}, \qquad y_2 = \frac{-x_2}{x_1^2 + x_2^2}.$$

Here $m = n = 2$, and D consists of V^2 minus the vector $(0, 0)$.   ■

We can use vector notation and replace $f_1(x_1, \dots, x_n)$ by $f_1(\mathbf{x})$, and so on. In fact we can replace (2.12) by one vector equation

$$\mathbf{y} = \mathbf{f}(\mathbf{x}). \tag{2.12'}$$

Here $\mathbf{x}$ is a vector of V^n, $\mathbf{y}$ is a vector of V^m, and $\mathbf{f}$ is a vector function: $\mathbf{f} = (f_1, \dots, f_m)$.

We call the mapping continuous at $\mathbf{x}^0$ in D if, for each $\epsilon > 0$, there is a $\delta > 0$ such that $|\mathbf{f}(\mathbf{x}) - \mathbf{f}(\mathbf{x}^0)| < \epsilon$ for each $\mathbf{x}$ in D satisfying $|\mathbf{x} - \mathbf{x}^0| < \delta$.

The previous discussion extends to such mappings. We observe also that the mapping (2.12') is continuous at $\mathbf{x}^0 = (x_1^0, \dots, x_n^0)$ if and only if all the functions $f_1(x_1, \dots, x_n), \dots, f_m(x_1, \dots, x_n)$ are continuous at $(x_1^0, \dots, x_n^0)$ (see Problem 10 which follows).

Instead of using vector notation, we can think of $(x_1, \dots, x_n)$ as a point in E^n

and $(y_1,\ldots,y_m)$ as a point in E^m. Then we write (2.12) as

$$Q = F(P),\tag{2.12''}$$

where P is a point of D and Q a point of E^m. The definition of continuity then reads: F is continuous at P_0 in D if for each $\epsilon > 0$ there is a $\delta > 0$ such that, for P in D,

$$d(F(P), F(P_0)) < \epsilon \quad \text{whenever} \quad d(P, P_0) < \delta.$$

Here $d(A, B)$ stands for the distance between A and B in the appropriate space: E^n for $d(P, P_0)$, E^m for $d(F(P), F(P_0))$. As in Section 1.13, in E^n, for example, one has the properties:

(i) $d(A, B) \geq 0, \quad d(A, B) = 0 \quad$ only for $A = B$,

(ii) $d(A, B) = d(B, A)$,

(iii) $d(A, C) \leq d(A, B) + d(B, C)$.

These properties of distance turn out to be crucial for discussions of continuity. Because of their importance and generality, one uses the term *metric space* for any set with a real distance function d satisfying (i), (ii), and (iii). Thus E^n or V^n can be regarded as metric spaces with distance $d(A, B) = |\mathbf{a} - \mathbf{b}|$ for $A = (a_1,\ldots,a_n)$, $B = (b_1,\ldots,b_n)$ regarded as points or vectors.

PROBLEMS

1. Give several examples of functions of several variables occurring in geometry (area and volume formulas, law of cosines, and so on).

2. Represent the following functions by first sketching a surface, and second, drawing level curves:

   a) $z = 3 - x - 3y$ b) $z = x^2 + y^2 + 1$

   c) $z = \sin(x + y)$ d) $z = e^{xy}$

3. Analyze the following functions by describing their level surfaces in space:

   a) $u = x^2 + y^2 + z^2$ b) $u = x + y + z$

   c) $w = x^2 + y^2 - z$ d) $w = x^2 + y^2$

4. Determine the values of the following limits, wherever the limit exists:

   a) $\displaystyle\lim_{\substack{x \to 0 \\ y \to 0}} \frac{x^2 - y^2}{1 + x^2 + y^2}$ b) $\displaystyle\lim_{\substack{x \to 0 \\ y \to 0}} \frac{x}{x^2 + y^2}$

   c) $\displaystyle\lim_{\substack{x \to 0 \\ y \to 0}} \frac{(1 + y^2)\sin x}{x}$ d) $\displaystyle\lim_{\substack{x \to 0 \\ y \to 0}} \frac{1 + x - y}{x^2 + y^2}$

5. Show that the following functions are discontinuous at $(0,0)$ and graph the corresponding surfaces:*

   a) $z = \dfrac{x}{x - y}$ b) $z = \log(x^2 + y^2)$

---

*In this book, $\log x$ denotes the natural logarithm of x.

6. Describe the sets in which the following functions are defined:

a) $z = e^{x-y}$

b) $z = \log(x^2 + y^2 - 1)$

c) $z = \sqrt{1 - x^2 - y^2}$

d) $u = \dfrac{xy}{z}$

7. Prove the theorem: Let $f(x, y)$ be defined in domain D and continuous at the point (x, y_1) of D. If $f(x_1, y_1) > 0$, then there is a neighborhood of (x_1, y_1) in which $f(x, y) > \frac{1}{2}f(x_1, y_1) > 0$. [Hint: Use $\epsilon = \frac{1}{2}f(x_1, y_1)$ in the definition of continuity.]

8. Let D be a domain in the plane. Show that D cannot consist of two open sets E_1, E_2 with no point in common. [Hint: Suppose the contrary and choose point P in E_1 and point Q in E_2; join these points by a broken line in D. Regard this line as a path from P to Q and let s be distance from P along the path, so that the path is given by continuous functions $x = x(s), y = y(s), 0 \le s \le L$, with $s = 0$ at P and $s = L$ at Q. Let $f(s) = -1$ if $(x(s), y(s))$ is in E_1 and let $f(s) = 1$ if $(x(s), y(s))$ is in E_2. Show that $f(s)$ is continuous for $0 \le s \le L$. Now apply the *intermediate value theorem*: If $f(x)$ is continuous for $a \le x \le b$ and $f(a) < 0, f(b) > 0$, then $f(x) = 0$ for some x between a and b (see Problem 5 following Section 2.23).]

9. Prove the theorem: Let $f(x, y)$ be continuous in domain D. Let $f(x, y)$ be positive for at least one point of D and negative for at least one point of D. Then $f(x, y) = 0$ for at least one point of D. [Hint: Use Problem 7 to conclude that the set A where $f(x, y) > 0$ and the set B where $f(x, y) < 0$ are open. If $f(x, y) \neq 0$ in D, then D is formed of the two nonoverlapping open sets A and B; this is not possible by Problem 8.]

Remark. This result extends the intermediate value theorem to functions of two variables.

10. In V^n, $|\mathbf{x}| = \sqrt{x_1^2 + \cdots + x_n^2}$, as in Section 1.13.

a) Show that if $|\mathbf{x}| < \epsilon$, then $|x_1| < \epsilon, \ldots, |x_n| < \epsilon$. Interpret the result geometrically for $n = 2$.

b) Show that if $|x_1| < \delta, \ldots, |x_n| < \delta$, then $|\mathbf{x}| < n\delta$.

c) In (2.12), let $f_1(x_1, \ldots, x_n), \ldots, f_m(x_1, \ldots, x_n)$ be defined in domain D and continuous at $(x_1^0, \ldots, x_n^0)$. Show that the corresponding mapping (2.12') is continuous at $\mathbf{x}^0$. [Hint: Given $\epsilon > 0$, choose $\delta > 0$ so small that $|f_1(x_1, \ldots, x_n) - f_1(x_1^0, \ldots, x_n^0)| < \epsilon/m, \ldots, |f_m(x_1, \ldots, x_n) - f_m(x_1^0, \ldots, x_n^0)| < \epsilon/m$ for $|\mathbf{x} - \mathbf{x}^0| < \delta$. Conclude from (b) that $|\mathbf{f}(\mathbf{x}) - \mathbf{f}(\mathbf{x}^0)| < \epsilon$ for $|\mathbf{x} - \mathbf{x}^0| < \delta$.]

d) Show that if the mapping (2.12') is continuous at $\mathbf{x}^0$, then each of the functions $f_1(x_1, \ldots, x_n), \ldots, f_m(x_1, \ldots, x_n)$ is continuous at $(x_1^0, \ldots, x_n^0)$.

11. An (infinite) sequence of points $P_1, \ldots, P_n, \ldots$ in the plane is said to converge and have limit P_0:

$$\lim_{n \to \infty} P_n = P_0 \quad \text{or} \quad P_n \to P_0$$

if for each $\epsilon > 0$ there is an integer N such that $d(P_n, P_0) < \epsilon$ for $n \ge N$. Show that the limit P_0 is unique. [Hint: If $P_n \to P_0$ and $P_n \to P_0'$, $P_0' \neq P_0$, then take $\epsilon = \frac{1}{3}d(P_0, P_0')$ and obtain a contradiction.]

12. Show that a set E in the plane is closed if and only if for every convergent sequence (Problem 11) of points $\{P_n\}$ in E, the limit of the sequence is in E. [Hint: Suppose E

is closed and $P_n \to P_0$, with P_n in E for all n. If P_0 is not in E, then choose $\epsilon > 0$ such that $d(P, P_0) < \epsilon$ implies that P is not in E (why is this possible?) and obtain a contradiction. Next suppose E is such that whenever $\{P_n\}$ is in E and $P_n \to P_0$, then P_0 is in E. To show that E is closed, suppose that P_0 is a point not in E and that no neighborhood of P_0 consists solely of points not in E. Then choose P_n such that P_n is in E and $d(P_n, P_0) < 1/n$. Show that $P_n \to P_0$ and obtain a contradiction.]

13. Let E be a set in the plane.

   a) Show that the set formed of E and its boundary is closed.

   b) Show that the boundary of E is a closed set. [Note: The boundary may be an empty set. When does this arise?]

2.5 PARTIAL DERIVATIVES

Let $z = f(x, y)$ be defined in a domain D of the xy-plane and let (x_1, y_1) be a point of D. The function $f(x, y_1)$ then depends on x alone and is defined in an interval about x_1. Hence its derivative with respect to x at $x = x_1$ may exist. If it does, its value is called the *partial derivative of $f(x, y)$ with respect to x at (x_1, y_1)*, and is denoted by

$$\frac{\partial f}{\partial x}(x_1, y_1) \quad \text{or by} \quad \frac{\partial z}{\partial x}\bigg|_{(x_1, y_1)}.$$

One has thus, by the definition of the derivative,

$$\frac{\partial f}{\partial x}(x_1, y_1) = \lim_{\Delta x \to 0} \frac{f(x_1 + \Delta x, y_1) - f(x_1, y_1)}{\Delta x}. \tag{2.13}$$

When the point (x_1, y_1) at which the derivative is being evaluated is evident, one can write simply $\partial z/\partial x$ or $\partial f/\partial x$ for the derivative. Other notations commonly used are z_x, f_x, f_1 or, more explicitly, $z_x(x_1, y_1), f_x(x_1, y_1), f_1(x_1, y_1)$. When subscripts are used, there can be confusion with the symbol for components of a vector; hence *when vectors and partial derivatives are being used together, a notation such as $\partial z/\partial x$ or $\partial f/\partial x$ is preferable for partial derivatives.*

The function $z = f(x, y)$ can be represented by a surface in space. The equation $y = y_1$ then represents a plane cutting the surface in a curve. The partial derivative $\partial z/\partial x$ at (x_1, y_1) can then be interpreted as the slope of the tangent to the curve, that is, as $\tan \alpha$, where α is the angle shown in Fig. 2.4. In this figure, $z = 5 + x^2 - y^2$, and the derivative is being computed at the point $x = 1$, $y = 2$. For $y = 2$, $z = 1 + x^2$, so that the derivative along the curve is $2x$; for $x = 1$, one finds $f_x(1, 2)$ to be 2.

The partial derivative $\dfrac{\partial z}{\partial y}\bigg|_{(x_1, y_1)}$ is defined similarly; one now holds x constant, equal to x_1, and differentiates $f(x_1, y)$ with respect to y. One has thus

$$\frac{\partial f}{\partial y}(x_1, y_1) = \frac{\partial z}{\partial y}\bigg|_{(x_1, y_1)} = \lim_{\Delta y \to 0} \frac{f(x_1, y_1 + \Delta y) - f(x_1, y_1)}{\Delta y}.$$

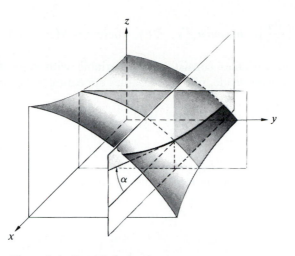

Figure 2.4   Partial derivatives.

This can also be interpreted as the slope of the tangent to the curve in which the plane $x = x_1$ cuts the surface $z = f(x, y)$. One also writes $f_y(x_1, y_1)$, $f_2(x_1, y_1)$ for this derivative.

If the point (x_1, y_1) is now varied, one obtains (wherever the derivative exists) a new function of two variables: the function $f_x(x, y)$. Similarly, the derivative $\partial z/\partial y$ at a variable point (x, y) is a function $f_y(x, y)$. For explicit functions $z = f(x, y)$, evaluation of these derivatives is carried out as in ordinary calculus, for one is always differentiating a function of one variable, the *other being treated as a constant*. For example, if $z = x^2 - y^2$, then

$$\frac{\partial z}{\partial x} = 2x, \qquad \frac{\partial z}{\partial y} = -2y.$$

The preceding definitions extend at once to functions of three or more variables. If $w = g(x, y, u, v)$, then one partial derivative at (x, y, u, v) is

$$\frac{\partial w}{\partial u} = \lim_{\Delta u \to 0} \frac{g(x, y, u + \Delta u, v) - g(x, y, u, v)}{\Delta u}. \qquad (2.14)$$

When only three variables x, y, z are involved in a discussion, the notation $\partial z/\partial x$ is self-explanatory: x and y are independent, and y is held constant. However, when four or more variables are involved, the partial derivative symbol by itself is ambiguous. Thus if x, y, u, v are involved, then $\partial u/\partial x$ may be interpreted as $f_x(x, y)$, where $u = f(x, y)$, $v = g(x, y)$; one may also interpret $\partial u/\partial x$ as $h_x(x, y, v)$, where $u = h(x, y, v)$. For this reason, when four or more variables are involved, it is advisable to supplement the partial derivative symbol by indicating the variables held constant. For example,

$$\left(\frac{\partial z}{\partial x}\right)_y \text{ means } f_x(x, y), \text{ where } z = f(x, y),$$

$$\left(\frac{\partial u}{\partial x}\right)_{yv} \text{ means } h_x(x, y, v), \text{ where } u = h(x, y, v).$$

The independent variables consist of the variable with respect to which the differentiation is being made, plus all variables appearing as subscripts.

EXAMPLE 1   If $w = xuv + u - 2v$, then

$$\frac{\partial w}{\partial x} = uv, \qquad \frac{\partial w}{\partial u} = xv + 1, \qquad \frac{\partial w}{\partial v} = xu - 2. \quad \blacksquare$$

EXAMPLE 2   If u, v, x, y are related by the equations

$$u = x - y, \qquad v = x + y,$$

then

$$\left(\frac{\partial u}{\partial x}\right)_y = 1, \qquad \left(\frac{\partial v}{\partial x}\right)_y = 1,$$

whereas

$$\left(\frac{\partial u}{\partial x}\right)_v = 2, \qquad \left(\frac{\partial v}{\partial x}\right)_u = 2,$$

since u can be expressed in terms of x and v by the equation

$$u = 2x - v,$$

from which one also obtains

$$v = 2x - u. \quad \blacksquare$$

EXAMPLE 3   If $x^2 + y^2 - z^2 = 1$, then

$$2x - 2z\frac{\partial z}{\partial x} = 0, \qquad 2y - 2z\frac{\partial z}{\partial y} = 0,$$

whence

$$\frac{\partial z}{\partial x} = \frac{x}{z}, \qquad \frac{\partial z}{\partial y} = \frac{y}{z} \quad (z \neq 0). \quad \blacksquare$$

2.6   TOTAL DIFFERENTIAL ▪ FUNDAMENTAL LEMMA

In forming the preceding partial derivatives $\partial z/\partial x$ and $\partial z/\partial y$, changes Δx and Δy in x and y were considered separately; we now consider the effect of changing x and y together. Let (x, y) be a fixed point of D and let $(x + \Delta x, y + \Delta y)$ be a second point of D. Then the function $z = f(x, y)$ changes by an amount Δz in going from (x, y) to $(x + \Delta x, y + \Delta y)$:

$$\Delta z = f(x + \Delta x, y + \Delta y) - f(x, y). \tag{2.15}$$

This defines Δz as a function of Δx and Δy (x and y being considered as constants), with the special property

$$\Delta z = 0 \quad \text{when} \quad \Delta x = 0 \quad \text{and} \quad \Delta y = 0.$$

For example, if $z = x^2 + xy + xy^2$, then

$$\Delta z = (x + \Delta x)^2 + (x + \Delta x)(y + \Delta y) + (x + \Delta x)(y + \Delta y)^2 - x^2 - xy - xy^2$$

$$= \Delta x(2x + y + y^2) + \Delta y(x + 2xy) + \overline{\Delta x}^2 + \Delta x \Delta y(1 + 2y)$$

$$+ \overline{\Delta y}^2 x + \Delta x \overline{\Delta y}^2.$$

Here Δz is of the form

$$\Delta z = a \Delta x + b \Delta y + c \overline{\Delta x}^2 + d \Delta x \Delta y + e \overline{\Delta y}^2 + f \Delta x \overline{\Delta y}^2,$$

that is, *a linear function of Δx and Δy plus terms of higher degree.*

In general, the function $z = f(x,y)$ is said to have a *total differential* or to be *differentiable* at the point (x,y) if, at this point,

$$\Delta z = a \Delta x + b \Delta y + \epsilon_1 \cdot \Delta x + \epsilon_2 \cdot \Delta y, \tag{2.16}$$

where a and b are independent of Δx, Δy and ϵ_1 and ϵ_2 are functions of Δx and Δy such that

$$\lim_{\substack{\Delta x \to 0 \\ \Delta y \to 0}} \epsilon_1 = 0, \qquad \lim_{\substack{\Delta x \to 0 \\ \Delta y \to 0}} \epsilon_2 = 0; \tag{2.17}$$

the linear function of Δx and Δy,

$$a \Delta x + b \Delta y$$

is then termed the *total differential* of z at the point (x, y) and is denoted by dz:

$$dz = a \Delta x + b \Delta y. \tag{2.18}$$

If Δx and Δy are sufficiently small, dz gives a close approximation to Δz. More precisely, one can write

$$\Delta z = \Delta x(a + \epsilon_1) + \Delta y(b + \epsilon_2),$$

where a and b are constants; by (2.17) the percentage error in each term caused by replacing ϵ_1 and ϵ_2 by 0 can be made as small as desired by choosing Δx and Δy sufficiently small. (This argument fails if a or b is 0.)

In the preceding example, Δz has a total differential at each point (x, y), with

$$a = 2x + y + y^2, \qquad b = x + 2xy,$$

and

$$\epsilon_1 = \Delta x + \Delta y(1 + 2y), \qquad \epsilon_2 = x \Delta y + \Delta x \Delta y.$$

THEOREM   If $z = f(x, y)$ has a total differential (2.18) at the point (x, y), then f is continuous at (x,y) and

$$a = \frac{\partial z}{\partial x}, \qquad b = \frac{\partial z}{\partial y}; \tag{2.19}$$

that is, the two partial derivatives exist at (x, y) and have the given values.

Proof. Set $\Delta y = 0$. Then, by (2.16) and (2.17),

$$\frac{\partial z}{\partial x} = \lim_{\Delta x \to 0} \frac{\Delta z}{\Delta x} = \lim_{\Delta x \to 0} \frac{\Delta x(a + \epsilon_1)}{\Delta x} = \lim_{\Delta x \to 0} (a + \epsilon_1) = a.$$

Similarly, one shows that $\partial z/\partial y = b$. The continuity of f follows from $f(x + \Delta x, y + \Delta y) - f(x,y) = \Delta z$, where by (2.16) and (2.17) Δz has limit 0 as $\Delta x \to 0$, $\Delta y \to 0$.□

Remark. The theorem shows that the differential, when it exists, is *unique.* One can describe the differential of $f(x,y)$ at a point (x_1,y_1) as the unique *linear function* $a(x - x_1) + b(y - y_1)$ which approximates $\Delta z = f(x,y) - f(x_1,y_1)$ as in Eq. (2.16), with $\Delta x = x - x_1$, $\Delta y = y - y_1$.

The existence of the partial derivatives at the point (x,y) is not sufficient to guarantee the existence of the total differential (see Problem 7 following this section); however, their continuity near the point is sufficient for this.

FUNDAMENTAL LEMMA If $z = f(x, y)$ has continuous first partial derivatives in D, then z has a differential

$$dz = \frac{\partial z}{\partial x}\Delta x + \frac{\partial z}{\partial y}\Delta y \tag{2.20}$$

at every point (x, y) of D.

Proof. Let (x, y) be a fixed point of D. If x alone changes, one obtains a change Δz in z:

$$\Delta z = f(x + \Delta x, y) - f(x, y);$$

this difference can be evaluated by the law of the mean for functions of one variable, for, with y held fixed, z is a function of x having a continuous derivative $f_x(x, y)$. Thus one concludes:

$$f(x + \Delta x, y) - f(x, y) = f_x(x_1, y) \Delta x,$$

where x_1 is between x and $x + \Delta x$. Since $f_x(x, y)$ is continuous, the difference

$$\epsilon_1 = f_x(x_1, y) - f_x(x, y)$$

approaches zero as Δx approaches 0. Thus

$$f(x + \Delta x, y) - f(x, y) = f_x(x, y) \Delta x + \epsilon_1 \Delta x. \tag{2.21}$$

Now if both x and y change, one obtains a change Δz in z:

$$\Delta z = f(x + \Delta x, y + \Delta y) - f(x, y).$$

This can be written as the sum of terms representing the effect of a change in x alone and a subsequent change in y alone:

$$\Delta z = [f(x + \Delta x, y) - f(x, y)] + [f(x + \Delta x, y + \Delta y) - f(x + \Delta x, y)]. \tag{2.22}$$

The first term can be evaluated by (2.21). The second is evaluated similarly, with z a function of y alone:

$$f(x + \Delta x, y + \Delta y) - f(x + \Delta x, y) = f_y(x + \Delta x, y_1) \Delta y,$$

where y_1 is between y and $y + \Delta y$. It follows from the continuity of $f_y(x, y)$ that the difference

$$\epsilon_2 = f_y(x + \Delta x, y_1) - f_y(x, y)$$

approaches 0 as *both* Δx and Δy approach 0. One has now

$$f(x + \Delta x, y + \Delta y) - f(x + \Delta x, y) = f_y(x, y) \Delta y + \epsilon_2 \Delta y. \qquad (2.23)$$

Equations (2.21) (2.22), and (2.23) now give

$$\Delta z = f_x(x, y) \Delta x + f_y(x, y) \Delta y + \epsilon_1 \Delta x + \epsilon_2 \Delta y,$$

where ϵ_1 and ϵ_2 satisfy (2.17). Thus z has a differential dz as stated in (2.20), and the Fundamental Lemma is proved. $\square$

For reasons to be explained, Δx and Δy can be replaced by dx and dy in (2.20). Thus one has

$$dz = \frac{\partial z}{\partial x} dx + \frac{\partial z}{\partial y} dy, \qquad (2.24)$$

which is the customary way of writing the differential.

The preceding analysis extends at once to functions of three or more variables. For example, if $w = f(x, y, u, v)$, then

$$dw = \frac{\partial w}{\partial x} dx + \frac{\partial w}{\partial y} dy + \frac{\partial w}{\partial u} du + \frac{\partial w}{\partial v} dv. \qquad (2.25)$$

EXAMPLE 1 If $z = x^2 - y^2$, then $dz = 2x\, dx - 2y\, dy$. ■

EXAMPLE 2 If $w = \dfrac{xy}{z}$, then $dw = \dfrac{y}{z} dx + \dfrac{x}{z} dy - \dfrac{xy}{z^2} dz$. ■

PROBLEMS

1. Evaluate $\dfrac{\partial z}{\partial x}$ and $\dfrac{\partial z}{\partial y}$ if

a) $z = \dfrac{y}{x^2 + y^2}$

b) $z = y \sin xy$

c) $x^3 + x^2 y - x^2 z + z^3 - 2 = 0$

d) $z = \sqrt{e^{x+2y} - y^2}$

e) $z = (x^2 + y^2)^{3/2}$

f) $z = \arcsin(x + 2y)$

g) $e^x + 2e^y - e^z - z = 0$

h) $xy^2 + yz^2 + xyz = 1$

2. A certain function $f(x, y)$ is known to have the following values: $f(0,0) = 0$, $f(1,0) = 1, f(2,0) = 4, f(0,1) = -2, f(1,1) = -1, f(2,1) = 2, f(0,2) = -4, f(1,2) = -3, f(2,2) = 0$. Compute approximately the derivatives $f_x(1,1)$ and $f_y(1,1)$.

3. Evaluate the indicated partial derivatives:

a) $\left(\dfrac{\partial u}{\partial x}\right)_y$ and $\left(\dfrac{\partial v}{\partial y}\right)_x$ if $u = x^2 - y^2, v = x - 2y$

b) $\left(\dfrac{\partial x}{\partial u}\right)_v$ and $\left(\dfrac{\partial y}{\partial v}\right)_u$ if $x = e^u \cos v, y = e^u \sin v$

c) $\left(\dfrac{\partial x}{\partial u}\right)_y$ and $\left(\dfrac{\partial y}{\partial v}\right)_u$ if $u = x - 2y, v = u - 2y$

d) $\left(\dfrac{\partial r}{\partial x}\right)_y$ and $\left(\dfrac{\partial r}{\partial \theta}\right)_x$ if $r = \sqrt{x^2 + y^2}, x = r\cos\theta$

4. Find the differentials of the following functions:

a) $z = \dfrac{x}{y}$

b) $z = \log\sqrt{x^2 + y^2}$

c) $z = \dfrac{xy}{1 - x - y}$

d) $z = (x - 2y)^5 e^{xy}$

e) $z = \arctan\dfrac{y}{x}$

f) $u = \dfrac{1}{\sqrt{x^2 + y^2 + z^2}}$

5. For the given function $z = f(x, y)$, find Δz and dz in terms of Δx and Δy at $x = 1, y = 1$. Compare these two functions for selected values of $\Delta x, \Delta y$ near 0.

a) $z = x^2 + 2xy$
b) $z = \dfrac{x}{x + y}$

6. A certain function $z = f(x, y)$ is known to have the value $f(1, 2) = 3$ and derivatives $f_x(1, 2) = 2, f_y(1, 2) = 5$. Make "reasonable" estimates of $f(1.1, 1.8), f(1.2, 1.8)$, and $f(1.3, 1.8)$.

7. Let $z = f(x, y) = xy/(x^2 + y^2)$ except at $(0, 0)$; let $f(0, 0) = 0$. Show that $\partial z/\partial x$ and $\partial z/\partial y$ exist for all (x, y) and are continuous except at $(0, 0)$. Show by the Fundamental Lemma that z has a differential for $(x, y) \neq (0, 0)$ but not at $(0, 0)$, since f is discontinuous at $(0, 0)$. [It is instructive to graph the level curves of f.]

2.7 DIFFERENTIAL OF FUNCTIONS OF n VARIABLES ▪ THE JACOBIAN MATRIX

For a function of n variables

$$y = f(x_1, \ldots, x_n) \tag{2.26}$$

the differential is obtained as in Section 2.6:

$$dy = f_{x_1} dx_1 + \cdots + f_{x_n} dx_n. \tag{2.27}$$

Thus it is a linear function of $dx_1,\ldots,dx_n$, whose coefficients $f_{x_1},\ldots,f_{x_n}$ are the partial derivatives of f at the point considered. This linear function is a close approximation to the increment Δy in the sense described in Section 2.6:

$$\Delta y = f(x_1 + dx_1,\ldots,x_n + dx_n) - f(x_1,\ldots,x_n)$$

$$= f_{x_1} dx_1 + \cdots + f_{x_n} dx_n + \epsilon_1 dx_1 + \cdots + \epsilon_n dx_n, \qquad (2.28)$$

where

$$\epsilon_1 \to 0,\ldots, \epsilon_n \to 0 \quad \text{as} \quad dx_1 \to 0,\ldots,dx_n \to 0.$$

On occasion, one has to deal with several functions of n variables:

$$\begin{cases} y_1 = f_1(x_1,\ldots,x_n), \\ \vdots \\ y_m = f_m(x_1,\ldots,x_n). \end{cases} \qquad (2.29)$$

If these functions have continuous partial derivatives in a domain D of E^n, then all have differentials:

$$\begin{cases} dy_1 = \dfrac{\partial f_1}{\partial x_1} dx_1 + \cdots + \dfrac{\partial f_1}{\partial x_n} dx_n, \\ \vdots \\ dy_m = \dfrac{\partial f_m}{\partial x_1} dx_1 + \cdots + \dfrac{\partial f_m}{\partial x_n} dx_n. \end{cases} \qquad (2.30)$$

These equations can be written in matrix form:

$$\begin{bmatrix} dy_1 \\ \vdots \\ dy_m \end{bmatrix} = \begin{bmatrix} \dfrac{\partial f_1}{\partial x_1} & \cdots & \dfrac{\partial f_1}{\partial x_n} \\ \vdots & & \vdots \\ \dfrac{\partial f_m}{\partial x_1} & \cdots & \dfrac{\partial f_m}{\partial x_n} \end{bmatrix} \begin{bmatrix} dx_1 \\ \vdots \\ dx_n \end{bmatrix}. \qquad (2.31)$$

Thus the *vector* col $(dy_1,\ldots,dy_m)$ is obtained from the *vector*

$$\text{col}\,(dx_1,\ldots,dx_n)$$

by multiplication by the *matrix*

$$\left(\frac{\partial f_i}{\partial x_j}\right) = \begin{bmatrix} \dfrac{\partial f_1}{\partial x_1} & \cdots & \dfrac{\partial f_1}{\partial x_n} \\ \vdots & & \vdots \\ \dfrac{\partial f_m}{\partial x_1} & \cdots & \dfrac{\partial f_m}{\partial x_n} \end{bmatrix}. \qquad (2.32)$$

This matrix is called the *Jacobian matrix* of the set of functions (2.29); its entries are partial derivatives of the functions (2.29), evaluated at a chosen point of D.

Equations (2.29) assign a point $(y_1,\ldots,y_m)$ in E^m to each point $(x_1,\ldots,x_n)$ of D. Thus they describe a *mapping* of D into E^m (see Fig. 2.5, in which $n = 3$, $m = 2$).

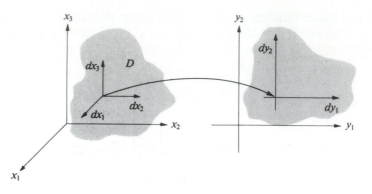

Figure 2.5 The differential as a linear mapping approximating a given mapping.

The linear equations (2.30) or (2.31) describe a *linear mapping* (Section 1.15) that approximates the given mapping near a chosen point; the linear mapping is expressed in terms of coordinates $(dx_1,\ldots,dx_n)$, with origin at the chosen point in D and axes parallel to the given axes, and coordinates $(dy_1,\ldots,dy_m)$ related similarly to the corresponding point in E^m.

We can simplify further by regarding (2.29) as a vector function of the vector $\mathbf{x} = (x_1,\ldots,x_n)$:

$$\mathbf{y} = \mathbf{f}(\mathbf{x}). \tag{2.29'}$$

Here $\mathbf{y} = (y_1,\ldots,y_m)$, and $\mathbf{f}$ is a *vector function* $(f_1,\ldots,f_m)$. We then write (2.31) in the concise form:

$$d\mathbf{y} = \mathbf{f_x}\, d\mathbf{x}. \tag{2.31'}$$

Here $d\mathbf{x} = \mathrm{col}\,(dx_1,\ldots,dx_n)$, $d\mathbf{y} = \mathrm{col}\,(dy_1,\ldots,dy_m)$, and $\mathbf{f_x}$ is an abbreviation for the Jacobian matrix $(\partial f_i/\partial x_j)$. We can also write $\partial y_i/\partial x_j$ for $\partial f_i/\partial x_j$ and are then led to write (2.31') in the form

$$d\mathbf{y} = \mathbf{y_x}\, d\mathbf{x},$$

which is much like the formula $dy = y'dx$ for functions of one variable.

EXAMPLE 1   The function $\mathbf{f}$ is defined by the equations

$$\begin{cases} y_1 = x_1^2 + x_2^2 - x_3^2, \\ y_2 = x_1^2 - x_2^2 + x_3^2, \\ y_3 = -x_1^2 + x_2^2 + x_3^2. \end{cases}$$

Hence

$$d\mathbf{y} = \begin{bmatrix} dy_1 \\ dy_2 \\ dy_3 \end{bmatrix} = \begin{bmatrix} 2x_1\,dx_1 + 2x_2\,dx_2 - 2x_3\,dx_3 \\ 2x_1\,dx_1 - 2x_2\,dx_2 + 2x_3\,dx_3 \\ -2x_1\,dx_1 + 2x_2\,dx_2 + 2x_3\,dx_3 \end{bmatrix}$$

$$= \begin{bmatrix} 2x_1 & 2x_2 & -2x_3 \\ 2x_1 & -2x_2 & 2x_3 \\ -2x_1 & 2x_2 & 2x_3 \end{bmatrix} \begin{bmatrix} dx_1 \\ dx_2 \\ dx_3 \end{bmatrix}.$$

At the point $(x_1, x_2, x_3) = (2, 1, 1)$ we find $(y_1, y_2, y_3) = (4, 4, -2)$ and

$$d\mathbf{y} = \begin{bmatrix} 4 & 2 & -2 \\ 4 & -2 & 2 \\ -4 & 2 & 2 \end{bmatrix} d\mathbf{x}.$$

If $\mathbf{x} = (2.01, 1.03, 1.02)$, then $d\mathbf{x} = (0.01, 0.03, 0.02)$, and the last equation gives

$$d\mathbf{y} = \begin{bmatrix} 4 & 2 & -2 \\ 4 & -2 & 2 \\ -4 & 2 & 2 \end{bmatrix} \begin{bmatrix} 0.01 \\ 0.03 \\ 0.02 \end{bmatrix} = \begin{bmatrix} 0.06 \\ 0.02 \\ 0.06 \end{bmatrix},$$

so that, approximately, $\mathbf{y} = (4.06, 4.02, -1.94)$; the exact value is $(4.0606, 4.0196, -1.9388)$. ▪

EXAMPLE 2 $u = x^2 - xy$, $v = xy + y^2$. Here the independent variable vector is (x, y), and the dependent variable vector is (u, v). We have

$$\begin{bmatrix} du \\ dv \end{bmatrix} = \begin{bmatrix} (2x - y)\,dx - x\,dy \\ y\,dx + (x + 2y)\,dy \end{bmatrix} = \begin{bmatrix} 2x - y & -x \\ y & x + 2y \end{bmatrix} \begin{bmatrix} dx \\ dy \end{bmatrix}.$$

At $(x, y) = (2, 1)$, $(u, v) = (2, 3)$, and the approximating linear mapping is

$$\begin{bmatrix} du \\ dv \end{bmatrix} = \begin{bmatrix} 3 & -2 \\ 1 & 4 \end{bmatrix} \begin{bmatrix} dx \\ dy \end{bmatrix}$$

(see Fig. 2.6). We study this linear mapping in more detail. For $dy = 0$ we have $du = 3\,dx$ and $dv = dx$, so that (du, dv) follows a line of slope $\frac{1}{3}$; for $dx = 0$ we have $du = -2\,dy$ and $dv = 4\,dy$, so that (du, dv) follows a line of slope -2. Similarly, the linear mapping can be studied along other lines. In particular, we

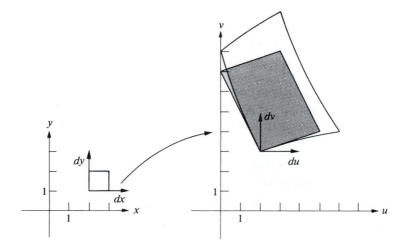

Figure 2.6 Mapping and approximating linear mapping for Example 2.

verify that the points of the square

$$0 \le dx \le 1, \qquad 0 \le dy \le 1$$

correspond to the points of the shaded parallelogram in the (du, dv) diagram. The area of the square is 1, and the area of the parallelogram is

$$|(3\mathbf{i} + \mathbf{j}) \times (-2\mathbf{i} + 4\mathbf{j})| = \begin{vmatrix} 3 & -2 \\ 1 & 4 \end{vmatrix} = 14 \text{ sq. units.}$$

For the given nonlinear mapping $u = x^2 - xy$, $v = xy + y^2$ the lines $y = 1$ and $x = 2$ (on which $dy = 0$ and $dx = 0$, respectively) correspond to parabolas through $(2, 3)$ in the uv-plane, as shown. The square $2 \le x \le 3$, $1 \le y \le 2$ (the same square as above) corresponds to a "curved parallelogram" as in the figure. Thus we see in a geometric way how the linear mapping approximates the given mapping. We observe that the linear mapping takes the line $dy = 0$ or the line $dx = 0$ to a line *tangent* to the curve obtained from the nonlinear mapping. ∎

Remark.   We have seen that for the square $0 \le dx \le 1$, $0 \le dy \le 1$, the corresponding region for the linear mapping is a parallelogram of area $14 = \begin{vmatrix} 3 & -2 \\ 1 & 4 \end{vmatrix}$. A similar calculation applies to an arbitrary square, and we find that the area is always multiplied by 14. Since arbitrary figures can be approximated by unions of squares, one concludes by a limiting process that all areas are multiplied by 14 under the linear mapping.

EXAMPLE 3   The mapping

$$x = \cos u \cos v, \qquad y = \cos u \sin v, \qquad z = \sin u$$

has the Jacobian matrix

$$\begin{bmatrix} \dfrac{\partial x}{\partial u} & \dfrac{\partial x}{\partial v} \\[2mm] \dfrac{\partial y}{\partial u} & \dfrac{\partial y}{\partial v} \\[2mm] \dfrac{\partial z}{\partial u} & \dfrac{\partial z}{\partial v} \end{bmatrix} = \begin{bmatrix} -\sin u \cos v & -\cos u \sin v \\[2mm] -\sin u \sin v & \cos u \cos v \\[2mm] \cos u & 0 \end{bmatrix}. \quad \blacksquare$$

EXAMPLE 4   Let $w = F(x, y, z)$. Then the Jacobian matrix of F is the *row vector* $(\partial F/\partial x, \partial F/\partial y, \partial F/\partial z)$. We call this vector the *gradient vector* of F and denote it by ∇F or grad F. This is discussed in Section 2.13 below. Similarly, $F(x_1, \ldots, x_n)$ has as Jacobian matrix the row vector $(F_{x_1}, \ldots, F_{x_n})$, called the gradient vector of F. ∎

We return to the general mapping $\mathbf{y} = \mathbf{f}(\mathbf{x})$ and assume $m = n$, so that the mapping is given by equations

$$\begin{cases} y_1 = f_1(x_1, \ldots, x_n), \\ \vdots \\ y_n = f_n(x_1, \ldots, x_n). \end{cases}$$

In this case the Jacobian matrix $\mathbf{y_x} = (\partial f_i/\partial x_j)$ is *square*, and we can form its determinant:

$$J = \det\left(\frac{\partial y_i}{\partial x_j}\right) = \begin{vmatrix} \dfrac{\partial y_1}{\partial x_1} & \cdots & \dfrac{\partial y_1}{\partial x_n} \\ \vdots & & \vdots \\ \dfrac{\partial y_n}{\partial x_1} & \cdots & \dfrac{\partial y_n}{\partial x_n} \end{vmatrix}.$$

We call this determinant the *Jacobian determinant* (or simply, the *Jacobian*) of the mapping. For the corresponding linear mapping $d\mathbf{y} = \mathbf{y_x}\,d\mathbf{x}$, J is the determinant of the matrix $\mathbf{y_x}$ of the mapping; this determinant measures the ratio of n-dimensional volumes. Symbolically,

$$\Delta V_y = |\det \mathbf{y_x}|\Delta V_x.$$

This is justified, as in the preceding Example 2, on the basis of the determinant formula for volume of an n-dimensional parallelotope (Eq. (1.107) in Section 1.13). For $n = 2$, $|J|$ measures the ratio of areas, as in Example 2 (see Problems 5 and 6 below).

Since the linear mapping approximates the nonlinear one, we can say that the absolute value of the Jacobian determinant J measures the ratio of corresponding volumes for small regions near the chosen point. This relationship is studied further in Sections 2.9, 4.6, and 5.14. An illustration is provided by the preceding Example 2, in which $n = 2$ and $J = 14$ at the point; this is precisely the ratio of the *area* of the parallelogram to the area of the square to which the parallelogram corresponds. For $n = 1$, J becomes the derivative dy/dx, and its absolute value does measure the ratio of corresponding *lengths* Δy, Δx: $|dy/dx| \sim |\Delta y|/|\Delta x|$ for small Δx, since $dy/dx = \lim (\Delta y/\Delta x)$ as $\Delta x \to 0$.

The Jacobian determinant is also denoted as follows:

$$J = \frac{\partial(y_1,\ldots,y_n)}{\partial(x_1,\ldots,x_n)} = \frac{\partial(f_1,\ldots,f_n)}{\partial(x_1,\ldots,x_n)}.$$

The concept of Jacobian determinant and these notations can also be applied to n functions of more than n variables. One simply forms the indicated partial derivatives, holding all other variables constant. For example, for $f(u, v, w)$, $g(u, v, w)$, one has

$$\frac{\partial(f, g)}{\partial(u, v)} = \begin{vmatrix} f_u & f_v \\ g_u & g_v \end{vmatrix},$$

$$\frac{\partial(f, g)}{\partial(u, w)} = \begin{vmatrix} f_u & f_w \\ g_u & g_w \end{vmatrix}.$$

PROBLEMS

1. Obtain the Jacobian matrix for each of the following mappings:

    a) $y_1 = 5x_1 + 2x_2$, $y_2 = 2x_1 + 3x_2$.

    b) $y_1 = 2x_1^2 + x_2^2$, $y_2 = 3x_1 x_2$.

    c) $y_1 = x_1 x_2 x_3$, $y_2 = x_1^2 x_3$.

    d) $u = x \cos y$, $v = x \sin y$, $w = x^2$.

    e) $w = x^2 yz$.

    f) $w = x^2 + y^2 - z^2$.

    g) $x = t^2$, $y = t^3$, $z = t^4$.

2. Obtain the linear mapping $dy = f_x \, dx$ approximating the given mapping $y = f(x)$ near the specified point and use the linear mapping to obtain an approximation to the value $f(x)$ specified.

    a) $y_1 = x_1^2 + x_2^2$, $y_2 = x_1 x_2$ at $(2, 1)$, approximate $f(2.04, 1.01)$.

    b) $y_1 = x_1 x_2 - x_3^2$, $y_2 = x_1 x_2 + x_1 x_3$ at $(3, 2, 1)$, approximate $f(3.01, 1.99, 1.03)$.

    c) $u = e^x \cos y$, $v = e^x \sin y$, $w = 2e^x$ at $(0, \pi/2)$, approximate value of (u, v, w) for $(x, y) = (0.1, 1.6)$.

    d) $y_1 = x_2^2 + \cdots + x_n^2$, $y_2 = x_1^2 + x_3^2 + \cdots + x_n^2, \ldots, y_n = x_1^2 + \cdots + x_{n-1}^2$ at $(1, 0, \ldots, 0)$, approximate $f(1, 0.1, \ldots, 0.1)$.

3. Obtain the Jacobian determinant requested:

    a) $\dfrac{\partial(u, v)}{\partial(x, y)}$ for $u = x^3 - 3xy^2$, $v = 3x^2 y - y^3$.

    b) $\dfrac{\partial(u, v, w)}{\partial(x, y, z)}$ for $u = xe^y \cos z$, $v = xe^y \sin z$, $w = xe^y$.

    c) $\dfrac{\partial(f, g)}{\partial(u, v)}$ for $f(u, v, w) = u^2 vw$, $g(u, v, w) = u^2 v^2 - w^4$.

    d) $\dfrac{\partial(f, g, h)}{\partial(x, y, z)}$ for $f(x, y, z, t) = x^2 + 2y + z^2 - t^2$, $g(x, y, z, t) = xyz + t^2$, $h(x, y, z, t) = z^2 - t^2$.

4. For the mapping $u = e^x \cos y$, $v = e^x \sin y$ from the xy-plane to the uv-plane, carry out the following steps:

    a) Evaluate the Jacobian determinant at $(1, 0)$.

    b) Show that the square R_{xy}: $0.9 \leq x \leq 1.1$, $-0.1 \leq y \leq 0.1$ corresponds to the region R_{uv} bounded by arcs of the circles $u^2 + v^2 = e^{1.8}$, $u^2 + v^2 = e^{2.2}$ and the rays $v = \pm(\tan 0.1)u$, $u \geq 0$, and find the ratio of the area of R_{uv} to that of R_{xy}. Compare with the result of (a).

    c) Obtain the approximating linear mapping at $(1, 0)$ and find the region R'_{uv} corresponding to the square R_{xy} of part (b) under this linear mapping. Find the ratio of the area of R'_{uv} to that of R_{xy} and compare with the results of parts (a) and (b).

5. **a)** Let $\mathbf{u}, \mathbf{v}$ be linearly independent vectors in V^2. Show by geometric reasoning that the points P of the plane for which

$$\mathbf{x} = \overrightarrow{OP} = a\mathbf{u} + b\mathbf{v}, \qquad 0 \le a \le 1, \quad 0 \le b \le 1,$$

fill a parallelogram whose edges, properly directed, represent $\mathbf{u}$ and $\mathbf{v}$.

b) With $\mathbf{u}, \mathbf{v}$ as in (a), let A be a nonsingular 2×2 matrix, so that $A\mathbf{u}, A\mathbf{v}$ are also linearly independent (Problem 13 following Section 1.15) and under the linear mapping $\mathbf{y} = A\mathbf{x}$ the parallelogram of part (a) is mapped onto a parallelogram given by

$$\mathbf{y} = \overrightarrow{OQ} = A(a\mathbf{u} + b\mathbf{v}) = aA\mathbf{u} + bA\mathbf{v}, \qquad 0 \le a \le 1, \quad 0 \le b \le 1,$$

in the plane.

Show that the area of the second parallelogram is $|\det A|$ times the area of the first. [Hint: Let B be the matrix whose column vectors are $\mathbf{u}, \mathbf{v}$ and let C be the matrix whose column vectors are $A\mathbf{u}, A\mathbf{v}$. Show that the areas in question are $|\det B|$ and $|\det C|$.]

6. Generalize the results of Problem 5 to 3-dimensional space. Thus in (a) use three vectors $\mathbf{u}, \mathbf{v}, \mathbf{w}$ and consider the corresponding parallelepiped and in (b) take A to be a nonsingular 3×3 matrix and consider volumes.

2.8 DERIVATIVES AND DIFFERENTIALS OF COMPOSITE FUNCTIONS

The functions to be considered in the following will be assumed to be defined in appropriate domains and to have continuous first partial derivatives, so that the corresponding differentials can be formed.

THEOREM If $z = f(x, y)$ and $x = g(t)$, $y = h(t)$, then

$$\frac{dz}{dt} = \frac{\partial z}{\partial x}\frac{dx}{dt} + \frac{\partial z}{\partial y}\frac{dy}{dt}. \tag{2.33}$$

If $z = f(x, y)$ and $x = g(u, v)$, $y = h(u, v)$, then

$$\frac{\partial z}{\partial u} = \frac{\partial z}{\partial x}\frac{\partial x}{\partial u} + \frac{\partial z}{\partial y}\frac{\partial y}{\partial u}, \qquad \frac{\partial z}{\partial v} = \frac{\partial z}{\partial x}\frac{\partial x}{\partial v} + \frac{\partial z}{\partial y}\frac{\partial y}{\partial v}. \tag{2.34}$$

In general, if $z = f(x, y, t, \ldots)$ and $x = g(u, v, w, \ldots)$, $y = h(u, v, w, \ldots)$, $t = p(u, v, w, \ldots), \ldots$, then

$$\frac{\partial z}{\partial u} = \frac{\partial z}{\partial x}\frac{\partial x}{\partial u} + \frac{\partial z}{\partial y}\frac{\partial y}{\partial u} + \frac{\partial z}{\partial t}\frac{\partial t}{\partial u} + \cdots,$$

$$\frac{\partial z}{\partial v} = \frac{\partial z}{\partial x}\frac{\partial x}{\partial v} + \frac{\partial z}{\partial y}\frac{\partial y}{\partial v} + \frac{\partial z}{\partial t}\frac{\partial t}{\partial v} + \cdots,$$

$$\frac{\partial z}{\partial w} = \frac{\partial z}{\partial x}\frac{\partial x}{\partial w} + \frac{\partial z}{\partial y}\frac{\partial y}{\partial w} + \frac{\partial z}{\partial t}\frac{\partial t}{\partial w} + \cdots. \tag{2.35}$$

These rules, known as "chain rules," are basic for computation of derivatives of composite functions. Equations (2.33), (2.34), and (2.35) are concise statements of the relations between the derivatives involved. Thus in (2.33),

$$z = f[g(t), h(t)]$$

is the function of t whose derivative is denoted by dz/dt, while dx/dt and dy/dt stand for $g'(t)$ and $h'(t)$, respectively. The derivatives $\partial z/\partial x$ and $\partial z/\partial y$, which could be written $(\partial z/\partial x)_y$ and $(\partial z/\partial y)_x$, stand for $f_x(x, y)$ and $f_y(x, y)$. In (2.34),

$$z = f[g(u, v), h(u, v)]$$

is the function whose derivative with respect to u is denoted by $\partial z/\partial u$, which should be understood as $(\partial z/\partial u)_v$. A more precise statement of the first equation in (2.34) would be as follows:

$$\left(\frac{\partial z}{\partial u}\right)_v = \left(\frac{\partial z}{\partial x}\right)_y \left(\frac{\partial x}{\partial u}\right)_v + \left(\frac{\partial z}{\partial y}\right)_x \left(\frac{\partial y}{\partial u}\right)_v,$$

and similar remarks apply to the other equations.

The proof of (2.33) will be given as a sample; the other rules are proved in the same way. Let t be a fixed value and let x, y, z be the corresponding values of the functions g, h, and f. Then, for given Δt, Δx and Δy are determined as

$$\Delta x = g(t + \Delta t) - g(t), \qquad \Delta y = h(t + \Delta t) - h(t),$$

while Δz is then determined as

$$\Delta z = f(x + \Delta x, y + \Delta y) - f(x, y).$$

By the Fundamental Lemma, one has

$$\Delta z = \frac{\partial z}{\partial x}\Delta x + \frac{\partial z}{\partial y}\Delta y + \epsilon_1 \Delta x + \epsilon_2 \Delta y.$$

Hence

$$\frac{\Delta z}{\Delta t} = \frac{\partial z}{\partial x}\frac{\Delta x}{\Delta t} + \frac{\partial z}{\partial y}\frac{\Delta y}{\Delta t} + \epsilon_1 \frac{\Delta x}{\Delta t} + \epsilon_2 \frac{\Delta y}{\Delta t}.$$

As Δt approaches 0, $\Delta x/\Delta t$ and $\Delta y/\Delta t$ approach the derivatives dx/dt and dy/dt, respectively, while ϵ_1 and ϵ_2 approach 0, since Δx and Δy approach 0. Hence

$$\lim_{\Delta t \to 0} \frac{\Delta z}{\Delta t} = \frac{\partial z}{\partial x} \cdot \frac{dx}{dt} + \frac{\partial z}{\partial y} \cdot \frac{dy}{dt} + 0 \cdot \frac{dx}{dt} + 0 \cdot \frac{dy}{dt};$$

that is,

$$\frac{dz}{dt} = \frac{\partial z}{\partial x}\frac{dx}{dt} + \frac{\partial z}{\partial y}\frac{dy}{dt},$$

as was to be proved.

The three functions of t considered here—$x = g(t)$, $y = h(t)$, $z = f[g(t), h(t)]$—have differentials

$$dx = \frac{dx}{dt}\Delta t, \qquad dy = \frac{dy}{dt}\Delta t, \qquad dz = \frac{dz}{dt}\Delta t.$$

From (2.33), one concludes that

$$\frac{dz}{dt}\Delta t = \frac{\partial z}{\partial x}\left(\frac{dx}{dt}\Delta t\right) + \frac{\partial z}{\partial y}\left(\frac{dy}{dt}\Delta t\right),$$

that is, that

$$dz = \frac{\partial z}{\partial x}dx + \frac{\partial z}{\partial y}dy.$$

But this is the same as (2.24), in which dx and dy are Δx and Δy, arbitrary increments of independent variables. Thus (2.24) holds whether x and y are independent and dz is the corresponding differential or whether x and y, and hence z, depend on t, so that dx, dy, dz are the differentials of these variables in terms of t.

Similar reasoning applies to (2.34). Here u and v are the independent variables on which x, y, and z depend. The corresponding differentials are

$$dx = \frac{\partial x}{\partial u}\Delta u + \frac{\partial x}{\partial v}\Delta v, \qquad dy = \frac{\partial y}{\partial u}\Delta u + \frac{\partial y}{\partial v}\Delta v, \qquad dz = \frac{\partial z}{\partial u}\Delta u + \frac{\partial z}{\partial v}\Delta v.$$

But (2.34) gives

$$
\begin{aligned}
dz &= \left(\frac{\partial z}{\partial x}\frac{\partial x}{\partial u} + \frac{\partial z}{\partial y}\frac{\partial y}{\partial u}\right)\Delta u + \left(\frac{\partial z}{\partial x}\frac{\partial x}{\partial v} + \frac{\partial z}{\partial y}\frac{\partial y}{\partial v}\right)\Delta v \\
&= \frac{\partial z}{\partial x}\left(\frac{\partial x}{\partial u}\Delta u + \frac{\partial x}{\partial v}\Delta v\right) + \frac{\partial z}{\partial y}\left(\frac{\partial y}{\partial u}\Delta u + \frac{\partial y}{\partial v}\Delta v\right) \\
&= \frac{\partial z}{\partial x}dx + \frac{\partial z}{\partial y}dy.
\end{aligned}
$$

Again (2.24) holds. Generalization of this to (2.35) permits one to conclude:

THEOREM   The differential formula

$$dz = \frac{\partial z}{\partial x}dx + \frac{\partial z}{\partial y}dy + \frac{\partial z}{\partial t}dt + \cdots, \tag{2.36}$$

which holds when $z = f(x, y, t, \ldots)$ and $dx = \Delta x$, $dy = \Delta y$, $dt = \Delta t, \ldots$, remains true when x, y, $t, \ldots$, and hence z, are all functions of other independent variables and dx, dy, $dt, \ldots, dz$ are the corresponding differentials.

As a consequence of this theorem, one can conclude: *Any equation in differentials that is correct for one choice of independent and dependent variables remains true for any other choice.* Another way of saying this is that any equation in differentials treats all variables on an equal basis. Thus if

$$dz = 2dx - 3dy$$

at a given point, then

$$dx = \tfrac{1}{2}dz + \tfrac{3}{2}dy$$

is the corresponding differential of x in terms of y and z.

An important practical application of the theorem is that in order to compute partial derivatives, one can first compute differentials, pretending that all variables are functions of a hypothetical single variable (for example, t), so that *all the rules of ordinary differential calculus apply*. From the resulting differential formula, one can at once obtain all partial derivatives desired.

EXAMPLE 1 If $z = \dfrac{x^2 - 1}{y}$, then

$$dz = \frac{2xy\,dx - (x^2 - 1)\,dy}{y^2}$$

by the quotient rule. Hence

$$\frac{\partial z}{\partial x} = \frac{2x}{y}, \qquad \frac{\partial z}{\partial y} = \frac{1 - x^2}{y^2}. \quad \blacksquare$$

EXAMPLE 2 If $r^2 = x^2 + y^2$, then $r\,dr = x\,dx + y\,dy$, whence

$$\left(\frac{\partial r}{\partial x}\right)_y = \frac{x}{r}, \qquad \left(\frac{\partial r}{\partial y}\right)_x = \frac{y}{r}, \qquad \left(\frac{\partial x}{\partial r}\right)_y = \frac{r}{x}, \qquad \text{and so on.} \quad \blacksquare$$

EXAMPLE 3 If $z = \arctan y/x \ (x \neq 0)$, then

$$dz = \frac{1}{1 + \left(\dfrac{y}{x}\right)^2}\,d\!\left(\frac{y}{x}\right) = \frac{x\,dy - y\,dx}{x^2 + y^2}$$

and hence

$$\frac{\partial z}{\partial x} = \frac{-y}{x^2 + y^2}, \qquad \frac{\partial z}{\partial y} = \frac{x}{x^2 + y^2}. \quad \blacksquare$$

PROBLEMS

1. If (a) $y = u + v$, (b) $y = u \cdot v$, (c) $y = u/v$, where u and v are functions of x, then apply (2.33) to find dy/dx.

2. If $y = u^v$, where u and v are functions of x, then find dy/dx by (2.33). [Hint: $(a^x)' = a^x \log a$, $(x^a)' = ax^{a-1}$.]

3. If $y = \log_u v$, where u and v are functions of x, then find dy/dx by (2.33). [Hint: $(\log_a x)' = 1/[x \log a]$, $\log_x a = 1/\log_a x$.]

4. If $z = e^x \cos y$, while x and y are implicit functions of t defined by the equations

$$x^3 + e^x - t^2 - t = 1, \qquad yt^2 + y^2t - t + y = 0,$$

then find dz/dt for $t = 0$. [Note that $x = 0$ and $y = 0$ for $t = 0$.]

5. Let $z = x^3 - 3x^2y$, where x and y are functions of t such that for $t = 5$, $x = 7$, $y = 2$, $dx/dt = 3$, and $dy/dt = -1$. Find dz/dt for $t = 3$.

6. Let $z = f(x, y)$, where $f_x(4, 4) = 7$, $f_y(4, 4) = 9$, $x = 2e^{3t} + t^2 - t + 2$, $y = 5e^{3t} + 3t - 1$. Find dz/dt for $t = 0$.

7. If $u = f(x, y)$ and $x = r\cos\theta$, $y = r\sin\theta$, then show that

$$\left(\frac{\partial u}{\partial x}\right)^2 + \left(\frac{\partial u}{\partial y}\right)^2 = \left(\frac{\partial u}{\partial r}\right)^2 + \frac{1}{r^2}\left(\frac{\partial u}{\partial \theta}\right)^2.$$

[Hint: Use the chain rules to evaluate the derivatives on the *right*-hand side.]

8. If $w = f(x, y)$ and $x = u\cosh v$, $y = u\sinh v$, then show that

$$\left(\frac{\partial w}{\partial x}\right)^2 - \left(\frac{\partial w}{\partial y}\right)^2 = \left(\frac{\partial w}{\partial u}\right)^2 - \frac{1}{u^2}\left(\frac{\partial w}{\partial v}\right)^2.$$

[Cf. hint for Problem 7.]

9. If $z = f(ax + by)$, show that

$$b\frac{\partial z}{\partial x} - a\frac{\partial z}{\partial y} = 0.$$

10. Find $\partial z/\partial x$ and $\partial z/\partial y$ by first obtaining dz:

a) $z = \log\sin(x^2y^2 - 1)$

b) $z = x^2y^2\sqrt{1 - x^2 - y^2}$

c) $x^2 + 2y^2 - z^2 = 1$

11. If $f(x, y)$ satisfies the identity

$$f(tx, ty) = t^n f(x, y)$$

for a fixed n, f is called *homogeneous* of degree n. Show that one then has the relation

$$x\frac{\partial f}{\partial x} + y\frac{\partial f}{\partial y} = nf(x, y).$$

This is *Euler's theorem on homogeneous functions*. [Hint: Differentiate both sides of the identity with respect to t and then set $t = 1$.]

12. (*The Stokes total time derivative in hydrodynamics*) Let $w = F(x, y, z, t)$, where $x = f(t)$, $y = g(t)$, $z = h(t)$, so that w can be expressed in terms of t alone. Show that

$$\frac{dw}{dt} = \frac{\partial w}{\partial x}\frac{dx}{dt} + \frac{\partial w}{\partial y}\frac{dy}{dt} + \frac{\partial w}{\partial z}\frac{dz}{dt} + \frac{\partial w}{\partial t}.$$

Here both dw/dt and $\partial w/\partial t = F_t(x, y, z, t)$ have meaning and are in general unequal. In hydrodynamics, dx/dt, dy/dt, dz/dt are the velocity components of a moving fluid particle, and dw/dt describes the variation of w "following the motion of the fluid." It is customary, following Stokes, to write Dw/Dt for dw/dt. [See H. Lamb, *Hydrodynamics*, 6th ed. (Cambridge University Press, 1932), p. 3.]

2.9   THE GENERAL CHAIN RULE

On occasion, one deals with two sets of functions:

$$y_1 = f_1(u_1,\ldots,u_p),$$
$$\vdots$$
$$y_m = f_m(u_1,\ldots,u_p), \tag{2.37}$$

and

$$u_1 = g_1(x_1,\ldots,x_n),$$
$$\vdots$$
$$u_p = g_p(x_1,\ldots,x_n). \tag{2.38}$$

If one substitutes the functions (2.38) in the functions (2.37), one obtains composite functions

$$y_1 = f_1\big(g_1(x_1,\ldots,x_n),\ldots, \quad g_p(x_1,\ldots,x_n)\big) = F_1(x_1,\ldots,x_n),$$
$$\vdots \tag{2.39}$$
$$y_m = f_m\big(g_1(x_1,\ldots,x_n),\ldots, \quad g_p(x_1,\ldots,x_n)\big) = F_m(x_1,\ldots,x_n).$$

Under the appropriate hypotheses, one can obtain the partial derivatives of these composite functions by chain rules, as in the previous section:

$$\frac{\partial y_i}{\partial x_j} = \frac{\partial y_i}{\partial u_1}\frac{\partial u_1}{\partial x_j} + \cdots + \frac{\partial y_i}{\partial u_p}\frac{\partial u_p}{\partial x_j} \qquad (i = 1,\ldots,m, \quad j = 1,\ldots,n). \tag{2.40}$$

The formulas (2.40) can be expressed concisely in matrix language. The partial derivatives $\partial y_i/\partial x_j$ are the entries in the $m \times n$ matrix

$$\left(\frac{\partial y_i}{\partial x_j}\right) = \begin{bmatrix} \dfrac{\partial y_1}{\partial x_1} & \dfrac{\partial y_1}{\partial x_2} & \cdots & \dfrac{\partial y_1}{\partial x_n} \\ \vdots & \vdots & & \vdots \\ \dfrac{\partial y_m}{\partial x_1} & \dfrac{\partial y_m}{\partial x_2} & \cdots & \dfrac{\partial y_m}{\partial x_n} \end{bmatrix}. \tag{2.41}$$

This is the Jacobian matrix of the mapping (2.39) (see Section 2.7). The formulas (2.40) involve two other Jacobian matrices:

$$\left(\frac{\partial y_i}{\partial u_j}\right) = \begin{bmatrix} \dfrac{\partial y_1}{\partial u_1} & \cdots & \dfrac{\partial y_1}{\partial u_p} \\ \vdots & & \vdots \\ \dfrac{\partial y_m}{\partial u_1} & \cdots & \dfrac{\partial y_m}{\partial u_p} \end{bmatrix}, \qquad \left(\frac{\partial u_i}{\partial x_j}\right) = \begin{bmatrix} \dfrac{\partial u_1}{\partial x_1} & \cdots & \dfrac{\partial u_1}{\partial x_n} \\ \vdots & & \vdots \\ \dfrac{\partial u_p}{\partial x_1} & \cdots & \dfrac{\partial u_p}{\partial x_n} \end{bmatrix}. \tag{2.42}$$

The chain rules (2.40) state that the product of the last two Jacobian matrices equals the previous one:

$$\left(\frac{\partial y_i}{\partial x_j}\right) = \left(\frac{\partial y_i}{\partial u_j}\right)\left(\frac{\partial u_i}{\partial x_j}\right). \tag{2.43}$$

This equation is called the *general chain rule*. It includes all the rules of the previous section.

EXAMPLE 1 Let $y_1 = u_1 u_2 - u_1 u_3$, $y_2 = u_1 u_3 + u_2^2$, $u_1 = x_1 \cos x_2 + (x_1 - x_2)^2$, $u_2 = x_1 \sin x_2 + x_1 x_2$, $u_3 = x_1^2 - x_1 x_2 + x_2^2$. Then, by (2.43),

$$\left(\frac{\partial y_i}{\partial x_j}\right) = \begin{bmatrix} u_2 - u_3 & u_1 & -u_1 \\ u_3 & 2u_2 & u_1 \end{bmatrix}$$

$$\cdot \begin{bmatrix} \cos x_2 + 2(x_1 - x_2) & -x_1 \sin x_2 - 2(x_1 - x_2) \\ \sin x_2 + x_2 & x_1 \cos x_2 + x_1 \\ 2x_1 - x_2 & 2x_2 - x_1 \end{bmatrix}.$$

On the right-hand side, u_1, u_2, u_3 can be expressed in terms of x_1, x_2, and the two matrices can be multiplied. However, for many purposes it is sufficient to leave the result in indicated form. In particular, to obtain numerical values, one can substitute the appropriate values and multiply the matrices only as a last step. For example, for $x_1 = 1$, $x_2 = 0$, we obtain $u_1 = 2$, $u_2 = 0$, $u_3 = 1$ and hence

$$\left(\frac{\partial y_i}{\partial x_j}\right) = \begin{bmatrix} -1 & 2 & -2 \\ 1 & 0 & 2 \end{bmatrix} \begin{bmatrix} 3 & -2 \\ 0 & 2 \\ 2 & -1 \end{bmatrix} = \begin{bmatrix} -7 & 8 \\ 7 & -4 \end{bmatrix};$$

that is to say, $\partial y_1/\partial x_1 = -7$, $\partial y_1/\partial x_2 = 8$, $\partial y_2/\partial x_1 = 7$, and $\partial y_2/\partial x_2 = -4$. ∎

Differentials and the chain rule.   If we take differentials in Eqs. (2.37) and (2.38), we obtain the equations

$$dy_1 = \frac{\partial y_1}{\partial u_1} du_1 + \cdots + \frac{\partial y_1}{\partial u_p} du_p,$$

$$\vdots$$

$$dy_m = \frac{\partial y_m}{\partial u_1} du_1 + \cdots + \frac{\partial y_m}{\partial u_p} du_p, \tag{2.44}$$

and

$$du_1 = \frac{\partial u_1}{\partial x_1} dx_1 + \cdots + \frac{\partial u_1}{\partial x_n} dx_n,$$

$$\vdots$$

$$du_p = \frac{\partial u_p}{\partial x_1} dx_1 + \cdots + \frac{\partial u_p}{\partial x_n} dx_n. \tag{2.45}$$

In (2.44), $du_1, \ldots, du_p$ are arbitrary increments $\Delta u_1, \ldots, \Delta u_p$, whereas in (2.45) they are functions of the arbitrary increments $dx_1(=\Delta x_1), \ldots, dx_n(=\Delta x_n)$. However, we know from Section 2.8 that the relationships are the same no matter how we interpret the differentials. We can write these equations in matrix form:

$$
\begin{bmatrix} dy_1 \\ \vdots \\ dy_m \end{bmatrix} = \left(\frac{\partial y_i}{\partial u_j} \right) \begin{bmatrix} du_1 \\ \vdots \\ du_p \end{bmatrix}, \qquad \begin{bmatrix} du_1 \\ \vdots \\ du_p \end{bmatrix} = \left(\frac{\partial u_i}{\partial x_j} \right) \begin{bmatrix} dx_1 \\ \vdots \\ dx_n \end{bmatrix}. \tag{2.46}
$$

If we eliminate the vector col $(du_1, \ldots, du_p)$ in these equations, we obtain

$$
\begin{bmatrix} dy_1 \\ \vdots \\ dy_m \end{bmatrix} = \left(\frac{\partial y_i}{\partial u_j} \right)\left(\frac{\partial u_i}{\partial x_j} \right) \begin{bmatrix} dx_1 \\ \vdots \\ dx_n \end{bmatrix} \tag{2.47}
$$

or

$$
dy_1 = \left(\frac{\partial y_1}{\partial u_1}\frac{\partial u_1}{\partial x_1} + \cdots + \frac{\partial y_1}{\partial u_p}\frac{\partial u_p}{\partial x_1} \right) dx_1 + \left(\frac{\partial y_1}{\partial u_1}\frac{\partial u_1}{\partial x_2} + \cdots \right) dx_2 + \cdots
$$

and so on. From these equations we can read off $\partial y_1/\partial x_1$, $\partial y_1/\partial x_2, \ldots$ Clearly, the results are the same as (2.40) or (2.43). Thus (2.47) can be termed the *general chain rule in differential form*.

The preceding development can be carried out even more concisely in terms of the notations of Section 2.7: The given functions are really vector functions

$$
\mathbf{y} = \mathbf{f(u)}, \qquad \mathbf{u} = \mathbf{g(x)}.
$$

If we take differentials, we obtain

$$
d\mathbf{y} = \mathbf{y_u}\, d\mathbf{u}, \qquad d\mathbf{u} = \mathbf{u_x}\, d\mathbf{x}
$$

and hence

$$
d\mathbf{y} = \mathbf{y_u u_x}\, d\mathbf{x}, \tag{2.48}
$$

so that

$$
\mathbf{y_x} = \mathbf{y_u u_x}. \tag{2.49}
$$

This last equation is the same as (2.43); the previous one is the same as (2.47).

We saw in Section 2.7 that a Jacobian matrix such as $\mathbf{y_x}$ is the matrix of the linear transformation approximating the given, in general nonlinear, mapping $\mathbf{y} = \mathbf{f(x)}$. Thus Eq. (2.43) asserts that if we have successive mappings $\mathbf{u} = \mathbf{g(x)}$, $\mathbf{y} = \mathbf{f(u)}$ and hence obtain a composite mapping $\mathbf{y} = \mathbf{f(g(x))}$, then the matrix of the linear approximation of the composite mapping is obtained by *multiplying* the approximating matrices of the two stages. In the special case when $\mathbf{f}$ and $\mathbf{g}$ are linear, then we have $\mathbf{u} = B\mathbf{x}, \mathbf{y} = A\mathbf{u}$ for appropriate matrices A and B ($\mathbf{u_x} = B, \mathbf{y_u} = A$), and the composite mapping is $\mathbf{y} = A(B\mathbf{x}) = AB\mathbf{x}$, as in Section 1.7; thus $\mathbf{y_x} = AB = \mathbf{y_u u_x}$.

Case of square matrices. In the preceding analysis, let $m = n = p$, so that all the Jacobian matrices appearing are *square* and each has a determinant—the *Jacobian determinant* of the corresponding mapping, as in Section 2.7. For example,

$$\det \mathbf{y_u} = \begin{vmatrix} \dfrac{\partial y_1}{\partial u_1} & \cdots & \dfrac{\partial y_1}{\partial u_n} \\ \vdots & \vdots & \vdots \\ \dfrac{\partial y_n}{\partial u_1} & \cdots & \dfrac{\partial y_n}{\partial u_n} \end{vmatrix} = \frac{\partial(y_1,\ldots,y_n)}{\partial(u_1,\ldots,u_n)}.$$

To the equation (2.49) we can apply the rule $\det AB = \det A \det B$ (Eq. (1.60) in Section 1.8) to obtain the following very useful rule:

$$\det \mathbf{y_x} = \det \mathbf{y_u} \det \mathbf{u_x}; \tag{2.50}$$

that is,

$$\frac{\partial(y_1,\ldots,y_n)}{\partial(x_1,\ldots,x_n)} = \frac{\partial(y_1,\ldots,y_n)}{\partial(u_1,\ldots,u_n)} \frac{\partial(u_1,\ldots,u_n)}{\partial(x_1,\ldots,x_n)}. \tag{2.51}$$

If, for example, $n = 2$, then each determinant here can be interpreted as in Section 2.7 as plus or minus the ratio of small corresponding areas, and (2.51) states roughly that

$$\frac{\Delta A_y}{\Delta A_x} = \frac{\Delta A_y}{\Delta A_u} \frac{\Delta A_u}{\Delta A_x},$$

where we have written ΔA_x for an "area element" in the $x_1 x_2$-plane, and similarly for ΔA_y, ΔA_u. There is a similar interpretation for $n = 3$, in terms of volumes, and for higher n in terms of higher-dimensional volume.

PROBLEMS

1. Find the Jacobian matrix $(\partial y_i/\partial x_j)$ in the form of a product of two matrices and evaluate the matrix for the given values of $x_1, x_2, \ldots$.

   a) $y_1 = u_1 u_2 - 3u_1$, $y_2 = u_2^2 + 2u_1 u_2 + 2u_1 - u_2$; $\quad u_1 = x_1 \cos 3x_2$, $u_2 = x_1 \sin 3x_2$; $x_1 = 0$, $x_2 = 0$.

   b) $y_1 = u_1^2 + u_2^2 - 3u_1 + u_3$, $y_2 = u_1^2 - u_2^2 + 2u_1 - 3u_3$; $\quad u_1 = x_1 x_2 x_3^2$, $u_2 = x_1 x_2^2 x_3$, $u_3 = x_1^2 x_2 x_3$; $x_1 = 1$, $x_2 = 1$, $x_3 = 1$.

   c) $y_1 = u_1 e^{u_2}$, $y_2 = u_1 e^{-u_2}$, $y_3 = u_1^2$; $\quad u_1 = x_1^2 + x_2$, $u_2 = 2x_1^2 - x_2$; $x_1 = 1$, $x_2 = 0$.

   d) $y_1 = u_1^2 + \cdots + u_n^2 - u_1^2$, $y_2 = u_1^2 + \cdots + u_n^2 - u_2^2, \ldots, y_n = u_1^2 + \cdots + u_n^2 - u_n^2$; $\quad u_1 = x_1^2 + x_1 x_2$, $u_2 = x_1^2 + 2x_1 x_2, \ldots, u_n = x_1^2 + nx_1 x_2$; $x_1 = 1$, $x_2 = 0$.

2. **a)** Find $\partial(z, w)/\partial(x, y)$ for $x = 1$, $y = 0$ if $z = u^3 + 3u^2 v - v^3 + u^2 - v^2$, $w = u^3 + v^3 - 2u^2$; $u = x \cos xy$, $v = x \sin xy + x^2 - y^2$.

   b) Find $\partial(x, y)/\partial(s, t)$ for $s = 0$, $t = 0$ if $x = (z^2 + w^2)^{1/2}$, $y = w(z^2 + w^2)^{-1/2}$; $z = (s + t + 1)^{-1}$, $w = (2s - t + 1)^{-1}$.

3. Justify the rules, under appropriate hypotheses:

   a) If $y = f(u)$, $u = g(v)$, $v = h(x)$, then $y_x = y_u u_v v_x$.

   b) $\dfrac{\partial(z, w)}{\partial(x, y)} = \dfrac{\partial(z, w)}{\partial(u, v)} \dfrac{\partial(u, v)}{\partial(s, t)} \dfrac{\partial(s, t)}{\partial(x, y)}$.

4. For certain functions $f(x, y)$, $g(x, y)$, $p(u, v)$, $q(u, v)$ it is known that $f(x_0, y_0) = u_0$, $g(x_0, y_0) = v_0$ and that $f_x(x_0, y_0) = 2$, $f_y(x_0, y_0) = 3$, $g_x(x_0, y_0) = -1$, $g_y(x_0, y_0) = 5$, $p_u(u_0, v_0) = 7$, $p_v(u_0, v_0) = 1$, $q_u(u_0, v_0) = -3$, $q_v(u_0, v_0) = 2$. Let $z = F(x, y) = p(f(x, y), g(x, y))$, $w = G(x, y) = q(f(x, y), g(x, y))$ and find the Jacobian matrix of $z(x, y)$, $w(x, y)$ at (x_0, y_0).

5. Let $u_1 = x_1 - 3x_2 + 2x_1x_2$, $u_2 = 2x_1 + 5x_2 - 3x_1x_2$. Let $\mathbf{w} = (w_1, w_2)$ be a vector function of $\mathbf{u} = (u_1, u_2)$ such that $\mathbf{w_u} = \begin{bmatrix} 2 & 11 \\ 7 & 5 \end{bmatrix}$ for $\mathbf{u} = (3, 3)$. Find the Jacobian matrix at $\mathbf{x} = (2, 1)$ for the composite function $\mathbf{w}[\mathbf{u}(\mathbf{x})]$.

6. Let $\mathbf{u} = \mathbf{f}(\mathbf{x})$ and $\mathbf{v} = \mathbf{g}(\mathbf{x})$ be differentiable mappings from a domain D in 3-dimensional space to 3-dimensional space. Let a and b be constant scalars. Let A be a constant 3×3 matrix. Show:

   a) $d(\mathbf{u} + \mathbf{v}) = d\mathbf{u} + d\mathbf{v}$

   b) $d(a\mathbf{u} + b\mathbf{v}) = a\,d\mathbf{u} + b\,d\mathbf{v}$

   c) $d(A\mathbf{u}) = A\,d\mathbf{u}$

   d) $d(\mathbf{u} \cdot \mathbf{v}) = \mathbf{u} \cdot d\mathbf{v} + \mathbf{v} \cdot d\mathbf{u}$

   e) $d(\mathbf{u} \times \mathbf{v}) = \mathbf{u} \times d\mathbf{v} + d\mathbf{u} \times \mathbf{v}$

2.10   IMPLICIT FUNCTIONS

If $F(x, y, z)$ is a given function of x, y, and z, then the equation

$$F(x, y, z) = 0 \qquad (2.52)$$

is a relation that may describe one or several functions z of x and y. Thus if $x^2 + y^2 + z^2 - 1 = 0$, then

$$z = \sqrt{1 - x^2 - y^2} \quad \text{or} \quad z = -\sqrt{1 - x^2 - y^2},$$

both functions being defined for $x^2 + y^2 \leq 1$. Either function is said to be *implicitly defined* by the equation $x^2 + y^2 + z^2 - 1 = 0$.

Similarly, an equation

$$F(x, y, z, w) = 0 \qquad (2.53)$$

may define one or more implicit functions w of x, y, z. If two such equations are given:

$$F(x, y, z, w) = 0, \qquad G(x, y, z, w) = 0, \qquad (2.54)$$

it is in general possible (at least in theory) to reduce the equations by elimination to the form

$$w = f(x, y), \qquad z = g(x, y), \qquad (2.55)$$

that is, to obtain two functions of two variables. In general, if m equations in n unknowns are given ($m < n$), it is possible to solve for m of the variables in terms of the remaining $n - m$ variables; the *number of dependent variables equals the number of equations*.

The main question to be considered here is of the following type. Suppose a particular solution of the m simultaneous equations is known, for example, a quadruple (x_1, y_1, z_1, w_1) of values satisfying (2.54); then one seeks to determine the behavior of the m dependent variables as functions of the independent variables near the given point; for example, for (2.54), one wishes to study $f(x, y)$ and $g(x, y)$ near (x_1, y_1), given that $f(x_1, y_1) = w_1$ and $g(x_1, y_1) = z_1$. Determining behavior of a function near a point will consist here of finding the first partial derivatives at the point; when the first derivatives are known, the total differential can be found, and hence a *linear approximation* to the function is known.

It will in fact be seen that the essential step to be taken consists of a *linearization* of the given simultaneous equations. This has its counterpart in the formulation of laws of physics; one seeks to describe natural phenomena by means of the simplest possible equations. These are usually linear in form and are usually valid only when the variables are restricted to a narrow range. The complete description of the phenomena in general involves simultaneous nonlinear equations, which are far more difficult to grasp. A typical example is Hooke's law for a spring: $F = k^2 x$; the linear dependence of force F on displacement x is valid only as a first approximation.

To analyze an equation of form (2.52), we assume that $z = f(x, y)$ is a differentiable function that satisfies the equation, so that

$$F(x, y, f(x, y)) = 0. \tag{2.56}$$

We assume that this relation holds for (x, y) in a domain D and that the points (x, y, z), for (x, y) in D and $z = f(x, y)$, all lie in a domain in which F is differentiable. Then from (2.52) or (2.56) we obtain:

$$F_x \, dx + F_y \, dy + F_z \, dz = 0. \tag{2.57}$$

Here z is considered to be the function $f(x, y)$, so that $dz = f_x \, dx + f_y \, dy$. From (2.57) we deduce that

$$dz = -\frac{F_x}{F_z} dx - \frac{F_y}{F_z} dy, \tag{2.58}$$

so that (provided that $F_z \neq 0$ at the points considered)

$$f_x = \frac{\partial z}{\partial x} = -\frac{F_x}{F_z}, \qquad f_y = \frac{\partial z}{\partial y} = -\frac{F_y}{F_z}. \tag{2.59}$$

These are the desired expressions for the derivatives.

EXAMPLE 1 $x^2 + y^2 + z^2 = 1$ or $x^2 + y^2 + z^2 - 1 = 0$. We imitate the procedure of the preceding paragraph:

$$2x \, dx + 2y \, dy + 2z \, dz = 0,$$

$$dz = -\frac{x}{z} dx - \frac{y}{z} dy,$$

$$\frac{\partial z}{\partial x} = -\frac{x}{z}, \qquad \frac{\partial z}{\partial y} = -\frac{y}{z} \qquad (z \neq 0).$$

These equations apply to each differentiable function satisfying the given equation—in particular to the function $z = (1 - x^2 - y^2)^{1/2}$. At $x = \frac{1}{2}, y = \frac{1}{2}$ we find $z = 1/\sqrt{2}$, and hence

$$\frac{\partial z}{\partial x} = -\frac{x}{z} = -\frac{\sqrt{2}}{2}, \qquad \frac{\partial z}{\partial y} = -\frac{y}{z} = -\frac{\sqrt{2}}{2}. \quad \blacksquare$$

We remark that the same results can be obtained by taking *partial derivatives* instead of differentials in the given equation. From (2.52), with z considered as a function of x and y, we differentiate with respect to x and then with respect to y to obtain

$$F_x + F_z \frac{\partial z}{\partial x} = 0, \qquad F_y + F_z \frac{\partial z}{\partial y} = 0,$$

from which (2.59) again follows. In the case of Example 1 we obtain

$$2x + 2z \frac{\partial z}{\partial x} = 0, \qquad 2y + 2z \frac{\partial z}{\partial y} = 0.$$

For the pair of equations (2.54) we assume that differentiable functions $w = f(x, y), z = g(x, y)$ satisfy the equations and can then take differentials:

$$F_x \, dx + F_y \, dy + F_z \, dz + F_w \, dw = 0,$$

$$G_x \, dx + G_y \, dy + G_z \, dz + G_w \, dw = 0. \tag{2.60}$$

We consider these equations as simultaneous *linear* equations for dz and dw and solve by elimination or by determinants. Cramer's Rule yields

$$dz = \frac{\begin{vmatrix} -F_x \, dx - F_y \, dy & F_w \\ -G_x \, dx - G_y \, dy & G_w \end{vmatrix}}{\begin{vmatrix} F_z & F_w \\ G_z & G_w \end{vmatrix}},$$

$$dw = \frac{\begin{vmatrix} F_z & -F_x \, dx - F_y \, dy \\ G_z & -G_x \, dx - G_y \, dy \end{vmatrix}}{\begin{vmatrix} F_z & F_w \\ G_z & G_w \end{vmatrix}},$$

and hence

$$dz = -\frac{\begin{vmatrix} F_x & F_w \\ G_x & G_w \end{vmatrix}}{\begin{vmatrix} F_z & F_w \\ G_z & G_w \end{vmatrix}} dx - \frac{\begin{vmatrix} F_y & F_w \\ G_y & G_w \end{vmatrix}}{\begin{vmatrix} F_z & F_w \\ G_z & G_w \end{vmatrix}} dy,$$

$$dw = -\frac{\begin{vmatrix} F_z & F_x \\ G_z & G_x \end{vmatrix}}{\begin{vmatrix} F_z & F_w \\ G_z & G_w \end{vmatrix}} dx - \frac{\begin{vmatrix} F_z & F_y \\ G_z & G_y \end{vmatrix}}{\begin{vmatrix} F_z & F_w \\ G_z & G_w \end{vmatrix}} dy.$$

Thus we can read off partial derivatives. Since the determinants appearing are Jacobian determinants, we can write the derivatives in terms of Jacobians:

$$\frac{\partial z}{\partial x} = -\frac{\dfrac{\partial(F,G)}{\partial(x,w)}}{\dfrac{\partial(F,G)}{\partial(z,w)}}, \qquad \frac{\partial z}{\partial y} = -\frac{\dfrac{\partial(F,G)}{\partial(y,w)}}{\dfrac{\partial(F,G)}{\partial(z,w)}},$$

$$\frac{\partial w}{\partial x} = -\frac{\dfrac{\partial(F,G)}{\partial(z,x)}}{\dfrac{\partial(F,G)}{\partial(z,w)}}, \qquad \frac{\partial w}{\partial y} = -\frac{\dfrac{\partial(F,G)}{\partial(z,y)}}{\dfrac{\partial(F,G)}{\partial(z,w)}}. \qquad (2.61)$$

Here we must assume that the Jacobian $\partial(F,G)/\partial(z,w)$ in the denominator is different from 0 at the points considered.

EXAMPLE 2

$$2x^2 + y^2 + z^2 - zw = 0,$$
$$x^2 + y^2 + 2z^2 + zw - 8 = 0.$$

We observe that the equations are satisfied for $x = 1$, $y = 1$, $z = 1$, $w = 4$, and seek differentiable functions $z(x, y)$, $w(x, y)$ satisfying the equations near this point. If such functions exist, then

$$4x\,dx + 2y\,dy + (2z - w)\,dz - z\,dw = 0,$$
$$2x\,dx + 2y\,dy + (4z + w)\,dz + z\,dw = 0.$$

By elimination we find

$$6x\,dx + 4y\,dy + 6z\,dz = 0,$$
$$6x(2z + w)\,dx + 4y(z + w)\,dy - 6z^2\,dw = 0$$

and hence

$$dz = -\frac{x}{z}\,dx - \frac{2y}{3z}\,dy,$$

$$dw = \frac{x(2z + w)}{z^2}\,dx + \frac{2y(z + w)}{3z^2}\,dy,$$

from which we can read off partial derivatives $\partial z/\partial x = -x/z$ and so on. We could have applied (2.61) directly. For example,

$$\frac{\partial z}{\partial x} = -\frac{\begin{vmatrix} 4x & -z \\ 2x & z \end{vmatrix}}{\begin{vmatrix} 2z - w & -z \\ 4z + w & z \end{vmatrix}} = \frac{-6xz}{6z^2} = -\frac{x}{z}.$$

We observe that the determinant in the denominator equals $6z^2$, and this is different from 0 near $z = 1$.

We could also have taken partial derivatives. By differentiating with respect to x, we obtain

$$4x + (2z - w)\frac{\partial z}{\partial x} - z\frac{\partial w}{\partial x} = 0,$$

$$2x + (4z + w)\frac{\partial z}{\partial x} + z\frac{\partial w}{\partial x} = 0.$$

Elimination gives $\partial z/\partial x = -x/z$, $\partial w/\partial x = x(2z + w)/z^2$. Taking differentials saves time, since all partial derivatives are obtained at once. ∎

The reasoning generalizes to an arbitrary set of m equations in $m + n$ unknowns, say

$$F_1(y_1, \ldots, y_m, x_1, \ldots, x_n) = 0,$$

$$\vdots$$

$$F_m(y_1, \ldots, y_m, x_1, \ldots, x_n) = 0. \tag{2.62}$$

We seek m differentiable functions

$$y_1 = f_1(x_1, \ldots, x_n),$$

$$\vdots$$

$$y_m = f_m(x_1, \ldots, x_n) \tag{2.63}$$

satisfying the equations. Assuming differentiability as before, we obtain from (2.62)

$$F_{1y_1}\, dy_1 + \cdots + F_{1y_m}\, dy_m + F_{1x_1}\, dx_1 + \cdots + F_{1x_n}\, dx_n = 0,$$

$$\vdots$$

$$F_{my_1}\, dy_1 + \cdots + F_{my_m}\, dy_m + F_{mx_1}\, dx_1 + \cdots + F_{mx_n}\, dx_n = 0, \tag{2.64}$$

where $dy_1, \ldots, dy_m$ are the differentials of the functions sought. Equations (2.64) are m *linear* equations in the m unknowns $dy_1, \ldots, dy_m$, and we have in effect *linearized* our problem. If the appropriate determinant is not 0, we can solve Eqs. (2.64) for $dy_1, \ldots, dy_m$. That determinant is

$$\begin{vmatrix} F_{1y_1} & \cdots & F_{1y_m} \\ \vdots & & \vdots \\ F_{my_1} & \cdots & F_{my_m} \end{vmatrix} = \frac{\partial(F_1, \ldots, F_m)}{\partial(y_1, \ldots, y_m)}. \tag{2.65}$$

The equations can be solved by elimination or by determinants.

We can also use matrices here. Equations (2.64) can be written:

$$\mathbf{F_y}\, d\mathbf{y} + \mathbf{F_x}\, d\mathbf{x} = \mathbf{0}, \tag{2.66}$$

where $\mathbf{F_y}$ is the Jacobian matrix of the coefficients of $dy_1, \ldots, dy_m$ in (2.64); that is, it is the matrix whose determinant is given in (2.65). Similarly,

$$\mathbf{F_x} = \begin{bmatrix} F_{1x_1} & \cdots & F_{1x_n} \\ \vdots & & \vdots \\ F_{mx_1} & \cdots & F_{mx_n} \end{bmatrix}. \tag{2.67}$$

Finally, $d\mathbf{y} = \mathrm{col}(dy_1, \ldots, dy_m)$, and $d\mathbf{x} = \mathrm{col}(dx_1, \ldots, dx_n)$. If $\mathbf{F_y}$ is not singular at the point considered—that is, if the Jacobian determinant (2.65) is not $\mathbf{0}$—then $\mathbf{F_y}$ has an inverse and we can write

$$\mathbf{F_y}\, d\mathbf{y} = -\mathbf{F_x}\, d\mathbf{x},$$

$$d\mathbf{y} = -(\mathbf{F_y})^{-1}\mathbf{F_x}\, d\mathbf{x}. \tag{2.68}$$

This equation gives $dy_1, \ldots, dy_m$ and hence all partial derivatives of the unknown functions $f_1, \ldots, f_m$.

We can also take partial derivatives in (2.62). By differentiation with respect to x_j, we obtain

$$F_{1y_1}\frac{\partial y_1}{\partial x_j} + \cdots + F_{1y_m}\frac{\partial y_m}{\partial x_j} + F_{1x_j} = 0,$$

$$\vdots$$

$$F_{my_1}\frac{\partial y_1}{\partial x_j} + \cdots + F_{my_m}\frac{\partial y_m}{\partial x_j} + F_{mx_j} = 0.$$

These are m linear equations for the m unknowns $\partial y_1/\partial x_j, \ldots, \partial y_m/\partial x_j$. If the determinant (2.65) is not 0, we can solve for the unknowns. Cramer's Rule gives, for example,

$$\frac{\partial y_1}{\partial x_j} = - \begin{vmatrix} F_{1x_j} & F_{1y_2} & \cdots & F_{1y_m} \\ \vdots & \vdots & & \vdots \\ F_{mx_j} & F_{my_2} & \cdots & F_{my_m} \end{vmatrix} \div \begin{vmatrix} F_{1y_1} & F_{1y_2} & \cdots & F_{1y_m} \\ \vdots & \vdots & & \vdots \\ F_{my_1} & F_{my_2} & \cdots & F_{my_m} \end{vmatrix}$$

$$= -\frac{\dfrac{\partial(F_1, \ldots, F_m)}{\partial(x_j, y_2, \ldots, y_m)}}{\dfrac{\partial(F_1, \ldots, F_m)}{\partial(y_1, \ldots, y_m)}}.$$

Similar formulas are obtained for $\partial y_2/\partial x_j, \ldots, \partial y_m/\partial x_j$. In general,

$$\frac{\partial y_i}{\partial x_j} = - \frac{\dfrac{\partial(F_1,\ldots,F_m)}{\partial(y_1,\ldots,y_{i-1},\, x_j,\, y_{i+1},\ldots,y_m)}}{\dfrac{\partial(F_1,\ldots,F_m)}{\partial(y_1,\ldots,y_m)}} \tag{2.69}$$

$$(i = 1,\ldots,m, \qquad j = 1,\ldots,n).$$

The denominator here is the Jacobian (2.65). The numerator is obtained from the denominator by replacing the ith column by $\mathrm{col}(\partial F_1/\partial x_j,\ldots,\partial F_m/\partial x_j)$. An illustration is given by Eqs. (2.61), where $m = 2$ and $n = 2$ (two equations in four unknowns). The following are additional illustrations:

one equation in two unknowns: $F(x, y) = 0$. Here

$$\frac{dy}{dx} = - \frac{F_x}{F_y} \qquad (F_y \neq 0). \tag{2.70}$$

one equation in three unknowns: $F(x, y, z) = 0$. Here, as in (2.59),

$$\frac{\partial z}{\partial x} = - \frac{F_x}{F_z}, \frac{\partial z}{\partial y} = - \frac{F_y}{F_z} \qquad (F_z \neq 0). \tag{2.71}$$

two equations in three unknowns: $F(x, y, z) = 0$, $G(x, y, z) = 0$. Here

$$\frac{dz}{dx} = - \frac{\dfrac{\partial(F, G)}{\partial(y, x)}}{\dfrac{\partial(F, G)}{\partial(y, z)}}, \qquad \frac{dy}{dx} = - \frac{\dfrac{\partial(F, G)}{\partial(x, z)}}{\dfrac{\partial(F, G)}{\partial(y, z)}}, \tag{2.72}$$

where

$$\frac{\partial(F, G)}{\partial(y, z)} = \begin{vmatrix} F_y & F_z \\ G_y & G_z \end{vmatrix} \neq 0.$$

three equations in five unknowns: $F(x, y, z, u, v) = 0$, $G(x, y, z, u, v) = 0$, $H(x, y, z, u, v) = 0$. Here

$$\frac{\partial x}{\partial u} = - \frac{\dfrac{\partial(F, G, H)}{\partial(u, y, z)}}{\dfrac{\partial(F, G, H)}{\partial(x, y, z)}}, \qquad \frac{\partial x}{\partial v} = - \frac{\dfrac{\partial(F, G, H)}{\partial(v, y, z)}}{\dfrac{\partial(F, G, H)}{\partial(x, y, z)}}, \qquad \frac{\partial y}{\partial u} = - \frac{\dfrac{\partial(F, G, H)}{\partial(x, u, z)}}{\dfrac{\partial(F, G, H)}{\partial(x, y, z)}},$$

$$\frac{\partial y}{\partial v} = - \frac{\dfrac{\partial(F, G, H)}{\partial(x, v, z)}}{\dfrac{\partial(F, G, H)}{\partial(x, y, z)}}, \qquad \frac{\partial z}{\partial u} = - \frac{\dfrac{\partial(F, G, H)}{\partial(x, y, u)}}{\dfrac{\partial(F, G, H)}{\partial(x, y, z)}}, \qquad \frac{\partial z}{\partial v} = - \frac{\dfrac{\partial(F, G, H)}{\partial(x, y, v)}}{\dfrac{\partial(F, G, H)}{\partial(x, y, z)}}.$$

$$\tag{2.73}$$

For greater precision the partial derivatives appearing should have subscripts attached. For example, in (2.73), $\partial x / \partial u$ should be $(\partial x / \partial u)_v$.

The entire preceding discussion has been based on the assumption that implicit functions are in fact defined by the equations. This assumption is not always fulfilled. For example, the equations

$$x^2 + y^2 + u^2 + v^2 + 1 = 0, \qquad x^2 - y^2 + 2uv = 0$$

define no functions at all. It can be shown that if the Jacobian determinant in the denominator is not 0, then the implicit equations do define functions as before. One has the following fundamental result:

IMPLICIT FUNCTION THEOREM   For $i = 1, \ldots, m$, let the functions $F_i(y_1, \ldots, y_m, x_1, \ldots, x_n)$ all be defined in a neighborhood of the point P_0: $(y_1^0, \ldots, y_m^0, x_1^0, \ldots, x_n^0)$ and have continuous first partial derivatives in this neighborhood.

Let the equations

$$F_i(y_1, \ldots, y_m, x_1, \ldots, x_n) = 0, \qquad i = 1, \ldots, m,$$

be satisfied at P_0 and let

$$\frac{\partial(F_1, \ldots, F_m)}{\partial(y_1, \ldots, y_m)} \neq 0 \quad \text{at } P_0.$$

Then in an appropriate neighborhood of $(x_1^0, \ldots, x_n^0)$, there is a unique set of continuous functions

$$y_i = f_i(x_1, \ldots, x_n), \qquad i = 1, \ldots, m,$$

such that $y_i^0 = f_i(x_1^0, \ldots, x_n^0)$ for $i = 1, \ldots, m$ and for all i

$$F_i(f_1(x_1, \ldots, x_n), \ldots, f_m(x_1, \ldots, x_n), x_1, \ldots, x_n) \equiv 0$$

in the neighborhood. Furthermore, the f_i have continuous partial derivatives satisfying (2.69).

A proof for the case $m = 1$, $n = 1$ is given in the next section.

Remark.   If the Jacobian determinant happens to be 0 at the point of interest, a different choice of dependent variables (such as $x_1, y_2, \ldots, y_m$) may avoid the difficulty. Thus if the Jacobian matrix formed of *all* $m + n$ first partial derivatives of each of $F_1, \ldots, F_m$ has rank m at the point P_0 (Section 1.16), then one can solve for m of the variables in terms of the remaining ones.

*2.11   PROOF OF A CASE OF THE IMPLICIT FUNCTION THEOREM

The discussion of implicit equations in the preceding section is purely formal; it is assumed that we can solve them, and then it gives formulas for derivatives of the solutions. It is important to know when we can solve. In practice, the typical situation is that one initially knows just one point on the graph of the solution sought and then one tries to find a continuous solution through this point. The Implicit Function Theorem ensures that *if the relevant Jacobian is not zero at the point, then this is possible at least in a sufficiently small neighborhood of the initial point.* The basis of the theorem is the fact that near the point our equations can be approximated by linear equations (by taking differentials), so that, provided that the relevant matrix is nonsingular, one can always solve.

We here consider only the simplest case of one equation in two unknowns, which we write as

$$F(x, y) = 0. \tag{2.74}$$

For a discussion of the general case we refer the reader to Chapter 9 of the book by Rudin listed at the end of the chapter. (See also the Remarks at the end of this Section.)

THEOREM   Let $F(x, y)$ be defined in an open region D of the xy-plane, let (x_0, y_0) be in D, and let $F(x_0, y_0) = 0$; let the partial derivatives F_x, F_y be continuous in D and suppose that $F_y(x_0, y_0) \neq 0$. Then there is a function

$$y = f(x) \quad \text{with domain} \quad |x - x_0| < \delta, \qquad \delta > 0, \tag{2.75}$$

whose graph is in D, with $f(x_0) = y_0$, and that satisfies Eq. (2.74). Furthermore, a positive number η can be chosen so that the graph of (2.75) lies in the set

$$|x - x_0| < \delta, \qquad |y - y_0| < \eta$$

and provides all solutions of (2.74) in this set. The function f is differentiable, and

$$f'(x) = -\frac{F_x(x, f(x))}{F_y(x, f(x))}.$$

Proof.   We let

$$g(x, y) = -\frac{F_x(x, y)}{F_y(x, y)}$$

wherever $F_y \neq 0$, so that

$$F_x = -F_y g.$$

Now $F_y(x_0, y_0) \neq 0$. Let us suppose, to be specific, that $F_y(x_0, y_0) > 0$. Then, by continuity, $F_y(x, y) > 0$ in a neighborhood of (x_0, y_0). Therefore we can choose δ, η so small and positive that the closed rectangular region

$$E: |x - x_0| \leq \delta, \qquad |y - y_0| \leq \eta$$

is in D, and $F_y > 0$ in E. The function g is also continuous in E and hence has absolute minimum m and maximum M in E (see Sections 2.19 and 2.23 below):

$$m \leq g(x, y) \leq M \quad \text{in } E.$$

We replace δ by a smaller number, if necessary, to ensure that $|m|\delta < \eta$ and $|M|\delta < \eta$. This is done to ensure that the graphs of the linear functions

$$y - y_0 = M(x - x_0), \qquad y - y_0 = m(x - x_0), \qquad |x - x_0| \le \delta \quad (2.76)$$

lie in E (see Figure 2.7). We assume δ to be so chosen.

Figure 2.7   Proof of the
Implicit Function Theorem.

Now, since $F_y > 0$ in E and $F(x_0, y_0) = 0$, F is monotone strictly increasing along the line $x = x_0$ in E and hence is positive for $y > y_0$ and negative for $y < y_0$. Along a line

$$y - y_0 = \lambda(x - x_0),$$

F becomes a function of x, and

$$\frac{dF}{dx} = F_x + F_y\frac{dy}{dx} = F_x + \lambda F_y = -F_y g + \lambda F_y = F_y(\lambda - g).$$

Hence, if λ is greater than M, the maximum of g, dF/dx is positive along the line; thus F itself is positive for $x > x_0$, negative for $x < x_0$. There is similar reasoning, with reversal of signs, for $\lambda < m$. Thus the sign of F is as in Figure 2.7. Since $F_y > 0$, F is monotone strictly increasing in y on each line $x = \text{const}$ in E and hence must go from negative to positive. Therefore by the Intermediate Value Theorem (Problem 5 following Section 2.23) $F(x, y) = 0$ for exactly one y for each x. This value of y we denote by $f(x)$. Thus

$$F(x, f(x)) = 0,$$

and $y = f(x)$ provides all solutions of the implicit equation in E.

The graph of f is squeezed between the two lines (2.76). Hence f is continuous at x_0, with $f(x) \to y_0 = f(x_0)$ as $x \to x_0$. But the same argument applies to each point (x, y) on the graph of f; we can find a rectangle E centered at the point and so on, as for (x_0, y_0). Therefore f is continuous for all x in the interval $|x - x_0| < \delta$. (We can also define f at the endpoints of the interval, $x_0 \pm \delta$, but we ignore these values in the subsequent discussion.)

Finally, we seek the derivative of f at x_0. Along a line $y - y_0 = \lambda(x - x_0)$ with $\lambda = g(x_0, y_0) + \epsilon$, $\epsilon > 0$, we have, as earlier,

$$\frac{dF}{dx} = F_y \cdot (\lambda - g) = F_y \cdot \left[g(x_0, y_0) - g(x, y) + \epsilon \right].$$

Since g is continuous at (x_0, y_0), $|g(x, y) - g(x_0, y_0)| < \epsilon$ for (x, y) sufficiently close to (x_0, y_0); thus the quantity in brackets is positive, and $dF/dx > 0$. Therefore F itself is positive along the line, for $x > x_0$ and x sufficiently close to x_0. But this means that the graph of f must be below the line, that is,

$$f(x) < y_0 + \left[g(x_0, y_0) + \epsilon \right](x - x_0)$$

or

$$\frac{f(x) - f(x_0)}{x - x_0} < g(x_0, y_0) + \epsilon$$

for $x > x_0$ as earlier. Similarly,

$$\frac{f(x) - f(x_0)}{x - x_0} > g(x_0, y_0) - \epsilon$$

for $x > x_0$ and x sufficiently close to x_0. Thus

$$\lim_{x \to x_0^+} \frac{f(x) - f(x_0)}{x - x_0} = g(x_0, y_0).$$

In the same way the same limit is found as $x \to x_0^-$. Therefore

$$f'(x_0) = g(x_0, y_0).$$

Again the argument applies to every point (x, y) on the graph of f, and we conclude that at each such point

$$f'(x) = g(x, y) = -\frac{F_x(x, y)}{F_y(x, y)}.$$

Thus the theorem is proved. $\square$

Remarks.   The theorem proved here gives more information about uniqueness than the general theorem stated at the end of the previous section: In an appropriate neighborhood of P_0: (x_0, y_0), *all* points (x, y) satisfying the implicit equation are on the graph of the solution $y = f(x)$ found. The general theorem can be extended in analogous fashion.

The proof gives some information about the size of the x-interval in which a solution $y = f(x)$ can be found. Specifically, one must choose a rectangle $E : |x - x_0| \le \delta$, $|y - y_0| \le \eta$ in which $F_y > 0$ (or $F_y < 0$) and in which δ has been restricted so that $\delta K < \eta$, where $K = \max|g(x, y)|$ in E. One sometimes has more information about F that permits one to give a better estimate of the size of the interval. For example, if $F_y > 0$ in E and F is positive for $y = y_0 + \eta$ and negative for $y = y_0 - \eta$, then the solution is defined and unique for $|x - x_0| < \delta$. The restriction $|y - y_0| \leq \eta$, in general, is needed to ensure the uniqueness of the solution. For example, the equation

$$y^2 - y \sin x - e^x y + e^x \sin x = 0$$

satisfies the hypotheses of the theorem with $x_0 = 0$, $y_0 = 0$, and a solution is given by $y = \sin x$. However, another solution is $y = e^x$; we exclude this solution by restricting to a sufficiently small rectangle E about $(0,0)$.

The existence of a unique continuous solution $y = f(x)$ can also be shown as follows. With E chosen as above so that $F_y > 0$ in E, $F(x_0, y)$ is strictly increasing in y, so that $F(x_0, y_0 + \eta) > 0$, $F(x_0, y_0 - \eta) < 0$. By the continuity of F, for δ sufficiently small, $F(x, y_0 + \eta) > 0$ and $F(x, y_0 - \eta) < 0$ for $|x - x_0| < \delta$. By the Intermediate Value Theorem, as above, for each x with $|x - x_0| < \delta$ $F(x,y) = 0$ for a unique $y = f(x)$ between $y_0 - \eta$ and $y_0 + \eta$. Given $\epsilon > 0$, we can choose η so that $\eta < \epsilon$ and we have $|f(x) - f(x_0)| < \epsilon$ for $|x - x_0| < \delta$. Thus f is continuous at x_0. By applying the same reasoning to each point (x_1, y_1) on the graph of f we conclude that f is continuous for $|x - x_0| < \delta$.

We can extend the proof to cover the case of $G(x,y,z) = 0$ with $G = 0$ and $G_z \neq 0$ at $P_0 : (x_0, y_0, z_0)$, say $G_z > 0$ at P_0. The proof of the preceding paragraph extends at once to this case to yield a unique continuous solution $z = f(x,y)$ with graph in an appropriate set $E : |x - x_0| < \delta$, $|y - y_0| < \delta$, $|z - z_0| < \eta$. For the equation $G(x, y_0, z) = 0$ we apply the theorem for the equation $F(x,y) = 0$ to show that $f(x, y_0)$ has a continuous derivative $f_x(x, y_0)$ and the equation

$$f_x(x,y) = -G_x(x,y,f(x,y))/G_z(x,y,f(x,y))$$

is satisfied at (x_0, y_0). By applying the same reasoning to each point $(x_1, y_1, f(x_1, y_1))$ on the graph of f, we conclude that the last equation holds at each point (x,y) with $|x - x_0| < \delta$, $|y - y_0| < \delta$. Since f is continuous, the right hand side of the equation is continuous for $|x - x_0| < \delta$, $|y - y_0| < \delta$ and therefore so also is $f_x(x,y)$. Similarly f_y is continuous and satisfies the analogous equation with G_y replacing G_x.

A similar reasoning applies to the single equation $F(y, x_1, \ldots, x_n) = 0$, to yield a solution $y = f(x_1, \ldots, x_n)$. For the general case of m equations $F_i(y_1, \ldots, y_m, x_1, \ldots, x_n) = 0$, $i = 1, \ldots, m$, one can proceed by induction. The case $m = 1$ has been disposed of. If we have proved the theorem for $m - 1$ equations, then we prove it for m equations by observing that the hypothesis on the Jacobian $\partial(F_1, \ldots, F_m)/\partial(y_1, \ldots, y_m)$ implies that for some i $\partial F_i/\partial y_m \neq 0$ at P_0. For proper numbering, we can take $i = m$ and then solve the m-th equation for y_m (by applying the theorem for a single equation); we substitute the resulting function in the previous equations; one verifies that the new equations in $y_1, \ldots, y_{m-1}, x_1, \ldots, x_n$ still satisfy the appropriate hypotheses. Hence by the induction assumption they yield desired functions $y_i = f_i(x_1, \ldots, x_n)$ for $i = 1, \ldots, m - 1$. Finally,

$$\begin{aligned} y_m &= y_m(y_1, \ldots, y_{m-1}, \quad x_1, \ldots, x_n) \\ &= y_m(f_1, \ldots, f_{m-1}, \quad x_1, \ldots, x_n) = f_m(x_1, \ldots, x_n). \end{aligned}$$

PROBLEMS

1. Find $(\partial z/\partial x)_y$ and $(\partial z/\partial y)_x$ by Eq. (2.71):
   a) $2x^2 + y^2 - z^2 = 3$
   b) $xyz + 2x^2z + 3xz^2 = 1$
   c) $z^3 + xz + 2yz - 1 = 0$
   d) $e^{xz} + e^{yz} + z - 1 = 0$

2. Given that
$$2x + y - 3z - 2u = 0, \qquad x + 2y + z + u = 0,$$
find the following partial derivatives:
$$\left(\frac{\partial x}{\partial y}\right)_z, \quad \left(\frac{\partial y}{\partial x}\right)_u, \quad \left(\frac{\partial z}{\partial u}\right)_x, \quad \left(\frac{\partial y}{\partial z}\right)_x.$$

3. Find $(\partial u/\partial x)_y$ and $(\partial u/\partial y)_x$:
   a) $x^2 - y^2 + u^2 + 2v^2 = 1$, $x^2 + y^2 - u^2 - v^2 = 2$
   b) $e^u + xu - yv - 1 = 0$, $e^v - xv + yu - 2 = 0$
   c) $x^2 + xu - yv^2 + uv = 1$, $xu - 2yv = 1$

4. Given that
$$x^2 + y^2 + z^2 - u^2 + v^2 = 1, \qquad x^2 - y^2 + z^2 + u^2 + 2v^2 = 21,$$
   a) find du and dv in terms of dx, dy, and dz at the point $x = 1$, $y = 1$, $z = 2$, $u = 3$, $v = 2$;
   b) find $(\partial u/\partial x)_{y,z}$ and $(\partial v/\partial y)_{x,z}$ at this point;
   c) find approximately the values of u and v for $x = 1.1$, $y = 1.2$, $z = 1.8$.

5. If $xy + 2xu + 3xv + uv - 1 = 0$, $2xy + 3yu - 2xv + 2uv + 2 = 0$, find the Jacobian matrix $\begin{bmatrix} u_x & u_y \\ v_x & v_y \end{bmatrix}$.

6. Let equations $F_1(x_1, x_2, x_3, x_4) = 0$, $F_2(x_1, x_2, x_3, x_4) = 0$ be given. At a certain point where the equations are satisfied, it is known that
$$\left(\frac{\partial F_i}{\partial x_j}\right) = \begin{bmatrix} 3 & 1 & 0 & 2 \\ 5 & 1 & -1 & 4 \end{bmatrix}.$$
   a) Evaluate $(\partial x_1/\partial x_3)_{x_4}$ and $(\partial x_1/\partial x_4)_{x_3}$ at the point.
   b) Evaluate $(\partial x_1/\partial x_3)_{x_2}$ and $(\partial x_4/\partial x_3)_{x_2}$ at the point.
   c) Evaluate $\partial(x_1, x_2)/\partial(x_3, x_4)$ and $\partial(x_3, x_4)/\partial(x_1, x_2)$ at the point.

7. Prove: If $F(x, y, z) = 0$, then $\left(\frac{\partial z}{\partial x}\right)_y \left(\frac{\partial x}{\partial y}\right)_z \left(\frac{\partial y}{\partial z}\right)_x = -1$.

8. In *thermodynamics* the variables p (pressure), T (temperature), U (internal energy), and V (volume) occur. For each substance these are related by two equations, so that any two of the four variables can be chosen as independent, the other two then being dependent. In addition, the second law of thermodynamics implies the relation
   a) $\dfrac{\partial U}{\partial V} - T\dfrac{\partial p}{\partial T} + p = 0$,

when V and T are independent. Show that this relation can be written in each of the following forms:

b) $\dfrac{\partial T}{\partial V} + T\dfrac{\partial p}{\partial U} - p\dfrac{\partial T}{\partial U} = 0$ $(U, V \text{ indep.})$,

c) $T - p\dfrac{\partial T}{\partial p} + \dfrac{\partial(T, U)}{\partial(V, p)} = 0$ $(V, p \text{ indep.})$,

d) $\dfrac{\partial U}{\partial p} + T\dfrac{\partial V}{\partial T} + p\dfrac{\partial V}{\partial p} = 0$ $(p, T \text{ indep.})$,

e) $\dfrac{\partial T}{\partial p} - T\dfrac{\partial V}{\partial U} + p\dfrac{\partial(V, T)}{\partial(U, p)} = 0$ $(U, p \text{ indep.})$,

f) $T\dfrac{\partial(p, V)}{\partial(T, U)} - p\dfrac{\partial V}{\partial U} - 1 = 0$ $(T, U \text{ indep.})$.

[Hint: The relation (a) implies that if

$$dU = a\, dV + b\, dT, \qquad dp = c\, dV + e\, dT$$

are the expressions for dU and dp in terms of dV and dT, then $a - Te + p = 0$. To prove (b), for example, one assumes relations

$$dT = \alpha\, dV + \beta\, dU, \qquad dp = \gamma\, dV + \delta\, dU.$$

If these are solved for dU and dp in terms of dV and dT, then one obtains expressions for a and e in terms of α, β, γ, δ. If these expressions are substituted in the equation $a - Te + p = 0$, one has an equation in α, β, γ, δ. Since $\alpha = \partial T/\partial V$, etc., the relation of form (b) is obtained. The others are proved in the same way.]

2.12   INVERSE FUNCTIONS · CURVILINEAR COORDINATES

A pair of functions

$$x = f(u, v), \qquad y = g(u, v) \tag{2.77}$$

can be regarded as a mapping from the uv-plane to the xy-plane (see Section 2.7). Under appropriate conditions this mapping is a one-to-one correspondence between a domain D_{uv} in the uv-plane and a domain D_{xy} in the xy-plane (Fig. 2.8). One can then consider the *inverse mapping* that takes each point (x, y) in D_{xy} to the unique point (u, v) such that (2.77) holds. The inverse mapping is given by functions

$$u = \varphi(x, y), \qquad v = \psi(x, y). \tag{2.78}$$

However, we may have difficulty in solving (2.77) to obtain these functions explicitly. We may, nevertheless, consider Eqs. (2.77), in the form

$$f(u, v) - x = 0, \qquad g(u, v) - y = 0$$

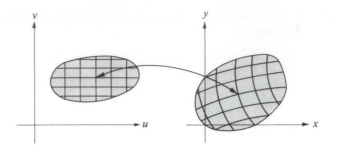

Figure 2.8 Mapping, inverse mapping, and curvilinear coordinates.

as implicit equations for the functions (2.78). We can then seek partial derivatives as in the preceding section. With $F(x, y, u, v) = f(u, v) - x$ and $G(x, y, u, v) = g(u, v) - y$ we have, for example,

$$\frac{\partial u}{\partial x} = -\frac{\dfrac{\partial(F, G)}{\partial(x, v)}}{\dfrac{\partial(F, G)}{\partial(u, v)}} = -\frac{\begin{vmatrix} -1 & f_v \\ 0 & g_v \end{vmatrix}}{\begin{vmatrix} f_u & f_v \\ g_u & g_v \end{vmatrix}} = \frac{g_v}{\begin{vmatrix} f_u & f_v \\ g_u & g_v \end{vmatrix}}.$$

The Jacobian determinant in the denominator is simply the Jacobian of the mapping (2.77). As earlier, we assume that it is not zero at the points considered. This condition, together with the continuity of the first partial derivatives appearing, ensures that the Implicit Function Theorem is applicable, so that the inverse mapping (2.78) is well defined.

The fact that the functions (2.78) are solutions of (2.77) means that

$$f[\varphi(x, y), \psi(x, y)] \equiv x, \qquad g[\varphi(x, y), \psi(x, y)] \equiv y.$$

(If we apply a mapping and then the inverse mapping, the composite is the identity mapping.) Hence by the rule (2.51),

$$\frac{\partial(x, y)}{\partial(u, v)} \frac{\partial(u, v)}{\partial(x, y)} = \begin{vmatrix} 1 & 0 \\ 0 & 1 \end{vmatrix} = 1. \tag{2.79}$$

Thus *the Jacobian of the inverse mapping is the reciprocal of the Jacobian of the mapping.*

Curvilinear coordinates. We can also regard (2.77) as equations describing a transformation from rectangular to curvilinear coordinates in the xy-plane. For fixed v, (x, y) traces a curve with u as parameter; for fixed u, (x, y) traces a curve with v as parameter. The curves mentioned serve as lines $v = $ const, $u = $ const in the curvilinear coordinate system (Fig. 2.8).

The most common example of curvilinear coordinates is provided by polar coordinates (Fig. 2.9). Here $x = r \cos \theta, y = r \sin \theta$.

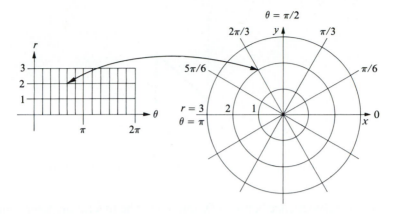

Figure 2.9   Polar coordinates as curvilinear coordinates.

The inverse functions (2.78) express the curvilinear coordinates in terms of rectangular coordinates. For polar coordinates the inverse functions are

$$r = \sqrt{x^2 + y^2}\,, \qquad \theta = \tan^{-1}\frac{y}{x}\,,$$

for an appropriate interpretation of the inverse tangent.

The preceding discussion extends to 3-dimensional space and generally to E^n. One considers a mapping

$$x_1 = f_1(u_1,\ldots,u_n),\ldots,x_n = f_n(u_1,\ldots,u_n) \tag{2.80}$$

or, in vector language, a vector function

$$\mathbf{x} = \mathbf{f}(\mathbf{u}).$$

The mapping is assumed to be defined in a domain D in E^n and the functions f_i are assumed to be continuous and have continuous first partial derivatives in D, with Jacobian $\partial(f_1,\ldots,f_n)/\partial(u_1,\ldots,u_n) \neq 0$ at the point $\mathbf{u}^0 = (u_1^0,\ldots,u_n^0)$ of D. Then there is a continuous inverse mapping

$$u_1 = g_1(x_1,\ldots,x_n),\ldots, \quad u_n = g_n(x_1,\ldots,x_n) \tag{2.81}$$

defined in a neighborhood (open set) D_0 of $\mathbf{x}^0 = \mathbf{f}(\mathbf{u}^0)$, whose range is a neighborhood D_1 of $\mathbf{u}^0$ contained in D, so that

$$\mathbf{f}\big(\mathbf{g}(\mathbf{x})\big) = \mathbf{x} \quad \text{in} \quad D_0, \qquad \mathbf{g}\big(\mathbf{f}(\mathbf{u})\big) = \mathbf{u} \quad \text{in} \quad D_1$$

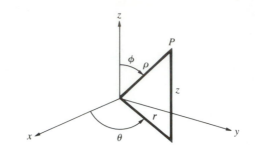

Figure 2.10 Cylindrical coordinates and spherical coordinates.

(one-to-one correspondence between D_0 and D_1). The inverse mapping functions g_i in (2.81) have continuous first partial derivatives in D_0. The partial derivatives of these functions can be obtained as earlier by the procedures for implicit functions. One can also reason that the differentials satisfy a matrix equation

$$d\mathbf{x} = \mathbf{f_u}d\mathbf{u}$$

and hence

$$d\mathbf{u} = (\mathbf{f_u})^{-1}d\mathbf{x}.$$

This shows that *the Jacobian matrix of the inverse mapping is simply the inverse of the Jacobian matrix of the mapping* (assumed to be nonsingular). In concise notation,

$$\mathbf{u_x} = (\mathbf{x_u})^{-1},$$

so that

$$\det \mathbf{u_x} = \frac{1}{\det \mathbf{x_u}} \quad \text{and} \quad \det \mathbf{u_x} \det \mathbf{x_u} = 1 \tag{2.82}$$

as in (2.79). The concept of curvilinear coordinates also applies to (2.80). Common examples, for $n = 3$, are the two systems

$$x = r\cos\theta, \qquad y = r\sin\theta, \qquad z = z, \tag{2.83}$$

$$x = \rho\sin\phi\cos\theta, \qquad y = \rho\sin\phi\sin\theta, \qquad z = \rho\cos\phi, \tag{2.84}$$

which are the equations of transformation from rectangular coordinates to cylindrical and spherical coordinates, respectively (Fig. 2.10).

PROBLEMS

1. For the transformation $x = r\cos\theta$, $y = r\sin\theta$ from rectangular to polar coordinates, verify the relations:

   a) $dx = \cos\theta\,dr - r\sin\theta\,d\theta$, $\quad dy = \sin\theta\,dr + r\cos\theta\,d\theta$

   b) $dr = \cos\theta\,dx + \sin\theta\,dy$, $\quad d\theta = -\dfrac{\sin\theta}{r}dx + \dfrac{\cos\theta}{r}dy$

   c) $\left(\dfrac{\partial x}{\partial r}\right)_\theta = \cos\theta$, $\quad \left(\dfrac{\partial x}{\partial r}\right)_y = \sec\theta$, $\quad \left(\dfrac{\partial r}{\partial x}\right)_y = \cos\theta$, $\quad \left(\dfrac{\partial r}{\partial x}\right)_\theta = \sec\theta$

d) $\dfrac{\partial(x,y)}{\partial(r,\theta)} = r, \qquad \dfrac{\partial(r,\theta)}{\partial(x,y)} = \dfrac{1}{r}$

2. Given the mapping

$$x = u - 2v, \qquad y = 2u + v,$$

a) write the equations of the inverse mapping,

b) evaluate the Jacobian of the mapping and that of the inverse mapping.

3. Given the mapping

$$x = u^2 - v^2, \qquad y = 2uv,$$

a) compute its Jacobian,

b) evaluate $\left(\dfrac{\partial u}{\partial x}\right)_y$ and $\left(\dfrac{\partial v}{\partial x}\right)_y$.

4. Given the mapping

$$x = f(u,v), \qquad y = g(u,v),$$

with Jacobian $J = \dfrac{\partial(x,y)}{\partial(u,v)}$, show that for the inverse functions one has

$$\dfrac{\partial u}{\partial x} = \dfrac{1}{J}\dfrac{\partial y}{\partial v}, \qquad \dfrac{\partial u}{\partial y} = -\dfrac{1}{J}\dfrac{\partial x}{\partial v}, \qquad \dfrac{\partial v}{\partial x} = -\dfrac{1}{J}\dfrac{\partial y}{\partial u}, \qquad \dfrac{\partial v}{\partial y} = \dfrac{1}{J}\dfrac{\partial x}{\partial u}.$$

Use these results to check Problem 1(b).

5. Given the mapping

$$x = f(u,v,w), \qquad y = g(u,v,w), \qquad z = h(u,v,w),$$

with Jacobian $J = \dfrac{\partial(x,y,z)}{\partial(u,v,w)}$, show that for the inverse functions one has

$$\dfrac{\partial u}{\partial x} = \dfrac{1}{J}\dfrac{\partial(y,z)}{\partial(v,w)}, \qquad \dfrac{\partial u}{\partial y} = \dfrac{1}{J}\dfrac{\partial(z,x)}{\partial(v,w)}, \qquad \dfrac{\partial u}{\partial z} = \dfrac{1}{J}\dfrac{\partial(x,y)}{\partial(v,w)},$$

$$\dfrac{\partial v}{\partial x} = \dfrac{1}{J}\dfrac{\partial(y,z)}{\partial(w,u)}, \qquad \dfrac{\partial v}{\partial y} = \dfrac{1}{J}\dfrac{\partial(z,x)}{\partial(w,u)}, \qquad \dfrac{\partial v}{\partial z} = \dfrac{1}{J}\dfrac{\partial(x,y)}{\partial(w,u)},$$

$$\dfrac{\partial w}{\partial x} = \dfrac{1}{J}\dfrac{\partial(y,z)}{\partial(u,v)}, \qquad \dfrac{\partial w}{\partial y} = \dfrac{1}{J}\dfrac{\partial(z,x)}{\partial(u,v)}, \qquad \dfrac{\partial w}{\partial z} = \dfrac{1}{J}\dfrac{\partial(x,y)}{\partial(u,v)}.$$

6. For the transformation (2.84) from rectangular to spherical coordinates, (a) compute the Jacobian $\dfrac{\partial(x,y,z)}{\partial(\rho,\phi,\theta)}$, (b) evaluate $\partial\rho/\partial y$, $\partial\phi/\partial z$, $\partial\theta/\partial x$ for the inverse transformation (cf. Problem 5).

7. Prove that if $x = f(u,v), y = g(u,v)$, then

$$\left(\dfrac{\partial x}{\partial u}\right)_v \left(\dfrac{\partial u}{\partial x}\right)_y = \left(\dfrac{\partial y}{\partial v}\right)_u \left(\dfrac{\partial v}{\partial y}\right)_x,$$

and

$$\left(\frac{\partial x}{\partial v}\right)_u \left(\frac{\partial v}{\partial x}\right)_y = \left(\frac{\partial u}{\partial y}\right)_x \left(\frac{\partial y}{\partial u}\right)_v,$$

also that

$$\left(\frac{\partial x}{\partial y}\right)_u \left(\frac{\partial y}{\partial x}\right)_u = 1.$$

8. *Polar coordinates in $E^n (n \geq 3)$ are introduced by the equations*

$$x_1 = r \cos \phi_1, \quad x_2 = r \sin \phi_1 \cos \phi_2, \ldots, \quad x_{n-2} = r \sin \phi_1 \cdots \sin \phi_{n-3} \cos \phi_{n-2},$$

$$x_{n-1} = r \sin \phi_1 \cdots \sin \phi_{n-2} \cos \phi_{n-1},$$

$$= x_n = r \sin \phi_1 \cdots \sin \phi_{n-2} \sin \phi_{n-1}.$$

a) Show that if $\mathbf{x} = (x_1, \ldots, x_n)$ and (x_{n-1}, x_n) is not $(0,0)$ or of the form $(c, 0)$ with $c > 0$, then unique polar coordinates are assigned to $\mathbf{x}$ by the requirements $r = (x_1^2 + \cdots + x_n^2)^{1/2} > 0, 0 < \phi_1 < \pi, \ldots, 0 < \phi_{n-2} < \pi, 0 < \phi_{n-1} < 2\pi$. (Hence the limits of integration for a multiple integral over the n-dimensional spherical region $x_1^2 + \cdots + x_n^2 \leq a^2$ are 0 to a for r, 0 to π for each ϕ_i except 0 to 2π for ϕ_{n-1}.) [Hint: $\mathbf{x}$ cannot be $\mathbf{0}$, so $x_1^2 + \cdots + x_n^2 > 0$ and $r > 0$. Choose ϕ_1 as $\cos^{-1}(x_1/r), 0 \leq \phi_1 \leq \pi$. Show that $\phi_1 = 0$ or $\phi_1 = \pi$ cannot occur, since they imply $|x_1| = r$ and hence $x_2 = x_3 = \cdots = 0$. Next choose $\phi_2 = \cos^{-1}(x_2/(r \sin \phi_1)), 0 \leq \phi_2 \leq \pi$, and show that $\phi_2 = 0$ or π cannot occur, etc.]

b) Show that the corresponding Jacobian is $J_n = r^{n-1} \sin^{n-2} \phi_1 \sin^{n-3} \phi_2 \cdots \sin \phi_{n-2}$. [Hint: Use induction. For the step from $n - 1$ to n, use the polar coordinates $\rho, \phi_2, \ldots, \phi_{n-1}$ in the subspace $x_1 = 0$, so that $x_2 = \rho \cos \phi_2, x_3 = \rho \sin \phi_2 \cos \phi_3, \ldots$, and $x_1, \rho, \phi_2, \ldots, \phi_{n-1}$ can be considered as cylindrical coordinates in E^n. Go from these cylindrical coordinates to full polar coordinates by the equations $x_1 = r \cos \phi_1$, $\rho = r \sin \phi_1$. Show that

$$\frac{\partial(x_1, \rho, \phi_2, \ldots, \phi_{n-1})}{\partial(r, \phi_1, \ldots, \phi_{n-1})} = r, \quad \frac{\partial(x_1, \ldots, x_n)}{\partial(x_1, \rho, \phi_2, \ldots, \phi_{n-1})} = J_{n-1},$$

where $J_{n-1} = \rho^{n-2} \sin^{n-3} \phi_2 \cdots \sin \phi_{n-2}$ by induction hypothesis. The conclusion now follows from (2.51).]

2.13   GEOMETRICAL APPLICATIONS

We recall the parametric representation for curves in the plane:

$$x = f(t), \quad y = g(t), \quad \alpha \leq t \leq \beta. \tag{2.85}$$

For each t the vector $\mathbf{v} = (f'(t), g'(t))$ can be considered as the *velocity vector* of the moving point $(x(t), y(t))$ in terms of time t. Its magnitude is

$$|\mathbf{v}| = \sqrt{\left(\frac{dx}{dt}\right)^2 + \left(\frac{dy}{dt}\right)^2} = \frac{ds}{dt}, \tag{2.86}$$

where s is distance along the path. One can interpret (2.85) as a vector function

$$\mathbf{r} = \mathbf{r}(t), \qquad \alpha \le t \le \beta,$$

where $\mathbf{r}$ is the position vector:

$$\mathbf{r} = x\mathbf{i} + y\mathbf{j}.$$

Then

$$\mathbf{v} = \frac{dx}{dt}\mathbf{i} + \frac{dy}{dt}\mathbf{j} = \mathbf{r}'(t).$$

The vector $\mathbf{v}$ is tangent to the path at the point $(x(t), y(t))$. Normally, one can use s as a parameter, instead of t, and then the velocity vector becomes

$$\mathbf{T} = \frac{dx}{ds}\mathbf{i} + \frac{dy}{ds}\mathbf{j}, \tag{2.87}$$

a *unit* tangent vector, since by (2.86),

$$\left(\frac{dx}{ds}\right)^2 + \left(\frac{dy}{ds}\right)^2 = \left[\left(\frac{dx}{dt}\right)^2 + \left(\frac{dy}{dt}\right)^2\right]\left(\frac{dt}{ds}\right)^2 = 1.$$

[For a detailed discussion, see CLA, pp. 329–336.]

The preceding analysis extends to curves in space with no significant change. A curve in space has parametric equations

$$x = f(t), \qquad y = g(t), \qquad z = h(t), \qquad \alpha \le t \le \beta, \tag{2.88}$$

and these are equivalent to a vector function $\mathbf{r} = \mathbf{r}(t)$, where now

$$\mathbf{r} = x\mathbf{i} + y\mathbf{j} + z\mathbf{k} \tag{2.89}$$

as in Fig. 2.11. The velocity vector

$$\mathbf{v} = \frac{dx}{dt}\mathbf{i} + \frac{dy}{dt}\mathbf{j} + \frac{dz}{dt}\mathbf{k} = \mathbf{r}'(t) \tag{2.90}$$

is again tangent to the path and has magnitude $|\mathbf{v}| = ds/dt$, where s is distance along the path, increasing with t. The corresponding unit tangent vector is

$$\mathbf{T} = \frac{1}{|\mathbf{v}|}\mathbf{v} = \frac{dx}{ds}\mathbf{i} + \frac{dy}{ds}\mathbf{j} + \frac{dz}{ds}\mathbf{k}. \tag{2.91}$$

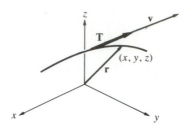

Figure 2.11   Curve in space, velocity vector **v**, and unit tangent vector **T**.

We remark that if $\mathbf{v} = \mathbf{0}$ at a point, then the tangent direction is not determined by $\mathbf{v}$; where this happens, it may be possible to determine a nonzero tangent vector by repeated differentiation (Problem 19 below).

At a value t_1 of t for which $\mathbf{v}$ is not $\mathbf{0}$, $\mathbf{v} = \mathbf{r}'(t_1)$ is a vector along the tangent line to the path at $(x_1, y_1, z_1) = \mathbf{r}(t_1)$, and hence the tangent line has parametric equations

$$x - x_1 = f'(t_1)(t - t_1), \qquad y - y_1 = g'(t_1)(t - t_1),$$
$$z - z_1 = h'(t_1)(t - t_1), \tag{2.92}$$

or, in vector form,

$$\overline{P_1 P} = (t - t_1)\mathbf{r}'(t_1) \tag{2.93}$$

(Section 1.2).

We observe that Eq. (2.92) can be obtained from Eqs. (2.88) by taking differentials

$$dx = f'(t)\, dt, \qquad dy = g'(t)\, dt, \qquad dz = h'(t)\, dt \tag{2.94}$$

and then interpreting dt as $t - t_1$, dx as $x - x_1$, dy as $y - y_1$, dz as $z - z_1$, while evaluating the derivatives $f'(t), \ldots$ at t_1. Thus dx, dy, dz are displacements from P_1 along the tangent line, as in Fig. 2.12. This result illustrates a basic principle, namely, that *taking differentials corresponds to forming tangents*. We can write Eqs. (2.94) in vector form:

$$d\mathbf{r} = f'(t)\, dt\,\mathbf{i} + g'(t)\, dt\,\mathbf{j} + h'(t)\, dt\,\mathbf{k}, \tag{2.95}$$

where $d\mathbf{r} = dx\,\mathbf{i} + dy\,\mathbf{j} + dz\,\mathbf{k}$, the "differential displacement vector."

Now let an equation

$$F(x, y, z) = 0 \tag{2.96}$$

be given; in general this represents a surface in space. Let (x_1, y_1, z_1) be a point on this surface, and let

$$x = f(t), \qquad y = g(t), \qquad z = h(t) \tag{2.97}$$

be a curve *in the surface*, passing through (x_1, y_1, z_1) when $t = t_1$. One has thus

$$F[f(t), g(t), h(t)] \equiv 0. \tag{2.98}$$

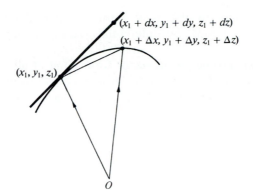

Figure 2.12 Displacements along the tangent line.

Taking differentials, one concludes

$$\frac{\partial F}{\partial x}\,dx + \frac{\partial F}{\partial y}\,dy + \frac{\partial F}{\partial z}\,dz = 0, \tag{2.99}$$

where $\partial F/\partial x$, $\partial F/\partial y$, $\partial F/\partial z$ are to be evaluated at (x_1, y_1, z_1) and $dx = f'(t_1)\,dt$, $dy = g'(t_1)\,dt$, $dz = h'(t_1)\,dt$; as earlier, these differentials can be replaced by $x - x_1$, $y - y_1$, $z - z_1$, where (x, y, z) is a point on the tangent to the given curve at (x_1, y_1, z_1). Accordingly, (2.99) can be written:

$$\frac{\partial F}{\partial x}\bigg|_{(x_1, y_1, z_1)}(x - x_1) + \frac{\partial F}{\partial y}\bigg|_{(x_1, y_1, z_1)}(y - y_1) + \frac{\partial F}{\partial z}\bigg|_{(x_1, y_1, z_1)}(z - z_1) = 0. \tag{2.100}$$

This is the equation of a plane containing the tangent line to the chosen curve. However, (2.100) no longer depends on the particular curve chosen; all tangent lines to curves in the surface through (x_1, y_1, z_1) lie in the one plane (2.100), which is termed the *tangent plane to the surface at* (x_1, y_1, z_1). [If all three partial derivatives of F are 0 at the chosen point, equation (2.100) fails to determine a plane, and the definition breaks down.]

It should be remarked that (2.99) is equivalent to (2.100), that is, that again *the operation of taking differentials yields the tangent*. As a further illustration of this, consider a curve determined by two intersecting surfaces:

$$F(x, y, z) = 0, \qquad G(x, y, z) = 0. \tag{2.101}$$

The corresponding differential relations

$$\frac{\partial F}{\partial x}\,dx + \frac{\partial F}{\partial y}\,dy + \frac{\partial F}{\partial z}\,dz = 0, \qquad \frac{\partial G}{\partial x}\,dx + \frac{\partial G}{\partial y}\,dy + \frac{\partial G}{\partial z}\,dz = 0 \tag{2.102}$$

represent two intersecting *tangent planes* at the point (x_1, y_1, z_1) at the point considered; the intersection of these planes is the *tangent line* to the curve (2.101).

To obtain the equations in the usual form, the partial derivatives must be evaluated at (x_1, y_1, z_1), and the differentials dx, dy, dz must be replaced by $x - x_1, y - y_1, z - z_1$.

These results can also be put in vector form. First of all, (2.100) shows that the vector $(\partial F/\partial x)\mathbf{i} + (\partial F/\partial y)\mathbf{j} + (\partial F/\partial z)\mathbf{k}$ is *normal* to the tangent plane at (x_1, y_1, z_1). This vector is known as the *gradient vector* of the function $F(x, y, z)$ (see Fig. 2.13). We write

$$\operatorname{grad} F = \frac{\partial F}{\partial x}\mathbf{i} + \frac{\partial F}{\partial y}\mathbf{j} + \frac{\partial F}{\partial z}\mathbf{k}. \tag{2.103}$$

The notation ∇F (read "del F") for grad F will also be used; this is discussed in the following section. There is a gradient vector of F at each point of the domain of definition at which the partial derivatives exist; in particular, there is a gradient vector at each point of the surface $F(x, y, z) = 0$ considered. Thus the gradient vector has a "point of application" and should be thought of as a *bound vector*.

Figure 2.13   Gradient vector of F and surface $F(x, y, z) = $ const.

The equation (2.99) for the tangent plane can now be written in the vector form

$$\operatorname{grad} F \cdot d\mathbf{r} = 0 \qquad (\mathbf{r} = x\mathbf{i} + y\mathbf{j} + z\mathbf{k}), \tag{2.104}$$

while the two equations (2.102) for the tangent line become

$$\operatorname{grad} F \cdot d\mathbf{r} = 0, \qquad \operatorname{grad} G \cdot d\mathbf{r} = 0. \tag{2.105}$$

Since (2.105) expresses the fact that $d\mathbf{r} = dx\mathbf{i} + dy\mathbf{j} + dz\mathbf{k}$ is perpendicular to both grad F and grad G, one concludes that

$$d\mathbf{r} \times (\operatorname{grad} F \times \operatorname{grad} G) = \mathbf{0}; \tag{2.106}$$

this equation again represents the tangent line. The vector grad $F \times$ grad G has components

$$
\begin{vmatrix} \dfrac{\partial F}{\partial y} & \dfrac{\partial F}{\partial z} \\[2mm] \dfrac{\partial G}{\partial y} & \dfrac{\partial G}{\partial z} \end{vmatrix}, \quad \begin{vmatrix} \dfrac{\partial F}{\partial z} & \dfrac{\partial F}{\partial x} \\[2mm] \dfrac{\partial G}{\partial z} & \dfrac{\partial G}{\partial x} \end{vmatrix}, \quad \begin{vmatrix} \dfrac{\partial F}{\partial x} & \dfrac{\partial F}{\partial y} \\[2mm] \dfrac{\partial G}{\partial x} & \dfrac{\partial G}{\partial y} \end{vmatrix}. \tag{2.107}
$$

Hence the tangent line can be written in the symmetric form

$$
\frac{x - x_1}{\begin{vmatrix} \dfrac{\partial F}{\partial y} & \dfrac{\partial F}{\partial z} \\[2mm] \dfrac{\partial G}{\partial y} & \dfrac{\partial G}{\partial z} \end{vmatrix}} = \frac{y - y_1}{\begin{vmatrix} \dfrac{\partial F}{\partial z} & \dfrac{\partial F}{\partial x} \\[2mm] \dfrac{\partial G}{\partial z} & \dfrac{\partial G}{\partial x} \end{vmatrix}} = \frac{z - z_1}{\begin{vmatrix} \dfrac{\partial F}{\partial x} & \dfrac{\partial F}{\partial y} \\[2mm] \dfrac{\partial G}{\partial x} & \dfrac{\partial G}{\partial y} \end{vmatrix}} \tag{2.108}
$$

or, in terms of Jacobians, in the form

$$
\frac{x - x_1}{\dfrac{\partial(F, G)}{\partial(y, z)}} = \frac{y - y_1}{\dfrac{\partial(F, G)}{\partial(z, x)}} = \frac{z - z_1}{\dfrac{\partial(F, G)}{\partial(x, y)}}. \tag{2.109}
$$

This discussion and that of the preceding section show the significance of the differential. Taking differentials in an equation or system of equations corresponds on the one hand to replacement of the equations by *linear* equations in the variables dx, dy, ... and on the other hand to replacement of curves and surfaces by *tangent* lines and planes.

PROBLEMS

1. Given the equations $x = \sin t$, $y = \cos t$, $z = \sin^2 t$ of a space curve,

   a) sketch the curve,

   b) find the equations of the tangent line at the point P for which $t = \dfrac{\pi}{3}$,

   c) find the equation of a plane cutting the curve at right angles at P.

2. a) Show that the curve of Problem 1 lies in the surface $x^2 + 2y^2 + z = 2$.

   b) Find the equation of the plane tangent to the surface at the point P of Problem 1(b).

   c) Show that the tangent line to the curve at P lies in the tangent plane to the surface.

3. Let $\mathbf{u} = \mathbf{u}(t)$ and $\mathbf{v} = \mathbf{v}(t)$ be vector functions of t, where $\mathbf{u}, \mathbf{v}$ are vectors in 3-dimensional space. Prove (assuming appropriate differentiability):

   a) $\dfrac{d}{dt}(\mathbf{u} + \mathbf{v}) = \dfrac{d\mathbf{u}}{dt} + \dfrac{d\mathbf{v}}{dt}$

   b) $\dfrac{d}{dt}[g(t)\mathbf{u}] = g(t)\dfrac{d\mathbf{u}}{dt} + g'(t)\mathbf{u}$

c) $\dfrac{d}{dt}(\mathbf{u} \cdot \mathbf{v}) = \mathbf{u} \cdot \dfrac{d\mathbf{v}}{dt} + \mathbf{v} \cdot \dfrac{d\mathbf{u}}{dt}$

d) $\dfrac{d}{dt}(\mathbf{u} \times \mathbf{v}) = \mathbf{u} \times \dfrac{d\mathbf{v}}{dt} + \dfrac{d\mathbf{u}}{dt} \times \mathbf{v}$ (watch the order!)

4. Let $\mathbf{u} = \mathbf{u}(t)$ have constant magnitude, $|\mathbf{u}(t)| \equiv a = \text{const}$. Show, assuming appropriate differentiability, that $\mathbf{u}$ is perpendicular to $d\mathbf{u}/dt$. What can be said of the locus of P such that $\overrightarrow{OP} = \mathbf{u}$?

5. Let a point P move in space, so that $\overrightarrow{OP} = \mathbf{r} = \mathbf{r}(t)$ and the velocity vector is $\mathbf{v} = d\mathbf{r}/dt$. The *acceleration vector* of P is

$$\mathbf{a} = \frac{d\mathbf{v}}{dt} = \frac{d^2\mathbf{r}}{dt^2}.$$

Assume further that the speed is 1, so that t can be identified with arc length s and

$$\mathbf{v} = \mathbf{T} = \frac{d\mathbf{r}}{ds}.$$

a) Show that $\mathbf{v}$ is perpendicular to the acceleration vector $\mathbf{a} = d\mathbf{v}/ds$ (cf. Problem 4). The plane through P determined by $\mathbf{v}$ and $\mathbf{a}$ is known as the *osculating plane* of the curve at P. It is given by the equation

$$\overrightarrow{PQ} \cdot \mathbf{v} \times \mathbf{a} = \mathbf{0},$$

where Q is an arbitrary point of the plane, provided that $\mathbf{v} \times \mathbf{a} \neq \mathbf{0}$. Show that the osculating plane is also given by

$$\overrightarrow{PQ} \cdot \frac{d\mathbf{r}}{dt} \times \frac{d^2\mathbf{r}}{dt^2} = 0,$$

where t is an arbitrary parameter along the curve, provided that $(d\mathbf{r}/dt) \times (d^2\mathbf{r}/dt^2) \neq \mathbf{0}$.

b) Find the osculating plane of the curve of Problem 1 at the point $t = \pi/3$ and graph. [It can be shown that this plane is the limiting position of a plane through three points P_1, P_2, P_3 on the curve, corresponding to parameter values t_1, t_2, t_3, as the values t_1, t_2, t_3 approach the value t at P.]

6. Let P move in space with speed 1 as in Problem 5. The *curvature* of the path is defined as κ, where

$$\kappa = |\mathbf{a}| = \left|\frac{d\mathbf{v}}{ds}\right| = \left|\frac{d\mathbf{T}}{ds}\right|.$$

Thus if $\kappa \neq 0$, $\mathbf{T}$ and $(1/\kappa)\,d\mathbf{T}/ds = \mathbf{N}$ are perpendicular unit vectors; $\mathbf{T}$ is tangent to the curve at P, $\mathbf{N}$ is normal to the curve at P and is termed the *principal normal*. The *radius of curvature* is $\rho = 1/\kappa$. The vector $\mathbf{B} = \mathbf{T} \times \mathbf{N}$ is called the *binormal*. Establish the following relations:

a) $\mathbf{B}$ is a unit vector and $(\mathbf{T}, \mathbf{N}, \mathbf{B})$ is a positive triple of unit vectors

b) $\frac{d\mathbf{B}}{ds} \cdot \mathbf{B} = 0$ and $\frac{d\mathbf{B}}{ds} \cdot \mathbf{T} = 0$

c) there is a scalar $-\tau$ such that $\frac{d\mathbf{B}}{ds} = -\tau\mathbf{N}$; τ is known as the *torsion*

d) $\frac{d\mathbf{N}}{ds} = -\kappa\mathbf{T} + \tau\mathbf{B}$. The equations

$$\frac{d\mathbf{T}}{ds} = \kappa\mathbf{N}, \qquad \frac{d\mathbf{N}}{ds} = -\kappa\mathbf{T} + \tau\mathbf{B}, \qquad \frac{d\mathbf{B}}{ds} = -\tau\mathbf{N}$$

are known as the *Frenet formulas.* For further properties of curves, see the book by Struik listed at the end of the chapter.

7. Let a point P move in space at speed $v = ds/dt = |\mathbf{v}| \neq 0$. Show:

a) $\mathbf{a} = \frac{d\mathbf{v}}{dt} = \frac{dv}{dt}\mathbf{T} + \frac{v^2}{\rho}\mathbf{N}$ $(\rho \neq 0)$. [Hint: Set $\mathbf{v} = v\mathbf{T}$, differentiate, and use the formula $d\mathbf{T}/ds = \kappa\mathbf{N} = (1/\rho)\mathbf{N}$ of Problem 6.]

b) $\kappa = v^{-3}|\mathbf{v} \times \mathbf{a}|$.

c) If $\mathbf{w} = d\mathbf{a}/dt$ and $\mathbf{v} \times \mathbf{a} \neq 0$, then $\tau = |\mathbf{v} \times \mathbf{a}|^{-2}\mathbf{v} \times \mathbf{a} \cdot \mathbf{w}$.

d) The path is a straight line if and only if $\kappa \equiv 0$.

e) The path lies in a plane if and only if $\tau \equiv 0$.

Note. Where $\mathbf{v} = \mathbf{0}$, the curvature κ is defined to be 0. If $\mathbf{v} \equiv \mathbf{0}$, the path reduces to a point.

8. For each of the following surfaces, find the tangent plane and normal line at the point indicated, verifying that the point is in the surface:

a) $x^2 + y^2 + z^2 = 9$ at $(2, 2, 1)$

b) $e^{x^2+y^2} - z^2 = 0$ at $(0, 0, 1)$

c) $x^3 - xy^2 + yz^2 - z^3 = 0$ at $(1, 1, 1)$

d) $x^2 + y^2 - z^2 = 0$ at $(0, 0, 0)$

   Why does the procedure break down in (d)? Show by graphing that a solution is impossible.

e) $xy - z = 0$ at (x_1, y_1, z_1), where $x_1 y_1 = z_1$

f) $xy + yz + xz = 1$ at (x_1, y_1, z_1), where $x_1 y_1 + y_1 z_1 + x_1 z_1 = 1$

9. Show that the tangent plane at (x_1, y_1, z_1) to a surface given by an equation $z = f(x, y)$ is as follows:

$$z - z_1 = \frac{\partial f}{\partial x}(x - x_1) + \frac{\partial f}{\partial y}(y - y_1).$$

Obtain the equations of the normal line.

10. Find the tangent plane and normal line to the following surfaces at the points shown (see Problem 9):

a) $z = x^2 + y^2$ at $(1, 1, 2)$,

b) $z = \sqrt{1 - x^2 - y^2}$ at $(\frac{2}{3}, \frac{2}{3}, \frac{1}{3})$,

c) $z = \dfrac{x}{y}$ at $(2, 1, 2)$,

d) $z = \log(x^2 + y^2)$ at $(\frac{3}{5}, \frac{4}{5}, 0)$.

11. For each of the following curves (represented by intersecting surfaces), find the equations of the tangent line at the point indicated, verifying that the point is on the curve:

a) $2x + y - z = 6$, $x + 2y + 2z = 7$ at $(3, 1, 1)$;

b) $x^2 + y^2 + z^2 = 9$, $x^2 + y^2 - 8z^2 = 0$ at $(2, 2, 1)$;

c) $x^2 + y^2 = 1$, $x + y + z = 0$ at $(1, 0, -1)$;

d) $x^2 + y^2 + z^2 = 9$, $x^2 + 2y^2 + 3z^2 = 9$ at $(3, 0, 0)$.

   Why does the procedure break down in (d)? Show that solution is impossible.

12. Show that the curve

$$x^2 - y^2 + z^2 = 1, \qquad xy + xz = 2$$

is tangent to the surface

$$xyz - x^2 - 6y = -6$$

at the point $(1,1,1)$.

13. Show that the equation of the plane normal to the curve

$$F(x, y, z) = 0, \qquad G(x, y, z) = 0$$

at the point (x_1, y_1, z_1) can be written in the vector form

$$d\mathbf{r} \cdot \operatorname{grad} F \times \operatorname{grad} G = 0,$$

and write out the equation in rectangular coordinates. Use the results to find the normal planes to the curves of Problem 11(b) and (c) at the points given.

14. Determine a plane normal to the curve $x = t^2$, $y = t$, $z = 2t$ and passing through the point $(1,0,0)$.

15. Find the gradient vectors of the following functions:

a) $F = x^2 + y^2 + z^2$ **b)** $F = 2x^2 + y^2$

Plot a level surface of each function and verify that the gradient vector is always normal to the level surface.

16. Three equations of form $x = f(u, v)$, $y = g(u, v)$, $z = h(u, v)$ can be considered as parametric equations of a surface, for elimination of u and v leads in general to a single equation $F(x, y, z) = 0$.

a) Show that the vector

$$\left(\frac{\partial f}{\partial u}\mathbf{i} + \frac{\partial g}{\partial u}\mathbf{j} + \frac{\partial h}{\partial u}\mathbf{k} \right) \times \left(\frac{\partial f}{\partial v}\mathbf{i} + \frac{\partial g}{\partial v}\mathbf{j} + \frac{\partial h}{\partial v}\mathbf{k} \right)$$

$$\equiv \frac{\partial(g,h)}{\partial(u,v)}\mathbf{i} + \frac{\partial(h,f)}{\partial(u,v)}\mathbf{j} + \frac{\partial(f,g)}{\partial(u,v)}\mathbf{k}$$

is normal to the surface at the point (x_1, y_1, z_1), where $x_1 = f(u_1, v_1)$, $y_1 = g(u_1, v_1)$, $z_1 = h(u_1, v_1)$ and the derivatives are evaluated at this point.

b) Write the equation of the tangent plane.

c) Apply the results to find the tangent plane to the surface

$$x = \cos u \cos v, \qquad y = \cos u \sin v, \qquad z = \sin u$$

at the point for which $u = \pi/4$, $v = \pi/4$.

d) Show that the surface (c) is a sphere.

17. Find the equations of a tangent line to a curve given by equations:

$$z = f(x, y), \qquad z = g(x, y).$$

18. Three equations

$$F(x, y, z, t) = 0, \qquad G(x, y, z, t) = 0, \qquad H(x, y, z, t) = 0$$

can be regarded as implicit parametric equations of a curve in terms of the parameter t. Find the equations of the tangent line.

19. Let a curve in space be given in parametric form by an equation: $\mathbf{r} = \mathbf{r}(t)$, $t_1 \leq t \leq t_2$, where $\mathbf{r} = \overrightarrow{OP}$ and O is fixed.

a) Show that if $d\mathbf{r}/dt \equiv 0$, then the curve degenerates into a single point.

b) Let $d\mathbf{r}/dt = 0$ at a point P_0 on the curve, at which $t = t_0$. Show that if $d^2\mathbf{r}/dt^2 \neq 0$ for $t = t_0$, then $d^2\mathbf{r}/dt^2$ represents a tangent vector to the curve at P_0 and, in general, that if

$$\frac{d\mathbf{r}}{dt} = 0, \quad \frac{d^2\mathbf{r}}{dt^2} = 0, \dots, \quad \frac{d^n\mathbf{r}}{dt^n} = 0, \quad \frac{d^{n+1}\mathbf{r}}{dt^{n+1}} \neq 0$$

for $t = t_0$, then

$$\left. \frac{d^{n+1}\mathbf{r}}{dt^{n+1}} \right|_{t=t_0}$$

represents a tangent vector to the curve at P_0. [Hint: For $t \neq t_0$ the vector

$$\frac{\mathbf{r}(t) - \mathbf{r}(t_0)}{(t - t_0)^{n+1}} = \frac{f(t) - f(t_0)}{(t - t_0)^{n+1}}\mathbf{i} + \frac{g(t) - g(t_0)}{(t - t_0)^{n+1}}\mathbf{j} + \frac{h(t) - h(t_0)}{(t - t_0)^{n+1}}\mathbf{k}$$

represents a secant to the curve, as in Fig. 2.11. Now let $t \to t_0$ and evaluate the limit of this vector with the aid of de l'Hôpital's Rule, noting that $f'(t_0) = 0$, $g'(t_0) = 0$, $h'(t_0) = 0$, but at least one of the three $(n + 1)$th derivatives is not 0.]

c) Show that if the parameter t is the arc length s, then $d\mathbf{r}/ds$ has magnitude 1 and hence is always a tangent vector.

20. Let a curve be given as in Problems 5–7 above, with $\mathbf{v} \neq 0$ and u as parameter. Then the equation

$$\mathbf{v} = \mathbf{r}(u) + w\mathbf{v}(u)$$

gives the tangent line to the path at the point $\mathbf{r}(u)$ with w as parameter along the line. The equation also defines a surface, with u and w as parameters. The surface is called the *tangential developable* of the curve. It is a ruled surface whose rulings are the tangent lines to the path.

a) Sketch the tangential developable for each of the following curves:

(i) a circular helix;

(ii) the curve $x = u$, $y = u^2$, $z = u^3$.

b) Show that along a chosen ruling of a tangential developable all points of the surface have the same tangent plane, which is the osculating plane to the curve at the point $\mathbf{r}(u)$.

Remark. A *developable surface* is one which can be obtained from a plane by bending without stretching. Examples are cones, cylinders, and the surfaces of this problem. (See pp. 66–73 of the book by Struik listed at the end of the chapter.)

2.14 THE DIRECTIONAL DERIVATIVE

Let $F(x, y, z)$ be given in a domain D of space. To compute the partial derivative $\partial F/\partial x$ at a point (x, y, z) of this domain, one considers the ratio of the change ΔF in the function F from (x, y, z) to $(x + \Delta x, y, z)$ to the change Δx in x. Thus only the values of F along a line parallel to the x axis are considered. Similarly, $\partial F/\partial y$ and $\partial F/\partial z$ involve a consideration of how F changes along parallels to the y and z axes, respectively. It appears unnatural to restrict attention to these three directions. Accordingly, one defines the *directional derivative* of F in a given direction as the limit of the ratio

$$\frac{\Delta F}{\Delta s} \qquad (2.110)$$

of the change in F to the distance Δs moved in the given direction, as Δs approaches 0.

Let the direction in question be given by a nonzero vector $\mathbf{v}$. The directional derivative of F in direction $\mathbf{v}$ at the point (x, y, z) is then denoted by $\nabla_v F(x, y, z)$ or, more concisely, by $\nabla_v F$. A displacement from (x, y, z) in direction $\mathbf{v}$ corresponds to changes Δx, Δy, Δz proportional to the components v_x, v_y, v_z; that is,

$$\Delta x = hv_x, \qquad \Delta y = hv_y, \qquad \Delta z = hv_z, \qquad (2.111)$$

where h is a positive scalar. The displacement is thus simply the vector $h\mathbf{v}$, and its magnitude Δs is $h|\mathbf{v}|$. The directional derivative is now by definition the limit:

$$\nabla_v F = \lim_{h \to 0} \frac{F(x + hv_x, y + hv_y, z + hv_z) - F(x, y, z)}{h|\mathbf{v}|}. \qquad (2.112)$$

In this definition we have allowed h to approach 0 through positive or negative values, just as in the definition of the derivative dy/dx as the limit of $\Delta y/\Delta x$ as Δx approaches 0 through positive or negative values. One can think of the directional derivative of F at (x,y,z) in direction $\mathbf{v}$ as the derivative of a function of one variable obtained by considering F as a function of a coordinate along an axis through (x,y,z) in the direction of $\mathbf{v}$ with the same unit of distance as along the $x,y,$ and z axes.

If F has a total differential at (x, y, z), then, as in Section 2.6,

$$\Delta F = F(x + hv_x, y + hv_y, z + hv_z) - F(x, y, z)$$

$$= \frac{\partial F}{\partial x} hv_x + \frac{\partial F}{\partial y} hv_y + \frac{\partial F}{\partial z} hv_z + \epsilon_1 hv_x + \epsilon_2 hv_y + \epsilon_3 hv_z.$$

Thus

$$\frac{\Delta F}{h|\mathbf{v}|} = \frac{\partial F}{\partial x}\frac{v_x}{|\mathbf{v}|} + \frac{\partial F}{\partial y}\frac{v_y}{|\mathbf{v}|} + \frac{\partial F}{\partial z}\frac{v_z}{|\mathbf{v}|} + \epsilon_1\frac{v_x}{|\mathbf{v}|} + \epsilon_2\frac{v_y}{|\mathbf{v}|} + \epsilon_3\frac{v_z}{|\mathbf{v}|}.$$

If h approaches zero, the last three terms approach 0, while the left-hand side approaches $\nabla_v F$. One has thus the equation

$$\nabla_v F = \frac{\partial F}{\partial x}\frac{v_x}{|\mathbf{v}|} + \frac{\partial F}{\partial y}\frac{v_y}{|\mathbf{v}|} + \frac{\partial F}{\partial z}\frac{v_z}{|\mathbf{v}|}. \tag{2.113}$$

Now $v_x/|\mathbf{v}|, v_y/|\mathbf{v}|, v_z/|\mathbf{v}|$ are simply the components of a unit vector $\mathbf{u}$ in the direction of $\mathbf{v}$ and are thus, by Section 1.1, the direction cosines of $\mathbf{v}$:

$$\mathbf{u} = \frac{1}{|\mathbf{v}|}(v_x\mathbf{i} + v_y\mathbf{j} + v_z\mathbf{k}) = \cos\alpha\,\mathbf{i} + \cos\beta\,\mathbf{j} + \cos\gamma\,\mathbf{k}.$$

Accordingly, (2.113) can be written in the following form:

$$\nabla_v F = \frac{\partial F}{\partial x}\cos\alpha + \frac{\partial F}{\partial y}\cos\beta + \frac{\partial F}{\partial z}\cos\gamma. \tag{2.114}$$

One has thus the fundamental rule:

The directional derivative of a function $F(x, y, z)$ is given by

$$\frac{\partial F}{\partial x}\cos\alpha + \frac{\partial F}{\partial y}\cos\beta + \frac{\partial F}{\partial z}\cos\gamma,$$

where α, β, γ are the direction angles of the direction chosen.

The right-hand side of (2.114) can be interpreted as the scalar product of the vector grad $F = \nabla F = (\partial F/\partial x)\mathbf{i} + (\partial F/\partial y)\mathbf{j} + (\partial F/\partial z)\mathbf{k}$ and the unit vector $\mathbf{u}$. Thus by Section 1.1 the *directional derivative equals the component of grad F in the direction of* $\mathbf{v}$:

$$\nabla_v F = \nabla F \cdot \frac{\mathbf{v}}{|\mathbf{v}|} = \text{comp}_v \nabla F. \tag{2.115}$$

It is for this reason that the notation $\nabla_v F$ is used for the directional derivative. In the special case when $\mathbf{v} = \mathbf{i}$, so that one is computing the directional derivative in the x direction, one writes

$$\nabla_v F = \nabla_x F = \text{comp}_x \, \nabla F = \frac{\partial F}{\partial x}; \tag{2.116}$$

similarly, the directional derivatives in the y and z directions are

$$\nabla_y F = \frac{\partial F}{\partial y}, \qquad \nabla_z F = \frac{\partial F}{\partial z}.$$

There are other situations in which a partial derivative notation is used for the directional derivative. A common one is that in which one is computing the directional derivative at a point (x, y, z) of a surface S along a given direction normal to S. If $\mathbf{n}$ is a unit normal vector in the chosen direction, one writes

$$\nabla_n F = \nabla F \cdot \mathbf{n} = \frac{\partial F}{\partial n}. \tag{2.117}$$

Equation (2.115) gives further information about the vector $\nabla F = \text{grad } F$. For from (2.115), one concludes that the directional derivative at a given point is a maximum when it is in the direction of ∇F; this maximum value is simply

$$|\nabla F| = \sqrt{\left(\frac{\partial F}{\partial x}\right)^2 + \left(\frac{\partial F}{\partial y}\right)^2 + \left(\frac{\partial F}{\partial z}\right)^2}. \tag{2.118}$$

Thus *the gradient vector points in the direction in which F increases most rapidly, and its length is the rate of increase in that direction.* If $\mathbf{v}$ makes an angle θ with ∇F, then the directional derivative in the direction of $\mathbf{v}$ is

$$\nabla_v F = |\nabla F| \cos \theta. \tag{2.119}$$

Thus if $\mathbf{v}$ is tangent to a level surface

$$F(x, y, z) = \text{const}$$

at (x, y, z), then $\nabla_v F = 0$; for ∇F is normal to such a level surface.

One sometimes refers to a "directional derivative along a curve." By this is meant the directional derivative along a direction tangent to the curve. Let the curve be given in terms of arc length as parameter:

$$x = f(s), \qquad y = g(s), \qquad z = h(s)$$

and let the direction chosen be that of increasing s. The vector

$$\mathbf{u} = \frac{dx}{ds}\mathbf{i} + \frac{dy}{ds}\mathbf{j} + \frac{dz}{ds}\mathbf{k}$$

is then tangent to the curve and has length 1 (cf. Section 2.13). Hence the directional derivative along the curve is

$$\nabla_u F = \frac{\partial F}{\partial x}\frac{dx}{ds} + \frac{\partial F}{\partial y}\frac{dy}{ds} + \frac{\partial F}{\partial z}\frac{dz}{ds} = \frac{dF}{ds}; \tag{2.120}$$

that is, it is the rate of change of F with respect to arc length on the curve.

The meaning of the directional derivative can be clarified by visualizing the following experiment. A traveler in a balloon carries a thermometer. At intervals he records the temperature. If his reading at position A is $42°$ and that at the position B is 44°, he would estimate that the directional derivative of the temperature in the direction $\overrightarrow{AB}$ is positive and has a value given approximately by 2° ÷ d, where d is the distance $|\overrightarrow{AB}|$; if $|\overrightarrow{AB}|$ = 5000 ft, the estimate would be 0.0004° per foot. If he continued traveling past B in the same direction, he would expect the temperature to rise at about the same rate. If the balloon moves on a curved path at a known constant speed, the traveler can compute the directional derivative *along the path* without looking out of the balloon.

The foregoing discussion has been given for three dimensions. It can be specialized to two dimensions without difficulty. Thus one considers a function $F(x, y)$ and its rate of change along a given direction $\mathbf{v}$ in the xy-plane. If $\mathbf{v}$ makes an angle α with the positive x axis, one can write

$$\nabla_v F = \nabla_\alpha F = \frac{\partial F}{\partial x} \cos \alpha + \frac{\partial F}{\partial y} \sin \alpha; \tag{2.121}$$

for $\mathbf{v}$ has direction cosines $\cos \alpha$ and $\cos \beta = \cos(\frac{1}{2}\pi - \alpha) = \sin \alpha$. Again the directional derivative is the component of grad $F = (\partial F/\partial x)\mathbf{i} + (\partial F/\partial y)\mathbf{j}$ in the given direction; the directional derivative at a given point has its maximum in the direction of grad F, the value being

$$|\nabla F| = \sqrt{\left(\frac{\partial F}{\partial x}\right)^2 + \left(\frac{\partial F}{\partial y}\right)^2}. \tag{2.122}$$

The directional derivative is zero along a level curve of F, as suggested in the accompanying Fig. 2.14.

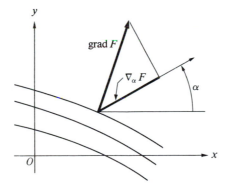

Figure 2.14   Gradient of $F(x, y)$ and curves $F(x, y) = $ const.

If the level curves are interpreted as contour lines of a landscape, that is, of the surface $z = F(x, y)$, then the directional derivative means simply the rate of climb in the given direction. The rate of climb in the direction of steepest ascent is the "gradient," precisely the term in common use. The bicyclist zigzagging up a hill is taking advantage of the component rule to reduce the directional derivative.

PROBLEMS

1. Evaluate the directional derivatives of the following functions for the points and directions given:

   a) $F(x, y, z) = 2x^2 - y^2 + z^2$ at $(1, 2, 3)$ in the direction of the line from $(1, 2, 3)$ to $(3, 5, 0)$;

   b) $F(x, y, z) = x^2 + y^2$ at $(0, 0, 0)$ in the direction of the vector $\mathbf{u} = a\mathbf{i} + b\mathbf{j} + c\mathbf{k}$; discuss the significance of the result;

   c) $F(x, y) = e^x \cos y$ at $(0, 0)$ in a direction making an angle of $60°$ with the x axis;

   d) $F(x, y) = 2x - 3y$ at $(1, 1)$ along the curve $y = x^2$ in the direction of increasing x;

   e) $F(x, y, z) = 3x - 5y + 2z$ at $(2, 2, 1)$ in the direction of the outer normal of the surface $x^2 + y^2 + z^2 = 9$;

   f) $F(x, y, z) = x^2 + y^2 - z^2$ at $(3, 4, 5)$ along the curve $x^2 + y^2 - z^2 = 0$, $2x^2 + 2y^2 - z^2 = 25$ in the direction of increasing x; explain the answer.

2. Evaluate $\partial F / \partial n$, where $\mathbf{n}$ is the outer normal to the surface given, at a general point (x, y, z) of the surface given:

   a) $F = x^2 - y^2$, surface $x^2 + y^2 + z^2 = 4$;

   b) $F = xyz$, surface $x^2 + 2y^2 + 4z^2 = 8$.

3. Prove that if $u = f(x, y)$ and $v = g(x, y)$ are functions such that

$$\frac{\partial u}{\partial x} = \frac{\partial v}{\partial y}, \qquad \frac{\partial u}{\partial y} = -\frac{\partial v}{\partial x},$$

then

$$\nabla_\alpha u = \nabla_{\alpha + \pi/2} v$$

for every angle α.

4. Show that if $u = f(x, y)$, then the directional derivatives of u along the line $\theta = $ const and circles $r = $ const (polar coordinates) are given respectively by

$$\nabla_\theta u = \frac{\partial u}{\partial r}, \qquad \nabla_{\theta + \pi/2} u = \frac{1}{r} \frac{\partial u}{\partial \theta}.$$

5. Show that under the hypotheses of Problem 3 one has

$$\frac{\partial u}{\partial r} = \frac{1}{r} \frac{\partial v}{\partial \theta}, \qquad \frac{1}{r} \frac{\partial u}{\partial \theta} = -\frac{\partial v}{\partial r}.$$

[Hint: Use Problem 4.]

6. Let arc length s be measured from the point $(2, 0)$ of the circle $x^2 + y^2 = 4$, starting in the direction of increasing y. If $u = x^2 - y^2$, evaluate du/ds on this circle. Check the result by using both the directional derivative and the explicit expression for u in terms of s. At what point of the circle does u have its smallest value?

7. Under the hypotheses of Problem 3, show that $\partial u / \partial s = \partial v / \partial n$ along each curve C of the domain in which u and v are given, for appropriate direction of the normal $\mathbf{n}$.

8. Determine the points (x, y) and directions for which the directional derivative of $u = 3x^2 + y^2$ has its largest value, if (x, y) is restricted to lie on the circle $x^2 + y^2 = 1$.

9. A function $F(x, y, z)$ is known to have the following values: $F(1, 1, 1) = 1$, $F(2, 1, 1)$ $= 4$, $F(2, 2, 1) = 8$, $F(2, 2, 2) = 16$. Compute approximately the directional derivatives:

$$\nabla_i F, \quad \nabla_{i+j} F, \quad \nabla_{i+j+k} F$$

at the point $(1, 1, 1)$.

2.15   PARTIAL DERIVATIVES OF HIGHER ORDER

Let a function $z = F(x, y)$ be given; its two partial derivatives $\partial z/\partial x$ and $\partial z/\partial y$ are themselves functions of x and y:

$$\frac{\partial z}{\partial x} = F_x(x, y), \qquad \frac{\partial z}{\partial y} = F_y(x, y).$$

Hence each can be differentiated with respect to x and y; one thus obtains the four *second partial derivatives*:

$$\frac{\partial^2 z}{\partial x^2} = F_{xx}(x, y), \qquad \frac{\partial^2 z}{\partial y\, \partial x} = F_{xy}(x, y),$$

$$\frac{\partial^2 z}{\partial x\, \partial y} = F_{yx}(x, y), \qquad \frac{\partial^2 z}{\partial y^2} = F_{yy}(x, y). \tag{2.123}$$

Thus $\partial^2 z/\partial x^2$ is the result of differentiating $\partial z/\partial x$ with respect to x, while $\partial^2 z/\partial y\, \partial x$ is the result of differentiation of $\partial z/\partial x$ with respect to y. Here a simplification is possible if all derivatives concerned are continuous in the domain considered, for one can prove that

$$\frac{\partial^2 z}{\partial y\, \partial x} = \frac{\partial^2 z}{\partial x\, \partial y}; \tag{2.124}$$

that is, the order of differentiation is immaterial. A proof of (2.124) is given at the end of this section.

Third- and higher-order partial derivatives are defined in a similar fashion, and again, under appropriate assumptions of continuity, the order of differentiation does not matter. Thus one obtains four third partial derivatives:

$$\frac{\partial^3 z}{\partial x^3}, \quad \frac{\partial^3 z}{\partial x^2\, \partial y} = \frac{\partial^3 z}{\partial x\, \partial y\, \partial x} = \frac{\partial^3 z}{\partial y\, \partial x^2},$$

$$\frac{\partial^3 z}{\partial x\, \partial y^2} = \frac{\partial^3 z}{\partial y\, \partial x\, \partial y} = \frac{\partial^3 z}{\partial y^2\, \partial x}, \quad \frac{\partial^3 z}{\partial y^3}.$$

A clue as to why the order of differentiation does not matter can be obtained by considering the mixed partial derivatives of $z = x^n y^m$. Here one has

$$\frac{\partial z}{\partial x} = nx^{n-1} y^m, \qquad \frac{\partial^2 z}{\partial y\, \partial x} = nmx^{n-1} y^{m-1},$$

$$\frac{\partial z}{\partial y} = mx^n y^{m-1}, \qquad \frac{\partial^2 z}{\partial x\, \partial y} = nmx^{n-1} y^{m-1}.$$

Thus the order does not matter in this case. A similar reasoning applies to a sum of such terms with constant factors, that is, a polynomial in x and y:

$$z = a_0 + a_1 x + a_2 y + a_3 x^2 + a_4 xy + a_5 y^2 + \cdots + a_s x^p y^q.$$

It is essentially because an "arbitrary" function can be approximated by such a polynomial near a given point that the rule holds.

Other notations for higher derivatives are illustrated by the following examples:

$$\frac{\partial^2 z}{\partial x^2} = z_{xx} = f_{11}(x, y), \qquad \frac{\partial^2 z}{\partial x \, \partial y} = z_{yx} = f_{21}(x, y),$$

$$\frac{\partial^3 w}{\partial x \, \partial y \, \partial z} = w_{zyx} = f_{321}(x, y, z), \qquad \frac{\partial^4 w}{\partial x^2 \, \partial z^2} = w_{zzxx} = f_{3311}(x, y, z).$$

If $z = f(x, y)$, the *Laplacian* of z, denoted by Δz or $\nabla^2 z$, is the expression

$$\Delta z = \nabla^2 z = \frac{\partial^2 z}{\partial x^2} + \frac{\partial^2 z}{\partial y^2}. \tag{2.125}$$

The Δ symbol here must not be confused with the symbol for increments; for this reason the $\nabla^2 z$ notation is preferable. If $w = f(x, y, z)$, the Laplacian of w is defined analogously:

$$\Delta w = \nabla^2 w = \frac{\partial^2 w}{\partial x^2} + \frac{\partial^2 w}{\partial y^2} + \frac{\partial^2 w}{\partial z^2}.$$

The origin of the ∇^2 symbol lies in the interpretation of ∇ as a "vector differential operator":

$$\nabla = \frac{\partial}{\partial x} \mathbf{i} + \frac{\partial}{\partial y} \mathbf{j} + \frac{\partial}{\partial z} \mathbf{k}.$$

One then has symbolically

$$\nabla^2 = \nabla \cdot \nabla = \frac{\partial^2}{\partial x^2} + \frac{\partial^2}{\partial y^2} + \frac{\partial^2}{\partial z^2}.$$

This point of view will be discussed further in Chapter 3.

If $z = f(x, y)$ has continuous second derivatives in a domain D and

$$\nabla^2 z = 0 \tag{2.126}$$

in D, then z is said to be *harmonic* in D. The same term is used for a function of three variables that has continuous second derivatives in a domain D in space and whose Laplacian is 0 in D. The two equations for harmonic functions:

$$\frac{\partial^2 z}{\partial x^2} + \frac{\partial^2 z}{\partial y^2} = 0, \qquad \frac{\partial^2 w}{\partial x^2} + \frac{\partial^2 w}{\partial y^2} + \frac{\partial^2 w}{\partial z^2} = 0, \tag{2.127}$$

are known as the *Laplace equations* in two and three dimensions, respectively.

Another important combination of derivatives occurs in the *biharmonic equation*:

$$\frac{\partial^4 z}{\partial x^4} + 2\frac{\partial^4 z}{\partial x^2 \, \partial y^2} + \frac{\partial^4 z}{\partial y^4} = 0, \tag{2.128}$$

which arises in the theory of elasticity. The combination that appears here can be expressed in terms of the Laplacian, for one has

$$\nabla^2(\nabla^2 z) = \frac{\partial^4 z}{\partial x^4} + 2\frac{\partial^4 z}{\partial x^2 \, \partial y^2} + \frac{\partial^4 z}{\partial y^4}.$$

If we write $\nabla^4 z = \nabla^2(\nabla^2 z)$, then the biharmonic equation can be written:

$$\nabla^4 z = 0. \tag{2.129}$$

Its solutions are termed *biharmonic* functions. This can again be generalized to functions of three variables, (2.129) suggesting the definition to be used.

Harmonic functions arise in the theory of electromagnetic fields, in fluid dynamics, in the theory of heat conduction, and in many other parts of physics; applications will be discussed in Chapters 5, 9, and 10. Biharmonic functions are used mainly in elasticity. (See Section 9.6.)

Proof of the rule (2.124).   We assume that $z = f(x, y)$ has continuous derivatives f_x, f_y, f_{xy}, f_{yx} in a domain D. Let R be a closed square region $x_0 \le x \le x_0 + h$, $y_0 \le y \le y_0 + h$ contained in D. Then by the rule $\int_a^b g'(x)\, dx = g(b) - g(a)$,

$$\int_R\!\!\int f_{xy}\, dx\, dy = \int_{x_0}^{x_0+h}\!\!\int_{y_0}^{y_0+h} f_{xy}\, dy\, dx$$

$$= \int_{x_0}^{x_0+h}[f_x(x, y_0 + h) - f_x(x, y_0)]\, dx$$

$$= \int_{x_0}^{x_0+h} f_x(x, y_0 + h)\, dx - \int_{x_0}^{x_0+h} f_x(x, y_0)\, dx$$

$$= f(x_0 + h, y_0 + h) - f(x_0, y_0 + h) - f(x_0 + h, y_0)$$
$$+ f(x_0, y_0).$$

We interchange the roles of x and y to obtain similarly

$$\int_R\!\!\int f_{yx}\, dx\, dy = f(x_0 + h, y_0 + h) - f(x_0 + h, y_0) - f(x_0, y_0 + h) + f(x_0, y_0).$$

Therefore the two double integrals are equal or

$$\int_R\!\!\int (f_{xy} - f_{yx})\, dx\, dy = 0.$$

Since this holds for *every* such square region R in D, we must have

$$f_{xy}(x, y) \equiv f_{yx}(x, y) \quad \text{in } D.$$

(See Problem 14 below.) Thus the proof is complete.   □

Remark.   This proof uses properties of double integrals which are usually developed in the first course in calculus; these properties are also considered in Chapter 4 (Sections 4.3 and 4.11).

2.16 HIGHER DERIVATIVES OF COMPOSITE FUNCTIONS

Let $z = f(x, y)$ and $x = g(t)$, $y = h(t)$, so that z can be expressed in terms of t alone. The derivative dz/dt can then be evaluated by the chain rule (2.33) of Section 2.8:

$$\frac{dz}{dt} = \frac{\partial z}{\partial x}\frac{dx}{dt} + \frac{\partial z}{\partial y}\frac{dy}{dt}. \tag{2.130}$$

By applying the product rule, one obtains the following expression for the second derivative:

$$\frac{d^2z}{dt^2} = \frac{d}{dt}\left(\frac{dz}{dt}\right) = \frac{\partial z}{\partial x}\frac{d^2x}{dt^2} + \frac{dx}{dt}\frac{d}{dt}\left(\frac{\partial z}{\partial x}\right) + \frac{\partial z}{\partial y}\frac{d^2y}{dt^2} + \frac{dy}{dt}\frac{d}{dt}\left(\frac{\partial z}{\partial y}\right).$$

To evaluate the expressions $(d/dt)(\partial z/\partial x)$ and $(d/dt)(\partial z/\partial y)$, one uses (2.130) again, this time applied to $\partial z/\partial x$ and $\partial z/\partial y$ rather than to z.

$$\frac{d}{dt}\left(\frac{\partial z}{\partial x}\right) = \frac{\partial^2 z}{\partial x^2}\frac{dx}{dt} + \frac{\partial^2 z}{\partial y\,\partial x}\frac{dy}{dt},$$

$$\frac{d}{dt}\left(\frac{\partial z}{\partial y}\right) = \frac{\partial^2 z}{\partial x\,\partial y}\frac{dx}{dt} + \frac{\partial^2 z}{\partial y^2}\frac{dy}{dt}.$$

One thus finds the rule:

$$\frac{d^2z}{dt^2} = \frac{\partial z}{\partial x}\frac{d^2x}{dt^2} + \frac{\partial^2 z}{\partial x^2}\left(\frac{dx}{dt}\right)^2 + 2\frac{\partial^2 z}{\partial x\,\partial y}\frac{dx}{dt}\frac{dy}{dt} + \frac{\partial^2 z}{\partial y^2}\left(\frac{dy}{dt}\right)^2 + \frac{\partial z}{\partial y}\frac{d^2y}{dt^2}. \tag{2.131}$$

This is a new chain rule.

Similarly, if $z = f(x, y)$, $x = g(u, v)$, $y = h(u, v)$, so that (2.34) holds, one has

$$\frac{\partial z}{\partial u} = \frac{\partial z}{\partial x}\frac{\partial x}{\partial u} + \frac{\partial z}{\partial y}\frac{\partial y}{\partial u},$$

$$\frac{\partial^2 z}{\partial u^2} = \frac{\partial z}{\partial x}\frac{\partial^2 x}{\partial u^2} + \frac{\partial}{\partial u}\left(\frac{\partial z}{\partial x}\right)\frac{\partial x}{\partial u} + \frac{\partial z}{\partial y}\frac{\partial^2 y}{\partial u^2} + \frac{\partial}{\partial u}\left(\frac{\partial z}{\partial y}\right)\frac{\partial y}{\partial u}. \tag{2.132}$$

Applying (2.132) again, one finds

$$\frac{\partial}{\partial u}\left(\frac{\partial z}{\partial x}\right) = \frac{\partial^2 z}{\partial x^2}\frac{\partial x}{\partial u} + \frac{\partial^2 z}{\partial y\,\partial x}\frac{\partial y}{\partial u}, \qquad \frac{\partial}{\partial u}\left(\frac{\partial z}{\partial y}\right) = \frac{\partial^2 z}{\partial x\,\partial y}\frac{\partial x}{\partial u} + \frac{\partial^2 z}{\partial y^2}\frac{\partial y}{\partial u},$$

so that

$$\frac{\partial^2 z}{\partial u^2} = \frac{\partial z}{\partial x}\frac{\partial^2 x}{\partial u^2} + \left(\frac{\partial^2 z}{\partial x^2}\right)\left(\frac{\partial x}{\partial u}\right)^2 + 2\frac{\partial^2 z}{\partial x\,\partial y}\frac{\partial x}{\partial u}\frac{\partial y}{\partial u} + \frac{\partial^2 z}{\partial y^2}\left(\frac{\partial y}{\partial u}\right)^2 + \frac{\partial z}{\partial y}\frac{\partial^2 y}{\partial u^2}. \tag{2.133}$$

It should be remarked that (2.133) is a special case of (2.131), since v is treated as a constant throughout.

Rules for $\partial^2 z / \partial u \, \partial v$, $\partial^2 z / \partial v^2$ and for higher derivatives can be formed, analogous to (2.131) and (2.133). These rules are of importance, but in most practical cases it is better to use only the chain rules (2.33), (2.34), (2.35), applying repeatedly if necessary. One reason for this is that simplifications are obtained if the derivatives occurring are expressed in terms of the right variables, and a complete description of all possible cases would be too involved to be useful.

The variations possible can be illustrated by the following example, which concerns only functions of one variable.

Let $y = f(x)$ and $x = e^t$. Then

$$\frac{dy}{dt} = \frac{dy}{dx}\frac{dx}{dt} = \frac{dy}{dx}e^t.$$

Hence

$$\frac{d^2 y}{dt^2} = \frac{d}{dt}\left(\frac{dy}{dx}\right)e^t + \frac{dy}{dx}e^t$$

$$= \frac{d^2 y}{dx^2}\frac{dx}{dt}e^t + \frac{dy}{dx}e^t$$

$$= \frac{d^2 y}{dx^2}e^{2t} + \frac{dy}{dx}e^t.$$

One could also write

$$\frac{dy}{dt} = \frac{dy}{dx}e^t = x\frac{dy}{dx}$$

and then

$$\frac{d^2 y}{dt^2} = \frac{d}{dt}\left(x\frac{dy}{dx}\right) = \frac{d}{dx}\left(x\frac{dy}{dx}\right)\frac{dx}{dt} = x\frac{d}{dx}\left(x\frac{dy}{dx}\right)$$

$$= \frac{d^2 y}{dx^2}x^2 + \frac{dy}{dx}x.$$

The second method is clearly simpler than the first; the answers obtained are equivalent because of the equation $x = e^t$.

2.17   THE LAPLACIAN IN POLAR, CYLINDRICAL, AND SPHERICAL COORDINATES

An important application of the method of the preceding section is the transformation of the Laplacian to its expression for other coordinate systems.

We consider first the 2-dimensional Laplacian

$$\nabla^2 w = \frac{\partial^2 w}{\partial x^2} + \frac{\partial^2 w}{\partial y^2}$$

and its expression in terms of polar coordinates r, θ. Thus we are given $w = f(x, y)$

and $x = r \cos \theta$, $y = r \sin \theta$, and we wish to express $\nabla^2 w$ in terms of r, θ, and derivatives of w with respect to r and θ. The solution is as follows. One has

$$\frac{\partial w}{\partial x} = \frac{\partial w}{\partial r}\frac{\partial r}{\partial x} + \frac{\partial w}{\partial \theta}\frac{\partial \theta}{\partial x}, \qquad \frac{\partial w}{\partial y} = \frac{\partial w}{\partial r}\frac{\partial r}{\partial y} + \frac{\partial w}{\partial \theta}\frac{\partial \theta}{\partial y} \qquad (2.134)$$

by the chain rule. To evaluate $\partial r/\partial x$, $\partial\theta/\partial x$, $\partial r/\partial y$, $\partial\theta/\partial y$ we use the equations

$$dx = \cos\theta \, dr - r \sin\theta \, d\theta, \qquad dy = \sin\theta \, dr + r\cos\theta \, d\theta.$$

These can be solved for dr and $d\theta$ by determinants or by elimination to give

$$dr = \cos\theta \, dx + \sin\theta \, dy, \qquad d\theta = -\frac{\sin\theta}{r}dx + \frac{\cos\theta}{r}dy.$$

Hence

$$\frac{\partial r}{\partial x} = \cos\theta, \qquad \frac{\partial r}{\partial y} = \sin\theta, \qquad \frac{\partial\theta}{\partial x} = -\frac{\sin\theta}{r}, \qquad \frac{\partial\theta}{\partial y} = \frac{\cos\theta}{r}.$$

Thus (2.134) can be written as follows:

$$\frac{\partial w}{\partial x} = \cos\theta\frac{\partial w}{\partial r} - \frac{\sin\theta}{r}\frac{\partial w}{\partial \theta}, \qquad \frac{\partial w}{\partial y} = \sin\theta\frac{\partial w}{\partial r} + \frac{\cos\theta}{r}\frac{\partial w}{\partial \theta}. \qquad (2.135)$$

These equations provide general rules for expressing derivatives with respect to x or y in terms of derivatives with respect to r and θ. By applying the first equation to the function $\partial w/\partial x$, one finds that

$$\frac{\partial^2 w}{\partial x^2} = \frac{\partial}{\partial x}\left(\frac{\partial w}{\partial x}\right) = \cos\theta\frac{\partial}{\partial r}\left(\frac{\partial w}{\partial x}\right) - \frac{\sin\theta}{r}\frac{\partial}{\partial \theta}\left(\frac{\partial w}{\partial x}\right);$$

by (2.135) this can be written as follows:

$$\frac{\partial^2 w}{\partial x^2} = \cos\theta\frac{\partial}{\partial r}\left(\cos\theta\frac{\partial w}{\partial r} - \frac{\sin\theta}{r}\frac{\partial w}{\partial \theta}\right) - \frac{\sin\theta}{r}\frac{\partial}{\partial \theta}\left(\cos\theta\frac{\partial w}{\partial r} - \frac{\sin\theta}{r}\frac{\partial w}{\partial \theta}\right).$$

The rule for differentiation of a product gives finally

$$\frac{\partial^2 w}{\partial x^2} = \cos^2\theta\frac{\partial^2 w}{\partial r^2} - \frac{2\sin\theta\cos\theta}{r}\frac{\partial^2 w}{\partial r\,\partial\theta} + \frac{\sin^2\theta}{r^2}\frac{\partial^2 w}{\partial\theta^2}$$

$$+ \frac{\sin^2\theta}{r}\frac{\partial w}{\partial r} + \frac{2\sin\theta\cos\theta}{r^2}\frac{\partial w}{\partial\theta}. \qquad (2.136)$$

In the same manner, one finds

$$\frac{\partial^2 w}{\partial y^2} = \frac{\partial}{\partial y}\left(\frac{\partial w}{\partial y}\right) = \sin\theta\frac{\partial}{\partial r}\left(\sin\theta\frac{\partial w}{\partial r} + \frac{\cos\theta}{r}\frac{\partial w}{\partial \theta}\right)$$

$$+ \frac{\cos\theta}{r}\frac{\partial}{\partial \theta}\left(\sin\theta\frac{\partial w}{\partial r} + \frac{\cos\theta}{r}\frac{\partial w}{\partial \theta}\right),$$

$$\frac{\partial^2 w}{\partial y^2} = \sin^2\theta\frac{\partial^2 w}{\partial r^2} + \frac{2\sin\theta\cos\theta}{r}\frac{\partial^2 w}{\partial r\,\partial\theta} + \frac{\cos^2\theta}{r^2}\frac{\partial^2 w}{\partial\theta^2}$$

$$+ \frac{\cos^2\theta}{r}\frac{\partial w}{\partial r} - \frac{2\sin\theta\cos\theta}{r^2}\frac{\partial w}{\partial\theta}. \qquad (2.137)$$

Adding (2.136) and (2.137), we conclude:

$$\nabla^2 w = \frac{\partial^2 w}{\partial x^2} + \frac{\partial^2 w}{\partial y^2} = \frac{\partial^2 w}{\partial r^2} + \frac{1}{r^2} \frac{\partial^2 w}{\partial \theta^2} + \frac{1}{r} \frac{\partial w}{\partial r}; \qquad (2.138)$$

this is the desired result.

Equation (2.138) at once permits one to write the expression for the 3-dimensional Laplacian in cylindrical coordinates; for the transformation of coordinates

$$x = r \cos \theta, \qquad y = r \sin \theta, \qquad z = z$$

involves only x and y and in the same way as earlier. One finds:

$$\nabla^2 w = \frac{\partial^2 w}{\partial x^2} + \frac{\partial^2 w}{\partial y^2} + \frac{\partial^2 w}{\partial z^2} = \frac{\partial^2 w}{\partial r^2} + \frac{1}{r^2} \frac{\partial^2 w}{\partial \theta^2} + \frac{1}{r} \frac{\partial w}{\partial r} + \frac{\partial^2 w}{\partial z^2}.$$

$$(2.139)$$

A procedure similar to the preceding gives the 3-dimensional Laplacian in spherical coordinates (Fig. 2.10):

$$\nabla^2 w = \frac{\partial^2 w}{\partial x^2} + \frac{\partial^2 w}{\partial y^2} + \frac{\partial^2 w}{\partial z^2} = \frac{\partial^2 w}{\partial \rho^2} + \frac{1}{\rho^2} \frac{\partial^2 w}{\partial \phi^2} + \frac{1}{\rho^2 \sin^2 \phi} \frac{\partial^2 w}{\partial \theta^2}$$

$$+ \frac{2}{\rho} \frac{\partial w}{\partial \rho} + \frac{\cot \phi}{\rho^2} \frac{\partial w}{\partial \phi}. \qquad (2.140)$$

(See Problem 8 below.)

2.18 HIGHER DERIVATIVES OF IMPLICIT FUNCTIONS

In Section 2.10, procedures are given for obtaining differentials or first partial derivatives of functions implicitly defined by simultaneous equations. Since the results are in the form of *explicit* expressions for the first partial derivatives [cf. (2.71), (2.72), (2.73), etc.], these derivatives can be differentiated explicitly. An example will illustrate the situation. Let x and y be defined as functions of u and v by the implicit equations:

$$x^2 + y^2 + u^2 + v^2 = 1, \qquad x^2 + 2y^2 - u^2 + v^2 = 1.$$

Then with $F(x, y, u, v) = x^2 + y^2 + u^2 + v^2 - 1$, $G(x, \ldots) = x^2 + 2y^2 - u^2 + v^2 - 1$,

$$\frac{\partial x}{\partial u} = -\frac{\dfrac{\partial(F, G)}{\partial(u, y)}}{\dfrac{\partial(F, G)}{\partial(x, y)}} = -\frac{\begin{vmatrix} 2u & 2y \\ -2u & 4y \end{vmatrix}}{\begin{vmatrix} 2x & 2y \\ 2x & 4y \end{vmatrix}} = -\frac{3u}{x},$$

$$\frac{\partial y}{\partial u} = -\frac{\dfrac{\partial(F, G)}{\partial(x, u)}}{\dfrac{\partial(F, G)}{\partial(x, y)}} = -\frac{\begin{vmatrix} 2x & 2u \\ 2x & -2u \end{vmatrix}}{\begin{vmatrix} 2x & 2y \\ 2x & 4y \end{vmatrix}} = \frac{2u}{y}.$$

Hence

$$\frac{\partial^2 x}{\partial u^2} = \frac{3u}{x^2}\frac{\partial x}{\partial u} - \frac{3}{x} = \frac{3u}{x^2}\left(\frac{-3u}{x}\right) - \frac{3}{x} = -\frac{9u^2}{x^3} - \frac{3}{x},$$

$$\frac{\partial^2 y}{\partial u^2} = -\frac{2u}{y^2}\frac{\partial y}{\partial u} + \frac{2}{y} = -\frac{2u}{y^2}\left(\frac{2u}{y}\right) + \frac{2}{y} = -\frac{4u^2}{y^3} + \frac{2}{y}.$$

PROBLEMS

1. Find the indicated partial derivatives:

  a) $\dfrac{\partial^2 w}{\partial x^2}$ and $\dfrac{\partial^2 w}{\partial y^2}$ if $w = \dfrac{1}{\sqrt{x^2 + y^2}}$,

  b) $\dfrac{\partial^2 w}{\partial x^2}$ and $\dfrac{\partial^2 w}{\partial y^2}$ if $w = \arctan\dfrac{y}{x}$,

  c) $\dfrac{\partial^3 w}{\partial x\,\partial y^2}$ and $\dfrac{\partial^3 w}{\partial x^2\,\partial y}$ if $w = e^{x^2 - y^2}$,

  d) $\dfrac{\partial^{m+n} w}{\partial x^m\,\partial y^n}$ if $w = x^m y^n$.

  Formula (2.138) can be used as a check for (a) and (b).

2. Verify that the mixed derivatives are identical for the following cases:

  a) $\dfrac{\partial^2 z}{\partial x\,\partial y}$, and $\dfrac{\partial^2 z}{\partial y\,\partial x}$ for $z = \dfrac{x}{x^2 + y^2}$,

  b) $\dfrac{\partial^3 w}{\partial x\,\partial y\,\partial z}$, $\dfrac{\partial^3 w}{\partial z\,\partial y\,\partial x}$, and $\dfrac{\partial^3 w}{\partial y\,\partial z\,\partial x}$ for $w = \sqrt{x^2 + y^2 + z^2}$.

3. Show that the following functions are harmonic in x and y:

  a) $e^x \cos y$,   **b)** $x^3 - 3xy^2$,   **c)** $\log\sqrt{x^2 + y^2}$.

4. **a)** Show that every harmonic function is biharmonic.

  b) Show that the following functions are biharmonic in x and y:

$$xe^x \cos y, \qquad x^4 - 3x^2 y^2.$$

  c) Choose the constants a, b, c so that $ax^2 + bxy + cy^2$ is harmonic.

  d) Choose the constants a, b, c, d so that $ax^3 + bx^2 y + cxy^2 + dy^3$ is harmonic.

5. **a)** Prove the identity:

$$\nabla^2(uv) = u\,\nabla^2 v + v\,\nabla^2 u + 2\nabla u \cdot \nabla v$$

  for functions u and v of x and y.

  b) Prove the identity of (a) for functions of x, y, and z.

  c) Prove that if u and v are harmonic in two or three dimensions, then

$$w = xu + v$$

  is biharmonic. [Hint: Use the identity of (a) and (b).]

d) Prove that if u and v are harmonic in two or three dimensions, then

$$w = r^2 u + v$$

is biharmonic, where $r^2 = x^2 + y^2$ for two dimensions and $r^2 = x^2 + y^2 + z^2$ for three dimensions.

6. Establish a chain rule analogous to (2.133) for $\partial^2 z / \partial u\, \partial v$.

7. Use the rule (2.133), applied to $\partial^2 w / \partial x^2$ and $\partial^2 w / \partial y^2$, to prove (2.138).

8. Prove (2.140). [Hint: Use (2.139) to express $\nabla^2 w$ in cylindrical coordinates; then note that the equations of transformation from (z, r) to (ρ, ϕ) are the same as those from (x, y) to (r, θ).]

9. Prove that the biharmonic equation in x and y becomes

$$w_{rrrr} + \frac{2}{r^2} w_{rr\theta\theta} + \frac{1}{r^4} w_{\theta\theta\theta\theta} + \frac{2}{r} w_{rrr} - \frac{2}{r^3} w_{r\theta\theta} - \frac{1}{r^2} w_{rr} + \frac{4}{r^4} w_{\theta\theta} + \frac{1}{r^3} w_r = 0$$

in polar coordinates (r, θ). [Hint: Use (2.138).]

10. If u and v are functions of x and y defined by the equations

$$xy + uv = 1, \qquad xu + yv = 1,$$

find $\partial^2 u / \partial x^2$.

11. If u and v are inverse functions of the system

$$x = u^2 - v^2, \qquad y = 2uv,$$

show that u is harmonic.

12. A *differential equation* is an equation relating one or more variables and their derivatives; the Laplace equation and biharmonic equation above are illustrations (see Chapters 8 and 10). A basic tool in the solution of differential equations, that is, in determining the functions that satisfy the equations, is that of introducing new variables by appropriate substitution formulas. The introduction of polar coordinates in the Laplace equation in Section 2.17 illustrates this. The substitutions can involve independent or dependent variables or both; in each case, one must indicate which of the new variables are to be treated as independent and which as dependent. Make the indicated substitutions in the following differential equations:

a) $\dfrac{dy}{dx} = \dfrac{2x}{y + x^2}$; new variables y (dep.) and $u = x^2$ (indep.);

b) $\dfrac{dy}{dx} = \dfrac{2x - y + 1}{x + y - 4}$; new variables $v = y - 3$ (dep.) and $u = x - 1$ (indep.);

c) $x^2 \dfrac{d^3 y}{dx^3} + 3x \dfrac{d^2 y}{dx^2} + \dfrac{dy}{dx} = 0$; new variables y (dep.) and $t = \log x$ (indep.);

d) $\dfrac{d^2 y}{dx^2} + \left(\dfrac{dy}{dx}\right)^3 = 0$; new variables x (dep.) and y (indep.);

e) $\dfrac{d^2 y}{dx^2} - 4x \dfrac{dy}{dx} + y(3x^2 - 2) = 0$; new variables $v = e^{-x^2} y$ (dep.) and x (indep.);

f) $a \dfrac{\partial u}{\partial x} + b \dfrac{\partial u}{\partial y} = 0$ (a, b constants); new variables u (dep.), $z = bx - ay$ (indep.), and $w = ax + by$ (indep.);

g) $\dfrac{\partial^2 u}{\partial x^2} - \dfrac{\partial^2 u}{\partial y^2} = 0$;   new variables u (dep.), $z = x + y$ (indep.), and $w = x - y$ (indep.).

13. *Transformation of the Sturm-Liouville differential equation.* The equation is

$$p(x)\frac{d^2 u}{dx^2} + p'(x)\frac{du}{dx} + [\lambda k(x) - q(x)]u = 0$$

on an interval $a \le x \le b$ on which $p(x) > 0$, $k(x) > 0$ and λ is a constant. One transforms this by the substitution $u = z(x)v$, $y = h(x)$, where $h'(x) \ne 0$ and $z(x) \ne 0$ on the interval, to obtain a new dependent variable v and new independent variable y.

a) Show that the new equation is

$$pzh'^2 \frac{d^2 v}{dy^2} + (pzh'' + 2pz'h' + p'zh)\frac{dv}{dy}$$
$$+ [pz'' + p'z' + z(\lambda k - q)]v = 0,$$

where in all coefficients x is expressed in terms of y through the inverse function $x = \phi(y)$ of $y = h(x)$. (Assume all functions and derivatives appearing are continuous.)

b) Show that for proper choice of $h(x)$ and $z(x)$ the new equation can be written thus:

$$\frac{d^2 v}{dy^2} + \left(\lambda - Q(y)\right)v = 0.$$

[Hint: To obtain the desired form, require that the coefficient of dv/dy is 0 and that the coefficient of λv equals the coefficient of $d^2 v/dy^2$; then divide by this latter coefficient. Show that the second condition gives $h' = (k/p)^{1/2}$, so that a suitable function h is $\int_{x_0}^{x} [k(t)/p(t)]^{1/2} dt$, where x_0 is on the given interval, and that the first condition gives $(d/dx) \log(h'z^2 p) = 0$, which is satisfied by a suitable function by making $h'z^2 p = 1$ and then choosing $z = (kp)^{-1/4}$. Verify that the $Q(y)$ obtained can be written as $(1/f)d^2 f/dy^2 + q/k$, where $f = 1/z$ as a function of y. This change of variables is known as the *Liouville substitution.*]

14. The proof at the end of Section 2.15 uses properties of the double integral. Justify the following:

a) If $g(x, y)$ is continuous on a square region R: $x_0 \le x \le x_0 + h$, $y_0 \le y \le y_0 + h$, and $g(x, y) > 0$ in R, then

$$\int_R\!\!\int g(x, y)\, dx\, dy > 0.$$

b) If $g(x, y)$ is continuous in domain D and

$$\int_R\!\!\int g(x, y)\, dx\, dy = 0$$

for *every* region R as in (a) contained in D, then $g(x, y) \equiv 0$. [Hint: Suppose $g(x_0, y_0) > 0$. Then by continuity, $g(x, y) > 0$ in some region R (Problem 7 following Section 2.4). Hence by (a) the double integral over R could not be 0. If $g(x_0, y_0) < 0$, then apply the same reasoning to $-g(x, y)$.]

2.19   MAXIMA AND MINIMA OF FUNCTIONS OF SEVERAL VARIABLES

We first recall the basic facts concerning maxima and minima of functions of one variable. Let $y = f(x)$ be defined and differentiable in a closed interval $a \leq x \leq b$, and let x_0 be a number between a and b: $a < x_0 < b$. The function $f(x)$ is said to have a *relative maximum* at x_0 if $f(x) \leq f(x_0)$ for x sufficiently close to x_0. It follows from the very definition of the derivative that if $f'(x_0) > 0$, then $f(x) > f(x_0)$ for all $x > x_0$ and sufficiently close to x_0; similarly, if $f'(x_0) < 0$, then $f(x) > f(x_0)$ for all $x < x_0$ and sufficiently close to x_0. Hence *at a relative maximum, necessarily $f'(x_0) = 0$*. A *relative minimum* of $f(x)$ is defined by the condition $f(x) \geq f(x_0)$ for all x sufficiently close to x_0. A reasoning such as the preceding enables one to conclude that *at a relative minimum of $f(x)$, necessarily $f'(x_0) = 0$*.

The points x_0 at which $f'(x_0) = 0$ are termed the *critical points* of $f(x)$. Although every relative maximum and minimum occurs at a critical point, a critical point need not give either maximum or minimum. This is illustrated by the function $y = x^3$, $-1 \leq x \leq 1$, at $x = 0$. This function has a critical point at $x = 0$, but the point is neither maximum nor minimum and is an example of what is termed a *horizontal inflection point*. This is illustrated in Fig. 2.15.

Let x_0 be a critical point, so that $f'(x_0) = 0$, and let it be assumed that $f''(x_0) > 0$. Then $f(x)$ *has a relative minimum at x_0*. For by the law of the mean, when $x > x_0$, $f(x) - f(x_0) = f'(x_1)(x - x_0)$, where $x_0 < x_1 < x$. Since $f''(x_0) > 0, f'(x_1) > 0$ for $x_1 > x_0$ and x_1 sufficiently close to x_0. Hence $f(x) - f(x_0) > 0$ for $x > x_0$ and x sufficiently close to x_0. Similarly, $f(x) - f(x_0) > 0$ for $x < x_0$ and x sufficiently close to x_0. Therefore $f(x)$ has a relative minimum at x_0. An analogous reasoning applies when $f''(x_0) < 0$. Accordingly, one can state the rules:

If $f'(x_0) = 0$ and $f''(x_0) > 0$, then $f(x)$ has a relative minimum at x_0; if $f'(x_0) = 0$ and $f''(x_0) < 0$, then $f(x)$ has a relative maximum at x_0.

These two rules cover most cases of interest. It can of course happen that $f''(x_0) = 0$ at a critical point, in which case one must consider the higher derivatives. A repeated application of the previous reasoning gives the following general rule:

Let $f'(x_0) = 0, f''(x_0) = 0, \ldots, f^{(n)}(x_0) = 0$, but $f^{(n+1)}(x_0) \neq 0$; then $f(x)$ has a relative maximum at x_0 if n is odd and $f^{(n+1)}(x_0) < 0$; $f(x)$ has a relative minimum at x_0 if n is odd and $f^{(n+1)}(x_0) > 0$; $f(x)$ has neither relative maximum nor relative minimum at x_0 but a horizontal inflection point at x_0 if n is even.

Figure 2.15   Critical points of function $y = f(x)$.

The preceding discussion has been restricted to points x_0 *within* the interval of definition of $f(x)$ and to *relative* maxima and minima. The notion of relative maximum and minimum can readily be extended to the endpoints a and b, and rules can be formulated in terms of derivatives. However, the interest in these points arises mainly in connection with the *absolute* maximum and minimum of $f(x)$ on the interval $a \leq x \leq b$. A function $f(x)$ is said to have an *absolute maximum* M for a certain set of values of x if $f(x_0) = M$ for some x_0 of the set given and $f(x) \leq M$ for all x of the set; an absolute minimum is defined similarly, with the condition $f(x) \geq M$ replacing the condition $f(x) \leq M$. The following theorem is then fundamental.

THEOREM   If $f(x)$ is continuous in the closed interval $a \leq x \leq b$, then $f(x)$ has an absolute minimum M_1 and an absolute maximum M_2 on this interval.

The proof of this theorem requires a more profound analysis of the real number system; see Section 2.23.

It should be remarked that the inclusion of the end-values is essential for this theorem, as the simple example $y = x$ for $0 < x < 1$ illustrates; here the function has no minimum or maximum *on the set given*. For the set $0 \leq x \leq 1$, the absolute minimum is 0, for $x = 0$, and the absolute maximum is 1, for $x = 1$. The example $y = \tan x$ for $-\pi/2 < x < \pi/2$ illustrates a function with no absolute minimum or maximum; in this case, adjoining values at the endpoints $x = \pm\pi/2$ does not help.

A function $f(x)$ is said to be *bounded* for a given set of x if there is a constant K such that $|f(x)| \leq K$ on this set. The theorem above implies that *if $f(x)$ is continuous in the closed interval $a \leq x \leq b$, then $f(x)$ is bounded in this interval*; for K can be chosen as the larger of $|M_1|, |M_2|$. For example, let $y = \sin x$ for $0 \leq x \leq \pi$. This function has the absolute minimum $M_1 = 0$ and the absolute maximum $M_2 = 1$ and is bounded, with $K = 1$. The example $y = x$ for $0 < x < 1$ illustrates a bounded function ($K = 1$) having neither absolute minimum nor absolute maximum; the example $y = \tan x$ for $-\frac{1}{2}\pi < x < \frac{1}{2}\pi$ illustrates an unbounded function.

To find the absolute maximum M of a function $y = f(x)$, differentiable for $a \leq x \leq b$, one can reason as follows: if $f(x_0) = M$ and $a < x_0 < b$, then $f(x)$ has necessarily a relative maximum at x_0; thus the absolute maximum occurs either at a critical point within the interval or at $x = a$ or $x = b$. One therefore first locates all critical points within the interval and compares the values of y at these points with the values at $x = a$ and $x = b$; the largest value of y is the maximum sought. It is thus not necessary to use the second derivative test described at the beginning of this section. The absolute minimum can be found in the same way.

The determination of critical points and their classification into maxima, minima, or neither are important also for graphing functions; from a knowledge of the critical points and the corresponding values of y, a very good first approximation to the graph of $y = f(x)$ can be obtained.

After these preliminaries we can consider the analogous questions for *functions of two or more variables*. Let $z = f(x, y)$ be defined and continuous in a domain D. This function is said to have a *relative maximum* at (x_0, y_0) if $f(x, y) \leq f(x_0, y_0)$ for (x, y) sufficiently close to (x_0, y_0) and to have a *relative minimum* at (x_0, y_0) if $f(x, y) \geq f(x_0, y_0)$ for (x, y) sufficiently close to (x_0, y_0). Let (x_0, y_0) give a relative maximum of $f(x, y)$; then the function $f(x, y_0)$, which depends on x alone, has a relative maximum at x_0, as is illustrated in Fig. 2.16. Hence if $f_x(x_0, y_0)$ exists, then $f_x(x_0, y_0) = 0$; similarly, if $f_y(x_0, y_0)$ exists, then $f_y(x_0, y_0) = 0$. Points (x, y) at which both partial derivatives are 0 are termed *critical points* of f. As before, one concludes that every relative maximum and every relative minimum occur at a critical point of f, if f_x and f_y exist in D.

One might expect that the nature of a critical point could be determined by studying the functions $f(x, y_0)$ and $f(x_0, y)$, with the aid of the second derivatives as previously. First of all, it should be remarked that one of these functions can have a maximum at (x_0, y_0), while the other has a minimum. This is illustrated by the function $z = 1 + x^2 - y^2$ at $(0, 0)$; this function has a minimum with respect to x at $x = 0$ for $y = 0$ and a maximum with respect to y at $y = 0$ for $x = 0$, as shown in Fig. 2.17. This critical point is an example of what is termed a *saddle point*; this will be discussed further. The level curves of z are shown in Fig. 2.17; the configuration shown is typical of that at a saddle point.

A further complication is the following: It can happen that $z = f(x, y_0)$ has a relative maximum for $x = x_0$, and $z = f(x_0, y)$ has also a relative maximum for $y = y_0$, but $z = f(x, y)$ does *not* have a relative maximum at (x_0, y_0). This is illustrated by the function $z = 1 - x^2 + 4xy - y^2$ at $(0, 0)$. For $y = 0$, $z = 1 - x^2$, with maximum for $x = 0$; for $x = 0$, $z = 1 - y^2$, with maximum for $y = 0$. On the other hand, for $y = x$, $z = 1 + 2x^2$, so that the section of the surface by the

Figure 2.16   Maximum of $z = f(x, y)$.

plane $y = x$ has a *minimum* at $x = 0$. This is shown in Fig. 2.18. The level curves, shown in the same figure, again reveal the presence of a saddle point.

A better understanding of this example and a clue to a general procedure can be obtained by introducing cylindrical coordinates (r, θ). One has then $z = 1 - r^2(1 - 2\sin 2\theta)$. If one sets $\theta = \text{const} = \alpha$, one obtains z as a function of r. For $\alpha = 0$, one obtains the xz-trace: $z = 1 - r^2$; for $\alpha = \pi/2$, one obtains the

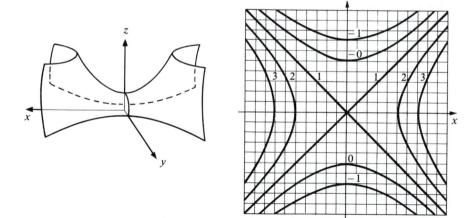

Figure 2.17   Saddle point ($z = 1 + x^2 - y^2$).

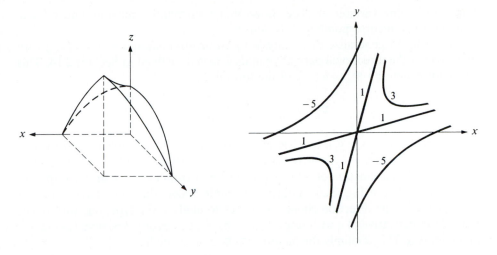

Figure 2.18 $z = 1 - x^2 + 4xy - y^2$.

yz-trace: $z = 1 - r^2$; for $\alpha = \pi/4$, one obtains the trace in the plane $y = x$: $z = 1 + r^2$. For a general α, one has the trace $z = 1 - r^2(1 - 2\sin 2\alpha)$; r can here be allowed both positive and negative values. On each trace, $\partial z/\partial r = -2r(1 - 2\sin 2\alpha)$; thus z, as a function of r, has a critical point for $r = 0$. However, $\partial^2 z/\partial r^2 = -2(1 - 2\sin 2\alpha)$. This second derivative takes on both positive and negative values, as shown graphically in Fig. 2.19. The second derivative test for

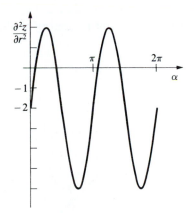

Figure 2.19 $\partial^2 z/\partial r^2$ for function of Fig. 2.18.

functions of one variable at once shows that both relative maxima and relative minima occur in corresponding directions α.

In order to generalize this analysis to an arbitrary function $z = f(x, y)$ we make use of the directional derivative in direction α, defined in Section 2.14. This we again denote by $\nabla_\alpha z$ and recall the formula:

$$\nabla_\alpha z = \frac{\partial z}{\partial x} \cos \alpha + \frac{\partial z}{\partial y} \sin \alpha.$$

At a critical point (x_0, y_0), one has necessarily $\nabla_\alpha z = 0$; this means that in each plane $(x - x_0) \sin \alpha - (y - y_0) \cos \alpha = 0$, z has a critical point at (x_0, y_0), when z is regarded as a function of "directed distance" s from (x_0, y_0), in the xy-plane; this is illustrated in Fig. 2.20. As in the example above, the type of critical point may vary with the direction chosen. In order to analyze the type, one introduces the second derivative of z with respect to s, that is, the *second directional derivative in direction* α. This is simply the quantity $\nabla_\alpha \nabla_\alpha z$, and one has

$$\nabla_\alpha \nabla_\alpha z = \nabla_\alpha \left(\frac{\partial z}{\partial x} \cos \alpha + \frac{\partial z}{\partial y} \sin \alpha \right)$$

$$= \frac{\partial^2 z}{\partial x^2} \cos^2 \alpha + 2 \frac{\partial^2 z}{\partial x \, \partial y} \sin \alpha \cos \alpha + \frac{\partial^2 z}{\partial y^2} \sin^2 \alpha. \qquad (2.141)$$

This is precisely the quantity $\partial^2 z / \partial r^2$ of the preceding example. In order to guarantee a relative maximum for z at (x_0, y_0), one can require that this second derivative be negative for all α, $0 \le \alpha \le 2\pi$; this ensures that z has a relative maximum as a function of s in each plane in direction α, and in fact (all second derivatives being assumed continuous) that z has a relative maximum at the point; cf. Problem 10 below. A similar reasoning applies to relative minima. One

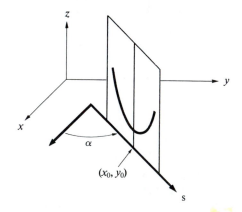

Figure 2.20 Critical point of $z = f(x, y)$ analyzed by vertical sections.

therefore obtains the rule:

If $\partial z/\partial x = 0$ and $\partial z/\partial y = 0$ at (x_0, y_0) and

$$\frac{\partial^2 z}{\partial x^2} \cos^2 \alpha + 2\frac{\partial^2 z}{\partial x \, \partial y} \sin \alpha \cos \alpha + \frac{\partial^2 z}{\partial y^2} \sin^2 \alpha < 0 \qquad (2.142)$$

for $x = x_0$, $y = y_0$, and all α: $0 \le \alpha \le 2\pi$, then $z = f(x, y)$ has a relative maximum at (x_0, y_0); if $\partial z/\partial x = 0$ and $\partial z/\partial y = 0$ and

$$\frac{\partial^2 z}{\partial x^2} \cos^2 \alpha + 2\frac{\partial^2 z}{\partial x \, \partial y} \sin \alpha \cos \alpha + \frac{\partial^2 z}{\partial y^2} \sin^2 \alpha > 0 \qquad (2.143)$$

for $x = x_0$, $y = y_0$, and all α: $0 \le \alpha \le 2\pi$, then $z = f(x, y)$ has a relative minimum at (x_0, y_0).

It thus appears that the study of the critical points is reduced to the analysis of the expression

$$A \cos^2 \alpha + 2B \sin \alpha \cos \alpha + C \sin^2 \alpha, \qquad (2.144)$$

where the abbreviations

$$A = \frac{\partial^2 z}{\partial x^2}(x_0, y_0), \qquad B = \frac{\partial^2 z}{\partial x \, \partial y}(x_0, y_0), \qquad C = \frac{\partial^2 z}{\partial y^2}(x_0, y_0)$$

$$(2.145)$$

are used. Here an algebraic analysis reduces the question to a simpler one, for one has the following theorem:

THEOREM   If $B^2 - AC < 0$ and $A + C < 0$, then the expression (2.144) is negative for all α; if $B^2 - AC < 0$ and $A + C > 0$, then the expression (2.144) is positive for all α.

Proof.   Let the expression (2.144) be denoted by $P(\alpha)$. Let $B^2 - AC < 0$ and $A + C < 0$. Then $P(\pm\pi/2) = C < 0$, for if $C \ge 0$, then $A + C < 0$ implies $A < 0$, so that $AC \le 0$; this contradicts the condition $B^2 - AC < 0$. Similarly, one shows $P(0) = A < 0$. For $\alpha \ne \pm\pi/2$, one has

$$P(\alpha) = \cos^2 \alpha (A + 2B \tan \alpha + C \tan^2 \alpha).$$

Thus $P(\alpha)$ is positive, negative, or 0, according to whether the quadratic expression

$$Q(u) = Cu^2 + 2Bu + A \qquad (u = \tan \alpha)$$

is positive, negative, or 0. Since $B^2 - AC < 0$, $Q(u)$ has no real roots; thus $Q(u)$ is always positive or always negative. For $u = 0$, $Q = A < 0$. Hence $Q(u)$ is always negative, and the same holds for $P(\alpha)$. Accordingly, the first assertion is proved. The second is proved in the same way. $\square$

If $B^2 - AC > 0$, the proof just given shows that $P(\alpha)$ will be positive for some values of α and negative for others, as in the example of Fig. 2.17 above. In this case the critical point is termed a *saddle point*. If $B^2 - AC = 0$, there will be some directions in which $P(\alpha) = 0$, and one must introduce higher derivatives in order to determine the nature of the critical point.

We summarize the results obtained:

THEOREM   Let $z = f(x, y)$ be defined and have continuous first and second partial derivatives in a domain D. Let (x_0, y_0) be a point of D for which $\partial z/\partial x$ and $\partial z/\partial y$ are 0. Let

$$A = \frac{\partial^2 z}{\partial x^2}(x_0, y_0), \qquad B = \frac{\partial^2 z}{\partial x \, \partial y}(x_0, y_0), \qquad C = \frac{\partial^2 z}{\partial y^2}(x_0, y_0).$$

Then one has the following cases:

$B^2 - AC < 0$ and $A + C < 0$, relative maximum at (x_0, y_0);   (2.146)

$B^2 - AC < 0$ and $A + C > 0$, relative minimum at (x_0, y_0);   (2.147)

$B^2 - AC > 0$, saddle point at (x_0, y_0);   (2.148)

$B^2 - AC = 0$, nature of critical point undetermined.   (2.149)

If A, B, and C are all 0, so that $P(\alpha) \equiv 0$, one can study the critical point with the aid of the third derivative: $\nabla_\alpha \nabla_\alpha \nabla_\alpha z$. This is illustrated by the function $z = x^3 - 3xy^2$, whose level curves are graphed in Fig. 2.21. If the third derivatives are all 0, one can go on to higher derivatives as for functions of one variable. In essence the whole problem is reduced to one for functions of one variable, and the directional derivative is the chief tool in this reduction.

In the accompanying Fig. 2.21, examples are given of various types of critical points; in each case the level curves of the function are shown.

Remark.   For the cases $B^2 - AC < 0$ and $B^2 - AC > 0$, one can show that the level curves of f have the qualitative appearance of Fig. 2.21(a), (b), and (c). To this end, one applies the implicit function theorem to the equation

$$f(x, y) - c = 0$$

to obtain a differential equation for these level curves:

$$y' = - \frac{f_x(x, y)}{f_y(x, y)}.$$

The differential equation has a *singular point* at the critical point, since the numerator and denominator are 0 there. By applying the theory of configurations at such singular points, one deduces that in the case of a maximum or minimum, the level curves are all closed curves around the singular point, whereas at a saddle point the curves resemble hyperbolas and asymptotes as in Fig. 2.21(c). For details, see Chapter 15 of the book by Coddington and Levinson listed at the end of the chapter. This book considers the problem for pairs of first-order

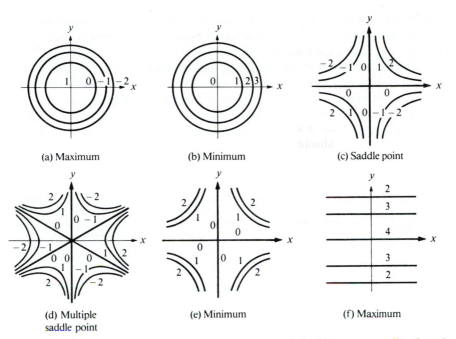

(a) Maximum   (b) Minimum   (c) Saddle point

(d) Multiple
saddle point   (e) Minimum   (f) Maximum

Figure 2.21   Examples of critical points of $f(x, y)$ at $(0,0)$. The corresponding functions
are as follows: (a) $z = 1 - x^2 - y^2$, (b) $z = x^2 + y^2$, (c) $z = xy$, (d) $z = x^3 - 3xy^2$,
(e) $z = x^2y^2$, (f) $z = 4 - y^2$.

equations. Our problem is replaced by such a pair by writing

$$\frac{dx}{dt} = -f_y(x, y), \qquad \frac{dy}{dt} = f_x(x, y),$$

which gives the solutions in parametric form.

The preceding discussion has concerned only relative maxima and minima.
Just as with functions of one variable, one can define the notions of absolute
maximum and absolute minimum; again their investigation calls for a study of
the function at the "ends" of the domain D of definition, that is, on the boundary.
In order to ensure the existence of absolute maximum and minimum, one must
demand that the domain D itself be *bounded* (cf. Section 2.2). One has then, as for
functions of one variable:

> **THEOREM**   Let D be a bounded domain of the xy-plane. Let $f(x, y)$ be
> defined and continuous in the closed region E formed of D plus its boundary.
> Then $f(x, y)$ has an absolute maximum and an absolute minimum in E.

The fact that D is a domain is actually inessential for this theorem; the
theorem remains valid for any bounded closed set E (Section 2.2). For a proof,
see Section 2.23.

Again one has the corollary that a function $f(x, y)$ that is continuous in a bounded closed region (or set) E is necessarily bounded on E; that is, $|f(x, y)| \leq M$ for (x, y) in E and for a suitable choice of M.

To determine the absolute maximum and minimum of a function $f(x, y)$ defined on a bounded closed region E, one proceeds as for functions of one variable. One determines the critical points of f inside E and the values of f at the critical points; one then determines the maximum and minimum of f on the boundary. Among these values the desired maximum and minimum will be found.

EXAMPLE Find the maximum and minimum of $z = x^2 + 2y^2 - x$ on the set $x^2 + y^2 \leq 1$.

One has

$$\frac{\partial z}{\partial x} = 2x - 1, \qquad \frac{\partial z}{\partial y} = 4y.$$

Thus the critical point is at $(\frac{1}{2}, 0)$, at which $z = -\frac{1}{4}$. On the boundary of E, one has $x^2 + y^2 = 1$, so that $z = 2 - x - x^2$, $-1 \leq x \leq 1$. For this function we find the absolute maximum to be $2\frac{1}{4}$, at the critical point $x = -\frac{1}{2}$; the absolute minimum is 0, at the end $x = 1$. Thus the absolute maximum is $2\frac{1}{4}$, occurring at $\left(-\frac{1}{2}, \pm\frac{\sqrt{3}}{2}\right)$, and the absolute minimum is $-\frac{1}{4}$, occurring at $(\frac{1}{2}, 0)$. Since the minimum occurs *inside* E, it is also a relative minimum, as can be checked by (2.146) through (2.149). One has here

$$A = \frac{\partial^2 z}{\partial x^2} = 2, \qquad B = \frac{\partial^2 z}{\partial x \partial y} = 0, \qquad C = \frac{\partial^2 z}{\partial y^2} = 4,$$

so that (2.147) is satisfied. ∎

The maximum and minimum on the boundary can also be analyzed by the method of Lagrange multipliers, to be explained in the following section.

All of the foregoing extends to functions of three or more variables without essential modification. Thus at a critical point (x_0, y_0, z_0) of $w = f(x, y, z)$, all three derivatives $\partial w/\partial x$, $\partial w/\partial y$, $\partial w/\partial z$ are 0, so that the directional derivative:

$$\nabla_u w = \frac{\partial w}{\partial x} \cos \alpha + \frac{\partial w}{\partial y} \cos \beta + \frac{\partial w}{\partial z} \cos \gamma$$

in the direction of an arbitrary unit vector $\mathbf{u} = \cos \alpha \mathbf{i} + \cos \beta \mathbf{j} + \cos \gamma \mathbf{k}$ is zero at the point (x_0, y_0, z_0). This directional derivative is the derivative dw/ds of a function of one variable s, serving as coordinate on a line through (x_0, y_0, z_0) in direction $\mathbf{u}$. To analyze the critical point, one uses the second derivative d^2w/ds^2

$= \nabla_u \nabla_u w$:

$$\nabla_u \nabla_u w = \frac{\partial}{\partial x}(\nabla_u w) \cos \alpha + \frac{\partial}{\partial y}(\nabla_u w) \cos \beta + \frac{\partial}{\partial z}(\nabla_u w) \cos \gamma$$

$$= \frac{\partial^2 w}{\partial x^2} \cos^2 \alpha + 2\frac{\partial^2 w}{\partial x \, \partial y} \cos \alpha \cos \beta + 2\frac{\partial^2 w}{\partial x \, \partial z} \cos \alpha \cos \gamma$$

$$+ \frac{\partial^2 w}{\partial y^2} \cos^2 \beta + 2\frac{\partial^2 w}{\partial y \, \partial z} \cos \beta \cos \gamma + \frac{\partial^2 w}{\partial z^2} \cos^2 \gamma. \qquad (2.150)$$

If this expression is positive for all **u**, then w has a relative minimum at (x_0, y_0, z_0). Algebraic criteria for the positiveness of this "quadratic form" can be obtained; see Section 2.21.

Remarks.   The tests for maxima or minima of $f(x)$ given at the beginning of this section can be deduced from Taylor's formula with remainder (Section 6.17). There is a similar formula for functions of several variables (Section 6.21), and it can also be used to derive the tests for maxima and minima.

*2.20   EXTREMA FOR FUNCTIONS WITH SIDE CONDITIONS ▪ LAGRANGE MULTIPLIERS

A problem of considerable importance for applications is that of maximizing or minimizing a function of several variables, where the variables are related by one or more equations, termed *side conditions*. Thus the problem of finding the radius of the largest sphere inscribable in the ellipsoid $x^2 + 2y^2 + 3z^2 = 6$ is equivalent to minimizing the function $w = x^2 + y^2 + z^2$, with the side condition $x^2 + 2y^2 + 3z^2 = 6$.

To handle such problems, one can, if possible, eliminate some of the variables by using the side conditions and eventually reduce the problem to an ordinary maximum and minimum problem such as that considered in the preceding section. This procedure is not always feasible, and the following procedure is often more convenient; it also treats the variables in a more symmetrical way, so that various simplifications may be possible.

To illustrate the method, we consider the problem of maximizing $w = f(x, y, z)$, where equations $g(x, y, z) = 0$ and $h(x, y, z) = 0$ are given. The equations $g = 0$ and $h = 0$ describe two surfaces in space, and the problem is thus one of maximizing $f(x, y, z)$ as (x, y, z) varies on the curve of intersection of these surfaces. At a maximum point the derivative of f along the curve, that is, the directional derivative along the tangent to the curve, must be 0. This directional derivative is the component of the vector ∇f along the tangent. It follows that ∇f must lie in a plane normal to the curve at the point. This plane also contains the vectors ∇g and ∇h (Section 2.13); that is, the vectors ∇f, ∇g, and ∇h are coplanar at the point. Hence (Section 1.2) there must exist scalars λ_1 and λ_2 such that

$$\nabla f + \lambda_1 \nabla g + \lambda_2 \nabla h = \mathbf{0} \qquad (2.151)$$

at the critical point. This is equivalent to three scalar equations:

$$\frac{\partial f}{\partial x} + \lambda_1 \frac{\partial g}{\partial x} + \lambda_2 \frac{\partial h}{\partial x} = 0, \qquad \frac{\partial f}{\partial y} + \lambda_1 \frac{\partial g}{\partial y} + \lambda_2 \frac{\partial h}{\partial y} = 0,$$

$$\frac{\partial f}{\partial z} + \lambda_1 \frac{\partial g}{\partial z} + \lambda_2 \frac{\partial h}{\partial z} = 0. \tag{2.152}$$

These three equations, together with the equations $g(x, y, z) = 0$, $h(x, y, z) = 0$, serve as five equations in the five unknowns $x, y, z, \lambda_1, \lambda_2$. By solving them for x, y, z, one locates the critical points on the curve.

It has been tacitly assumed here that the surfaces $g = 0$, $h = 0$ do actually intersect in a curve and that ∇g and ∇h are linearly independent. Cases in which these conditions fail are degenerate and require further investigation (cf. Section 2.22).

The method described applies quite generally. To find the critical points of $w = f(x, y, z, u,...)$, where the variables $x, y, z,...$ are related by equations $g(x, y, z, u,...) = 0$, $h(x, y, z, u,...) = 0,...$, one solves the system of equations

$$\frac{\partial f}{\partial x} + \lambda_1 \frac{\partial g}{\partial x} + \lambda_2 \frac{\partial h}{\partial x} + \cdots = 0, \qquad \frac{\partial f}{\partial y} + \lambda_1 \frac{\partial g}{\partial y} + \lambda_2 \frac{\partial h}{\partial y} + \cdots = 0,...,$$

$$g(x, y, z, u,...) = 0, \qquad h(x, y, z, u,...) = 0,... \tag{2.153}$$

for the unknowns $x, y, z, u,..., \lambda_1, \lambda_2,...$. The parameters $\lambda_1, \lambda_2,...$ are known as *Lagrange multipliers*.

EXAMPLE   To find the critical points of $w = xyz$, subject to the condition $x^2 + y^2 + z^2 = 1$, one forms the function

$$f + \lambda g = xyz + \lambda(x^2 + y^2 + z^2 - 1)$$

and then obtains four equations:

$$yz + 2\lambda x = 0, \qquad xz + 2\lambda y = 0, \qquad xy + 2\lambda z = 0, \qquad x^2 + y^2 + z^2 = 1.$$

Multiplying the first three by x, y, z respectively, adding, and using the fourth equation, one finds $\lambda = -\frac{1}{2}(3xyz)$. Using this relation, one easily finds the 14 critical points $(0, 0, \pm 1)$, $(0, \pm 1, 0)$, $(\pm 1, 0, 0)$, $(\pm \sqrt{3}/3, \pm \sqrt{3}/3, \pm \sqrt{3}/3)$. The first six listed are saddle points, whereas the remaining eight furnish four minima and four maxima, as simple considerations of signs show. The maxima and minima are absolute.   ∎

*2.21   MAXIMA AND MINIMA OF QUADRATIC FORMS ON THE UNIT SPHERE

One of the most important applications of the method of Lagrange multipliers is to the problem of maximizing or minimizing a quadratic form in n variables (Section 1.11):

$$w = f(x_1,...,x_n) = \sum_{i=1}^{n} \sum_{j=1}^{n} a_{ij} x_i x_j \tag{2.154}$$

subject to the side condition

$$x_1^2 + \cdots + x_n^2 = 1. \tag{2.155}$$

We can interpret (2.155) as the equation of the *unit sphere* in E^n, so that our problem is that of maximizing or minimizing f on the unit sphere. As in Section 1.11, we let $A = (a_{ij})$ be the $n \times n$ matrix of coefficients of the quadratic form and can always assume that f is written in a form which makes A *symmetric*: $a_{ij} = a_{ji}$ for all i and j, or $A = A'$. As in Section 1.11, we can then also write

$$f(x_1,\ldots,x_n) = \mathbf{x}'A\mathbf{x}, \qquad (\mathbf{x} = \operatorname{col}(x_1,\ldots,x_n)).$$

We proceed as in the preceding section and are led to the $n + 1$ equations:

$$\frac{\partial f}{\partial x_1} + \lambda\frac{\partial g}{\partial x_1} = 0, \qquad \frac{\partial f}{\partial x_2} + \lambda\frac{\partial g}{\partial x_2} = 0,\ldots, \frac{\partial f}{\partial x_n} + \lambda\frac{\partial g}{\partial x_n} = 0,$$

$$g(x_1,\ldots,x_n) = 0. \tag{2.156}$$

Here f is as previously, and $g(x_1,\ldots,x_n) = 1 - x_1^2 - \cdots - x_n^2$. Since A is symmetric,

$$f(x_1,\ldots,x_n) = a_{11}x_1^2 + a_{12}x_1x_2 + a_{12}x_2x_1 + a_{13}x_1x_3 + a_{13}x_3x_1 + \cdots$$

$$= a_{11}x_1^2 + 2a_{12}x_1x_2 + 2a_{13}x_1x_3 + \cdots$$

(we are suggesting only the terms in x_1), and accordingly,

$$\frac{\partial f}{\partial x_1} = 2a_{11}x_1 + 2a_{12}x_2 + 2a_{13}x_3 + \cdots.$$

Thus the equations (2.156) become

$$2a_{11}x_1 + 2a_{12}x_2 + \cdots + 2a_{1n}x_n - 2\lambda x_1 = 0,$$

$$\vdots$$

$$2a_{n1}x_1 + 2a_{n2}x_2 + \cdots + 2a_{nn}x_n - 2\lambda x_n = 0,$$

$$x_1^2 + \cdots + x_n^2 = 1.$$

Here the first n equations are equivalent to the equation

$$A\mathbf{x} = \lambda\mathbf{x}.$$

Therefore they state that $\mathbf{x}$ (if nonzero) is to be an *eigenvector* of A, associated with the *eigenvalue* λ (Section 1.10). The last equation states that $|\mathbf{x}| = 1$, so that $\mathbf{x}$ is a unit vector and cannot be $\mathbf{0}$. Thus we are seeking the unit eigenvectors of A. If $\mathbf{x}$ is such an eigenvector, then

$$f(x_1,\ldots,x_n) = \mathbf{x}'A\mathbf{x} = \mathbf{x}'\lambda\mathbf{x} = \lambda\mathbf{x}'\mathbf{x}$$

$$= \lambda(x_1^2 + \cdots + x_n^2) = \lambda.$$

Hence the eigenvalues of A provide the values of f at its critical points on the unit sphere. In particular, the absolute maximum of f on the unit sphere equals the largest eigenvalue of A, and the absolute minimum of f on the unit sphere equals the smallest eigenvalue of A. As in Section 1.12, all eigenvalues are real.

EXAMPLE 1   Let $n = 2$ and let

$$f(x, y) = ax^2 + 2bxy + cy^2,$$

so that $A = \begin{bmatrix} a & b \\ b & c \end{bmatrix}$. Then the eigenvalues of A are the solutions of

$$\begin{vmatrix} a - \lambda & b \\ b & c - \lambda \end{vmatrix} = 0 \quad \text{or} \quad \lambda^2 - (a + c)\lambda + ac - b^2 = 0.$$

Hence

$$\lambda = \frac{a + c \pm \sqrt{(a + c)^2 - 4(ac - b^2)}}{2} = \frac{a + c \pm \sqrt{(a - c)^2 + 4b^2}}{2}.$$

The last form shows that the roots are always real (see Problem 12 following Section 1.12). If $a = c$ and $b = 0$, the roots are equal and equal to $(a + c)/2$. Otherwise, the larger root is obtained from the plus sign, the smaller from the minus sign. Also, if $ac - b^2 > 0$, both roots have the same sign as $a + c$. Thus if $a + c < 0$. then both roots are negative. If $ac - b^2 < 0$, the roots have opposite signs; if $ac - b^2 = 0$, one root is 0, the other is $a + c$.

Since our side condition is $x^2 + y^2 = 1$, we can write $x = \cos \alpha$, $y = \sin \alpha$, and f becomes $a \cos^2 \alpha + 2b \sin \alpha \cos \alpha + c \sin^2 \alpha$. When both roots are positive, f has a positive minimum and remains positive for all α; when both roots are negative, f has a negative maximum and remains negative for all α. Thus the conditions found provide a new proof of the rules given in Section 2.19 for the function

$$A \cos^2 \alpha + 2B \sin \alpha \cos \alpha + C \sin^2 \alpha. \qquad \blacksquare$$

In this example we have thus far considered only the nature and significance of the eigenvalues. To find the eigenvectors, and hence to find (x, y) that maximizes or minimizes f, one proceeds as in Section 1.10.

For a general quadratic form f, one can imitate the example and ask for the conditions under which f has a positive minimum or a negative maximum on the unit sphere. We remark that every nonzero vector $\mathbf{x} = (x_1, \ldots, x_n)$ can be written as $k(u_1, \ldots, u_n)$, where $k = |\mathbf{x}|$ and $\mathbf{u} = (u_1, \ldots, u_n)$ is a unit vector. Hence

$$f(x_1, \ldots, x_n) = \sum \sum a_{ij} x_i x_j = k^2 \sum \sum a_{ij} u_i u_j = k^2 f(u_1, \ldots, u_n).$$

Accordingly, if f has a positive minimum on the unit sphere, then f is positive for all nonzero $\mathbf{x}$. Conversely, if f is positive for all nonzero $\mathbf{x}$, then f is positive on the unit sphere, in particular, and hence f has a positive minimum on the unit sphere. Similarly, f has a negative maximum on the unit sphere if and only if f is negative for all nonzero $\mathbf{x}$. A quadratic form that is positive (negative) for all nonzero $\mathbf{x}$ is said to be *positive definite* (*negative definite*). Hence we can say, for example: f is *positive definite if and only if all eigenvalues of A are positive.*

Another way to obtain these results is to choose new Cartesian coordinates $(y_1, \ldots, y_n)$ in E^n. As was pointed out in Section 1.12, for proper choice of the new coordinates, f becomes simply

$$\lambda_1 y_1^2 + \cdots + \lambda_n y_n^2,$$

where $\lambda_1,\ldots,\lambda_n$ are the (*necessarily real*) eigenvalues of A. It is thus clear that f is positive definite precisely when all the λ's are positive.

For further information on this topic, see Chapter 10 of Vol. 1 of the book by Gantmacher listed at the end of the chapter.

For a general function $F(x_1,\ldots,x_n)$, with critical point at $(x_1^0,\ldots,x_n^0)$, the method of Section 2.19 leads us to the quadratic form

$$\sum_{i,j} \frac{\partial^2 F}{\partial x_i\, \partial x_j} u_i u_j, \quad \text{\textbackslash}$$

where $(u_1,\ldots,u_n)$ is a unit vector and all derivatives are evaluated at the critical point. If this form is positive definite, then F has a minimum at the critical point. Hence *if all eigenvalues of the matrix* $(\partial^2 F/\partial x_i \partial x_j)$ *are positive, there is a minimum. Similarly, if all eigenvalues are negative, then there is a maximum.*

PROBLEMS

1. Locate the critical points of the following functions, classify them, and graph the functions:

   a) $y = x^3 - 3x$, **b)** $y = 2 \sin x + \sin 2x$, **c)** $y = e^{-x} - e^{-2x}$.

2. Determine the nature of the critical point of $y = x^n (n = 2, 3, \ldots)$ at $x = 0$.

3. Determine the absolute maximum and absolute minimum, if they exist, of the following functions:

   a) $y = \cos x$, $-\dfrac{\pi}{2} \le x \le \dfrac{\pi}{2}$,

   b) $y = \log x$, $0 < x \le 1$,

   c) $y = \tanh x$, all x,

   d) $y = \dfrac{x}{1 + x^2}$, all x.

4. Find the critical points of the following functions and test for maxima and minima:

   a) $z = \sqrt{1 - x^2 - y^2}$,

   b) $z = 1 + x^2 + y^2$,

   c) $z = 2x^2 - xy - 3y^2 - 3x + 7y$,

   d) $z = x^2 - 5xy - y^2$,

   e) $z = x^2 - 2xy + y^2$,

   f) $z = x^3 - 3xy^2 + y^3$,

   g) $z = x^2 - 2x(\sin y + \cos y) + 1$,

   h) $z = xy^2 + x^2y - xy$,

   i) $z = x^3 + y^3$,

j) $z = x^4 + 3x^2y^2 + y^4$,

k) $z = [x^2 + (y + 1)^2][x^2 + (y - 1)^2]$ (interpret geometrically).

5. Find the critical points of the following functions, classify, and graph the level curves of the functions:

a) $z = e^{-x^2-y^2}$,

b) $z = x^4 - y^4$,

c) $z = \sin x \cosh y$,

d) $z = \dfrac{x}{x^2 + y^2}$,

e) $z = x^2 - xy + y^2$,

f) $z = x + y + \sqrt{1 - x^2 - y^2}$.

6. Find the critical points of the following functions with given side conditions and test for maxima and minima:

a) $z = 3x + 4y$, where $x^2 + y^2 = 1$,

b) $z = x^2 + y^2$, where $x^4 + y^4 = 1$,

c) $z = x^2 + 24xy + 8y^2$, where $x^2 + y^2 = 25$,

d) $w = x + z$, where $x^2 + y^2 + z^2 = 1$,

e) $w = xyz$, where $x^2 + y^2 = 1$ and $x - z = 0$,

f) $w = x^2 + y^2 + z^2$, where $x + y + z = 1$ and $x^2 + y^2 - z^2 = 0$.

7. Find the point of the curve

$$x^2 - xy + y^2 - z^2 = 1, \qquad x^2 + y^2 = 1$$

nearest to the origin $(0, 0, 0)$.

8. Find the absolute minimum and maximum, if they exist, of the following functions:

a) $z = \dfrac{1}{1 + x^2 + y^2}$, all (x, y),

b) $z = xy$, $x^2 + y^2 \le 1$,

c) $w = x + y + z$, $x^2 + y^2 + z^2 \le 1$,

d) $w = e^{-x^2-y^2-z^2}$, all (x, y, z).

9. Determine whether the given quadratic form is positive definite:

a) $3x^2 + 2xy + y^2$,

b) $x^2 - xy - 2y^2$,

c) $\frac{5}{3}x_1^2 + \frac{4}{3}x_1x_2 + 2x_2^2 + \frac{4}{3}x_2x_3 + \frac{7}{3}x_3^2$.

10. Prove the validity of the criterion (2.143) for a minimum, under the conditions stated. [Hint: The function $\nabla_\alpha \nabla_\alpha f(x_0, y_0)$ is continuous in α for $0 \le \alpha \le 2\pi$ and has a minimum M_1 in this interval; by (2.143), $M_1 > 0$. By the Fundamental Lemma of Section 2.6, $\partial z/\partial x$ and $\partial z/\partial y$ have differentials at (x_0, y_0). Show that this implies that

$$\nabla_\alpha f(x, y) = \nabla_\alpha f(x_0, y_0) + s\nabla_\alpha \nabla_\alpha f(x_0, y_0) + \epsilon s = s\nabla_\alpha \nabla_\alpha f(x_0, y_0) + \epsilon s,$$

where $x = x_0 + s\cos\alpha$, $y = y_0 + s\sin\alpha$ $(s > 0)$ and $|\epsilon|$ can be made as small as desired by choosing s sufficiently small. Choose δ so that $|\epsilon| < \frac{1}{2}M_1$ for $0 < s < \delta$ and show that

$$\nabla_\alpha f(x, y) = s\left[\nabla_\alpha \nabla_\alpha f(x_0, y_0) + \epsilon\right] > 0 \qquad \text{for } 0 < s < \delta.$$

Accordingly, f increases steadily, as one recedes from (x_0, y_0) on a straight line in the neighborhood of radius δ of (x_0, y_0).]

11. (*The method of least squares*) Let five numbers e_1, e_2, e_3, e_4, e_5 be given. It is in general impossible to find a quadratic expression $f(x) = ax^2 + bx + c$ such that $f(-2) = e_1$, $f(-1) = e_2$, $f(0) = e_3$, $f(1) = e_4$, $f(2) = e_5$. However, one can try to make the "total square-error"

$$E = \left(f(-2) - e_1\right)^2 + \left(f(-1) - e_2\right)^2 + \left(f(0) - e_3\right)^2$$
$$+ \left(f(1) - e_4\right)^2 + \left(f(2) - e_5\right)^2$$

as small as possible. Determine the values of a, b, and c that make E a minimum. This is the method of *least squares*, which is basic in the theory of statistics and in curve-fitting (see Chapter 7).

12. Find the shortest distance d between the two skew lines

$$x = 1 + 2t, \quad y = 2 - 3t, \quad z = 5t \quad \text{and} \quad x = 2 + \tau, \quad y = 3 + 2\tau, \quad z = 1 - \tau$$

by minimizing the squared distance function for variable points on the two lines.

13. Let f be a positive definite quadratic form $\sum\sum a_{ij}x_ix_j$ in $(x_1, \ldots, x_n)$ $(a_{ij} = a_{ji})$. Show that each level set $E : f(x_1, \ldots, x_n) = c$ $(c > 0)$ is a bounded closed set [Hint: Let $\lambda_0 > 0$ be the smallest eigenvalue of $A = (a_{ij})$ and show that $f(\mathbf{x}) = c$ implies $|\mathbf{x}|^2 \le \lambda_0/c$. This shows that E is bounded. To show E is closed, proceed as in Problem 7 following Section 2.4 to show that if $\mathbf{x}_0$ is not in E, so that $\pm(f(\mathbf{x}_0) - c) > 0$, then some neighborhood of $\mathbf{x}_0$ has no point in E.]

14. Let A and B be symmetric $n \times n$ matrices and let $f(x_1, \ldots, x_n) = \mathbf{x}'A\mathbf{x}$, $g(x_1, \ldots, x_n) = \mathbf{x}'B\mathbf{x}$ be the corresponding quadratic forms. Let g be positive definite. Use Lagrange multipliers to seek critical points of f on the set $g(x_1, \ldots, x_n) = 1$ (a bounded closed set by Problem 13).

a) Show that, at each critical point $\mathbf{x}, A\mathbf{x} - \lambda B\mathbf{x} = \mathbf{0}$ and $\mathbf{x}'B\mathbf{x} = 1$, where λ is a root of the n-th degree equation

$$\det(A - \lambda B) = 0, \tag{*}$$

and that $f(\mathbf{x}) = \lambda$.

b) Show that if λ_1, λ_2 are distinct roots of $(*)$ and $\mathbf{x}_1, \mathbf{x}_2$ are corresponding choices of $\mathbf{x}$, then $\mathbf{x}_1'B\mathbf{x}_2 = 0$. (Compare Problem 15 following Section 1.12.)

15. Show that Problem 14 can be reduced to the problem of Section 2.21 by a suitable linear transformation $\mathbf{x} = E\mathbf{z}$, where E is a nonsingular $n \times n$ matrix. [Hint: First set $\mathbf{x} = C\mathbf{y}$, where C is an orthogonal matrix such that $C'BC = B_1 = \operatorname{diag}(\mu_1, \ldots, \mu_n)$, with all $\mu_j > 0$, so that

$$g(C\mathbf{y}) = \mu_1 y_1^2 + \cdots + \mu_n y_n^2 = g_1(\mathbf{y})$$

as in Section 1.12. Then set $\mathbf{y} = D\mathbf{z}$, where $D = \operatorname{diag}(1/\mu_1^{1/2}, \ldots, 1/\mu_n^{1/2})$, so that

$$g_1(D\mathbf{z}) = z_1^2 + \cdots + z_n^2 = g_2(\mathbf{z}).$$

Verify that x satisfies the critical point conditions $Ax - \lambda Bx = 0$, $x'Bx = 1$, if and only if z satisfies the conditions

$$A_2z - \lambda z = 0, \quad z'z = 1,$$

where $A_2 = D'A_1D$, $A_1 = C'AC$, so that A_2 is symmetric, and that these conditions are those for a critical point of $f_2(z) = z'A_2z$, with A_2 symmetric, on the unit sphere. Finally remark that $x = Ez$, where $E = CD$ and E is nonsingular.]

Remark.   A further orthogonal transformation $z = C_1w$ diagonalizes A_2. The combined nonsingular transformation $x = Fw$, with $F = CDC_1$, diagonalizes *both* coefficient matrices A, B of the given quadratic forms.

*2.22   FUNCTIONAL DEPENDENCE

Throughout this chapter the condition that a derivative or Jacobian be zero has played an important role. Thus in Section 2.10 the condition that the Jacobian *not* be zero was needed to obtain the derivatives of implicit functions; in Section 2.19 the condition that all partial derivatives be zero was used to locate critical points. In this section we consider these questions from a more general point of view, with emphasis on certain extreme cases that are of importance.

Let $w = f(x, y)$ be given in a domain D. If $\nabla w = (\partial f/\partial x)\mathbf{i} + (\partial f/\partial y)\mathbf{j}$ is not $\mathbf{0}$ in D, then the level curves $f(x, y) = \text{const}$ of w are well-defined curves, one through each point of D. This follows from the implicit function theorem of Section 2.11; at each point (x_1, y_1), either $\partial f/\partial x \neq 0$ or $\partial f/\partial y \neq 0$, so that (2.70) gives a definite derivative for the implicit function. The family of level curves is thus without singularity, as shown in Fig. 2.22.

Figure 2.22   Level curves of $f(x, y)$ with $\nabla f \neq \mathbf{0}$.

Singularities will be introduced if $\nabla w = \mathbf{0}$ at certain points of D. These are precisely the critical points considered in Section 2.19. Figure 2.21 illustrates some of the possible complications.

The extreme case is that in which $\nabla w \equiv \mathbf{0}$ in D, that is, every point of D is a critical point. Here $\partial f/\partial x \equiv 0$ and $\partial f/\partial y \equiv 0$ in D, and one concludes that f is constant in D:

THEOREM Let $f(x, y)$ be defined in domain D and let

$$\frac{\partial f}{\partial x} \equiv 0, \qquad \frac{\partial f}{\partial y} \equiv 0 \qquad\qquad (2.157)$$

in D. Then there is a constant c such that

$$f \equiv c \qquad\qquad (2.158)$$

in D.

Proof. Let P_1: (x_1, y_1) and P_2: (x_2, y_2) be two points of D such that the line segment $P_1 P_2$ lies in D. Let P: (x, y) vary on this segment and let s be the distance $P_1 P$. The directional derivative of f at P in the direction $\overrightarrow{P_1 P_2}$ is then equal to df/ds; since $\partial f/\partial x \equiv 0$, $\partial f/\partial y \equiv 0$, one has $df/ds \equiv 0$. Hence by the familiar theorem for functions of one variable, f, as a function of s, is constant on $P_1 P_2$. Thus $f(x_1, y_1) = f(x_2, y_2) = c$ for some c. Every point of D can be joined to P_1 by a broken line (cf. Section 2.2 and Fig. 2.1); by repetition of the argument just given, one concludes that $f(x, y) = f(x_1, y_1) = c$ for every point (x, y) of D. The theorem is now proved. $\square$

Now let two functions $u = f(x, y)$, $v = g(x, y)$ be given in D. Suppose that $\nabla f \neq \mathbf{0}$ in D and $\nabla g \neq \mathbf{0}$ in D, so that both functions have well-defined level curves, as in Fig. 2.22. "In general" the two families of level curves determine curvilinear coordinates in D and a mapping of D onto a *domain* in the uv-plane; this will be the case if the Jacobian $\partial(f, g)/\partial(x, y) \neq 0$ in D, as a study of the theorem on implicit functions shows. It should be noted that the condition $\partial(f, g)/\partial(x, y) = 0$ is the same as the condition $\nabla f \times \nabla g = \mathbf{0}$, that is, that ∇f and ∇g are collinear vectors. For

$$\nabla f \times \nabla g = \begin{vmatrix} \mathbf{i} & \mathbf{j} & \mathbf{k} \\ \dfrac{\partial f}{\partial x} & \dfrac{\partial f}{\partial y} & 0 \\ \dfrac{\partial g}{\partial x} & \dfrac{\partial g}{\partial y} & 0 \end{vmatrix} = \mathbf{k}\,\frac{\partial(f, g)}{\partial(x, y)}.$$

When these vectors are collinear, the level curves $f = $ const and $g = $ const are tangent, the curvilinear coordinates are interfered with, and the mapping to the uv-plane may degenerate.

The extreme case here is that for which $\nabla f \times \nabla g \equiv \mathbf{0}$ in D or, equivalently, $\partial(f, g)/\partial(x, y) \equiv 0$ in D. In this case, *each level curve of f is a level curve of g and vice versa.* For a tangent vector to a level curve $f = $ const is a vector perpendicular to the normal vector $(\partial f/\partial x)\mathbf{i} + (\partial f/\partial y)\mathbf{j}$; thus $\mathbf{T} = -(\partial f/\partial y)\mathbf{i} + (\partial f/\partial x)\mathbf{j}$

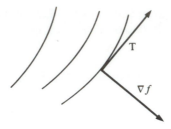

Figure 2.23 Tangent to level curves.

is such a tangent vector, as shown in Fig. 2.23. The component of ∇g in the direction of this tangent is 0, since

$$\nabla g \cdot \mathbf{T} = \frac{\partial g}{\partial x}\left(-\frac{\partial f}{\partial y}\right) + \frac{\partial g}{\partial y}\frac{\partial f}{\partial x} = \frac{\partial(f, g)}{\partial(x, y)} \tag{2.159}$$

and the Jacobian is 0 by assumption. Thus the directional derivative of g along the curve is 0, and g must be constant. Therefore *if $\partial(f, g)/\partial(x, y) \equiv 0$ in D, then the level curves of f and g coincide. Conversely, if f and g have the same level curves, then $\partial(f, g)/\partial(x, y) \equiv 0$ in D,* for the argument just given can be reversed.

EXAMPLE 1 Let $f(x,y) = e^x \sin y$ and $g(x,y) = x + \log \sin y$ for $0 < x < 1$ and $0 < y < \pi$. Then

$$\frac{\partial(f, g)}{\partial(x, y)} = \begin{vmatrix} e^x \sin y & e^x \cos y \\ 1 & \cot y \end{vmatrix} = e^x \cos y - e^x \cos y \equiv 0. \quad \blacksquare$$

In this example the functions f and g are related by an identity:

$$\log f(x, y) - g(x, y) \equiv 0$$

in the domain considered; that is, g is simply the function $\log f$, a "function of the function." From this fact it is clear that on a level curve $f = \text{const} = c$, one must also have $g = \text{const} = \log c$. In general, two functions f and g related by an identity

$$F[f(x, y), g(x, y)] \equiv 0 \tag{2.160}$$

in a given domain D, are termed *functionally dependent* in D. Here $F[u, v]$ is a function of the variables, u, v such that $F[f(x, y), g(x, y)]$ is defined in D, and in order to rule out degenerate cases we assume $\nabla F \neq \mathbf{0}$ for the range of u, v involved.

THEOREM If $f(x, y)$ and $g(x, y)$ are differentiable in the domain D and are functionally dependent in D, then

$$\frac{\partial(f, g)}{\partial(x, y)} \equiv 0, \tag{2.161}$$

so that the level curves of f and g coincide. Conversely, if (2.161) holds and $\nabla f \neq \mathbf{0}$, $\nabla g \neq \mathbf{0}$, then, in some neighborhood of each point of D, f and g are functionally dependent.

Proof. Let f and g be functionally dependent in D, so that (2.160) holds for a suitable $F[u, v]$. On differentiating (2.160) with respect to x and y and using the chain rules, one obtains the identities:

$$\frac{\partial F}{\partial u}\frac{\partial f}{\partial x} + \frac{\partial F}{\partial v}\frac{\partial g}{\partial x} \equiv 0, \qquad \frac{\partial F}{\partial u}\frac{\partial f}{\partial y} + \frac{\partial F}{\partial v}\frac{\partial g}{\partial y} \equiv 0. \qquad (2.162)$$

Since $\partial F/\partial u$ and $\partial F/\partial v$ are not both zero, these equations are consistent only if the "determinant of the coefficients" is zero (Section 1.4), that is, only if

$$\begin{vmatrix} \dfrac{\partial f}{\partial x} & \dfrac{\partial g}{\partial x} \\[2mm] \dfrac{\partial f}{\partial y} & \dfrac{\partial g}{\partial y} \end{vmatrix} \equiv 0. \qquad (2.163)$$

Thus (2.161) holds.

Conversely, let (2.161) hold, so that f and g have the same level curves (Fig. 2.24). The equations

$$u = f(x, y), \qquad v = g(x, y) \qquad (2.164)$$

then define a mapping from the xy-plane to the uv-plane. This mapping is degenerate, for along each level curve of f and g, u and v have constant values, so that the entire level curve maps into a *single point* in the uv-plane. If one considers a particular point (x_1, y_1) of D, then on proceeding from this point in a direction normal to the level curve, f must either increase or decrease, since $\nabla f \neq 0$; similarly, g must either increase or decrease. Thus a sufficiently small neighborhood in the xy-plane is mapped by (2.164) on a curve in the uv-plane expressible either as $u = \phi(v)$ or as $v = \psi(u)$. Thus $f(x,y) - \phi(g(x,y)) \equiv 0$ in the neighborhood. Along the normal, in terms of a suitable coordinate t, f and g are differentiable functions of t and at (x_1, y_1) $u'(t) = \partial f/\partial n = \nabla f \cdot \mathbf{n} \neq 0$ as in Eq. (2.117) and similarly $v'(t) \neq 0$, so that $du/dv = u'(t)/v'(t) = \phi'(v)$. Hence ϕ is differentiable, as is $F(u,v) = u - \phi(v)$ and f, g are functionally dependent in the neighborhood. $\square$

The proof just given brings out the significance of the condition $\partial(f, g)/\partial(x, y) \equiv 0$ for the mapping (2.164): This mapping maps D not onto a domain but onto a *curve* or several curves. For the functions $f = e^x \sin y$, $g = x + \log \sin y$ the corresponding curve is given by part of the graph of $\log u - v = 0$.

The results obtained can be generalized to the case of three functions of three variables or, in general, to n functions of n variables. Thus for three functions of three variables, functional dependence

$$F[f(x, y, z), g(x, y, z), h(x, y, z)] \equiv 0$$

is equivalent as previously to the condition

$f = a_4, g = b_4$

$f = a_3, g = b_3$

$f = a_2, g = b_2$

$f = a_1, g = b_1$

Figure 2.24 Level curves of functionally dependent functions.

$$\frac{\partial(f, g, h)}{\partial(x, y, z)} \equiv 0.$$

This in turn is equivalent to the statement that the three vectors $\nabla f, \nabla g, \nabla h$ are coplanar at each point, so that the three families of level surfaces have a common tangent direction at each point.

One can also consider the case of n functions of m variables: $f_1(x_1,\ldots,x_m),\ldots,$ $f_n(x_1,\ldots,x_m)$ and allow for several functional dependencies among them.

Linear case.   Here $f_i(x_1,\ldots) = a_{i\,1}x_1 + \cdots + a_{im}x_m$ for $i = 1,\ldots,n$ or, in vector notation, $\mathbf{f}(\mathbf{x}) = A\mathbf{x}$, with $A = (a_{ij})$ a constant $n \times m$ matrix. We restrict attention to linear dependencies, of form

$$z_1 f_1 + \cdots + z_n f_n \equiv 0, \tag{2.165}$$

where $z_1,\ldots,z_n$ are constants, not all 0. (A general functional dependence $F(f_1,\ldots,$ $f_n) \equiv 0$ implies a linear one; see Problem 11 below.) Equation (2.165) is equivalent to $\mathbf{z}'A\mathbf{x} = 0$ for all $\mathbf{x}$, where $\mathbf{z} = \mathrm{col}(z_1,\ldots,z_n)$, and hence to $\mathbf{z}'A = \mathbf{0}$ or

$$A'\mathbf{z} = \mathbf{0}, \tag{2.166}$$

If A has rank r, then as in Section 1.16 the solutions $\mathbf{z}$ of (2.166) form a subspace W of dimension $n - r$ of V^n. We choose a basis $\mathbf{z}_1,\ldots,\mathbf{z}_{n-r}$ for W and obtain $n - r$ independent linear dependencies:

$$z_{\alpha 1} f_1 + \cdots + z_{\alpha n} f_n \equiv 0, \qquad \alpha = 1,\ldots,n - r, \tag{2.167}$$

where $\mathbf{z}_\alpha = \mathrm{col}(z_{\alpha 1},\ldots,z_{\alpha n})$, and all other linear dependencies are linear combinations of these. If $n \le m$ and A has maximal rank n, then there are no such relations and the n functions f_i are functionally independent. If $m < n$, then $r \le m < n$ and the dependencies exist. In all cases A maps V^m onto an r-dimensional subspace of V^n and the mapping is one-to-one precisely when $r = m$.

General case.   The results for the linear case extend in a natural way to the general case of n functions $f_i(x_1,\ldots,x_m)(i = 1,\ldots,n)$, defined and differentiable in domain D. Functional dependence of $f_1, \ldots, f_n$ in D is understood to mean the existence of h functions $F_j(u_1, \ldots, u_n)$ $(j = 1, \ldots, h)$, defined and differentiable in a domain D_u of E^n, such that the composite functions

$$F_j\big(f_1(x_1, \ldots), \ldots, f_n(x_1, \ldots)\big), \quad j = 1, \ldots, h, \tag{2.168}$$

are defined in D and are identically 0 in D. Here it is assumed that the Jacobian matrix $(\partial F_j / \partial u_k)$ has maximum rank h in D_u, so that $h \le n$.

When these conditions hold, differentiation yields the equations

$$\sum_{k=1}^{n} \frac{\partial F_j}{\partial u_k} \frac{\partial f_k}{\partial x_\ell} = 0, \quad j = 1, \ldots, h, \quad \ell = 1, \ldots, m. \tag{2.169}$$

We denote the Jacobian matrix $(\partial f_k / \partial x_\ell)$ by A. Equation (2.169) then asserts that $A' \mathbf{z}_j = 0$ at each point $\mathbf{x}$ of D, where A is evaluated at $\mathbf{x}$ and $\mathbf{z}_j'$ is the j-th row of $(\partial F_j / \partial u_k)$, evaluated at the corresponding point $\mathbf{u}$ of D_u. Since the last matrix is assumed to have rank h, $\mathbf{z}_1, \ldots, \mathbf{z}_h$ are linearly independent; accordingly, $h \le n - r$, where r is the rank of A. Therefore, $r \le n - h$:

THEOREM Functional dependence:

$$F_j\big(f_1(x_1, \ldots, x_m), \ldots, f_n(x_1, \ldots, x_m)\big) \equiv 0 \quad \text{in} \quad D \tag{2.170}$$

for $j = 1, \ldots, h$, implies that the Jacobian matrix $A = (\partial f_i / \partial x_j)$ has rank $r \le n - h$ at each point of D.

To obtain a converse, we assume that A has constant rank $r < n$ in D. Then one can prove that, in an appropriate neighborhood D_1 of each point $\mathbf{x}^0$ of D, the functions $f_1, \ldots, f_n$ are functionally dependent, so that they satisfy relations of form (2.170) in D_1 with $h = n - r$, where $F_1, \ldots, F_h$ are as above. In fact, one can show that, if $f \le n$, then in terms of appropriate curvilinear coordinates $\xi_1, \ldots, \xi_m$ in D_1 and $\eta_1, \ldots, \eta_n$ in a neighborhood of the corresponding point $\mathbf{u}^0 = \mathbf{f}(\mathbf{x}^0)$, the mapping $\mathbf{u} = \mathbf{f}(\mathbf{x})$ becomes simply the linear mapping

$$\eta_1 = \xi_1, \ldots, \eta_r = \xi_r, \quad \eta_{r+1} = 0, \ldots, \eta_n = 0.$$

In the original coordinates $\eta_j = \eta_j(u_1, \ldots, u_n)$ and the functional dependencies are as in (2.170), with $F_1(u_1, \ldots, u_n) = \eta_{r+1}, \ldots, F_h(u_1, \ldots, u_n) = \eta_n(u_1, \ldots, u_n)$. If $r = n$, there are no functional dependencies.

For proofs of the results just stated one is referred to pp. 86–87 of Differential Geometry, Lie Groups, and Symmetric Spaces, by S. Helgason (Academic Press, Orlando, 1978).

The range of the mapping $\mathbf{u} = \mathbf{f}(\mathbf{x})$ just considered is often a smooth curve, surface or hypersurface in E^n. As the expression of the mapping in local coordinates shows, S has dimension r. One calls the mapping a *submersion* of D onto S.

EXAMPLE 2   Let $f_1(x_1,x_2,x_3) = f_2(x_1,x_2,x_3) = \sin(x_1 + x_2 + x + 3), f_3(x_1,x_2,x_3) = \cos(x_1 + x_2 + x_3)$. Then $A = (\partial f_i/\partial x_j)$ has rank 1 in E^3 and there are two dependencies $F_j(f_1,f_2,f_3) \equiv 0$; one choice with $(\partial F_j/\partial u_k)$ of rank 2 is

$$F_1(u_1,u_2,u_3) = u_1 - u_2, \qquad F_2(u_1,u_2,u_3) = u_1^2 + u_2^2 + 2u_3^2 - 2.$$

(See Problem 12 below.)   ■

PROBLEMS

1. A function $f(x, y)$, defined for all (x, y), satisfies the conditions

$$f(x,0) = \sin x, \qquad \frac{\partial f}{\partial y} \equiv 0.$$

   Evaluate $f(\pi/2, 2), f(\pi, 3), f(x, 1)$.

2. Two functions $f(x, y)$ and $g(x, y)$ are such that

$$\nabla f \equiv \nabla g$$

   in a domain D. Show that

$$f \equiv g + c$$

   for some constant c.

3. Determine all functions $f(x, y)$ whose second partial derivatives are identically 0.

4. A function $f(x, y)$, defined for all (x, y), is such that $\dfrac{\partial f}{\partial y} \equiv 0$. Show that there is a function $g(x)$ such that

$$f(x, y) \equiv g(x).$$

5. Determine all functions $f(x, y)$ such that $\dfrac{\partial^2 f}{\partial x \, \partial y} \equiv 0$ for all (x, y). [Hint: Cf. Problem 4.]

6. Show that the following sets of functions are functionally dependent:

   a) $f = \dfrac{y}{x}, \quad g = \dfrac{x-y}{x+y}$;

   b) $f = x^2 + 2xy + y^2 + 2x + 2y, \quad g = e^x e^y$;

   c) $f = x^2 y - xy^2 + xyz, \quad g = xy + x - y + z, \quad h = x^2 + y^2 + z^2 - 2yz + 2xz$;

   d) $f = u + v - x, \quad g = x - y + u, \quad h = u - 2v + 5x - 3y.$

7. Find an identity relating each of the sets of functions of Problem 6.

8. Plot the level curves of the functions f and g of Problem 6(a) and (b).

9. Let $f(x, y)$ and $g(x, y, u)$ be such that

$$\frac{\partial f}{\partial x}\frac{\partial g}{\partial y} - \frac{\partial f}{\partial y}\frac{\partial g}{\partial x} = 0$$

   when $u = f(x, y)$. Then show that

$$f(x, y) \quad \text{and} \quad g[x, y, f(x, y)]$$

   are functionally dependent.

10. Let $u(x, y)$ and $v(x, y)$ be harmonic in a domain D and have no critical points in D. Show that if u and v are functionally dependent, then they are "linearly" dependent: $u = av + b$ for suitable constants a and b. [Hint: Assume a relation of form $u = f(v)$ and take the Laplacian of both sides.]

11. a) Show that if linear functions $u = ax + by$, $v = cx + dy$ are functionally dependent then they are linearly dependent: $hu(x,y) + kv(x,y) \equiv 0$ for suitable constants h and k. [Hint: Differentiate the identity $F(ax + by, cx + dy) \equiv 0$ with respect to x and y.]

   b) Generalize the result of part (a) to n functions of m variables.

12. With reference to Example 2 in Section 2.22, verify that A has rank 1 in E^3, that the F_j can be chosen as stated, that the range of $\mathbf{u} = \mathbf{f}(\mathbf{x})$ is an ellipse and that $(\partial F_j / \partial u_k)$ has rank 2 in a domain D_u containing the ellipse. (Here $\mathbf{u} = \mathbf{f}(\mathbf{x})$ is a submersion of E^3 onto the ellipse.)

*2.23   REAL VARIABLE THEORY · THEOREM ON MAXIMUM AND MINIMUM

In this section we prove some basic theorems on sets and on functions of real variables. As will be seen, the concept of sequence is important for the development. (By "sequence" we here mean "infinite sequence.")

   We take the real number system itself as known and think of real numbers as those representable as decimal expressions such as 2.3175, $-1.3333\ldots$, and $3.14159\ldots$. These include the integers $0, \pm 1, \pm 2, \ldots$, and every real number x satisfies $k \le x < k + 1$ for a unique integer k. They also include the *rational numbers*: numbers representable as p/q, where p and q are integers with $q \ne 0$. If we represent the rational numbers as points on a line, the number axis, then these points do not fill the line: there are infinitely many points left out, such as those representing the real numbers $\sqrt{2}, \pi$ and e and, in general, all the *irrational numbers* (those which are not rational). For each irrational number x and each positive integer q, we can find a unique integer k such that $k/q < x < (k + 1)/q$. We need only choose k as above so that $k < qx < k + 1$ (equality is excluded here); thus x is approximated by rational numbers above and below within distance $\epsilon = 1/q$.

   The sequence s_n is said to converge to the real number s (written $s_n \to s$) or have limit s if for each $\epsilon > 0$ there is an integer N for which $|s_n - s| < \epsilon$ for $n > N$. The limit, when it exists, is unique, since if also $s_n \to s'$, then

$$|s' - s| = |s' - s_n + s_n - s| \le |s' - s_n| + |s_n - s| < 2\epsilon$$

for n sufficiently large; this holds for every $\epsilon > 0$, so that $s' = s$.

   A sequence s_n is said to be *monotone increasing* if $s_n \le s_{n+1}$ for all n and *monotone decreasing* if $s_n \ge s_{n+1}$ for all n; if the inequality is strict ($<$ or $>$), then these are said to be *strictly increasing* or *strictly decreasing*. In all cases the sequence is called monotone.

   A sequence s_n is said to be *bounded above* if for all n, $s_n \le B$ for some number B and *bounded below* if for all n, $s_n \ge A$ for some number A; it is said to be *bounded* if it is bounded above and bounded below. We observe that a monotone increasing (decreasing) sequence bounded above (below) is necessarily bounded.

   THEOREM A   Every bounded monotone sequence converges.

Proof.   Let the sequence be s_n, $n = 1, 2, \ldots$ and let it be increasing, with $s_n \le B$

Figure 2.25   Convergence of bounded monotone sequence.

for all n. We then try to find s by locating it to the nearest integer below s, then to the nearest tenth and so on, as in Fig. 2.25.

In detail we let $h \le B < h + 1$, $c \le s_1 < c + 1$ for integers c, h, so that necessarily $c \le h$. Also $c \le s_n < h + 1$ for all n. We consider the finite set of integers $c, c + 1, \ldots, h$ and let p be the largest of these that is attained or exceeded by some s_n. Thus $p \le s_n < p + 1$ for n sufficiently large. We consider the finite set of numbers

$$p, \quad p + \tfrac{1}{10}, \quad p + \tfrac{2}{10}, \ldots, p + \tfrac{9}{10}$$

and let $p + (k_1/10)$ be the largest number in this set attained or exceeded by some s_n, so that k_1 is one of the integers $0, 1, \ldots, 9$, and

$$p + \frac{k_1}{10} \le s_n < p + \frac{k_1 + 1}{10}$$

for n sufficiently large. Continuing in this way, we obtain a sequence $k_1, k_2, \ldots, k_m, \ldots$ of integers, each having one of the values $0, 1, \ldots, 9$. Let s be the real number $p + 0.k_1 k_2 \cdots k_m \cdots$. Then from our construction we have $s_n \le s$ for all n; for if $s_n > s$, then we would have failed to choose one of the k_i as large as possible. Also, for each m,

$$s_n \ge p + 0.k_1 k_2 \cdots k_m \ge s - 10^{-m}$$

for all n sufficiently large. Given $\epsilon > 0$, we choose m so large that $10^{-m} < \epsilon$ (m can be obtained from the decimal expression for ϵ). Thus for n sufficiently large, $s - \epsilon < s_n \le s$, so that $|s_n - s| < \epsilon$. Accordingly, $s_n \to s$.   □

A set E of real numbers is said to be *bounded above* if there is a number B such that $x \le B$ for all numbers x in E. We call B an *upper bound* of E. For example, if E is the set of numbers

$$\frac{1}{2}, \frac{2}{3}, \ldots, \frac{n}{n + 1}, \ldots,$$

then E is bounded above, and $B = 1$ or any number larger than 1 is an upper bound of E. Similarly, E is *bounded below* if $x \ge B$ for all x in E, for some number B; B is then a *lower bound* of E.

THEOREM B   Let E be a set of real numbers that is nonempty and bounded above. Then E has a least upper bound; that is, there is a number B_0 such that B_0 is an upper bound of E and $B_0 \le B$ for every upper bound of E.

Proof.   We can imitate the proof of Theorem A. There are integers that are not upper bounds of E—for we can choose an integer less than some x in E (E is nonempty). There is a largest such integer, s_1. For E has an upper bound B, and every integer that is not an upper bound must be less than B. Similarly, there is a largest integer $k_1, 0 \le k_1 \le 9$, such that $s_2 = s_1 + (k_1/10)$ is not an upper bound of E. Continuing, we obtain a monotone increasing sequence s_n such that

$$s_n = s_1 + \frac{k_1}{10} + \cdots + \frac{k_{n-1}}{10^{n-1}}, \qquad 0 \le k_i \le 9,$$

and s_n is not an upper bound of E but $s_n + (1/10^{n-1})$ is an upper bound. It follows that for every x in E, for all n,

$$s_n + \frac{1}{10^{n-1}} \ge x. \tag{2.171}$$

Now the sequence s_n is bounded above (by B, as earlier). Therefore by Theorem A, s_n converges to some B_0. By (2.171),

$$\lim_{n \to \infty} \left(s_n + \frac{1}{10^{n-1}} \right) = B_0 \ge x,$$

so that B_0 is an upper bound of E. On the other hand, $s_n < B$ for all n implies that $B_0 \le B$. Therefore B_0 is the least upper bound (and is clearly unique).   □

COROLLARY TO THEOREM B   If a set E of real numbers is bounded below, then E has a greatest lower bound A_0.

The proof is left as an exercise (Problem 1 below).

One abbreviates the least upper bound of E and the greatest lower bound of E by

$$\text{lub } E, \quad \text{glb } E.$$

A sequence t_n is said to be a subsequence of s_n if $t_1 = s_{n_1}, t_2 = s_{n_2}, \ldots, t_k = s_{n_k}, \ldots$, where $n_1 < n_2 < \cdots < n_k < n_{k+1} < \cdots$. We remark that if the sequence s_n converges, then the subsequence t_n also converges, with the same limit (Problem 2).

THEOREM C   (*Weierstrass-Bolzano Theorem*)   Let s_n be a bounded sequence of real numbers. Then s_n has a convergent subsequence.

Proof.   Let E be the set of all real numbers x such that $s_n \ge x$ for infinitely many values of n. Since the sequence s_n satisfies $A \le s_n \le B$ for all n, for some A and B, E contains at least the number A, but E contains no number larger than B. Therefore E is bounded above. Let $s_0 = \text{lub } E$. Then any number s less than s_0 is in E. Otherwise, $s_n \ge s$ for only finitely many values of n, and hence no number x of E could be greater than s. That is, s would be an upper bound of E less than s_0, and that is impossible. Also, any number s greater than s_0 is not in E, since $s_0 = \text{lub } E$. Now we consider the intervals

$$s_0 - \frac{1}{2^k} < x < s_0 + \frac{1}{2^k} \qquad (k = 1, 2, \dots).$$

For each k, by the two cases just considered, $s_n > s_0 - 2^{-k}$ for infinitely many values of n, but $s_n \geq s_0 + 2^{-k}$ for only a finite set of values of n. Therefore at least one of the s_n, s_{n_k}, can be chosen to lie in the interval, and n_k can be made as large as desired. In particular, we can successively choose s_{n_1}, $s_{n_2}, \dots$ so that

$$s_0 - \tfrac{1}{2} < s_{n_1} < s_0 + \tfrac{1}{2}, \qquad s_0 - \tfrac{1}{4} < s_{n_2} < s_0 + \tfrac{1}{4}, \dots$$

and $n_1 < n_2 < \cdots < n_k < \cdots$. Clearly, s_{n_k} is the desired subsequence, converging to s_0 as $k \to \infty$. $\square$

We now consider sequences of points in E^N. It will be convenient to represent each point $P = (x_1, \dots, x_N)$ by the vector $\mathbf{x} = (x_1, \dots, x_N)$ of V^N. As in Section 2.4, we write $d(P, Q)$ for the distance between P and Q, so that

$$d(P, Q) = |\mathbf{x} - \mathbf{y}|$$

if $\mathbf{x}, \mathbf{y}$ are the corresponding vectors. As in Section 2.4, we have the rules

(i) $d(P, Q) \geq 0$, $\qquad d(P, Q) = 0 \qquad$ only for $P = Q$,

(ii) $d(P, Q) = d(Q, P)$,

(iii) $d(P, S) \leq d(P, Q) + d(Q, S)$.

Furthermore, as in Problem 10 following Section 2.4, if $\mathbf{x} = (x_1, \dots, x_N)$, then

(iv) $|\mathbf{x}| < \epsilon \quad$ implies $\quad |x_1| < \epsilon, \dots, |x_N| < \epsilon$,

(v) $|x_1| < \delta, \dots, |x_N| < \delta \quad$ together imply $\quad |\mathbf{x}| < N\delta$.

A sequence P_n of points in E^N is said to converge and have limit P_0 if

$$\lim_{n \to \infty} d(P_n, P_0) = 0.$$

We write $P_n \to P_0$. As before, the limit is unique. For $N = 1$, this is consistent with the definition given earlier of convergence of sequences of real numbers.

THEOREM D Let $P_n = (x_1^n, x_2^n, \dots, x_N^n)$ be a sequence of points in E^N. Then P_n converges to $P_0 = (x_1^0, x_2^0, \dots, x_N^0)$ if and only if

$$x_1^n \to x_1^0, \qquad x_2^n \to x_2^0, \dots, x_N^n \to x_N^0.$$

The proof, which is similar to that of Problem 10(c) and (d) following Section 2.4, is left as an exercise (Problem 3).

A set G in E^N is said to be *bounded* if $|\mathbf{x}| \leq K$ for some constant K, for all $\mathbf{x} = (x_1, \dots, x_n)$ in G. This is again consistent with the definition for $N = 1$ of bounded sets of real numbers.

THEOREM E   (*Weierstrass-Bolzano Theorem for E^N*) Let P_n be a sequence of points in the bounded set G of E^N. Then P_n has a convergent subsequence.

Proof.   For simplicity we take $N = 2$, so that $P_n = (x_1^n, x_2^n) = \mathbf{x}^n$. Since $|\mathbf{x}^n| \le K$, we conclude from Rule (iv) that $|x_1^n| \le K, |x_2^n| \le K$ for all n. Thus x_1^n is a bounded sequence of real numbers. By Theorem C it has a convergent subsequence $x_1^{n_k}$. The sequence

$$P_{n_k} = (x_1^{n_k}, x_2^{n_k}) \qquad (k = 1, 2, \ldots)$$

is a subsequence of P_n, and $x_2^{n_k}$ is a bounded sequence of real numbers. By Theorem C again, this sequence has a convergent subsequence. By relabeling we can again denote this by $x_2^{n_k}$; it is still a subsequence of x_2^n, and the corresponding sequence $x_1^{n_k}$ is a subsequence of a convergent sequence and therefore also converges (Problem 2 below). Therefore by Theorem D the new sequence $P_{n_k} = (x_1^{n_k}, x_2^{n_k})$ converges and is the desired subsequence of P_n.   □

THEOREM F   A set G in E^N is closed if and only if, for each convergent sequence P_n in G, the limit of the sequence is in G.

For the proof, see Problem 12 following Section 2.4.

We consider a mapping f from a set G in E^N into E^M. This is said to be continuous at P_0 in G if for each $\epsilon > 0$, there is a $\delta > 0$ such that $d(f(P), f(P_0)) < \epsilon$ whenever $d(P, P_0) < \delta$. The mapping f is said to be continuous in G if it is continuous at every point of G.

THEOREM G   Let f be a mapping of a set G in E^N into E^M. Then f is continuous at P_0 in G if and only if $f(P_n) \to f(P_0)$ for every sequence P_n in G converging to P_0.

Proof.   Let f be continuous at P_0 and let $P_n \to P_0$, with $P_1, P_2, \ldots$ all in G. Given $\epsilon > 0$, we choose $\delta > 0$ as in the definition of continuity above. Then we choose n_ϵ so that $d(P_n, P_0) < \delta$ for $n > n_\epsilon$. Then $d(f(P_n), f(P_0)) < \epsilon$ for $n > n_\epsilon$, so that $f(P_n) \to f(P_0)$.

Conversely, let f be such that $f(P_n) \to f(P_0)$ wherever $P_n \to P_0$. Suppose f is not continuous at P_0. Then for some $\epsilon > 0$ for *every* $\delta > 0$ there is some P in G such that $d(P, P_0) < \delta$ but $d(f(P), f(P_0)) \ge \epsilon$. We take δ successively equal to $1/2, 1/4, \ldots, 1/2^n, \ldots$ to obtain a sequence P_n such that $d(P_n, P_0) < 2^{-n}$ and $d(f(P_n), f(P_0)) \ge \epsilon$. Then clearly, $P_n \to P_0$, but $f(P_n)$ does not converge to $f(P_0)$, contrary to assumption. Hence f must be continuous at P_0.   □

THEOREM H   Let G be a bounded closed set in E^N. Let f be a continuous mapping of G into E^M. Then the range of f is also bounded and closed.

Proof. We observe first that to say that G is bounded is equivalent to requiring that $d(O, P) \leq K$, for some constant K, for all P in G; here O is the origin of E^N.

If the range of f is not bounded, then we can find a sequence P_n in G such that $d(O', f(P_n)) > n$ for $n = 1, 2, \ldots,$; here O' is the origin of E^M. By Theorem E, P_n has a convergent subsequence P_{n_k}, $P_{n_k} \to P_0$. By Theorem F, P_0 is in G. By Theorem G, $f(P_{n_k}) \to f(P_0)$ as $k \to \infty$. But $n_k < d(O', f(P_{n_k})) \leq d(O', f(P_0)) + d(f(P_0), f(P_{n_k}))$. As $k \to \infty$, the last term has limit 0, but $n_k \to \infty$. This would imply $\infty \leq d(O', f(P_0))$, which is impossible. Hence the range of f is bounded.

By Theorem F we show that the range if closed by showing that if $Q_n = f(P_n)$ is a convergent sequence in the range, then its limit is also in the range. As earlier, we choose a subsequence P_{n_k} converging to P_0 in G and then $f(P_{n_k}) \to f(P_0)$ by continuity. But the sequences $f(P_n)$ and $f(P_{n_k})$ have the same limit. Therefore $Q_n \to f(P_0)$, which is in the range. Therefore the range is closed. $\square$

THEOREM I   Let E be a nonempty, bounded closed set of real numbers. Then lub E and glb E belong to E.

Proof. Let $x_0 = \text{lub } E$. For each $k = 1, 2, \ldots$ there must be some x_k in E such that

$$x_0 - \frac{1}{2^k} < x_k \leq x_0.$$

Otherwise, x_0 could not be the least upper bound of E. Clearly, $x_k \to x_0$ as $k \to \infty$. Since E is closed, x_0 is in E, as was asserted. The proof is similar for glb E. $\square$

THEOREM J   Let G be a bounded closed set in E^N and let f be a continuous real-valued function on G, hence mapping G into E^1. Then f takes on an absolute maximum and minimum on G.

Proof. By Theorem H the range of f is a bounded, closed set F of real numbers. By Theorem I, F contains $a = \text{glb } F$ and $b = \text{lub } F$. Therefore $a \leq f(P) \leq b$ for all P in G and $f(P_1) = a$ for at least one point P_1, and $f(P_2) = b$ for at least one point P_2. Thus f takes on its absolute minimum, a, at P_1 and its absolute maximum, b, at P_2. $\square$

This theorem includes as special cases the theorems stated in Section 2.19: That a continuous function $f(x)$ on the interval $a \leq x \leq b$ and a continuous function $f(x, y)$ on a bounded closed region in the plane take on their absolute minimum and maximum.

Remark.   A bounded closed set in E^N is also called a *compact* set in E^N.

PROBLEMS

1. Prove the corollary to Theorem B.

2. Prove: If s_n is a convergent sequence of real numbers with limit c, then every subsequence of the sequence s_n converges to c.

3. Prove Theorem D.

4. Prove the following:

   a) A closed interval $a \leq x \leq b$ is a bounded closed set.

   b) If $y_1(x)$ and $y_2(x)$ are continuous for $a \leq x \leq b$ and $y_1(x) \leq y_2(x)$ for $a \leq x \leq b$, then the set G: $a \leq x \leq b$, $y_1(x) < y \leq y_2(x)$ is a bounded closed set. Is G a closed region? Explain.

5. Prove the intermediate value theorem for functions of one variable: If $f(x)$ is continuous for $a \leq x \leq b$ and $f(a) < 0$, $f(b) > 0$, then there exists an x_0, $a < x_0 < b$, such that $f(y_0) = 0$. [Hint: Let E be the set of x on the interval for which $f(x) < 0$ and let x_0 be lub E.]

Suggested References

Apostol, Tom M., Mathematical Analysis: A Modern Approach to Advanced Calculus, 2nd ed. Reading, Mass.: Addison-Wesley, 1974.

Buck, R. C., Advanced Calculus, 3rd ed. New York: McGraw-Hill, 1978.

Coddington, Earl A., and Norman Levinson, Theory of Ordinary Differential Equations. New York: McGraw-Hill, 1955.

Courant, Richard J., Differential and Integral Calculus, transl. by E. J. McShane, 2 vols. New York: Interscience, 1947.

Gantmacher, F. R., Theory of Matrices, transl. by K. A. Hirsch, 2 vols. New York: Chelsea Publishing Co., 1960.

Rudin, W., Principles of Mathematical Analysis. New York: McGraw-Hill, 1964.

Struik, Dirk J., Lectures on Classical Differential Geometry. Cambridge, Mass.: Addison-Wesley Press, Inc., 1950.

3

Vector
Differential
Calculus

3.1 INTRODUCTION

In Section 2.13 the path of a moving point P is described by giving its position vector $\mathbf{r} = \overrightarrow{OP}$ as a function of time t. It is shown how this vector function can be differentiated to give the velocity vector of the moving point. This operation can properly be termed a part of *vector differential calculus*.

In this chapter there is again a natural physical model—a fluid in motion. At each point of the fluid, one has a velocity vector $\mathbf{v}$, the velocity of the "fluid particle" located at the given point. Such vectors are defined for all points of the fluid and together form what is termed a *vector field*. This is illustrated in Fig. 3.1. The field may change with time or remain the same (*stationary* flow).

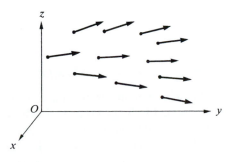

Figure 3.1   Vector field in space.

Figure 3.2   Vector field in the plane.

One can again trace the paths of the individual particles, the "stream lines," and determine the velocity vector $\mathbf{v} = d\mathbf{r}/dt$ for each. However, one can also consider the velocity field at a given time as describing a vector $\mathbf{v}$, which is a function of x, y, and z, that is, of position in space. The vector function $\mathbf{v}(x, y, z)$ can then be differentiated with respect to x, y, and z; that is, one can consider the rate and manner in which $\mathbf{v}$ varies from point to point in space.

The description of the variation of $\mathbf{v}$ turns out to require not merely partial derivatives, but special combinations of these, the *divergence* and *curl*. At each point of the field a scalar, div $\mathbf{v}$, the divergence of $\mathbf{v}$, and a vector, curl $\mathbf{v}$, will be defined. The divergence measures the net rate at which matter is being transported away from the neighborhood of each point, and the condition

$$\operatorname{div} \mathbf{v} \equiv 0$$

describes the *incompressible* flow of a fluid. The curl is essentially a measure of the *angular velocity* of the motion; in the case when the fluid is rotating as a rigid body with angular velocity ω about the z axis, the curl of $\mathbf{v}$ is everywhere equal to $2\omega\mathbf{k}$ (Problem 16 following Section 3.6).

There are other important physical examples of vector fields, such as force fields arising from gravitational attraction or from electromagnetic sources; the familiar experiment showing the effect of a magnet on iron filings illustrates the latter.

In many cases the vectors of the field are parallel to a fixed plane and form the same pattern in each plane parallel to this plane. In this case the study of the field can be reduced to a 2-dimensional problem. One is thus led to study *vector fields* in the plane, as illustrated in Fig. 3.2.

3.2   VECTOR FIELDS AND SCALAR FIELDS

If to each point (x, y, z) of a domain D in space a vector $\mathbf{v} = \mathbf{v}(x, y, z)$ is assigned, then a *vector field* is said to be given in D. Each vector $\mathbf{v}$ of the field can be regarded as a "*bound vector*" attached to the corresponding point (x, y, z). If $\mathbf{v}$ is expressed in terms of components $\mathbf{v} = v_x\mathbf{i} + v_y\mathbf{j} + v_z\mathbf{k}$, then these components will also vary from point to point, so that one has

$$\mathbf{v} = v_x(x, y, z)\mathbf{i} + v_y(x, y, z)\mathbf{j} + v_z(x, y, z)\mathbf{k}, \tag{3.1}$$

that is, each vector field is equivalent to a triple of scalar functions of the three variables x, y, z.

EXAMPLE 1   Let $\mathbf{v} = x\mathbf{i} + y\mathbf{j} + z\mathbf{k}$. Here $\mathbf{v} = \mathbf{0}$ at the origin; at other points, $\mathbf{v}$ is a vector pointing away from the origin, as illustrated in Fig. 3.3.   ■

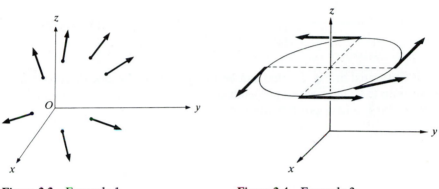

Figure 3.3   Example 1.   **Figure 3.4**   Example 2.

EXAMPLE 2   Let $\mathbf{v} = -y\mathbf{i} + x\mathbf{j}$. Here the vectors can be interpreted as the velocity vectors of a rigid rotation about the z axis. This is illustrated in Fig. 3.4.   ■

EXAMPLE 3   Let $\mathbf{F}$ denote the gravitational force exerted on a particle of mass m at $P: (x, y, z)$ by a mass M concentrated at the origin. Newton's law of gravitation gives

$$\mathbf{F} = -k\frac{Mm}{r^2}\frac{\mathbf{r}}{r}, \tag{3.2}$$

where $\mathbf{r} = \overrightarrow{OP}$ and k is a universal constant. Here $\mathbf{r}/r$ is a unit vector, so $\mathbf{F}$ has magnitude

$$F = \frac{kMm}{r^2},$$

and the force is inversely proportional to the square of the distance.   ■

The notion of vector field can be specialized to two dimensions. Thus a vector field $\mathbf{v}$ in a domain D of the xy-plane is given by

$$\mathbf{v} = v_x(x, y)\mathbf{i} + v_y(x, y)\mathbf{j}, \tag{3.3}$$

where $v_x(x, y)$ and $v_y(x, y)$ are two scalar functions of x and y defined in D. Two-dimensional fields arise in applications mainly in connection with *planar problems*, that is, problems concerning a vector field $\mathbf{v}$ in space such that $\mathbf{v}$ is always parallel to the xy-plane and $\mathbf{v}$ is independent of z; that is, $v_z = 0$ and v_x and v_y depend only on x and y, as in (3.3). From (3.3), one can construct the vectors $\mathbf{v}$ first in the xy-plane and then, using exactly the same vectors, in any plane parallel to the xy-plane.

EXAMPLE 4 Let $\mathbf{F}$ be the field

$$\mathbf{F} = \frac{1}{\left[(x+1)^2 + y^2\right]\left[(x-1)^2 + y^2\right]} \left[2(x^2 - y^2 - 1)\mathbf{i} + 4xy\mathbf{j}\right]. \quad (3.4)$$

This is illustrated in Fig. 3.5 and can be interpreted as the electric force field due to two infinite straight wires, perpendicular to the xy-plane at $(1, 0)$ and $(-1, 0)$, homogeneously and oppositely charged with electricity. ∎

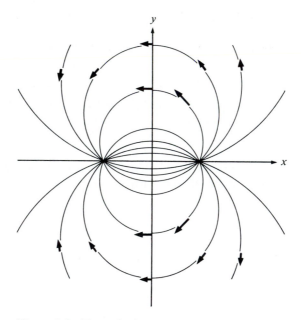

Figure 3.5 Example 4.

If we assign to each point of a domain D in space a scalar, rather than a vector, we obtain a *scalar field* in D. For example, the temperature at each point in a room defines a scalar field. If coordinates x, y, z are introduced, the scalar field determines a function $f(x, y, z)$ in D. One can also consider scalar fields in the plane; each such scalar field is described by a function $f(x, y)$.

It will be seen that scalar fields give rise to vector fields (for example, the field of the vector grad f) and that vector fields give rise to scalar fields (for example, the field of the scalar $|v|$).

3.3 THE GRADIENT FIELD

Let a scalar field f be given in space and a coordinate system chosen so that $f = f(x,y,z)$ and is defined and differentiable in a certain domain of space; thus the first partial derivatives of f exist in this domain. They form the components of the vector grad f, the *gradient* of the scalar f. Thus one has

$$\operatorname{grad} f = \frac{\partial f}{\partial x}\mathbf{i} + \frac{\partial f}{\partial y}\mathbf{j} + \frac{\partial f}{\partial z}\mathbf{k}. \tag{3.5}$$

For example, if $f = x^2 y - z^2$, then

$$\operatorname{grad} f = 2xy\mathbf{i} + x^2\mathbf{j} - 2z\mathbf{k}.$$

Formula (3.5) can be written in the following symbolic form:

$$\operatorname{grad} f = \left(\frac{\partial}{\partial x}\mathbf{i} + \frac{\partial}{\partial y}\mathbf{j} + \frac{\partial}{\partial z}\mathbf{k} \right) f, \tag{3.6}$$

where the suggested multiplication actually leads to a differentiation. The expression in parentheses is denoted by the symbol ∇ and is called "del" or "nabla." Thus

$$\nabla \equiv \frac{\partial}{\partial x}\mathbf{i} + \frac{\partial}{\partial y}\mathbf{j} + \frac{\partial}{\partial z}\mathbf{k}; \tag{3.7}$$

∇ is a "vector differential operator." By itself, the ∇ has no numerical significance; it takes on such significance when it is applied to a function, that is, in forming

$$\nabla f \equiv \operatorname{grad} f \equiv \frac{\partial f}{\partial x}\mathbf{i} + \frac{\partial f}{\partial y}\mathbf{j} + \frac{\partial f}{\partial z}\mathbf{k}. \tag{3.8}$$

The operator ∇ will be shown to be exceedingly useful.

It was shown in Section 2.14 that the directional derivative of the scalar f in the direction of the unit vector $\mathbf{u} = \cos \alpha \mathbf{i} + \cos \beta \mathbf{j} + \cos \gamma \mathbf{k}$ is given by

$$\nabla_u f = \nabla f \cdot \mathbf{u} = \frac{\partial f}{\partial x}\cos \alpha + \frac{\partial f}{\partial y}\cos \beta + \frac{\partial f}{\partial z}\cos \gamma. \tag{3.9}$$

This shows that grad f has a meaning independent of the coordinate system chosen: *Its component in a given direction represents the rate of change of f in that direction.* In particular, grad f points in the direction of maximum increase of f.

The gradient obeys the following laws:

$$\operatorname{grad}(f + g) = \operatorname{grad} f + \operatorname{grad} g, \tag{3.10}$$

$$\operatorname{grad}(fg) = f\operatorname{grad} g + g\operatorname{grad} f; \tag{3.11}$$

that is, with the ∇ symbol,

$$\nabla(f + g) = \nabla f + \nabla g, \qquad \nabla(fg) = f\nabla g + g\nabla f. \tag{3.12}$$

These hold, provided $\operatorname{grad} f$ and $\operatorname{grad} g$ exist in the domain considered. The proofs are left for the problems.

If f is a constant c, (3.11) reduces to the simpler condition:

$$\operatorname{grad}(cg) = c\operatorname{grad} g \qquad (c = \text{const}). \tag{3.13}$$

If the terms in z are dropped, the preceding discussion specializes at once to two dimensions. Thus for $f = f(x, y)$, one has

$$\operatorname{grad} f \equiv \nabla f \equiv \frac{\partial f}{\partial x}\mathbf{i} + \frac{\partial f}{\partial y}\mathbf{j},$$

$$\nabla \equiv \frac{\partial}{\partial x}\mathbf{i} + \frac{\partial}{\partial y}\mathbf{j}. \tag{3.14}$$

PROBLEMS

1. Sketch the following vector fields:

   a) $\mathbf{v} = (x^2 - y^2)\mathbf{i} + 2xy\mathbf{j}$, b) $\mathbf{u} = (x - y)\mathbf{i} + (x + y)\mathbf{j}$,

   c) $\mathbf{v} = -y\mathbf{i} + x\mathbf{j} + \mathbf{k}$, d) $\mathbf{v} = -x\mathbf{i} - y\mathbf{j} - z\mathbf{k}$.

2. Sketch the level curves or surfaces of the following scalar fields:

   a) $f = xy$, b) $f = x^2 + y^2 - z^2$, c) $f = e^{x+y-z}$.

3. Determine $\operatorname{grad} f$ for the scalar fields of Problem 2 and sketch several of the corresponding vectors.

4. Show that the gravitational field (3.2) is the gradient of the scalar

$$f = \frac{kMm}{r}.$$

5. Show that the force field (3.4) is the gradient of the scalar

$$f = \log \frac{\sqrt{(x - 1)^2 + y^2}}{\sqrt{(x + 1)^2 + y^2}}.$$

6. Prove (3.10) and (3.11).

7. Prove: If $f(x, y, z)$ is a composite function $F(u)$, where $u = g(x, y, z)$, then $\operatorname{grad} f = F'(u)\operatorname{grad} g$.

8. Prove: $\operatorname{grad} \dfrac{f}{g} = \dfrac{1}{g^2}[g\operatorname{grad} f - f\operatorname{grad} g]$.

9. If $f = f(x_1, \ldots, x_n)$, then the *Hessian matrix* of f is the matrix

$$H = \left(\frac{\partial^2 f}{\partial x_i \partial x_j}\right).$$

a) Find this matrix for the functions $w = x^3 y - y^3 z$ and $w = x_1^2 + 2x_1 x_2 + 5x_1 x_3 + 2x_2 x_1 + 4x_2^2 + x_2 x_3 + 5x_3 x_1 + x_3 x_2 + 2x_3^2$.

b) Show, under appropriate assumptions, that H is symmetric.

c) If $z = f(x, y)$, show that $\nabla_\alpha \nabla_\beta z = [\cos \alpha \ \cos \beta] H [\cos \alpha \ \cos \beta]'$ (see Section 2.21).

3.4   THE DIVERGENCE OF A VECTOR FIELD

Given a vector field v in a domain D of space, one has (for a given coordinate system) three scalar functions v_x, v_y, v_z. If these possess first partial derivatives in D, we can form in all nine partial derivatives, which we arrange to form a matrix:

$$\begin{bmatrix} \dfrac{\partial v_x}{\partial x} & \dfrac{\partial v_x}{\partial y} & \dfrac{\partial v_x}{\partial z} \\[2mm] \dfrac{\partial v_y}{\partial x} & \dfrac{\partial v_y}{\partial y} & \dfrac{\partial v_y}{\partial z} \\[2mm] \dfrac{\partial v_z}{\partial x} & \dfrac{\partial v_z}{\partial y} & \dfrac{\partial v_z}{\partial z} \end{bmatrix}.$$

From three of these the scalar $\operatorname{div} v$, the *divergence* of v, is constructed by the formula:

$$\operatorname{div} v = \frac{\partial v_x}{\partial x} + \frac{\partial v_y}{\partial y} + \frac{\partial v_z}{\partial z}. \tag{3.15}$$

It will be noted that the derivatives used form a diagonal (*principal* diagonal) of the matrix.

For example, if $v = x^2 i - xy j + xyz k$, then

$$\operatorname{div} v = 2x - x + xy = x + xy.$$

Formula (3.15) can be written in the symbolic form:

$$\operatorname{div} v = \nabla \cdot v; \tag{3.16}$$

for, treating ∇ as a vector, one has

$$\nabla \cdot v = \left(\frac{\partial}{\partial x} i + \frac{\partial}{\partial y} j + \frac{\partial}{\partial z} k \right) \cdot (v_x i + v_y j + v_z k)$$

$$= \frac{\partial v_x}{\partial x} + \frac{\partial v_y}{\partial y} + \frac{\partial v_z}{\partial z} = \operatorname{div} v.$$

The definition of the divergence at first appears to be quite arbitrary and to depend on the choice of axes in space. It will be seen in Section 3.8 that this is not the case. The divergence has in fact a definite physical significance. In fluid dynamics it appears as a measure of the rate of decrease of density at a point. More precisely, let $u = u(x, y, z, t)$ denote the velocity vector of a fluid motion and let $\rho = \rho(x, y, z, t)$ denote the density. Then $v = \rho u$ is a vector whose divergence satisfies the equation

$$\text{div}\,\mathbf{v} = -\frac{\partial \rho}{\partial t}. \qquad (3.17)$$

This is in fact the "continuity equation" of fluid mechanics. If the fluid is incompressible, this reduces to the simpler equation

$$\text{div}\,\mathbf{u} = 0. \qquad (3.18)$$

The law (3.17) will be established in Chapter 5; the derivation of (3.18) from (3.17) is considered in Problem 2 following Section 3.6.

The divergence also plays an important part in the theory of electromagnetic fields. Here the divergence of the electric force vector $\mathbf{E}$ satisfies the equation

$$\text{div}\,\mathbf{E} = 4\pi \rho, \qquad (3.19)$$

where ρ is the charge density. Thus where there is no charge, one has

$$\text{div}\,\mathbf{E} = 0. \qquad (3.20)$$

The divergence has the basic properties:

$$\text{div}\,(\mathbf{u} + \mathbf{v}) = \text{div}\,\mathbf{u} + \text{div}\,\mathbf{v}, \qquad (3.21)$$

$$\text{div}\,(f\mathbf{u}) = f\,\text{div}\,\mathbf{u} + \text{grad}\,f \cdot \mathbf{u}; \qquad (3.22)$$

that is, with the nabla symbol:

$$\nabla \cdot (\mathbf{u} + \mathbf{v}) = \nabla \cdot \mathbf{u} + \nabla \cdot \mathbf{v}, \qquad \nabla \cdot (f\mathbf{u}) = f(\nabla \cdot \mathbf{u}) + (\nabla f \cdot \mathbf{u}).$$

The proofs are left to the problems.

3.5 THE CURL OF A VECTOR FIELD

From the remaining six partial derivatives of the square array of Section 3.4, one constructs a new vector field, $\text{curl}\,\mathbf{v}$, by the definition:

$$\text{curl}\,\mathbf{v} = \left(\frac{\partial v_z}{\partial y} - \frac{\partial v_y}{\partial z}\right)\mathbf{i} + \left(\frac{\partial v_x}{\partial z} - \frac{\partial v_z}{\partial x}\right)\mathbf{j} + \left(\frac{\partial v_y}{\partial x} - \frac{\partial v_x}{\partial y}\right)\mathbf{k}. \qquad (3.23)$$

It will be noted that each component is formed of elements symmetrically placed relative to the principal diagonal. The curl can be expressed in terms of ∇, for one has

$$\text{curl}\,\mathbf{v} = \nabla \times \mathbf{v} = \begin{vmatrix} \mathbf{i} & \mathbf{j} & \mathbf{k} \\ \dfrac{\partial}{\partial x} & \dfrac{\partial}{\partial y} & \dfrac{\partial}{\partial z} \\ v_x & v_y & v_z \end{vmatrix}. \qquad (3.24)$$

The determinant must be expanded by minors of the first row, that is, so as to yield (3.23).

The fact that this vector field has a meaning independent of the choice of axes will also be shown in Section 3.8. The curl is important in the analysis of the velocity fields of fluid dynamics and in the analysis of electromagnetic force fields. The curl can be interpreted as measuring angular motion of a fluid (see Problem 16 following Section 3.6) and the condition

$$\text{curl } \mathbf{v} = \mathbf{0} \tag{3.25}$$

for a velocity field $\mathbf{v}$ characterizes what are termed *irrotational flows*. The analogous equation

$$\text{curl } \mathbf{E} = \mathbf{0} \tag{3.26}$$

for the electric force vector $\mathbf{E}$ holds when only electrostatic forces are present.

The curl satisfies the basic laws:

$$\text{curl } (\mathbf{u} + \mathbf{v}) = \text{curl } \mathbf{u} + \text{curl } \mathbf{v}, \tag{3.27}$$

$$\text{curl } (f\mathbf{u}) = f\text{curl } \mathbf{u} + \text{grad } f \times \mathbf{u}. \tag{3.28}$$

The proofs are left to the problems.

3.6 COMBINED OPERATIONS

As a result of the new definitions, we now have at our disposal the operations listed in Table 3.1.

The theory of vector algebra, discussed in Section 1.1, concerns the properties of the algebraic operations and their combinations. The theory of vector differential calculus concerns the theory of the differential operations (g) to (k) and their combinations with each other and with the algebraic operations (a) to (f).

TABLE 3.1

Operation		Symbols	Where Discussed (Section No.)
Algebraic Operations	(a) sum of scalars	$f + g$	
	(b) product of scalars	fg	
	(c) sum of vectors	$\mathbf{u} + \mathbf{v}$	1.1
	(d) scalar times vector	$f\mathbf{u}$	1.1
	(e) scalar product	$\mathbf{u} \cdot \mathbf{v}$	1.1
	(f) vector product	$\mathbf{u} \times \mathbf{v}$	1.1
Differential Operations	(g) derivative of scalar	$\dfrac{df}{dt}, \dfrac{\partial f}{\partial x}$	2.5
	(h) derivative of vector	$\dfrac{d\mathbf{v}}{dt}$	2.13
	(i) gradient of scalar	$\nabla f \equiv \text{grad } f$	2.13, 3.3
	(j) divergence of vector	$\nabla \cdot \mathbf{v} \equiv \text{div } \mathbf{v}$	3.4
	(k) curl of vector	$\nabla \times \mathbf{v} \equiv \text{curl } \mathbf{v}$	3.5

The combinations of (g) and (h) with the algebraic operations are discussed in Problem 3 following Section 2.13.

The combinations of (i), (j), and (k) with (a) and (c) are discussed in Sections 3.3, 3.4, and 3.5. The results can be summarized in the one rule:

$$\text{operator on sum} = \text{sum of operators on terms}. \qquad (3.29)$$

The combinations of (i), (j), (k) with (b) and (d) are also discussed in Sections 3.3 to 3.5. The results include the important case of scalar constant times scalar or vector. Here we have the general rule:

$$\text{operator on scalar constant factor} = \text{scalar factor times operator}; \qquad (3.30)$$

that is, a scalar constant can be factored out. Thus $\nabla(cf) = c\nabla f$, $\nabla \cdot (c\mathbf{u}) = c\nabla \cdot \mathbf{u}$, and so on. The rules (3.29) and (3.30) characterize what are called *linear operators*; thus grad, div, and curl are linear operators.

If one considers the other possible combinations, one obtains a long list of identities, some of which will be considered here. The proofs are left to the problems. All derivatives occurring are to be assumed continuous.

Curl of a gradient. Here one has the rule:

$$\text{curl grad } f = \mathbf{0}. \qquad (3.31)$$

This relation is suggested by the fact that curl grad $f = \nabla \times (\nabla f)$, that is, has the appearance of the vector product of collinear vectors. There is an important converse:

$$\text{if curl } \mathbf{v} = \mathbf{0}, \quad \text{then} \quad \mathbf{v} = \text{grad } f \text{ for some } f; \qquad (3.32)$$

further assumptions are needed here, and the rule (3.32) must be used with caution. A proof and full discussion are given in Chapter 5. A vector field $\mathbf{v}$ such that curl $\mathbf{v} = \mathbf{0}$ is often termed *irrotational*.

Divergence of a curl. Here one concludes that

$$\text{div curl } \mathbf{v} = 0; \qquad (3.33)$$

this relation is again suggested by a vector identity, for div curl $\mathbf{v} = \nabla \cdot (\nabla \times \mathbf{v})$, so that one has an expression resembling a scalar triple product of coplanar vectors (Section 1.3). Again there is a converse:

$$\text{if div } \mathbf{u} = 0, \quad \text{then} \quad \mathbf{u} = \text{curl } \mathbf{v} \text{ for some } \mathbf{v}; \qquad (3.34)$$

as with (3.32) there are restrictions on the use of (3.34), and one is again referred to Chapter 5. A vector field $\mathbf{u}$ such that div $\mathbf{u} = 0$ is often termed *solenoidal*.

Divergence of a vector product. Here one has

$$\text{div} (\mathbf{u} \times \mathbf{v}) = \mathbf{v} \cdot \text{curl } \mathbf{u} - \mathbf{u} \cdot \text{curl } \mathbf{v}. \qquad (3.35)$$

Divergence of a gradient. If one expands in terms of components, one finds that

$$\text{div grad } f = \frac{\partial^2 f}{\partial x^2} + \frac{\partial^2 f}{\partial y^2} + \frac{\partial^2 f}{\partial z^2}. \qquad (3.36)$$

The expression on the right-hand side is known as the Laplacian of f and is also denoted by Δf or by $\nabla^2 f$, since div grad $f = \nabla \cdot (\nabla f)$. A function f (having continuous second derivatives) such that div grad $f = 0$ in a domain is called *harmonic* in that domain. The equation satisfied by f:

$$\frac{\partial^2 f}{\partial x^2} + \frac{\partial^2 f}{\partial y^2} + \frac{\partial^2 f}{\partial z^2} = 0 \qquad (3.37)$$

is called *Laplace's equation* (see Sections 2.15 and 2.17).

Curl of a curl. Here an expansion into components yields the relation:

$$\operatorname{curl} \operatorname{curl} \mathbf{u} = \operatorname{grad} \operatorname{div} \mathbf{u} - \left(\nabla^2 u_x \mathbf{i} + \nabla^2 u_y \mathbf{j} + \nabla^2 u_z \mathbf{k} \right). \qquad (3.38)$$

If one defines the Laplacian of a vector $\mathbf{u}$ to be the vector

$$\nabla^2 \mathbf{u} = \nabla^2 u_x \mathbf{i} + \nabla^2 u_y \mathbf{j} + \nabla^2 u_z \mathbf{k}, \qquad (3.39)$$

then (3.38) becomes

$$\operatorname{curl} \operatorname{curl} \mathbf{u} = \operatorname{grad} \operatorname{div} \mathbf{u} - \nabla^2 \mathbf{u}. \qquad (3.40)$$

This identity can be written as an expression for the *gradient of a divergence*:

$$\operatorname{grad} \operatorname{div} \mathbf{u} = \operatorname{curl} \operatorname{curl} \mathbf{u} + \nabla^2 \mathbf{u}. \qquad (3.41)$$

The identities listed here, together with those previously obtained, cover all of interest except for those for *gradient of a scalar product* and *curl of a vector product*; these two are considered in Problems 13 and 14.

PROBLEMS

1. Prove (3.21) and (3.22).

2. Prove that the continuity equation (3.17) can be written in the form

$$\frac{\partial \rho}{\partial t} + \operatorname{grad} \rho \cdot \mathbf{u} + \rho \operatorname{div} \mathbf{u} = 0$$

or, in terms of the Stokes derivative (Problem 10 following Section 2.8), thus:

$$\frac{D\rho}{Dt} + \rho \operatorname{div} \mathbf{u} = 0.$$

Prove that (3.17) reduces to (3.18) when $\rho \equiv$ const. It will be shown in Chapter 5 that the same simplification can be made when ρ is variable, provided that the fluid is incompressible. This follows from the fact that $D\rho/Dt$ measures the variation in density at a point moving with the fluid; for an incompressible fluid this local density cannot vary.

3. Prove (3.27) and (3.28).

4. Prove (3.31). Verify by applying to $f = \dfrac{1}{\sqrt{x^2 + y^2 + z^2}}$.

5. For the given vector field $\mathbf{v}$, verify that $\operatorname{curl} \mathbf{v} = \mathbf{0}$ and find all functions f such that $\operatorname{grad} f = \mathbf{v}$.

   a) $\mathbf{v} = 2xyz\mathbf{i} + x^2z\mathbf{j} + x^2y\mathbf{k}$.

   b) $\mathbf{v} = e^{xy}[(2y^2 + yz^2)\mathbf{i} + (2xy + xz^2 + 2)\mathbf{j} + 2z\mathbf{k}]$.

6. Prove (3.33). Verify by applying to $\mathbf{v} = x^2yz\mathbf{i} - x^3y^3\mathbf{j} + xyz^2\mathbf{k}$.

7. a) Given the vector field $\mathbf{v} = 2x\mathbf{i} + y\mathbf{j} - 3z\mathbf{k}$, verify that $\operatorname{div} \mathbf{v} = 0$. Find all vectors $\mathbf{u}$ such that $\operatorname{curl} \mathbf{u} = \mathbf{v}$. [Hint: First remark that on the basis of (3.32), all solutions of the equation $\operatorname{curl} \mathbf{u} = \mathbf{v}$ are given by $\mathbf{u} = \mathbf{u}_0 + \operatorname{grad} f$, where f is an arbitrary scalar and $\mathbf{u}_0$ is any one vector whose curl is $\mathbf{v}$. To find $\mathbf{u}_0$, assume $\mathbf{u}_0 \cdot \mathbf{k} = 0$.]

   b) Proceed as in (a) for $\mathbf{v} = y\mathbf{i} + z\mathbf{j} + x\mathbf{k}$.

8. Prove (3.36). Verify that the function f of Problem 4 is harmonic in space (except at the origin). [This function, which represents the electrostatic potential of a charge of $+1$ at the origin, is in a sense the fundamental harmonic function in space, for every harmonic function in space can be represented as a sum, or limit of a sum, of such functions.]

9. Prove (3.35).

10. Prove (3.38).

11. Prove the following identities:

   a) $\operatorname{div}[\mathbf{u} \times (\mathbf{v} \times \mathbf{w})] = (\mathbf{u} \cdot \mathbf{w})\operatorname{div} \mathbf{v} - (\mathbf{u} \cdot \mathbf{v})\operatorname{div} \mathbf{w} + \operatorname{grad}(\mathbf{u} \cdot \mathbf{w}) \cdot \mathbf{v} - \operatorname{grad}(\mathbf{u} \cdot \mathbf{v}) \cdot \mathbf{w}$,

   b) $\operatorname{div}(\operatorname{grad} f \times f \operatorname{grad} g) = 0$,

   c) $\operatorname{curl}(\operatorname{curl} \mathbf{v} + \operatorname{grad} f) = \operatorname{curl} \operatorname{curl} \mathbf{v}$,

   d) $\nabla^2 f = \operatorname{div}(\operatorname{curl} \mathbf{v} + \operatorname{grad} f)$.

   These should be established by means of the identities already found in this chapter and not by expanding into components.

12. One defines the scalar product $\mathbf{u} \cdot \nabla$, with $\mathbf{u}$ on the *left* of the operator ∇, as the operator

$$\mathbf{u} \cdot \nabla = u_x \frac{\partial}{\partial x} + u_y \frac{\partial}{\partial y} + u_z \frac{\partial}{\partial z}.$$

This is thus quite unrelated to $\nabla \cdot \mathbf{u} = \operatorname{div} \mathbf{u}$. The operator $\mathbf{u} \cdot \nabla$ can be applied to a scalar f:

$$(\mathbf{u} \cdot \nabla)f = u_x \frac{\partial f}{\partial x} + u_y \frac{\partial f}{\partial y} + u_z \frac{\partial f}{\partial z} = \mathbf{u} \cdot (\nabla f);$$

thus an associative law holds. The operator $\mathbf{u} \cdot \nabla$ can also be applied to a vector $\mathbf{v}$:

$$(\mathbf{u} \cdot \nabla)\mathbf{v} = u_x \frac{\partial \mathbf{v}}{\partial x} + u_y \frac{\partial \mathbf{v}}{\partial y} + u_z \frac{\partial \mathbf{v}}{\partial z},$$

where the partial derivatives $\partial \mathbf{v}/\partial x, \ldots$ are defined just as is $d\mathbf{r}/dt$ in Section 2.13; thus one has

$$\frac{\partial \mathbf{v}}{\partial x} = \frac{\partial v_x}{\partial x}\mathbf{i} + \frac{\partial v_y}{\partial x}\mathbf{j} + \frac{\partial v_z}{\partial x}\mathbf{k}.$$

   a) Show that if $\mathbf{u}$ is a unit vector, then $(\mathbf{u} \cdot \nabla)f = \nabla_u f$.

b) Evaluate $[(\mathbf{i} - \mathbf{j}) \cdot \nabla] f$.

c) Evaluate $[(x\mathbf{i} - y\mathbf{j}) \cdot \nabla](x^2\mathbf{i} - y^2\mathbf{j} + z^2\mathbf{k})$.

13. Prove the identity (cf. Problem 12):

$$\text{grad}\,(\mathbf{u} \cdot \mathbf{v}) = (\mathbf{u} \cdot \nabla)\mathbf{v} + (\mathbf{v} \cdot \nabla)\mathbf{u} + (\mathbf{u} \times \text{curl}\,\mathbf{v}) + (\mathbf{v} \times \text{curl}\,\mathbf{u}).$$

14. Prove the identity (cf. Problem 12):

$$\text{curl}\,(\mathbf{u} \times \mathbf{v}) = \mathbf{u}\,\text{div}\,\mathbf{v} - \mathbf{v}\,\text{div}\,\mathbf{u} + (\mathbf{v} \cdot \nabla)\mathbf{u} - (\mathbf{u} \cdot \nabla)\mathbf{v}.$$

15. Let $\mathbf{n}$ be the unit outer normal vector to the sphere $x^2 + y^2 + z^2 = 9$ and let $\mathbf{u}$ be the vector $(x^2 - z^2)(\mathbf{i} - \mathbf{j} + 3\mathbf{k})$. Evaluate $\partial/\partial n\,(\text{div}\,\mathbf{u})$ at $(2, 2, 1)$.

16. A rigid body is rotating about the z-axis with angular velocity ω. Show that a typical particle of the body follows a path

$$\overrightarrow{OP} = r\cos\,(\omega t + \alpha)\mathbf{i} + r\sin\,(\omega t + \alpha)\mathbf{j} + z\mathbf{k},$$

where α, r, and z are constant, and that at each instant the velocity is

$$\mathbf{v} = \boldsymbol{\omega} \times \overrightarrow{OP},$$

where $\boldsymbol{\omega} = \omega\mathbf{k}$ (the *angular velocity vector* of the motion). Evaluate $\text{div}\,\mathbf{v}$ and $\text{curl}\,\mathbf{v}$.

17. A steady fluid motion has velocity $\mathbf{u} = y\mathbf{i}$. Show that all points that move do so on straight lines and that the flow is incompressible. Determine the volume occupied at time $t = 1$ by the points which at time $t = 0$ fill the cube bounded by the coordinate planes and the planes $x = 1$, $y = 1$, $z = 1$.

18. A steady fluid motion has velocity $\mathbf{u} = x\mathbf{i}$. Show that all points either do not move or else move on straight lines. Determine the volume occupied at time $t = 1$ by the points which at time $t = 0$ fill the cube of Problem 17. [Hint: Show that the paths of the individual points are given by $x = c_1 e^t$, $y = c_2$, $z = c_3$, where c_1, c_2, c_3 are constants.] Is the flow incompressible?

*3.7   CURVILINEAR COORDINATES IN SPACE ▪ ORTHOGONAL COORDINATES

The discussion of the preceding sections has been confined to a fixed rectangular coordinate system in space. We now consider the extension of the theory to curvilinear coordinates, such as cylindrical or spherical coordinates.

New coordinates u, v, w can be introduced in space by equations

$$x = f(u, v, w), \qquad y = g(u, v, w), \qquad z = h(u, v, w). \qquad (3.42)$$

It will be assumed that f, g and h are defined and have continuous first partial derivatives in a domain D_1 of uvw-space and that Equations (3.42) can be solved uniquely for u, v and w as similar functions of x, y, z:

$$u = F(x, y, z), \qquad v = G(x, y, z), \qquad w = H(x, y, z). \qquad (3.43)$$

The inverse functions will be defined in a domain D of xyz-space, and we term

u, v, w *curvilinear coordinates* in D. We assume that the Jacobian

$$J = \frac{\partial(x, y, z)}{\partial(u, v, w)} \tag{3.44}$$

is positive in D_1.

If v and w are assigned constant values v_0 and w_0 while u is allowed to vary, equations (3.42) define a curve in D, on which u is a parameter. As the numbers v_0 and w_0 are varied, one obtains a family of curves in D, like the parallels to one of the axes in rectangular coordinates. For example, in spherical coordinates ρ, ϕ, θ, the curves $\phi = \phi_0$, $\theta = \theta_0$ (ρ variable) are rays through the origin.

If we write $\mathbf{r} = x\mathbf{i} + y\mathbf{j} + z\mathbf{k}$ for the position vector of a point (x, y, z), the equations (3.42) can be interpreted as defining a vector function $\mathbf{r} = \mathbf{r}(u, v, w)$. When $v = v_0$, $w = w_0$, this is the vector representation $\mathbf{r} = \mathbf{r}(u, v_0, w_0)$ of the curve of the preceding paragraph. The tangent vector to this curve is defined as in Section 2.13 to be the derivative of $\mathbf{r}$ with respect to the parameter u. Here we must write $\partial \mathbf{r} / \partial u$ for the derivative, to indicate that v and w are held constant. Similarly, $\partial \mathbf{r} / \partial v$ is tangent to a curve $u = \text{const}$, $w = \text{const}$, and $\partial \mathbf{r} / \partial w$ is tangent to a curve $u = \text{const}$, $v = \text{const}$. This is illustrated in Fig. 3.6.

We write:

$$\alpha = \left| \frac{\partial \mathbf{r}}{\partial u} \right|, \qquad \beta = \left| \frac{\partial \mathbf{r}}{\partial v} \right|, \qquad \gamma = \left| \frac{\partial \mathbf{r}}{\partial w} \right|. \tag{3.45}$$

Accordingly,

$$\alpha = \sqrt{ \left(\frac{\partial x}{\partial u} \right)^2 + \left(\frac{\partial y}{\partial u} \right)^2 + \left(\frac{\partial z}{\partial u} \right)^2 }$$

is the "speed," in terms of time u, with which a curve $v = v_0$, $w = w_0$ is traced, and $ds = \alpha \, du$ is the element of distance.

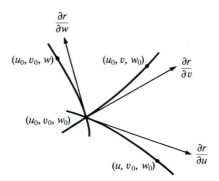

Figure 3.6   Curvilinear coordinates in space.

The tangent vectors $\partial\mathbf{r}/\partial u$, $\partial\mathbf{r}/\partial v$, $\partial\mathbf{r}/\partial w$ and Jacobian J can be expressed in terms of x, y, and z by the equations

$$\frac{\partial\mathbf{r}}{\partial u} = J(\nabla G \times \nabla H), \qquad \frac{\partial\mathbf{r}}{\partial v} = J(\nabla H \times \nabla F), \qquad \frac{\partial\mathbf{r}}{\partial w} = J(\nabla F \times \nabla G),$$

$$(3.46)$$

$$J = \frac{1}{\dfrac{\partial(u, v, w)}{\partial(x, y, z)}} = \frac{1}{\nabla F \cdot \nabla G \times \nabla H}. \qquad (3.47)$$

The gradients $\nabla F, \nabla G, \nabla H$ can be expressed in terms of u, v, w by the equations

$$\nabla F = \frac{\dfrac{\partial\mathbf{r}}{\partial v} \times \dfrac{\partial\mathbf{r}}{\partial w}}{J}, \qquad \nabla G = \frac{\dfrac{\partial\mathbf{r}}{\partial w} \times \dfrac{\partial\mathbf{r}}{\partial u}}{J}, \qquad \nabla H = \frac{\dfrac{\partial\mathbf{r}}{\partial u} \times \dfrac{\partial\mathbf{r}}{\partial v}}{J}. \qquad (3.48)$$

The proofs are left to Problems 1 to 3 following Section 3.8. Because of the assumption $J > 0$, the vectors $\partial\mathbf{r}/\partial u$, $\partial\mathbf{r}/\partial v$, $\partial\mathbf{r}/\partial w$ form a positive triple; because of (3.47), $\nabla F, \nabla G, \nabla H$ also form a positive triple.

We note two further sets of identities:

$$\frac{1}{J}\frac{\partial\mathbf{r}}{\partial u} = \text{curl}\,(G\nabla H), \qquad \frac{1}{J}\frac{\partial\mathbf{r}}{\partial v} = \text{curl}\,(H\nabla F), \qquad \frac{1}{J}\frac{\partial\mathbf{r}}{\partial w} = \text{curl}\,(F\nabla G);$$

$$(3.49)$$

$$\text{div}\left(\frac{1}{J}\frac{\partial\mathbf{r}}{\partial u}\right) = 0, \qquad \text{div}\left(\frac{1}{J}\frac{\partial\mathbf{r}}{\partial v}\right) = 0, \qquad \text{div}\left(\frac{1}{J}\frac{\partial\mathbf{r}}{\partial w}\right) = 0. \qquad (3.50)$$

The first reduces to (3.46) on application of the identity (3.28). The second then follows from (3.33).

The curvilinear coordinate system defined by (3.42) and (3.43) is said to be *orthogonal* if the tangent vectors $\partial\mathbf{r}/\partial u$, $\partial\mathbf{r}/\partial v$, $\partial\mathbf{r}/\partial w$ at each point of D form a triple of mutually perpendicular vectors. It will be seen that when this is the case, important simplifications in the formulas are possible. The most commonly used curvilinear coordinates are orthogonal, and for this reason we shall confine attention to this case. *In this and the following section the coordinates will be assumed to be orthogonal.*

As a first consequence of orthogonality, we remark that $(1/\alpha)\partial\mathbf{r}/\partial u$, $(1/\beta)\partial\mathbf{r}/\partial v$, $(1/\gamma)\partial\mathbf{r}/\partial w$ are a positive triple of mutually perpendicular unit vectors. Hence

$$\left(\frac{1}{\alpha}\frac{\partial\mathbf{r}}{\partial u}\right) \cdot \left(\frac{1}{\beta}\frac{\partial\mathbf{r}}{\partial v}\right) \times \left(\frac{1}{\gamma}\frac{\partial\mathbf{r}}{\partial w}\right) = 1.$$

Accordingly, by (3.44),

$$J = \frac{\partial\mathbf{r}}{\partial u} \cdot \frac{\partial\mathbf{r}}{\partial v} \times \frac{\partial\mathbf{r}}{\partial w} = \alpha\beta\gamma. \qquad (3.51)$$

By (3.48),

$$\alpha \nabla F = \frac{\alpha}{J}\left(\frac{\partial \mathbf{r}}{\partial v} \times \frac{\partial \mathbf{r}}{\partial w}\right) = \left(\frac{1}{\beta}\frac{\partial \mathbf{r}}{\partial v}\right) \times \left(\frac{1}{\gamma}\frac{\partial \mathbf{r}}{\partial w}\right) = \frac{1}{\alpha}\frac{\partial \mathbf{r}}{\partial u}.$$

A similar reasoning applies to ∇G and ∇H; we conclude that $\alpha \nabla F, \beta \nabla G, \gamma \nabla H$ are also mutually perpendicular unit vectors and

$$\alpha \nabla F = \frac{1}{\alpha}\frac{\partial \mathbf{r}}{\partial u}, \qquad \beta \nabla G = \frac{1}{\beta}\frac{\partial \mathbf{r}}{\partial v}, \qquad \gamma \nabla H = \frac{1}{\gamma}\frac{\partial \mathbf{r}}{\partial w}. \tag{3.52}$$

The surfaces $F = $ const, $G = $ const, $H = $ const must hence meet at right angles; they form what is called a *triply orthogonal* family of surfaces. Conversely, when the vectors $\nabla F, \nabla G, \nabla H$ are mutually perpendicular throughout D, the coordinates must be orthogonal (Problem 4 following Section 3.8).

A curve in D can be described by equations: $x = x(t), y = y(t), z = z(t)$, or, by (3.43), in terms of the curvilinear coordinates by the equations $u = u(t)$, $v = v(t), w = w(t)$. The element of arc ds on such a curve is defined by the equation

$$ds^2 = dx^2 + dy^2 + dz^2. \tag{3.53}$$

Hence

$$ds^2 = \left(\frac{\partial x}{\partial u}du + \frac{\partial x}{\partial v}dv + \frac{\partial x}{\partial w}dw\right)^2 + \left(\frac{\partial y}{\partial u}du + \cdots\right)^2 + \left(\frac{\partial z}{\partial u}du + \cdots\right)^2$$

$$= \left|\frac{\partial \mathbf{r}}{\partial u}\right|^2 du^2 + \left|\frac{\partial \mathbf{r}}{\partial v}\right|^2 dv^2 + \left|\frac{\partial \mathbf{r}}{\partial w}\right|^2 dw^2$$

$$+ 2\left(\frac{\partial \mathbf{r}}{\partial u} \cdot \frac{\partial \mathbf{r}}{\partial v}\right)du\,dv + 2\left(\frac{\partial \mathbf{r}}{\partial v} \cdot \frac{\partial \mathbf{r}}{\partial w}\right)dv\,dw + 2\left(\frac{\partial \mathbf{r}}{\partial w} \cdot \frac{\partial \mathbf{r}}{\partial u}\right)dw\,du.$$

Since the coordinates are orthogonal, we conclude:

$$ds^2 = \alpha^2\,du^2 + \beta^2\,dv^2 + \gamma^2\,dw^2. \tag{3.54}$$

Now $\alpha\,du, \beta\,dv, \gamma\,dw$ are the elements of arc on the u curves, v curves, and w curves, respectively. The expression (3.54) is built up as a sum of squares of the elements in three coordinate directions just as in (3.53). This is one of the basic properties of orthogonal coordinates. The above derivation shows that (3.54) can hold only when $\partial \mathbf{r}/\partial u, \partial \mathbf{r}/\partial v, \partial \mathbf{r}/\partial w$ are mutually perpendicular, so that (3.54) itself can be used to define what is meant by orthogonal coordinates.

We note also that $\alpha\beta\gamma\,du\,dv\,dw$ can be interpreted as the volume dV of an "elementary rectangular parallelepiped." Hence by (3.44) and (3.51),

$$dV = \alpha\beta\gamma\,du\,dv\,dw = J\,du\,dv\,dw = \frac{\partial(x, y, z)}{\partial(u, v, w)}du\,dv\,dw.$$

This formula will be discussed in Chapter 4.

From Eqs. (3.52) and the identity (3.31) we deduce the important rule:

$$\text{curl}\left(\frac{1}{\alpha^2}\frac{\partial \mathbf{r}}{\partial u}\right) = \mathbf{0}, \qquad \text{curl}\left(\frac{1}{\beta^2}\frac{\partial \mathbf{r}}{\partial v}\right) = \mathbf{0}, \qquad \text{curl}\left(\frac{1}{\gamma^2}\frac{\partial \mathbf{r}}{\partial w}\right) = \mathbf{0}. \tag{3.55}$$

*3.8   VECTOR OPERATIONS IN ORTHOGONAL CURVILINEAR COORDINATES

Now let a vector field $\mathbf{p}$ be given in D. The vector $\mathbf{p}$ can be described by its components p_x, p_y, p_z in terms of the given rectangular system. However, at each point of D, the vectors $(1/\alpha)\,\partial\mathbf{r}/\partial u$, $(1/\beta)\,\partial\mathbf{r}/\partial v$, $(1/\gamma)\,\partial\mathbf{r}/\partial w$ are a triple of mutually perpendicular unit vectors. Hence we can write

$$\mathbf{p} = p_u \frac{1}{\alpha}\frac{\partial\mathbf{r}}{\partial u} + p_v \frac{1}{\beta}\frac{\partial\mathbf{r}}{\partial v} + p_w \frac{1}{\gamma}\frac{\partial\mathbf{r}}{\partial w} \tag{3.56}$$

in terms of the components p_u, p_v, p_w in the directions of the three unit vectors. It should be emphasized that the triple of unit vectors in general *varies from point to point* in D. By (3.52) we can also write

$$\mathbf{p} = p_u \alpha \nabla F + p_v \beta \nabla G + p_w \gamma \nabla H. \tag{3.56$'$}$$

The components p_u, p_v, p_w can be computed from the components p_x, p_y, p_z in the rectangular system. For example,

$$p_u = \mathbf{p} \cdot \frac{1}{\alpha}\frac{\partial\mathbf{r}}{\partial u} = \frac{1}{\alpha}\left(p_x \frac{\partial x}{\partial u} + p_y \frac{\partial y}{\partial u} + p_z \frac{\partial z}{\partial u} \right). \tag{3.57}$$

Similarly, the components p_x, p_y, p_z can be computed from p_u, p_v, p_w:

$$p_x = \mathbf{p} \cdot \mathbf{i} = p_u \frac{1}{\alpha}\frac{\partial\mathbf{r}}{\partial u} \cdot \mathbf{i} + p_v \frac{1}{\beta}\frac{\partial\mathbf{r}}{\partial v} \cdot \mathbf{i} + p_w \frac{1}{\gamma}\frac{\partial\mathbf{r}}{\partial w} \cdot \mathbf{i},$$

$$p_x = \frac{1}{\alpha} p_u \frac{\partial x}{\partial u} + \frac{1}{\beta} p_v \frac{\partial x}{\partial v} + \frac{1}{\gamma} p_w \frac{\partial x}{\partial w}. \tag{3.58}$$

If the representation (3.56$'$) is used, one finds:

$$p_u = \alpha\left(p_x \frac{\partial u}{\partial x} + p_y \frac{\partial u}{\partial y} + p_z \frac{\partial u}{\partial z} \right), \tag{3.57$'$}$$

$$p_x = \alpha p_u \frac{\partial u}{\partial x} + \beta p_v \frac{\partial v}{\partial x} + \gamma p_w \frac{\partial w}{\partial x}. \tag{3.58$'$}$$

Since the vectors $(1/\alpha)\,\partial\mathbf{r}/\partial u$, $(1/\beta)\,\partial\mathbf{r}/\partial v$, $(1/\gamma)\,\partial\mathbf{r}/\partial w$ form a positive triple of mutually perpendicular unit vectors, the operations scalar times vector, sum of vectors, scalar product, and vector product can be carried out in terms of components in the directions of these unit vectors just as in terms of x, y, and z components. In particular,

$$[\mathbf{p} + \mathbf{q}]_u = p_u + q_u, \qquad [\mathbf{p} + \mathbf{q}]_v = p_v + q_v, \dots,$$

$$[\phi\mathbf{p}]_u = \phi p_u, \qquad [\phi\mathbf{p}]_v = \phi p_v, \dots,$$

$$\mathbf{p} \cdot \mathbf{q} = p_u q_u + p_v q_v + p_w q_w,$$

$$[\mathbf{p} \times \mathbf{q}]_u = p_v q_w - p_w q_v, \qquad [\mathbf{p} \times \mathbf{q}]_v = p_w q_u - p_u q_w, \dots. \tag{3.59}$$

On the other hand, because the base vectors vary from point to point, the differential operations become more complicated:

$$[\text{grad }\phi]_u = \frac{1}{\alpha}\frac{\partial\phi}{\partial u}, \qquad [\text{grad }\phi]_v = \frac{1}{\beta}\frac{\partial\phi}{\partial v}, \qquad [\text{grad }\phi]_w = \frac{1}{\gamma}\frac{\partial\phi}{\partial w}, \quad (3.60)$$

$$\text{div }\mathbf{p} = \frac{1}{\alpha\beta\gamma}\left[\frac{\partial}{\partial u}(\beta\gamma p_u) + \frac{\partial}{\partial v}(\gamma\alpha p_v) + \frac{\partial}{\partial w}(\alpha\beta p_w)\right], \qquad (3.61)$$

$$[\text{curl }\mathbf{p}]_u = \frac{1}{\beta\gamma}\left[\frac{\partial}{\partial v}(\gamma p_w) - \frac{\partial}{\partial w}(\beta p_v)\right],$$

$$[\text{curl }\mathbf{p}]_v = \frac{1}{\gamma\alpha}\left[\frac{\partial}{\partial w}(\alpha p_u) - \frac{\partial}{\partial u}(\gamma p_w)\right], \qquad (3.62)$$

$$[\text{curl }\mathbf{p}]_w = \frac{1}{\alpha\beta}\left[\frac{\partial}{\partial u}(\beta p_v) - \frac{\partial}{\partial v}(\alpha p_u)\right].$$

To prove (3.60), we use (3.57):

$$[\text{grad }\phi]_u = \frac{1}{\alpha}\left(\frac{\partial\phi}{\partial x}\frac{\partial x}{\partial u} + \frac{\partial\phi}{\partial y}\frac{\partial y}{\partial u} + \frac{\partial\phi}{\partial z}\frac{\partial z}{\partial u}\right) = \frac{1}{\alpha}\frac{\partial\phi}{\partial u}.$$

The other components are found in the same way. We remark that $[\text{grad }\phi]_u$ is the directional derivative of ϕ along a u curve, that is, $d\phi/ds$ in terms of arc length s on the curve. Since $ds = \alpha\,du$, the result (3.60) follows at once.

To prove (3.61), we use (3.51) and (3.56) to write:

$$\mathbf{p} = (\beta\gamma p_u)\left(\frac{1}{J}\frac{\partial\mathbf{r}}{\partial u}\right) + (\gamma\alpha p_v)\left(\frac{1}{J}\frac{\partial\mathbf{r}}{\partial v}\right) + (\alpha\beta p_w)\left(\frac{1}{J}\frac{\partial\mathbf{r}}{\partial w}\right).$$

The divergence of $\mathbf{p}$ is the sum of the divergences of the terms on the right-hand side. By (3.22) and (3.50) the divergence of the first term is

$$\text{grad }(\beta\gamma p_u)\cdot\left(\frac{1}{J}\frac{\partial\mathbf{r}}{\partial u}\right) + \beta\gamma p_u\,\text{div}\left(\frac{1}{J}\frac{\partial\mathbf{r}}{\partial u}\right) = (\text{grad }\beta\gamma p_u)\cdot\left(\frac{1}{J}\frac{\partial\mathbf{r}}{\partial u}\right).$$

By (3.60) this can be written as

$$\frac{\alpha}{J}(\text{grad }\beta\gamma p_u)\cdot\frac{1}{\alpha}\frac{\partial\mathbf{r}}{\partial u} = \frac{\alpha}{J}[\text{grad }\beta\gamma p_u]_u = \frac{1}{J}\frac{\partial}{\partial u}(\beta\gamma p_u) = \frac{1}{\alpha\beta\gamma}\frac{\partial}{\partial u}(\beta\gamma p_u).$$

This gives the first term on the right-hand side of (3.61); the others are found in the same way.

The proof of (3.62) is left to Problem 5 below.

From (3.60) and (3.61) we obtain an expression for the Laplacian in orthogonal curvilinear coordinates:

$$\nabla^2\phi = \text{div grad }\phi = \frac{1}{\alpha\beta\gamma}\left[\frac{\partial}{\partial u}\left(\frac{\beta\gamma}{\alpha}\frac{\partial\phi}{\partial u}\right) + \frac{\partial}{\partial v}\left(\frac{\gamma\alpha}{\beta}\frac{\partial\phi}{\partial v}\right) + \frac{\partial}{\partial w}\left(\frac{\alpha\beta}{\gamma}\frac{\partial\phi}{\partial w}\right)\right].$$

$$(3.63)$$

Remark. If the curvilinear coordinates are not orthogonal, the very notion of vector components must be generalized. This leads to *tensor analysis* (see Section 3.9).

The formulas (3.60), (3.61), (3.62) can be applied to the special case in which the new coordinates are obtained simply by choosing new rectangular coordinates u, v, w in space. This can be accomplished by choosing a positive triple $\mathbf{i}_1, \mathbf{j}_1, \mathbf{k}_1$ of unit vectors and then choosing axes u, v, w through an origin $O_1: (x_1, y_1, z_1)$ having the directions $\mathbf{i}_1, \mathbf{j}_1, \mathbf{k}_1$, respectively (Problems 13 and 14 following Section 1.14). The new coordinates (u, v, w) of a point $P: (x, y, z)$ are defined by the equation:

$$\overline{O_1P} = u\mathbf{i}_1 + v\mathbf{j}_1 + w\mathbf{k}_1.$$

One then finds

$$u = \overline{O_1P} \cdot \mathbf{i}_1 = \left[(x - x_1)\mathbf{i} + (y - y_1)\mathbf{j} + (z - z_1)\mathbf{k}\right] \cdot \mathbf{i}_1$$
$$= (x - x_1)(\mathbf{i} \cdot \mathbf{i}_1) + (y - y_1)(\mathbf{j} \cdot \mathbf{i}_1) + (z - z_1)(\mathbf{k} \cdot \mathbf{i}_1);$$

similar expressions are found for v and w. Also,

$$x = \overline{OP} \cdot \mathbf{i} = \left(\overline{OO_1} + \overline{O_1P}\right) \cdot \mathbf{i} = x_1 + u(\mathbf{i}_1 \cdot \mathbf{i}) + v(\mathbf{j}_1 \cdot \mathbf{i}) + w(\mathbf{k}_1 \cdot \mathbf{i}),$$

and similar expressions are found for y and z. These two sets of equations correspond to (3.43) and (3.42). We note that in all cases the functions involved are *linear*.

The quantities α, β, γ can be evaluated for this case without computation. For, since the new coordinates are rectangular and *no change of scale is made*, one must have

$$ds^2 = du^2 + dv^2 + dw^2 \tag{3.64}$$

for element of arc on a general curve. Hence

$$\alpha = \beta = \gamma = 1. \tag{3.65}$$

If we substitute these values in (3.60), (3.61), (3.62), we find that the basic formulas (3.5), (3.15), (3.23) reappear, with x replaced by u, y by v, and z by w. For example,

$$\operatorname{div}\mathbf{p} = \frac{\partial p_u}{\partial u} + \frac{\partial p_v}{\partial v} + \frac{\partial p_w}{\partial w}.$$

This shows that *the fundamental definitions* (3.5), (3.15), (3.23) *of grad, div, curl are not dependent on the particular coordinate system chosen.*

If a change of scale is made, the formulas will be altered. Physically, this corresponds to a change in unit of length (for example, from inches to feet), and one could hardly expect a temperature gradient, for example, to have the same value in degrees per inch as in degrees per foot. For this reason, in practice one must always specify the units used.

It has been assumed throughout that the Jacobian J is positive; this implies, in particular, that in the case just considered the vectors $\mathbf{i}_1, \mathbf{j}_1, \mathbf{k}_1$ form a positive triple. If J is negative, one finds that the only change in (3.60), (3.61), (3.62) is a reversal of sign in the components of the curl. In particular, if new rectangular coordinates are chosen, based on a negative triple $\mathbf{i}_1, \mathbf{j}_1, \mathbf{k}_1$, the curl as defined by

(3.23) relative to these axes is the negative of the curl relative to the original x, y, and z axes. *If the orientation of the axes is reversed, the curl reverses direction.* This is not surprising if one thinks of the curl as an angular velocity vector. On the other hand, the gradient vector and divergence do not depend on the orientation. The vector product components in (3.59) also change sign if the orientation is reversed. This could be predicted from the very definition of the vector product in Section 1.1.

For further discussion of change of coordinates in space, see Chapter 4 of *Classical Mechanics*, by H. Goldstein (Cambridge, Mass.: Addison-Wesley Press, 1950).

If the Jacobian J is 0 at a point P, it is in general impossible to use the curvilinear coordinates at P; in particular, the vectors $\partial \mathbf{r}/\partial u$, $\partial \mathbf{r}/\partial v$, $\partial \mathbf{r}/\partial w$ are coplanar, so that one cannot express an arbitrary vector $\mathbf{p}$ in terms of components p_u, p_v, p_w. Furthermore, the point P will in general correspond to many values of the coordinates (u, v, w); that is, the inverse functions (3.43) become ambiguous at P.

However, let us assume that the functions (3.42) remain continuous and differentiable in the domain D_1 of the coordinates (u, v, w) and have values in a domain D; then the chain rules (2.35) imply that every function $U(x, y, z)$ that is differentiable in D becomes a function

$$U[f(u, v, w), \quad g(u, v, w), \quad h(u, v, w)]$$

that is defined and differentiable in D_1. Accordingly, even though the inverse transformation (3.43) may fail to exist, *scalar* functions of (x, y, z) can be transformed into functions of (u, v, w) without difficulty.

Precisely this situation arises in cylindrical and spherical coordinates. For example, in cylindrical coordinates the equations (3.42) are as follows:

$$x = r\cos\theta, \quad y = r\sin\theta, \quad z = z;$$

these functions are defined and differentiable as often as desired for all real values of r, θ, z; as (r, θ, z) ranges over all possible combinations, (x, y, z) ranges over all of space. An inverse transformation:

$$r = \sqrt{x^2 + y^2}, \quad \theta = \arctan\frac{y}{x}, \quad z = z$$

can be defined by suitably restricting θ; but in no way can this be defined as a triple of continuous functions when (x, y, z) ranges over a domain D including points of the z axis. It is precisely when $r = 0$ that the Jacobian $J = r$ (Problem 6) is zero.

It is shown in Problem 6 that the Laplacian in cylindrical coordinates has the expression:

$$\nabla^2 U = \frac{1}{r^2}\left[r\frac{\partial}{\partial r}\left(r\frac{\partial U}{\partial r}\right) + \frac{\partial^2 U}{\partial \theta^2} + r^2\frac{\partial^2 U}{\partial z^2}\right]. \tag{3.66}$$

This expression becomes meaningless when $r = 0$. However, if we know that U, when expressed in rectangular coordinates, has continuous first and second derivatives on the z axis, then $\nabla^2 U$ must also be a continuous function of (r, θ, z) for $r = 0$. Under these assumptions, $\nabla^2 U$ can be obtained from (3.66) for $r = 0$ by a limit process. For example, if

$$U = x^2 + x^2 y = r^2 \cos^2 \theta + r^3 \cos^2 \theta \sin \theta,$$

then

$$\nabla^2 U = 2 + 2y = 2 + 2r \sin \theta;$$

(3.66) gives the indeterminate expression

$$\nabla^2 U = \frac{1}{r^2} [2r^2 + 2r^3 \sin \theta].$$

While this function is indeterminate for $r = 0$, it has a definite limit as $r \to 0$; the limit is 2 and is independent of θ. By cancelling r^2 from numerator and denominator we automatically remove the indeterminacy and assign the correct limiting value for $r = 0$:

$$\nabla^2 U = 2 + 2r \sin \theta.$$

A similar discussion applies to spherical coordinates ρ, ϕ, θ. In Problem 7 it is shown that $J = \rho^2 \sin \phi$, so that $J = 0$ on the z axis. The Laplacian is found to be

$$\nabla^2 U = \frac{1}{\rho^2 \sin^2 \phi} \left[\sin^2 \phi \frac{\partial}{\partial \rho} \left(\rho^2 \frac{\partial U}{\partial \rho} \right) + \sin \phi \frac{\partial}{\partial \phi} \left(\sin \phi \frac{\partial U}{\partial \phi} \right) + \frac{\partial^2 U}{\partial \theta^2} \right]. \quad (3.67)$$

This is indeterminate on the z axis ($\rho = 0$ or $\sin \phi = 0$); when $\nabla^2 U$ is known to be continuous on the z axis, the values of $\nabla^2 U$ on this line can be found from (3.67) by a limit process.

It is also possible to obtain the values of the Laplacian at the troublesome points directly in terms of derivatives. For example, $\nabla^2 U$ can be computed at the origin ($\rho = 0$) by the formula:

$$\nabla^2 U|_{\rho=0} = \frac{\partial^2 U}{\partial \rho^2} (0, \tfrac{1}{2}\pi, 0) + \frac{\partial^2 U}{\partial \rho^2} (0, \tfrac{1}{2}\pi, \tfrac{1}{2}\pi) + \frac{\partial^2 U}{\partial \rho^2} (0, 0, 0). \quad (3.68)$$

The three terms on the right are simply the three terms of the Laplacian

$$\frac{\partial^2 U}{\partial x^2} + \frac{\partial^2 U}{\partial y^2} + \frac{\partial^2 U}{\partial z^2}$$

at the origin.

PROBLEMS

1. Prove (3.48). [Hint: Use the result of Problem 5 following Section 2.12.]
2. Prove (3.47). [Hint: Use (3.48) and the rules (1.19) of Section 1.1.]
3. Prove (3.46). [Hint: Use (3.48) and the rules (1.19) of Section 1.1.]
4. Prove that when the vectors $\nabla F, \nabla G, \nabla H$ are mutually perpendicular in D, the coordinates are orthogonal.

5. Prove (3.62). [Hint: Use (3.56) to write:

$$\mathbf{p} = (\alpha p_u)\left(\frac{1}{\alpha^2}\frac{\partial \mathbf{r}}{\partial u}\right) + \beta p_v\left(\frac{1}{\beta^2}\frac{\partial \mathbf{r}}{\partial v}\right) + (\gamma p_w)\left(\frac{1}{\gamma^2}\frac{\partial \mathbf{r}}{\partial w}\right).$$

Use (3.28) and (3.55) to show that

$$\text{curl } \mathbf{p} = \text{grad}(\alpha p_u) \times \left(\frac{1}{\alpha^2}\frac{\partial \mathbf{r}}{\partial u}\right) + \text{grad}(\beta p_v) \times \frac{1}{\beta^2}\left(\frac{\partial \mathbf{r}}{\partial v}\right) + \cdots .$$

To compute $[\text{curl } \mathbf{p}]_u$, take the scalar product of both sides with $\dfrac{1}{\alpha}\dfrac{\partial \mathbf{r}}{\partial u}$ and use (3.59) and (3.60) to compute the scalar triple products.]

6. Verify the following relations for cylindrical coordinates $u = r$, $v = \theta$, $w = z$:

a) the surfaces $r = \text{const}$, $\theta = \text{const}$, $z = \text{const}$ form a triply orthogonal family, and

$$J = \frac{\partial(x, y, z)}{\partial(r, \theta, z)} = r;$$

b) the element of arc length is given by

$$ds^2 = dr^2 + r^2\, d\theta^2 + dz^2;$$

c) the components of a vector $\mathbf{p}$ are given by

$$p_r = p_x \cos\theta + p_y \sin\theta, \qquad p_\theta = -p_x \sin\theta + p_y \cos\theta, \qquad p_z = p_z;$$

d) grad U has components: $\dfrac{\partial U}{\partial r}$, $\dfrac{1}{r}\dfrac{\partial U}{\partial \theta}$, $\dfrac{\partial U}{\partial z}$;

e) $\text{div }\mathbf{p} = \dfrac{1}{r}\left[\dfrac{\partial}{\partial r}(rp_r) + \dfrac{\partial p_\theta}{\partial \theta} + r\dfrac{\partial p_z}{\partial z}\right];$

f) curl $\mathbf{p}$ has components:

$$\frac{1}{r}\left[\frac{\partial p_z}{\partial \theta} - r\frac{\partial p_\theta}{\partial z}\right], \quad \left[\frac{\partial p_r}{\partial z} - \frac{\partial p_z}{\partial r}\right], \quad \frac{1}{r}\left[\frac{\partial}{\partial r}(rp_\theta) - \frac{\partial p_r}{\partial \theta}\right];$$

g) $\nabla^2 U$ is given by (3.66).

7. Verify the following relations for spherical coordinates $u = \rho$, $v = \phi$, $w = \theta$:

a) the surfaces $\rho = \text{const}$, $\phi = \text{const}$, $\theta = \text{const}$ form a triply orthogonal family, and

$$J = \frac{\partial(x, y, z)}{\partial(\rho, \phi, \theta)} = \rho^2 \sin\phi;$$

b) the element of arc length is given by

$$ds^2 = d\rho^2 + \rho^2\, d\phi^2 + \rho^2 \sin^2\phi\, d\theta^2;$$

c) the components of a vector $\mathbf{p}$ are given by

$$p_\rho = p_x \sin\phi\cos\theta + p_y \sin\phi\sin\theta + p_z \cos\phi,$$
$$p_\phi = p_x \cos\phi\cos\theta + p_y \cos\phi\sin\theta - p_z \sin\phi,$$
$$p_\theta = -p_x \sin\theta + p_y \cos\theta;$$

d) grad U has components $\dfrac{\partial U}{\partial \rho}$, $\dfrac{1}{\rho}\dfrac{\partial U}{\partial \phi}$, $\dfrac{1}{\rho \sin\phi}\dfrac{\partial U}{\partial \theta}$;

e) $\operatorname{div} \mathbf{p} = \dfrac{1}{\rho^2 \sin \phi} \left[\sin \phi \dfrac{\partial}{\partial \rho} \left(\rho^2 p_\rho \right) + \rho \dfrac{\partial}{\partial \phi} \left(p_\phi \sin \phi \right) + \rho \dfrac{\partial p_\theta}{\partial \theta} \right];$

f) curl $\mathbf{p}$ has components:

$$\dfrac{1}{\rho \sin \phi} \left[\dfrac{\partial}{\partial \phi} \left(p_\theta \sin \phi \right) - \dfrac{\partial p_\phi}{\partial \theta} \right],$$

$$\dfrac{1}{\rho \sin \phi} \left[\dfrac{\partial p_\rho}{\partial \theta} - \sin \phi \dfrac{\partial}{\partial \rho} \left(\rho p_\theta \right) \right], \quad \dfrac{1}{\rho} \left[\dfrac{\partial}{\partial \rho} \left(\rho p_\phi \right) - \dfrac{\partial p_\rho}{\partial \phi} \right];$$

g) $\nabla^2 U$ is given by (3.67).

8. (*Curvilinear coordinates on a surface*) Equations

$$x = f(u, v), \qquad y = g(u, v), \qquad z = h(u, v)$$

can be interpreted as parametric equations of a surface S in space. They can be considered as a special case of (3.42), in which w is restricted to a constant value, while (u, v) varies over a domain D of the uv plane; the surface S then corresponds to a surface $w = $ const for (3.42). We consider u, v as *curvilinear coordinates* on S. The two sets of curves $u = $ const and $v = $ const on S form families like the parallels to the axes in the xy-plane. Graph the surface and the lines $u = $ const, $v = $ const, for the following cases:

a) sphere: $x = \sin u \cos v$, $y = \sin u \sin v$, $z = \cos u$;

b) cylinder: $x = \cos u$, $y = \sin u$, $z = v$;

c) cone: $x = \sinh u \sin v$, $y = \sinh u \cos v$, $z = \sinh u$.

9. For the surface of Problem 8, assume that f, g, h have continuous first partial derivatives in D and that the Jacobian matrix

$$\begin{pmatrix} f_u & g_u & h_u \\ f_v & g_v & h_v \end{pmatrix}'$$

has rank 2 in D (Section 1.16). Show that each point (u_0, v_0) of D has a neighborhood D_0 in which the correspondence between points (u, v) and points of the surface is one-to-one. [Hint: If, for example, $\partial(f, g)/\partial(u, v) \neq 0$ at (u_0, v_0), then apply the Implicit Function Theorem of Section 2.10, as in Section 2.12, to obtain inverse functions $u = \phi(x, y)$, $v = \psi(x, y)$ of $x = f(u, v)$, $y = g(u, v)$ in D_0. Show that each point (x, y, z) of the surface, for (u, v) in D_0, comes from a unique (u, v) in D_0.]

10. Let a surface S be given as in Problems 8 and 9 and let $\mathbf{r} = x\mathbf{i} + y\mathbf{j} + z\mathbf{k}$.

a) Show that $\partial \mathbf{r}/\partial u$ and $\partial \mathbf{r}/\partial v$ are vectors tangent to the lines $v = $ const, $u = $ const on the surface.

b) Show that the curves $v = $ const, $u = $ const intersect at right angles, so that the coordinates are *orthogonal*, if and only if

$$\dfrac{\partial x}{\partial u} \dfrac{\partial x}{\partial v} + \dfrac{\partial y}{\partial u} \dfrac{\partial y}{\partial v} + \dfrac{\partial z}{\partial u} \dfrac{\partial z}{\partial v} = 0.$$

c) Show that the element of arc on a curve $u = u(t)$, $v = v(t)$ on S is given by

$$ds^2 = E\, du^2 + 2F\, du\, dv + G\, dv^2,$$

$$E = \left| \frac{\partial \mathbf{r}}{\partial u} \right|^2 = \left(\frac{\partial x}{\partial u} \right)^2 + \left(\frac{\partial y}{\partial u} \right)^2 + \left(\frac{\partial z}{\partial u} \right)^2,$$

$$G = \left| \frac{\partial \mathbf{r}}{\partial v} \right|^2 = \left(\frac{\partial x}{\partial v} \right)^2 + \left(\frac{\partial y}{\partial v} \right)^2 + \left(\frac{\partial z}{\partial v} \right)^2,$$

$$F = \frac{\partial \mathbf{r}}{\partial u} \cdot \frac{\partial \mathbf{r}}{\partial v} = \frac{\partial x}{\partial u} \frac{\partial x}{\partial v} + \frac{\partial y}{\partial u} \frac{\partial y}{\partial v} + \frac{\partial z}{\partial u} \frac{\partial z}{\partial v}.$$

d) Show that the coordinates are *orthogonal* precisely when

$$ds^2 = E \, du^2 + G \, dv^2.$$

e) Let $u = u(t)$, $v = v(t)$ and $u = U(\tau)$, $v = V(\tau)$ be two curves on S meeting at a point P_0 of S for $t = t_0$, $\tau = \tau_0$, so that $u(t_0) = u_0 = U(\tau_0)$, $v(t_0) = v_0 = V(\tau_0)$. Show that the angle θ between the corresponding velocity vectors $d\mathbf{r}/dt$ at t_0, $d\mathbf{r}/d\tau$ at τ_0 (assumed to be nonzero vectors) satisfies

$$\cos \theta = \frac{Eu'U' + F(u'V' + v'U') + Gv'V'}{(Eu'^2 + 2Fu'v' + Gv'^2)^{1/2} (EU'^2 + 2FU'V' + GV'^2)^{1/2}}$$

where E, F, G are evaluated at (u_0, v_0) and $u' = u'(t_0)$, $v' = v'(t_0)$, $U' = U'(\tau_0)$, $V' = V'(\tau_0)$.

f) Show that if the paths of part **(e)** are coordinate lines:

$$u(t) = u_0 + t - t_0, \quad v(t) = v_0, \quad U(\tau) = u_0, \quad V(\tau) = v_0 + \tau - \tau_0,$$

then $\cos \theta = F(E \, G)^{-1/2}$ at (u_0, v_0).

g) Show from the hypotheses of Problem 9 that $\partial \mathbf{r}/\partial u$ and $\partial \mathbf{r}/\partial v$ are linearly independent in D.

h) Show that $E > 0$, $G > 0$, $EG - F^2 > 0$. [Hint: $E > 0$, $G > 0$ follow from (g). By Problem 12 (a) following Section 1.4,

$$EG - F^2 = |\partial \mathbf{r}/\partial u \times \partial \mathbf{r}/\partial v|^2;$$

show from (g) that this is > 0.]

i) Show from the meaning of ds^2 in terms of arc length that $Edu^2 + 2F \, du \, dv + G \, dv^2$ is a positive definite quadratic form and hence $E > 0$, $G > 0$, $EG - F^2 > 0$. [Hint: See Example 1 in Section 2.21.]

For further theory of surfaces, see the book by Struik listed at the end of this chapter.

*3.9   TENSORS

When nonorthogonal curvilinear coordinates are introduced, the methods of the preceding section are no longer adequate for the analysis of the basic vector operations. The desired analysis can be carried out with the aid of tensors, which we proceed to develop briefly here.

We first consider the very simple case of a change of scale in space. Let

(x, y, z) be given Cartesian coordinates and let $(\bar{x}, \bar{y}, \bar{z})$ be new coordinates, where

$$\bar{x} = \lambda x, \qquad \bar{y} = \lambda y, \qquad \bar{z} = \lambda z, \tag{3.69}$$

λ being a positive constant scalar. Thus we have changed scale in the ratio $\lambda : 1$, and $1/\lambda$ is our new unit of distance. For a point moving on a path we have hitherto assigned a velocity vector $\mathbf{v} = (v_x, v_y, v_z)$ and thought of this as a definite geometric object, represented by a directed line segment. But the components of $\mathbf{v}$, as a velocity vector, depend on the coordinate system; that is, here, where we are considering a change of scale, these components depend on the unit of distance chosen. In the (x, y, z) coordinates we would assign

$$v_x = \frac{dx}{dt}, \qquad v_y = \frac{dy}{dt}, \qquad v_z = \frac{dz}{dt},$$

but in the $(\bar{x}, \bar{y}, \bar{z})$ coordinates we would assign

$$\bar{v}_x = \frac{d\bar{x}}{dt}, \qquad \bar{v}_y = \frac{d\bar{y}}{dt}, \qquad \bar{v}_z = \frac{d\bar{z}}{dt}.$$

By virtue of Eqs. (3.69), $d\bar{x}/dt = \lambda\, dx/dt$, so that $\bar{v}_x = \lambda v_x$, and in general,

$$\bar{v}_x = \lambda v_x, \qquad \bar{v}_y = \lambda v_y, \qquad \bar{v}_z = \lambda v_z. \tag{3.70}$$

Thus in the new coordinates $(\bar{x}, \bar{y}, \bar{z})$ we assign to $\mathbf{v}$ the new components $(\bar{v}_x, \bar{v}_y, \bar{v}_z)$, which are λ times the previous components.

Now a vector $\mathbf{v}$ can also be obtained as the gradient vector of a function f: $\mathbf{v} = \operatorname{grad} f$, so that

$$v_x = \frac{\partial f}{\partial x}, \qquad v_y = \frac{\partial f}{\partial y}, \qquad v_z = \frac{\partial f}{\partial z}.$$

If we change scale by (3.69), we would still like $\mathbf{v}$ to be the gradient of f. But in the new coordinates, f becomes $f(\bar{x}/\lambda, \bar{y}/\lambda, \bar{z}/\lambda) = \bar{f}(\bar{x}, \bar{y}, \bar{z})$, and this has gradient

$$\bar{v}_x = \frac{\partial \bar{f}}{\partial \bar{x}} = \frac{1}{\lambda} f_x, \qquad \bar{v}_y = \frac{\partial \bar{f}}{\partial \bar{y}} = \frac{1}{\lambda} f_y, \qquad \bar{v}_z = \frac{\partial \bar{f}}{\partial \bar{z}} = \frac{1}{\lambda} f_z.$$

Hence now

$$\bar{v}_x = \frac{1}{\lambda} v_x, \qquad \bar{v}_y = \frac{1}{\lambda} v_y, \qquad \bar{v}_z = \frac{1}{\lambda} v_z. \tag{3.71}$$

Thus when we change scale by (3.69), the three components of the gradient vector are *divided* by λ. This result is to be expected, since the gradient vector measures the rate of change of f with respect to distance (in various directions), and we are changing the unit of distance. If for example, $\lambda = 2$, then the new unit of distance is half the old one, and the amount of change in f per unit of distance is half as much as before, whereas velocity components are twice as large as before, since one covers twice as many units of distance per unit of time.

The two different rules (3.70) and (3.71) show that we really have two different ways of assigning components to vectors. In the case of (3.70), one is dealing with a "contravariant vector," and in the case of (3.71), one is dealing with a "covariant vector." In obtaining the two types of components we have emphasized the change in unit of distance. However, in changing coordinates we can always keep in mind the original unit of distance (as a standard of reference), so that all distances can ultimately be restated in the original units. This concept of a standard unit of distance will be important in the development to follow.

In tensor analysis, one allows changes of coordinates that are much more general than changes of scale. It is more convenient to number our coordinates as (x^1, x^2, x^3) and simply to refer to the (x^i) coordinates. We can in fact reason generally about n-dimensional space and allow i to go from 1 to n. The use of superscripts rather than subscripts is required for a certain consistency of all the tensor notations. If we introduce new coordinates $(\bar{x}^i)$, then we have equations

$$x^i = x^i(\bar{x}^1, \ldots, \bar{x}^n), \qquad i = 1, \ldots, n, \tag{3.72}$$

relating new and old coordinates. We consider these equations only in a neighborhood D of a certain point and assume that they define a one-to-one mapping with inverse

$$\bar{x}^i = \bar{x}^i(x^1, \ldots, x^n), \qquad i = 1, \ldots, n, \tag{3.73}$$

and that all functions in (3.72) and (3.73) are differentiable, with continuous second partial derivatives and nonzero Jacobian

$$\frac{\partial(x^1, \ldots, x^n)}{\partial(\bar{x}^1, \ldots, \bar{x}^n)}.$$

It is essential for the theory of tensors that we allow *all* such changes of coordinates.

In the following discussion it will be convenient to denote by $(\xi^1, \ldots, \xi^n)$ a *fixed* Cartesian coordinate system in E^n, in terms of which distance and angle are measured as usual. We denote by $(x^1, \ldots, x^n)$ and $(\bar{x}^1, \ldots, \bar{x}^n)$ two other, generally curvilinear, coordinate systems introduced as before in the neighborhood D of a point, and related by Eqs. (3.72) and (3.73) to each other and related in similar fashion to the (ξ^i). We shall refer to the (ξ^i) as *standard coordinates*.

Now let a vector field be given in the chosen neighborhood. Then the vectors all have sets of components *in each coordinate system*. For a *contravariant vector field* (or briefly, *contravariant vector*) the components will be denoted in the (x^i) coordinates by $(u^1, \ldots, u^n)$ (upper indices) and in the $(\bar{x}^i)$ coordinates by $(\bar{u}^1, \ldots, \bar{u}^n)$. Furthermore, for any two such coordinate systems the components are to be related by the rule

$$\bar{u}^i = \sum_{j=1}^{n} \frac{\partial \bar{x}^i}{\partial x^j} u^j. \tag{3.74}$$

This rule is chosen to fit the case when the vector in question is a velocity vector:

$$u^i = \frac{dx^i}{dt}, \qquad i = 1,\ldots,n.$$

For then

$$\bar{u}^i = \frac{d\bar{x}^i}{dt} = \sum_{j=1}^{n} \frac{\partial \bar{x}^i}{\partial x^j} \frac{dx^j}{dt} = \sum_{j=1}^{n} \frac{\partial \bar{x}^i}{\partial x^j} u^j.$$

For a *covariant vector* we denote the components in the two systems of coordinates by $(u_1,\ldots,u_n)$ and $(\bar{u}_1,\ldots,\bar{u}_n)$, respectively, and require that

$$\bar{u}_i = \sum_{j=1}^{n} \frac{\partial x^j}{\partial \bar{x}^i} u_j. \tag{3.75}$$

This rule is chosen to fit the case when the vector in question is a gradient vector:

$$u_i = \frac{\partial f}{\partial x^i}, \qquad i = 1,\ldots,n.$$

For then

$$\bar{u}_i = \frac{\partial}{\partial \bar{x}^i} f\left(x^1(\bar{x}^1,\ldots),\ldots,x^n(\bar{x}^1,\ldots)\right)$$

$$= \sum_{j=1}^{n} \frac{\partial f}{\partial x^j} \frac{\partial x^j}{\partial \bar{x}^i} = \sum_{j=1}^{n} \frac{\partial x^j}{\partial \bar{x}^i} u_j.$$

Each contravariant vector field in the chosen neighborhood D can be obtained in the following way. One first selects a vector field U^i in the (ξ^i) coordinates—that is, n functions $U^1(\xi^1,\ldots,\xi^n),\ldots,U^n(\xi^1,\ldots,\xi^n)$ defined in D. Then to each other coordinate system, say (x^i), one assigns components by the equations analogous to (3.74):

$$u^i = \sum_{j=1}^{n} \frac{\partial x^i}{\partial \xi^j} U^j. \tag{3.76}$$

Thus in $(\bar{x}^i)$ one has similarly

$$\bar{u}^i = \sum_{j=1}^{n} \frac{\partial \bar{x}^i}{\partial \xi^j} U^j. \tag{3.77}$$

Now the inverse of the matrix $(\partial x^i / \partial \xi^j)$ is the matrix $(\partial \xi^i / \partial x^j)$. Hence we can solve (3.76) to obtain

$$U^j = \sum_{k=1}^{n} \frac{\partial \xi^j}{\partial x^k} u^k.$$

If we substitute in (3.77), we obtain

$$\bar{u}^i = \sum_{j=1}^{n} \sum_{k=1}^{n} \frac{\partial \bar{x}^i}{\partial \xi^j} \frac{\partial \xi^j}{\partial x^k} u^k = \sum_{k=1}^{n} \sum_{j=1}^{n} \frac{\partial \bar{x}^i}{\partial \xi^j} \frac{\partial \xi^j}{\partial x^k} u^k = \sum_{k=1}^{n} \frac{\partial \bar{x}^i}{\partial x^k} u^k,$$

so that (3.74) holds. Accordingly, once we have assigned components in the standard coordinates, we automatically obtain components in all other coordinate systems, related by (3.74), and a contravariant vector is obtained.

A similar reasoning applies to covariant vector fields (Problem 5).

From this discussion it follows that starting with n functions $f_i(\xi^1,\ldots,\xi^n)$, defined in D, we can generate either a contravariant vector field u^i or a covariant field u_i. For the contravariant vector we assign components $U^i = f_i(\xi^1,\ldots,\xi^n)$ in standard coordinates and hence obtain the u^i as above in all other allowed coordinates (x^i); for the covariant vector we assign components $U_i = f_i(\xi^1,\ldots,\xi^n)$ in standard coordinates and hence obtain the u_i in all other coordinate systems (x^i). The contravariant and covariant vector fields thus obtained are related in a special way—namely, in that in standard coordinates, corresponding components are equal:

$$U^i(\xi^1,\ldots,\xi^n) = f_i(\xi^1,\ldots,\xi^n) = U_i(\xi^1,\ldots,\xi^n).$$

In such a case we say that the contravariant vector u^i and covariant vector u_i are *associated*. We regard the u^i and u_i as different aspects of one underlying geometric object **u**, a vector field abstracted from its components. Thus we speak of u^i as contravariant components of **u**, u_i as covariant components of **u**. In standard coordinates the two types of components coincide and can be identified with our usual vector components.

The contravariant and covariant vectors are the tensors of *first order*. We also introduce tensors of *order zero*, as scalar functions whose values are not changed by coordinate transformations. Thus in the two coordinate systems (x^i) and $(\bar{x}^i)$, such a function is given by $f(x^1,\ldots,x^n)$ and $\bar{f}(\bar{x}^1,\ldots,\bar{x}^n)$, respectively, where

$$f(x^1,\ldots,x^n) = \bar{f}(\bar{x}^1,\ldots,\bar{x}^n)$$

whenever (x^i) and $(\bar{x}^i)$ refer to the same point; that is,

$$f(x^1(\bar{x}^1,\ldots),\ldots,x^n(\bar{x}^1,\ldots)) \equiv \bar{f}(\bar{x}^1,\ldots,\bar{x}^n). \tag{3.78}$$

We also call such a tensor an *invariant*.

We can also introduce tensors of higher order. A tensor of order two requires two indices and hence can be arranged as a matrix. For example, we denote by v_{ij} a tensor of order two that is *covariant* in both indices; here v_{ij} $(i = 1,\ldots,n, j = 1,\ldots,n)$ are the n^2 components of the tensor in the (x^i) coordinate system. In the second coordinate system $(\bar{x}^i)$ the tensor has components $\bar{v}_{ij}$, and we require that

$$\bar{v}_{ij} = \sum_{k=1}^{n} \sum_{l=1}^{n} \frac{\partial x^k}{\partial \bar{x}^i} \frac{\partial x^l}{\partial \bar{x}^j} v_{kl}. \tag{3.79}$$

Because of the many such sums appearing in tensor analysis, one agrees to drop the sigma signs and to write (3.79) simply as

$$\bar{v}_{ij} = \frac{\partial x^k}{\partial \bar{x}^i} \frac{\partial x^l}{\partial \bar{x}^j} v_{kl},$$

with the understanding that we sum over each index which appears more than once (k and l in this case). This notational rule is called the *summation convention*.

By v_j^i we denote a second-order tensor, called *mixed*, which is contravariant in the index i and covariant in the index j, and require that for a change of coordinates,

$$\bar{v}_j^i = \frac{\partial \bar{x}^i}{\partial x^k} \frac{\partial x^l}{\partial \bar{x}^j} v_l^k. \tag{3.80}$$

By v^{ij} we denote a second-order tensor that is contravariant in both indices and require that

$$\bar{v}^{ij} = \frac{\partial \bar{x}^i}{\partial x^k} \frac{\partial \bar{x}^j}{\partial x^l} v^{kl}. \tag{3.81}$$

Similar definitions are given for tensors of third, fourth, and higher orders. For example, w_k^{ij} denotes a tensor contravariant in i and j and covariant in k, whereby

$$\bar{w}_k^{ij} = \frac{\partial \bar{x}^i}{\partial x^p} \frac{\partial \bar{x}^j}{\partial x^r} \frac{\partial x^s}{\partial \bar{x}^k} w_s^{pr}. \tag{3.82}$$

The tensors of higher order appear in many geometrical and physical theories. In particular, they are needed for the basic operations on vector fields, as we shall illustrate.

Each tensor of higher order can also be obtained by first assigning components in standard coordinates (ξ^i) and then using the appropriate rule to obtain components in an arbitrary coordinate system (x^i). This is proved just as for contravariant and covariant vectors earlier. In particular, starting with a set of functions $f_{ijklm}(\xi^1,\ldots,\xi^n)$, where each index $i, j,\ldots$ runs from 1 to n, one can choose some indices as contravariant and others as covariant, in some chosen order in each case, and then introduce corresponding tensor components in D, for example,

$$W_{klj}^{im}(\xi^1,\ldots,\xi^n) = f_{ijklm}(\xi^1,\ldots,\xi^n).$$

Then the rules tell us how to define the components w_{klj}^{im} in (x^i), to obtain a tensor of the type indicated. All these tensors, arising from the same set of functions in standard coordinates, are said to be *associated*, and we regard them as different aspects of one geometric object **w**. From $f_{ij}(\xi^1,\ldots,\xi^n)$ we obtain in this way six associated tensors

$$u_{ij}, \quad u_{ji}, \quad u_j^i, \quad u_i^j, \quad u^{ij}, \quad u^{ji}.$$

A very important tensor of second order arises in discussing distance in curvilinear coordinates. We start with standard coordinates (ξ^i) in D. For a point moving along a path in D the usual reasoning leads us to the formula

$$\frac{ds}{dt} = \left(\sum_{i=1}^{n} \left(\frac{d\xi^i}{dt} \right)^2 \right)^{\frac{1}{2}}$$

for the speed. Correspondingly, we write

$$ds^2 = d\xi^i \, d\xi^i$$

(summation convention!) for the square of the "element of distance" ds. If we now introduce curvilinear coordinates (x^i) but require that *distance be unchanged*, then we find

$$ds^2 = d\xi^i \, d\xi^i = \left(\frac{\partial \xi^i}{\partial x^k} dx^k \right) \left(\frac{\partial \xi^i}{\partial x^l} dx^l \right) = \frac{\partial \xi^i}{\partial x^k} \frac{\partial \xi^i}{\partial x^l} dx^k \, dx^l.$$

Hence in curvilinear coordinates, ds^2 appears as a "quadratic differential form"

$$ds^2 = g_{kl} \, dx^k \, dx^l, \tag{3.83}$$

where

$$g_{kl} = \frac{\partial \xi^i}{\partial x^k} \frac{\partial \xi^i}{\partial x^l} = g_{lk} \qquad (k = 1, \ldots, n, \quad l = 1, \ldots, n). \tag{3.84}$$

(Thus the matrix (g_{kl}) is symmetric.) By this procedure we assign sets of components g_{kl} in every curvilinear coordinate system. For two sets of coordinates (x^i) and $(\bar{x}^i)$ we have

$$g_{ij} = \frac{\partial \xi^r}{\partial x^i} \frac{\partial \xi^r}{\partial x^j}, \qquad \bar{g}_{ij} = \frac{\partial \xi^r}{\partial \bar{x}^i} \frac{\partial \xi^r}{\partial \bar{x}^j},$$

and hence

$$\bar{g}_{ij} = \left(\frac{\partial \xi^r}{\partial x^k} \frac{\partial x^k}{\partial \bar{x}^i} \right) \left(\frac{\partial \xi^r}{\partial x^l} \frac{\partial x^l}{\partial \bar{x}^j} \right) = \frac{\partial x^k}{\partial \bar{x}^i} \frac{\partial x^l}{\partial \bar{x}^j} \frac{\partial \xi^r}{\partial x^k} \frac{\partial \xi^r}{\partial x^l}$$

$$= \frac{\partial x^k}{\partial \bar{x}^i} \frac{\partial x^l}{\partial \bar{x}^j} g_{kl}.$$

Therefore g_{ij} is indeed a second-order tensor, covariant in both indices. We call g_{ij} the *fundamental metric tensor*.

We observe that the g_{ij} reduce to δ_{ij} (Kronecker delta) in standard coordinates, since in the (ξ^i),

$$ds^2 = d\xi^i \, d\xi^i = \delta_{ij} \, d\xi^i \, d\xi^j.$$

Hence we can regard the g_{ij} as the covariant tensor obtained from δ_{ij} (constant functions) in standard coordinates. From these functions we obtain six associated tensors as earlier. However, since $\delta_{ji} = \delta_{ij}$, $g_{ji} = g_{ij}$ (as noted before), so that these two associated tensors are the same. By setting $G^i_j = \delta_{ij}$ in standard coordinates (ξ^i) we obtain a mixed second-order tensor g^i_j. In the (x^i) coordinates we obtain

$$g^i_j = \frac{\partial x^i}{\partial \xi^k} \frac{\partial \xi^l}{\partial x^j} \delta_{kl} = \frac{\partial x^i}{\partial \xi^k} \frac{\partial \xi^k}{\partial x^j} = \delta_{ij}$$

(since the matrices $(\partial x^i/\partial \xi^j)$ and $(\partial \xi^i/\partial x^j)$ are inverses of each other). Hence

$$g^i_j = g^j_i = \delta_{ij} \qquad \text{(constant functions)} \tag{3.85}$$

in all coordinates. We often write δ^i_j for this tensor. Next, by setting $G^{ij} = \delta_{ij}$ in standard coordinates we obtain a contravariant second-order tensor g^{ij}; as earlier,

$$g^{ij} = \frac{\partial x^i}{\partial \xi^k} \frac{\partial x^j}{\partial \xi^l} \delta_{kl} = \frac{\partial x^i}{\partial \xi^k} \frac{\partial x^j}{\partial \xi^k} = g^{ji}. \tag{3.86}$$

Comparison with (3.84) suggests that (g^{ij}) is the inverse matrix of (g_{ij}), that is, that

$$g_{i\alpha} g^{\alpha j} = \delta_{ij}. \tag{3.87}$$

This can be directly verified (Problem 6). It follows that both matrices are nonsingular. We denote by g the determinant of (g_{ij}):

$$g = \det(g_{ij}). \tag{3.88}$$

Thus g is a scalar function, depending on the coordinate system; it is not an invariant. One can show that g *must be positive* (Problem 7).

We thus see that our fundamental metric tensor has three aspects:

$$g_{ij} = g_{ji}, \qquad g^i_j = g^j_i = \delta^j_i, \qquad g^{ij} = g^{ji} \qquad \text{(inverse matrix of the } g_{ij}). \tag{3.89}$$

Tensor algebra. Tensors can be combined in certain ways to yield new tensors. We can *add* two tensors of the same type, to obtain another of the same type (same number of contravariant and covariant indices)—for example,

$$v^i_j + w^i_j = u^i_j \tag{3.90}$$

defines the sum of the tensors v^i_j and w^i_j. Here we are giving the rule for a typical (x^i) coordinate system and assume that v^i_j and w^i_j are both given in the neighborhood of a point. Thus we understand that in an $(\bar{x}^i)$ coordinate system,

$$\bar{v}^i_j + \bar{w}^i_j = \bar{u}^i_j.$$

We can now verify that the sum is indeed a tensor (Problem 8).

We can *multiply* a tensor by an *invariant*; for example, from u_{ij} and f we form

$$f u_{ij} = v_{ij} \tag{3.91}$$

and verify that this again leads to a tensor v_{ij} of the type indicated (Problem 9).

We can multiply two tensors to form their *tensor product*—for example, from first-order tensors u_i, v_j, w^k, z^l we can form

$$s_{ij} = u_i v_j, \qquad t_i^k = u_i w^k, \qquad q^{kl} = w^k z^l$$

and can verify that these are tensors of second order of the types indicated (Problem 10).

We can *contract* a tensor by summing over a pair of indices, one contravariant, one covariant. For example, from u_{kl}^{ij} we can form

$$w_k^j = u_{ki}^{ij} \quad \text{and} \quad v_k^i = u_{kj}^{ij}$$

(summed over i and j, respectively). Each such contraction lowers the order by two. We can verify that the procedure described always leads to a tensor (Problem 11). In particular, if a_{ij}, u^i and v^j are tensors, then

$$a_{ij} u^i v^j = f$$

is an invariant, for $a_{ij} u^k v^l$ is a product of tensors and hence is a tensor of order four; f is obtained from this tensor by contracting twice.

We can *raise or lower indices* of a given tensor $u_{kl\ldots}^{ij\ldots}$ by the following procedure. We multiply by the metric tensor g_{pq} or g^{pq} to first increase the order by two. Then we use contraction, based on q and one selected index of $u_{kl\ldots}^{ij\ldots}$, to lower the order by 2. For example,

$$g_{i\alpha} u_{kl}^{\alpha j} \quad \text{is a new tensor} \quad u_{kli}^{j}; \qquad g^{k\alpha} u_{\alpha l}^{ij} \quad \text{is a new tensor} \quad u_l^{ijk}.$$

We have some freedom here in where we locate the new index. However, this freedom corresponds to a choice among several associated tensors. In fact, all tensors obtained by this process are associated with the initial tensor $u_{kl\ldots}^{ij\ldots}$. To see this, we carry out the first process in standard coordinates (ξ^i). In these coordinates, $g_{ij} = \delta_{ij}$, $g^{ij} = \delta_{ij}$, $u_{kl\ldots}^{ij\ldots}$ becomes $U_{kl\ldots}^{ij\ldots}$ and $g_{i\alpha} u_{kl\ldots}^{\alpha i\ldots}$ becomes

$$\delta_{i\alpha} U_{kl\ldots}^{\alpha j\ldots} = U_{kl\ldots i}^{j\ldots}.$$

But the left-hand side equals $U_{kl\ldots}^{ij\ldots}$. Hence in standard coordinates we have merely moved an index to a new position. This shows that we are dealing with associated tensors.

Covariant and contravariant derivatives.   We seek a process, analogous to forming partial derivatives, that we can apply to tensor components to yield a new tensor. In doing so we shall add one index. We can make this covariant or contravariant and obtain corresponding covariant and contravariant derivatives. We make the process definite by simply requiring that in standard coordinates (ξ^i) we are in fact computing partial derivatives.

For example, for a (differentiable) contravariant vector u^i, in (x^i), we obtain the corresponding U^i in (ξ^i), then form $W_j^i = \partial U^i / \partial \xi^j$. The W_j^i determine a mixed second-order tensor, becoming w_j^i in (x^i), and since we have added a covariant index, we call w_j^i the *covariant derivative* of u^i and denote it by $\Delta_j u^i$. It remains to calculate the w_j^i. We have

$$U^i = \frac{\partial \xi^i}{\partial x^l} u^l,$$

$$W_j^i = \frac{\partial U^i}{\partial \xi^j} = \frac{\partial U^i}{\partial x^k} \frac{\partial x^k}{\partial \xi^j} = \left(\frac{\partial^2 \xi^i}{\partial x^k \partial x^l} u^l + \frac{\partial \xi^i}{\partial x^l} \frac{\partial u^l}{\partial x^k} \right) \frac{\partial x^k}{\partial \xi^j},$$

$$w_j^i = W_q^p \frac{\partial x^i}{\partial \xi^p} \frac{\partial \xi^q}{\partial x^j} = \frac{\partial^2 \xi^p}{\partial x^k \partial x^l} \frac{\partial x^k}{\partial \xi^q} \frac{\partial x^i}{\partial \xi^p} \frac{\partial \xi^q}{\partial x^j} u^l + \frac{\partial \xi^p}{\partial x^l} \frac{\partial x^k}{\partial \xi^q} \frac{\partial x^i}{\partial \xi^p} \frac{\partial \xi^q}{\partial x^j} \frac{\partial u^l}{\partial x^k}.$$

Here we can simplify by noting various products of inverse matrices appearing and can write our result as

$$\Delta_j u^i = w_j^i = \frac{\partial u^i}{\partial x^j} + \Gamma_{jl}^i u^l, \tag{3.92}$$

where

$$\Gamma_{jl}^i = \frac{\partial^2 \xi^p}{\partial x^j \partial x^l} \frac{\partial x^i}{\partial \xi^p} = \Gamma_{lj}^i. \tag{3.93}$$

The Γ_{jl}^i are called *Christoffel symbols*; they are not the components of a tensor. One can express them in terms of the components of the fundamental metric tensor (Problem 12):

$$\Gamma_{jl}^i = \tfrac{1}{2} g^{is} \left(\frac{\partial g_{js}}{\partial x^l} + \frac{\partial g_{ls}}{\partial x^j} - \frac{\partial g_{jl}}{\partial x^s} \right). \tag{3.94}$$

By proceeding in this way we obtain covariant derivatives of other types of tensors. For example, we find

$$\Delta_j u_i = \frac{\partial u_i}{\partial x^j} - \Gamma_{ij}^l u_l, \tag{3.95}$$

$$\Delta_k u_j^i = \frac{\partial u_j^i}{\partial x^k} + \Gamma_{kl}^i u_j^l - \Gamma_{kj}^l u_l^i, \tag{3.96}$$

$$\Delta_k u_{jh}^i = \frac{\partial u_{jh}^i}{\partial x^k} + \Gamma_{kl}^i u_{jh}^l - \Gamma_{kj}^l u_{lh}^i - \Gamma_{kh}^l u_{jl}^i. \tag{3.97}$$

From these results the general pattern should be clear.

Contravariant derivatives are found in similar fashion. In fact one has the following simple rule: For every tensor A,

$$\Delta^k A = g^{k\alpha} \Delta_\alpha A. \tag{3.98}$$

To justify this, we observe that the right-hand side is a tensor of the type desired and that in standard coordinates (ξ^i) it reduces to the set of partial derivatives of the components of A with respect to x^k. Equation (3.98) expresses contravariant derivatives in terms of covariant derivatives and shows that $\Delta_k A$ and $\Delta^k A$ are associated tensors.

For an invariant f, one finds

$$\Delta_k f = \frac{\partial f}{\partial x^k}, \qquad \Delta^k f = g^{k\alpha} \frac{\partial f}{\partial x^k}. \tag{3.99}$$

Thus the covariant derivative of an invariant f is simply the *gradient* of f:

$$\Delta_k f = \operatorname{grad} f = \frac{\partial f}{\partial x^k}. \tag{3.100}$$

The contravariant derivative of f is the associated contravariant vector.

We can now define the *divergence* of a vector field. We want this to reduce to the usual divergence in standard coordinates. If $\mathbf{u}$ has contravariant components u^i, then we define

$$\operatorname{div} \mathbf{u} = \operatorname{div} u^i = \Delta_\alpha u^\alpha, \tag{3.101}$$

since this reduces to the expected expression

$$\frac{\partial U^1}{\partial \xi^1} + \cdots + \frac{\partial U^n}{\partial \xi^n} \tag{3.102}$$

in standard coordinates. Since $\Delta_\alpha u^\alpha$ is obtained by contraction of the tensor $\Delta_i u^j$, it is a tensor of order 0, an invariant. We can obtain the same invariant from the covariant components of $\mathbf{u}$:

$$\operatorname{div} \mathbf{u} = \operatorname{div} u_i = \Delta^\alpha u_\alpha, \tag{3.103}$$

for in standard coordinates the components become $U_i = U^i$, and again (3.102) is obtained.

One can also obtain (Problem 13) the following useful expression for the divergence:

$$\operatorname{div} \mathbf{u} = \frac{1}{\sqrt{g}} \frac{\partial}{\partial x^\lambda} \left(\sqrt{g}\, g^{\alpha\lambda} u_\alpha \right). \tag{3.104}$$

From the divergence and gradient we obtain the *Laplacian* of an invariant:

$$\nabla^2 f = \operatorname{div} \operatorname{grad} f = \Delta^\alpha \Delta_\alpha f. \tag{3.105}$$

We can also obtain this from the contravariant components of grad f:

$$\nabla^2 f = \Delta_\alpha \Delta^\alpha f. \tag{3.106}$$

Corresponding to (3.104), one has also the useful expression:

$$\nabla^2 f = \frac{1}{\sqrt{g}} \frac{\partial}{\partial x^\lambda} \left(\sqrt{g}\, g^{\alpha\lambda} \frac{\partial f}{\partial x^\alpha} \right). \tag{3.107}$$

For the curl of a vector field our usual definition is specially adapted to 3-dimensional space. The appropriate generalization turns out to be a *second-order* tensor. We define the *curl* of $\mathbf{u}$ to be the second-order tensor

$$b_{ij} = \Delta_i u_j - \Delta_j u_i. \tag{3.108}$$

However, we verify (Problem 14) that this is equivalent to

$$\operatorname{curl} \mathbf{u} = b_{ij} = \frac{\partial u_j}{\partial x^i} - \frac{\partial u_i}{\partial x^j}. \tag{3.109}$$

Thus the form that (3.108) takes in standard coordinates is the form that it takes in arbitrary coordinates. (The same is true for the gradient operation.) By (3.109) we have

$$b_{ij} = -b_{ji}, \quad b_{ij} = 0 \quad \text{if} \quad i = j. \tag{3.110}$$

In 3-dimensional space the curl has nine components. By (3.110), three of these are 0, and the other six come in pairs of opposite sign:

$$b_{23} = -b_{32} = \frac{\partial u_3}{\partial x^2} - \frac{\partial u_2}{\partial x^3}, \qquad b_{31} = -b_{13} = \frac{\partial u_1}{\partial x^3} - \frac{\partial u_3}{\partial x^1},$$

$$b_{12} = -b_{21} = \frac{\partial u_2}{\partial x^1} - \frac{\partial u_1}{\partial x^2}. \tag{3.111}$$

These three values are the familiar components of the curl vector in space: $\partial u_z/\partial y - \partial u_y/\partial z, \ldots$. It is because these *three* quantities determine the curl that the curl can be interpreted as a vector in 3-dimensional space. In 4-dimensional space, six components are needed, and one cannot associate these with a vector. For $n = 2$, one has essentially one component

$$b_{12} = -b_{21} = \frac{\partial u_2}{\partial x^1} - \frac{\partial u_1}{\partial x^2}$$

(which is the $\partial Q/\partial x - \partial P/\partial y$ of Green's theorem—see Chapter 5).

Case of orthogonal curvilinear coordinates.   We consider here only the case of 3-dimensional space. In Section 3.7 it is shown that for orthogonal coordinates, one has $ds^2 = \alpha^2\,dx^2 + \beta^2\,dy^2 + \gamma^2\,dz^2$—that is, in the present notations,

$$ds^2 = g_{11}\,dx^1\,dx^1 + g_{22}\,dx^2\,dx^2 + g_{33}\,dx^3\,dx^3.$$

Thus (g_{ij}) is a diagonal matrix; its inverse is the diagonal matrix $\operatorname{diag}(1/g_{11}, 1/g_{22}, 1/g_{33})$, and its determinant $g = g_{11}g_{22}g_{33}$. From (3.107) the Laplacian of f is

$$\frac{1}{\sqrt{g}}\frac{\partial}{\partial x^1}\left(\frac{\sqrt{g}}{g_{11}}\frac{\partial f}{\partial x^1}\right) + \frac{1}{\sqrt{g}}\frac{\partial}{\partial x^2}\left(\frac{\sqrt{g}}{g_{22}}\frac{\partial f}{\partial x^2}\right) + \frac{1}{\sqrt{g}}\frac{\partial}{\partial x^3}\left(\frac{\sqrt{g}}{g_{33}}\frac{\partial f}{\partial x^3}\right).$$

With $g_{11} = \alpha^2$, $g_{22} = \beta^2$, $g_{33} = \gamma^2$, $g = \alpha^2\beta^2\gamma^2$, this reduces to (3.63).

One can discuss $\operatorname{div} \mathbf{u}$ in the same way for the case of orthogonal coordinates. One appears to obtain a formula differing from (3.61). However, as (3.57′) shows, the components p_u, p_v, p_w of Section 3.8 are neither contravariant nor covariant components of the vector $\mathbf{p}$; these components are obtained geometrically by perpendicular projection. From (3.57′) we see that p_u/α, p_v/β, p_w/γ are contravariant components of **p** and can now verify that (3.61) is in agreement with (3.104) for the case of orthogonal coordinates.

For further information on tensors, one is referred to the books by Brand, Coburn, Hay, Levi-Civita, Rainich, and Weyl listed at the end of this chapter.

*3.10   TENSORS ON A SURFACE OR HYPERSURFACE

A surface S in space can be represented by parametric equations $x = f(u,v), y = g(u,v), z = h(u,v)$. For example,

$$x = \sin u \cos v, \quad y = \sin u \sin v, \quad z = \cos u$$

are parametric equations of a sphere (see Problems 8–10 following Section 3.8). There are many parametrizations of a chosen surface; a second one for a sphere is $x = u, y = v, z = (1 - u^2 - v^2)^{1/2}$.

It is natural to allows *arbitrary parametrizations,* as we have done for coordinates in Section 3.9. If we are concentrating on a neighborhood D of a point on the surface, this leads us to consider all changes of coordinates from (u,v) to $(\bar{u},\bar{v})$, $(\bar{\bar{u}},\bar{\bar{v}})$, ... Here for example,

$$\bar{u} = \bar{u}(u,v), \quad \bar{v} = \bar{v}(u,v)$$

in a neighborhood of (u_0, v_0), with inverse

$$u = u(\bar{u},\bar{v}), \quad v = v(\bar{u},\bar{v})$$

just as in Eqs. (3.72), (3.73).

Indeed even though $(u,v),\dots$ are regarded as parameters, we are here dealing with the case $n = 2$ of Section 3.9: that is, we have a collection of 2-dimensional coordinate systems, all referring to the points of a neighborhood D of a point on the surface and related by appropriately differentiable functions. The only difference is that there are no natural *standard* coordinates.

The absence of standard coordinates does not prevent our developing the theory of tensors. We can again do this for the n-dimensional case: that is, we consider a hypersurface of n-dimensions. This is given, for a neighborhood of a point, by a set of (parameter) coordinate systems $(x^1,\dots,x^n)$, $(\bar{x}^1,\dots,\bar{x}^n),\dots$ related by Eqs. (3.72), (3.73), as in Section 3.9. Contravariant vectors, covariant vectors and tensors in general are defined as in Section 3.9, except that one has no preferred set of standard coordinates.

Typically one can measure distance on a hypersurface and hence has again $ds^2 = g_{ij}dx^i dx^j$, where g_{ij} is a covariant tensor. However, in the absence of standard coordinates, g_{ij} in general never reduces to δ_{ij}. We can nevertheless define g_j^i to be δ_{ij} and g^{ij} to be the inverse matrix of g_{ij}. We verify that g_j^i and g^{ij} are tensors of the types indicated. With the aid of these tensors, we can again raise and lower indices to obtain 'associated tensors.'

With the aid of the g_{ij} and g^{ij} we can *define* covariant derivatives by Eqs. (3.92), (3.94)–(3.97) and verify that these are tensors of the appropriate types.

Thus it appears that, although the standard coordinates were the basis of many of the definitions in Section 3.9, we can wholly dispense with them to obtain a satisfactory theory of tensors on a surface or n-dimensional hypersurface.

*3.11 ALTERNATING TENSORS • EXTERIOR PRODUCT

We restrict attention to invariants and *covariant* tensors $u_i, u_{ij}, \ldots$. Each such tensor is called *alternating* if interchange of two subscripts changes the sign: for example, u_{ij} is alternating if

$$u_{ij} = -u_{ji} \quad (i = 1, \ldots, n, \quad j = 1, \ldots, n) \tag{3.112}$$

and u_{ijk} is alternating if for $1 \le i, j, k \le n$

$$u_{ijk} = -u_{jik} = -u_{ikj} = -u_{kji}. \tag{3.113}$$

In general, the sign is reversed (unchanged) after an odd (even) number of interchanges (cf. Section 1.3 and Problem 16 following Section 1.4). Because of (3.112), $u_{ij} = 0$ if $i = j$ and, in general, each component having two subscript equal is 0. We remark that, if a covariant tensor is alternating in one coordinate system, then it is alternating in each allowed coordinate system (Problem 20 following Section 3.11). Thus we are dealing with a true tensor property.

For convenience we include invariants and first order covariant tensors as alternating tensors. For a tensor $u_{ij\ldots}$ of order n, each component with n different subscripts equals $\pm u_{12\ldots n}$. For tensors of order $m > n$, all components are 0, since of the m subscripts at least two must be equal.

It is easily seen that addition of two alternating tensors of the same order produces an alternating tensor of that order, as does multiplication of such a tensor by an invariant. However, the tensor product of two alternating tensors of order at least 1 is in general not alternating (Problem 21 below).

To remedy this deficiency, we observe that from each covariant tensor (order at least 2) we can obtain an alternating tensor by forming a suitable "average" of the tensors obtained from the given one by permuting indices. Thus from u_{ij} we form $v_{ij} = \frac{1}{2}(u_{ij} - u_{ji})$ and from u_{ijk} we form

$$v_{ijk} = \frac{1}{6}(u_{ijk} + u_{kij} + u_{jki} - u_{jik} - u_{ikj} - u_{kji}).$$

For a tensor $u_{i_1 \ldots i_m}$ of order m we form

$$v_{i_1 \ldots i_m} = \frac{1}{m!} \sum \eta_{j_1 j_2 \ldots j_m} u_{j_1 j_2 \ldots j_m}, \tag{3.114}$$

where the sum is over all permutations of $i_1, \ldots, i_m$ and $\eta_{j_1 j_2 \ldots j_m}$ is $+1$ for an even permutation, -1 for an odd one (cf. Problem 16 following Section 1.4); when two of $i_1, \ldots, i_m$ are equal, both sides of (3.114) are 0.

We write $\mathcal{A}(u_{ij})$, $\mathcal{A}(u_{ijk})$, $\ldots$ for the tensors formed in this way from u_{ij}, u_{ijk}, $\ldots$ and call $\mathcal{A}$ the *alternation operator*. We also write $\mathcal{A}(u_i) = u_i$, $\mathcal{A}(f) = f$ for orders 1 and 0. It is left as an exercise to verify that the new objects are indeed alternating tensors (Problem 22 following Section 3.11).

With the aid of the operators $\mathcal{A}$ we can now obtain an alternating product of two alternating tensors. We can simply apply $\mathcal{A}$ to the tensor product of the two tensors. The result, slightly modified by a constant factor, is called the exterior product (or *wedge product*) of the two alternating tensors: If $u_{i_1 \ldots i_r}$ and $v_{j_1 \ldots j_s}$ are alternating tensors, the exterior product is written as $u_{i_1 \ldots i_r} \wedge v_{j_1 \ldots j_s}$ and is defined as

$$\frac{(r+s)!}{r!\,s!} \mathcal{A}(u_{i_1 \ldots i_r} v_{j_1 \ldots j_s}).$$

Thus $u_i \wedge v_{jk}$ equals

$$\frac{3!}{1!\,2!} \frac{1}{3!} \left(u_i v_{jk} + u_k v_{ij} + u_j v_{ki} - u_j v_{ik} - u_i v_{kj} - u_k v_{ji} \right)$$
$$= u_i v_{jk} + u_k v_{ij} + u_j v_{ki}.$$

This is a new alternating tensor w_{ijk} of order 3.

For more information on alternating tensors and references, see Sections 5.19 to 5.21.

PROBLEMS

1. In E^2 let (ξ^1, ξ^2) be standard coordinates and let (x^1, x^2) be new coordinates given by $x^1 = 3\xi^1 + 2\xi^2$, $x^2 = 4\xi^1 + 3\xi^2$. Find the (x^i) components of the following tensors, for which the components in standard coordinates are given.

   a) u_i, where $U_1 = \xi^1 \xi^2$, $U_2 = \xi^1 - \xi^2$.

   b) v^i, where $V^1 = \xi^1 \cos \xi^2$, $V^2 = \xi^1 \sin \xi^2$.

   c) w_{ij}, where $W_{11} = 0$, $W_{12} = \xi^1 \xi^2$, $W_{21} = -\xi^1 \xi^2$, $W_{22} = 0$.

   d) z^i_j, where $Z^1_1 = \xi^1 + \xi^2$, $Z^1_2 = Z^2_1 = 3\xi^1 + 2\xi^2$, $Z^2_2 = \xi^1 - \xi^2$.

2. For the coordinates of Problem 1, find (a) the components of the fundamental metric tensor g_{ij} in standard coordinates and in the (x^i) and (b) the g^i_j and g^{ij} in both sets of coordinates.

3. With reference to the coordinates and tensors of Problem 1 and with the aid of the results of Problem 2, find the following tensor components in (x^i):

   a) u^i, associated to u_i;   b) v_i, associated to v^i;   c) w^{ij}, associated to w_{ij}.

4. a) Show that if a tensor has all components 0 at a point in one coordinate system, then all components are 0 at the point in every allowed coordinate system.

b) Show that if two tensors of the same type have corresponding components equal (identically) in one coordinate system, then corresponding components are equal in every allowed coordinate system. (One then says that the tensors are equal.)

c) Show that if a tensor u_{ij} is such that $u_{ij} = u_{ji}$ in one coordinate system (x^i), then $\bar{u}_{ij} = \bar{u}_{ji}$ in every other coordinate system $(\bar{x}^i)$. (One then calls the tensor *symmetric*.)

d) Show that if a tensor u_{ij} is such that $u_{ij} = -u_{ji}$ in one coordinate system (x^i), then $\bar{u}_{ij} = -\bar{u}_{ji}$ in every other coordinate system $(\bar{x}^i)$.
(One calls such a tensor *alternating*, as in Section 3.11.)

5. Prove: If components U_i are defined in (ξ^i), and corresponding components u_i are defined in every other allowed coordinate system (x^i) by the equation $u_i = (\partial \xi^j / \partial x^i) U_j$, then the components $\bar{u}_i$ in $(\bar{x}^i)$ satisfy (3.75), so that a covariant vector has been defined.

6. Prove (3.87) from (3.84) and (3.86).

7. Prove that $g = \det(g_{ij})$ is positive. [Hint: Show from (3.84) that (g_{ij}) is the product of two matrices and then take determinants.]

8. a) Show that, the tensors v^i_j and w^i_j being given in a neighborhood D, Eq. (3.90) defines u^i_j as components of a tensor.

b) Generalize the result of part (a) to addition of two tensors of arbitrary type.

9. Show that multiplication of all tensor components by the same invariant, as in (3.91), defines a new tensor.

10. Let $u_i, v_j, w^k, z^l, p_{mh}$ be tensors. Show that each of the following tensor products is a tensor:

a) $u_i v_j = s_{ij}$ **b)** $u_i w^k = t^k_i$ **c)** $w^k z^l = q^{kl}$ **d)** $w^k p_{mh} = d^k_{mh}$

11. a) Let u^{ij}_{kl} be a tensor of the type indicated. Show that each of the following is a tensor: $w^j_k = u^{ij}_{ki}, v^i_k = u^{ij}_{kj}$.

b) Prove generally that contraction of a tensor yields a tensor.

12. Show from (3.93) that the Γ^i_{ji} can be expressed as in (3.94). [Hint: Show first that $\Gamma^i_{jl} g_{is} = (\partial^2 \xi^\alpha / \partial x^j \partial x^l)(\partial \xi^\alpha / \partial x^s)$. Next, with the aid of (3.84), compute the quantity in parentheses in (3.94) and show that it equals $2\Gamma^i_{jl} g_{is}$. Now multiply both sides by g^{st}.]

13. Derive the formula (3.104). [Hint: First show that

$$\operatorname{div} \mathbf{u} = (\partial u^\alpha / \partial x^\alpha) + \Gamma^\alpha_{\alpha l} u^l.$$

Next put $i = j = \alpha$ in (3.94), multiply out, and show that the last two terms cancel, so that $\Gamma^\alpha_{\alpha l} = \frac{1}{2} g^{\alpha s} (\partial g_{\alpha s} / \partial x^l)$. Now obtain $\partial g / \partial x^l$, from the determinant g, as a sum of n determinants, the αth of which is obtained from the given determinant by differentiating each element in the αth row with respect to x^l. Expand each of these determinants by minors of the αth row to obtain

$$\partial g / \partial x^l = A_{\alpha s} \partial g_{\alpha s} / \partial x^l,$$

where $A_{\alpha s}$ is the cofactor of $g_{\alpha s}$ in the determinant g. Now $g^{\alpha s} = g^{s\alpha} = A_{\alpha s}/g$ (see Problem 7 following Section 1.8). Thus conclude that $\Gamma^\alpha_{\alpha l} = (2g)^{-1}(\partial g / \partial x^l)$ and show that this permits one to write $\operatorname{div} \mathbf{u}$ in the form asserted.]

14. Derive (3.109) from (3.108), using (3.95).

15. We define the *norm* of a vector **u** to be its norm or length in standard coordinates. Thus for components U^i or U_i the length is

$$|\mathbf{u}| = (U^i U^i)^{\frac{1}{2}} = (U_i U_i)^{\frac{1}{2}}.$$

Show that |**u**| can be written in the forms:

$$|\mathbf{u}| = \left(u_i u^i\right)^{\frac{1}{2}}, \qquad |\mathbf{u}| = \left(g_{ij} u^i u^j\right)^{\frac{1}{2}}, \qquad |\mathbf{u}| = \left(g^{ij} u_i u_j\right)^{\frac{1}{2}},$$

each of which shows that the norm is an invariant.

16. The *inner product* of two vectors **u**, **v** is defined to be the invariant

$$(\mathbf{u}, \mathbf{v}) = u_i v^i.$$

Show that this can be written in the forms:

$$(\mathbf{u}, \mathbf{v}) = g_{ij} u^i v^j, \qquad (\mathbf{u}, \mathbf{v}) = g^{ij} u_i v_j$$

and that

$$(\mathbf{u}, \mathbf{u}) = |\mathbf{u}|^2.$$

17. **a)** Prove: If u^i, v^j are contravariant vectors and $b_{ij} u^i v^j$ is an invariant, then b_{ij} is a covariant tensor.

b) Prove generally: If $u_i, v_j, \ldots, w^k, z^l, \ldots$ are covariant and contravariant vectors and

$$b_{kl\ldots}^{ij\cdots} u_i v_j \ldots w^k z^l \ldots$$

is an invariant, then $b_{kl\ldots}^{ij\cdots}$ is a tensor of the type indicated.

18. **a)** Show that the associated tensor b^{ij} to the curl b_{ij} of a covariant vector u_i (as in Eq. (3.109)) is given by

$$b^{ij} = g^{ik} g^{jl} \left(\frac{\partial u_\ell}{\partial x^k} - \frac{\partial u_k}{\partial x^\ell} \right).$$

b) Show that for $n = 3$ in orthogonal coordinate as in Section 3.8, the three nonzero components b^{23}, b^{31}, b^{12} of the tensor b^{ij} are the same as (3.62). [Note that the covariant components of **p** are αp_u, βp_v, γp_w by (3.58').]

c) Verify that b_{ij} is an alternating tensor.

19. Show that in 3-dimensional space the exterior product $w_{ij} = u_i \wedge v_j$ of two covariant vectors u_i, v_j can be interpreted in terms of the vector product. In detail, show that, in standard coordinates x, y, z with usual orientation and $u_1 = u_x, \ldots, v_1 = v_x, \ldots$, the three components of the vector product $(u_x \mathbf{i} + \cdots) \times (v_x \mathbf{i} + \cdots)$ appear with + or − signs as the nonzero components of w_{ij}.

20. Let $u_{i_1 \ldots i_r}$ be a covariant tensor in E^n ($r \geq 2$). Let this tensor be alternating in one coordinate system. Show that it is alternating in each allowed coordinate system.

21. Let u_i, v_i, w_{ij} be alternating covariant tensors in E^3.

a) Show that $u_i v_j$ fails to be alternating if $u_1 v_1 \neq 0$.

b) Show that $u_i w_{jk}$ fails to be alternating if $u_1 w_{21} \neq 0$.

c) In one coordinate system let $u_i = 1$ for $i = 1, 2, 3$. Show that if $u_i v_j$ is alternating, then $v_j = 0$ for $j = 1, 2, 3$.

22. Show that if $u_{i_1 \ldots i_m}$ is a covariant tensor, then Eq. (3.114) defines a new covariant tensor $v_{i_1 \ldots i_m}$ which is alternating.

Differential geometry of surfaces in space. In the following problem let S be a surface in space as in Problems 8–10 following Section 3.8. Thus one has curvilinear coordinates (u, v) on S and a positive definite quadratic differential form $ds^2 = E \, du^2 + 2F \, du \, dv + G \, dv^2$, called the *first fundamental form* of S. If we write $u^1 = u$, $u^2 = v$, $g_{11} = E$, $g_{12} = g_{21} = F$, $g_{22} = G$, then $ds^2 = g_{ij} du^i du^j$ and, as in Section 3.10, g_{ij} is a covariant second order tensor. Also $g = \det(g_{ij}) = EG - F^2 > 0$.

As in Problem 16 following Section 2.13 a unit normal vector for S can be obtained as $\mathbf{n} = \mathbf{w}/|\mathbf{w}|$, where $\mathbf{w} = \partial \mathbf{r}/\partial u \times \partial \mathbf{r}/\partial v$. The choice of normal $\mathbf{n}$ (varying from point to point) will be assumed *fixed* and unaffected by changes of coordinate on S. (We are here dealing with an *oriented surface* as in Section 5.9)

23. Let a path C on S be given by parametric equations $u = u(t)$, $v = v(t)$ in curvilinear coordinates, so that $\mathbf{r} = \mathbf{r}(u(t), v(t))$ is the vector equation of the path. We assume that t is arc length s on C and that appropriate derivatives exist, so that the Frenet formulas of Problem 6 following Section 2.13 apply. We write $\mathbf{r}_u, \mathbf{r}_v, \mathbf{r}_{uu}, \mathbf{r}_{uv}, \mathbf{r}_{vv}$ for partial derivatives, all assumed to be continuous.

a) Show that the unit tangent vector $\mathbf{T} = d\mathbf{r}/ds$ satisfies the equation $\mathbf{T} = (du/ds)\mathbf{r}_u + (dv/ds)/\mathbf{r}_v$, where

$$E \left(\frac{du}{ds} \right)^2 + 2F \frac{du}{ds} \frac{dv}{ds} + G \left(\frac{dv}{ds} \right)^2 = 1.$$

b) Show that the acceleration vector $\mathbf{a} = d\mathbf{T}/ds$ satisfies the equation

$$\mathbf{a} = \left(\frac{du}{ds} \right)^2 \mathbf{r}_{uu} + 2 \frac{du}{ds} \frac{dv}{ds} \mathbf{r}_{uv} + \left(\frac{dv}{ds} \right)^2 \mathbf{r}_{vv} + \frac{d^2 u}{ds^2} \mathbf{r} + \frac{d^2 v}{ds^2} \mathbf{r}_v$$

c) Show that the curvature κ satisfies the equation

$$\kappa \mathbf{N} \cdot \mathbf{n} = L \left(\frac{du}{ds} \right)^2 + 2M \frac{du}{ds} \frac{dv}{ds} + N \left(\frac{dv}{ds} \right)^2,$$

where $L = \mathbf{r}_{uu} \cdot \mathbf{n}$, $M = \mathbf{r}_{uv} \cdot \mathbf{n}$, $N = \mathbf{r}_{vv} \cdot \mathbf{n}$. The quadratic differential form $L \, du^2 + 2M \, du \, dv + N \, dv^2$ is called the *second fundamental form* of S.

d) Write $L_{11} = L$, $L_{12} = L_{21} = M$, $L_{22} = N$ and show that L_{ij} is a tensor as indicated. [Hint: Fix a point $P_0 : (u_0, v_0)$ on S and choose a special basis $\mathbf{i}_1, \mathbf{j}_1, \mathbf{k}_1$ for vectors in space with $\mathbf{k}_1 = \mathbf{n}$ at P_0. Then write $\mathbf{r} = \mathbf{r}_0 + \xi \mathbf{i}_1 + \eta \mathbf{j}_1 + \zeta \mathbf{k}_1$, with $(\xi, \eta, \zeta) = (0, 0, 0)$ at P_0. Show that at P_0 $\mathbf{r}_u \cdot \mathbf{n} = 0$ and $\mathbf{r}_v \cdot \mathbf{n} = 0$, so that at (u_0, v_0) $\zeta_u = 0$ and $\zeta_v = 0$ and $L = \zeta_{uu}, M = \zeta_{uv}, N = \zeta_{vv}$. These conditions must hold at P_0 for each allowable coordinate system (u, v). For new coordinates $(\bar{u}, \bar{v})$ with $\bar{u} = \bar{u}_0, \bar{v} = \bar{v}_0$ at P_0, show by the rules of Section 2.16 (Eq. (2.133) and Problem 6 following that section) that at P_0

$$\zeta_{uu} = \zeta_{\bar{u}\bar{u}} (\bar{u}_u)^2 + 2\zeta_{\bar{u}\bar{v}} \bar{u}_u \bar{v}_u + \zeta_{\bar{v}\bar{v}} (\bar{v}_u)^2$$

and obtain analogous formulas for ζ_{uv} and ζ_{vv}. Show that in tensor notation these become

$$L_{ij} = \bar{L}_{k\ell} \frac{\partial u^k}{\partial \bar{u}^i} \frac{\partial u^\ell}{\partial \bar{u}^j} \cdot]$$

24. *Normal curvature on S.* The result of Problem 19(c) show that all paths C on S through a chosen point P_0 with fixed tangent vector $\mathbf{T}$ and principal normal $\mathbf{N}$ at P_0, have the same curvature κ at P_0. Hence it suffices to consider plane sections of S. We consider such sections by planes through P_0, other than the tangent plane to S at P_0, and let $\theta = \measuredangle (\mathbf{N}, \mathbf{n})$ for each path C obtained as such a plane section.

   a) Show that for fixed $\mathbf{T}$ and variable $\mathbf{N}$ one has $\kappa \cos \theta = $ const. at P_0. Here the constant has absolute value κ_n, where κ_n, called the *normal curvature* of S, is the curvature at P_0 of the section of the surface containing the tangent line to the path at P_0 and the normal line to the surface at P_0. [Thus $\kappa = \pm \kappa_n$; the $\pm$ sign follows from the fact that either $0 \le \theta < \pi/2$ or $\pi/2 < \theta \le \pi$ for all $\mathbf{N}$.]

   b) Show that

$$\kappa_n = \left| L \left(\frac{du}{ds} \right)^2 + 2M \frac{du}{ds} \frac{dv}{ds} + N \left(\frac{dv}{ds} \right)^2 \right|.$$

25. *Mean and Gaussian curvatures.* Seek the maximum and minimum of the normal curvature κ_n for variable tangent vector $\mathbf{T}$ at a fixed point P_0. [Hint: Let $\xi = du/ds$, $\eta = dv/ds$. Then by Problem 24(b) $\kappa_n = |L\xi^2 + 2M\xi\eta + N\eta^2|$ and one is led to seek the critical points of $f(\xi, \eta) = L\xi^2 + 2M\xi\eta + N\eta^2$, where $E\xi^2 + 2F\xi\eta + G\eta^2 = 1$. Use Lagrange multipliers as in Problems 14, 15 following Section 2.21 to obtain the quadratic equation

$$\begin{vmatrix} L - \lambda E & M - \lambda F \\ M - \lambda F & N - \lambda G \end{vmatrix} = 0$$

with real roots $\lambda_1 \le \lambda_2$ and corresponding eigenvectors $\mathbf{v}_1 = (\xi_1, \eta_1)$, $\mathbf{v}_2 = (\xi_2, \eta_2)$. Conclude that f has an absolute minimum λ_1 and absolute maximum λ_2 under the side condition. One terms $H = \frac{1}{2}(\lambda_1 + \lambda_2)$ the *mean curvature* and $K = \lambda_1 \lambda_2$ the *Gaussian curvature* of the surface at P_0. By Problem 14(b) following Section 2.21, if $\lambda_1 \ne \lambda_2$, then

$$\mathbf{v}_1' \begin{bmatrix} E & F \\ F & G \end{bmatrix} \mathbf{v}_2 = 0$$

and hence the corresponding tangent vectors are orthogonal (Problem 10(e) following Section 3.8). The corresponding directions, or their opposites, are called *directions of principal curvature*. If $\lambda_1 = \lambda_2$, $f(\xi, \eta)$ has a constant value λ_1 on the ellipse $E\xi^2 + 2F\xi\eta + G\eta^2 = 1$ and hence $f = \lambda_1(E\xi^2 + 2F\xi\eta + G\eta^2)$ for all (ξ, η); if further $\lambda_1 \ne 0$, then the point P_0 is called an *umbilical point* of S—here all plane sections of S through P_0 have the same curvature $|\lambda_1|$; if $\lambda_1 = \lambda_2 = 0$, then P_0 is called a *parabolic umbilic* of S.]

Suggested References

Brand, Louis, Vector and Tensor Analysis. New York: John Wiley and Sons, Inc., 1947.
Coburn, N., Vector and Tensor Analysis. New York: Macmillan, 1955.

Gibbs, J. Willard, Vector Analysis. New Haven: Yale University Press, 1913.

Hay, G. E., Vector and Tensor Analysis. New York: Dover, 1954.

Levi-Civita, T., The Absolute Differential Calculus, transl. by M. Long. London: Blackie and Son, 1927.

Phillips, H. B., Vector Analysis. New York: John Wiley and Sons, Inc., 1933.

Rainich, G. Y., Mathematics of Relativity. New York: John Wiley and Sons, Inc., 1950. Especially Chapters 1, 2, and 4.

Struik, Dirk J., Lectures on Classical Differential Geometry. Cambridge, Mass.: Addison-Wesley Press, Inc., 1950.

Weyl, Hermann, Space, Time, Matter, transl. by H. L. Brose. London. Methuen, 1922.

4

Integral Calculus of Functions of Several Variables

4.1 THE DEFINITE INTEGRAL

We review here the properties of the integral for functions of one variable.

Let $f(x)$ be defined for $a \leq x \leq b$. The definite integral

$$\int_a^b f(x)\, dx$$

is defined as a limit:

$$\lim_{h \to 0} \sum_{i=1}^n f(x_i^*)\, \Delta_i x. \tag{4.1}$$

Here one is considering subdivisions of the interval $a \leq x \leq b$ by values $a = x_0 < x_1 < x_2 \cdots < x_n = b$, and $\Delta_i x = x_i - x_{i-1}$, $x_{i-1} \leq x_i^* \leq x_i$, h is the largest of $\Delta_1 x, \ldots, \Delta_n x$; we call h the *mesh* of the subdivision. The limit (4.1) is said to exist and have value c if for every $\epsilon > 0$, one can choose $\delta > 0$ so small that for *every* such subdivision of mesh h less than δ and no matter how the x_i^* are chosen in the intervals $x_{i-1} \leq x \leq x_i$, one has

$$\left| \sum_{i=1}^n f(x_i^*)\, \Delta_i x - c \right| < \epsilon. \tag{4.2}$$

The integral we have defined is called the *Riemann* integral. It can be shown to exist if, for example, $f(x)$ is continuous for $a \leq x \leq b$; see Section 4.10. Further, it exists if f is continuous except at a finite number of points, provided that f is *bounded*: $|f(x)| \leq M$ for some constant M, for $a \leq x \leq b$.

From the definition we deduce some basic properties of the integral (see CLA, Chapters 4 and 13):

$$\int_a^b [f(x) + g(x)]\, dx = \int_a^b f(x)\, dx + \int_a^b g(x)\, dx; \tag{4.3}$$

$$c\int_a^b f(x)\, dx = \int_a^b cf(x)\, dx \ (c = \text{const}); \tag{4.4}$$

$$\left|\int_a^b f(x)\, dx\right| \leq \int_a^b |f(x)|\, dx; \tag{4.5}$$

$$\left|\int_a^b f(x)\, dx\right| \leq M(b - a) \quad \text{if} \quad |f(x)| \leq M, \qquad a \leq x \leq b; \tag{4.6}$$

$$A(b - a) \leq \int_a^b f(x)\, dx \leq B(b - a)$$
$$\text{if} \quad A \leq f(x) \leq B, \qquad a \leq x \leq b. \tag{4.7}$$

These are all valid if, for example, f is continuous for $a \leq x \leq b$.

We call $F(x)$ a *primitive* or *antiderivative* of $f(x)$ or an *indefinite integral* of $f(x)$ if $F'(x) = f(x)$ on the interval. All such primitives are expressible as $F(x) + C$, where $F(x)$ is one primitive and C is an arbitrary constant. One writes

$$\int f(x)\, dx = F(x) + C \tag{4.8}$$

to describe the relationship.

If f is continuous on the interval, then one primitive of f is given by

$$F(x) = \int_a^x f(u)\, du. \tag{4.9}$$

Thus one has the rule, often called the *fundamental theorem of calculus*,

$$\frac{d}{dx}\int_a^x f(u)\, du = f(x), \qquad a \leq x \leq b. \tag{4.10}$$

The function $F(x)$ can be interpreted as the *area* under the graph of f from a to x, as in Fig. 4.1, provided that f is positive.

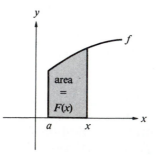

Figure 4.1 Function $F(x)$ of Eq. (4.9) as area.

If a primitive F of f is known for $a \leq x \leq b$, then

$$\int_a^b f(x)\, dx = \int_a^b F'(x)\, dx = F(x)\Big|_a^b = F(b) - F(a). \qquad (4.11)$$

This rule is often considered to be part of the fundamental theorem of calculus. It is the basis for the familiar procedures of elementary calculus; for example,

$$\int_0^{\pi/2} \sin 2x\, dx = -\frac{\cos 2x}{2}\Big|_0^{\pi/2} = -\frac{\cos \pi}{2} + \frac{1}{2} = 1.$$

The *average value* or *mean value* of f on the interval $a \leq x \leq b$ is

$$\frac{1}{b - a}\int_a^b f(x)\, dx. \qquad (4.12)$$

If $f(x)$ is continuous on the interval, then for some x^* on the interval, $f(x^*)$ equals the mean value; that is,

$$\int_a^b f(x)\, dx = f(x^*)(b - a). \qquad (4.13)$$

This is the *mean value theorem* for integrals.

If a continuous function f is such that

$$\int_{a_1}^{b_1} f(x)\, dx = 0 \qquad (4.14)$$

for every interval $a_1 \leq x \leq b_1$ contained in the given interval $a \leq x \leq b$, then $f(x) \equiv 0$. This follows from the mean value theorem (Problem 11 below). As a consequence of this result, we deduce that if f is continuous and $f(x) \geq 0$ for $a \leq x \leq b$, then

$$\int_a^b f(x)\, dx = 0$$

implies that $f(x) \equiv 0$ (see Problem 12 below).

For $b < a$, one defines:

$$\int_a^b f(x)\, dx = -\int_b^a f(x)\, dx, \qquad (4.15)$$

and for $a = b$ the integral is defined to be 0. One then verifies that for f continuous on an interval containing a, b, c,

$$\int_a^c f(x)\, dx = \int_a^b f(x)\, dx + \int_b^c f(x)\, dx. \qquad (4.16)$$

The rule $(uv)' = uv' + u'v$ leads to a procedure for integrating by parts, which we state for definite integrals:

$$\int_a^b u(x)v'(x)\, dx = u(x)v(x)\Big|_a^b - \int_a^b v(x)u'(x)\, dx. \qquad (4.17)$$

Here we assume that $u(x)$, $v(x)$, $u'(x)$, $v'(x)$ are continuous for $a \leq x \leq b$.

If the Riemann integral of f fails to exist, one may still be able to obtain a value by treating the integral as an *improper* integral. For example,

$$\int_0^1 \frac{1}{x^p}\, dx \qquad (p = \text{const})$$

does not exist as a Riemann integral for $p > 0$, since the integrand is unbounded, and the integral is improper because of the discontinuity at $x = 0$. One seeks a value by a limit process

$$\lim_{b \to 0+} \int_b^1 \frac{1}{x^p}\, dx.$$

For $p = 1$ the integral from b to 1 has the value $-\log b$, so that there is no limit as $b \to 0$. Otherwise, the integral has value

$$\frac{1 - b^{1-p}}{1 - p}$$

and this has limit $1/(1 - p)$ for $0 < p < 1$ but has no limit for $p > 1$. Hence one writes

$$\int_0^1 \frac{dx}{x^p} = \frac{1}{1 - p}, \qquad 0 < p < 1 \tag{4.18}$$

and assigns no value for $p \geq 1$. Here the improper integral is said to converge for $0 < p < 1$ and to diverge for $p \geq 1$.

If there are several discontinuities (a finite number) on the interval, one breaks the integral up into integrals with just one discontinuity; if each of these converges, then the given integral is termed convergent and assigned as value the sum of the values of the partial integrals; otherwise, it is termed *divergent*. For example, for an integral from 0 to 5 with discontinuities at 1 and 4, one considers the integrals from 0 to 1, 1 to 3, 3 to 4, and 4 to 5.

Integrals with infinite limits are also improper, and one evaluates them by a limit process. For example,

$$\int_0^\infty \frac{1}{1 + x^2}\, dx = \lim_{b \to \infty} \int_0^b \frac{1}{1 + x^2}\, dx = \lim_{b \to \infty} \arctan b = \frac{\pi}{2},$$

$$\int_{-\infty}^\infty x e^{-x^2}\, dx = \int_{-\infty}^0 x e^{-x^2}\, dx + \int_0^\infty x e^{-x^2}\, dx$$

$$= \lim_{b \to -\infty} \frac{e^{-x^2}}{-2}\Big|_b^0 + \lim_{b \to \infty} \frac{e^{-x^2}}{-2}\Big|_0^b$$

$$= -\tfrac{1}{2} + \tfrac{1}{2} = 0.$$

Improper integrals are discussed further in Sections 6.22 to 6.24.

Numerical evaluation of definite integrals.   The formula (4.11) is often of no help, since the primitive F is hard to find (or may not exist in usable form). Hence in practice, many definite integrals are evaluated numerically with the aid of

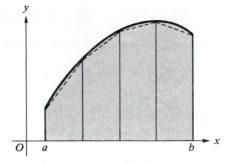

Figure 4.2 Trapezoidal rule
for $\int_a^b f(x)\, dx$.

computers. Sophisticated procedures to this end have been developed. We mention here only one simple rule, which is often sufficiently accurate, the trapezoidal
rule:

$$\int_a^b f(x)\, dx \sim \Delta_1 x \frac{f(a) + f(x_1)}{2} + \cdots + \Delta_n x \frac{f(x_{n-1}) + f(b)}{2}. \quad (4.19)$$

Here one has a subdivision of the interval as above. The right-hand side can be
interpreted as replacing the integrals from x_{i-1} to x_i by areas of trapezoids, as in
Fig. 4.2. When all $\Delta_i x$ are equal, the rule becomes

$$\int_a^b f(x)\, dx \sim \frac{b - a}{2n} [f(a) + 2f(x_1) + 2f(x_2) + \cdots + 2f(x_{n-1}) + f(b)].$$

$$(4.20)$$

If $|f''(x)| \le L = $ const on the interval, then one can show that the sum on the
right-hand side of (4.20) differs from the integral by at most $L(b - a)^3/(12n^2)$.
(See CLA, pp. 561–562.)

PROBLEMS

1. Evaluate the following indefinite integrals:

a) $\int x^2 \sin x\, dx$

b) $\int \dfrac{x\, dx}{1 + x^4}$

c) $\int \dfrac{dx}{(x - 1)(x - 2)}, 1 < x < 2$

d) $\int \dfrac{dx}{1 + \sqrt{x - 1}}, x \ge 1$

2. Evaluate the following definite integrals:

a) $\int_0^1 \sqrt{1 - x^2}\, dx$

b) $\int_0^\pi \sin 2x \sin 3x\, dx$

c) $\int_0^1 (2x^2 - 3x + 1)e^x \, dx$ **d)** $\int_0^1 \arctan x \, dx$

3. Show that the following integrals converge by evaluating the corresponding limits:

a) $\int_0^1 \dfrac{dx}{\sqrt{1 - x^2}}$,   **b)** $\int_0^\infty e^{-x} \, dx$,   **c)** $\int_0^1 \log x \, dx$,

d) $\int_1^\infty \dfrac{dx}{x\sqrt{1 + x^2}}$,   **e)** $\int_0^\infty x^2 e^{-x} \, dx$,   **f)** $\int_1^\infty \dfrac{\log x}{x^2} \, dx$

4. Evaluate where possible:

a) $\int_{-1}^1 \dfrac{dx}{x^{\frac{1}{3}}}$,   **b)** $\int_{-1}^1 \dfrac{dx}{x^3}$,   **c)** $\int_0^\infty \dfrac{dx}{1 + x^2}$,

d) $\int_0^\infty \dfrac{x^2 - x - 1}{x(x^3 + 1)} \, dx$,   **e)** $\int_0^\infty \sin x \, dx$,   **f)** $\int_0^\infty (1 - \tanh x) \, dx$

5. Find the area between the curves:

a) $y = 0$ and $y = 1 - x^2$,

b) $y = x^3$ and $y = x^{\frac{1}{3}}$,

c) $y = 6 \arc \sin x$ and $y = \pi \sin \pi x$.

6. Find the average value of the function given over the interval indicated:

a) $f(x) = \sin x$, $0 \le x \le \dfrac{\pi}{2}$;

b) $f(x) = \sin x$, $-\dfrac{\pi}{2} \le x \le 0$;

c) $f(x) = \sin^2 x$, $0 \le x \le \dfrac{\pi}{2}$;

d) $f(x) = ax + b$, $x_1 \le x \le x_2$.

7. Prove: If $f(x)$ and $g(x)$ are continuous for $a \le x \le b$ and $|g(x) - f(x)| \le \epsilon$ for $a \le x \le b$, then

$$\left| \int_a^b g(x) \, dx - \int_a^b f(x) \, dx \right| \le \epsilon(b - a).$$

[Hint: Use (4.6).]

8. On the basis of Problem 7, use the suggested approximation to evaluate each integral and estimate the error:

a) $\int_0^1 \sin(x^2) \, dx$, approximate $\sin(x^2)$ by $x^2 - (x^6/6)$;

b) $\int_0^1 e^{-x^2} \, dx$, approximate e^{-x^2} by $1 - x^2 + \dfrac{x^4}{2}$.

9. Show that if $f(x)$ is continuous for $0 \le x \le 1$, then

$$\lim_{n \to \infty} \frac{1}{n}\left[f\!\left(\frac{1}{n}\right) + f\!\left(\frac{2}{n}\right) + \cdots + f\!\left(\frac{n-1}{n}\right) + f\!\left(\frac{n}{n}\right) \right] = \int_0^1 f(x) \, dx.$$

10. Using the result of Problem 9, show that

a) $\lim_{n \to \infty} \dfrac{1 + 2 + \cdots + n}{n^2} = \int_0^1 x \, dx = \dfrac{1}{2}$

b) $\lim_{n \to \infty} \dfrac{1^2 + 2^2 + \cdots + n^2}{n^3} = \dfrac{1}{3}$

c) $\displaystyle\lim_{n\to\infty} \frac{1^P + 2^P + \cdots + n^P}{n^{P+1}} = \frac{1}{P+1}, P \geq 0$

d) $\displaystyle\lim_{n\to\infty} \frac{1}{n}\{(n+1)(n+2)\cdots(2n)\}^{\frac{1}{n}} = \frac{4}{e}$ [Hint: Use $f(x) = \log(1+x)$.]

11. Prove: If $f(x)$ is continuous for $a \leq x \leq b$ and

$$\int_{a_1}^{b_1} f(x)\, dx = 0$$

for every interval $a_1 \leq x \leq b_1$ contained in the interval $a \leq x \leq b$, then $f(x) \equiv 0$.
[Hint: Fix x_0 and apply the mean value theorem (4.13) to the interval $x_0 < x \leq x_0 + \delta$
(or $x_0 - \delta \leq x \leq x_0$) for small positive δ. Now let $\delta \to 0$.]

12. Prove: If $f(x)$ is continuous for $a \leq x \leq b$, $f(x) \geq 0$ on the interval, and

$$\int_a^b f(x)\, dx = 0,$$

then $f(x) \equiv 0$. [Hint: Take c, $a < c < b$, and observe that the integrals from a to c
and from c to b must be positive or 0, but their sum is 0, so both must be 0. Conclude
that $\int_{a_1}^{b_1} f(x)\, dx = 0$ for every choice of a_1, b_1 on the interval $a \leq x \leq b$ and hence,
by Problem 11, $f(x) \equiv 0$.]

4.2 NUMERICAL EVALUATION OF INDEFINITE INTEGRALS ▪ ELLIPTIC INTEGRALS

It is shown in advanced mathematics that the indefinite integral

$$\int e^{-x^2}\, dx$$

cannot be expressed in terms of elementary functions. It is, however, possible to
evaluate the definite integral

$$\int_a^b e^{-x^2}\, dx$$

as accurately as desired, by the methods mentioned in Section 4.1. We shall now
see that a similar statement holds for the indefinite integral.

The procedure to be given is an application of the fundamental theorem

$$\frac{d}{dx}\int_a^x f(t)\, dt = f(x) \tag{4.21}$$

of Eq. (4.10). As was pointed out in Section 4.1, the theorem implies that the
equation

$$F(x) = \int_a^x f(t)\, dt$$

defines a function $F(x)$ such that $F'(x) = f(x)$; that is, $F(x)$ is an indefinite
integral of $f(x)$; the complete indefinite integral is then given by

$$\int f(x)\, dx = F(x) + C.$$

The function $F(x)$ can be numerically evaluated, for each *fixed* x, by the procedures mentioned at the end of the preceding section. If this is done for a sequence of values of x covering the range $a \le x \le b$, $F(x)$ will be known in *tabular* form and can be graphed. It is in this sense that the indefinite integral is being numerically evaluated.

EXAMPLE   To evaluate

$$\int e^{\sin x}\, dx \tag{4.22}$$

for $0 \le x \le 1$, one computes

$$F(x) = \int_0^x e^{\sin t}\, dt$$

for $x = 0, 0.1, 0.2, \ldots, 1$. The work is shown in Table 4.1. The integrals are computed by the trapezoidal rule. The fourth column, which shows $F(x)$, is the "cumulative sum" of the third column. Thus the third column gives the differences ΔF, and if these are divided by $\Delta x = 0.1$, one should obtain $f(x)$ again approximately. For example,

$$\frac{F(0.6) - F(0.5)}{0.1} = \frac{0.816 - 0.647}{0.1} = \frac{0.169}{0.1} = 1.69.$$

This gives $f(x)$ for x about equal to 0.55.

The complete indefinite integral (4.22) is now given by $F(x) + C$, where $F(x)$ is given by Table 4.1, for $0 \le x \le 1$. The range of x can clearly be extended as far as desired and the accuracy improved as required, so that the method should meet all practical needs. ■

TABLE 4.1

x	$e^{\sin x}$	$\int_x^{x+0.1} e^{\sin t}\, dt$	$\int_0^x e^{\sin t}\, dt = F(x)$
0	1.00	0.105	0
0.1	1.11	0.116	0.105
0.2	1.22	0.129	0.221
0.3	1.35	0.142	0.350
0.4	1.48	0.155	0.492
0.5	1.62	0.169	0.647
0.6	1.75	0.182	0.816
0.7	1.90	0.197	0.998
0.8	2.05	0.212	1.195
0.9	2.18	0.225	1.407
1	2.32		1.632

Elliptic integrals. An integral of the form

$$\int \frac{dx}{\sqrt{1 - k^2 \sin^2 x}}, \qquad 0 < k^2 < 1 \tag{4.23}$$

is an example of an *elliptic* integral. It can be shown that this integral cannot be evaluated in terms of elementary functions. Accordingly, some numerical method is called for. Because of the importance of this integral in applications, elaborate tables have been computed of the function

$$y = F(x) = \int_0^x \frac{dt}{\sqrt{1 - k^2 \sin^2 t}} \tag{4.24}$$

for various values of the constant k. Such tables are given in *A Short Table of Integrals*, 3rd ed., by B. O. Peirce (Boston: Ginn, 1929) and *Tables of Higher Functions* 6th ed. revised by F. Losch, E. Jahnke and F. Emde (New York: McGraw-Hill, 1960).

The integral (4.24) is called an elliptic integral of first kind. Integrals of the second and third kinds are illustrated by

$$y = E(x) = \int_0^x \sqrt{1 - k^2 \sin^2 t} \, dt, \tag{4.25}$$

$$y = \int_0^x \frac{dt}{\sqrt{1 - k^2 \sin^2 t} \, (1 + a^2 \sin^2 t)}, \tag{4.26}$$

where $0 < k^2 < 1$, $a \neq 0$, $a^2 \neq k^2$. The integral (4.25) arises in computing the length of an arc of an ellipse (Problem 4), and this is the basis of the term *elliptic integral.*

It can be shown that if $R(x, y)$ is a rational function of x and y (Section 2.4) and $g(x)$ is a polynomial in x of degree 3 or 4, then the integral

$$\int R\left[x, \sqrt{g(x)}\right] dx \tag{4.27}$$

can be expressed as an elementary function plus elliptic integrals of the first, second, or third kind. Accordingly, numerical evaluation of a few integrals permits precise evaluation of a broad class of integrals. This puts additional emphasis on the fact that every indefinite integral determines a function of x. For each new indefinite integral that we study and evaluate, a broad class of new functions becomes both numerically useful and open to a thorough analysis. Indeed, the trigonometric functions could have been *defined* in such a manner:

$$y = \int_0^x \frac{dt}{\sqrt{1 - t^2}} \tag{4.28}$$

defines $y = $ arc sin x and hence $x = $ sin y. This is more awkward than the usual geometric definition, but all properties can be deduced from such a definition (cf. Problem 3).

For further information on elliptic integrals we refer to the books by von Kármán and Biot (Chapter 4) and Whittaker and Watson (Chapters 20–22) listed at the end of the chapter. Some properties are obtained in Problems 5 to 7.

PROBLEMS

1. Evaluate numerically:

   a) $\int x \, dx$ for $0 \le x \le 10$, using $x = 0, 1, 2, \ldots$;

   b) $\int e^{-x^2} \, dx$ for $0 \le x \le 1$, using $x = 0, 0.1, 0.2, \ldots$.

   c) $\int \cos x \, dx$, $0 \le x \le 1$, using $x = 0, 0.1, 0.2, \ldots$;

   d) $\int \dfrac{1}{1 + x^3} \, dx$, $0 \le x \le 1$, using $x = 0, 0.1, 0.2, \ldots$;

   e) $\int \sqrt{1 - x^3} \, dx$, $0 \le x \le 0.5$, using $x = 0, 0.1, 0.2, \ldots$.

   The answers should be graphed and checked by differentiation.

2. Let $f(x)$ be continuous for $a \le x \le b$.

   a) Prove that

   $$\frac{d}{dx} \int_x^b f(t) \, dt = -f(x), \qquad a \le x \le b.$$

   b) Prove that

   $$\frac{d}{dx} \int_a^{x^2} f(t) \, dt = 2xf(x^2), \qquad a \le x^2 \le b.$$

   [Hint: Let $u = x^2$, so that the integral becomes a function $F(u)$. Then, by the chain rule, $\dfrac{d}{dx} F(u) = F'(u) \dfrac{du}{dx}$.]

   c) Prove that

   $$\frac{d}{dx} \int_{x^2}^b f(t) \, dt = -2xf(x^2), \qquad a \le x^2 \le b.$$

   d) Prove that

   $$\frac{d}{dx} \int_{x^2}^{x^3} f(t) \, dt = 3x^2 f(x^3) - 2xf(x^2), \qquad a \le x^2 \le b, \qquad a \le x^3 \le b.$$

   [Hint: Let $u = x^2$, $v = x^3$. Then the integral is $F(u, v)$. By the chain rule,

   $$\frac{d}{dx} F(u, v) = \frac{\partial F}{\partial u} \frac{du}{dx} + \frac{\partial F}{\partial v} \frac{dv}{dx}.]$$

3. The function $\log x$ (base e always understood) can be *defined* by the equation

   $$\log x = \int_1^x \frac{dt}{t}, \qquad x > 0. \tag{a}$$

   a) Use this equation to evaluate $\log 1$, $\log 2$, and $\log 0.5$ approximately.

   b) Prove, from Eq. (a), that $\log x$ is defined and continuous for $0 < x < \infty$ and that

   $$\frac{d \log x}{dx} = \frac{1}{x}.$$

c) Prove that $\log(ax) = \log a + \log x$ for $a > 0$ and $x > 0$. [Hint: Let $F(x) = \log(ax) - \log x$. Use the result of (b) to show that $F'(x) \equiv 0$, so that $F(x) \equiv$ const. Take $x = 1$ to evaluate the constant.]

4. Let an ellipse be given by parametric equations: $x = a\cos\phi$, $y = b\sin\phi$, $b > a > 0$. Show that the length of arc from $\phi = 0$ to $\phi = \alpha$ is given by

$$s = b\int_0^\alpha \sqrt{1 - k^2\sin^2\phi}\, d\phi, \qquad k^2 = \frac{b^2 - a^2}{b^2}.$$

5. Show that the function $F(x)$ defined by (4.24) has the following properties:

a) $F(x)$ is defined and continuous for all x;

b) as x increases, $F(x)$ increases;

c) $F(x + \pi) - F(x) = 2K$, where K is a constant;

d) $\lim\limits_{x\to\infty} F(x) = \infty$, $\quad \lim\limits_{x\to-\infty} F(x) = -\infty$.

6. The function $x = am(y)$ is defined as the inverse of the function $y = F(x)$ of (4.24). Using the results of Problem 5, show that $am(y)$ is defined and continuous for all y and has the following properties:

a) as y increases, $am(y)$ increases:

b) $am(y + 2K) = am(y) + \pi$;

c) $\dfrac{dx}{dy} = \sqrt{1 - k^2\sin^2 x}$.

7. The functions $sn(y)$, $cn(y)$, $dn(y)$ are defined in terms of the function of Problem 6 by the equations:

$$sn(y) = \sin[am(y)], \qquad cn(y) = \cos[am(y)], \qquad dn(y) = \sqrt{1 - k^2 sn^2(y)}.$$

Prove the following identities:

a) $sn^2(y) + cn^2(y) = 1$;

b) $\dfrac{d}{dy} sn(y) = cn(y)\, dn(y)$;

c) $\dfrac{d}{dy} cn(y) = -sn(y)\, dn(y)$;

d) $sn(y + 4K) = sn(y)$;

e) $cn(y + 4K) = cn(y)$;

f) $dn(y + 2K) = dn(y)$.

The functions $sn(y)$, $cn(y)$, $dn(y)$ are called *elliptic functions*. It should be emphasized that they depend on k, in addition to y.

8. The error function $y = erf(x)$ is defined by the equation

$$y = erf(x) = \int_0^x e^{-t^2}\, dt.$$

This function is of great importance in probability and statistics and is tabulated in the books mentioned after (4.24). Establish the following properties:

a) $erf(x)$ is defined and continuous for all x;

b) $erf(-x) = -erf(x)$;

c) $-1 < erf(x) < 1$ for all x.

4.3 DOUBLE INTEGRALS

The definite integral $\int_a^b f(x)\, dx$ is defined with respect to a function $f(x)$ defined over an interval $a \leq x \leq b$. The double integral

$$\int_R \int f(x, y)\, dx\, dy$$

will be defined with reference to a function $f(x, y)$ defined over a closed region R of the xy-plane. It is necessary further to assume that R is *bounded*, that is, that R can be enclosed in a circle of sufficiently large radius; otherwise, just as in the case when a or b is infinite, the integral is improper.

The definition of the double integral parallels that of the definite integral. One subdivides the region R by drawing parallels to the x and y axes, as in Fig. 4.3. One considers only those rectangles that are within R and numbers them from 1 to n, denoting by $\Delta_i A$ the area of the ith rectangle and by h the maximum diagonal of the rectangles; h is called the *mesh*. One then forms the sum

$$\sum_{i=1}^{n} f(x_i^*, y_i^*)\, \Delta_i A, \tag{4.29}$$

where (x_i^*, y_i^*) is chosen arbitrarily in the ith rectangle. If this sum approaches a unique limit as the maximum diagonal h approaches 0, then the double integral is defined as that limit:

$$\int_R \int f(x, y)\, dx\, dy = \lim_{h \to 0} \sum_{i=1}^{n} f(x_i^*, y_i^*)\, \Delta_i A. \tag{4.30}$$

The existence of the double integral can be demonstrated when f is continuous and R satisfies simple conditions, in particular when R can be split into a finite number of pieces, each of which is described by inequalities of the form

$$y_1(x) \leq y \leq y_2(x), \qquad x_1 \leq x \leq x_2 \tag{4.31}$$

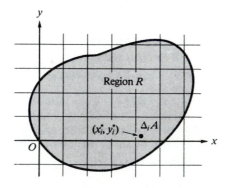

Figure 4.3   The double integral.

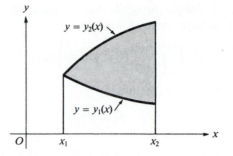

Figure 4.4 Reduction of double integral to iterated integral.

or of the form

$$x_1(y) \leqq x \leqq x_2(y), \qquad y_1 \leqq y \leqq y_2, \tag{4.32}$$

where the functions $y_1(x)$, $y_2(x)$, $x_1(y)$, $x_2(y)$ are continuous. Fig. 4.4 illustrates the first form.

Just as the definite integral of a function $f(x)$ can be interpreted in terms of area, so can the double integral be interpreted in terms of volume, as suggested in Fig. 4.5. Here $f(x, y)$ is assumed to be positive over the region R (the same type as in Fig. 4.4), and the volume being computed is that beneath the surface $z = f(x, y)$ and above the region R in the xy-plane. Each term $f(x_i^*, y_i^*) \Delta_i A$ in the sum (4.30) represents the volume of a rectangular parallelepiped with base $\Delta_i A$ and altitude $f(x_i^*, y_i^*)$, as shown in Fig. 4.5. It is clear that the sum of these volumes can be regarded as an approximation to the volume beneath the surface and that the double integral (4.30) can be used as definition of the volume.

Figure 4.5 The double integral as volume.

The evaluation of double integrals for a region like that of Fig. 4.4 is facilitated by reduction to an *iterated integral*:

$$\int_{x_1}^{x_2} \left[\int_{y_1(x)}^{y_2(x)} f(x, y) \, dy \right] dx.$$

For each fixed value of x the inner integral

$$\int_{y_1(x)}^{y_2(x)} f(x, y) \, dy$$

is simply a definite integral with respect to y of the continuous function $f(x, y)$. This integral can be interpreted as the area of a cross section, perpendicular to the x axis, of the volume being computed; this is suggested in Fig. 4.5. If the area of the cross section is denoted by $A(x)$, then the iterated integral gives

$$\int_{x_1}^{x_2} A(x) \, dx.$$

If this integral is interpreted as a limit of a sum of terms $\Sigma A(x) \Delta x$, then it is geometrically evident that the integral does represent the volume.

THEOREM   If $f(x, y)$ is continuous in a closed region R described by the inequalities (4.31), then, for $x_1 \le x \le x_2$,

$$\int_{y_1(x)}^{y_2(x)} f(x, y) \, dy$$

is a continuous function of x, and

$$\int_R \int f(x, y) \, dx \, dy = \int_{x_1}^{x_2} \int_{y_1(x)}^{y_2(x)} f(x, y) \, dy \, dx. \qquad (4.33)$$

Similarly, if R is described by (4.32),

$$\int_R \int f(x, y) \, dx \, dy = \int_{y_1}^{y_2} \int_{x_1(y)}^{x_2(y)} f(x, y) \, dx \, dy. \qquad (4.34)$$

For a proof of existence of the double integral of a continuous function and of the reduction to an iterated integral, the reader is referred to Section 4.11.

EXAMPLE   Let R be the quarter-circle, $0 \le y \le \sqrt{1 - x^2}$, $0 \le x \le 1$, and let $f(x, y) = x^2 + y^2$. Then one has

$$\int_R \int (x^2 + y^2) \, dx \, dy = \int_0^1 \int_0^{\sqrt{1-x^2}} (x^2 + y^2) \, dy \, dx$$

$$= \int_0^1 \left(x^2 \sqrt{1 - x^2} + \tfrac{1}{3}(1 - x^2)^{\frac{3}{2}} \right) dx$$

$$= \int_0^{\frac{\pi}{2}} (\sin^2 \theta \cos^2 \theta + \tfrac{1}{3} \cos^4 \theta) \, d\theta = \frac{\pi}{8}. \quad \blacksquare$$

The double integral can be interpreted and applied in a variety of ways. The following are illustrations:

I. *Volume.* If $z = f(x, y)$ is the equation of a surface, then

$$V = \int_R \int f(x, y) \, dx \, dy \tag{4.35}$$

gives the volume between the surface and the xy-plane, volumes above the xy-plane being counted positively and those below the xy-plane being counted negatively.

II. *Area.* If one takes $f(x, y) \equiv 1$, one obtains

$$A = \text{area of } R = \int_R \int dx \, dy. \tag{4.36}$$

III. *Mass.* If f is interpreted as density, that is, as mass per unit area, then

$$M = \text{mass of } R = \int_R \int f(x, y) \, dx \, dy. \tag{4.37}$$

IV. *Center of mass.* If f is density, then the center of mass $(\bar{x}, \bar{y})$ of the *thin plate* represented by R is located by the equations

$$M\bar{x} = \int_R \int xf(x, y) \, dx \, dy, \qquad M\bar{y} = \int_R \int yf(x, y) \, dx \, dy, \tag{4.38}$$

where M is given by (4.37).

V. *Moment of inertia.* The moments of inertia of the thin plate about the x- and y-axes are given by the equations

$$I_x = \int_R \int y^2 f(x, y) \, dx \, dy, \qquad I_y = \int_R \int x^2 f(x, y) \, dx \, dy, \tag{4.39}$$

while the polar moment of inertia about the origin O is given by

$$I_O = I_x + I_y = \int_R \int (x^2 + y^2) f(x, y) \, dx \, dy. \tag{4.40}$$

The basic properties of the double integral are essentially the same as those for the definite integral:

$$\int_R \int [f(x, y) + g(x, y)] \, dx \, dy = \int_R \int f(x, y) \, dx \, dy + \int_R \int g(x, y) \, dx \, dy; \tag{4.41}$$

$$\int_R \int cf(x, y) \, dx \, dy = c \int_R \int f(x, y) \, dx \, dy \qquad (c = \text{const}); \tag{4.42}$$

$$\int_R \int f(x, y) \, dx \, dy = \int_{R_1} \int f(x, y) \, dx \, dy + \int_{R_2} \int f(x, y) \, dx \, dy, \tag{4.43}$$

where R is composed of two pieces R_1 and R_2 overlapping only at boundary points;

$$\int_R\int f(x, y)\, dx\, dy = f(x_1, y_1) \cdot A, \tag{4.44}$$

where A is the area of R, as in (4.36), and (x_1, y_1) is a suitably chosen point of R;

$$\left| \int_R\int f(x, y)\, dx\, dy \right| \leq \int_R\int |f(x, y)|\, dx\, dy; \tag{4.45}$$

if $M_1 \leq f(x, y) \leq M_2$ for (x, y) in R, then

$$M_1 A \leq \int_R\int f(x, y)\, dx\, dy \leq M_2 A, \tag{4.46}$$

where A is the area of R. The functions $f(x, y)$ and $g(x, y)$ are here assumed to be continuous in R. The proofs of these properties are essentially the same as for functions of one variable and will not be reviewed here.

From (4.45) one can prove the inequality:

$$\left| \int_R\int f(x, y)\, dx\, dy \right| \leq M \cdot A, \tag{4.47}$$

where $|f(x, y)| \leq M$ in R.

Equation (4.44) is the *law of the mean for double integrals*. If both sides are divided by A, one obtains

$$f(x_1, y_1) = \frac{1}{A} \int_R\int f(x, y)\, dx\, dy. \tag{4.48}$$

The right-hand side can properly be termed the *average value of f in R* [cf. Eq. (4.12)]. The law of the mean thus asserts that a function continuous in R takes on its average value somewhere in R.

If R is a circular region R_r of radius r with center at (x_0, y_0), then one can consider the effect of letting r approach 0 in (4.48). The region R_r is thus being shrunk down to the point (x_0, y_0). Since (x_1, y_1) is always in R_r, it must approach (x_0, y_0); since f is assumed to be continuous, it must approach $f(x_0, y_0)$. Accordingly,

$$\lim_{r \to 0} \frac{1}{A_r} \int_{R_r}\int f(x, y)\, dx\, dy = f(x_0, y_0), \tag{4.49}$$

where $A_r = \pi r^2 = $ area of R. The operation on the left-hand side suggests what might be called "differentiation of the double integral with respect to area." Thus the derivative, in this sense, of the integral reproduces the function being integrated. Accordingly, (4.49) parallels the theorem

$$\frac{d}{dx} \int_a^x f(t)\, dt = f(x)$$

for functions of one variable. It should be remarked that the "incremental area" R_r of (4.49) could also be chosen as a rectangle, ellipse, and so on.

Equation (4.49) has a physical interpretation, namely, the determination of density from mass. Thus if a thin plate of variable density is given, one can measure the mass of the plate or any portion of it directly, simply by weighing the portion concerned. To determine the density at a particular point of the plate, one chooses a small area about this point, measures its mass, and divides by its area. One is thus carrying out experimentally a stage of the limit process (4.49) in the form

$$\lim \frac{\text{mass}}{\text{area}} = \text{density}.$$

The following property of double integrals is established as for single integrals (Problems 11 and 12 following Section 4.1; see also Problem 14 following Section 2.18):

Let $f(x, y)$ be continuous in the bounded closed region R. If

$$\int_{R_1}\int f(x, y)\, dx\, dy = 0$$

for every bounded closed region R_1 contained in R, then $f(x, y) \equiv 0$ in R. If $f(x, y) \geq 0$ in R and

$$\int_{R}\int f(x, y)\, dx\, dy = 0,$$

then $f(x, y) \equiv 0$ in R.

Remarks on the definition of the double integral.   For a subdivision such as that of Fig. 4.3, one can include in the sum (4.29) a term for each rectangle partly included in R, using a point (x_i^*, y_i^*) of R in the rectangle and taking as $\Delta_i A$ the area of the irregular piece. For the type of region considered here, the total contribution of such terms can be shown to be negligible in the sense that for h sufficiently small, their sum has an absolute value that is less than a prescribed positive ϵ. The same assertion holds if one includes the boundary pieces and in each case uses as $\Delta_i A$ the area of the whole rectangle.

One can also subdivide R by curves other than straight lines, as in Fig. 4.6.

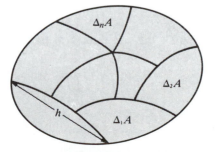

Figure 4.6   General subdivision for a double integral.

Under some natural restrictions on these curves and on R the areas $\Delta_1 A$, $\Delta_2 A$, ... are well determined, and the sum (4.29) can be formed. The mesh h is now the maximum "diameter" of the n pieces (see Fig. 4.6). If f is continuous in R, then Eq. (4.30) remains valid. Because of the freedom to use arbitrary shapes, one also writes the double integral as

$$\iint_R f(x, y)\, dA.$$

4.4 TRIPLE INTEGRALS AND MULTIPLE INTEGRALS IN GENERAL

The notion of double integral generalizes to integrals of functions of three, four, or more variables:

$$\iiint_R f(x, y, z)\, dx\, dy\, dz, \qquad \iiiint_R f(x, y, z, w)\, dx\, dy\, dz\, dw, \ldots .$$

These are called *triple, quadruple,* ... integrals. In general, one terms them *multiple* integrals, the double integral being the simplest multiple integral.

For the triple integral, for example, one considers a function $f(x, y, z)$ defined in a bounded closed region R of space. One subdivides R into rectangular parallelepipeds by planes parallel to the coordinate planes, numbers the parallelepipeds inside R from 1 to n, and denotes the ith volume by $\Delta_i V$. The triple integral is then obtained as the limit of a sum:

$$\iiint_R f(x, y, z)\, dx\, dy\, dz = \lim_{h \to 0} \sum_{i=1}^n f(x_i^*, y_i^*, z_i^*)\Delta_i V;$$

here h is the mesh, the maximum diagonal of the $\Delta_i V$. The point (x_i^*, y_i^*, z_i^*) is arbitrarily chosen in the ith parallelepiped. The existence of a unique limit can be shown if $f(x, y, z)$ is continuous in R.

The simplest theory holds for a region R described by inequalities such as the following:

$$x_1 \leqq x \leqq x_2, \qquad y_1(x) \leqq y \leqq y_2(x), \qquad z_1(x, y) \leqq z \leqq z_2(x, y). \quad (4.50)$$

For this region, one can reduce the triple integral to an iterated integral by the equation

$$\iiint_R f(x, y, z)\, dx\, dy\, dz = \int_{x_1}^{x_2} \int_{y_1(x)}^{y_2(x)} \int_{z_1(x, y)}^{z_2(x, y)} f(x, y, z)\, dz\, dy\, dx, \quad (4.51)$$

as in the theorem of Section 4.3.

EXAMPLE If R is described by the inequalities

$$0 \leqq x \leqq 1, \qquad 0 \leqq y \leqq x^2, \qquad 0 \leqq z \leqq x + y$$

and $f = 2x - y - z$, one has

$$\int\int_R\int f \, dx \, dy \, dz = \int_0^1 \int_0^{x^2} \int_0^{x+y} (2x - y - z) \, dz \, dy \, dx$$

$$= \frac{3}{2} \int_0^1 \int_0^{x^2} (x^2 - y^2) \, dy \, dx$$

$$= \frac{3}{2} \int_0^1 \left(x^4 - \frac{x^6}{3} \right) dx = \frac{8}{35}. \quad \blacksquare$$

The limits of integration can be determined by the following procedure. Let the order chosen be $dz \, dy \, dx$ as in the preceding example. Then we determine the x limits x_1, x_2 first as the smallest and largest values of x on the figure R. We then consider a cross section: $x = $ const of the figure, taken between x_1 and x_2. The y limits are now determined as the minimum and maximum values of y in the cross section; these depend on the x chosen and are hence $y_1(x)$ and $y_2(x)$. Finally, within the cross section we find the minimum and maximum values of z for each fixed y; these are the z limits; they depend on the constant values of x and y chosen and are hence $z_1(x, y)$ and $z_2(x, y)$. Although the minima and maxima referred to can often be read off a sketch of R as a 3-dimensional figure, it is usually easier to plot only typical cross sections as 2-dimensional figures.

Since the definite integral can be interpreted as area and the double integral as volume, one would expect the triple integral to be interpreted as "hyper-volume," or volume in a 4-dimensional space. Although such an interpretation has some value, it is simpler to think of mass; for example,

$$\int\int_R\int f(x, y, z) \, dx \, dy \, dz = \text{mass of solid of density } f. \qquad (4.52)$$

The other interpretations of the double integral also generalize:

$$\int\int_R\int dx \, dy \, dz = \text{volume of } R; \qquad (4.53)$$

$$M\bar{x} = \int\int_R\int xf(x, y, z) \, dx \, dy \, dz, \ldots, \qquad (4.54)$$

where $(\bar{x}, \bar{y}, \bar{z})$ is the center of mass;

$$I_x = \int\int_R\int (y^2 + z^2) f(x, y, z) \, dx \, dy \, dz, \qquad (4.55)$$

the moment of inertia about Ox.

The basic properties (4.41) to (4.47) also generalize to all multiple integrals. Thus, in particular, one has the law of the mean:

$$\int\int_R\int f(x, y, z) \, dx \, dy \, dz = f(x_1, y_1, z_1) \cdot V, \qquad (4.56)$$

where V is the volume of R. This can also be made the basis for "differentiating a triple integral with respect to volume," as in (4.49).

4.5 INTEGRALS OF VECTOR FUNCTIONS

Let

$$\mathbf{F}(t) = f(t)\mathbf{i} + g(t)\mathbf{j} + h(t)\mathbf{k}, \qquad a \le t \le b,$$

be a continuous vector function of t, so that $f(t)$, $g(t)$, $h(t)$ are all continuous for $a \le t \le b$. We then define:

$$\int_a^b \mathbf{F}(t)\,dt = \int_a^b f(t)\,dt\,\mathbf{i} + \int_a^b g(t)\,dt\,\mathbf{j} + \int_a^b h(t)\,dt\,\mathbf{k}. \tag{4.57}$$

This integral can be interpreted as the limit of a sum:

$$\lim_{h \to 0} \sum_{i=1}^n \Delta_i t\,\mathbf{F}(t_i^*)$$

as in (4.1). The limit is a *vector* $\mathbf{c}$ and existence of the limit means that given $\epsilon > 0$, there is a $\delta > 0$ such that, for $0 < h < \delta$,

$$\left| \sum_{i=1}^n \Delta_i t\,\mathbf{F}(t_i^*) - \mathbf{c} \right| < \epsilon.$$

The properties (4.3),...,(4.6), (4.10), (4.16) all extend at once to the integral of $\mathbf{F}(t)$, where the absolute value $|f(x)|$ is replaced by the vector norm $|\mathbf{F}(t)|$.

One can also integrate vector functions of several variables. For example, if

$$\mathbf{F}(x, y, z) = f(x, y, z)\mathbf{i} + g(x, y, z)\mathbf{j} + h(x, y, z)\mathbf{k}$$

is continuous on the bounded closed region R, suitable for triple integrals, then

$$\iiint_R \mathbf{F}(x, y, z)\,dV = \iiint_R f(x, y, z)\,dV\,\mathbf{i} + \iiint_R g(x, y, z)\,dV\,\mathbf{j}$$

$$+ \iiint_R h(x, y, z)\,dV\,\mathbf{k}. \tag{4.58}$$

This integral can also be interpreted as the limit of a sum, and the familiar properties of triple integrals carry over.

PROBLEMS

1. Evaluate the following integrals:

a) $\iint_R (x^2 + y^2)\,dx\,dy$, where R is the triangle with vertices $(0,0), (1,0), (1,1)$;

b) $\iiint_R u^2 v^2 w\,du\,dv\,dw$, where R is the region: $u^2 + v^2 \le 1, 0 \le w \le 1$;

c) $\iint_R r^3 \cos\theta\,dr\,d\theta$, where R is the region: $1 \le r \le 2, \dfrac{\pi}{4} \le \theta \le \pi$;

d) $\iiint_R (x + z) dV$, where R is the tetrahedron with vertices $(0,0,0)$, $(1,0,0)$, $(0,2,0)$, $(0,0,3)$.

2. Find the volume of the solid region below the given surface $z = f(x,y)$ for (x,y) in the region R define by the given inequalities:

a) $z = e^x \cos y$, $0 \le x \le 1$, $0 \le y \le \pi/2$

b) $z = x^2 e^{-x-y}$, $0 \le x \le 1$, $0 \le y \le 2$

c) $z = x^2 y$, $0 \le x \le 1$, $x + 1 \le y \le x + 2$

d) $z = \sqrt{x^2 - y^2}$, $x^2 - y^2 \ge 0$, $0 \le x \le 1$

3. For each of the following choice of R, represent $\int\int f(x,y) dA$ over R as an iterated integral of both forms (4.33) and (4.34):

a) $1 \le x \le 2$, $1 - x \le y \le 1 + x$

b) $y^2 + x(x-1) \le 0$

4. Evaluate $\iiint_R f(x,y,z) dV$ for the following choices of f and R:

a) $f(x,y,z) = \sqrt{x+y+z}$, R the cube of vertices $(0,0,0)$, $(1,0,0)$, $(0,1,0)$, $(0,0,1)$, $(1,1,0)$, $(1,0,1)$, $(0,1,1)$, $(1,1,1)$.

b) $f(x,y,z) = x^2 + z^2$, R the pyramid of vertices $(\pm 1, \pm 1, 0)$ and $(0,0,1)$.

5. For each of the following iterated integrals, find the region R and write the integral in the other form (interchanging the order of integration):

a) $\int_{1/2}^{1} \int_{0}^{1-x} f(x, y) \, dy \, dx$

b) $\int_{0}^{1} \int_{0}^{\sqrt{1-x^2}} f(x, y) \, dy \, dx$

c) $\int_{0}^{1} \int_{y-1}^{0} f(x, y) \, dx \, dy$

d) $\int_{0}^{1} \int_{1-x}^{1+x} f(x, y) \, dy \, dx$

6. Express the following in terms of multiple integrals and reduce to iterated integrals, but do not evaluate:

a) the mass of a sphere whose density is proportional to the distance from one diametral plane;

b) the coordinates of the center of mass of the sphere of part (a);

c) the moment of inertia about the x-axis of the solid filling the region $0 \le z \le 1 - x^2 - y^2$, $0 \le x \le 1$, $0 \le y \le 1 - x$ and having density proportional to xy.

7. The moment of inertia of a solid about an arbitrary line L is defined as

$$I_L = \iiint_R d^2 f(x, y, z) \, dx \, dy \, dz,$$

where f is density and d is the distance from a general point (x, y, z) of the solid to the line L. Prove the *parallel axis theorem*:

$$I_L = I_{\bar{L}} + Mh^2,$$

where $\bar{L}$ is a line parallel to L through the center of mass, M is the mass, and h is the distance between L and $\bar{L}$. (Hint: Take $\bar{L}$ to be the z axis.)

8. Let L be a line through the origin O with direction cosines l, m, n. Prove that

$$I_L = I_x l^2 + I_y m^2 + I_z n^2 - 2I_{xy} lm - 2I_{yz} mn - 2I_{zx} ln,$$

where

$$I_{xy} = \iiint_R xyf(x, y, z) \, dx \, dy \, dz, \qquad I_{yz} = \iiint_R yzf \ldots .$$

The new integrals are called *products of inertia*. The locus

$$I_x x^2 + I_y y^2 + I_z z^2 - 2(I_{xy} xy + I_{yz} yz + I_{zx} zx) = 1$$

is an ellipsoid called the *ellipsoid of inertia*.

9. It is shown in geometry that the medians of a triangle meet at a point, which is the centroid of the triangle, and that the lines from the vertices of a tetrahedron to the centroids of the opposite faces meet at a point which is 3/4 of the way from each vertex to the opposite face along the lines described. Show that this last point is the centroid of the tetrahedron. [Hint: Take the base of the tetrahedron to be in the xy-plane and show that $\bar{z} = h/4$, if h is the z-coordinate of the vertex not in the xy-plane.]

10. Evaluate the integrals:

a) $\int_0^1 \mathbf{F}(t) \, dt$, if $\mathbf{F}(t) = t^2 \mathbf{i} - e^t \mathbf{j} + \dfrac{1}{1+t} \mathbf{k}$.

b) $\iint_R \mathbf{F}(x, y) \, dA$, if R is the triangular region enclosed by the triangle of vertices $(0, 0)$, $(1, 0)$, and $(0, 1)$ and $\mathbf{F}(x, y) = x^2 y \mathbf{i} + xy^2 \mathbf{j}$.

11. Let $\mathbf{F}(t)$ be continuous for $a \le t \le b$ and let $\mathbf{q}$ be a constant vector. Prove:

a) $\int_a^b \mathbf{q} \cdot \mathbf{F}(t)\, dt = \mathbf{q} \cdot \int_a^b \mathbf{F}(t)\, dt$

b) $\int_a^b \mathbf{q} \times \mathbf{F}(t)\, dt = \mathbf{q} \times \int_a^b \mathbf{F}(t)\, dt$

4.6   CHANGE OF VARIABLES IN INTEGRALS

For functions of one variable the chain rule

$$\frac{dF}{du} = \frac{dF}{dx}\frac{dx}{du} \tag{4.59}$$

at once gives the rule for change of variable in a definite integral:

$$\int_{x_1}^{x_2} f(x)\, dx = \int_{u_1}^{u_2} f[x(u)]\frac{dx}{du}\, du. \tag{4.60}$$

Here $f(x)$ is assumed to be continuous at least for $x_1 \le x \le x_2$, $x = x(u)$ is defined for $u_1 \le u \le u_2$ and has a continuous derivative, with $x_1 = x(u_1)$, $x_2 = x(u_2)$, and $f[x(u)]$ is continuous for $u_1 \le u \le u_2$.

Proof.   If $F(x)$ is an indefinite integral of $f(x)$, then

$$\int_{x_1}^{x_2} f(x)\, dx = F(x_2) - F(x_1).$$

But $F[x(u)]$ is then an indefinite integral of $f[x(u)]\dfrac{dx}{du}$, for (4.59) gives

$$\frac{dF}{du} = \frac{dF}{dx}\frac{dx}{du} = f(x)\frac{dx}{du} = f[x(u)]\frac{dx}{du},$$

when x is expressed in terms of u. Thus the integral on the right of (4.60) is

$$F[x(u_2)] - F[x(u_1)] = F(x_2) - F(x_1).$$

Since this is the same as the value of the left-hand side of (4.60), the rule is established. $\square$

It is worth noting that the emphasis in (4.60) is on the function $x(u)$ rather than on its inverse $u = u(x)$. Such an inverse will exist only when x is a steadily increasing function of u or a steadily decreasing function of u. This is not required for (4.60). In fact, the function $x(u)$ can take on values outside the interval $x_1 \le x \le x_2$, as illustrated in Fig. 4.7. However, $f[x(u)]$ must remain continuous for $u_1 \le u \le u_2$.

There is a formula analogous to (4.60) for double integrals:

$$\iint_{R_{xy}} f(x, y)\, dx\, dy = \iint_{R_{uv}} f[x(u, v), y(u, v)]\left|\frac{\partial(x, y)}{\partial(u, v)}\right|\, du\, dv. \tag{4.61}$$

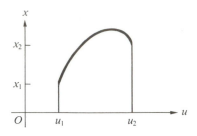

Figure 4.7 The substitution $x = x(u)$ in a definite integral.

Here the functions

$$x = x(u, v), \qquad y = y(u, v) \tag{4.62}$$

are assumed to be defined and have continuous derivatives in a region R_{uv} of the uv-plane. The corresponding points (x, y) lie in the region R_{xy} of the xy-plane, and it is assumed that the inverse functions

$$u = u(x, y), \qquad v = v(x, y) \tag{4.63}$$

are defined and continuous in R_{xy}, so that the correspondence between R_{xy} and R_{uv} is *one-to-one*, as suggested in Fig. 4.8. The function $f(x, y)$ is assumed to be continuous in R_{xy}, so that $f[x(u, v), y(u, v)]$ is continuous in R_{uv}. Finally, it is assumed that the Jacobian

$$J = \frac{\partial(x, y)}{\partial(u, v)}$$

is either positive throughout R_{uv} or negative throughout R_{uv}. It should be noted that it is the *absolute value* of J that is used in (4.61).

A proof of (4.61) will be given in the next chapter, with the aid of line integrals. Here we discuss the significance of (4.61) and its applications.

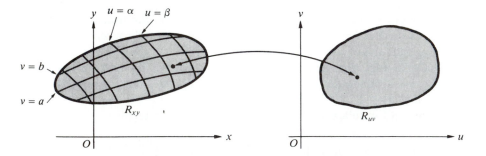

Figure 4.8 Curvilinear coordinates for change of variables in a double integral.

Equations (4.62) can be interpreted as an introduction of *curvilinear* coordinates in the xy-plane, as suggested in Fig. 4.8. The lines $u =$ const and $v =$ const in R_{xy} form a system of curves like the parallels to the axes. It is natural to use them to cut the region R_{xy} into elements of area ΔA for formation of the double integral. With such curvilinear elements the volume "beneath the surface $z = f(x, y)$" will still be approximated by $f(x, y)\Delta A$, where ΔA denotes the area of one of the curvilinear elements. If ΔA can be expressed as a multiple k of $\Delta u\,\Delta v$ and f is expressed in terms of u and v, one obtains a sum

$$\Sigma f[x(u, v), y(u, v)]\,k\,\Delta u\,\Delta v,$$

which approaches a double integral

$$\int\int_{R_{uv}} f[x(u, v), y(u, v)]\,k\,du\,dv$$

as limit. The crucial question is thus the evaluation of the factor k. As (4.61) shows, one must prove that

$$k = \left|\frac{\partial(x, y)}{\partial(u, v)}\right|.$$

The number k can also be interpreted as the ratio of an element of area ΔA_{xy} in the xy-plane to the element $\Delta A_{uv} = \Delta u\,\Delta v$ in the uv-plane. Thus one must show (cf. Section 2.7) that

$$\left|\frac{\partial(x, y)}{\partial(u, v)}\right| = \lim \frac{\Delta A_{xy}}{\Delta A_{uv}}.$$

EXAMPLE 1   Let

$$x = r\cos\theta, \qquad y = r\sin\theta,$$

so that the curvilinear coordinates are polar coordinates. The element of area is approximately a rectangle with sides $r\,\Delta\theta$ and Δr, as shown in Fig. 4.9. Thus

$$\Delta A \sim r\,\Delta\theta\,\Delta r,$$

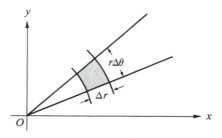

Figure 4.9   Element of area in polar coordinates.

and one expects the formula

$$\iint_{R_{xy}} f(x, y)\, dx\, dy = \iint_{R_{r\theta}} f(r\cos\theta, r\sin\theta)\, r\, d\theta\, dr. \tag{4.64}$$

Now the Jacobian J in this case is

$$J = \frac{\partial(x, y)}{\partial(r, \theta)} = \begin{vmatrix} \cos\theta & -r\sin\theta \\ \sin\theta & r\cos\theta \end{vmatrix} = r.$$

Thus (4.64) is correct. The region $R_{r\theta}$ can be pictured in an $r\theta$-plane or, more simply, can be described by inequalities such as the following:

$$\alpha \le \theta \le \beta, \qquad r_1(\theta) \le r \le r_2(\theta), \tag{4.65}$$

which can be read off the figure in the xy-plane. From (4.65), one finds

$$\iint_{R_{r\theta}} f(r\cos\theta, r\sin\theta)\, r\, d\theta\, dr = \int_\alpha^\beta \int_{r_1(\theta)}^{r_2(\theta)} f(r\cos\theta, r\sin\theta)\, r\, dr\, d\theta,$$

so that the integral has been reduced to an iterated integral in r and θ. It may be necessary to decompose the region $R_{r\theta}$ into several parts and obtain the integral as a sum of iterated integrals of the form given. For some problems it is simpler to integrate in the order $d\theta, dr$; the region $R_{r\theta}$ must then be described by inequalities

$$a \le r \le b, \qquad \theta_1(r) \le \theta \le \theta_2(r), \tag{4.65'}$$

and the integral becomes

$$\int_a^b \int_{\theta_1(r)}^{\theta_2(r)} f(r\cos\theta, r\sin\theta)\, r\, d\theta\, dr. \quad \blacksquare$$

EXAMPLE 2   Let it be required to evaluate

$$\iint_{R_{xy}} (x + y)^3\, dx\, dy,$$

where R_{xy} is the parallelogram shown in Fig. 4.10. The sides of R_{xy} are straight lines having equations of form

$$x + y = c_1, \qquad x - 2y = c_2$$

for appropriate choices of c_1, c_2. It is therefore natural to introduce as new coordinates

$$u = x + y, \qquad v = x - 2y.$$

The region R_{xy} then corresponds to the rectangle $1 \le u \le 4$, $-2 \le v \le 1$. The correspondence is clearly one-to-one. The Jacobian is

$$\frac{\partial(x, y)}{\partial(u, v)} = \frac{1}{\dfrac{\partial(u, v)}{\partial(x, y)}} = \frac{1}{\begin{vmatrix} 1 & 1 \\ 1 & -2 \end{vmatrix}} = -\frac{1}{3}.$$

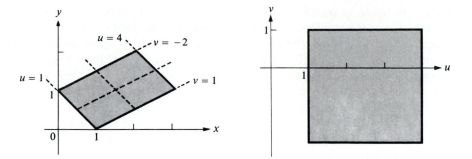

Figure 4.10 Curvilinear coordinates $u = x + y$, $v = x - 2y$.

Hence

$$\iint_{R_{xy}} (x + y)^3 \, dx \, dy = \iint_{R_{uv}} \frac{u^3}{3} \, du \, dv = \int_{-2}^{1} \int_{1}^{4} \frac{u^3}{3} \, du \, dv = 63\tfrac{3}{4}.$$

It should be noted that the limits of integration for the uv-integral are determined from the figure and are not directly related to the limits that would be assigned to the corresponding iterated integral in the xy-plane. ∎

The fundamental formula (4.61) generalizes to triple integrals and multiple integrals of any order. Thus

$$\iiint_{R_{xyz}} f(x, y, z) \, dx \, dy \, dz = \iiint_{R_{uvw}} F(u, v, w) \left| \frac{\partial(x, y, z)}{\partial(u, v, w)} \right| du \, dv \, dw,$$

$$(4.66)$$

where $F(u, v, w) = f[x(u, v, w), y(u, v, w), z(u, v, w)]$, under corresponding assumptions. Two important special cases are those of *cylindrical coordinates*:

$$\iiint_{R_{xyz}} f(x, y, z) \, dx \, dy \, dz = \iiint_{R_{r\theta z}} F(r, \theta, z) r \, dr \, d\theta \, dz,$$

$$F(r, \theta, z) = f(r \cos \theta, r \sin \theta, z), \qquad (4.67)$$

and *spherical coordinates*:

$$\iiint_{R_{xyz}} f(x, y, z) \, dx \, dy \, dz = \iiint_{R_{\rho\phi\theta}} F(\rho, \phi, \theta) \rho^2 \sin \phi \, d\rho \, d\phi \, d\theta,$$

$$F(\rho, \phi, \theta) = f(\rho \sin \phi \cos \theta, \rho \sin \phi \sin \theta, \rho \cos \phi). \quad (4.68)$$

These are discussed in Problems 8 and 9 below.

Remark.   In order to determine the region R_{uv} for (4.61) and, in particular, to verify that the correspondence between R_{uv} and R_{xy} is one-to-one a variety of techniques are available. In (4.62) or (4.63), one can set $u = $ const and plot the resulting level curves of u in R_{xy}. The same can be done for v. If these level curves have the property that a curve $u = c_1$ meets a curve $v = c_2$ in at most one point in R_{xy}, then the correspondence must be one-to-one. From the level curves, one can follow the variation of u and v on the boundary of R_{xy} and thereby determine the boundary of R_{uv}. It can be shown that if R_{xy} and R_{uv} are each bounded by a single closed curve, as in Fig. 4.8, if the correspondence between (x, y) and (u, v) is one-to-one on these boundary curves, and $J \neq 0$ in R_{uv}, then the correspondence is necessarily one-to-one in all of R_{xy} and R_{uv}. For a further discussion of this point, the reader is referred to Section 5.14. Actually, the conditions that the correspondence be one-to-one and that $J \neq 0$ are not vital for the theorem. It is shown in Section 5.14 that (4.61) can be written in a different form that covers the more general cases.

PROBLEMS

1. Evaluate with the aid of the substitution indicated:

a) $\int_0^1 (1 - x^2)^{3/2} \, dx, \ x = \sin \theta$

b) $\int_0^1 \frac{1}{1 + \sqrt{1 + x}} \, dx, \ x = u^2 - 1$

c) $\int_0^{\pi/2} \frac{1}{\sin x + \cos x + 2} \, dx, \quad t = \tan(x/2)$

d) $\int_0^{\pi/4} \frac{x \cos x (x \sin x - \cos x)}{1 + x \cos x} \, dx, \quad t = 1 + x \cos x$

2. Prove the formula

$$\int_{u_1}^{u_2} \phi'(u) \, du = \phi(u_2) - \phi(u_1)$$

as a special case of (4.60).

3. a) Prove that (4.60) remains valid for improper integrals, that is, if $f(x)$ is continuous for $x_1 \leq x < x_2$, $x(u)$ is defined and has a continuous derivative for $u_1 \leq u < u_2$, with $x(u_1) = x_1$, $\displaystyle\lim_{u \to u_2} x(u) = x_2$, and $f[x(u)]$ is continuous for $u_1 \leq u < u_2$. [Hint: Use the fact that (4.60) holds with u_2 and x_2 replaced by u_0 and $x_0 = x(u_0)$, $u_1 < u_0 < u_2$. Then let u_0 approach u_2. One concludes that if either side of the equation has a limit, then the other side has a limit also and the limits are equal. Note that x_2 or u_2 or both may be ∞.]

b) Evaluate $\displaystyle\int_1^\infty \frac{1}{x^2} \sinh \frac{1}{x} \, dx$ by setting $u = \frac{1}{x}$.

c) Evaluate $\displaystyle\int_0^\infty (1 - \tanh x) \, dx$ by setting $u = \tanh x$.

4. Evaluate the following integrals with the aid of the substitution suggested:

a) $\iint\limits_{R_{xy}} (1 - x^2 - y^2)\, dx\, dy$, where R_{xy} is the region $x^2 + y^2 \leq 1$, using $x = r\cos\theta$, $y = r\sin\theta$;

b) $\iint\limits_{R} \dfrac{y\sqrt{x^2 + y^2}}{x}\, dx\, dy$, where R is the region $1 \leq x \leq 2,\ 0 \leq y \leq x,$ using $x = r\cos\theta,\ y = r\sin\theta$;

c) $\iint\limits_{R_{xy}} (x - y)^2 \sin^2(x + y)\, dx\, dy$, where R_{xy} is the parallelogram with successive vertices $(\pi, 0), (2\pi, \pi), (\pi, 2\pi), (0, \pi)$, using $u = x - y,\ v = x + y$;

d) $\iint\limits_{R} \dfrac{(x - y)^2}{1 + x + y}\, dx\, dy$, where R is the trapezoidal region bounded by the lines $x + y = 1,\ x + y = 2$ in the first quadrant, using $u = 1 + x + y,\ v = x - y$;

e) $\iint\limits_{R} \sqrt{5x^2 + 2xy + 2y^2}\, dx\, dy$ over the region R bounded by the ellipse $5x^2 + 2xy + 2y^2 = 1$, using $x = u + v,\ y = -2u + v$.

5. Verify that the transformation $u = e^x \cos y,\ v = e^x \sin y$ defines a one-to-one mapping of the rectangle $R_{xy}: 0 \leq x \leq 1,\ 0 \leq y \leq \pi/2$ onto a region of the uv-plane and express as an iterated integral in u, v the integral

$$\iint\limits_{R_{xy}} \dfrac{e^{2x}}{1 + e^{4x}\cos^2 y \sin^2 y}\, dx\, dy.$$

6. Verify that the transformation

$$u = 2xy, \qquad v = x^2 - y^2$$

defines a one-to-one mapping of the square $0 \leq x \leq 1,\ 0 \leq y \leq 1$ onto a region of the uv-plane. Express the integral

$$\iint\limits_{R_{xy}} \sqrt[3]{x^4 - 6x^2y^2 + y^4}\, dx\, dy$$

over the square as an iterated integral in u and v.

7. Transform the integrals given, using the substitutions indicated:

a) $\displaystyle\int_0^1 \int_0^x \log(1 + x^2 + y^2)\, dy\, dx, \quad x = u + v,\ \ y = u - v$

b) $\displaystyle\int_0^1 \int_{1-x}^{1+x} \sqrt{1 + x^2 y^2}\, dy\, dx, \quad x = u,\ \ y = u + v$

8. Let R_{uv} be the square $0 \leq u \leq 1,\ 0 \leq v \leq 1$. Show that the given equations define a one-to-one mapping of R_{uv} onto R_{xy} and graph R_{xy}:

a) $x = u + u^2,\ \ y = e^v$ b) $x = ue^v,\ \ y = e^v$

c) $x = 2u - v^2,\ \ y = v + uv$ d) $x = 5u - u^2 + v^2,\ \ y = 5v + 10uv$

9. Verify the correctness of (4.67) as a special case of (4.66). Show the geometric meaning of the volume element $r\Delta r\Delta\theta\,\Delta z$.

10. Verify the correctness of (4.68) as a special case of (4.66). Show the geometric meaning of the volume element $\rho^2 \sin\phi\,\Delta\rho\,\Delta\phi\,\Delta\theta$.

11. Transform to cylindrical coordinates but do not evaluate:

a) $\iiint\limits_{R_{xyz}} x^2 y \, dx \, dy \, dz$, where R_{xyz} is the region $x^2 + y^2 \leq 1$, $0 \leq z \leq 1$;

b) $\int_0^1 \int_0^{\sqrt{1-x^2}} \int_0^{1+x+y} (x^2 - y^2) \, dz \, dy \, dx$.

12. Transform to spherical coordinates but do not evaluate:

a) $\iiint\limits_{R_{xyz}} x^2 y \, dx \, dy \, dz$, where R_{xyz} is the sphere: $x^2 + y^2 + z^2 \leq a^2$;

b) $\int_{-1}^1 \int_{-\sqrt{1-x^2}}^{\sqrt{1-x^2}} \int_{\sqrt{x^2+y^2}}^1 (x^2 + y^2 + z^2) \, dz \, dy \, dx$.

4.7 ARC LENGTH AND SURFACE AREA

In elementary calculus it is shown that a curve $y = f(x)$, $a \leq x \leq b$, has length

$$s = \int_a^b \sqrt{1 + \left(\frac{dy}{dx}\right)^2} \, dx \qquad (4.69)$$

and that if the curve is given parametrically by equations $x = x(t)$, $y = y(t)$ for $t_1 \leq t \leq t_2$, then it has length

$$s = \int_{t_1}^{t_2} \sqrt{\left(\frac{dx}{dt}\right)^2 + \left(\frac{dy}{dt}\right)^2} \, dt. \qquad (4.70)$$

Furthermore, a curve $x = x(t)$, $y = y(t)$, $z = z(t)$ in space has length

$$s = \int_{t_1}^{t_2} \sqrt{\left(\frac{dx}{dt}\right)^2 + \left(\frac{dy}{dt}\right)^2 + \left(\frac{dz}{dt}\right)^2} \, dt. \qquad (4.71)$$

The length is here defined as a limit of the lengths of inscribed polygons; the functions are assumed to have continuous derivatives over the intervals concerned.

The main question here is the generalization of (4.69) and (4.71) to the area of surfaces in space. For a surface $z = f(x, y)$ it will be seen that the area is given by

$$S = \iint\limits_{R_{xy}} \sqrt{1 + \left(\frac{\partial z}{\partial x}\right)^2 + \left(\frac{\partial z}{\partial y}\right)^2} \, dx \, dy. \qquad (4.72)$$

This parallels (4.69). A surface in space can be represented parametrically by equations

$$x = x(u, v), \qquad y = y(u, v), \qquad z = z(u, v), \qquad (4.73)$$

where u and v vary in a region R_{uv} of the uv-plane. The area of the surface (4.73) is given by

$$S = \iint\limits_{R_{uv}} \sqrt{EG - F^2} \, du \, dv, \qquad (4.74)$$

where

$$E = \left(\frac{\partial x}{\partial u}\right)^2 + \left(\frac{\partial y}{\partial u}\right)^2 + \left(\frac{\partial z}{\partial u}\right)^2,$$

$$F = \frac{\partial x}{\partial u}\frac{\partial x}{\partial v} + \frac{\partial y}{\partial u}\frac{\partial y}{\partial v} + \frac{\partial z}{\partial u}\frac{\partial z}{\partial v}, \tag{4.75}$$

$$G = \left(\frac{\partial x}{\partial v}\right)^2 + \left(\frac{\partial y}{\partial v}\right)^2 + \left(\frac{\partial z}{\partial v}\right)^2.$$

A complete justification of (4.72) and (4.74) turns out to involve a very delicate analysis, much more difficult than that of arc length. In particular, one cannot define surface area S simply as the limit of the areas of inscribed polyhedra. For a full discussion the reader is referred to P. Franklin's *Treatise on Advanced Calculus*, pages 371–378 (New York: John Wiley, 1940). Here we give an intuitive discussion of why (4.72) and (4.74) are to be expected.

We first note that the arc length s can be defined in a somewhat different way, using tangents rather than an inscribed polygon, as suggested in Fig. 4.11. Thus for a curve $y = f(x)$, one subdivides the interval $a \le x \le b$ as for integrating $f(x)$. At a point x_i^* between x_{i-1} and x_i, one draws the corresponding tangent line to the curve:

$$y - y_i^* = f'(x_i^*)(x - x_i^*),$$

and denotes by $\Delta_i T$ the length of the segment of this line between x_{i-1} and x_i. It is natural to expect that

$$s = \lim \sum_{i=1}^{n} \Delta_i T$$

as n becomes infinite and max $\Delta_i x$ approaches 0. Now if α_i^* is the angle of inclination of $\Delta_i T$, so that $f'(x_i^*) = \tan \alpha_i^*$, one has

$$\Delta_i x = \Delta_i T \cos \alpha_i^*$$

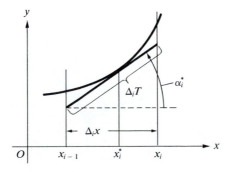

Figure 4.11   Definition of arc length.

or

$$\Delta_i T = \Delta_i x \sec \alpha_i^*. \tag{4.76}$$

The sum $\Sigma \Delta_i T$ thus becomes

$$\Sigma \sec \alpha_i^* \, \Delta_i x = \Sigma \sqrt{1 + f'(x_i^*)^2} \, \Delta_i x.$$

If $f'(x)$ is continuous for $a \le x \le b$, this sum approaches as limit the desired expression

$$\int_a^b \sec \alpha \, dx = \int_a^b \sqrt{1 + f'(x)^2} \, dx = s.$$

Now let a surface $z = f(x, y)$ be given, where $f(x, y)$ is defined and has continuous partial derivatives in a domain D. To find the area of the part of the surface above a bounded closed region R_{xy}, contained in D, we subdivide R_{xy} as in Fig. 4.3. Let (x_i^*, y_i^*) be a point of the ith rectangle; we then construct the tangent plane to the surface at the corresponding point:

$$z - z_i^* = f_x(x_i^*, y_i^*)(x - x_i^*) + f_y(x_i^*, y_i^*)(y - y_i^*). \tag{4.77}$$

Let $\Delta_i S^*$ be the area of the part of this tangent plane above the ith rectangle in the xy-plane. Thus $\Delta_i S^*$ is the area of a certain parallelogram whose projection on the xy-plane is a rectangle of area $\Delta_i A$, as in Fig. 4.12. Now let $\mathbf{n}^*$ be the normal vector to the surface at the point of tangency:

$$\mathbf{n}^* = -\frac{\partial z}{\partial x}\mathbf{i} - \frac{\partial z}{\partial y}\mathbf{j} + \mathbf{k}. \tag{4.78}$$

Then it follows readily from geometry that

$$\Delta_i S^* = \sec \gamma_i^* \, \Delta_i A, \tag{4.79}$$

where γ_i^* is the angle between $\mathbf{n}^*$ and $\mathbf{k}$. This is analogous to (4.76) (cf. Problem 5

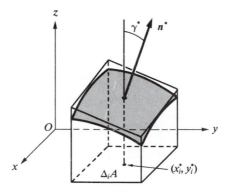

Figure 4.12   Definition of surface area.

below). Now from (4.78),

$$\cos \gamma_i^* = \frac{1}{\sqrt{1 + \left(\dfrac{\partial z}{\partial x}\right)^2 + \left(\dfrac{\partial z}{\partial y}\right)^2}} = \frac{\mathbf{n}^* \cdot \mathbf{k}}{|\mathbf{n}^*|},$$

so that

$$\sec \gamma_i^* = \sqrt{1 + \left(\frac{\partial z}{\partial x}\right)^2 + \left(\frac{\partial z}{\partial y}\right)^2},$$

all derivatives being evaluated at (x_i^*, y_i^*). Paralleling the procedure for arc length given earlier, it is natural to expect the surface area S to be obtained as the limit of a sum

$$\sum_{i=1}^{n} \Delta_i S^* = \sum_{i=1}^{n} \sec \gamma_i^* \, \Delta_i A$$

$$= \sum_{i=1}^{n} \sqrt{1 + \left(\frac{\partial z}{\partial x}\right)^2 + \left(\frac{\partial z}{\partial y}\right)^2} \, \Delta_i A,$$

as the number n of subdivisions approaches ∞, while the maximum diagonal approaches 0. This limit is precisely the double integral

$$S = \iint_{R_{xy}} \sec \gamma \, dx \, dy = \iint_{R_{xy}} \sqrt{1 + \left(\frac{\partial z}{\partial x}\right)^2 + \left(\frac{\partial z}{\partial y}\right)^2} \, dx \, dy$$

as desired.

Surfaces in parametric form.   The parametric equations

$$x = x(u, v), \qquad y = y(u, v), \qquad z = z(u, v), \qquad (u, v) \text{ in } R_{uv}, \qquad (4.80)$$

of a surface can be regarded as a mapping of R_{uv} onto a "curved region" in space, as suggested in Fig. 4.13. The lines $u = $ const, $v = $ const on the surface determine

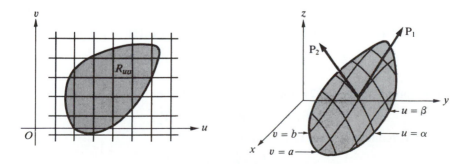

Figure 4.13   Curvilinear coordinates on a surface.

curvilinear coordinates on the surface. When $v = $ const, (4.80) can be regarded as parametric equations of a curve; the tangent vector to such a curve is the vector

$$\mathbf{P}_1 = \frac{\partial x}{\partial u}\mathbf{i} + \frac{\partial y}{\partial u}\mathbf{j} + \frac{\partial z}{\partial u}\mathbf{k}, \tag{4.81}$$

in accordance with Section 2.13. Similarly, a curve $u = $ const has tangent vector

$$\mathbf{P}_2 = \frac{\partial x}{\partial v}\mathbf{i} + \frac{\partial y}{\partial v}\mathbf{j} + \frac{\partial z}{\partial v}\mathbf{k}. \tag{4.82}$$

The lines $u = $ const, $v = $ const in R_{uv} can be used to subdivide R_{uv} as in the formation of the double integral. Each rectangle, of area $\Delta A = \Delta u\,\Delta v$, of this subdivision corresponds to a curved area element on the surface. To first approximation this area element is a parallelogram with sides $|\mathbf{P}_1|\,\Delta u$ and $|\mathbf{P}_2|\,\Delta v$, for $\mathbf{P}_1$ can be interpreted as a velocity vector in terms of time u, so that

$$\frac{ds_1}{du} = |\mathbf{P}_1|,$$

where s_1 is distance along the line $v = $ const chosen. For a small change Δu in the time u the distance moved is approximately $ds_1 = |\mathbf{P}_1|\,\Delta u$. Similarly, the other side of the parallelogram is approximately $|\mathbf{P}_2|\,\Delta v$. The vectors $\mathbf{P}_1$ and $\mathbf{P}_2$ are here evaluated at one corner of the parallelogram, as suggested in Fig. 4.13. The parallelogram has then as sides segments representing the vectors

$$\mathbf{P}_1\,\Delta u \quad \text{and} \quad \mathbf{P}_2\,\Delta v.$$

Its area is thus given by

$$|(\mathbf{P}_1\,\Delta u) \times (\mathbf{P}_2\,\Delta v)| = |\mathbf{P}_1 \times \mathbf{P}_2|\,\Delta u\,\Delta v.$$

If we grant that the sum of the areas of these parallelograms over the surface approaches the area of the surface as limit as the number of subdivisions of R_{uv} is increased, as in forming the double integral, then we conclude that the surface area is given by

$$S = \iint_{R_{uv}} |\mathbf{P}_1 \times \mathbf{P}_2|\,du\,dv. \tag{4.83}$$

By means of the vector identity

$$|\mathbf{P}_1 \times \mathbf{P}_2|^2 = |\mathbf{P}_1|^2|\mathbf{P}_2|^2 - (\mathbf{P}_1 \cdot \mathbf{P}_2)^2 \tag{4.84}$$

(see Problem 12(a) following Section 1.4) this can be written in the form

$$S = \iint_{R_{uv}} \sqrt{|\mathbf{P}_1|^2|\mathbf{P}_2|^2 - (\mathbf{P}_1 \cdot \mathbf{P}_2)^2}\,du\,dv. \tag{4.85}$$

Comparison of (4.81), (4.82), and (4.75) shows that one has

$$|\mathbf{P}_1|^2 = E, \quad \mathbf{P}_1 \cdot \mathbf{P}_2 = F, \quad |\mathbf{P}_2|^2 = G. \tag{4.86}$$

Thus (4.85) reduces at once to the formula desired:

$$S = \iint_{R_{uv}} \sqrt{EG - F^2}\,du\,dv. \tag{4.87}$$

It should be noted that if the point (x, y, z) traces a general curve on the surface, the differential

$$d\mathbf{r} = dx\mathbf{i} + dy\mathbf{j} + dz\mathbf{k}$$

can be expressed as follows:

$$d\mathbf{r} = \left(\frac{\partial x}{\partial u}du + \frac{\partial x}{\partial v}dv\right)\mathbf{i} + \left(\frac{\partial y}{\partial u}du + \frac{\partial y}{\partial v}dv\right)\mathbf{j} + \left(\frac{\partial z}{\partial u}du + \frac{\partial z}{\partial v}dv\right)\mathbf{k}$$

$$= \mathbf{P}_1\,du + \mathbf{P}_2\,dv = \frac{\partial \mathbf{r}}{\partial u}du + \frac{\partial \mathbf{r}}{\partial v}dv.$$

The element of arc on such a curve is defined by

$$ds^2 = dx^2 + dy^2 + dz^2 = |d\mathbf{r}|^2 = (\mathbf{P}_1\,du + \mathbf{P}_2\,dv)\cdot(\mathbf{P}_1\,du + \mathbf{P}_2\,dv).$$

Expanding the last product, one finds

$$ds^2 = |\mathbf{P}_1|^2\,du^2 + 2(\mathbf{P}_1\cdot\mathbf{P}_2)\,du\,dv + |\mathbf{P}_2|^2\,dv^2. \tag{4.88}$$

Accordingly, by (4.86),

$$ds^2 = E\,du^2 + 2F\,du\,dv + G\,dv^2. \tag{4.89}$$

This shows the significance of E, F, G for the geometry on the surface.

PROBLEMS

1. Find the length of the circumference of a circle
   a) using the parametric representation
   $$x = a\cos\theta, \qquad y = a\sin\theta,$$
   b) using the parametric representation
   $$x = a\frac{1 - t^2}{1 + t^2}, \qquad y = a\frac{2t}{1 + t^2}.$$

2. Find the area of the surface of a sphere
   a) using the equation
   $$z = \pm\sqrt{a^2 - x^2 - y^2},$$
   b) using the parametric equations
   $$x = a\sin\phi\cos\theta, \qquad y = a\sin\phi\sin\theta, \qquad z = a\cos\phi.$$

3. Find the surface area of the surface having the given parametric equations:
   a) $x = (b + a\cos v)\cos u$, $y = (b + a\cos v)\sin u$, $z = a\sin v$, $0 \le u \le 2\pi$, $0 \le v \le 2\pi$, where a and b are constants, $0 < a < b$ (torus).
   b) $x = u\cos v$, $y = u\sin v$, $z = u^2\sin 2v$, $0 \le u \le 1$, $0 \le v \le \pi/2$ (portion of saddle surface $z = 2xy$).

4. Using the parametric equations of Problem 2(b), set up a double integral for the area of a portion of the earth's surface bounded by two parallels of latitude and two

meridians of longitude. Apply this to find the area of the United States of America, approximating this by the "rectangle" between parallels 30° N and 47° N and meridians 75° W and 122° W. Take the radius of the earth to be 4000 miles.

5. Let a parallelogram be given in space whose sides represent the vectors **a** and **b**. Let **c** be a unit vector perpendicular to a plane C.

   a) Show that $\mathbf{a} \times \mathbf{b} \cdot \mathbf{c}$ equals plus or minus the area of the projection of the parallelogram on C.

   b) Show that this can also be written as $S \cos \gamma$, where S is the area of the parallelogram and γ is the angle between $\mathbf{a} \times \mathbf{b}$ and $\mathbf{c}$.

   c) Show that one has

$$S = \sqrt{S_{yz}^2 + S_{zx}^2 + S_{xy}^2},$$

   where S_{yz}, S_{zx}, S_{xy} are the areas of the projections of the parallelogram on the yz-plane, zx-plane, xy-plane.

6. A surface of revolution is obtained by rotating a curve $z = f(x)$, $y = 0$ in the xz-plane about the z-axis.

   a) Show that this surface has the equation $z = f(r)$ in cylindrical coordinates.

   b) Show that the area of the surface is

$$S = \int_0^{2\pi} \int_a^b \sqrt{1 + f'(r)^2}\, r\, dr\, d\theta = 2\pi \int_a^b \sqrt{1 + f'(r)^2}\, r\, dr.$$

7. Prove (4.74) by assuming that the surface in question can also be represented as $z = f(x, y)$ and hence has an area given by (4.72). The parametric equations (4.73) are to be regarded as a transformation of variables in the double integral (4.72), as in Section 4.6.

8. Show that (4.74) reduces to

$$S = \iint_{R_{uv}} \left| \frac{\partial(x, y)}{\partial(u, v)} \right| du\, dv = \iint_{R_{xy}} dx\, dy,$$

   when u, v are curvilinear coordinates in a plane area R_{xy}.

9. Prove that if a surface $z = f(x, y)$ is given in implicit form $F(x, y, z) = 0$, then the surface area (4.72) becomes

$$\iint_{R_{xy}} \frac{\sqrt{F_x^2 + F_y^2 + F_z^2}}{|F_z|} dx\, dy.$$

4.8   IMPROPER MULTIPLE INTEGRALS

Since the discontinuities of functions of several variables can be much more complicated than those of functions of one variable, the discussion of improper multiple integrals is not as simple as that for ordinary definite integrals. However, in principle, the analysis of Section 4.1 carries over to multiple integrals.

The definition of boundedness of a function of several variables is the same as that of a function of one variable. Thus the function $\log(x^2 + y^2)$ is bounded for $1 \leq x \leq 2, 1 \leq y \leq 2$ but is not bounded for $0 < x^2 + y^2 \leq 1$. If a function $f(x, y)$ is defined and continuous throughout a bounded closed region R such as (4.31), except for a finite number of points of discontinuity, and if f is bounded, then the double integral

$$\int_R \int f(x, y)\, dx\, dy$$

continues to exist as a limit of a sum, for the same reason as for functions of one variable. The discontinuities can even take the form of whole curves, finite in number and composed of "smooth curves," for the similar reason that these together form a set of *zero area*. An important case of this is when $f(x, y)$ is continuous and bounded only in a bounded domain D, nothing being known about values of f on the boundary of D (assumed to consist of such smooth curves). In such a case the integral

$$\int_D \int f(x, y)\, dx\, dy$$

continues to exist as a limit of a sum, provided that one evaluates f only where it is given, that is, inside D. The same result is obtained if one assigns f arbitrary values, for example, 0, on the boundary of D.

EXAMPLE   The integral

$$\int_D \int \sin \frac{y}{x}\, dx\, dy,$$

where D is the square domain $0 < x < 1$, $0 < y < 1$, exists, even though the function is badly discontinuous on the y axis, since $\left|\sin \dfrac{y}{x}\right| \leq 1$ in D. ∎

For any such bounded function f, one can approximate the integral over the whole region arbitrarily closely by the integral over a smaller region, avoiding the discontinuities. For if R is split into two regions R_1, R_2, overlapping only on boundary points, then

$$\int_R \int f(x, y)\, dx\, dy = \int_{R_1} \int f(x, y)\, dx\, dy + \int_{R_2} \int f(x, y)\, dx\, dy.$$

Further, if $|f| \leq M$ in R, then

$$\left| \int_{R_2} \int f(x, y)\, dx\, dy \right| \leq M \cdot A_2, \tag{4.90}$$

where A_2 is the area of R_2, as follows from (4.47). If A_2 is sufficiently small, the integral over R_1 will approximate the integral over R as closely as desired.

True improper integrals arise, as for functions of one variable, when $f(x, y)$ is unbounded in R. The most important case of this is when $f(x, y)$ is defined and continuous but unbounded in a bounded domain D, with no information given about values of f on the boundary of D. In this case the limit of the sum $\Sigma f(x, y) \Delta A$ will fail to exist because f is unbounded. The integral of f over D is termed improper and is assigned a value by the limit process

$$\int_D\!\!\int f(x, y)\, dx\, dy = \lim_{R \to D} \int_R\!\!\int f(x, y)\, dx\, dy, \tag{4.91}$$

provided that the limit exists. Here R denotes a closed region contained, with its boundary, in D, as suggested in Fig. 4.14. The limit process is understood as follows: The limit exists and has value K if, given $\epsilon > 0$, a particular region R_1 can be found such that

$$\left| \int_R\!\!\int f(x, y)\, dx\, dy - K \right| < \epsilon$$

for all regions R containing R_1 and lying in D.

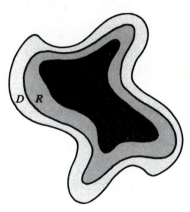

Figure 4.14   The limit process $R \to D$.

What amounts to a special case of the preceding is the case of a function having a *point discontinuity*, that is, a function continuous in a domain D except at a single point P of D. It would be natural in this case to isolate the trouble at P by integrating up to a small circle of radius h about P and then letting h approach 0 (Fig. 4.15). This definition of the improper integral is equivalent to the preceding, provided that $f(x, y)$ has the same sign ($+$ or $-$) near P. This is the most frequent case. Thus the integral

$$\int_R\!\!\int \frac{1}{r^p}\, dx\, dy, \qquad p > 0,$$

where R is the circle $x^2 + y^2 \leq 1$, is improper because of a point discontinuity at the origin. The limit process gives, in polar coordinates,

$$\iint_R \frac{1}{r^p}\, dx\, dy = \lim_{h \to 0} \int_0^{2\pi} \int_h^1 \frac{1}{r^p} r\, dr\, d\theta$$

$$= \lim_{h \to 0} 2\pi \left(\frac{1}{2-p} - \frac{h^{2-p}}{2-p} \right), \qquad (p \neq 2). \qquad (4.92)$$

Thus if $p < 2$, the integral converges to the value $2\pi/(2-p)$; if $p > 2$, the integral diverges. For $p = 2$, one obtains a logarithm and the integral again diverges. It should be noted that the critical value here is $p = 2$, as compared with $p = 1$ for single integrals.

Figure 4.15   Integral improper at a point P.

A second type of improper integral, generalizing the definite integrals with infinite limits, is an integral

$$\iint_R f(x, y)\, dx\, dy$$

where R is an *unbounded* closed region. Here one obtains a value by a limit process just like that of (4.91). The most important case of this is that of a function continuous outside and on a circle $x^2 + y^2 = a^2$. If $f(x, y)$ is of one sign, the integral over this region R can be defined as the limit

$$\lim_{k \to \infty} \iint_{R_k} f(x, y)\, dx\, dy$$

where R_k is the region $a^2 \leq x^2 + y^2 \leq k^2$. Thus the improper integral

$$\iint_R \frac{1}{r^p}\, dx\, dy$$

has the value

$$\lim_{k \to \infty} \int_a^k \int_0^{2\pi} \frac{1}{r^p} \, d\theta \, r \, dr = \lim_{k \to \infty} 2\pi \frac{k^{2-p} - a^{2-p}}{2 - p},$$

which equals $2\pi a^{2-p}/(p - 2)$, for $p > 2$. For $p \leq 2$ the integral diverges.

While the emphasis here has been on double integrals, the statements hold with minor changes [affecting in particular the critical value of p for the integral (4.92)] for triple and other multiple integrals.

For further discussion of this topic, see Section 6.26.

PROBLEMS

1. One way of evaluating the *error integral*

$$\int_0^\infty e^{-x^2} \, dx$$

is to use the equations

$$\left(\int_0^\infty e^{-x^2} \, dx \right)^2 = \int_0^\infty e^{-x^2} \, dx \int_0^\infty e^{-y^2} \, dy = \int_0^\infty \int_0^\infty e^{-x^2 - y^2} \, dx \, dy$$

and to evaluate the double integral by polar coordinates. Carry out this evaluation, showing that the integral equals $\frac{1}{2}\sqrt{\pi}$; also discuss the significance of the above equations in terms of the limit definitions of the improper integrals.

2. Show that the integral

$$\int_R \int \log \sqrt{x^2 + y^2} \, dx \, dy$$

converges, where R is the region $x^2 + y^2 \leq 1$, and find its value. This can be interpreted as minus the *logarithmic potential*, at the origin, of a uniform mass distribution over the circle.

3. **a)** Show that the integral

$$\int_R \int \int \frac{1}{r^p} \, dx \, dy \, dz, \qquad r = \sqrt{x^2 + y^2 + z^2},$$

over the spherical region $x^2 + y^2 + z^2 \leq 1$ converges for $p < 3$ and find its value. For $p = 1$ this is the *Newtonian potential* of a uniform mass distribution over the solid sphere, evaluated at the origin.

b) For the integral of part (a), let R be the *exterior* region $x^2 + y^2 + z^2 \geq 1$. Show that the integral converges for $p > 3$ and find its value.

4. Test for convergence or divergence:

a) $\int_R \int \frac{1}{x^2 + y^2} \, dx \, dy$, over the square $|x| < 1, |y| < 1$;

b) $\int_R \int \frac{\log (x^2 + y^2)}{\sqrt{x^2 + y^2}} \, dx \, dy$ over the circle $x^2 + y^2 \leq 1$;

c) $\displaystyle\int_R\!\!\int \log (x^2 + y^2)\, dx\, dy$ over the region $x^2 + y^2 \geq 1$;

d) $\displaystyle\int_R\!\!\int \frac{\sqrt{x^2 + xy + y^2}}{x^2 + y^2}\, dx\, dy$ over the region $x^2 + y^2 \leq 1$;

e) $\displaystyle\int_R\!\!\int\!\!\int \log (x^2 + y^2 + z^2)\, dx\, dy\, dz$ over the solid $x^2 + y^2 + z^2 \leq 1$.

4.9   INTEGRALS DEPENDING ON A PARAMETER—LEIBNITZ'S RULE

A definite integral

$$\int_a^b f(x, t)\, dx$$

of a continuous function $f(x, t)$ has a value that depends on the choice of t, so that one can write

$$\int_a^b f(x, t)\, dx = F(t). \tag{4.93}$$

One calls such an expression an *integral depending on a parameter*, and t is termed the parameter. Thus

$$\int_0^{\frac{\pi}{2}} \frac{dx}{\sqrt{1 - k^2 \sin^2 x}}$$

is an integral depending on the parameter k; this example happens to be a *complete elliptic integral* (Section 4.2).

If an integral depending on a parameter can be evaluated in terms of familiar functions, it becomes simply an explicit function of one variable. Thus, for example,

$$\int_0^{\pi} \sin (xt)\, dx = \frac{1}{t} - \frac{\cos (\pi t)}{t} \qquad (t \neq 0).$$

However, it can easily happen, as the preceding elliptic integral illustrates, that the integral cannot be expressed in terms of elementary functions. In such a case the function of the parameter is nevertheless well defined. It can be evaluated as accurately as desired for each particular parameter value and then tabulated; precisely this has been done for the preceding elliptic integral, and tables thereof are easily accessible (Section 4.2).

The question to be studied here is the evaluation of the *derivative* of a function $F(t)$ defined by an integral as in Eq. (4.93):

LEIBNITZ'S RULE   Let $f(x, t)$ be continuous and have a continuous derivative $\partial f/\partial t$ in a domain of the xt-plane that includes the rectangle $a \leq x \leq b$, $t_1 \leq t \leq t_2$. Then for $t_1 \leq t \leq t_2$,

$$\frac{d}{dt} \int_a^b f(x, t)\, dx = \int_a^b \frac{\partial f}{\partial t}(x, t)\, dx. \tag{4.94}$$

In other words, differentiation and integration can be interchanged; for example,

$$\frac{d}{dt} \int_0^\pi \sin(xt)\, dx = \int_0^\pi x \cos(xt)\, dx.$$

Here both sides can be fully evaluated so that the result can be checked.

Proof of Leibnitz's Rule.   Let

$$g(t) = \int_a^b \frac{\partial f}{\partial t}(x, t)\, dx \qquad (t_1 \leqq t \leqq t_2).$$

Since $\partial f/\partial t$ is continuous, one concludes from the theorem of Section 4.3 that $g(t)$ is continuous for $t_1 \leq t \leq t_2$. Now for $t_1 \leq t_3 \leq t_2$,

$$\int_{t_1}^{t_3} g(t)\, dt = \int_{t_1}^{t_3} \int_a^b \frac{\partial f}{\partial t}(x, t)\, dx\, dt;$$

by the theorem referred to one can interchange the order of integration:

$$\int_{t_1}^{t_3} g(t)\, dt = \int_a^b \int_{t_1}^{t_3} \frac{\partial f}{\partial t}(x, t)\, dt\, dx = \int_a^b [f(x, t_3) - f(x, t_1)]\, dx$$

$$= \int_a^b f(x, t_3)\, dx - \int_a^b f(x, t_1)\, dx = F(t_3) - F(t_1),$$

where $F(t)$ is defined by (4.93). If we now let t_3 be simply a variable t, we have

$$F(t) - F(t_1) = \int_{t_1}^t g(u)\, du.$$

Both sides can now be differentiated with respect to t. By the fundamental theorem (4.10), one obtains

$$F'(t) = g(t) = \int_a^b \frac{\partial f}{\partial t}(x, t)\, dx.$$

Thus the rule is proved.   □

The notion of integral depending on a parameter extends at once to multiple integrals, and Leibnitz's Rule also generalizes. Thus

$$\frac{d}{d\alpha} \int_1^2 \int_1^2 \sqrt{x^\alpha + y^\alpha}\, dx\, dy = \int_1^2 \int_1^2 \frac{x^\alpha \log x + y^\alpha \log y}{2\sqrt{x^\alpha + y^\alpha}}\, dx\, dy.$$

It can be extended to improper integrals, but with complications; this is discussed in Chapter 6.

One has at times to consider expressions like (4.93) in which the limits of integration a and b themselves depend on the parameter t. For example, one might have

$$F(t) = \int_{t^2}^{t^3} e^{-x^2 t} \, dx.$$

Here again, one has a method for finding the derivative in terms of an integral.

THEOREM   Let $f(x, t)$ satisfy the condition stated previously for Leibnitz's Rule. In addition, let $a(t)$ and $b(t)$ be defined and have continuous derivatives for $t_1 \le t \le t_2$. Then for $t_1 \le t \le t_2$,

$$\frac{d}{dt} \int_{a(t)}^{b(t)} f(x, t) \, dx = f[b(t), t] b'(t) - f[a(t), t] a'(t) + \int_{a(t)}^{b(t)} \frac{\partial f}{\partial t}(x, t) \, dx.$$

$$(4.95)$$

Thus, for example,

$$\frac{d}{dt} \int_{t^2}^{t^3} e^{-x^2 t} \, dx = e^{-t^7}(3t^2) - e^{-t^5}(2t) + \int_{t^2}^{t^3} e^{-x^2 t}(-x^2) \, dx.$$

Proof.   Let $u = b(t)$, $v = a(t)$, $w = t$, so that the integral $F(t)$ can be written as follows:

$$F(t) = \int_v^u f(x, w) \, dx = G(u, v, w),$$

where u, v, w all depend on t. Hence by the chain rule,

$$\frac{dF}{dt} = \frac{\partial G}{\partial u} \frac{du}{dt} + \frac{\partial G}{\partial v} \frac{dv}{dt} + \frac{\partial G}{\partial w} \frac{dw}{dt}.$$

It will be seen that the three terms here correspond to the three terms on the right of (4.95). Indeed, one has

$$\frac{\partial G}{\partial u} = \frac{\partial}{\partial u} \int_v^u f(x, w) \, dx = f(u, w),$$

by the fundamental theorem (4.10). Since $u = b(t)$, $du/dt = b'(t)$ and

$$\frac{\partial G}{\partial u} \frac{du}{dt} = f[b(t), t] b'(t).$$

The second term is accounted for similarly, the minus sign appearing because

$$\frac{\partial}{\partial v} \int_v^u f(x, w) \, dx = \frac{\partial}{\partial v} \left\{ -\int_u^v f(x, w) \, dx \right\} = -f(v, w).$$

Finally,

$$\frac{\partial G}{\partial w} = \frac{\partial}{\partial w} \int_v^u f(x, w) \, dx = \int_v^u \frac{\partial f}{\partial w}(x, w) \, dx$$

by Leibnitz's Rule. Since $w = t$, $dw/dt = 1$ and the third term is accounted for.   ◻

PROBLEMS

1. Obtain the indicated derivatives in the form of integrals:

a) $\dfrac{d}{dt} \displaystyle\int_{\frac{\pi}{2}}^{\pi} \dfrac{\cos(xt)}{x}\, dx$

b) $\dfrac{d}{dt} \displaystyle\int_{1}^{2} \dfrac{x^2}{(1-tx)^2}\, dx$

c) $\dfrac{d}{du} \displaystyle\int_{1}^{2} \log(xu)\, dx$

d) $\dfrac{d^n}{dy^n} \displaystyle\int_{1}^{2} \dfrac{\sin x}{x-y}\, dx$

2. Obtain the indicated derivatives:

a) $\dfrac{d}{dx} \displaystyle\int_{1}^{x} t^2\, dt$

b) $\dfrac{d}{dt} \displaystyle\int_{1}^{t^2} \sin(x^2)\, dx$

c) $\dfrac{d}{dt} \displaystyle\int_{t^3}^{2} \log(1 + x^2)\, dx$

d) $\dfrac{d}{dx} \displaystyle\int_{x}^{\tan x} e^{-t^2}\, dt$

3. Prove the following:

a) $\dfrac{d}{d\alpha} \displaystyle\int_{\sin\alpha}^{\cos\alpha} \log(x+\alpha)\, dx = \log\dfrac{\cos\alpha + \alpha}{\sin\alpha + \alpha} - [\sin\alpha \log(\cos\alpha + \alpha)$

$\qquad\qquad\qquad\qquad\qquad\qquad\qquad\qquad + \cos\alpha \log(\sin\alpha + \alpha)];$

b) $\dfrac{d}{du} \displaystyle\int_{0}^{\frac{\pi}{2u}} u \sin ux\, dx = 0;$

c) $\dfrac{d}{dy} \displaystyle\int_{y}^{y^2} e^{-x^2 y^2}\, dx = 2 y e^{-y^6} - e^{-y^4} - 2y \displaystyle\int_{y}^{y^2} x^2 e^{-x^2 y^2}\, dx.$

4. a) Evaluate $\displaystyle\int_{0}^{1} x^n \log x\, dx$ by differentiating both sides of the equation $\displaystyle\int_{0}^{1} x^n\, dx = \dfrac{1}{n+1}$ with respect to n $(n > -1)$.

b) Evaluate $\displaystyle\int_{0}^{\infty} x^n e^{-ax}\, dx$ by repeated differentiation of $\displaystyle\int_{0}^{\infty} e^{-ax}\, dx$ $(a > 0)$.

c) Evaluate $\displaystyle\int_{0}^{\infty} \dfrac{dy}{(x^2+y^2)^n}$ by repeated differentiation of $\displaystyle\int_{0}^{\infty} \dfrac{dy}{x^2+y^2}$.

[In (b) and (c) the improper integrals are of a type to which Leibnitz's Rule is applicable, as is shown in Chapter 6. The result of (a) can be explicitly verified.]

5. Leibnitz's Rule extends to indefinite integrals in the form:

$$\frac{\partial}{\partial t} \int f(x,t)\, dx + C = \int \frac{\partial}{\partial t} f(x,t)\, dx. \tag{a}$$

There is still an arbitrary constant in the equation because we are evaluating an *indefinite* integral. Thus from the equation

$$\int e^{tx}\, dx = \frac{e^{tx}}{t} + C,$$

one deduces that

$$\int x e^{tx}\, dx = e^{tx}\left(\frac{x}{t} - \frac{1}{t^2}\right) + C_1.$$

a) By differentiating n times, prove that

$$\int \frac{dx}{(x^2 + a)^n} = \frac{(-1)^{n-1}}{(n-1)!}\frac{\partial^{n-1}}{\partial a^{n-1}}\left(\frac{1}{\sqrt{a}}\ \arctan\frac{x}{\sqrt{a}}\right) + C \qquad (a > 0).$$

b) Prove $\int x^n \cos ax\, dx = \dfrac{\partial^n}{\partial a^n}\left(\dfrac{\sin ax}{a}\right) + C,\ n = 4, 8, 12, \ldots\,.$

c) Let $\int f(x, t)\, dx = F(x, t) + C$, so that $\partial F/\partial x = f(x, t)$. Show that Eq. (a) is equivalent to the statement

$$\frac{\partial^2 F}{\partial x\, \partial t} = \frac{\partial^2 F}{\partial t\, \partial x}.$$

6. It is known that

$$\int_0^{2\pi} \frac{\cos\theta}{1 - a\cos\theta}\, d\theta = 2\pi\frac{1 - \sqrt{1 - a^2}}{a\sqrt{1 - a^2}},$$

where a is a constant, $0 < a < 1$. (This can be established as in elementary calculus with the aid of the substitution $t = \tan(\theta/2)$.) Use this result to prove that

$$\int_0^{2\pi} \log(1 - a\cos\theta)\, d\theta = 2\pi\log\frac{1 + \sqrt{1 - a^2}}{2}.$$

[Hint: Call the left-hand side of the new equation $g(a)$, find $g'(a)$, and integrate to find $g(a) = 2\pi\log(1 + \sqrt{1 - a^2}) + C$. Use continuity of g for $a = 0$ and $g(0) = 0$ to find C.]

7. Consider a 1-dimensional fluid motion, the flow taking place along the x axis. Let $v = v(x, t)$ be the velocity at position x at time t, so that if x is the coordinate of a fluid particle at time t, one has $dx/dt = v$. If $f(x, t)$ is any scalar associated with the flow (velocity, acceleration, density, ...), one can study the variation of f following the flow with the aid of the Stokes derivative:

$$\frac{Df}{Dt} = \frac{\partial f}{\partial x}\frac{dx}{dt} + \frac{\partial f}{\partial t}$$

[see Problem 12 following Section 2.8]. A piece of the fluid occupying an interval $a_0 \le x \le b_0$ when $t = 0$ will occupy an interval $a(t) \le x \le b(t)$ at time t, where $\dfrac{da}{dt} = v(a, t)$, $\dfrac{db}{dt} = v(b, t)$. The integral

$$F(t) = \int_{a(t)}^{b(t)} f(x, t)\, dx$$

is then an integral of f over a definite piece of the fluid, whose position varies with time; if f is density, this is the mass of the piece. Show that

$$\frac{dF}{dt} = \int_{a(t)}^{b(t)}\left[\frac{\partial f}{\partial t}(x, t) + \frac{\partial}{\partial x}(fv)\right] dx = \int_{a(t)}^{b(t)}\left(\frac{Df}{Dt} + f\frac{dv}{dx}\right) dx.$$

This is generalized to arbitrary 3-dimensional flows in Section 5.15.

8. Let $f(\alpha)$ be continuous for $0 \leq \alpha \leq 2\pi$. Let

$$u(r, \theta) = \frac{1}{2\pi} \int_0^{2\pi} f(\alpha) \frac{1 - r^2}{1 + r^2 - 2r \cos(\theta - \alpha)} d\alpha$$

for $r < 1$, r and θ being polar coordinates. Show that u is harmonic for $r < 1$. This is the *Poisson integral formula*. [Hint: Write $w = 1 + r^2 - 2r \cos(\theta - \alpha)$ and $v(r, \theta, \alpha) = (1 - r^2)w^{-1}$. Use Leibnitz's rule to conclude that $\nabla^2 u = (2\pi)^{-1} \int_0^{2\pi} f(\alpha) \nabla^2 v \, d\theta$, where $\nabla^2 v = v_{rr} + r^{-2}v_{\theta\theta} + r^{-1}v_r$ as in Eq. (2.138) in Section 2.17. Show that $v_r = -2rw^{-1} - (1 - r^2)w^{-2}w_r$, etc. and finally

$$\nabla^2 v = -4w^{-1} + (5r - r^{-1})ww_r + (r^2 - 1)(w^{-2}w_{rr} - 2w^{-3}w_r^2)$$
$$+ r^{-2}w^{-2}w_{\theta\theta} - 2r^{-2}w^{-3}w_\theta^2).$$

Multiply both sides by r^2w^3, insert the proper expressions for $w, w_r, \ldots$ on the right and collect terms in powers of r ($r^6, r^5, \ldots$) to verify that $r^2w^3\nabla^2 v = 0$ and hence $\nabla^2 u = 0$.]

*4.10 UNIFORM CONTINUITY · EXISTENCE OF THE RIEMANN INTEGRAL

In this section and the following one we provide additional real variable theory and prove the existence of single and double integrals. The theory can be regarded as a continuation of that developed in Section 2.23; hence we continue naming theorems by letters (Theorem K, and so on) and refer to similar theorems in Section 2.23.

DEFINITION Let f be a mapping of the set G in E^N into E^M. Then f is said to be *uniformly continuous* if for each $\epsilon > 0$ there is a $\delta > 0$ such that whenever P, Q are in G and $d(P, Q) < \delta$, one has $d(f(P), f(Q)) < \epsilon$.

A uniformly continuous mapping is clearly continuous at every point. However, δ now depends only on ϵ and not on the point. For example, $y = f(x) = x^2$, $0 \leq x < \infty$, is a continuous mapping of the set G: $0 \leq x < \infty$ into E^1, but this mapping is not uniformly continuous. Thus let us take $\epsilon = 1$ and ask how close x_1, x_2 must be so that $|f(x_1) - f(x_2)| < 1$. If $|x_2 - x_1| = c$, then $|f(x_1) - f(x_2)| = |x_1^2 - x_2^2| = |x_1 + x_2| |x_2 - x_1| = c|x_1 + x_2|$. If this is to be less than 1, then we must have

$$c|x_1 + x_2| < 1 \quad \text{or} \quad c < \frac{1}{|x_1 + x_2|}$$

But we can choose x_1, x_2 as close together as we wish while $|x_1 + x_2|$ is arbitrarily large. Thus there is no $\delta > 0$ such that $c < \delta$ implies $|f(x_1) - f(x_2)| < 1$.

We can illustrate this numerically by trying to calculate $y = x^2$ with error less than 0.001 by approximating x by a nearby value with small error. For $x = 1$ we approximate by 1.0004 and get

$$y = (1.0004)^2 = 1.00080016,$$

which is within the allowed error. For $x = 10$, use of the same accuracy in x leads us to 10.0004 and $y = 100.0080016$, so that the error in y is now greater than

0.001. The larger the x, the smaller the error in x allowed to ensure the same accuracy in y; the δ depends on ϵ *and* x.

THEOREM K   Let f be a continuous mapping of a bounded closed set G in E^N into E^M. Then f is uniformly continuous.

Proof. Suppose f is not uniformly continuous. Then for some $\epsilon > 0$ and every $\delta > 0$ we can find a pair of points P, Q in G such that $d(P, Q) < \delta$ but $d(f(P), f(Q)) \leq \delta$. In particular, we take δ successively equal to $1/2, 1/4, \ldots, 1/2^n, \ldots$ to obtain sequences P_n, Q_n such that

$$d(P_n, Q_n) < \frac{1}{2^n} \quad \text{and} \quad f(P_n, Q_n) \geq \epsilon.$$

By Theorems E and F of Section 2.23 we can choose a subsequence P_{n_k} converging to P_0 in G. Then

$$d(Q_{n_k}, P_0) \leq d(Q_{n_k}, P_{n_k}) + d(P_{n_k}, P_0).$$

Since both sequences on the right have limit 0, we conclude that $Q_{n_k} \to P_0$ also. Hence $f(P_{n_k}) \to f(P_0), f(Q_{n_k}) \to f(P_0)$. But

$$0 < \epsilon \leq d\big(f(P_{n_k}), f(Q_{n_k})\big) \leq d\big(f(P_{n_k}), f(P_0)\big) + d\big(f(P_0), f(Q_{n_k})\big).$$

As $k \to \infty$, both terms on the right have limit 0. That is a contradiction. Hence f is uniformly continuous.   □

THEOREM L   Let $y = f(x)$ be a continuous real function of the real variable x for $a \leq x \leq b$. Then the Riemann integral

$$\int_a^b f(x)\, dx$$

exists.

Proof. We first observe that the interval $a \leq x \leq b$ is a bounded closed set (Problem 4 following Section 2.23). Hence by Theorem K, f is uniformly continuous.

As in Section 4.1, we now consider subdivisions of the interval: $a = x_0 < x_1 < \cdots < x_n = b$. We let h be the mesh of the subdivision, the maximum of $\Delta_1 x = x_1 - x_0, \ldots, \Delta_n x = x_n - x_{n-1}$. We must show that there is a unique number c such that for each $\epsilon > 0$ there is a $\delta > 0$ such that

$$\left| \sum_{i=1}^{n} f(x_i^*) \Delta_i x - c \right| < \epsilon$$

for all subdivisions of mesh h less than δ, no matter how $x_1^*, \ldots, x_n^*$ are chosen in the successive subintervals: $x_{i-1} \leq x_i^* \leq x_i$ for $i = 1, \ldots, n$.

To this end, we first observe that by Theorem J of Section 2.23, f has an absolute maximum B and minimum B, so that $A \leq f(x) \leq B$ for $a \leq x \leq b$. Hence

$$A \sum_{i=1}^{n} \Delta_i x \leq \sum_{i=1}^{n} f(x_i^*) \Delta_i x \leq B \sum_{i=1}^{n} \Delta_i x$$

or

$$A(b-a) \le \sum_{i=1}^{n} f(x_i^*) \Delta_i x \le B(b-a)$$

for all sums $\Sigma f(x_i^*) \Delta_i x$. Thus the sums form a bounded set of real numbers.

In particular, we consider E_δ, the set of sums arising only from those subdivisions of mesh h less than δ, where $\delta > 0$. We let

$$\alpha_\delta = \text{glb } E_\delta, \qquad \beta_\delta = \text{lub } E_\delta.$$

Thus

$$\alpha_\delta \le \sum_{i=1}^{n} f(x_i^*) \Delta_i x \le \beta_\delta$$

for all subdivisions of mesh h less than δ.

By uniform continuity we now choose $\eta > 0$ so that

$$|f(x') - f(x'')| < \frac{\epsilon}{2(b-a)},$$

whenever x', x'' are on the interval and $|x' - x''| < \eta$. We let $\delta = \eta/2$ and consider two sums in E_δ. The two sums may come from different subdivisions of the interval. However, by combining the points of the two subdivisions we obtain a new subdivision of mesh h less than δ and can write each sum in terms of the new subdivision. If a point $\bar{x}$ is a new subdivision point between x_{i-1} and x_i, we write

$$f(x_i^*)(x_i - x_{i-1}) = f(x_i^*)(x_i - \bar{x}) + f(x_i^*)(\bar{x} - x_{i-1})$$

and can proceed similarly for all new subdivision points. Thus both sums now have the form, in terms of the new subdivision,

$$\sum_{i=1}^{n} f(x_i^*)(x_i - x_{i-1}), \qquad \sum_{i=1}^{n} f(x_i^{**})(x_i - x_{i-1}).$$

However, the points x_i^*, x_i^{**} are no longer necessarily between x_{i-1} and x_i. We can say that in any case, x_i^* and x_i^{**} are at most $2\delta = \eta$ apart. Accordingly, $|f(x_i^*) - f(x_i^{**})| < \epsilon/[2(b-a)]$ and hence

$$\left| \sum_{i=1}^{n} f(x_i^*) \Delta_i x - \sum_{i=1}^{n} f(x_i^{**}) \Delta_i x \right| = \left| \sum_{i=1}^{n} [f(x_i^*) - f(x_i^{**})] \Delta_i x \right|$$

$$\le \sum_{i=1}^{n} |f(x_i^*) - f(x_i^{**})| \Delta_i x$$

$$< \frac{\epsilon}{2(b-a)} \sum_{i=1}^{n} \Delta_i x = \frac{\epsilon}{2}.$$

Hence any two numbers in E_δ are less than $\epsilon/2$ apart and therefore $\beta_\delta - \alpha_\delta \le \epsilon/2$.

We now choose the sequence $\delta_n = 2^{-n}$ $(n = 1, 2, \ldots)$ converging to 0 and let s_n be the corresponding sequence α_{δ_n} of greatest lower bounds of E_{δ_n}. Since $E_{\delta_{n+1}}$ is a subset of E_{δ_n} (every subdivision of mesh less than 2^{-n-1} is a subdivision of mesh less than 2^{-n}), $s_n \leq s_{n+1}$ for all n, so that s_n is a monotone increasing sequence; this sequence is also bounded, since $s_n \leq B(b - a)$ for all n. Therefore, by Theorem A of Section 2.23, s_n converges to some number c, so that $|s_n - c| < \epsilon/2$ for n sufficiently large. But for n sufficiently large, δ_n is less than δ as above, so that $|\alpha_{\delta_n} - \beta_{\delta_n}| \leq \epsilon/2$ and every sum s in E_{δ_n} satisfies $|s - s_n| \leq \epsilon/2$. Therefore for n sufficiently large,

$$|s - c| = |s - s_n + s_n - c| \leq |s - s_n| + |s_n - c|$$

$$< \frac{\epsilon}{2} + \frac{\epsilon}{2} = \epsilon$$

for every sum s in E_{δ_n}; that is, for δ sufficiently small and $0 < h < \delta$, we have $|s - c| < \epsilon$ for every sum s obtained from a subdivision of mesh h. Therefore the theorem is proved. $\square$

*4.11 THEORY OF DOUBLE INTEGRALS

We now prove two theorems that show the existence of the double integral and justify its representation as an iterated integral. We confine attention to regions G of form $a \leq x \leq b, y_1(x) \leq y \leq y_2(x)$.

THEOREM M   Let $f(x, y)$ be continuous on the set $G: a \leq x \leq b, y_1(x) \leq y \leq y_2(x)$, where $y_1(x)$ and $y_2(x)$ are continuous for $a \leq x \leq b$ and $y_1(x) \leq y_2(x)$ for $a \leq x \leq b$. Then

$$g(x) = \int_{y_1(x)}^{y_2(x)} f(x, y)\, dy \tag{4.96}$$

is continuous for $a \leq x \leq b$.

Proof. We first consider the case when $y_1(x)$ and $y_2(x)$ are unequal constant functions so that

$$g(x) = \int_c^d f(x, y)\, dy; \tag{4.97}$$

the function f is now continuous on the bounded closed set $a \leq x \leq b, c \leq y \leq d$. Hence it is uniformly continuous: Given $\epsilon > 0$, we choose $\delta > 0$ so that $|f(x_1, y_2) - f(x_2, y_2)| < \epsilon/(d - c)$ when $d((x_1, y_1), (x_2, y_2)) < \delta$. Then for $|x_1 - x_2| < \delta$,

$$g(x_1) - g(x_2) = \int_c^d [f(x_1, y) - f(x_2, y)]\, dy,$$

$$|g(x_1) - g(x_2)| \leq \int_c^d |f(x_1, y) - f(x_2, y)|\, dy$$

$$\leq \int_c^d \frac{\epsilon}{d - c}\, dy = \epsilon,$$

since $d((x_1, y), (x_2, y)) = |x_1 - x_2| < \delta$. Therefore g is continuous for $a \leq x \leq b$.

For the general case (4.96) we make the substitution, for fixed x,

$$y = y_1(x) + u[y_2(x) - y_1(x)], \qquad 0 \leq u \leq 1.$$

As in Section 4.6, we obtain

$$g(x) = [y_2(x) - y_1(x)] \int_0^1 f(x, y_1(x) + u[y_2(x) - y_1(x)]) \, du.$$

As a composite function of continuous functions, the integrand $f(x, y_1(x) + \cdots)$ is continuous for $a \leq x \leq b$, $0 \leq u \leq 1$. The integral now has the form (4.97), with $c = 0$, $d = 1$. Therefore by the case considered previously it defines a continuous function of x for $a \leq x \leq b$; multiplication by the continuous function $y_2(x) - y_1(x)$ produces a continuous function $g(x)$. Therefore the proof is complete. $\square$

THEOREM N   Let $f(x, y)$ be continuous on the set $G: a \leq x \leq b$, $y_1(x) \leq y \leq y_2(x)$, where $y_1(x)$ and $y_2(x)$ are continuous for $a \leq x \leq b$ and $y_1(x) < y_2(x)$ for $a < x < b$. Then the double integral of f over G exists and equals the iterated integral

$$\int_a^b \int_{y_1(x)}^{y_2(x)} f(x, y) \, dy \, dx. \tag{4.98}$$

Proof. As in Section 4.3, we subdivide the region by parallels to the axes: Lines $x = x_i = $ const for $a = x_0 < x_1 < \cdots < x_n = b$, lines $y = y_j = $ const for $y_0 < y_1 < \cdots < y_m$. Here y_0 must be less than or equal to the minimum of $y_1(x)$, and y_m must be greater than or equal to the maximum of $y_2(x)$. These lines form rectangles, and we let h be the mesh of the subdivision, the largest diagonal of the rectangles.

By Theorem M the iterated integral (4.98) exists and has a value c. We let $g(x)$ denote the inner integral, as in (4.96), so that $g(x)$ is continuous. Now

$$c = \int_a^b g(x) \, dx = \int_{x_0}^{x_1} g(x) \, dx + \cdots + \int_{x_{n-1}}^{x_n} g(x) \, dx$$

$$= g(x_1^*) \Delta_1 x + \cdots + g(x_n^*) \Delta_n x, \tag{4.99}$$

by the mean value theorem, with $x_{i-1} \leq x_i^* \leq x_i$ and $\Delta_i x = x_i - x_{i-1}$ for $i = 1, \ldots, n$. Similarly,

$$g(x_i^*) = \int_{y_1(x_i^*)}^{y_2(x_i^*)} f(x_i^*, y) \, dy$$

$$= f(x_i^*, y_{i1}^*) \Delta_{i1} y + \cdots + f(x_i^*, y_{im_i}^*) \Delta_{im_i} y. \tag{4.100}$$

Here we use the values y_{ij} to subdivide the interval $y_1(x_i^*) \leq y \leq y_2(x_i^*)$, letting y_{i1} be the first of the values $y_0, y_1, \ldots$ to exceed $y_1(x_i^*)$ and y_{i,m_i-1} be the last to fall below $y_2(x_i^*)$. We also write

$$y_{i0} = y_1(x_i^*), \qquad y_{im_i} = y_2(x_i^*)$$

Figure 4.16   Formation of sum (4.101).

(see Fig. 4.16). We write

$$\Delta_{ik} y = y_{ik} - y_{i,k-1}.$$

Also the values y_{i1}^*, y_{i2}^*,... satisfy

$$y_{i,k-1} \le y_{ik}^* \le y_{ik}.$$

If we combine (4.99) and (4.100), we obtain

$$c = \sum_{i=1}^{n} \sum_{k=1}^{m_i} f(x_i^*, y_{ik}^*) \Delta_i x \Delta_{ik} y. \tag{4.101}$$

On the right-hand side we have a sum like that for the double integral; on the left-hand side we have a *constant* c. It follows that as the mesh $h \to 0$, the sums on the right-hand side have limit c.

However, the sum in (4.101) includes terms for rectangles not wholly contained in G. This is clear from Fig. 4.16, where the top and bottom rectangles extend beyond G. We show below that for given ϵ, a $\delta > 0$ can be found so that for $0 < h < \delta$ the absolute value of the sum of such terms is less than $\epsilon/2$. Therefore we can write, for $0 < k < \delta$,

$$c = \text{a sum for double integral} + \text{term of absolute value less than } \epsilon/2. \tag{4.102}$$

Granting this point, we now consider an arbitrary sum for the double integral, using the same subdivision by lines parallel to the axes; this allows us to vary the choice of the points at which f is evaluated in each rectangle. We obtain a number

$$S = \sum \sum f(x_i^{**}, y_{ik}^{**}) \Delta_i x \Delta_{ik} y, \tag{4.103}$$

where the sum is as in (4.101), but we exclude the contribution from rectangles extending beyond G, as in the first term on the right of (4.102). Hence we can write

$$c - S = \sum\sum\left[f(x_i^*, y_{ik}^*) - f(x_i^{**}, y_{ik}^{**})\right]\Delta_i x \Delta_{ik} y$$

$$+ \text{ term of absolute value} < \epsilon/2. \tag{4.104}$$

Now G is a bounded closed set. Hence f is uniformly continuous on G. Therefore we can choose δ so small that the values of f at points less than δ apart differ by less than $\epsilon/(2A)$, where A is the area of G. For subdivisions of mesh h, with $h < \delta$ and δ thus restricted (and restricted as earlier), we have from (4.104)

$$|c - S| < \sum\sum \frac{\epsilon}{2A}\Delta_i x \Delta_{ik} y + \frac{\epsilon}{2} < \frac{\epsilon}{2A} \cdot A + \frac{\epsilon}{2} = \epsilon.$$

For here $\sum\sum \Delta_i x \Delta_{ik} y$ is the total area of the rectangles contained in G, and this area is at most A. Accordingly, by definition of the double integral,

$$\int_G\int f(x, y)\, dA = c = \int_a^b \int_{y_1(x)}^{y_2(x)} f(x, y)\, dy\, dx,$$

as asserted.

It remains to show that for δ sufficiently small the rectangles extending beyond G contribute a sum of absolute value less than $\epsilon/2$ to the whole sum (4.101).

Now $f(x, y)$ is continuous on G, and therefore $|f(x, y)| \leq K$ for some constant K, for all (x, y) in G. This follows from Theorem J of Section 2.23. Accordingly, the absolute value of the sum we are considering is at most K times the sum of the areas of the rectangles. Thus it suffices to show that the total *area* of the rectangles can be made arbitrarily small by choosing h sufficiently small.

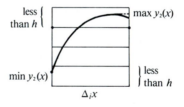

Figure 4.17 Contribution of area due to boundary curve.

But, as Fig. 4.17 shows, the extruding rectangles in the interval $x_{i-1} \leq x \leq x_i$ provide along the upper boundary a total area less than

$$(\max y_2(x) - \min y_2(x))\Delta_i x + 2h\Delta_i x. \tag{4.105}$$

By uniform continuity, for h sufficiently small, $\max y_2(x) - \min y_2(x)$ is less than a given $\eta > 0$, and if h is also less than η, the whole expression (4.105) is less than $3\eta\Delta_i x$, so that the sum of the areas from the upper boundary is at most $3\eta(b - a)$. If $\eta = \epsilon/[12(b - a)]$, the upper boundary contributes less than $\epsilon/4$ to the area of extruding rectangles. Similarly, the lower one contributes less than $\epsilon/4$ to this area. Hence for h sufficiently small the total unwanted area is at most $\epsilon/2$. Thus our conclusion follows. $\square$

Remarks.   The argument about the extruding rectangles shows why, in finding the double integral as a limit, one can include or exclude such terms. For h sufficiently small their contribution is as small as desired. Roughly stated, the boundary curves here have *zero area*. There are bounded regions whose boundary sets do *not* have zero area, and for these our limit process for the double integral leads to difficulty.

To handle such problems and, in general, to provide a more satisfactory theory of integration, it is better to use the Lebesgue integral, which is developed in many advanced texts including that of Rudin listed at the end of the chapter.

PROBLEMS

1. **a)** Prove the rule for vectors in V^N:

$$|\,|\mathbf{u}| - |\mathbf{v}|\,| \le |\mathbf{u} - \mathbf{v}|.$$

   b) Prove: The function $f(\mathbf{x}) = |\mathbf{x}|$ is a continuous mapping of E^N into E^1. Is it uniformly continuous?

2. Determine whether the function is uniformly continuous:

   a) $y = e^x, 0 \le x \le 1$;

   b) $y = \log x, 0 < x \le 1$;

   c) $y = \log x, 1 \le x < \infty$;

   d) $y = \sin x, -\infty < x < \infty$.

3. Prove: If $f(x)$ is defined for $a < x < b$, $f'(x)$ is continuous for $a < x < b$ and $f'(x)$ is bounded, $|f'(x)| \le K$, then f is uniformly continuous for $a < x < b$.

4. Let $f(x, y)$ have continuous partial derivatives in the domain D in the xy-plane. Further let $|\nabla f| \le K$ in D, where K is a constant. In each of the following cases, determine whether this implies that f is uniformly continuous:

   a) D is the domain $x^2 + y^2 < 1$. [Hint: If s is distance along a line segment from (x_1, y_1) to (x_2, y_2), then f has directional derivative $df/ds = \nabla f \cdot \mathbf{u}$ along the line segment, where $\mathbf{u}$ is an appropriate unit vector.]

   b) D is the domain $|x| < 1, |y| < 1$, excluding the points $(x, 0)$ for $0 \le x < 1$.

5. Show that if f and f' are continuous for $a \le x \le b$ and $|f'(x)| \le K = $ const for $a \le x \le b$, then for each subdivision of mesh less than $\delta = \epsilon/[2K(b - a)]$, each sum $\Sigma f(x_i^*)\Delta_i x$ differs from $\int_a^b f(x)\, dx$ by less than ϵ.

Suggested References

Courant, Richard J., Differential and Integral Calculus, transl. by E. J. McShane, 2 vols. New York: Interscience, 1947.

Franklin, Philip, A Treatise on Advanced Calculus. New York: John Wiley and Sons, Inc., 1940.

Goursat, Edouard, A Course in Mathematical Analysis, Vol. 1, transl. by E. R. Hedrick. New York: Dover Publications, 1970.

Henrici, Peter K., Elements of Numerical Analysis. New York: John Wiley and Sons, Inc., 1964.

Kaplan, Wilfred, and Donald J. Lewis, Calculus and Linear Algebra, 2 vols. New York: John Wiley and Sons, Inc., 1970–1971.

Ralston, Anthony, and Rabinowitz, Philip, First Course in Numerical Analysis, 2nd ed. New York: McGraw-Hill, 1978.

Rudin, Walter, Principles of Mathematical Analysis, 2nd ed. New York: McGraw-Hill, 1964.

Scarborough, James B., Numerical Mathematical Analysis, 6th ed. Baltimore: Johns Hopkins Press, 1966.

von Kármán, Theodore, and Maurice A. Biot, Mathematical Methods in Engineering. New York: McGraw-Hill, 1940.

Whittaker, E. T., and G. N. Watson, Modern Analysis, 4th ed. Cambridge: Cambridge University Press, 1940.

Widder, David V., Advanced Calculus, 2nd ed. New York: Prentice-Hall, Inc., 1961.

5

Vector Integral Calculus

Part I. Two-Dimensional Theory

5.1 INTRODUCTION

The topic of this chapter is *line and surface integrals*. It will be seen that these can both be regarded as integrals of vectors and that the principal theorems can be most simply stated in terms of vectors; hence the title "vector integral calculus."

A familiar line integral is that of arc length: $\int_C ds$. The subscript C indicates that one is measuring the length of a curve C, as in Fig. 5.1. If C is given in parametric form $x = x(t)$, $y = y(t)$, the line integral reduces to the ordinary definite integral:

$$\int_C ds = \int_{t_1}^{t_2} \sqrt{\left(\frac{dx}{dt}\right)^2 + \left(\frac{dy}{dt}\right)^2}\, dt.$$

If the curve C represents a wire whose density (mass per unit length) varies along C, then the wire has a total mass

$$M = \int_C f(x, y)\, ds,$$

where $f(x, y)$ is the density at the point (x, y) of the wire. The new integral can be expressed in terms of a parameter as previously or can be thought of simply as a limit of a sum

$$\int_C f(x, y)\, ds = \lim \sum_{i=1}^{n} f(x_i^*, y_i^*)\, \Delta_i s.$$

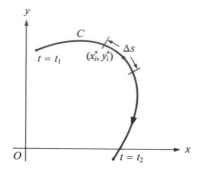

Figure 5.1   Line integral.

Here the curve has been subdivided into n pieces of lengths $\Delta_1 s, \Delta_2 s, \ldots, \Delta_n s$, and the point (x_i^*, y_i^*) lies on the ith piece. The limit is taken as n becomes infinite, while the maximum $\Delta_i s$ approaches 0.

A third example of a line integral is that of *work*. If a particle moves from one end of C to the other under the influence of a force $\mathbf{F}$, the work done by this force is defined as

where F_T denotes the component of $\mathbf{F}$ on the tangent $\mathbf{T}$ in the direction of motion. This integral can be thought of as a limit of a sum as previously. However, another interpretation is possible. We first remark that the work done by a constant force $\mathbf{F}$ in moving a particle from A to B on the line segment AB is $\mathbf{F} \cdot \overrightarrow{AB}$; for this scalar product is equal to $|\mathbf{F}| \cdot \cos \alpha \cdot |\overrightarrow{AB}|$, α being the angle between $\mathbf{F}$ and $\overrightarrow{AB}$, and hence to the product of force component in direction of motion by the distance moved. Now the motion of the particle along C can be thought of as the sum of many small displacements along line segments, as suggested in Fig. 5.2. If these displacements are denoted by $\Delta_1 \mathbf{r}, \Delta_2 \mathbf{r}, \ldots, \Delta_n \mathbf{r}$, the work done would be approximated by a sum of form

$$\sum_{i=1}^{n} \mathbf{F}_i \cdot \Delta_i \mathbf{r},$$

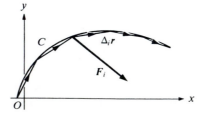

Figure 5.2   Work $= \displaystyle\int_C \mathbf{F} \cdot d\mathbf{r}.$

where $\mathbf{F}_i$ is the force acting for the ith displacement. The limiting form of this is again equal to the line integral $\int F_T \, ds$, but because of the way the limit is obtained, we can also write it as

$$\int_C \mathbf{F} \cdot d\mathbf{r}.$$

One can thus write

$$\text{work} = \int_C F_T \, ds = \int_C \mathbf{F} \cdot d\mathbf{r}.$$

If the displacement vector $\Delta \mathbf{r}$ and force $\mathbf{F}$ are expressed in components,

$$\mathbf{F} = F_x \mathbf{i} + F_y \mathbf{j}, \qquad \Delta \mathbf{r} = \Delta x \mathbf{i} + \Delta y \mathbf{j},$$

the element of work $\mathbf{F} \cdot \Delta \mathbf{r}$ becomes

$$\mathbf{F} \cdot \Delta \mathbf{r} = F_x \, \Delta x + F_y \, \Delta y.$$

The total amount of work done is then approximated by a sum of form

$$\sum (F_x \, \Delta x + F_y \, \Delta y) = \sum F_x \, \Delta x + \sum F_y \, \Delta y.$$

The limiting form of this is a sum of two integrals:

$$\int_C F_x \, dx + \int_C F_y \, dy.$$

The first integral represents the work done by the x-component of the force; the second integral represents the work done by the y-component of the force.

It thus appears that one has three types of line integrals to consider, namely, the types

$$\int_C f(x, y) \, ds, \qquad \int_C P(x, y) \, dx, \qquad \int_C Q(x, y) \, dy,$$

which are limits of sums

$$\sum f(x, y) \, \Delta s, \qquad \sum P(x, y) \, \Delta x, \qquad \sum Q(x, y) \, \Delta y.$$

The foregoing gives the basis for the theory of line integrals in the plane. A very slight extension of these ideas leads to line integrals in space:

$$\int_C f(x, y, z) \, ds, \qquad \int_C f(x, y, z) \, dx, \qquad \ldots .$$

Surface integrals appear as a natural generalization, with the surface area element $d\sigma$ replacing the arc element ds:

$$\int\int_S f(x, y, z) \, d\sigma = \lim \sum f(x, y, z) \, \Delta \sigma.$$

There are corresponding component integrals

$$\int\int_S f(x, y, z) \, dx \, dy, \qquad \int\int_S f(x, y, z) \, dy \, dz, \qquad \ldots$$

and a vector surface integral

$$\int_S\int \mathbf{F} \cdot d\boldsymbol{\sigma} = \int_S\int (\mathbf{F} \cdot \mathbf{n})\, d\sigma,$$

where $d\boldsymbol{\sigma} = \mathbf{n}\, d\sigma$ is the "area element vector," $\mathbf{n}$ being a unit normal vector to the surface.

It will be seen that the basic theorems—those of Green, Gauss, and Stokes—concern the relations between line, surface, and volume (triple) integrals. These correspond to fundamental physical relations between such quantities as flux, circulation, divergence, and curl. The applications will be considered at the end of the chapter.

5.2   LINE INTEGRALS IN THE PLANE

We now state in precise form the definitions outlined in the preceding section.

By a *smooth curve* C in the xy-plane will be meant a curve representable in the form:

$$x = \phi(t), \qquad y = \psi(t), \qquad h \le t \le k, \tag{5.1}$$

where x and y are continuous and have continuous derivatives for $h \le t \le k$. The curve C can be assigned a direction, which will usually be that of increasing t. If A denotes the point $[\phi(h), \psi(h)]$ and B denotes the point $[\phi(k), \psi(k)]$, then C can be thought of as the path of a point moving continuously from A to B. This path may cross itself, as for the curve C_1 of Fig. 5.3. If the initial point A and terminal point B coincide, C is termed a *closed* curve; if, in addition, (x, y) moves from A to $B = A$ without retracing any other point, C is called a *simple closed* curve (curve C_2 of Fig. 5.3).

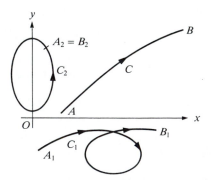

Figure 5.3   Paths of integration.

Let C be a smooth curve as previously, with positive direction that of increasing t. Let $f(x, y)$ be a function defined at least when (x, y) is on C. The line integral $\int_C f(x, y)\, dx$ is defined as a limit:

$$\int_C f(x, y)\, dx = \lim \sum_{i=1}^{n} f(x_i^*, y_i^*)\, \Delta_i x. \tag{5.2}$$

The limit refers to a subdivision of C as indicated in Fig. 5.4. The successive subdivision points are $A: (x_0, y_0), (x_1, y_1), \ldots, B: (x_n, y_n)$. These correspond to parameter values: $h = t_0 < t_1 < \cdots < t_n = k$. The point (x_i^*, y_i^*) is some point of C between (x_{i-1}, y_{i-1}) and (x_i, y_i); that is, (x_i^*, y_i^*) corresponds to a parameter value t_i^*, where $t_{i-1} \leq t_i^* \leq t_i$. $\Delta_i x$ denotes the difference $x_i - x_{i-1}$. The limit is taken as n becomes infinite and the largest $\Delta_i t$ approaches 0, where $\Delta_i t = t_i - t_{i-1}$. Similarly,

$$\int_C f(x, y)\, dy = \lim \sum f(x_i^*, y_i^*)\, \Delta_i y, \tag{5.3}$$

where $\Delta_i y = y_i - y_{i-1}$.

The significance of these definitions is guaranteed by the following basic theorems:

I If $f(x, y)$ is continuous on C, then

$$\int_C f(x, y)\, dx \quad \text{and} \quad \int_C f(x, y)\, dy \quad \text{exist.}$$

II If $f(x, y)$ is continuous on C, then

$$\int_C f(x, y)\, dx = \int_h^k f[\phi(t), \psi(t)]\, \phi'(t)\, dt, \tag{5.4}$$

$$\int_C f(x, y)\, dy = \int_h^k f[\phi(t), \psi(t)]\, \psi'(t)\, dt. \tag{5.5}$$

Formulas (5.4) and (5.5) reduce the integrals to ordinary definite integrals and are thus essential for computation of particular integrals. Thus let C be the

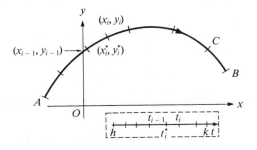

Figure 5.4   Definition of line integral.

path $x = 1 + t$, $y = t^2$, $0 \le t \le 1$, directed with increasing t. Then

$$\int_C (x^2 - y^2)\, dx = \int_0^1 \left[(1 + t)^2 - t^4 \right] dt = \tfrac{32}{15},$$

$$\int_C (x^2 - y^2)\, dy = \int_0^1 \left[(1 + t)^2 - t^4 \right] 2t\, dt = 2\tfrac{1}{2}.$$

It is logically easier to prove II first, for I is an immediate consequence of II. To prove II, one notes that the sum $\Sigma f(x_i^*, y_i^*)\, \Delta_i x$ can be written as

$$\sum_{i=1}^n f\left[\phi(t_i^*), \psi(t_i^*)\right] \frac{\Delta_i x}{\Delta_i t} \Delta_i t.$$

Now $\Delta_i x = x_i - x_{i-1} = \phi'(t_i^{**})\, \Delta_i t$ by the Law of the Mean. Hence the sum can be written as

$$\sum_{i=1}^n F(t_i^*) \phi'(t_i^{**})\, \Delta_i t,$$

where $F(t) = f[\phi(t), \psi(t)]$ and t_i^* and t_i^{**} are both between t_{i-1} and t_i. It is easily shown [see CLA, Section 12-25] that this sum approaches as limit the integral

$$\int_h^k F(t)\phi'(t)\, dt = \int_h^k f[\phi(t), \psi(t)]\phi'(t)\, dt$$

as required. Formula (5.5) is proved in the same way.

We remark that the value of a line integral on C does not depend on the particular parametrization of C, but only on the order in which the points of C are traced. (See Problem 5.)

In many applications the path C is not itself smooth but is composed of a finite number of arcs, each of which is smooth. Thus C might be a broken line. In this case, C is termed *piecewise* smooth. The line integral along C is simply, by definition, the sum of the integrals along the pieces. One verifies at once that (5.2), (5.3), and the theorems I and II continue to hold. In (5.4) and (5.5) the functions $\phi'(t)$ and $\psi'(t)$ will have jump discontinuities, which will not interfere with the existence of the integral (cf. Section 4.1). *Throughout this book all paths of integration for line integrals will be piecewise smooth unless otherwise specified.*

If the curve C is represented in the form

$$y = g(x), \qquad a \le x \le b,$$

then one can regard x itself as parameter, replacing t; that is, C is given by the equations

$$x = x, \qquad y = g(x), \qquad a \le x \le b$$

in terms of the parameter x. If the direction of C is that of increasing x, (5.4) and

(5.5) become

$$\int_C f(x, y)\, dx = \int_a^b f[x, g(x)]\, dx,$$

(5.6)

$$\int_C f(x, y)\, dy = \int_a^b f[x, g(x)]\, g'(x)\, dx.$$

(5.7)

The ordinary definite integral $\int_a^b y\, dx$, where $y = g(x)$, is a special case of (5.6). Similarly, if C is represented in the form

$$x = F(y), \qquad c \le y \le d,$$

and the direction of C is that of increasing y, then

$$\int_C f(x, y)\, dx = \int_c^d f[F(y), y]\, F'(y)\, dy,$$

(5.8)

$$\int_C f(x, y)\, dy = \int_c^d f[F(y), y]\, dy.$$

(5.9)

In most applications the line integrals appear as a combination,

$$\int_C P(x, y)\, dx + \int_C Q(x, y)\, dy,$$

which is abbreviated as follows:

$$\int_C [P(x, y)\, dx + Q(x, y)\, dy] \quad \text{or} \quad \int_C P(x, y)\, dx + Q(x, y)\, dy,$$

the brackets being used only when necessary.

In the formulas thus far the direction of C has been that of increasing parameter. If the opposite direction is chosen, upper and lower limits are reversed on all integrals. Thus (5.4) becomes

$$\int_C f(x, y)\, dx = \int_k^h f[\phi(t), \psi(t)]\, \phi'(t)\, dt.$$

(5.4′)

The line integral is therefore multiplied by -1. Often it is convenient to specify the path by its equations in some form and to indicate the direction by using the initial and terminal points as lower and upper limits:

$$\int_{C\,A}^{B} P\, dx + Q\, dy \quad \text{or} \quad \int_{C\,(x_1, y_1)}^{(x_2, y_2)} P\, dx + Q\, dy.$$

It will be seen later that under certain conditions, one needs only prescribe initial and terminal points:

$$\int_A^B P\, dx + Q\, dy.$$

EXAMPLE 1   To evaluate

$$\int_{C\,(1,0)}^{(-1,0)} (x^3 - y^3)\, dy,$$

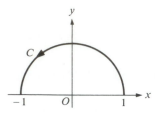

Figure 5.5 Example 1.

where C is the semicircle $y = \sqrt{1 - x^2}$ shown in Fig. 5.5, one can represent C parametrically:

$$x = \cos t, \qquad y = \sin t, \qquad 0 \le t \le \pi,$$

and the integral becomes

$$\int_0^\pi (\cos^3 t - \sin^3 t) \cos t \, dt = \frac{3\pi}{8}.$$

One can use x as parameter, and the integral becomes

$$\int_1^{-1} \left[x^3 (1 - x^2)^{\frac{3}{2}} \right] \frac{-x}{\sqrt{1 - x^2}} \, dx;$$

this is clearly in a more awkward form for integration. The substitution $x = \cos t$ brings one back to the parametric form. One can use y as parameter but has then to split the integral into two parts, from $(1,0)$ to $(0,1)$ and from $(0,1)$ to $(-1,0)$:

$$\int_0^1 \left[(1 - y^2)^{\frac{3}{2}} - y^3 \right] dy + \int_1^0 \left[-(1 - y^2)^{\frac{3}{2}} - y^3 \right] dy = 2 \int_0^1 (1 - y^2)^{\frac{3}{2}} dy.$$

Note that $x = \sqrt{1 - y^2}$ on the first part of the path and $x = -\sqrt{1 - y^2}$ on the second part. ■

EXAMPLE 2 Let C be the parabolic arc $y = x^2$ from $(0,0)$ to $(-1,1)$. Then

$$\int_C xy^2 \, dx + x^2 y \, dy = \int_0^{-1} \left(xy^2 + x^2 y \frac{dy}{dx} \right) dx = \int_0^{-1} (x^5 + 2x^5) \, dx = \tfrac{1}{2}. \quad ■$$

If C is a *closed* curve, then there is no need to specify initial and terminal point, though the direction must be indicated. If C is a simple closed curve (traced just once), then one need only specify which of the two possible directions is chosen. The notations

$$\text{(a)} \oint P \, dx + Q \, dy, \qquad \text{(b)} \oint P \, dx + Q \, dy$$

refer to the two cases of Figs. 5.6(a) and 5.6(b). The counterclockwise arrow refers to what is roughly a counterclockwise direction on C; this will be termed the *positive* direction (as for angular measure); the clockwise direction will be called

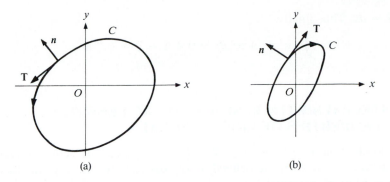

Figure 5.6   (a) Positive direction and (b) negative direction on a simple closed path.

the *negative* direction. It should be noted that the direction can be specified by reference to the unit tangent vector **T** in the direction of integration and the unit normal vector **n** that points to the outside of the region bounded by C; for the positive direction, **n** is $90°$ behind **T**, as in Fig. 5.6(a); for the negative direction, **n** is $90°$ ahead of **T** as in Fig. 5.6(b).

EXAMPLE 3   To evaluate

$$\oint_C y^2\, dx + x^2\, dy,$$

where C is the triangle with vertices $(1,0)$, $(1,1)$, $(0,0)$, shown in Fig. 5.7, one has to compute three integrals. The first is the integral from $(0,0)$ to $(1,0)$; along this path, $y = 0$ and, if x is the parameter, $dy = 0$. Hence the first integral is 0. The second integral is that from $(1,0)$ to $(1,1)$; if y is used as parameter, this reduces to

$$\int_0^1 dy = 1,$$

since $dx = 0$. For the third integral, from $(1,1)$ to $(0,0)$, x can be used as parameter, so that the integral is

$$\int_1^0 2x^2\, dx = -\tfrac{2}{3},$$

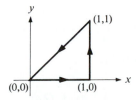

Figure 5.7   Example 3.

since $dy = dx$. Thus finally

$$\oint_C y^2 \, dx + x^2 \, dy = 0 + 1 - \tfrac{2}{3} = \tfrac{1}{3}. \quad \blacksquare$$

5.3 INTEGRALS WITH RESPECT TO ARC LENGTH · BASIC PROPERTIES OF LINE INTEGRALS

For a smooth or piecewise smooth path C, as in the preceding section, arc length s is well defined. Thus s can be defined as the distance traversed from the initial point $(t = h)$ up to a general t:

$$s = \int_h^t \sqrt{\left(\frac{dx}{dt}\right)^2 + \left(\frac{dy}{dt}\right)^2} \, dt. \tag{5.10}$$

If the curve C is directed with increasing t, then s also increases in the direction of motion, going from 0 up to the length L of C. Let C be subdivided as in Fig. 5.4 and let $\Delta_i s$ denote the increment in s from t_{i-1} to t_i, that is, the distance moved in this interval. One then makes the definition

$$\int_C f(x, y) \, ds = \lim_{\substack{n \to \infty \\ \max \Delta_i s \to 0}} \sum_{i=1}^{n} f(x_i^*, y_i^*) \, \Delta_i s. \tag{5.11}$$

If f is continuous on C, this integral will exist and can be evaluated as follows:

$$\int_C f(x, y) \, ds = \int_h^k f[\phi(t), \psi(t)] \sqrt{\phi'(t)^2 + \psi'(t)^2} \, dt. \tag{5.12}$$

This is proved in the same way as (5.4) and (5.5), with the aid of the formula

$$\frac{ds}{dt} = \sqrt{\left(\frac{dx}{dt}\right)^2 + \left(\frac{dy}{dt}\right)^2} = \sqrt{\phi'(t)^2 + \psi'(t)^2}.$$

One can in principle use s itself as the parameter on the curve C; if this is done, x and y become functions of s: $x = x(s)$, $y = y(s)$. The point $[x(s), y(s)]$ is then the position of the moving point after a distance s has been traversed. In this case, (5.11) reduces to a definite integral with respect to s:

$$\int_C f(x, y) \, ds = \int_0^L f[x(s), y(s)] \, ds. \tag{5.13}$$

If x is used as parameter, one has

$$\int_C f(x, y) \, ds = \int_a^b f[x, y(x)] \sqrt{1 + \left(\frac{dy}{dx}\right)^2} \, dx; \tag{5.14}$$

there is an analogous formula for y.

The basic combination

$$\int_C P \, dx + Q \, dy$$

Figure 5.8   Element of arc.

referred to above can be written as an integral with respect to s as follows:

$$\int_C P\,dx + Q\,dy = \int_C (P\cos\alpha + Q\sin\alpha)\,ds, \qquad (5.15)$$

where α is the angle between the positive x-axis and a tangent vector in direction of increasing s. For, as shown in Fig. 5.8 (cf. Section 2.13), one has

$$\frac{dx}{ds} = \cos\alpha, \qquad \frac{dy}{ds} = \sin\alpha \qquad (5.16)$$

and

$$\int_C P\,dx + Q\,dy = \int_0^L \left(P\frac{dx}{ds} + Q\frac{dy}{ds} \right) ds = \int_C (P\cos\alpha + Q\sin\alpha)\,ds.$$

$$(5.17)$$

The basic properties of the line integral are analogous to those of ordinary definite integrals; the properties to be listed here can in fact be demonstrated by reducing the line integral to a definite integral by use of a parameter:

$$\int_{C_1}^{A_2}(P\,dx + Q\,dy) + \int_{C_2}^{A_3}(P\,dx + Q\,dy) = \int_{C_3}^{A_3}(P\,dx + Q\,dy), \quad (5.18)$$

where C_3 is the combined path: from A_1 to A_2 via C_1 and from A_2 to A_3 via C_2;

$$\int_{C}^{B}{}_{A} P\,dx + Q\,dy = -\int_{C'}^{A}{}_{B} P\,dx + Q\,dy, \qquad (5.19)$$

where C' denotes C traced in the reverse direction;

$$\int_C (P_1\,dx + Q_1\,dy) + \int_C (P_2\,dx + Q_2\,dy) = \int_C (P_1 + P_2)\,dx + (Q_1 + Q_2)\,dy;$$

$$(5.20)$$

$$K\int_C (P\,dx + Q\,dy) = \int_C (KP)\,dx + (KQ)\,dy, \qquad K = \text{const}, \quad (5.21)$$

$$\int_C ds = L = \text{length of } C; \qquad (5.22)$$

if $|f(x, y)| \leq M$ on C, then

$$\left| \int_C f(x, y) \, ds \right| \leq M \cdot L;$$ (5.23)

if C is a simple closed curve, as in Fig. 5.6(a) then

$$\oint_C x \, dy = - \oint_C y \, dx$$

$$= \text{area enclosed by } C.$$ (5.24)

These theorems all hold under the assumptions that the paths are piecewise smooth and that the functions being integrated are continuous. All except the formula (5.24) follow at once from appropriate parametric representations; (5.24) will be proved in Section 5.5.

PROBLEMS

1. Evaluate the following integrals along the straight-line paths joining the end points:

a) $\int_{(0,0)}^{(2,2)} y^2 \, dx,$

b) $\int_{(2,1)}^{(1,2)} y \, dx,$

c) $\int_{(1,1)}^{(2,1)} x \, dy.$

2. Evaluate the following line integrals:

a) $\int_{C}^{(0,1)}{}_{(0,-1)} y^2 \, dx + x^2 \, dy$, where C is the semicircle $x = \sqrt{1 - y^2}$;

b) $\int_{C}^{(2,4)}{}_{(0,0)} y \, dx + x \, dy$, where C is the parabola $y = x^2$;

c) $\int_{C}^{(0,1)}{}_{(1,0)} \dfrac{y \, dx - x \, dy}{x^2 + y^2}$, where C is the curve $x = \cos^3 t, y = \sin^3 t, 0 \leq t \leq \dfrac{\pi}{2}$.

   (Hint: Set $u = \tan^3 t$ in the integral for t.)

3. Evaluate the following line integrals:

a) $\oint_C y^2 \, dx + xy \, dy$, where C is the square with vertices $(1,1), (-1,1), (-1,-1),$ $(1,-1)$;

b) $\oint_C y \, dx - x \, dy$, where C is the circle $x^2 + y^2 = 1$ (cf. (5.24));

c) $\oint_C x^2 y^2 \, dx - xy^3 \, dy$, where C is the triangle with vertices $(0,0), (1,0), (1,1)$.

4. Evaluate the following line integrals:

   a) $\oint (x^2 - y^2)\, ds$, where C is the circle $x^2 + y^2 = 4$;

   b) $\int_{C\,(0,0)}^{(1,1)} x\, ds$, where C is the line $y = x$;

   c) $\int_{C\,(0,0)}^{(1,1)} ds$, where C is the parabola $y = x^2$.

5. Let a path (5.1) be given and let a *change of parameter* be made by an equation $t = g(\tau)$, $\alpha \le \tau \le \beta$, where $g'(\tau)$ is continuous and positive in the interval and $g(\alpha) = h$, $g(\beta) = k$. As in (5.4) the line integral $\int f(x, y)\, dx$ on the path $x = \phi(g(\tau))$, $y = \psi(g(\tau))$ is given by

$$\int_{\alpha}^{\beta} f[\phi(g(\tau)), \psi(g(\tau))] \frac{d}{d\tau} \phi(g(\tau))\, d\tau.$$

Show that this equals the integral in (5.4), so that such a change of parameter does not affect the value of the line integral.

6. (*Numerical evaluation of line integrals*) If the parametric equations of the path C are explicitly known, Eqs. (5.4) and (5.5) reduce the evaluation of line integrals to a problem in ordinary definite integrals, to which the methods of Section 4.1 apply. One can in any case evaluate the integral directly from its definition as limit of a sum. Thus

$$\int_C P\, dx + Q\, dy \sim \sum_{i=1}^{n} (P(x_i^*, y_i^*)\, \Delta_i x + Q(x_i^*, y_i^*)\, \Delta_i y).$$

The points (x_i^*, y_i^*) can be chosen as the subdivision points (x_{i-1}, y_{i-1}) or as (x_i, y_i); one can also use a trapezoidal rule to obtain the sum:

$$\sum_{i=1}^{n} \{\tfrac{1}{2}[P(x_{i-1}, y_{i-1}) + P(x_i, y_i)]\, \Delta_i x + \tfrac{1}{2}[Q(x_{i-1}, y_{i-1}) + Q(x_i, y_i)]\, \Delta_i y\}.$$

(a)

For example,

$$\int_{C\,(0,0)}^{(2,2)} y^2\, dx + x^2\, dy \sim [\tfrac{1}{2}(0 + 1) \cdot 1 + \tfrac{1}{2}(0 + 1) \cdot 1]$$

$$+ [\tfrac{1}{2}(1 + 4) \cdot 1 + \tfrac{1}{2}(1 + 4) \cdot 1] = 6,$$

where C is the straight line segment from $(0, 0)$ to $(2, 2)$ and the subdivision points are $(0, 0)$, $(1, 1)$, $(2, 2)$.

Let the functions P and Q be given by the following table at the points $A \ldots S$.

	A	B	C	D	E	F	G	H	I	J	K	L	M	N	O	S
x	1	2	3	4	1	2	3	4	1	2	3	4	1	2	3	4
y	1	1	1	1	2	2	2	2	3	3	3	3	4	4	4	4
P	0	3	8	5	3	0	5	2	8	5	0	1	2	7	3	4
Q	1	2	3	4	2	4	6	8	3	6	9	2	4	8	2	6

Evaluate the integral $\int P\,dx + Q\,dy$ approximately by the trapezoidal rule, Eq. (a) above, on the following broken line paths: (a) *ABFG* (b) *AFGKH* (c) *ABCDHL-SONMIEA* (d) *AFJNMIJFA* (e) *ABFEAEFBA*.

7. Show that if $f(x, y) > 0$ on C, the integral $\int\limits_{C} f(x, y)\,ds$ can be interpreted as the area of the cylindrical surface $0 \le z \le f(x, y), (x, y)$ on C.

5.4   LINE INTEGRALS AS INTEGRALS OF VECTORS

The functions $P(x, y)$ and $Q(x, y)$ appearing above can be regarded as the components of a vector **u**:

$$\mathbf{u} = P(x, y)\mathbf{i} + Q(x, y)\mathbf{j}, \qquad u_x = P(x, y), \qquad u_y = Q(x, y). \quad (5.25)$$

The line integral

$$\int\limits_{C} P\,dx + Q\,dy$$

has then a simple vector interpretation as follows:

$$\int\limits_{C} P\,dx + Q\,dy = \int\limits_{C} u_T\,ds, \qquad (5.26)$$

where u_T denotes the tangential component of **u**, that is, the component of **u** in the direction of the unit tangent vector **T** in the direction of increasing s. For as Fig. 5.9 shows (cf. Section 2.13), **T** has components dx/ds, dy/ds:

$$\mathbf{T} = \frac{dx}{ds}\mathbf{i} + \frac{dy}{ds}\mathbf{j} = \cos\alpha\,\mathbf{i} + \sin\alpha\,\mathbf{j}. \qquad (5.27)$$

Hence

$$u_T = \mathbf{u} \cdot \mathbf{T} = P\cos\alpha + Q\sin\alpha, \qquad (5.28)$$

so that

$$\int\limits_{C} u_T\,ds = \int\limits_{C} (P\cos\alpha + Q\sin\alpha)\,ds = \int\limits_{C} P\,dx + Q\,dy \qquad (5.29)$$

by (5.15).

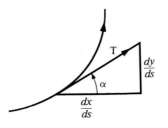

Figure 5.9   Unit tangent vector.

In the case in which $\mathbf{u} = P\mathbf{i} + Q\mathbf{j}$ is a *force* field, the integral $\int u_T \, ds = \int P \, dx + Q \, dy$ represents the *work* done by this force in moving a particle from one end of C to the other. For the work done is defined in mechanics as "force times distance" or, more precisely, as follows:

$$\text{work} = \int_C (\text{tangential force component}) \, ds, \tag{5.30}$$

and u_T is precisely this tangential force component.

The line integral $\int P \, dx + Q \, dy$ can be defined directly as a vector integral as follows:

$$\int_C P \, dx + Q \, dy = \int_C \mathbf{u} \cdot d\mathbf{r}, \qquad d\mathbf{r} = dx\mathbf{i} + dy\mathbf{j}, \tag{5.31}$$

where

$$\int_C \mathbf{u} \cdot d\mathbf{r} = \lim_{\substack{n \to \infty \\ \max \Delta_i s \to 0}} \sum_{i=1}^{n} \mathbf{u}(x_i^*, y_i^*) \cdot \Delta_i \mathbf{r} \tag{5.32}$$

and $\Delta_i \mathbf{r} = \Delta_i x \mathbf{i} + \Delta_i y \mathbf{j}$. Equation (5.17) can then be written in the vector form:

$$\int_C \mathbf{u} \cdot d\mathbf{r} = \int_0^L \left(\mathbf{u} \cdot \frac{d\mathbf{r}}{ds} \right) ds = \int_C (\mathbf{u} \cdot \mathbf{T}) \, ds. \tag{5.33}$$

If C is represented in terms of a parameter t, then

$$\int_C \mathbf{u} \cdot d\mathbf{r} = \int_C P \, dx + Q \, dy = \int_h^k \left(P \frac{dx}{dt} + Q \frac{dy}{dt} \right) dt = \int_h^k \left(\mathbf{u} \cdot \frac{d\mathbf{r}}{dt} \right) dt. \tag{5.34}$$

If $\mathbf{r}$ is the position vector of a particle of mass m moving on C and $\mathbf{u}$ is the force applied, then by Newton's Second Law,

$$\mathbf{u} = m \frac{d^2 \mathbf{r}}{dt^2} = m \frac{d\mathbf{v}}{dt}. \tag{5.35}$$

Hence, if $|\mathbf{v}| = v$,

$$\int_C \mathbf{u} \cdot d\mathbf{r} = \int_h^k \left(\mathbf{u} \cdot \frac{d\mathbf{r}}{dt} \right) dt = \int_h^k \left(m \frac{d\mathbf{v}}{dt} \cdot \mathbf{v} \right) dt = \int_h^k \frac{d}{dt} \left(\tfrac{1}{2} m \mathbf{v} \cdot \mathbf{v} \right) dt$$

$$= \int_h^k \frac{d}{dt} \left(\tfrac{1}{2} m v^2 \right) dt.$$

One thus concludes that

$$\int_C u_T \, ds = \int_C \mathbf{u} \cdot d\mathbf{r} = \tfrac{1}{2} m v^2 \bigg|_{t=h}^{t=k}; \tag{5.36}$$

that is, the *work done equals the gain in kinetic energy*. This is a basic law of mechanics.

Figure 5.10 Tangent and normal.

The line integral $\int_C P\,dx + Q\,dy$ can be interpreted as a vector integral in another way, namely, as

$$\int_C \mathbf{v} \cdot \mathbf{n}\,ds = \int_C v_n\,ds,$$

where $\mathbf{v}$ is the vector $Q\mathbf{i} - P\mathbf{j}$ and $\mathbf{n}$ is the unit normal vector $90°$ behind $\mathbf{T}$, so that v_n is the normal component of $\mathbf{v}$. One has then (cf. Fig. 5.10)

$$\mathbf{n} = \mathbf{T} \times \mathbf{k} = \left(\frac{dx}{ds}\mathbf{i} + \frac{dy}{ds}\mathbf{j}\right) \times \mathbf{k} = \frac{dy}{ds}\mathbf{i} - \frac{dx}{ds}\mathbf{j}, \tag{5.37}$$

since the vector product $\mathbf{a} \times \mathbf{k}$ of a vector $\mathbf{a}$ in the xy-plane with $\mathbf{k}$ is a vector $\mathbf{b}$ having the same magnitude as $\mathbf{a}$ and $90°$ behind $\mathbf{a}$. Accordingly,

$$v_n = \mathbf{v} \cdot \mathbf{n} = (Q\mathbf{i} - P\mathbf{j}) \cdot \left(\frac{dy}{ds}\mathbf{i} - \frac{dx}{ds}\mathbf{j}\right) = Q\frac{dy}{ds} + P\frac{dx}{ds}.$$

Thus one has

$$\int_C v_n\,ds = \int_C \left(Q\frac{dy}{ds} + P\frac{dx}{ds}\right) ds = \int_C P\,dx + Q\,dy,$$
$$\mathbf{v} = Q\mathbf{i} - P\mathbf{j}, \qquad \mathbf{n} = \mathbf{T} \times \mathbf{k}, \tag{5.38}$$

as asserted. It should be noted that for $\mathbf{u} = P\mathbf{i} + Q\mathbf{j}$, one has then

$$\int_C u_n\,ds = \int_C \left(P\frac{dy}{ds} - Q\frac{dx}{ds}\right) ds = \int_C -Q\,dx + P\,dy,$$
$$\mathbf{u} = P\mathbf{i} + Q\mathbf{j}, \qquad \mathbf{n} = \mathbf{T} \times \mathbf{k}. \tag{5.39}$$

5.5 GREEN'S THEOREM

The following theorem and its generalizations are fundamental in the theory of line integrals.

GREEN'S THEOREM Let D be a domain of the xy-plane and let C be a piecewise smooth simple closed curve in D whose interior is also in D. Let $P(x, y)$ and $Q(x, y)$ be functions defined and continuous and having continuous first partial derivatives in D. Then

$$\oint_C P\,dx + Q\,dy = \int\!\!\int_R \left(\frac{\partial Q}{\partial x} - \frac{\partial P}{\partial y} \right) dx\,dy, \qquad (5.40)$$

where R is the closed region bounded by C.

The theorem will be proved first for the case in which R is representable in both of the forms:

$$a \leq x \leq b, \qquad f_1(x) \leq y \leq f_2(x), \qquad (5.41)$$
$$c \leq y \leq d, \qquad g_1(y) \leq x \leq g_2(y), \qquad (5.42)$$

as in Fig. 5.11.

The double integral

$$\int\!\!\int_R \frac{\partial P}{\partial y}\,dx\,dy$$

can by (5.41) be written as an iterated integral:

$$\int\!\!\int_R \frac{\partial P}{\partial y}\,dx\,dy = \int_a^b \int_{f_1(x)}^{f_2(x)} \frac{\partial P}{\partial y}\,dy\,dx.$$

One can now integrate out:

$$\int\!\!\int_R \frac{\partial P}{\partial y}\,dx\,dy = \int_a^b \{ P[x, f_2(x)] - P[x, f_1(x)] \}\,dx$$

$$= - \int_b^a P[x, f_2(x)]\,dx - \int_a^b P[x, f_1(x)]\,dx$$

$$= - \oint_C P(x, y)\,dx.$$

Figure 5.11 Special region for Green's theorem.

In the same way, $\int_R\int \frac{\partial Q}{\partial x}\,dx\,dy$ can be written as an iterated integral, with the aid of (5.42), and one concludes that

$$\int_R\int \frac{\partial Q}{\partial x}\,dx\,dy = \oint_C Q\,dy.$$

Adding the two double integrals, one finds:

$$\int_R\int \left(\frac{\partial Q}{\partial x} - \frac{\partial P}{\partial y}\right)\,dx\,dy = \oint_C P\,dx + Q\,dy.$$

The theorem is thus proved for the special type of region R.

Suppose next that R is not itself of this form but can be decomposed into a finite number of such regions: $R_1, R_2, \ldots, R_n$ by suitable lines or arcs, as in Fig. 5.12. Let $C_1, C_2, \ldots, C_n$ denote the corresponding boundaries. Then Eq. (5.40) can be applied to each region separately. Adding, one obtains the equation:

$$\oint_{C_1} (P\,dx + Q\,dy) + \oint_{C_2} (\quad) + \cdots \oint_{C_n} (\quad)$$

$$= \int_{R_1}\int \left(\frac{\partial Q}{\partial x} - \frac{\partial P}{\partial y}\right)\,dx\,dy + \cdots + \int_{R_n}\int (\quad)\,dx\,dy.$$

But the sum of the integrals on the left is just

$$\oint_C P\,dx + Q\,dy.$$

For the integrals along the added arcs are taken once in each direction and hence cancel each other; the remaining integrals add up to precisely the integral around C in the positive direction. The integrals on the right add up to

$$\int_R\int \left(\frac{\partial Q}{\partial x} - \frac{\partial P}{\partial y}\right)\,dx\,dy$$

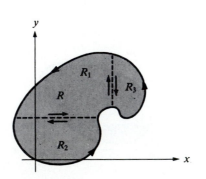

Figure 5.12   Decomposition of region into special regions.

and hence

$$\oint_C P\,dx + Q\,dy = \int\int_R \left(\frac{\partial Q}{\partial x} - \frac{\partial P}{\partial y} \right) dx\,dy.$$

The proof thus far covers all regions R of interest in practical problems. To prove the theorem for the most general region R, it is necessary to approximate this region by those of the special type just considered and then to use a limiting process. For details, see the book by Kellogg cited at the end of the chapter.

EXAMPLE 1   Let C be the circle: $x^2 + y^2 = 1$. Then

$$\oint_C 4xy^3\,dx + 6x^2y^2\,dy = \int\int_R (12xy^2 - 12xy^2)\,dx\,dy = 0. \quad \blacksquare$$

EXAMPLE 2   Let C be the ellipse $x^2 + 4y^2 = 4$. Then

$$\oint_C (2x - y)\,dx + (x + 3y)\,dy = \int\int_R (1 + 1)\,dx\,dy = 2A,$$

where A is the area of R. Since the ellipse has semi-axes $a = 2$, $b = 1$, the area is $\pi ab = 2\pi$, and the value of the line integral is 4π. $\quad \blacksquare$

EXAMPLE 3   Let C be the ellipse $4x^2 + y^2 = 4$. Then Green's theorem is not applicable to the integral

$$\oint_C \frac{y}{x^2 + y^2}\,dx - \frac{x}{x^2 + y^2}\,dy,$$

since P and Q fail to be continuous at the origin. $\quad \blacksquare$

EXAMPLE 4   Let C be the square with vertices $(1,1)$, $(1, -1)$, $(-1, -1)$, $(-1, 1)$. Then

$$\oint_C (x^2 + 2y^2)\,dx = -\int\int_R 4y\,dx\,dy = -4A\bar{y},$$

where $(\bar{x}, \bar{y})$ is the centroid of the square. Hence the value of the line integral is 0. $\quad \blacksquare$

Vector interpretation of Green's theorem.   If $\mathbf{u} = P(x, y)\mathbf{i} + Q(x, y)\mathbf{j}$, then, as previously,

$$\oint_C P\,dx + Q\,dy = \oint_C u_T\,ds,$$

where $u_T = \mathbf{u} \cdot \mathbf{T}$ is the tangential component of $\mathbf{u}$. The integrand in the right-hand member of Green's theorem is the z component of curl $\mathbf{u}$; that is,

$$\operatorname{curl}_z \mathbf{u} = \frac{\partial Q}{\partial x} - \frac{\partial P}{\partial y}.$$

Hence Green's theorem states:

$$\oint_C u_T \, ds = \int\int_R \mathrm{curl}_z \, \mathbf{u} \, dx \, dy. \tag{5.43}$$

This result is a special case of Stokes's theorem, to be considered later.

The integral $\oint_C P \, dx + Q \, dy$ can also be interpreted, as above, as the integral

$$\oint_C \mathbf{v} \cdot \mathbf{n} \, ds = \oint_C v_n \, ds,$$

where $\mathbf{v}$ is the vector $Q\mathbf{i} - P\mathbf{j}$. The right-hand member of Green's theorem is then the double integral of $\mathrm{div}\,\mathbf{v}$. Thus one has, for an arbitrary vector field,

$$\oint_C v_n \, ds = \int\int_R \mathrm{div}\,\mathbf{v} \, dx \, dy, \tag{5.44}$$

where $\mathbf{n}$ is the outer normal on C, as in Fig. 5.6(a). This result is the 2-dimensional form of Gauss's theorem, to be considered later.

Application to area.   Equation (5.24) asserts that

$$\oint_C x \, dy = - \oint_C y \, dx = \text{area of } R.$$

This follows at once from Green's theorem since

$$\oint_C x \, dy = - \oint_C y \, dx = \int\int_R 1 \cdot dx \, dy.$$

If these two equal line integrals are averaged, another expression for area is obtained:

$$\tfrac{1}{2} \oint_C (-y \, dx + x \, dy) = \text{area of } R. \tag{5.45}$$

This can also be checked by Green's theorem.

PROBLEMS

1. If $\mathbf{v} = (x^2 + y^2)\mathbf{i} + (2xy)\mathbf{j}$, evaluate $\int_C v_T \, ds$ for the following paths:

   a) from $(0,0)$ to $(1,1)$ on the line $y = x$,

   b) from $(0,0)$ to $(1,1)$ on the line $y = x^2$;

   c) from $(0,0)$ to $(1,1)$ on the broken line with corner at $(1,0)$.

2. Evaluate $\int_C v_n \, ds$ for the vector $\mathbf{v}$ given in Problem 1 on the paths (a), (b), and (c) of Problem 1, $\mathbf{n}$ being chosen as the normal 90° behind $\mathbf{T}$.

3. The gravitational force near a point on the earth's surface is represented approximately by the vector $-mg\mathbf{j}$, where the y axis points upwards. Show that the work

done by this force on a body moving in a vertical plane from height h_1 to height h_2 along any path is equal to $mg(h_1 - h_2)$.

4. Show that the earth's gravitational potential $U = -kMm/r$ is equal to the negative of the work done by the gravitational force $\mathbf{F} = -(kMm/r^2)(\mathbf{r}/r)$ in bringing the particle to its present position from infinite distance along the ray through the earth's center.

5. Evaluate by Green's theorem:

a) $\oint_C ay\,dx + bx\,dy$ on any path;

b) $\oint_C e^x \sin y\,dx + e^x \cos y\,dy$ around the rectangle with vertices $(0,0)$, $(1,0)$, $(1, \frac{1}{2}\pi)$, $(0, \frac{1}{2}\pi)$;

c) $\oint_C (2x^3 - y^3)\,dx + (x^3 + y^3)\,dy$ around the circle $x^2 + y^2 = 1$;

d) $\oint_C u_T\,ds$, where $\mathbf{u} = \text{grad}\,(x^2 y)$ and C is the circle $x^2 + y^2 = 1$;

e) $\oint_C v_n\,ds$, where $\mathbf{v} = (x^2 + y^2)\mathbf{i} - 2xy\mathbf{j}$, and C is the circle $x^2 + y^2 = 1$, $\mathbf{n}$ being the outer normal;

f) $\oint_C \frac{\partial}{\partial n}[(x - 2)^2 + y^2]\,ds$, where C is the circle $x^2 + y^2 = 1$, $\mathbf{n}$ is the outer normal;

g) $\oint_C \frac{\partial}{\partial n} \log \frac{1}{\left[(x - 2)^2 + y^2\right]}\,ds$, where C and $\mathbf{n}$ are as in (f);

h) $\oint_C f(x)\,dx + g(y)\,dy$ on any path.

6. If $\mathbf{r} = x\mathbf{i} + y\mathbf{j}$ is the position vector of an arbitrary point (x, y), show that

$$\frac{1}{2} \oint_C r_n\,ds = \text{area enclosed by } C,$$

$\mathbf{n}$ being the outer normal to C.

7. Check the answers to Problems 2(a), 3(a), (b), (c), and 4(a) following Section 5.3 by Green's theorem.

5.6 INDEPENDENCE OF PATH ▪ SIMPLY CONNECTED DOMAINS

Let the functions $P(x, y)$ and $Q(x, y)$ be defined and continuous in a domain D. Then the line integral $\int P\,dx + Q\,dy$ is said to be *independent of path in D* if, for every pair of endpoints A and B in D, the value of the line integral

$$\int_A^B P\,dx + Q\,dy$$

Figure 5.13 Independence of path.

is the same for all paths C from A to B. The value of the integral will then in general depend on the choice of A and B, but not on the choice of the path joining them. Thus as in Fig. 5.13, the integrals on C_1, C_2, C_3 have the same value.

THEOREM I If the integral $\int P\,dx + Q\,dy$ is independent of path in D, then there is a function $F(x, y)$ defined in D such that

$$\frac{\partial F}{\partial x} = P(x, y), \qquad \frac{\partial F}{\partial y} = Q(x, y) \tag{5.46}$$

holds throughout D. Conversely, if a function $F(x, y)$ can be found such that (5.46) holds in D, then $\int P\,dx + Q\,dy$ is independent of path in D.

Proof. Suppose first that the integral is independent of path in D. Then choose a point (x_0, y_0) in D and let $F(x, y)$ be defined as follows:

$$F(x, y) = \int_{(x_0, y_0)}^{(x, y)} P\,dx + Q\,dy, \tag{5.47}$$

where the integral is taken on an arbitrary path in D joining (x_0, y_0) to (x, y). Since the integral is independent of path, the integral in (5.47) does indeed depend only on the endpoint (x, y) and defines a function $F(x, y)$. It remains to show that $\partial F/\partial x = P(x, y)$, $\partial F/\partial y = Q(x, y)$ in D.

For a particular (x, y) in D, choose (x_1, y) so that $x_1 \neq x$ and the line segment from (x_1, y) to (x, y) is in D, as shown in Fig. 5.14. Then, because of

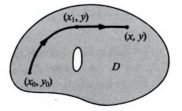

Figure 5.14 Construction of function F such that $dF = P\,dx + Q\,dy$.

independence of path, one has

$$F(x, y) = \int_{(x_0, y_0)}^{(x_1, y)} (P\,dx + Q\,dy) + \int_{(x_1, y)}^{(x, y)} (P\,dx + Q\,dy).$$

Here x_1 and y are thought of as fixed, while (x, y) may vary along the line segment. Thus y has been restricted to a constant value, and $F(x, y)$ is being considered as a function of x near a particular choice of x. The first integral on the right is then independent of x, while the second can be integrated along the line segment. Hence for this fixed y,

$$F(x, y) = \text{const} + \int_{x_1}^{x} P(x, y)\,dx, \qquad (y = \text{const})$$

or, with the dummy variable x replaced by a t,

$$F(x, y) = \text{const} + \int_{x_1}^{x} P(t, y)\,dt.$$

The fundamental theorem of calculus (Section 4.1) now gives

$$\frac{\partial F}{\partial x} = P(x, y).$$

Similarly, $\partial F/\partial y = Q(x, y)$, and the first part of the theorem is proved.

Conversely, let (5.46) hold in D for some F. Then, in terms of a parameter t,

$$\int_{C}^{(x_2, y_2)}_{(x_1, y_1)} P\,dx + Q\,dy = \int_{t_1}^{t_2} \left(\frac{\partial F}{\partial x}\frac{dx}{dt} + \frac{\partial F}{\partial y}\frac{dy}{dt} \right) dt$$

$$= \int_{t_1}^{t_2} \frac{dF}{dt}\,dt = F\big|_{t=t_2} - F\big|_{t=t_1}$$

$$= F(x_2, y_2) - F(x_1, y_1).$$

In other words, in simplified notation,

$$\int_{C}^{B}_{A} P\,dx + Q\,dy = \int_{A}^{B} dF = F(B) - F(A). \qquad (5.48)$$

The value of the integral is simply the difference of the values of F at the two ends and is therefore independent of the path C. □

Remark 1.   Formula (5.48) is the analogue of the usual formula

$$\int_{a}^{b} f(x)\,dx = F(x)\Big|_{a}^{b} = F(b) - F(a), \qquad F'(x) = f(x)$$

for evaluating definite integrals and should be used whenever the function F can be found. Thus

$$\int_{(1,2)}^{(5,6)} y\,dx + x\,dy = \int_{(1,2)}^{(5,6)} d(xy) = xy\Big|_{(1,2)}^{(5,6)} = 30 - 2 = 28.$$

Remark 2.   Suppose $\partial F/\partial x = P \equiv 0$ in D, $\partial F/\partial y = Q \equiv 0$ in D. Then (5.48)
shows that

$$F(B) - F(A) = \int_C{}_A^B 0 \, dx + 0 \, dy = 0,$$

that is, that $F(B) = F(A)$ for every two points of D. Hence $F \equiv$ const in D (cf.
Section 2.22).

Remark 3.   If P and Q are given such that $\int P \, dx + Q \, dy$ is independent of path,
then the function F such that $dF = P \, dx + Q \, dy$ is determined only up to an
additive constant:

$$F = \int_{(x_0, y_0)}^{(x, y)} P \, dx + Q \, dy + \text{const}. \tag{5.49}$$

For if G is a second function such that $\partial G/\partial x = P$, $\partial G/\partial y = Q$, then $(\partial/\partial x)$
$(G - F) = P - P \equiv 0$, $(\partial/\partial y)(G - F) = Q - Q \equiv 0$; hence by Remark 2, $G -$
F is a constant, or $G = F + \text{const}$.
    It often happens that $P = \partial F/\partial x$ and $Q = \partial F/\partial y$ are known (perhaps in
tabular form), while F is not; in this case, (5.49) permits evaluation of F at any
point in D, once its value at (x_0, y_0) has been chosen. For (5.49) can be written:

$$F = \int_{(x_0, y_0)}^{(x, y)} P \, dx + Q \, dy + F(x_0, y_0), \tag{5.50}$$

as the replacement of the upper limit (x, y) by (x_0, y_0) shows. In (5.50), any
convenient path of integration can be used, for example, one formed of lines
parallel to the axes. This is very important in thermodynamics, in which the
partial derivatives of such functions as internal energy, entropy, and free energy
are the measured quantities, rather than the functions themselves (cf. Section
5.15).

Remark 4.   It is to be emphasized that the function $F(x, y)$ in (5.46) must be
defined and continuous in a domain containing the path. A function such as
arc tan (y/x) is "many-valued" and hence useless for the theorem unless it is
unambiguously defined and has the required continuity; these requirements
cannot be satisfied when C is the circular path $x = \cos t$, $y = \sin t$, $0 \le t \le 2\pi$.
This point is considered further in connection with Example 2.

THEOREM II   If the integral $\int P \, dx + Q \, dy$ is independent of path in D,
then

$$\oint_C P \, dx + Q \, dy = 0 \tag{5.51}$$

on every simple closed path in D. Conversely, if (5.51) holds for every simple
closed path in D, then $\int P \, dx + Q \, dy$ is independent of path in D.

Proof.   Suppose first that the integral is independent of path. Let C be a simple
closed path in D and divide C into arcs AB and BA by points A, B as in Fig. 5.15.

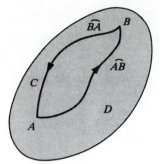

Figure 5.15 Proof of first part of theorem II.

Then

$$\oint_C P\,dx + Q\,dy = \int_{\widehat{AB}} (P\,dx + Q\,dy) + \int_{\widehat{BA}} (P\,dx + Q\,dy).$$

Because of independence of path, one can write this equation in the form

$$\oint_C P\,dx + Q\,dy = \int_A^B (P\,dx + Q\,dy) + \int_B^A (P\,dx + Q\,dy)$$

$$= \int_A^B (P\,dx + Q\,dy) - \int_A^B (P\,dx + Q\,dy) = 0.$$

Conversely, suppose $\int P\,dx + Q\,dy = 0$ on every simple closed path in D. Let A and B be two points of D and let C_1 and C_2 be two paths from A to B as in Fig. 5.16. We must show that

$$\int_{C_1}\!\!\!\!{}_A^B P\,dx + Q\,dy = \int_{C_2}\!\!\!\!{}_A^B P\,dx + Q\,dy$$

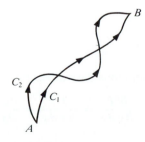

Figure 5.16 Proof of second part of theorem II.

or, if C_2' denotes C_2 traced in the reverse direction,

$$\int_{C_1}^{B}{}_{A} (P \, dx + Q \, dy) + \int_{C_2'}^{A}{}_{B} P \, dx + Q \, dy = 0.$$

If the paths C_1 and C_2' do not cross, together they form a simple closed path C, and

$$\int_{C_1'}^{B}{}_{A} (P \, dx + Q \, dy) + \int_{C_2'}^{A}{}_{B} (P \, dx + Q \, dy) = \int_{C} P \, dx + Q \, dy = 0,$$

regardless of the direction of integration.

If the paths cross a finite number of times, as in Fig. 5.16, a repetition of the same argument gives the proof; this reasoning covers, in particular, the case in which C_1 and C_2 are broken lines.

To cover the most general case, we can now reason as follows. We consider the function (5.47), using as paths only broken lines from (x_0, y_0) to (x, y). We have just shown that the value is independent of the path on such paths, so that $F(x, y)$ is well defined. Furthermore, we deduce exactly as previously that $\partial F/\partial x = P(x, y)$, $\partial G/\partial y = Q(x, y)$, using only broken line paths. Hence by Theorem I the line integral is independent of path for all types of paths in D. $\square$

Remark.   The first part of the preceding proof shows that, if $\int P \, dx + Q \, dy$ is independent of path, then

$$\int_{C} P \, dx + Q \, dy = 0$$

on every *closed path* C in D, regardless of whether or not C crosses itself.

THEOREM III   If $P(x, y)$ and $Q(x, y)$ have continuous partial derivatives in D and $\int P \, dx + Q \, dy$ is independent of path in D, then

$$\frac{\partial P}{\partial y} = \frac{\partial Q}{\partial x} \qquad (5.52)$$

in D.

Proof.   By Theorem I, one has

$$P = \frac{\partial F}{\partial x}, \qquad Q = \frac{\partial F}{\partial y}$$

and, since P and Q have continuous derivatives in D,

$$\frac{\partial P}{\partial y} = \frac{\partial^2 F}{\partial y \, \partial x} = \frac{\partial^2 F}{\partial x \, \partial y} = \frac{\partial Q}{\partial x}. \quad \square$$

Simply connected domains.   The converse of Theorem III does not hold without a further restriction, namely, that the domain D be *simply connected*. In plain terms,

Figure 5.17 Simply connected
domain.

Figure 5.18 Doubly connected
domain.

a domain is simply connected if it has no "holes." Thus the domains of Figs. 5.13
and 5.14 are not simply connected, but rather are, as one says, *multiply connected*.
More precisely, D is simply connected if for every simple closed curve C in D the
region R formed of C plus interior lies wholly in D (cf. Fig. 5.17).

Examples of simply connected domains are the following: The interior of a
circle, the interior of a square, a sector, a quadrant, and the whole xy-plane. The
annular region between two circles is not simply connected, nor is the interior of a
circle minus the center. Thus the holes may reduce to single points.

One can distinguish between types of multiply connected domains as follows:
A domain with just one hole, as in Fig. 5.18, is doubly connected; one with two
holes is triply connected; one with $n - 1$ holes is n-tuply connected. One can even
have infinitely many holes, in which case one speaks of a domain that is infinitely
multiply connected.

After this preparation we can state the converse of Theorem III:

THEOREM IV   Let $P(x, y)$ and $Q(x, y)$ have continuous derivatives in D
and let D be simply connected. If

$$\frac{\partial P}{\partial y} = \frac{\partial Q}{\partial x} \tag{5.53}$$

in D, then $\int P\, dx + Q\, dy$ is independent of path in D.

This theorem and its consequences are fundamental in physical applications.

Proof.   Suppose (5.53) to be satisfied. Then choose any simple closed curve C in
D. By Green's theorem,

$$\oint_C P\, dx + Q\, dy = \int_R\!\!\int \left(\frac{\partial Q}{\partial x} - \frac{\partial P}{\partial y} \right) dx\, dy = 0,$$

where R is the region formed of C plus interior. It should be remarked that
Green's theorem is applicable because D is simply connected, so that R is in D
and, in particular, $\partial Q/\partial x$ and $\partial P/\partial y$ are continuous in R. Since we have shown

that

$$\oint_C P\,dx + Q\,dy = 0$$

for every simple closed path C in D, it follows from Theorem II that $\int P\,dx + Q\,dy$ is independent of path in D. $\square$

Vector interpretation. Theorem I asserts that $\int P\,dx + Q\,dy$ is independent of path precisely when $P = \partial F/\partial x$, $Q = \partial F/\partial y$, that is,

$$P\mathbf{i} + Q\mathbf{j} = \text{grad } F.$$

Thus $\int u_T\,ds$ is independent of path precisely when $\mathbf{u}$ is the gradient vector of some scalar F: $\mathbf{u} = \nabla F$. Equation (5.48) then states that

$$\int_A^B \nabla F \cdot d\mathbf{r} = F(B) - F(A). \tag{5.54}$$

Theorem III asserts that if $\int P\,dx + Q\,dy$ is independent of path, then $\partial P/\partial y = \partial Q/\partial x$. If $\mathbf{u} = P\mathbf{i} + Q\mathbf{j}$, this is the statement that

$$\text{curl } \mathbf{u} = \mathbf{0},$$

for

$$\text{curl } \mathbf{u} = \nabla \times \mathbf{u} = \begin{vmatrix} \mathbf{i} & \mathbf{j} & \mathbf{k} \\ \dfrac{\partial}{\partial x} & \dfrac{\partial}{\partial y} & \dfrac{\partial}{\partial z} \\ P & Q & 0 \end{vmatrix} = \mathbf{k}\left(\frac{\partial Q}{\partial x} - \frac{\partial P}{\partial y} \right).$$

Since we know by Theorem I that $\mathbf{u}$ is a gradient, Theorem III is equivalent to the statement

$$\text{curl grad } F = \mathbf{0}.$$

This fundamental identity was pointed out in Chapter 3 [Eq. (3.31)].

Theorem IV then provides the converse: *If* $\text{curl } \mathbf{u} = \mathbf{0}$ *in* D, *then* $\mathbf{u} = \text{grad } F$ *for some* F, *provided that* D *is simply connected.* The analogous statement for three dimensions will be proved in Section 5.13.

EXAMPLE 1 The integral

$$\int 2xy^3\,dx + 3x^2y^2\,dy$$

is independent of path in the whole plane, since

$$\frac{\partial P}{\partial y} - \frac{\partial Q}{\partial x} = 6xy^2 - 6xy^2 = 0.$$

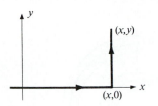

Figure 5.19   Example 1.

To find a function F whose gradient is $P\mathbf{i} + Q\mathbf{j}$, set

$$F = \int_{(0,0)}^{(x,\,y)} 2xy^3\, dx + 3x^2y^2\, dy$$

$$= \int_{(0,0)}^{(x,0)} 2xy^3\, dx + 3x^2y^2\, dy$$

$$+ \int_{(x,0)}^{(x,\,y)} 2xy^3\, dx + 3x^2y^2\, dy,$$

using a broken line path, as in Fig. 5.19. Along the first part, $dy = 0$ and $y = 0$, so that the integral reduces to 0. Along the second part, $dx = 0$, and x is constant, so that

$$F = \int_0^y 3x^2y^2\, dy = x^2y^3.$$

The general solution is then $x^2y^3 + C$. To evaluate the integral on any path from $(1, 2)$ to $(3, -2)$, one then writes

$$\int_{(1,2)}^{(3,\,-2)} 2xy^3\, dx + 3x^2y^2\, dy = \int_{(1,2)}^{(3,\,-2)} d(x^2y^3) = x^2y^3 \Big|_{(1,2)}^{(3,\,-2)} = -80. \quad \blacksquare$$

EXAMPLE 2   The integral

$$\int \frac{-y\, dx + x\, dy}{x^2 + y^2}$$

is independent of path in any simply connected domain D not containing the origin, for

$$\frac{\partial P}{\partial y} = \frac{\partial Q}{\partial x} = \frac{y^2 - x^2}{(x^2 + y^2)^2}$$

except at the origin. To find the function F whose differential is $P\, dx + Q\, dy$, one can proceed as in Example 1, using as path a broken line from $(1, 0)$ to (x, y). However, $P\, dx + Q\, dy$ is here a familiar differential (cf. Example 3, Section 2.8), namely, that of the polar coordinate angle θ:

$$d\theta = d\left(\arctan \frac{y}{x}\right) = \frac{-y\, dx + x\, dy}{x^2 + y^2}.$$

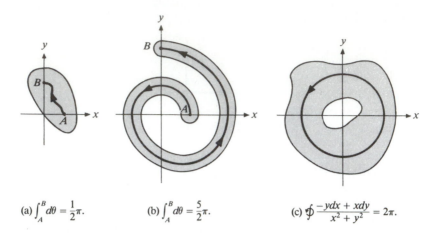

(a) $\int_A^B d\theta = \dfrac{1}{2}\pi.$ (b) $\int_A^B d\theta = \dfrac{5}{2}\pi.$ (c) $\oint \dfrac{-y\,dx + x\,dy}{x^2 + y^2} = 2\pi.$

Figure 5.20   Example 2.

The arc tangent is rather to be avoided if possible, because one cannot in general restrict to principal values and also because of the awkwardness of arc tan ∞ for the y-axis. One would therefore write

$$\int_A^B \frac{-y\,dx + x\,dy}{x^2 + y^2} = \int_A^B d\theta = \theta_B - \theta_A.$$

However, in order for θ to be a well-defined continuous function in a domain including the path (which in particular requires that θ be *single-valued*), so that Theorem I can be applied, it is necessary to restrict to a simply connected domain not containing the origin. Two such domains are shown in Figs. 5.20(a) and (b). In Fig. 5.20(c) the path is the circle $x^2 + y^2 = 1$, and the domain is doubly connected. Since $x^2 + y^2 = 1$ on the path,

$$\oint \frac{-y\,dx + x\,dy}{x^2 + y^2} = \oint -y\,dx + x\,dy = \int_R\!\!\int 2\,dx\,dy = 2\pi;$$

here Green's theorem, which is not applicable to the given integral, could be applied after the line integral had been simplified by using the relation $x^2 + y^2 = 1$ on the path; one can also regard the line integral as a sum of integrals from $\theta = 0$ to $\theta = \pi$ and from $\theta = \pi$ to $\theta = 2\pi$, giving $\pi + \pi = 2\pi$. A similar procedure can be followed in general, and one concludes that for any path C not through $(0,0)$,

$$\int_{\substack{A \\ C}}^B \frac{-y\,dx + x\,dy}{x^2 + y^2} = \text{total increase in } \theta \text{ from } A \text{ to } B,$$

as θ varies *continuously* on the path C. The integral is not independent of path but depends on the number of times C goes around the origin. ∎

EXAMPLE 3   The integral

$$\int \frac{x\,dx + y\,dy}{x^2 + y^2}$$

is independent of path in the *doubly connected* domain consisting of the xy-plane minus the origin. First of all,

$$\frac{\partial P}{\partial y} = \frac{\partial Q}{\partial x} = \frac{-2xy}{\left(x^2 + y^2\right)^2},$$

but, as Example 2 shows, this would guarantee independence of path only in a *simply connected* domain. However, for this example, one has additional information:

$$\frac{x\,dx + y\,dy}{x^2 + y^2} = d\log\sqrt{x^2 + y^2},$$

and the function $\log\sqrt{x^2 + y^2} = \log r$ is well defined (that is, *single valued*), with continuous derivatives except at the origin. Hence

$$\int_A^B \frac{x\,dx + y\,dy}{x^2 + y^2} = \int_A^B d\log r = \log r\Big|_A^B = \log\frac{r_B}{r_A}$$

for any path from A to B not through the origin—in particular, then, for the paths of Fig. 5.20(a) and (b). On the circle of Fig. 5.20(c) the integral is 0, in accordance with Theorem II.   ■

5.7   EXTENSION OF RESULTS TO MULTIPLY CONNECTED DOMAINS

If the closed curve C of Green's theorem encloses a point at which $\partial P/\partial y$ or $\partial Q/\partial x$ fails to exist, then the theorem cannot be applied. The following theorem shows that even in this case the reduction of a properly chosen line integral to a double integral is possible.

THEOREM   Let $P(x, y)$ and $Q(x, y)$ be continuous and have continuous derivatives in a domain D of the plane. Let R be a closed region in D whose boundary consists of n distinct simple closed curves $C_1, C_2, \ldots, C_n$, where C_1 includes $C_2, \ldots, C_n$ in its interior. Then

$$\oint_{C_1} (P\,dx + Q\,dy) + \oint_{C_2} (P\,dx + Q\,dy) + \cdots + \oint_{C_n} (P\,dx + Q\,dy)$$

$$= \int\int_R \left(\frac{\partial Q}{\partial x} - \frac{\partial P}{\partial y}\right) dx\,dy. \quad (5.55)$$

In particular, if $\partial Q/\partial x = \partial P/\partial y$ in D, then

$$\oint_{C_1}(P\,dx + Q\,dy) + \oint_{C_2}(P\,dx + Q\,dy) + \cdots + \oint_{C_n}(P\,dx + Q\,dy) = 0.$$

$$(5.55')$$

To prove the theorem, one first introduces auxiliary arcs from C_1 to C_2, C_1 to $C_3, \ldots$, two to each, as illustrated in Fig. 5.21. These decompose the region R into smaller regions, each of which is simply connected; there are four subregions in Fig. 5.21. If one integrates in a positive direction around the boundary of each subregion and then adds the results, one finds that the integrals along the auxiliary arcs cancel out, leaving just the integral around C_1 in the positive direction plus the integrals around $C_2, C_3, \ldots$, in the negative direction. On the other hand, the line integral around the boundary of each subregion can be expressed as a double integral

$$\iint \left(\frac{\partial Q}{\partial x} - \frac{\partial P}{\partial y} \right) dx\,dy$$

over the subregion by Green's theorem. Hence the sum of the line integrals is equal to the double integral over R. This gives (5.55); Eq. (5.55′) then follows as a special case.

If one denotes by B_R the directed boundary of R, that is, the curves $C_1, C_2, \ldots, C_n$ with the given directions, then Eqs. (5.55) and (5.55′) can be written in the concise form:

$$\int_{B_R} P\,dx + Q\,dy = \iint_R \left(\frac{\partial Q}{\partial x} - \frac{\partial P}{\partial y} \right) dx\,dy; \qquad (5.56)$$

$$\int_{B_R} P\,dx + Q\,dy = 0 \left(\frac{\partial Q}{\partial x} = \frac{\partial P}{\partial y} \text{ in } D \right). \qquad (5.56')$$

It is to be noted that the correct direction of integration always keeps the region to the left; that is, the normal pointing away from the region is 90° behind the tangent.

Figure 5.21   Green's theorem for multiply connected domains.

Figure 5.22   Doubly connected
domain (point A excluded).

As an application of the theorem just proved, let us consider the case of a
doubly connected region D with a hole which, for the sake of simplicity, we take
to be a point A, as in Fig. 5.22. Let $\partial P/\partial y = \partial Q/\partial x$ in D. What are the possible
values of an integral

$$\oint_C P\,dx + Q\,dy$$

about a simple closed path in D? We assert that there are only two possible
values, namely, the value 0 when C does not enclose the hole A (the curve C_0 of
Fig. 5.22) and a value k, *which is the same for all curves C enclosing the hole A.*
That the value is 0 for a curve such as C_0 follows from the fact that C_0 lies in a
simply connected part of D, so that Theorem IV of the preceding section is
applicable. It remains to show that for paths C_1 and C_2 which both enclose A, one
has

$$\oint_{C_1} P\,dx + Q\,dy = \oint_{C_2} P\,dx + Q\,dy. \tag{5.57}$$

To prove this, let us assume first that C_1 and C_2 do not meet and that, for
example, C_2 lies within C_1. Then C_1 and C_2 form the boundary of a region R to
which (5.55′) is applicable:

$$\oint_{C_1} P\,dx + Q\,dy + \oint_{C_2} P\,dx + Q\,dy = 0$$

or

$$\oint_{C_1} P\,dx + Q\,dy = \oint_{C_2} P\,dx + Q\,dy,$$

as was to be shown. If C_1 and C_2 do intersect each other, a sufficiently small circle
C_3 about A will meet neither of them. One then has

$$\oint_{C_1} P\,dx + Q\,dy = \oint_{C_3} P\,dx + Q\,dy, \qquad \oint_{C_2} P\,dx + Q\,dy = \oint_{C_3} P\,dx + Q\,dy$$

as before. Thus again

$$\oint_{C_1} P\,dx + Q\,dy = \oint_{C_2} P\,dx + Q\,dy,$$

and (5.57) is completely proved.

EXAMPLE 1   The integral

$$\oint_C \frac{-y\,dx + x\,dy}{x^2 + y^2}$$

of Example 2 of the preceding section is of the type considered, with the hole A at the origin. Hence the integral is 0 when C does not enclose the origin and has a certain value k when C does enclose the origin. To find k, we choose C to be the circle $x^2 + y^2 = 1$ and find that

$$k = \oint_C -y\,dx + x\,dy = \int_R\int 2\,dx\,dy = 2\pi.$$

This is in agreement with the previous results. ■

EXAMPLE 2   Example 3 of the preceding section,

$$\oint_C \frac{x\,dx + y\,dy}{x^2 + y^2},$$

illustrates a case in which the constant value k is 0. Thus the integral is 0 on all simple closed paths not through the origin. ■

EXAMPLE 3   To evaluate

$$\oint_C \frac{y^3\,dx - xy^2\,dy}{(x^2 + y^2)^2}$$

around the ellipse $x^2 + 3y^2 = 1$, one verifies that $\partial P/\partial y = \partial Q/\partial x$ except at $(0,0)$, where both functions are discontinuous. Then, since the ellipse is awkward for integration, one replaces it by the circle $x^2 + y^2 = 1$, which also encloses the point of discontinuity. With the parametrization $x = \cos t$, $y = \sin t$, the integral on the circle is

$$\int_0^{2\pi}(-\sin^3 t \sin t - \cos t \sin^2 t \cos t)\,dt = -\int_0^{2\pi} \sin^2 t\,dt = -\pi.$$

Thus the integral on the given ellipse equals $-\pi$. ■

The results (5.55) or (5.56) can be put in vector form. Thus as in (5.43) and (5.44) (Section 5.5), one has

$$\int_{B_R} u_T\,ds = \int_R\int \operatorname{curl}_z \mathbf{u}\,dx\,dy, \tag{5.58}$$

$$\int_{B_R} v_n\,ds = \int_R\int \operatorname{div}\mathbf{v}\,dx\,dy, \tag{5.59}$$

where **T** is the unit tangent vector in the direction of integration and **n** is 90° behind **T**, so that **n** is an *outer normal*, that is, points away from R.

Remark.   It can happen that the path C passes *through* a discontinuity of P or Q. If the line integral is reduced to a definite integral $\int_a^b f(t) \, dt$ in the parameter t, one has then the problem discussed in Section 4.1. If $f(t)$ is discontinuous only at a finite number of points and is bounded, then the integral continues to exist; if $f(t)$ is unbounded, the integral is improper and may or may not exist.

PROBLEMS

1. Determine by inspection a function $F(x, y)$ whose differential has the given value and integrate the corresponding line integral:

   a) $dF = 2xy \, dx + x^2 \, dy$, $\int_{C \, (0,0)}^{(1,1)} 2xy \, dx + x^2 \, dy$, where C is the curve $y = x^{\frac{3}{2}}$;

   b) $dF = ye^{xy} \, dx + xe^{xy} \, dy$, $\int_{C \, (0,0)}^{(\pi,0)} ye^{xy} \, dx + xe^{xy} \, dy$, where C is the curve $y = \sin^3 x$;

   c) $dF = \dfrac{x \, dx + y \, dy}{(x^2 + y^2)^{\frac{3}{2}}}$, $\int_{C \, (1,0)}^{(e^{2\pi},0)} \dfrac{x \, dx + y \, dy}{(x^2 + y^2)^{\frac{3}{2}}}$, where C is the curve $x = e^t \cos t$, $y = e^t \sin t$.

2. Test for independence of path and evaluate the following integrals:

   a) $\int_{(1,-2)}^{(3,4)} \dfrac{y \, dx - x \, dy}{x^2}$ on the line $y = 3x - 5$;

   b) $\int_{(0,2)}^{(1,3)} \dfrac{3x^2}{y} dx - \dfrac{x^3}{y^2} dy$ on the parabola $y = 2 + x^2$.

   c) $\int_{(1,0)}^{(-1,0)} (2xy - 1) \, dx + (x^2 + 6y) \, dy$ on the circular arc $y = \sqrt{1 - x^2}$, $-1 \le x \le 1$;

   d) $\int_{(0,0)}^{(\frac{\pi}{4}, \frac{\pi}{4})} \sec^2 x \tan y \, dx + \sec^2 y \tan x \, dy$ on the curve $y = 16x^3/\pi^2$.

3. Evaluate the following integrals:

   a) $\oint [\sin(xy) + xy \cos(xy)] \, dx + x^2 \cos(xy) \, dy$ on the circle $x^2 + y^2 = 1$;

   b) $\oint \dfrac{y \, dx - (x - 1) \, dy}{(x - 1)^2 + y^2}$ on the circle $x^2 + y^2 = 4$;

   c) $\oint y^3 \, dx - x^3 \, dy$ on the square $|x| + |y| = 1$;

   d) $\oint xy^6 \, dx + (3x^2y^5 + 6x) \, dy$ on the ellipse $x^2 + 4y^2 = 4$.

e) $\oint (7x - 3y + 2)\,dx + (4y - 3x - 5)\,dy$ on the ellipse $2x^2 + 3y^2 = 1$;

f) $\oint \dfrac{(e^x \cos y - 1)\,dx + e^x \sin y\,dy}{e^{2x} - 2e^x \cos y + 1}$ on the circle $x^2 + y^2 = 1$. [Hint: First show

that the denominator is 0 only for $x = 0$ and $\cos y = 1$ by writing it as $(e^x - 1)^2 + 2e^x(1 - \cos y)$.]

4. Determine all values of the integral

$$\int_{(1,0)}^{(2,2)} \frac{-y\,dx + x\,dy}{x^2 + y^2}$$

on a path not passing through the origin.

5. On a path $C : x = f(t), y = g(t), a \leq t \leq b$ not passing through the origin, for which $f(a) = f(b), g(a) = g(b)$, the analysis of Section 5.6 shows that

$$\int_C \frac{-y\,dx + x\,dy}{x^2 + y^2}$$

equals $n \cdot 2\pi$, where n is the number of times C encircles the origin (counted positive or negative according as θ has a net increase or decrease on the path); n is called the *winding number* of C. The value of n can be determined from a sketch of the path. Evaluate the line integral for the following paths:

a) $x = 5 + \cos^3 t, y = 8 + \sin^3 t, 0 \leq t \leq 2\pi$

b) $x = \cos + t \sin t, y = \sin t, 0 \leq t \leq 2\pi$

c) $x = 2 \cos 2t - \cos t, y = 2 \sin 2t - \sin t, 0 \leq t \leq 2\pi$

d) $x = e^{\cos^3 t}, y = \sin^4 t, 0 \leq t \leq 2\pi$

Remark. In this way, some difficult definite integrals are evaluated with ease. For example, b) integrates $\cos^2 t / (1 + 2t \sin t \cos t + t^2 \sin^2 t)$ from 0 to 2π.

6. Show that the following functions are independent of path in the xy-plane and evaluate them:

a) $\int_{(1,1)}^{(x,y)} 2xy\,dx + (x^2 - y^2)\,dy$

b) $\int_{(0,0)}^{(x,y)} \sin y\,dx + x \cos y\,dy$.

7. Evaluate

$$\oint \frac{x^2y \, dx - x^3 \, dy}{\left(x^2 + y^2\right)^2}$$

around the square with vertices $(\pm 1, \pm 1)$ (note that $\partial P/\partial y = \partial Q/\partial x$).

8. Let D be a domain that has a finite number of "holes" at points $A_1, A_2, \ldots, A_k$, so that D is $(k + 1)$-tuply connected; cf. Fig. 5.23. Let P and Q be continuous and have continuous derivatives in D, with $\partial P/\partial y = \partial Q/\partial x$ in D. Let C_1 denote a circle about A_1 in D, enclosing none of the other A's. Let C_2 be chosen similarly for A_2, and so on. Let

$$\oint_{C_1} P \, dx + Q \, dy = \alpha_1, \quad \oint_{C_2} P \, dx + Q \, dy = \alpha_2, \ldots, \quad \oint_{C_k} P \, dx + Q \, dy = \alpha_k.$$

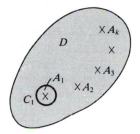

Figure 5.23 Problem 7.

a) Show that if C is an arbitrary simple closed path in D enclosing $A_1, A_2, \ldots, A_k$, then

$$\oint_C P \, dx + Q \, dy = \alpha_1 + \alpha_2 + \cdots + \alpha_k.$$

b) Determine all possible values of the integral

$$\int_{(x_1, \, y_1)}^{(x_2, \, y_2)} P \, dx + Q \, dy$$

between two fixed points of D, if it is known that this integral has the value K for one particular path.

9. Let P and Q be continuous and have continuous derivatives, with $\partial P/\partial y = \partial Q/\partial x$, except at the points $(4,0)$, $(0,0)$, $(-4,0)$. Let C_1 denote the circle $(x-2)^2 + y^2 = 9$; let C_2 denote the circle $(x+2)^2 + y^2 = 9$; let C_3 denote the circle $x^2 + y^2 = 25$. Given that

$$\oint_{C_1} P\,dx + Q\,dy = 11, \qquad \oint_{C_2} P\,dx + Q\,dy = 9, \qquad \oint_{C_3} P\,dx + Q\,dy = 13,$$

find

$$\oint_{C_4} P\,dx + Q\,dy,$$

where C_4 is the circle $x^2 + y^2 = 1$. [Hint: Use the result of Problem 8(a).]

10. Let $F(x, y) = x^2 - y^2$. Evaluate

a) $\int_{(0,0)}^{(2,8)} \nabla F \cdot d\mathbf{r}$ on the curve $y = x^3$;

b) $\oint \dfrac{\partial F}{\partial n}\,ds$ on the circle $x^2 + y^2 = 1$, if $\mathbf{n}$ is the outer normal and $\dfrac{\partial F}{\partial n} = \nabla F \cdot \mathbf{n}$ is the directional derivative of F in the direction of $\mathbf{n}$ (Section 2.14).

11. Let $F(x, y)$ and $G(x, y)$ be continuous and have continuous derivatives in a domain D. Let R be a closed region in D with directed boundary B_R consisting of closed curves $C_1, \ldots, C_n$ as in Fig. 5.21. Let $\mathbf{n}$ be the unit outer normal of R and let $\partial F/\partial n$, $\partial G/\partial n$ denote the directional derivatives of F and G in the direction of $\mathbf{n}$: $\partial F/\partial n = \nabla F \cdot \mathbf{n}$, $\partial G/\partial n = \nabla G \cdot \mathbf{n}$.

a) Show that $\displaystyle\int_{B_R} \dfrac{\partial F}{\partial n}\,ds = \int_R \int \nabla^2 F\,dx\,dy$.

b) Show that $\displaystyle\int_{B_R} \nabla F \cdot d\mathbf{r} = 0$.

c) Show that if F is harmonic in D, then $\displaystyle\int_{B_R} \dfrac{\partial F}{\partial n}\,ds = 0$.

d) Show that

$$\int_{B_R} F\dfrac{\partial G}{\partial n}\,ds = \int_R\int F\,\nabla^2 G\,dx\,dy + \int_R\int (\nabla F \cdot \nabla G)\,dx\,dy.$$

[Hint: Use the identity $\operatorname{div}(f\mathbf{u}) = f\operatorname{div}\mathbf{u} + \operatorname{grad} f \cdot \mathbf{u}$ of Section 3.6.]

12. Under the assumptions of Problem 11, prove the identities:

a) $\displaystyle\int_{B_R} \left(F\dfrac{\partial G}{\partial n} - G\dfrac{\partial F}{\partial n} \right) ds = \int_R\int (F\,\nabla^2 G - G\,\nabla^2 F)\,dx\,dy$

[Hint: Use Problem 11(d), applied to F and G and then with F and G interchanged];

b) if F and G are harmonic in R, then

$$\int_{B_R} \left(F \frac{\partial G}{\partial n} - G \frac{\partial F}{\partial n} \right) ds = 0.$$

Remark.   Problems 11(d) and 12(a) are known as *identities of Green.* Generalization to space and further applications are considered in Problem 4 following Section 5.11.

Part II.   Three-Dimensional Theory and Applications

5.8   LINE INTEGRALS IN SPACE

The basic definitions (Sections 5.2 and 5.3) of the line integrals

$$\int F(x, y)\, dx, \qquad \int F(x, y)\, dy, \qquad \int F(x, y)\, ds$$

can be generalized almost without change to give corresponding line integrals along a curve C in xyz-space. Let C be given parametrically by equations:

$$x = \phi(t), \qquad y = \psi(t), \qquad z = \omega(t), \qquad h \le t \le k, \qquad (5.60)$$

and let C be directed by increasing t. We assume that $\phi(t), \psi(t), \omega(t)$ are continuous and have continuous derivatives over the interval, so that C is "smooth." The interval $h \le t \le k$ is now subdivided by the successive points $t_0 = h, t_1, \ldots, t_i, \ldots, t_n = k$. The value t_i^* is chosen between t_{i-1} and t_i, and one sets $x_i^* = \phi(t_i^*), y_i^* = \psi(t_i^*), z_i^* = \omega(t_i^*)$. One denotes by $\Delta_i t$ the difference $t_i - t_{i-1}$ and by $\Delta_i s$ the length of the corresponding piece of C. One defines

$$\Delta_i x = \phi(t_i) - \phi(t_{i-1}), \qquad \Delta_i y = \psi(t_i) - \psi(t_{i-1}), \qquad \Delta_i z = \omega(t_i) - \omega(t_{i-1}).$$

These quantities are indicated in Fig. 5.24. One can now make the definition

$$\int_C X(x, y, z)\, dx = \lim_{\substack{n \to \infty \\ \max \Delta_i t \to 0}} \sum_{i=1}^{n} X(x_i^*, y_i^*, z_i^*)\, \Delta_i x. \qquad (5.61)$$

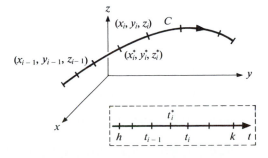

Figure 5.24   Line integral in space.

The integrals

$$\int_C Y(x, y, z)\, dy, \qquad \int_C Z(x, y, z)\, dz$$

are defined in similar fashion. Finally,

$$\int_C F(x, y, z)\, ds = \lim_{\substack{n \to \infty \\ \max \Delta_i s \to 0}} \sum_{i=1}^{n} F(x_i^*, y_i^*, z_i^*)\, \Delta_i s. \tag{5.62}$$

These definitions can again be shown to be meaningful if the functions X, Y, Z are continuous on C. The integrals can be reduced to ordinary definite integrals as in the plane:

$$\int_C X(x, y, z)\, dx = \int_h^k X[\phi(t), \psi(t), \omega(t)]\, \phi'(t)\, dt. \tag{5.63}$$

One can also extend the definitions and results to arbitrary "piecewise smooth" curves C, that is, curves made up of a finite number of smooth pieces. The x-, y-, and z-integrals will usually be considered together:

$$\int_C X\, dx + Y\, dy + Z\, dz,$$

and this combination is related to an s-integral by the equation:

$$\int_C X\, dx + Y\, dy + Z\, dz = \int_C (X \cos \alpha + Y \cos \beta + Z \cos \gamma)\, ds, \tag{5.64}$$

where

$$\mathbf{T} = \cos \alpha \mathbf{i} + \cos \beta \mathbf{j} + \cos \gamma \mathbf{k}$$

is the unit tangent vector in the direction of increasing s. Equation (5.64) has a vector interpretation:

$$\int_C \mathbf{u} \cdot d\mathbf{r} = \int_C u_T\, ds, \tag{5.65}$$

where the vector line integral on the left is the limit of a sum

$$\sum \mathbf{u}(x_i^*, y_i^*, z_i^*) \cdot \Delta_i \mathbf{r}$$

and $u_T = \mathbf{u} \cdot \mathbf{T}$ is the tangential component of the vector

$$\mathbf{u} = X\mathbf{i} + Y\mathbf{j} + Z\mathbf{k}.$$

The integrals in (5.64) and (5.65) all measure the *work* done by the *force* $\mathbf{u}$ on the given path.

The interpretation $\int v_n\, ds$, in terms of a normal component, does not generalize to line integrals in space but does to surface integrals, to be defined later.

5.9   SURFACES IN SPACE · ORIENTABILITY

Let a surface S be given in space. The surface may be represented in the form:

$$z = f(x, y), \qquad (x, y) \text{ in region } R_{xy}, \tag{5.66}$$

in the form

$$F(x, y, z) = 0, \tag{5.67}$$

or even in a parametric form (Section 4.7):

$$x = f(u, v), \qquad y = g(u, v), \qquad z = h(u, v), \qquad (u, v) \text{ in } R_{uv}. \tag{5.68}$$

It will be assumed that S is "smooth," that is, that in (5.66), for example, $f(x, y)$ has continuous derivatives in a domain containing the region R_{xy}. It will be assumed that R_{xy} is bounded and closed. It is then possible to evaluate the area of S or of a part of S.

   If the surface is given in form (5.66), the area of S is given by the formula (Section 4.7):

$$\iint\limits_{R_{xy}} \sec \gamma \, dx \, dy = \iint\limits_{R_{xy}} \sqrt{1 + \left(\frac{\partial z}{\partial x}\right)^2 + \left(\frac{\partial z}{\partial y}\right)^2} \, dx \, dy, \tag{5.69}$$

where γ is the angle between the *upper* normal to S and the z direction, as in Fig. 5.25. Thus $0 \leq \gamma < \frac{\pi}{2}$ and $\sec \gamma > 0$. We can regard formula (5.69) as stating that the surface *area element* $d\sigma$ for a surface $z = f(x, y)$ is

$$d\sigma = \sec \gamma \, dx \, dy = \sqrt{1 + \left(\frac{\partial z}{\partial x}\right)^2 + \left(\frac{\partial z}{\partial y}\right)^2} \, dx \, dy. \tag{5.70}$$

   If the surface is given in form (5.68), the area is given by

$$\iint\limits_{R_{uv}} \sqrt{EG - F^2} \, du \, dv, \tag{5.71}$$

$$E = x_u^2 + y_u^2 + z_u^2, \qquad F = x_u x_v + y_u y_v + z_u z_v, \qquad G = x_v^2 + y_v^2 + z_v^2,$$

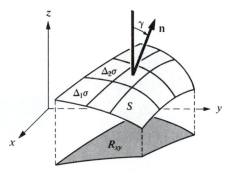

Figure 5.25   Surface integral.

where the subscripts denote partial derivatives (Section 4.7). It is assumed here that the correspondence between parameter points (u, v) and surface points (x, y, z) is one-to-one and that $EG - F^2 \neq 0$. Corresponding to (5.71), one has the *surface area element*

$$d\sigma = \sqrt{EG - F^2}\, du\, dv. \tag{5.72}$$

The preceding formulas can be extended to surfaces S that are *piecewise smooth*, that is, pieced together of a finite number of smooth parts. One can compute the total area by adding the areas of the parts. For surfaces in implicit form $F(x, y, z) = 0$, one can in general decompose the surface into smooth pieces in each of which a representation $z = z(x, y)$ or $x = x(y, z)$ or $y = y(x, z)$ is available, so that the area can be computed by several applications of the formula (5.69) (cf. Problem 9 following Section 4.7).

In order to define a surface integral analogous to the line integral $\int X\, dx + Y\, dy + Z\, dz$ it is necessary to assume that the surface S has a "direction assigned." One does not ordinarily consider the possibility of assigning a direction to a surface, but something equivalent to this is not unfamiliar, namely, the choice of a positive direction for angles, as in the xy-plane. For a general smooth surface S in space a simple way of achieving this is by choosing a unit normal vector $\mathbf{n}$ at each point of S in such a manner that $\mathbf{n}$ *varies continuously* on S. If this field of normal vectors has been chosen we term the surface *oriented*. One can then define a positive direction for angles at each point of the surface, as suggested in Fig. 5.26(a). When this has been done, one can also assign positive directions to simple closed curves forming the boundary of S. For each boundary curve C, one can choose a tangent $\mathbf{T}$ in the direction chosen and an *inner* normal $\mathbf{N}$, in a plane tangent to S, as in Fig. 5.26(a). The vectors $\mathbf{T}, \mathbf{N}, \mathbf{n}$ must then form a positive triple at each point of C.

If the surface S is only piecewise smooth, as is, for example, the surface of a cylinder, one cannot choose a continuously varying normal vector for all of S. In this case we say that S has been oriented if an orientation has been chosen in each

(a)   (b)

Figure 5.26 (a) Oriented smooth surface. (b) Möbius strip, a nonorientable surface.

smooth piece of S, in such a manner that along each curve that is a common boundary of two pieces the positive direction relative to one piece is the opposite of the positive direction relative to the other piece. This is illustrated for a cylindrical surface in Fig. 5.27. If a piecewise smooth surface S can be so oriented, we term the surface S *orientable*.

It should be remarked that not every surface is orientable. Fig. 5.26(b) suggests a Möbius strip, which is nonorientable. For this surface S, one easily convinces one's self that the normal **n** cannot be chosen to vary continuously on S and that the surface has the peculiar property of having only "one side." One-sidedness and nonorientability go together for surfaces in space.

For surfaces $z = f(x, y)$ there is no difficulty in this regard, since one can always choose **n** as the upper normal or, alternatively, as the lower normal. For surfaces in parametric form there is also no difficulty. It follows from the discussion of Section 4.7 that the vector

$$\mathbf{P}_1 \times \mathbf{P}_2 = (x_u\mathbf{i} + y_u\mathbf{j} + z_u\mathbf{k}) \times (x_v\mathbf{i} + y_v\mathbf{j} + z_v\mathbf{k})$$

is a normal vector having magnitude $\sqrt{EG - F^2}$. One can hence use

$$\mathbf{n} = \frac{\mathbf{P}_1 \times \mathbf{P}_2}{|\mathbf{P}_1 \times \mathbf{P}_2|} \tag{5.73}$$

throughout or the negative of this vector throughout, provided that $EG - F^2 \neq 0$. For a surface defined by an implicit equation $F(x, y, z) = 0$, one can choose **n** as $\nabla F / |\nabla F|$ (or its negative), provided that $\nabla F \neq \mathbf{0}$.

Figure 5.27   Oriented piecewise smooth surface (cylinder).

5.10   SURFACE INTEGRALS

Let S be a smooth surface as previously and let $H(x, y, z)$ be a function defined and continuous on S. Then the surface integral of H over S is defined as follows:

$$\int_S\int H(x, y, z)\, d\sigma = \lim_{n\to\infty} \sum_{i=1}^n H(x_i^*, y_i^*, z_i^*)\, \Delta_i\sigma. \tag{5.74}$$

The surface S is here assumed cut into n pieces as in Fig. 5.25, $\Delta_i\sigma$ denotes the area of the ith piece, and it is assumed that the ith piece shrinks to a point as $n \to \infty$ in an appropriate manner. The subdivision and the limit process are best described in terms of a representation (5.66) or (5.68), in which case they are precisely the same as the procedures used for the double integral in Section 4.3.

If the representation $z = f(x, y)$ is used, the integral is reducible to a double integral in the xy-plane:

$$\int_S\int H\, d\sigma = \int\int_{R_{xy}} H[x, y, f(x, y)]\sec\gamma\, dx\, dy, \tag{5.75}$$

where γ is the angle between the upper normal and the z axis. This follows from the area formula (5.69) or, more concisely, from the formula $d\sigma = \sec\gamma\, dx\, dy$.

If the parametric representation (5.68) is used, the integral has the expression:

$$\int_S\int H\, d\sigma = \int\int_{R_{uv}} H[f(u, v), g(u, v), h(u, v)]\sqrt{EG - F^2}\, du\, dv. \tag{5.76}$$

This follows similarly from the expression $d\sigma = \sqrt{EG - F^2}\, du\, dv$ for the area element.

The existence of the surface integral (5.74) and the expressions (5.75) and (5.76) can be established when H is continuous. As was pointed out in Section 4.7, the notion of surface area is an awkward one. Once that has been settled and the formula for the area element justified, the proof of (5.75) and (5.76) parallels that for line integrals.

The integral $\int\int H\, d\sigma$ is analogous to the line integral $\int F\, ds$. In order to obtain a surface integral analogous to the line integral $\int X\, dx + Y\, dy + Z\, dz$ we consider an oriented smooth surface S, with continuous unit normal vector $\mathbf{n}$:

$$\mathbf{n} = \cos\alpha\,\mathbf{i} + \cos\beta\,\mathbf{j} + \cos\gamma\,\mathbf{k}.$$

Let $L(x, y, z)$, $M(x, y, z)$, $N(x, y, z)$ be functions defined and continuous on S. Then by definition

$$\int_S\int L\, dy\, dz = \int_S\int L\cos\alpha\, d\sigma,$$

$$\int_S\int M\, dz\, dx = \int_S\int M\cos\beta\, d\sigma, \tag{5.77}$$

$$\int_S\int N\, dx\, dy = \int_S\int N\cos\gamma\, d\sigma.$$

If these are added, we obtain the general surface integral

$$\int_S\int L\, dy\, dz + M\, dz\, dx + N\, dx\, dy = \int_S\int (L\cos\alpha + M\cos\beta + N\cos\gamma)\, d\sigma.$$

$$(5.78)$$

This can at once be written in terms of vectors:

$$\int_S\int L\, dy\, dz + M\, dz\, dx + N\, dx\, dy = \int_S\int (\mathbf{v}\cdot\mathbf{n})\, d\sigma,$$

$$\mathbf{v} = L\mathbf{i} + M\mathbf{j} + N\mathbf{k}; \qquad (5.79)$$

thus the surface integral $\int\int L\, dy\, dz + M\, dz\, dx + N\, dx\, dy$ is the integral over the surface of the *normal* component of the vector $\mathbf{v} = L\mathbf{i} + M\mathbf{j} + N\mathbf{k}$.

For evaluation of the surface integral, one has the following formulas:

I If S is given in the form

$$z = f(x, y), \qquad (x, y) \text{ in } R_{xy},$$

with normal vector $\mathbf{n}$, then

$$\int_S\int L\, dy\, dz + M\, dz\, dx + N\, dx\, dy = \pm \int_{R_{xy}}\int \left(-L\frac{\partial z}{\partial x} - M\frac{\partial z}{\partial y} + N\right) dx\, dy,$$

$$(5.80)$$

with the $+$ sign when $\mathbf{n}$ is the upper normal and the $-$ sign when $\mathbf{n}$ is the lower normal.

II If S is given in parametric form

$$x = f(u, v), \qquad y = g(u, v), \qquad z = h(u, v), \qquad (u, v) \text{ in } R_{uv},$$

with normal vector $\mathbf{n}$, then

$$\int_S\int L\, dy\, dz + M\, dz\, dx + N\, dx\, dy$$

$$= \pm \int_{R_{uv}}\int \left[L\frac{\partial(y, z)}{\partial(u, v)} + M\frac{\partial(z, x)}{\partial(u, v)} + N\frac{\partial(x, y)}{\partial(u, v)} \right] du\, dv, \quad (5.81)$$

with the $+$ or $-$ sign according as

$$\mathbf{n} = \pm \frac{\mathbf{P}_1 \times \mathbf{P}_2}{|\mathbf{P}_1 \times \mathbf{P}_2|},$$

$$\mathbf{P}_1 = x_u\mathbf{i} + y_u\mathbf{j} + z_u\mathbf{k}, \qquad \mathbf{P}_2 = x_v\mathbf{i} + y_v\mathbf{j} + z_v\mathbf{k}. \qquad (5.82)$$

The proof of (5.80) is given here, the proof of (5.81) being left as an exercise [Problem 8(c)].

For a surface $z = f(x, y)$ the methods of Section 2.13 show that the tangent plane at (x_1, y_1, z_1) is

$$z - z_1 = \frac{\partial z}{\partial x}(x - x_1) + \frac{\partial z}{\partial y}(y - y_1)$$

and hence that the unit normal vector is

$$\mathbf{n} = \pm \frac{-\dfrac{\partial z}{\partial x}\mathbf{i} - \dfrac{\partial z}{\partial y}\mathbf{j} + \mathbf{k}}{\sqrt{1 + \left(\dfrac{\partial z}{\partial x}\right)^2 + \left(\dfrac{\partial z}{\partial y}\right)^2}},$$

with the $+$ or $-$ sign according to whether $\mathbf{n}$ is upper or lower. The denominator here is $\sec \gamma'$, where γ' is the angle between the *upper* normal and $\mathbf{k}$. Thus

$$\mathbf{n} = \pm \frac{-\dfrac{\partial z}{\partial x}\mathbf{i} - \dfrac{\partial z}{\partial y}\mathbf{j} + \mathbf{k}}{\sec \gamma'}.$$

By (5.79),

$$\int_S\int L\,dy\,dz + M\,dz\,dx + N\,dx\,dy = \int_S\int [(L\mathbf{i} + M\mathbf{j} + N\mathbf{k}) \cdot \mathbf{n}]\,d\sigma$$

$$= \pm \int_S\int \frac{-L\dfrac{\partial z}{\partial x} - M\dfrac{\partial z}{\partial y} + N}{\sec \gamma'}\,d\sigma$$

$$= \pm \int_{R_{xy}}\int \left(-L\frac{\partial z}{\partial x} - M\frac{\partial z}{\partial y} + N\right)dx\,dy,$$

since $d\sigma = \sec \gamma'\,dx\,dy$. Accordingly, (5.80) is proved.

When $L = 0$ and $M = 0$, (5.80) reduces to the formula

$$\int_S\int N(x, y, z)\,dx\,dy = \pm \int_{R_{xy}}\int N[x, y, f(x, y)]\,dx\,dy.$$

Thus the surface integral $\int\int N\,dx\,dy$ over S is the same as plus or minus the double integral of N over the projection of S on the xy-plane, with plus or minus sign according to whether the chosen normal on S is upper or lower. In particular, S can be chosen as a region R in the xy-plane, and $N = N(x, y)$. Thus the surface integral $\int_S\int N(x, y)\,dx\,dy$ and the double integral $\int_R\int N(x, y)\,dx\,dy$ are not the same; they are equal when S has the orientation determined by $\mathbf{k}$ and are otherwise negatives of each other, as illustrated in Fig. 5.28. This situation indicates how one can enlarge the concept of double integral to permit a choice of direction of integration. If this is done, Green's theorem can be stated in the form

$$\int_C P\,dx + Q\,dy = \int_S\int \left(\frac{\partial Q}{\partial x} - \frac{\partial P}{\partial y}\right)dx\,dy,$$

without a direction of integration indicated on C but with the understanding that the "directions" on C and S match, that is, that when C is positively directed the normal associated with S is $\mathbf{k}$, but when C is negatively directed the normal associated with S is $-\mathbf{k}$. These two cases are shown in Fig. 5.28.

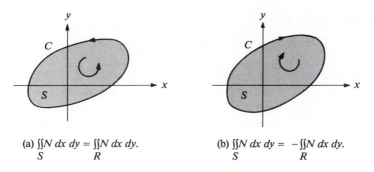

(a) $\iint_S N\,dx\,dy = \iint_R N\,dx\,dy.$

(b) $\iint_S N\,dx\,dy = -\iint_R N\,dx\,dy.$

Figure 5.28   Double integral over R versus surface integral over S, $S = R$ with orientation.

The surface integral $\iint L\,dy\,dz + \cdots$ can also be defined in a manner paralleling the vector line integral $\int \mathbf{u} \cdot d\mathbf{r}$ of (5.65). One introduces a *differential area vector* $d\boldsymbol{\sigma}$ by the equation

$$d\boldsymbol{\sigma} = \mathbf{n}\,d\sigma,$$

so that

$$d\boldsymbol{\sigma} = \cos\alpha\,d\sigma\,\mathbf{i} + \cos\beta\,d\sigma\,\mathbf{j} + \cos\gamma\,d\sigma\,\mathbf{k}$$

and

$$\iint_S L\,dy\,dz + M\,dz\,dx + N\,dx\,dy = \iint_S \mathbf{v}\cdot\mathbf{n}\,d\sigma = \iint_S \mathbf{v}\cdot d\boldsymbol{\sigma}.$$

One can also define this as the limit of a sum, as for the line integral. For the surface $z = f(x, y)$, one finds

$$d\boldsymbol{\sigma} = \pm\left(-\frac{\partial z}{\partial x}\mathbf{i} - \frac{\partial z}{\partial y}\mathbf{j} + \mathbf{k}\right)dx\,dy.$$

This can also be written:

$$d\boldsymbol{\sigma} = \mathbf{n}|\sec\gamma|\,dx\,dy = \frac{\mathbf{n}}{|\mathbf{n}\cdot\mathbf{k}|}dx\,dy.$$

From this and the corresponding expressions for surfaces $x = g(y, z), y = h(x, z)$, one obtains the formulas

$$\iint_S (\mathbf{v}\cdot\mathbf{n})\,d\sigma = \iint_{R_{xy}} \frac{\mathbf{v}\cdot\mathbf{n}}{|\mathbf{n}\cdot\mathbf{k}|}dx\,dy = \iint_{R_{yz}} \frac{\mathbf{v}\cdot\mathbf{n}}{|\mathbf{n}\cdot\mathbf{i}|}dy\,dz = \iint_{R_{zx}} \frac{\mathbf{v}\cdot\mathbf{n}}{|\mathbf{n}\cdot\mathbf{j}|}dz\,dx.$$

$$(5.83)$$

For the surfaces in parametric form we have

$$d\boldsymbol{\sigma} = \pm\left(\frac{\partial(y, z)}{\partial(u, v)}\mathbf{i} + \frac{\partial(z, x)}{\partial(u, v)}\mathbf{j} + \frac{\partial(x, y)}{\partial(u, v)}\mathbf{k}\right)du\,dv$$

or, more concisely, by (5.82),

$$d\boldsymbol{\sigma} = \pm(\mathbf{P}_1 \times \mathbf{P}_2)\, du\, dv.$$

The formal properties of line and surface integrals are analogous to those for line integrals in the plane (Section 5.3) and need no special discussion here. Furthermore, the definitions and properties of line and surface integrals carry over without change to piecewise smooth curves and surfaces, provided that the surfaces are orientable.

PROBLEMS

1. Evaluate the line integrals:

a) $\displaystyle\int_{C}^{(1,0,2\pi)}_{(1,0,0)} z\, dx + x\, dy + y\, dz$, where C is the curve $x = \cos t$, $y = \sin t$, $z = t$,

   $0 \le t \le 2\pi$;

b) $\displaystyle\int_{(1,0,1)}^{(2,3,2)} x^2\, dx - xz\, dy + y^2\, dz$ on the straight line joining the two points;

c) $\displaystyle\int_{(1,1,0)}^{(0,0,\sqrt{2})} x^2 yz\, ds$ on the curve $x = \cos t$, $y = \cos t$, $z = \sqrt{2}\,\sin t$, $0 \le t \le \dfrac{\pi}{2}$;

d) $\displaystyle\int_{C} u_T\, ds$, where $\mathbf{u} = 2xy^2 z\mathbf{i} + 2x^2 yz\mathbf{j} + x^2 y^2 \mathbf{k}$ and C is the circle $x = \cos t$,

   $y = \sin t$, $z = 2$, directed by increasing t;

e) $\displaystyle\int_{C} u_T\, ds$, where $\mathbf{u} = \operatorname{curl} \mathbf{v}$, $\mathbf{v} = y^2\mathbf{i} + z^2\mathbf{j} + x^2\mathbf{k}$, and C is the path $x = 2t + 1$,

   $y = t^2$, $z = 1 + t^3$, $0 \le t \le 1$, directed by increasing t.

2. If $\mathbf{u} = \operatorname{grad} F$ in a domain D, then show that

a) $\displaystyle\int_{(x_1, y_1, z_1)}^{(x_2, y_2, z_2)} u_T\, ds = F(x_2, y_2, z_2) - F(x_1, y_1, z_1)$, where the integral is along any

   path in D joining the two points;

b) $\displaystyle\int_{C} u_T\, ds = 0$ on any closed path in D.

3. Let a wire be given as a curve C in space. Let its density (mass per unit length) be $\delta = \delta(x, y, z)$, where (x, y, z) is a variable point in C. Justify the following formulas:

a) length of wire $= \displaystyle\int_{C} ds = L$;

b) mass of wire $= \displaystyle\int_{C} \delta\, ds = M$;

c) center of mass of the wire is $(\bar{x}, \bar{y}, \bar{z})$, where

$$M\bar{x} = \int_{C} x\,\delta\, ds, \qquad M\bar{y} = \int_{C} y\,\delta\, ds, \qquad M\bar{z} = \int_{C} z\,\delta\, ds;$$

d) moment of inertia of the wire about the z axis is

$$I_z = \int_C (x^2 + y^2)\, \delta\, ds.$$

4. Formulate and justify the formulas analogous to those of Problem 3 for the surface area, mass, center of mass, and moment of inertia of a thin curved sheet of metal forming a surface S in space.

5. Evaluate the following surface integrals:

a) $\iint_S x\, dy\, dz + y\, dz\, dx + z\, dx\, dy$, where S is the triangle with vertices $(1, 0, 0)$, $(0, 1, 0)$, $(0, 0, 1)$ and the normal points away from $(0, 0, 0)$;

b) $\iint_S dy\, dz + dz\, dx + dx\, dy$, where S is the hemisphere $z = \sqrt{1 - x^2 - y^2}$, $x^2 + y^2 \le 1$, and the normal is the upper normal;

c) $\iint_S (x \cos \alpha + y \cos \beta + z \cos \gamma)\, d\sigma$ for the surface of part (b);

d) $\iint_S x^2 z\, d\sigma$, where S is the cylindrical surface $x^2 + y^2 = 1$, $0 \le z \le 1$.

6. Evaluate the surface integrals of Problem 5, using the parametric representation:

a) $x = u + v$, $y = u - v$, $z = 1 - 2u$

b) $x = \sin u \cos v$, $y = \sin u \sin v$, $z = \cos u$

c) same as (b)

d) $x = \cos u$, $y = \sin u$, $z = v$

7. Evaluate the surface integrals:

a) $\iint_S \mathbf{w} \cdot \mathbf{n}\, d\sigma$, if $\mathbf{w} = xy^2 z \mathbf{i} - 2x^3 \mathbf{j} + yz^2 \mathbf{k}$, S is the surface $z = 1 - x^2 - y^2$, $x^2 + y^2 \le 1$, and $\mathbf{n}$ is upper;

b) $\iint_S \mathbf{w} \cdot \mathbf{n}\, d\sigma$, if $\mathbf{w} = \mathbf{i} + 2\mathbf{j} - 3\mathbf{k}$, S is the surface $x = e^u \cos v$, $y = e^u \sin v$, $z = \cos v \sin v$, $0 \le u \le 1$, $0 \le v \le \pi/2$, and $\mathbf{n}$ is given by (5.82) with the $+$ sign;

c) $\iint_S \dfrac{\partial w}{\partial n}\, d\sigma$ if $w = x^2 y^2 z$ and S and $\mathbf{n}$ are as in (a);

d) $\iint_S \dfrac{\partial w}{\partial n}\, d\sigma$ if $w = x^2 - y^2 + z^2$ and S and $\mathbf{n}$ are as in (b);

e) $\iint_S \operatorname{curl} \mathbf{u} \cdot \mathbf{n}\, d\sigma$ if $\mathbf{u} = yz\mathbf{i} - xz\mathbf{j} + xz\mathbf{k}$, S is the triangle with vertices $(1, 2, 8)$, $(3, 1, 9)$, $(2, 1, 7)$ and $\mathbf{n}$ is upper.

8. **a)** Let a surface S: $z = f(x, y)$ be defined by an implicit equation $F(x, y, z) = 0$. Show that the surface integral $\iint_S H\, d\sigma$ over S becomes

$$\iint_{R_{xy}} \sqrt{\left(\frac{\partial F}{\partial x}\right)^2 + \left(\frac{\partial F}{\partial y}\right)^2 + \left(\frac{\partial F}{\partial z}\right)^2}\; \frac{H}{\left|\dfrac{\partial F}{\partial z}\right|}\, dx\, dy,$$

provided that $\dfrac{\partial F}{\partial z} \ne 0$.

b) Prove that for the surface of part (a) with $\mathbf{n} = \nabla F / |\nabla F|$,

$$\iint_S (\mathbf{v} \cdot \mathbf{n})\, d\sigma = \iint_{R_{xy}} (\mathbf{v} \cdot \nabla F) \frac{1}{\left| \dfrac{\partial F}{\partial z} \right|}\, dx\, dy.$$

c) Prove (5.81).

d) Prove that (5.81) reduces to (5.80) when $x = u$, $y = v$, $z = f(u, v)$.

9. Let S be an oriented surface in space that is planar; that is, S lies in a plane. With S one can associate the vector $\mathbf{S}$, which has the direction of the normal chosen on S and has a length equal to the area of S.

a) Show that if S_1, S_2, S_3, S_4 are the faces of a tetrahedron, oriented so that the normal is the exterior normal, then

$$\mathbf{S}_1 + \mathbf{S}_2 + \mathbf{S}_3 + \mathbf{S}_4 = \mathbf{0}.$$

[Hint: Let $\mathbf{S}_i = A_i \mathbf{n}_i (A_i > 0)$ for $i = 1, \ldots, 4$ and let $\mathbf{S}_1 + \cdots + \mathbf{S}_4 = \mathbf{b}$. Let p_1 be the foot of the altitude on face S_1 and join p_1 to the vertices of S_1 to form three triangles of areas $A_{12}, \ldots, A_{14}$. Show that, for proper numbering, $A_{1j} = \pm A_j \mathbf{n}_j \cdot \mathbf{n}_1$, with $+$ or $-$ according as $\mathbf{n}_j \cdot \mathbf{n}_1 > 0$ or < 0, and $A_{1j} = 0$ if $\mathbf{n}_j \cdot \mathbf{n}_1 = 0$ $(j = 2, 3, 4)$. Hence deduce that $\mathbf{b} \cdot \mathbf{n}_j = 0$ for $j = 2, 3, 4$ and thus $\mathbf{b} \cdot \mathbf{b} = 0$.]

b) Show that the result of (a) extends to an arbitrary convex polyhedron with faces $S_1, \ldots, S_n$, that is, that

$$\mathbf{S}_1 + \mathbf{S}_2 + \cdots + \mathbf{S}_n = \mathbf{0},$$

when the orientation is that of the exterior normal.

c) Using the result of (b), indicating a reasoning to justify the relation

$$\iint_S \mathbf{v} \cdot d\boldsymbol{\sigma} = 0$$

for any convex closed surface S (such as the surface of a sphere or ellipsoid), provided that $\mathbf{v}$ is a constant vector.

d) Apply the result of (b) to a triangular prism whose edges represent the vectors $\mathbf{a}, \mathbf{b}, \mathbf{a} + \mathbf{b}, \mathbf{c}$ to prove the *distributive law* (Equation (1.19))

$$\mathbf{c} \times (\mathbf{a} + \mathbf{b}) = \mathbf{c} \times \mathbf{a} + \mathbf{c} \times \mathbf{b}$$

for the vector product. This is the method used by Gibbs (cf. the book of Gibbs listed at the end of this chapter).

5.11   THE DIVERGENCE THEOREM

It was pointed out in Section 5.5 that Green's theorem can be written in the form

$$\int_C v_n\, ds = \iint_R \operatorname{div} \mathbf{v}\, dx\, dy.$$

The following generalization thus appears natural:

$$\iint_S v_n\, d\sigma = \iiint_R \operatorname{div} \mathbf{v}\, dx\, dy\, dz,$$

where S is a surface forming the complete boundary of a bounded closed region R in space and $\mathbf{n}$ is the outer normal of S, that is, the one pointing away from R.

The theorem, known as the *divergence theorem* or *Gauss's theorem*, plays a role like that of Green's theorem for line integrals, in showing that certain surface integrals are 0 and that others are "independent of path." The theorem has a basic physical significance, which will be considered later. If one considers a continuous flow of matter in space, the left-hand side can be interpreted as the flux across the boundary S, that is, the total mass leaving R per unit time; the right-hand side measures the rate at which the density is decreasing throughout R, that is, the total loss of mass per unit time. Under the assumptions made, the only way mass can be lost is by crossing the boundary S. For this reason the flux equals the total divergence.

DIVERGENCE THEOREM (*Gauss's theorem*)   Let $\mathbf{v} = L\mathbf{i} + M\mathbf{j} + N\mathbf{k}$ be a vector field in a domain D of space; let L, M, N be continuous and have continuous derivatives in D. Let S be a piecewise smooth surface in D that forms the complete boundary of a bounded closed region R in D. Let $\mathbf{n}$ be the outer normal of S with respect to R. Then

$$\iint_S v_n \, d\sigma = \iiint_R \operatorname{div} \mathbf{v} \, dx \, dy \, dz; \qquad (5.84)$$

that is,

$$\iint_S L \, dy \, dz + M \, dz \, dx + N \, dx \, dy = \iiint_R \left(\frac{\partial L}{\partial x} + \frac{\partial M}{\partial y} + \frac{\partial N}{\partial z} \right) dx \, dy \, dz. \qquad (5.85)$$

Proof.   It will be proved that

$$\iint_S N \, dx \, dy = \iiint_R \frac{\partial N}{\partial z} \, dx \, dy \, dz, \qquad (5.86)$$

the proofs of the two equations

$$\iint_S L \, dy \, dz = \iiint_R \frac{\partial L}{\partial x} \, dx \, dy \, dz, \qquad (5.87)$$

$$\iint_S M \, dz \, dx = \iiint_R \frac{\partial M}{\partial y} \, dx \, dy \, dz \qquad (5.88)$$

being exactly the same.

It should first be remarked that the normal $\mathbf{n}$, defined as the outer normal of S with respect to R, necessarily varies continuously on each smooth part of S, so that the surface integral in (5.85) is well defined. The orientation defined by $\mathbf{n}$ on each part of S in fact determines an orientation of all of S, so that S must be orientable.

We assume now that R is representable in the form:

$$f_1(x, y) \leqq z \leqq f_2(x, y), \qquad (x, y) \text{ in } R_{xy}, \qquad (5.89)$$

Figure 5.29   Proof of divergence theorem.

where R_{xy} is a bounded closed region in the xy-plane (cf. Fig. 5.29) bounded by a simple closed curve C. The surface S is then composed of three parts:

$$S_1: z = f_1(x, y), \qquad (x, y) \text{ in } R_{xy},$$
$$S_2: z = f_2(x, y), \qquad (x, y) \text{ in } R_{xy},$$
$$S_3: f_1(x, y) \leqq z \leqq f_2(x, y), \qquad (x, y) \text{ on } C.$$

S_2 forms the "top" of S, S_1 forms the "bottom," and S_3 forms the "sides" (the portion S_3 may degenerate into a curve, as for a sphere). Now by definition (5.77),

$$\int_S \int N \, dx \, dy = \int_S \int N \cos \gamma \, d\sigma,$$

where $\gamma = \sphericalangle(\mathbf{n}, \mathbf{k})$. Along S_3, $\gamma = \pi/2$, so that $\cos \gamma = 0$; along S_2, $\gamma = \gamma'$ where γ' is the angle between the *upper* normal and $\mathbf{k}$; along S_1, $\gamma = \pi - \gamma'$. Since $d\sigma = \sec \gamma' \, dx \, dy$ on S_1 and S_2, one has

$$\int_S \int N \, dx \, dy = \int_{S_1} \int N \, dx \, dy + \int_{S_2} \int N \, dx \, dy + \int_{S_3} \int N \, dx \, dy$$

$$= - \int_{R_{xy}} \int N \cos \gamma' \sec \gamma' \, dx \, dy + \int_{R_{xy}} \int N \cos \gamma' \sec \gamma' \, dx \, dy,$$

where $z = f_1(x, y)$ in the first integral and $z = f_2(x, y)$ in the second:

$$\int_S \int N \, dx \, dy = \int_{R_{xy}} \int \{N[x, y, f_2(x, y)] - N[x, y, f_1(x, y)]\} \, dx \, dy.$$

$$(5.90)$$

Now the triple integral on the right-hand side of (5.86) can be evaluated as follows:

$$\iint_R \int \frac{\partial N}{\partial z} dx\,dy\,dz = \int \int_{R_{xy}} \left[\int_{f_1(x,y)}^{f_2(x,y)} \frac{\partial N}{\partial z} dz \right] dx\,dy$$

$$= \int \int_{R_{xy}} \{ N[x, y, f_2(x, y)] - N[x, y, f_1(x, y)] \}\, dx\,dy.$$

This is the same as (5.90); hence

$$\int_S \int N\,dx\,dy = \iint_R \int \frac{\partial N}{\partial z} dx\,dy\,dz.$$

This is the result desired for the particular R assumed. For any region R that can be cut up into a finite number of pieces of this type by means of piecewise smooth auxiliary surfaces, the theorem then follows by adding the result for each part separately (cf. the proof of Green's theorem in Section 5.5). The surface integrals over the auxiliary surfaces cancel in pairs, and the sum of the surface integrals is precisely that over the complete boundary S of R; the volume integrals over the pieces add up to that over R. The result can finally be extended to the most general R envisaged in the theorem by a limit process [see the book by Kellogg, listed at the end of the chapter].

Once (5.86) has been established, (5.87) and (5.88) then follow by merely relabeling the variables. Adding these three equations, one obtains Eq. (5.85).   □

As a first application of the divergence theorem, a new interpretation of the divergence of a vector field $\mathbf{v}$ will be given. Let S_r be a sphere with center (x_1, y_1, z_1) and radius r. Let R_r denote S_r plus interior. We recall that the law of the mean for integrals asserts that

$$\iint_R \int f(x, y, z)\, dx\,dy\,dz = f(x^*, y^*, z^*) \cdot V,$$

where (x^*, y^*, z^*) is a point of R and V is the volume of R, provided that f is continuous in R (cf. the formulation for double integrals in Section 4.3). If this is applied to the integral

$$\iint_{R_r} \int \text{div}\,\mathbf{v}\,dx\,dy\,dz,$$

one concludes that

$$\iint_{R_r} \int \text{div}\,\mathbf{v}\,dx\,dy\,dz = \text{div}\,\mathbf{v}(x^*, y^*, z^*) \cdot V_r,$$

for some (x^*, y^*, z^*) in R_r, V_r denoting the volume $(\frac{4}{3}\pi r^3)$ of R_r. If the divergence theorem is applied, one concludes that

$$\text{div}\,\mathbf{v}(x^*, y^*, z^*) = \frac{1}{V_r} \int_{S_r} \int v_n\,d\sigma.$$

If now we let r approach 0, the point (x^*, y^*, z^*) must approach the center (x_1, y_1, z_1). Thus

$$\text{div } \mathbf{v}(x_1, y_1, z_1) = \lim_{r \to 0} \frac{1}{V_r} \int\int_{S_r} v_n \, d\sigma. \tag{5.91}$$

The surface integral $\int\int v_n \, d\sigma$ can be interpreted as *the flux* of the vector field $\mathbf{v}$ across S_r. Hence the *divergence of a vector field at a point equals the limiting value of the flux across a sphere about the point divided by the volume of the sphere, as the radius of the sphere approaches* 0. The spherical surface and solid used here can be replaced by a more general one, provided that the "shrinking to zero" takes place in a suitable manner. Thus in concise form, divergence equals flux per unit volume.

This result has fundamental significance, for it assigns a meaning to the divergence that is independent of any coordinate system. It also provides considerable insight into the meaning of positive, negative, and zero divergence in a physical problem. One can use (5.91) as the *definition* of divergence, and this is often done; one must then verify, with the aid of the divergence theorem in the form (5.85), that one obtains the usual formula

$$\frac{\partial v_x}{\partial x} + \frac{\partial v_y}{\partial y} + \frac{\partial v_z}{\partial z}$$

in each coordinate system.

One can apply (5.91) to the case in which $\mathbf{v} = \mathbf{u}$, where $\mathbf{u}$ is the velocity vector of a fluid motion. The expression $u_n \, d\sigma$ can be interpreted as the volume filled up per unit time by the fluid crossing the surface element $d\sigma$ in the direction of $\mathbf{n}$ (see Fig. 5.30). The flux integral $\int\int u_n \, d\sigma$ over the surface S_r then measures the total rate of filling volume (outgoing minus incoming volume per unit time). By (5.91),

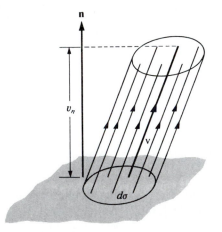

Figure 5.30   Flux $= \int\int v_n \, d\sigma$.

div $\mathbf{u}$ then measures the limiting value of this rate, per unit volume, at a point (x_1, y_1, z_1). If the fluid is *incompressible*, the outgoing volume precisely equals the incoming volume, so that the flux across S_r is always 0. By (5.91) this implies that div $\mathbf{u} = 0$; conversely, by (5.84), if div $\mathbf{u} = 0$ in D, then the flow is incompressible. Another interpretation of div $\mathbf{u}$ is given in Problem 7 following Section 5.15.

One can also let $\mathbf{v} = \rho\mathbf{u}$, where ρ is the density and $\mathbf{u}$ is the velocity of the fluid motion; the flux integral

$$\int_S\!\!\int v_n \, d\sigma = \int_S\!\!\int \rho u_n \, d\sigma$$

then measures the rate at which *mass* is leaving the solid R via the surface S. For $u_n \, d\sigma$ measures the volume of fluid crossing a surface element per unit of time, and $\rho u_n \, d\sigma$ measures its mass. The divergence div $\mathbf{v}$ measures this rate of loss of mass per unit volume, by (5.91). But this is precisely the rate at which the density is *decreasing* at the point (x_1, y_1, z_1). Hence

$$\frac{\partial \rho}{\partial t} = -\text{div}\,\mathbf{v} = -\text{div}\,(\rho\mathbf{u})$$

or

$$\frac{\partial \rho}{\partial t} + \text{div}\,(\rho\mathbf{u}) = 0. \tag{5.92}$$

This is the *continuity equation* of hydrodynamics. It expresses the *conservation of mass*. Another derivation is given in Problem 9 following Section 5.15.

PROBLEMS

1. Evaluate by the divergence theorem:

   a) $\int_S\!\!\int x \, dy \, dz + y \, dz \, dx + z \, dx \, dy$, where S is the sphere $x^2 + y^2 + z^2 = 1$ and $\mathbf{n}$ is the outer normal;

   b) $\int_S\!\!\int v_n \, d\sigma$, where $\mathbf{v} = x^2\mathbf{i} + y^2\mathbf{j} + z^2\mathbf{k}$, $\mathbf{n}$ is the outer normal and S is the surface of the cube $0 \le x \le 1, 0 \le y \le 1, 0 \le z \le 1$;

   c) $\int_S\!\!\int e^y \cos z \, dy \, dz + e^x \sin z \, dz \, dx + e^x \cos y \, dx \, dy$, with S and $\mathbf{n}$ as in (a);

   d) $\int_S\!\!\int \nabla F \cdot \mathbf{n} \, d\sigma$ if $F = x^2 + y^2 + z^2$, $\mathbf{n}$ is the exterior normal, and S bounds a solid region R;

   e) $\int_S\!\!\int \nabla F \cdot \mathbf{n} \, d\sigma$ if $F = 2x^2 - y^2 - z^2$, with $\mathbf{n}$ and S as in (d);

   f) $\int_S\!\!\int \nabla F \cdot \mathbf{n} \, d\sigma$ if $F = [(x-2)^2 + y^2 + z^2]^{-1/2}$ and S and $\mathbf{n}$ are as in (a).

2. Let S be the boundary surface of a region R in space and let $\mathbf{n}$ be its outer normal.

Prove the formulas:

a) $V = \int_S \int x \, dy \, dz = \int_S \int y \, dz \, dx = \int_S \int z \, dx \, dy$

$= \frac{1}{3} \int_S \int x \, dy \, dz + y \, dz \, dx + z \, dx \, dy,$

where V is the volume of R;

b) $\int_S \int x^2 \, dy \, dz + 2xy \, dz \, dx + 2xz \, dx \, dy = 6V\bar{x},$

where $(\bar{x}, \bar{y}, \bar{z})$ is the centroid of R;

c) $\int_S \int \operatorname{curl} \mathbf{v} \cdot \mathbf{n} \, d\sigma = 0$, where $\mathbf{v}$ is an arbitrary vector field.

3. Deduce the results of Problem 9(a), (b), (c) following Section 5.10 by proving (c) first, using the incompressibility of the flow of constant velocity $\mathbf{v}$.

4. Let S be the boundary surface of a region R, with outer normal $\mathbf{n}$, as in the divergence theorem. Let $f(x, y, z)$ and $g(x, y, z)$ be functions defined and continuous, with continuous first and second derivatives, in a domain D containing R. Prove the following relations:

a) $\int_S \int f \partial g / \partial n \, d\sigma = \int \int_R \int f \nabla^2 g \, dx \, dy \, dz + \int \int_R \int (\nabla f \cdot \nabla g) \, dx \, dy \, dz;$

[Hint: use the identity $\nabla \cdot (f\mathbf{u}) = \nabla f \cdot \mathbf{u} + f(\nabla \cdot \mathbf{u}).$]

b) if g is harmonic in D, then

$$\int_S \int \frac{\partial g}{\partial n} \, d\sigma = 0;$$

[Hint: Put $f = 1$ in (a).]

c) if f is harmonic in D, then

$$\int_S \int f \frac{\partial f}{\partial n} \, d\sigma = \int \int_R \int |\nabla f|^2 \, dx \, dy \, dz;$$

d) if f is harmonic in D and $f \equiv 0$ on S, then $f \equiv 0$ in R [cf. the last paragraph before the remarks at the end of Section 4.3];

e) if f and g are harmonic in D and $f \equiv g$ on S, then $f \equiv g$ in R; [Hint: Use (d).]

f) if f is harmonic in D and $\partial f / \partial n = 0$ on S, then f is constant in R;

g) if f and g are harmonic in D and $\partial f / \partial n = \partial g / \partial n$ on S, then $f = g + \text{const}$ in R;

h) if f and g are harmonic in R, and

$$\frac{\partial f}{\partial n} = -f + h, \qquad \frac{\partial g}{\partial n} = -g + h \text{ on } S, \qquad h = h(x, y, z),$$

then

$$f \equiv g \text{ in } R;$$

i) if f and g both satisfy the same *Poisson equation* in R,

$$\nabla^2 f = -4\pi h, \qquad \nabla^2 g = -4\pi h, \qquad h = h(x, y, z),$$

and $f = g$ on S, then

$$f \equiv g \text{ in } R;$$

j) $\int_S \int \left(f \frac{\partial g}{\partial n} - g \frac{\partial f}{\partial n} \right) d\sigma = \int \int_R \int (f \nabla^2 g - g \nabla^2 f) \, dx \, dy \, dz;$

[Hint: Use (a).]

k) if f and g are harmonic in R, then

$$\int_S\int\left(f\frac{\partial g}{\partial n} - g\frac{\partial f}{\partial n}\right)d\sigma = 0;$$

l) if f and g satisfy the equations:

$$\nabla^2 f = hf, \qquad \nabla^2 g = hg, \qquad h = h(x, y, z),$$

in R, then

$$\int_S\int\left(f\frac{\partial g}{\partial n} - g\frac{\partial f}{\partial n}\right)d\sigma = 0.$$

Remark. Parts (a) and (j) are known as *Green's first and second identities*, respectively.

5. Let S and R be as in Problem 4. Prove, under appropriate continuity assumptions:

a) $\int_S\int fn \cdot i \, d\sigma = \int\int\int_R \frac{\partial f}{\partial x}dV.$

[Hint: Apply the divergence theorem.]

b) $\int_S\int fn \, d\sigma = \int\int\int_R \nabla f \, dV.$

[Hint: These are integrals of vectors as in Section 4.5. Use (a) to show that the x-components of both sides are equal and, similarly, that the y- and z-components are equal.]

c) $\int_S\int v \times i \cdot n \, d\sigma = \int\int\int_R \text{curl } v \cdot i \, dV.$

[Hint: Apply the divergence theorem and then evaluate div $(v \times i)$ by (3.35).]

d) $\int_S\int n \times v \, d\sigma = \int\int\int_R \text{curl } v \, dV.$

[Hint: These are vector integrals. Use (c) to show that the x-components of both sides are equal and, similarly, that the y- and z-components are equal.]

5.12 STOKES'S THEOREM

It was seen in Section 5.5 that Green's theorem can be written in the form

$$\oint_C u_T \, ds = \int_R\int \text{curl}_z \, u \, dx \, dy.$$

This suggests that for any simple closed plane curve C in space (Fig. 5.31),

$$\int_C u_T \, ds = \int_S\int \text{curl}_n \, u \, d\sigma, \tag{5.93}$$

where n is normal to the plane in which C lies, S is the planar surface bounded by C, and the direction of C is positive in terms of the orientation of S determined by n. It was pointed out in Section 5.9 that the choice of a continuously varying normal vector on a surface determines a positive sense for each simple closed curve C, when regarded as bounding a portion of the surface.

Figure 5.31   Case of Stokes's
theorem.

Equation (5.93) can be generalized further, namely, by replacing S by an arbitrary smooth orientable surface whose boundary is a simple closed curve C, not necessarily a plane curve. Again the choice of a continuous normal determines the direction on C, as in Fig. 5.32. It should be stressed that this relation between **n** and the direction of C, as defined in Section 5.9, depends on the notion of *positive triple* of vectors, hence on a particular orientation of space (Section 1.1). If the orientation were reversed, the given normal **n** would correspond to the opposite direction on C.

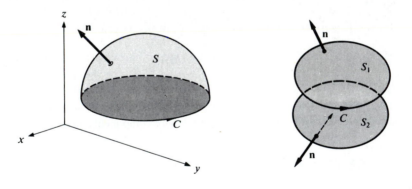

Figure 5.32   Another case of
Stokes's theorem.

Figure 5.33   Proof of (5.94).

Equation (5.93) is Stokes's theorem, which will be shown to be correct under very general assumptions. An additional reason for expecting such a relation as this is the fact that if two surfaces S_1 and S_2 have the common boundary C and no other points in common, as in Fig. 5.33, then

$$\int_{S_1}\int \operatorname{curl}_n \mathbf{u}\, d\sigma = \int_{S_2}\int \operatorname{curl}_n \mathbf{u}\, d\sigma, \tag{5.94}$$

provided that S_1 and S_2 are oriented so as to determine the same direction on C.

For

$$\int_{S_1}\int \text{curl}_n \mathbf{u} \, d\sigma - \int_{S_2}\int \text{curl}_n \mathbf{u} \, d\sigma = \int_S\int \text{curl}_n \mathbf{u} \, d\sigma,$$

where S is the oriented surface formed of S_1 and of S_2 with its normal reversed. S then bounds a certain solid region R and by the divergence theorem

$$\int_S\int \text{curl}_n \mathbf{u} \, d\sigma = \pm \int\int\int_R \text{div curl } \mathbf{u} \, dx \, dy \, dz = 0.$$

(The $+$ or $-$ depends on whether $\mathbf{n}$ is outer or inner normal.) Thus the surface integral $\int\int \text{curl}_n \mathbf{u} \, dA$ has the same value for all surfaces with boundary C, and it is natural to expect the surface integrals to be expressible in terms of a line integral of $\mathbf{u}$ on C.

THEOREM OF STOKES   Let S be a piecewise smooth oriented surface in space, whose boundary C is a piecewise smooth simple closed curve, directed in accordance with the given orientation in S. Let $\mathbf{u} = L\mathbf{i} + M\mathbf{j} + N\mathbf{k}$ be a vector field, with continuous and differentiable components, in a domain D of space including S. Then

$$\int_C u_T \, ds = \int_S\int (\text{curl } \mathbf{u} \cdot \mathbf{n}) \, d\sigma, \tag{5.95}$$

where $\mathbf{n}$ is the chosen unit normal vector on S; that is,

$$\int_C L \, dx + M \, dy + N \, dz = \int_S\int \left(\frac{\partial N}{\partial y} - \frac{\partial M}{\partial z} \right) dy \, dz + \left(\frac{\partial L}{\partial z} - \frac{\partial N}{\partial x} \right) dz \, dx$$

$$+ \left(\frac{\partial M}{\partial x} - \frac{\partial L}{\partial y} \right) dx \, dy. \tag{5.96}$$

Proof.   Just as with the divergence theorem, it is sufficient to prove three separate equations:

$$\int_C L \, dx = \int_S\int \frac{\partial L}{\partial z} dz \, dx - \frac{\partial L}{\partial y} dx \, dy, \qquad \int M \, dy = \cdots, \cdots.$$

One can further restrict attention to the case in which S is representable in the form

$$z = f(x, y), \qquad (x, y) \text{ in } R_{xy},$$

as shown in Fig. 5.34; as in the proof of the divergence theorem, the general case is handled by decomposing into a finite number of such pieces, when possible, and by a limit process.

If S has the form $z = f(x, y)$, the curve C has as projection on the xy-plane a curve C_{xy} as shown in Fig. 5.34; as (x, y, z) goes around C once in the given direction, its projection $(x, y, 0)$ goes once around C_{xy} in a corresponding

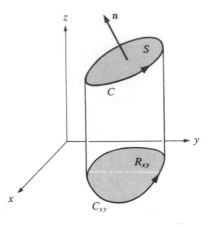

Figure 5.34   Proof of Stokes's theorem.

direction. If the normal **n** chosen on S is the upper normal, the direction on C_{xy} is the positive direction, and by Green's theorem,

$$\int_C L(x, y, z)\, dx = \int_{C_{xy}} L[x, y, f(x, y)]\, dx = -\int\int_{R_{xy}} \left(\frac{\partial L}{\partial y} + \frac{\partial L}{\partial z}\frac{\partial f}{\partial y} \right) dx\, dy.$$

Under the same assumption about the normal,

$$\int\int_S \frac{\partial L}{\partial z} dz\, dx - \frac{\partial L}{\partial y} dx\, dy = \int\int_{R_{xy}} \left(-\frac{\partial L}{\partial z}\frac{\partial f}{\partial y} - \frac{\partial L}{\partial y} \right) dx\, dy,$$

by (5.80). It at once follows that

$$\int_C L\, dx = \int\int_S \frac{\partial L}{\partial z} dz\, dx - \frac{\partial L}{\partial y} dx\, dy. \tag{5.97}$$

If the direction of **n** is reversed, both sides change sign, so that (5.97) holds in general. By the reasoning described above, one then finds that (5.97) holds for a general orientable S. In the same way, equations analogous to (5.97) are established for M and N for a general S. Upon adding the equations for L, M, and N, one obtains Stokes's theorem in full generality. □

Just as the divergence theorem gives a new interpretation for the divergence of a vector, so does the Stokes theorem give a new interpretation for the curl of a vector. To obtain this, we take S_r to be a circular disk in space of radius r and center (x_1, y_1, z_1) bounded by the circle C_r (see Fig 5.35). By the Stokes theorem and the law of the mean for integrals,

$$\int_{C_r} u_T\, ds = \int\int_{S_r} \operatorname{curl}_n \mathbf{u}\, d\sigma = \operatorname{curl}_n \mathbf{u}(x^*, y^*, z^*) \cdot A_r,$$

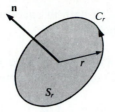

Figure 5.35   Meaning
of curl.

where A_r is the area (πr^2) of S_r and (x^*, y^*, z^*) is a suitably chosen point of S_r.
One can now write

$$\text{curl}_n \, \mathbf{u}(x^*, y^*, z^*) = \frac{1}{A_r} \int_{C_r} u_T \, ds.$$

In the case of a fluid motion with velocity $\mathbf{u}$ the integral $\int_{C_r} u_T \, ds$ is termed the
circulation around C_r; it measures the extent to which the corresponding fluid
motion is a rotation around the circle C_r in the given direction. If r is now allowed
to approach 0, we find

$$\text{curl}_n \, \mathbf{u}(x_1, y_1, z_1) = \lim_{r \to 0} \frac{1}{A_r} \int_{C_r} u_T \, ds; \qquad (5.98)$$

that is, the component of curl $\mathbf{u}$ at (x_1, y_1, z_1) in the direction of $\mathbf{n}$ is the limiting
ratio of circulation to area for a circle about (x_1, y_1, z_1) with $\mathbf{n}$ as normal. Briefly,
curl equals circulation per unit area. In the limit process the circular disk can be
replaced by a more general surface with normal $\mathbf{n}$, provided that the shrinking to
zero is properly carried out. If $\mathbf{n}$ is taken as $\mathbf{i}$, $\mathbf{j}$, and $\mathbf{k}$ successively, one obtains the
three components of curl $\mathbf{u}$ along the axes.

Since (5.98) has a significance independent of the coordinate system chosen,
this equation proves that the curl of a vector field has a meaning independent of
the particular (right-handed) coordinate system chosen in space. One could in fact
use (5.98) to define the curl. [If the orientation of space is reversed, the direction
of the curl will also be reversed (cf. also Section 3.8).]

5.13   INTEGRALS INDEPENDENT OF PATH ·
IRROTATIONAL AND SOLENOIDAL FIELDS

Since the generalization of Green's theorem to space takes two different forms,
the divergence theorem and Stokes's theorem, one can generalize the discussions
of Sections 5.6 and 5.7 in two directions: to surface integrals and to line integrals.
In the case of line integrals the results for two dimensions carry over with minor
modifications. The surface integrals require a somewhat different treatment.

Line integrals independent of path in space are defined just as in the plane. The following theorems then hold.

THEOREM I   Let $\mathbf{u} = X\mathbf{i} + Y\mathbf{j} + Z\mathbf{k}$ be a vector field with continuous components in a domain D of space. The line integral

$$\int u_T \, ds = \int X \, dx + Y \, dy + Z \, dz$$

is independent of path in D if and only if there is a function $F(x, y, z)$, defined in D such that

$$\frac{\partial F}{\partial x} = X, \qquad \frac{\partial F}{\partial y} = Y, \qquad \frac{\partial F}{\partial z} = Z$$

throughout D. In other words, the line integral is independent of path if and only if $\mathbf{u}$ is a gradient vector:

$$\mathbf{u} = \operatorname{grad} F.$$

The proof for two dimensions in Section 5.6 can be repeated without essential change. When the integral is independent of path, $X \, dx + Y \, dy + Z \, dz = dF$ for some F, and

$$\int_A^B X \, dx + Y \, dy + Z \, dz = \int_A^B dF = F(B) - F(A)$$

as in the plane.

THEOREM II   Let X, Y, Z be continuous in a domain D of space. The line integral

$$\int X \, dx + Y \, dy + Z \, dz$$

is independent of path in D if and only if

$$\int_C X \, dx + Y \, dy + Z \, dz = 0$$

on every simple closed curve C in D.

This is proved as in the plane.

THEOREM III   Let $\mathbf{u} = X\mathbf{i} + Y\mathbf{j} + Z\mathbf{k}$ be a vector field in a domain D of space; let X, Y, and Z have continuous partial derivatives in D. If

$$\int u_T \, ds = \int X \, dx + Y \, dy + Z \, dz$$

is independent of path in D, then $\operatorname{curl} \mathbf{u} \equiv \mathbf{0}$ in D; that is,

$$\frac{\partial Z}{\partial y} \equiv \frac{\partial Y}{\partial z}, \qquad \frac{\partial X}{\partial z} \equiv \frac{\partial Z}{\partial x}, \qquad \frac{\partial Y}{\partial x} \equiv \frac{\partial X}{\partial y}. \qquad (5.99)$$

Conversely, if D is simply connected and (5.99) holds, then $\int X\,dx + Y\,dy + Z\,dz$ is independent of the path in D; that is, if D is simply connected and curl $\mathbf{u} \equiv \mathbf{0}$ in D, then

$$\mathbf{u} = \text{grad } F$$

for some F.

A domain D of space is called simply connected if every simple closed curve in D forms the boundary of a smooth orientable surface in D. Thus the interior of a sphere is simply connected whereas the interior of a torus is not. The domain between two concentric spheres is simply connected, as is the interior of a sphere with a finite number of points removed. (This definition, which is adequate for practical applications, differs from the standard one of advanced mathematics.)

In the first part of the theorem we assume that $\int u_T\,ds$ is independent of path. Hence by Theorem I, $\mathbf{u} = \text{grad } F$. Accordingly,

$$\text{curl } \mathbf{u} = \text{curl grad } F \equiv \mathbf{0}$$

by the identity of Section 3.6.

In the second part of the theorem we assume that D is simply connected and curl $\mathbf{u} \equiv \mathbf{0}$. To show independence of path, it is sufficient, by Theorem II, to show that $\int_C u_T\,ds = 0$ on each simple closed curve C in D. By assumption, C forms the boundary of a piecewise smooth oriented surface S in D. Stokes's theorem is applicable, and one finds

$$\int_C u_T\,ds = \iint_S \text{curl}_n \mathbf{u}\,d\sigma = 0$$

for proper direction on C and normal $\mathbf{n}$ on S.

We remark that the Stokes theorem can be extended to an arbitrary oriented surface S whose boundary is formed of distinct simple closed curves $C_1,\ldots,C_n$. If B_S denotes this boundary, with proper directions, one has

$$\int_{B_S} u_T\,ds = \iint_S \text{curl}_n \mathbf{u}\,d\sigma.$$

The proof is like that of Section 5.7. In particular, if curl $\mathbf{u} \equiv \mathbf{0}$ in D,

$$\int_{B_S} u_T\,ds = 0.$$

This can be applied to evaluate line integrals in "multiply connected" domains, as in Section 5.7.

A vector field $\mathbf{u}$ (whose components have continuous derivatives) such that

$$\text{curl } \mathbf{u} \equiv \mathbf{0}$$

in a domain D, is called *irrotational* in D. By virtue of the above theorems, irrotationality in a simply connected domain is equivalent to each of the properties:

$$\int_C u_T \, ds = 0 \text{ for every simple closed curve in } D;$$

$$\int u_T \, ds \text{ is independent of the path in } D;$$

$$u = \operatorname{grad} F \text{ in } D.$$

A theory similar to the preceding holds for surface integrals. Rather than give a full discussion here, we confine our attention to the counterpart of the last statement in Theorem III; for more details we refer the reader to the books of Brand and Kellogg listed at the end of the chapter.

THEOREM IV Let $\mathbf{u} = L\mathbf{i} + M\mathbf{j} + N\mathbf{k}$ be a vector field whose components have continuous partial derivatives in a spherical domain D. If $\operatorname{div} \mathbf{u} \equiv 0$ in D:

$$\frac{\partial L}{\partial x} + \frac{\partial M}{\partial y} + \frac{\partial N}{\partial z} \equiv 0 \text{ in } D,$$

then a vector field $\mathbf{v} = X\mathbf{i} + Y\mathbf{j} + Z\mathbf{k}$ in D can be found such that

$$\operatorname{curl} \mathbf{v} \equiv \mathbf{u} \text{ in } D;$$

that is,

$$\frac{\partial Z}{\partial y} - \frac{\partial Y}{\partial z} = L, \qquad \frac{\partial X}{\partial z} - \frac{\partial Z}{\partial x} = M, \qquad \frac{\partial Y}{\partial x} - \frac{\partial X}{\partial y} = N \quad (5.100)$$

in D.

Remark. The theorem provides a converse to the theorem of Section 3.6:

$$\operatorname{div} \operatorname{curl} \mathbf{u} \equiv 0,$$

for it asserts that if $\operatorname{div} \mathbf{u} \equiv 0$, then $\mathbf{u} \equiv \operatorname{curl} \mathbf{v}$ for some $\mathbf{v}$. However, while $\operatorname{div} \mathbf{u}$ may be 0 in an arbitrary domain D_1, the theorem provides $\mathbf{v}$ whose curl is $\mathbf{u}$ only in each spherical domain D contained in D_1, that is, no one $\mathbf{v}$ serves for all of D_1. Actually, the proof to follow gives one $\mathbf{v}$ for all of D_1 when D_1 is the interior of a cube or of an ellipsoid or of any "convex" surface. The existence of $\mathbf{v}$ for all of D_1 for more general cases can be established (see, for example, pages 203 ff. in the book by Lamb listed at the end of the chapter).

Vector fields $\mathbf{u}$ satisfying the condition $\operatorname{div} \mathbf{u} \equiv 0$ are termed *solenoidal*. The theorem amounts to the assertion that solenoidal fields are (in suitable domains) fields of the form $\operatorname{curl} \mathbf{v}$, provided that the components of $\mathbf{u}$ have continuous partial derivatives. The field $\mathbf{v}$ is not unique (Problem 5).

Proof of the theorem. Let the spherical domain D have center at P_0; for simplicity we assume P_0 to be the origin $(0, 0, 0)$. If $P_1(x_1, y_1, z_1)$ is an arbitrary point of D, we set

$$X(x_1, y_1, z_1) = \int_0^1 [zM(x, y, z) - yN(x, y, z)] \, dt,$$

$$Y(x_1, y_1, z_1) = \int_0^1 [xN(x, y, z) - zL(x, y, z)] \, dt,$$

$$Z(x_1, y_1, z_1) = \int_0^1 [yL(x, y, z) - xM(x, y, z)] \, dt, \qquad (5.101)$$

where, on the right-hand side, x, y, and z are the following functions of t:

$$x = x_1 t, \qquad y = y_1 t, \qquad z = z_1 t. \qquad (5.102)$$

As t varies from 0 to 1, the point (x, y, z) varies from P_0 to P_1 on the line segment $P_0 P_1$; hence (x, y, z) remains in D. We have now by Leibnitz's Rule and the chain rule (Sections 4.9 and 2.8):

$$\frac{\partial Z}{\partial y_1} = \int_0^1 \left[y \frac{\partial L}{\partial y} \frac{\partial y}{\partial y_1} + \frac{\partial y}{\partial y_1} L - x \frac{\partial M}{\partial y} \frac{\partial y}{\partial y_1} \right] dt$$

$$= \int_0^1 \left[yt \frac{\partial L}{\partial y} + tL - xt \frac{\partial M}{\partial y} \right] dt$$

and, similarly,

$$\frac{\partial Y}{\partial z_1} = \int_0^1 \left[xt \frac{\partial N}{\partial z} - zt \frac{\partial L}{\partial z} - tL \right] dt.$$

Accordingly,

$$\frac{\partial Z}{\partial y_1} - \frac{\partial Y}{\partial z_1} = \int_0^1 \left[2tL - xt \left(\frac{\partial M}{\partial y} + \frac{\partial N}{\partial z} \right) + yt \frac{\partial L}{\partial y} + zt \frac{\partial L}{\partial z} \right] dt.$$

Since $\mathrm{div}\,(L\mathbf{i} + M\mathbf{j} + N\mathbf{k}) = 0$, this can be written as follows:

$$\frac{\partial Z}{\partial y_1} - \frac{\partial Y}{\partial z_1} = \int_0^1 \left[2tL + xt \frac{\partial L}{\partial x} + yt \frac{\partial L}{\partial y} + zt \frac{\partial L}{\partial z} \right] dt.$$

Now

$$t \frac{\partial L}{\partial t} = t \left(\frac{\partial L}{\partial x} \frac{\partial x}{\partial t} + \frac{\partial L}{\partial y} \frac{\partial y}{\partial t} + \frac{\partial L}{\partial z} \frac{\partial z}{\partial t} \right)$$

$$= t \left(x_1 \frac{\partial L}{\partial x} + y_1 \frac{\partial L}{\partial y} + z_1 \frac{\partial L}{\partial z} \right)$$

$$= x \frac{\partial L}{\partial x} + y \frac{\partial L}{\partial y} + z \frac{\partial L}{\partial z}.$$

Accordingly,

$$\frac{\partial Z}{\partial y_1} - \frac{\partial Y}{\partial z_1} = \int_0^1 \left(t^2 \frac{\partial L}{\partial t} + 2tL \right) dt = \int_0^1 \frac{\partial}{\partial t} (t^2 L) \, dt$$

$$= t^2 L \big|_{t=0}^{t=1} = L(x_1, y_1, z_1).$$

This gives the first of (5.100). The other two equations are proved in the same way.   □

The solution $\mathbf{v}$ can be expressed in the compact form:

$$\mathbf{v}(x, y, z) = \int_0^1 t\mathbf{u}(xt, yt, zt) \times (x\mathbf{i} + y\mathbf{j} + z\mathbf{k})\, dt. \tag{5.103}$$

If the vector field $\mathbf{u}$ is *homogeneous of degree n*, that is,

$$\mathbf{u}(xt, yt, zt) = t^n \mathbf{u}(x, y, z)$$

(Problem 11 following Section 2.8), the formula can be simplified further:

$$\mathbf{v} = \int_0^1 t^{n+1} \mathbf{u}(x, y, z) \times (x\mathbf{i} + y\mathbf{j} + z\mathbf{k})\, dt$$

$$= \frac{1}{n+2}(\mathbf{u} \times \mathbf{r}), \qquad \mathbf{r} = x\mathbf{i} + y\mathbf{j} + z\mathbf{k}.$$

For an interesting discussion of this topic the reader is referred to pages 487–489 of Vol. 58 (1951) of the *American Mathematical Monthly*.

PROBLEMS

1. Evaluate by Stokes's theorem:

   a) $\int_C u_T\, ds$, where C is the circle $x^2 + y^2 = 1$, $z = 2$, directed so that y increases for positive x, and $\mathbf{u}$ is the vector $-3y\mathbf{i} + 3x\mathbf{j} + \mathbf{k}$;

   b) $\int_C 2xy^2z\, dx + 2x^2yz\, dy + (x^2y^2 - 2z)\, dz$ around the curve $x = \cos t$, $y = \sin t$, $z = \sin t$, $0 \le t \le 2\pi$, directed with increasing t.

2. By showing that the integrand is an exact differential, evaluate

   a) $\int_{(1,1,2)}^{(3,5,0)} yz\, dx + xz\, dy + xy\, dz$ on any path;

   b) $\int_{(1,0,0)}^{(1,0,2\pi)} \sin yz\, dx + xz \cos yz\, dy + xy \cos yz\, dz$ on the helix $x = \cos t$, $y = \sin t$, $z = t$.

3. Let C be a simple closed *plane* curve in space. Let $\mathbf{n} = a\mathbf{i} + b\mathbf{j} + c\mathbf{k}$ be a unit vector normal to the plane of C and let the direction on C match that of $\mathbf{n}$. Prove that

   $$\tfrac{1}{2} \int_C (bz - cy)\, dx + (cx - az)\, dy + (ay - bx)\, dz$$

   equals the plane area enclosed by C. What does the integral reduce to when C is in the xy-plane?

4. Let $\mathbf{u} = \dfrac{-y}{x^2 + y^2}\mathbf{i} + \dfrac{x}{x^2 + y^2}\mathbf{j} + z\mathbf{k}$ and let D be the interior of the torus obtained by rotating the circle $(x - 2)^2 + z^2 = 1$, $y = 0$ about the z-axis. Show that $\operatorname{curl} \mathbf{u} = \mathbf{0}$ in D but $\int_C u_T\, ds$ is not zero when C is the circle $x^2 + y^2 = 4$, $z = 0$. Determine the possible values of the integral $\int_{(2,0,0)}^{(0,2,0)} u_T\, ds$ on a path in D.

5. **a)** Show that if **v** is one solution of the equation curl $\mathbf{v} = \mathbf{u}$ for given **u** in a simply connected domain D, then all solutions are given by $\mathbf{v} + \operatorname{grad} f$, where f is an arbitrary differentiable scalar in D.

   b) Find all vectors **v** such that curl $\mathbf{v} = \mathbf{u}$ if

   $$\mathbf{u} = (2xyz^2 + xy^3)\mathbf{i} + (x^2y^2 - y^2z^2)\mathbf{j} - (y^3z + 2x^2yz)\mathbf{k}.$$

6. Show that if f and g are scalars having continuous second partial derivatives in a domain D, then

   $$\mathbf{u} = \nabla f \times \nabla g$$

   is solenoidal in D. (It can be shown that every solenoidal vector has such a representation, at least in a suitably restricted domain.)

7. Show that if $\iint_S u_n \, d\sigma = 0$ for every oriented spherical surface S in a domain D and the components of **u** have continuous derivatives in D, then **u** is solenoidal in D. Does the converse hold?

8. Let C and S be as in Stokes's theorem. Prove, under appropriate assumptions:

   a) $\displaystyle\int_C f\, \mathbf{T} \cdot \mathbf{i}\, ds = \iint_S \mathbf{n} \times \nabla f \cdot \mathbf{i}\, d\sigma;$

   [Hint: Apply Stokes's theorem, taking $\mathbf{u} = f\mathbf{i}$. Evaluate curl **u** by (3.28).]

   b) $\displaystyle\int_C f\, \mathbf{T}\, ds = \iint_S \mathbf{n} \times \nabla f\, d\sigma.$

   [Hint: These are vector integrals, as in Section 4.5. Show by (a) that the x-components of both sides are equal and, similarly, that the y- and z-components are equal.]

9. The operator $\mathbf{v} \times \nabla$ is defined formally as $(v_x\mathbf{i} + v_y\mathbf{j} + v_z\mathbf{k}) \times (\nabla_x\mathbf{i} + \nabla_y\mathbf{j} + \nabla_z\mathbf{k})$
   $= (v_y\nabla_z - v_z\nabla_y)\mathbf{i} + \cdots$
   Show that, formally:

   a) $(\mathbf{v} \times \nabla) \cdot \mathbf{u} = \mathbf{v} \cdot \nabla \times \mathbf{u} = \mathbf{v} \cdot \operatorname{curl} \mathbf{u};$

   b) $(\mathbf{v} \times \nabla) \times \mathbf{u} = \nabla_u(\mathbf{v} \cdot \mathbf{u}) - (\nabla \cdot \mathbf{u})\mathbf{v}$, where $\nabla_u(\mathbf{v} \cdot \mathbf{u})$ indicates that **v** is treated as constant: $\nabla_u(\mathbf{v} \cdot \mathbf{u}) = v_x\nabla u_x + v_y\nabla u_y + v_z\nabla u_z.$

10. Let C and S be as in Stokes's theorem. Show with the aid of Problem 9:

    a) $\displaystyle\int_C \mathbf{T} \times \mathbf{u} \cdot \mathbf{i}\, ds = \iint_S (\mathbf{n} \times \nabla) \times \mathbf{u} \cdot \mathbf{i}\, d\sigma;$

    b) $\displaystyle\int_C \mathbf{T} \times \mathbf{u}\, ds = \iint_S (\mathbf{n} \times \nabla) \times \mathbf{u}\, d\sigma.$

*5.14   CHANGE OF VARIABLES IN A MULTIPLE INTEGRAL

The formula for change of variables in a double integral:

$$\iint_{R_{xy}} F(x, y)\, dx\, dy = \iint_{R_{uv}} F[f(u, v), g(u, v)] \left| \frac{\partial(x, y)}{\partial(u, v)} \right| du\, dv, \quad (5.104)$$

is given in Section 4.6. In this section we shall give a proof of this formula under appropriate assumptions. We shall also indicate how widely the formula is applicable and shall explain the more general formula:

$$\delta \int_{R_{xy}} \int F(x, y)\, dx\, dy = \int_{R_{uv}} \int F[f(u, v), g(u, v)] \frac{\partial(x, y)}{\partial(u, v)}\, du\, dv, \quad (5.105)$$

where δ is the "degree" of the mapping of the boundary of R_{uv} into the boundary of R_{xy}.

THEOREM I   The formula

$$\int_{R_{xy}} \int F(x, y)\, dx\, dy = \pm \int_{R_{uv}} \int F[f(u, v), g(u, v)] \frac{\partial(x, y)}{\partial(u, v)}\, du\, dv$$

$$(5.106)$$

is valid under the following assumptions:

a) R_{xy} and R_{uv} are bounded closed regions in the xy- and uv-planes bounded by piecewise smooth simple closed curves C_{xy} and C_{uv}, respectively.

b) The function $F(x, y)$ is defined and has continuous first derivatives in a circular domain D_{xy} containing R_{xy}.

c) The functions $x = f(u, v)$, $y = g(u, v)$ are defined and have continuous second derivatives in a domain D_{uv} including R_{uv}; when (u, v) is in D_{uv}, the point (x, y) is in D_{xy}.

d) When (u, v) is on C_{uv}, the corresponding point (x, y), $x = f(u, v)$, $y = g(u, v)$, is on C_{xy}; as (u, v) traces C_{uv} once in the positive direction, (x, y) traces C_{xy} once in the positive direction [corresponding to the $+$ sign in (5.106)] or negative direction [corresponding to the $-$ sign in (5.106)].

The situation is illustrated in Fig. 5.36. It is to be stressed that the mapping from the uv-plane to the xy-plane is not assumed to be one-to-one except on the boundary, that when (u, v) is in R_{uv} the corresponding point (x, y) need not be in

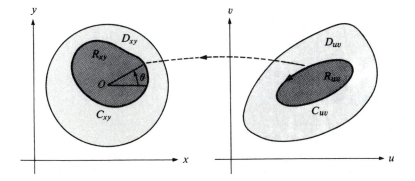

Figure 5.36   Transformation of double integrals.

R_{xy}, and that no assumptions are made about the sign of the Jacobian $\partial(x, y)/\partial(u, v)$.

To prove the theorem, we first remark that one can represent $F(x, y)$ as follows:

$$F(x, y) = \frac{\partial Q}{\partial x}$$

in D_{xy}, where Q has continuous first derivatives. One need only take

$$Q = \int_{x_0}^{x} F(t, y) \, dt,$$

where (x_0, y_0) is the center of D_{xy}.

Green's theorem now gives

$$\int_{R_{xy}} \int F(x, y) \, dx \, dy = \int_{R_{xy}} \int \frac{\partial Q}{\partial x} \, dx \, dy = \oint_{C_{xy}} Q \, dy.$$

The line integral $\int Q \, dy$ can at once be written as a line integral in the uv-plane:

$$\oint_{C_{xy}} Q(x, y) \, dy = \pm \oint_{C_{uv}} Q[f(u, v), g(u, v)] \left(\frac{\partial g}{\partial u} \, du + \frac{\partial g}{\partial v} \, dv \right).$$

The differentials here should all be thought of as expressed in terms of a parameter t: $dy = (dy/dt) \, dt$, $du = (du/dt) \, dt$, $dv = (dv/dt) \, dt$. As t goes from h to k, C_{uv} is to be traced just once in the positive direction, so that C_{xy} is traced just once in the positive or negative direction. The $\pm$ sign corresponds to these two cases.

The line integral on the right is of form

$$\oint_{C_{uv}} P_1 \, du + Q_1 \, dv, \qquad P_1 = Q \frac{\partial g}{\partial u}, \qquad Q_1 = Q \frac{\partial g}{\partial v},$$

where P_1 and Q_1 are defined in D_{uv} and have continuous derivatives there. Hence Green's theorem is applicable:

$$\oint_{C_{uv}} P_1 \, du + Q_1 \, dv = \int_{R_{uv}} \int \left(\frac{\partial Q_1}{\partial u} - \frac{\partial P_1}{\partial v} \right) du \, dv$$

$$= \int_{R_{uv}} \int \left(\frac{\partial Q}{\partial u} \frac{\partial g}{\partial v} + Q \frac{\partial^2 g}{\partial u \, \partial v} - \frac{\partial Q}{\partial v} \frac{\partial g}{\partial u} - Q \frac{\partial^2 g}{\partial v \, \partial u} \right) du \, dv$$

$$= \int_{R_{uv}} \int \left[\left(\frac{\partial Q}{\partial x} \frac{\partial f}{\partial u} + \frac{\partial Q}{\partial y} \frac{\partial g}{\partial u} \right) \frac{\partial g}{\partial v} - \left(\frac{\partial Q}{\partial x} \frac{\partial f}{\partial v} + \frac{\partial Q}{\partial y} \frac{\partial g}{\partial v} \right) \frac{\partial g}{\partial u} \right] du \, dv$$

$$= \int_{R_{uv}} \int \frac{\partial Q}{\partial x} \left(\frac{\partial f}{\partial u} \frac{\partial g}{\partial v} - \frac{\partial f}{\partial v} \frac{\partial g}{\partial u} \right) du \, dv$$

$$= \int_{R_{uv}} \int F[f(u, v), g(u, v)] \frac{\partial(x, y)}{\partial(u, v)} \, du \, dv,$$

since $\partial Q/\partial x = F(x, y)$. One thus concludes

$$\int_{R_{xy}} \int F(x, y) \, dx \, dy = \pm \int_{R_{uv}} \int F[f(u, v), g(u, v)] \frac{\partial(x, y)}{\partial(u, v)} \, du \, dv,$$

as was to be shown.   □

The restrictions made in the statement of the theorem can be relaxed considerably. First of all, it is not necessary to assume that, as (u, v) makes a circuit of C_{uv}, the point (x, y) moves steadily around C_{xy}; the point (x, y) can move ahead, then move backwards, then ahead again, and so on, as long as one complete circuit is achieved. One can make clear what is allowed and, in fact, get even greater generality as follows: Let O be a point inside C_{xy} and let θ be a polar coordinate angle measured relative to O, as in Fig. 5.36. As the point (u, v) goes around C_{uv} in the positive direction, the angle θ for the corresponding point (x, y) can be chosen to vary *continuously*. When (u, v) has made a circuit of C_{uv}, θ will have increased by a certain multiple δ of 2π. This integer δ is known as the *degree* of the mapping of C_{uv} into C_{xy}. With the theorem as stated previously, $\delta = \pm 1$; for a general δ, the theorem must be restated:

THEOREM II   The formula

$$\delta \int_{R_{xy}} \int F(x, y) \, dx \, dy = \int_{R_{uv}} \int F[f(u, v), g(u, v)] \frac{\partial(x, y)}{\partial(u, v)} \, dx \, dv \quad (5.107)$$

is valid under the assumptions (a), (b), (c), of Theorem I and the following assumption:

d′) When (u, v) is on C_{uv}, the corresponding point (x, y) is on C_{xy}; the degree of the mapping of C_{uv} into C_{xy} is δ.

The extension of the previous proof to this case causes no great difficulty. One need only verify that

$$\delta \oint_{C_{xy}} Q(x, y) \, dy = \oint_{C_{uv}} Q\left(\frac{\partial g}{\partial u} \, du + \frac{\partial g}{\partial v} \, dv \right);$$

this relation follows from the fact that, as (u, v) makes one circuit of C_{uv} in the positive direction, the point (x, y) effectively traces the curve C_{xy} δ times. If δ is negative, this means that (x, y) traces C_{xy} $|\delta|$ times in the negative direction. It can happen that $\delta = 0$, in which case both sides of (5.107) are 0.   □

It is of interest to note that a careful treatment of the transformation from rectangular to polar coordinates requires a formula as general as (5.107) (see Problem 4).

One can also consider the case of a multiply connected region R_{xy}, bounded by curves $C_{xy}^{(1)}, C_{xy}^{(2)}, \ldots, C_{xy}^{(k)}$. If the boundary of R_{uv} is a similar set of curves

$C_{uv}^{(1)}, C_{uv}^{(2)}, \ldots, C_{uv}^{(k)}$, then the formula (5.107) continues to hold, provided that each $C_{uv}^{(i)}$ is mapped on the corresponding $C_{xy}^{(i)}$ with the same degree δ. In particular, the points on these boundaries can be in one-to-one correspondence, with δ always equal to 1; one then obtains formula (5.106) again, with the $+$ sign.

The requirement that $F(x, y)$ be defined in a *circular* domain D_{xy} was needed to show that a Q could be found such that $\partial Q/\partial x = F$. A more general domain would still permit this, but it is actually unnecessary to make any restriction whatsoever on the nature of D_{xy}.

The preceding proof clearly used the differentiability conditions on F, f, and g in an unavoidable manner. An entirely different method of proof can be devised that requires merely that F be continuous and that $f(u, v)$ and $g(u, v)$ have continuous first derivatives. Thus (5.107) remains valid if assumptions (b) and (c) are replaced by the following:

b′) The function $F(x, y)$ is defined and continuous in a domain D_{xy} containing R_{xy}.

c′) The functions $x = f(u, v)$, $y = g(u, v)$ are defined and have continuous derivatives in a domain D_{uv} including R_{uv}; when (u, v) is in D_{uv}, (x, y) is in D_{xy}.

For proofs the reader is referred to an article, "The Transformation of Double Integrals," by R. G. Helsel and T. Radó, in *Transactions of the American Mathematical Society*, Vol. 54 (1943), pages 83–102. This article is quite advanced; the author regrets that he cannot cite a more elementary treatment.

When $F(x, y)$ is chosen identically equal to 1 in (5.106), the left-hand side gives the area A of R_{xy}. Thus

$$A = \pm \int \int_{R_{uv}} \frac{\partial(x, y)}{\partial(u, v)}\, du\, dv.$$

If the Jacobian $J = \partial(x, y)/\partial(u, v)$ is always positive or zero, the integral on the right is positive; hence only the $+$ sign can hold. We therefore conclude, on the basis of (a), (b), (c), (d):

If the Jacobian J is always positive, then, as (u, v) traces C_{uv} once in the positive direction, the point (x, y) traces C_{xy} once in the positive direction.

A similar result holds when J is negative, the positive direction on C_{xy} being replaced by the negative direction. Combining the two cases, one obtains the following theorem:

THEOREM III   The formula

$$\int \int_{R_{xy}} F(x, y)\, dx\, dy = \int \int_{R_{uv}} F[f(u, v), g(u, v)] \left| \frac{\partial(x, y)}{\partial(u, v)} \right| du\, dv$$

$$(5.108)$$

holds under the assumptions (a), (b), (c), and (d), and

e) The Jacobian $\partial(x, y)/\partial(u, v)$ does not change sign in R_{uv}.

It can be shown that if $J \neq 0$, the assumptions made for Theorem III imply that the mapping from the uv-plane to the xy-plane is one-to-one. Theorem III is thus essentially the standard form of the transformation theorem, as given in the books of Courant, Goursat, and Franklin listed at the end of the chapter.

As was stated in Section 4.6, a formula analogous to (5.106) holds in three (or more) dimensions. The method of proof used here generalizes in a natural manner, the line integrals being replaced by surface integrals and Green's theorem by the divergence theorem. One can also define a *degree* δ for the correspondence of boundaries, and a formula analogous to (5.107) then holds without any assumption about one-to-one correspondence on boundaries:

$$\delta \iiint_{R_{xyz}} F(x, y, z)\, dx\, dy\, dz = \iiint_{R_{uvw}} F[x(u, v, w),\dots] \frac{\partial(x, y, z)}{\partial(u, v, w)}\, du\, dv\, dw.$$

$$(5.109)$$

The degree δ measures the effective number of times the bounding surface S_{xyz} is traced by (x, y, z) as the point (u, v, w) traces the bounding surface S_{uvw} of R_{uvw}. The surfaces S_{xyz} and S_{uvw} are both considered oriented, the normal being the outer normal for both; in measuring δ, one counts *negatively* the parts of S_{uvw} that are mapped into S_{xyz} with *reversal* of orientation. Thus for the mapping

$$x = -u, \qquad y = -v, \qquad z = -w$$

of the sphere $u^2 + v^2 + w^2 = 1$ onto the sphere $x^2 + y^2 + z^2 = 1$, one has $\delta = -1$. Formula (5.109) can also be extended to the case in which R_{uvw} and R_{xyz} are each bounded by several surfaces, as in two dimensions. In the case in which R_{xyz} is simply connected and is bounded by a single surface S_{xyz} (which has then the structure of a sphere), the degree δ can be computed, by analogy with the planar case, by reference to a "solid angle" (see Problem 6).

PROBLEMS

1. Transform the integrals, using the substitution given:

a) $\int_0^1 \int_0^y (x^2 + y^2)\, dx\, dy$, $u = y$, $v = x$;

b) $\iint_{R_{xy}} (x - y)\, dx\, dy$, where R_{xy} is the region $x^2 + y^2 \leq 1$, and $x = u +$
$(1 - u^2 - v^2)$, $y = v + (1 - u^2 - v^2)$; (Hint: Use as R_{uv} the region $u^2 + v^2 \leq 1$.)

c) $\iint_{R_{xy}} xy\, dx\, dy$, where R_{xy} is the region $x^2 + y^2 \leq 1$ and $x = u^2 - v^2$, $y = 2uv$.
[Hint: Choose R_{uv} as in (b).]

2. Let $x = f(u, v)$, $y = g(u, v)$ be given as in Theorem II and let C_{xy} enclose the origin O. Show that the degree δ can be evaluated by the formula

$$2\pi\delta = \oint_{C_{uv}} \frac{-y\,dx + x\,dy}{x^2 + y^2},$$

where x, y, dx, dy are expressed in terms of u, v:

$$x = f(u, v), \qquad dx = \frac{\partial x}{\partial u}\,du + \frac{\partial x}{\partial v}\,dv.\dots.$$

(Hint: The line integral measures the change in $\theta = \arctan y/x$, as shown in Section 5.6.)

3. Apply the formula of Problem 2 to evaluate the degree for the following mappings of the circle $u^2 + v^2 = 1$ into the circle $x^2 + y^2 = 1$:

a) $x = \dfrac{3u + 4v}{5}, \qquad y = \dfrac{4u - 3v}{5}$

b) $x = u^2 - v^2, \qquad y = 2uv$

c) $x = u^3 - 3uv^2, \qquad y = 3u^2v - v^3$

4. (*Polar coordinates*) Prove the validity of the transformation formula

$$\int_R\int F(x, y)\,dx\,dy = \int_0^{2\pi}\int_0^1 F(r\cos\theta, r\sin\theta)\,r\,dr\,d\theta,$$

where R is the circular region $x^2 + y^2 \le 1$ and $x = r\cos\theta$, $y = r\sin\theta$. [Hint: Consider first the semicircular region R_1: $x^2 + y^2 \le 1$, $y \ge 0$, and the corresponding *rectangle*: $0 \le \theta \le \pi$, $0 \le r \le 1$, in the $r\theta$ plane. Show that the conditions of Theorem II are met, with $\delta = 1$. Note that the correspondence between the boundary of the rectangle and that of the semicircle is not one-to-one. Obtain a similar result for the semicircular region R_2: $x^2 + y^2 \le 1$, $y \le 0$ and add the results for R_1 and R_2.]

5. (*The solid angle*) Let S be a plane surface, oriented in accordance with a unit normal $\mathbf{n}$. The solid angle Ω of S with respect to a point O not in S is defined as

$$\Omega(O, S) = \pm\text{area of projection of } S \text{ on } S_1,$$

where S_1 is the sphere of radius 1 about O and the $+$ or $-$ sign is chosen according to whether $\mathbf{n}$ points away from or toward the side of S on which O lies. This is suggested in Fig. 5.37.

a) Show that if O lies in the plane of S but not in S, then $\Omega(O, S) = 0$.

b) Show that if S is a complete (that is, infinite) plane, then $\Omega(O, S) = \pm 2\pi$.

c) For a general oriented surface S the surface can be thought of as made up of small elements, each of which is approximately planar and has a normal $\mathbf{n}$. Justify the following definition of *element of solid angle* for such a surface element:

$$d\Omega = \frac{\mathbf{r}\cdot\mathbf{n}}{r^3}\,d\sigma,$$

where $\mathbf{r}$ is the vector from O to the element.

d) On the basis of the formula of (c), one obtains as solid angle for a general oriented surface S the integral

$$\Omega(O, S) = \int_S\int \frac{\mathbf{r}\cdot\mathbf{n}}{r^3}\,d\sigma.$$

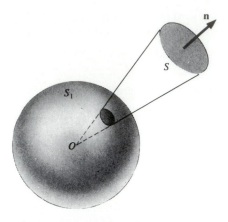

Figure 5.37 Solid angle.

Show that for surfaces in parametric form, if O is the origin,

$$\Omega(O, S) = \int\int_{R_{uv}} \begin{vmatrix} x & y & z \\ \dfrac{\partial x}{\partial u} & \dfrac{\partial y}{\partial u} & \dfrac{\partial z}{\partial u} \\ \dfrac{\partial x}{\partial v} & \dfrac{\partial y}{\partial v} & \dfrac{\partial z}{\partial v} \end{vmatrix} \frac{1}{(x^2 + y^2 + z^2)^{\frac{3}{2}}} \, du \, dv.$$

This formula permits one to define a solid angle for complicated surfaces that intersect themselves.

e) Show that if the normal of S_1 is the outer one, then $\Omega(O, S_1) = 4\pi$.

f) Show that if S forms the boundary of a bounded, closed, simply connected region R, then $\Omega(O, S) = \pm 4\pi$ when O is inside S and $\Omega(O, S) = 0$ when O is outside S.

g) If S is a fixed circular disk and O is variable, show that $-2\pi \le \Omega(O, S) \le 2\pi$ and that $\Omega(O, S)$ jumps by 4π as O crosses S.

6. (*Degree of mapping of one surface into another*) Let S_{uvw} and S_{xyz} be surfaces forming the boundaries of regions R_{uvw} and R_{xyz}, respectively; it is assumed that R_{uvw} and R_{xyz} are bounded and closed and that R_{xyz} is simply connected. Let S_{uvw} and S_{xyz} be oriented by the outer normal. Let s, t be parameters for S_{uvw}:

$$u = u(s, t), \qquad v = v(s, t), \qquad w = w(s, t), \tag{a}$$

the normal having the direction of

$$(u_s \mathbf{i} + v_s \mathbf{j} + w_s \mathbf{k}) \times (u_t \mathbf{i} + v_t \mathbf{j} + w_t \mathbf{k}).$$

Let

$$x = x(u, v, w), \qquad y = y(u, v, w), \qquad z = z(u, v, w) \tag{b}$$

be functions defined and having continuous derivatives in a domain containing S_{uvw}, and let these equations define a mapping of S_{uvw} into S_{xyz}. The degree δ of this mapping is defined as $1/4\pi$ times the solid angle $\Omega(O, S)$ of the image S of S_{uvw} with respect to a point O interior to S_{xyz}. If O is the origin, the degree is hence given by the integral (Kronecker integral)

$$\delta = \frac{1}{4\pi} \int\int_{R_{st}} \begin{vmatrix} x & y & z \\ \dfrac{\partial x}{\partial s} & \dfrac{\partial y}{\partial s} & \dfrac{\partial z}{\partial s} \\ \dfrac{\partial x}{\partial t} & \dfrac{\partial y}{\partial t} & \dfrac{\partial z}{\partial t} \end{vmatrix} \frac{1}{\left(x^2 + y^2 + z^2\right)^{\frac{3}{2}}} \, ds \, dt,$$

where x, y, z are expressed in terms of s, t by (a) and (b). It can be shown that δ, as thus defined, is independent of the choice of the interior point O, that δ is a positive or negative integer or zero, and that δ does measure the effective number of times that S_{xyz} is covered.

Let S_{uvw} be the sphere $u = \sin s \cos t$, $v = \sin s \sin t$, $w = \cos s$, $0 \le s \le \pi$, $0 \le t \le 2\pi$. Let S_{xyz} be the sphere $x^2 + y^2 + z^2 = 1$. Evaluate the degree for the following mappings of S_{uvv} into S_{xyz}:

a) $x = v$, $y = -w$, $z = u$

b) $x = u^2 - v^2$, $y = 2uv$, $z = w\sqrt{2 - w^2}$

*5.15   PHYSICAL APPLICATIONS

The following is a brief discussion of some of the important applications of the divergence and curl and of line and surface integrals.

a) *Dynamics.* If **F** is a force field, then, as shown previously, the work done by **F** on an arbitrary path C is

$$W = \int_C F_T \, ds. \tag{5.110}$$

In general, this will be dependent on the path. If, however, **F** is the gradient of a scalar, the work done will be independent of the path. If this holds, the scalar will be denoted by $-U$, so that

$$\mathbf{F} = -\operatorname{grad} U, \qquad U = U(x, y, z), \tag{5.111}$$

and **F** is said to be derived from the potential U; U is also termed the potential energy of the force field. U is determined uniquely up to an additive constant:

$$U = -\int_{(x_1, y_1, z_1)}^{(x, y, z)} F_T \, ds + \text{const} \tag{5.112}$$

as in Section 5.6. In practice the constant is often chosen so that U approaches 0 as $x^2 + y^2 + z^2$ becomes infinite.

The work done on a particle moving from A to B is then expressed in terms of U as follows:

$$W = \int_A^B F_T \, ds = U(A) - U(B); \tag{5.113}$$

that is, the *work done equals the loss in potential energy.*

The theorem *work done equals gain in kinetic energy*, proved in Section 5.4, holds for a general force field. Combining the two results, one concludes:

gain in kinetic energy = loss in potential energy

or

$$\text{(gain in kinetic energy)} + \text{(gain in potential energy)} = 0,$$

since the gain in potential energy equals the negative of the loss. If one now defines the *total energy* of the particle to be E, where

$$E = \text{(kinetic energy)} + \text{(potential energy)} = \frac{mv^2}{2} + U, \qquad (5.114)$$

then one concludes that for an arbitrary motion of the particle under the given force,

$$E = \text{const}; \qquad (5.115)$$

that is, the total energy is conserved. This is the law of *conservation of energy* for a particle. This was established here under the assumption that $\mathbf{F}$ was a gradient vector; it can be shown that such a conservation law can hold only when $\mathbf{F}$ is a gradient vector. For this reason, force fields that are gradients are termed *conservative* force fields.

b) *Fluid dynamics.* If $\mathbf{u}$ is the velocity vector and ρ is the density for a fluid motion, then, as in Section 5.11, the *equation of continuity*

$$\frac{\partial \rho}{\partial t} + \text{div} \, (\rho \mathbf{u}) = 0 \qquad (5.116)$$

holds. This can also be written, by virtue of an identity of Section 3.6, as follows:

$$\frac{\partial \rho}{\partial t} + \text{grad} \, \rho \cdot \mathbf{u} + \rho \, \text{div} \, \mathbf{u} = 0.$$

The first two terms are the Stokes total derivative of ρ:

$$\frac{D\rho}{Dt} = \frac{\partial \rho}{\partial t} + \frac{\partial \rho}{\partial x} \frac{dx}{dt} + \frac{\partial \rho}{\partial y} \frac{dy}{dt} + \frac{\partial \rho}{\partial z} \frac{dz}{dt} = \frac{\partial \rho}{\partial t} + \text{grad} \, \rho \cdot \mathbf{u},$$

and describe the rate of change of ρ as one stays with a particular particle of fluid in the motion. For an *incompressible* fluid, $D\rho/Dt = 0$, so that the equation of continuity becomes

$$\text{div} \, \mathbf{u} = 0, \qquad (5.117)$$

and $\mathbf{u}$ is solenoidal.

Another interpretation of $\text{div} \, \mathbf{u}$ and a new proof of the continuity equation are given in the problems that follow this section.

The integral $\int_C u_T \, ds$ about a closed curve C has already been introduced as the *circulation* of the velocity field. If this is zero for every closed path C, then by Theorem III of Section 5.13,

$$\text{curl} \, \mathbf{u} = \mathbf{0}. \qquad (5.118)$$

If (5.118) holds, the flow is called *irrotational*. This implies, by Theorem III of Section 5.13, that the circulation is zero on every closed path, provided that attention is restricted to a simply connected domain D. If this last assumption is made, then $\mathbf{u} = \text{grad} \, \phi$ for some scalar ϕ, termed the *velocity potential*.

If the flow is both irrotational and incompressible, then ϕ must satisfy the equation

$$\text{div grad } \phi = 0;$$

that is,

$$\frac{\partial^2 \phi}{\partial x^2} + \frac{\partial^2 \phi}{\partial y^2} + \frac{\partial^2 \phi}{\partial z^2} = 0, \tag{5.119}$$

and ϕ is *harmonic* in D.

c) *Electromagnetism.* An electromagnetic field is described, in accordance with Maxwell's theory, by two vector fields **E** and **H**, where **E** is the electric force and **H** is the magnetic field strength. Both **E** and **H** in general vary with time t. In the absence of conductors, **E** and **H** satisfy the *Maxwell equations*:

$$\begin{array}{ll} \textbf{a) div } \mathbf{E} = 4\pi\rho, & \textbf{b) div } \mathbf{H} = 0, \\[2mm] \textbf{c) curl } \mathbf{E} = -\dfrac{1}{c}\dfrac{\partial \mathbf{H}}{\partial t}, & \textbf{d) curl } \mathbf{H} = \dfrac{1}{c}\dfrac{\partial \mathbf{E}}{\partial t}, \end{array} \tag{5.120}$$

where ρ is the charge density and c is a universal constant.

We now apply (a) ,..., (d) of (5.120) and our rules of vector calculus to show that **E** and **H** are expressible in terms of suitable potentials, which in turn satisfy certain partial differential equations. We assume that all derivatives appearing here are continuous, so that the order of differentiation can be changed and that the discussion is confined to a simply connected domain to which Theorem IV of Section 5.13 applies, so that each irrotational field is a gradient and each solenoidal field is a curl.

From (b) of (5.120), we can choose a vector field **A** (the *vector potential*) such that

$$\mathbf{H} = \text{curl } \mathbf{A}. \tag{5.121}$$

Here both **H** and **A** depend on x,y,z and t, as do all functions appearing in this analysis. As in Problem 5 following Section 5.13, **A** is not unique; to any choice of **A** one can add grad ϕ, where ϕ is an arbitrary function of x,y,z and t. From (c) and (5.121) we deduce that

$$\text{curl}\left(\mathbf{E} + \frac{1}{c}\frac{\partial \mathbf{A}}{\partial t}\right) = \mathbf{0} \tag{5.122}$$

and hence that Φ (the *scalar potential*) can be chosen so that

$$\mathbf{E} + \frac{1}{c}\frac{\partial \mathbf{A}}{\partial t} = -\nabla\Phi. \tag{5.123}$$

If we take the divergence of both sides and use (a), we obtain

$$\nabla^2\Phi + \frac{1}{c}\frac{\partial}{\partial t}(\nabla \cdot \mathbf{A}) = -4\pi\rho. \tag{5.124}$$

From (d) and (5.121) we obtain:

$$\text{curl curl } \mathbf{A} - \frac{1}{c}\frac{\partial \mathbf{E}}{\partial t} = \mathbf{0}. \tag{5.125}$$

From (5.123), by differentiation,

$$\frac{1}{c}\frac{\partial \mathbf{E}}{\partial t} + \frac{1}{c^2}\frac{\partial^2 \mathbf{A}}{\partial t} + \frac{1}{c}\nabla\frac{\partial \Phi}{\partial t} = \mathbf{0}. \tag{5.126}$$

Addition of the last two equations and application of the identity (3.40) of Section 3.6 for curl curl $\mathbf{A}$ gives

$$\nabla^2 \mathbf{A} - \frac{1}{c^2}\frac{\partial^2 \mathbf{A}}{\partial t} - \nabla(\nabla \cdot \mathbf{A} + \frac{1}{c}\frac{\partial \Phi}{\partial t}) = \mathbf{0}. \tag{5.127}$$

We now use the freedom in the choice of $\mathbf{A}$ to achieve the *Lorentz condition*:

$$\nabla \cdot \mathbf{A} + \frac{1}{c}\frac{\partial \Phi}{\partial t} \equiv 0. \tag{5.128}$$

We return to this question below. Now from (5.124), (5.127) and (5.128) we obtain the desired partial differential equations for the scalar and vector potentials:

$$\nabla^2 \Phi - \frac{1}{c^2}\frac{\partial^2 \Phi}{\partial t^2} = -4\pi\rho, \tag{5.129}$$

$$\nabla^2 \mathbf{A} - \frac{1}{c^2}\frac{\partial^2 \mathbf{A}}{\partial t^2} = \mathbf{0}. \tag{5.130}$$

In each region free of charge ($\rho = 0$), $u = \Phi(x,y,z,t)$ satisfies the *wave equation*

$$\nabla^2 u - \frac{1}{c^2}\frac{\partial^2 u}{\partial t^2} = 0. \tag{5.131}$$

Each component of $\mathbf{A}$ also satisfies the wave equation and hence from (5.123) so does each component of the electric force field in each region free of charge.

We now consider the Lorenz condition (5.128). We can write $\mathbf{A} = \mathbf{A}_0 + \text{grad }\phi$, where ϕ is at our disposal. From $\mathbf{A}_0$ and (5.122) we now can choose a scalar potential Φ_0 such that

$$\mathbf{E} + \frac{1}{c}\frac{\partial \mathbf{A}_0}{\partial t} = -\nabla\Phi_0. \tag{5.132}$$

From (5.123) and (5.132),

$$\nabla\Phi - \nabla\Phi_0 = \frac{1}{c}\frac{\partial \mathbf{A}_0}{\partial t} - \frac{1}{c}\frac{\partial \mathbf{A}}{\partial t} = -\frac{1}{c}\text{grad}\frac{\partial \phi}{\partial t}.$$

Hence we satisfy (5.123) by taking

$$\Phi = \Phi_0 - \frac{1}{c}\frac{\partial \phi}{\partial t}.$$

Substitution of this expression and that for **A** in (5.128) and rearrangement gives

$$\nabla^2 \phi - \frac{1}{c^2}\frac{\partial^2 \phi}{\partial t^2} = -\nabla \cdot \mathbf{A}_0 - \frac{1}{c}\frac{\partial \Phi_0}{\partial t}. \tag{5.133}$$

Thus to achieve (5.128), we must choose ϕ to satisfy this "non-homogenous wave equation." This equation is studied in Chapter 10, as an equation for forced vibrations of a continuous medium; the general theory shows that solutions exist. Accordingly the Lorentz condition (5.128) can be satisfied.

In the electrostatic case, **H** is taken to be **0** and **E** and Φ are independent of t. Eqs. (5.123) and (5.124) give

$$\mathbf{E} = -\nabla \Phi, \tag{5.134}$$

where Φ satisfies the *Poisson equation*

$$\nabla^2 \Phi = -4\pi\rho. \tag{5.135}$$

Φ is termed the *electrostatic potential.* In any domain free of charge, Φ is therefore a *harmonic* function.

The function Φ can be computed by Coulomb's law for given charge distributions. Thus for a point charge e at the origin,

$$\Phi = \frac{e}{r} + \text{const}, \quad r = \sqrt{x^2 + y^2 + z^2}, \tag{5.136}$$

and Φ is obtained by simple addition for a sum of point charges. If the charge is distributed along a wire C and ρ_s is the density (charge per unit length), then

$$\Phi(x_1,y_1,z_1) = \int_C \frac{\rho_s \, ds}{r_1} + \text{const}, \tag{5.137}$$

where $r_1 = \sqrt{(x-x_1)^2 + (y-y_1)^2 + (z-z_1)^2}$. If the charge is spread out over a surface S, then Φ is given by the surface integral

$$\Phi(x_1,y_1,z_1) = \int_S \int \frac{\rho_a}{r_1} \, d\sigma + \text{const}, \tag{5.138}$$

where ρ_a is the charge density (charge per unit area) (see Section 5.18).

d) *Heat conduction.* Let $T(x, y, z, t)$ be the temperature at the point (x, y, z) of a body at time t. If heat is being conducted in the body, the flow of heat can be represented by a vector $\mathbf{u}$ such that the flux integral

$$\int_S \int u_n \, d\sigma$$

for each oriented surface S represents the number of calories crossing S in the direction of the given normal per unit of time. The simplest law of thermal conduction postulates that

$$\mathbf{u} = -k \operatorname{grad} T, \tag{5.139}$$

with $k > 0$; k is usually treated as a constant. Equation (5.139) implies that heat flows in the direction of decreasing temperature and the rate of flow is proportional to the temperature gradient: $|\operatorname{grad} T|$.

If S is a closed surface, forming the boundary of a region R in the body, then

$$\int_S \int u_n \, d\sigma = \int \int_R \int \operatorname{div} \mathbf{u} \, dx \, dy \, dz, \tag{5.140}$$

by the divergence theorem. Hence the total amount of heat *entering* R is

$$-\int_S \int u_n \, d\sigma = \int \int_R \int k \operatorname{div} \operatorname{grad} T \, dx \, dy \, dz. \tag{5.141}$$

On the other hand the rate at which heat is being absorbed per unit mass can also be measured by $c\dfrac{\partial T}{\partial t}$, where c is the specific heat; the rate at which R is receiving heat is then

$$\int \int_R \int c\rho \frac{\partial T}{\partial t} \, dx \, dy \, dz, \tag{5.142}$$

where ρ is the density. Equating the two expressions, one finds

$$\int \int_R \int \left(c\rho \frac{\partial T}{\partial t} - k \operatorname{div} \operatorname{grad} T \right) dx \, dy \, dz = 0. \tag{5.143}$$

Since this must hold for an *arbitrary* solid region R, the function integrated (if continuous) must be zero everywhere. Hence

$$c\rho \frac{\partial T}{\partial t} - k \operatorname{div} \operatorname{grad} T = 0. \tag{5.144}$$

This is the fundamental equation for heat conduction. If the body is in temperature equilibrium, $\partial T / \partial t = 0$, and one concludes that

$$\operatorname{div} \operatorname{grad} T = 0; \tag{5.145}$$

that is, T is *harmonic*.

e) *Thermodynamics.* Let a certain volume V of a gas be given, enclosed in a container, subject to a pressure p. It is known from experiment that for each kind of gas there is an "equation of state"

$$f(p, V, T) = 0, \tag{5.146}$$

connecting pressure, volume, and temperature T. For an "ideal gas" (low density and high temperature), Eq. (5.146) takes the special form:

$$pV = RT, \tag{5.147}$$

where R is constant (the same for all gases, if one mole of gas is used).

With each gas is also associated a scalar U, the total internal energy; this is analogous to the kinetic energy plus potential energy considered above. For each gas U is given as a definite function of the "state," hence of p and V:

$$U = U(p, V). \tag{5.148}$$

The particular equation (5.148) depends on the gas considered. For an ideal gas, (5.148) takes the form:

$$U = c_V \frac{pV}{R} = c_V T, \tag{5.149}$$

where c_V is constant, the *specific heat* at constant volume.

A particular *process* gone through by a gas is a succession of changes in the state, so that p and V, and hence T and U, become functions of time t. The state at time t can be represented by a point (p, V) on a pV diagram (Fig. 5.38) and the process by a curve C, with t as parameter. During such a process it is possible to measure the *amount of heat*, Q, received by the gas. The first law of thermodynamics is equivalent to the statement that

$$\frac{dQ}{dt} = \frac{dU}{dt} + p\frac{dV}{dt}, \tag{5.150}$$

where $Q(t)$ is the amount of heat received up to time t.

Hence the amount of heat introduced in a particular process is given by an integral:

$$Q(h) - Q(0) = \int_0^h \left(\frac{dU}{dt} + p\frac{dV}{dt} \right) dt. \tag{5.151}$$

Now by (5.148), dU is expressible in terms of dp and dV:

$$dU = \left(\frac{\partial U}{\partial p} \right)_V dp + \left(\frac{\partial U}{\partial V} \right)_p dV. \tag{5.152}$$

Hence (5.151) can be written as a line integral:

$$\begin{aligned} Q(h) - Q(0) &= \int_0^h \left\{ \left(\frac{\partial U}{\partial p} \right)_V \frac{dp}{dt} + \left[\left(\frac{\partial U}{\partial V} \right)_p + p \right] \frac{dV}{dt} \right\} dt \\ &= \int_C \left(\frac{\partial U}{\partial p} \right)_V dp + \left[\left(\frac{\partial U}{\partial V} \right)_p + p \right] dV, \end{aligned} \tag{5.153}$$

or, with (5.152) understood, simply thus:

$$Q(h) - Q(0) = \int_C dU + p\, dV. \tag{5.154}$$

For (5.153) or (5.154) to be independent of the path C, one must have

$$\frac{\partial}{\partial V}\left(\frac{\partial U}{\partial p} \right) = \frac{\partial}{\partial p}\left(\frac{\partial U}{\partial V} + p \right);$$

Figure 5.38   Thermodynamic processes.

that is,

$$\frac{\partial^2 U}{\partial V \, \partial p} = \frac{\partial^2 U}{\partial p \, \partial V} + 1.$$

Since this is impossible (when U has continuous second derivatives), the heat introduced is dependent on the path. For a simple closed path C_1, as in Fig. 5.38, the heat introduced is

$$\oint_{C_1} dU + p \, dV.$$

Since U is a given function of p and V, $\oint dU = 0$; thus the heat introduced reduces to

$$\int_{C_1} p \, dV.$$

This integral is precisely the area integral $\int y \, dx$ considered in Section 5.5, with p replacing y and V replacing x. Hence for such a counterclockwise cycle the heat introduced is *negative*; there is a heat *loss*, equal to the area enclosed (a unit of area corresponding to a unit of *energy*). The integral $\int p \, dV$ can also be interpreted as the mechanical *work* done by the gas on the surrounding medium or as the negative of the work done on the gas by the surrounding medium. For the process of the curve C_1 considered above, the heat loss equals the work done on the gas; the total energy remains unchanged, in agreement with the law of conservation of energy expressed by the thermodynamic law (5.150).

While the integral $\int dU + p \, dV$ is not independent of path, it is an experimental law that the integral

$$\int \frac{1}{T} dU + \frac{p}{T} dV$$

is independent of path. One can accordingly introduce a scalar S whose differential is the expression being integrated:

$$dS = \frac{1}{T} dU + \frac{p}{T} dV$$

$$= \frac{1}{T} \frac{\partial U}{\partial p} dp + \frac{1}{T} \left(\frac{\partial U}{\partial V} + p \right) dV; \qquad (5.155)$$

S is termed the *entropy*. In the first equation here, one can consider U and V as independent variables; in the second, p and V can be considered independent. Thus the first equation gives

$$\left(\frac{\partial S}{\partial U}\right)_V = \frac{1}{T}, \qquad \left(\frac{\partial S}{\partial V}\right)_U = \frac{p}{T}$$

and hence

$$\frac{\partial^2 S}{\partial V \, \partial U} = \frac{\partial}{\partial V}\left(\frac{1}{T}\right) = \frac{\partial}{\partial U}\left(\frac{p}{T}\right) = \frac{\partial^2 S}{\partial U \, \partial V}.$$

Accordingly, one finds

$$T\frac{\partial p}{\partial U} - p\frac{\partial T}{\partial U} + \frac{\partial T}{\partial V} = 0 \qquad (U, V \text{ indep.}). \tag{5.156}$$

A similar relation is obtainable from the second equation (5.155), and other are obtainable by varying the choice of independent variables. All these equations are simply different forms of the condition $\partial P/\partial y = \partial Q/\partial x$ for independence of path (see Problem 8 following Section 2.11).

The second law of thermodynamics states first the existence of the entropy S and second the fact that for any closed system,

$$\frac{dS}{dt} \geq 0;$$

that is, the entropy can never decrease.

PROBLEMS

1. **a)** A particle of mass m moves on a straight line, the x-axis, subject to a force $-k^2 x$. Find the potential energy and determine the law of conservation of energy for this motion. Does the law hold if a resistance $-c\dfrac{dx}{dt}$ is added?

   b) A particle of mass m moves in the xy-plane subject to a force $\mathbf{F} = -a^2 x\mathbf{i} - b^2 y\mathbf{j}$. Find the potential energy and determine the law of conservation of energy for this motion.

2. Let D be a simply connected domain in the xy-plane and let $\mathbf{w} = u\mathbf{i} - v\mathbf{j}$ be the velocity vector of an irrotational incompressible flow in D. (This is the same as an irrotational incompressible flow in a 3-dimensional domain whose projection is D and for which the z-component of velocity is 0 whereas the x- and y-components of velocity are independent of z.) Show that the following properties hold:

   a) u and v satisfy the Cauchy-Riemann equations:

$$\frac{\partial u}{\partial x} = \frac{\partial v}{\partial y}, \qquad \frac{\partial u}{\partial y} = -\frac{\partial v}{\partial x} \quad \text{in } D;$$

   b) u and v are harmonic in D;

   c) $\int u \, dx - v \, dy$ and $\int v \, dx + u \, dy$ are independent of the path in D;

d) there is a vector $\mathbf{F} = \phi\mathbf{i} - \psi\mathbf{j}$ in D such that

$$\frac{\partial \phi}{\partial x} = u = \frac{\partial \psi}{\partial y}, \qquad \frac{\partial \phi}{\partial y} = -v = -\frac{\partial \psi}{\partial x};$$

e) $\operatorname{div} \mathbf{F} = 0$ and $\operatorname{curl} \mathbf{F} = \mathbf{0}$ in D;

f) ϕ and ψ are harmonic in D;

g) $\operatorname{grad} \phi = \mathbf{w}$, ψ is constant on each stream line.
   The function ϕ is the *velocity potential*; ψ is the *stream function*.

3. Let a wire occupying the line segment from $(0, -c)$ to $(0, c)$ in the xy-plane have a constant charge density equal to ρ. Show that the electrostatic potential due to this wire at a point (x_1, y_1) of the xy-plane is given by

$$\Phi = \rho \log \frac{\sqrt{x_1^2 + (c - y_1)^2} + c - y_1}{\sqrt{x_1^2 + (c + y_1)^2} - c - y_1} + k,$$

where k is an arbitrary constant. Show that if k is chosen so that $\Phi(1,0) = 0$, then, as c becomes infinite, Φ approaches the limiting value $-2\rho \log |x_1|$. This is the potential of an infinite wire with uniform charge.

4. a) Write out in detail the steps leading from the Maxwell equations (5.120) to the partial differential equations (5.129), (5.130).

   b) Show from Eqs. (5.120),..., (5.130) that each component of $\mathbf{E}$ satisfies the wave equation in each domain free of charge.

5. Find the temperature distribution in a solid whose boundaries are two parallel planes, d units apart, kept at temperatures T_1, T_2, respectively. (Hint: Take the boundaries to be the planes $x = 0$, $x = d$ and note that, by symmetry, T must be independent of y and z.)

6. Show that on the basis of the laws of thermodynamics, the line integral

$$\int S\, dT + p\, dV$$

is independent of the path in the TV plane. The integrand is minus the differential of the *free energy* F.

7. Consider a fluid motion in space. A particle occupying position (x_0, y_0, z_0) at time $t = 0$ occupies position (x, y, z) at time t. Thus x, y, z become functions of x_0, y_0, z_0, t:

$$x = \phi(x_0, y_0, z_0, t),$$
$$y = \psi(x_0, y_0, z_0, t), \qquad (*)$$
$$z = \chi(x_0, y_0, z_0, t).$$

Let the ∇ symbol be used as follows:

$$\nabla = \frac{\partial}{\partial x_0}\mathbf{i} + \frac{\partial}{\partial y_0}\mathbf{j} + \frac{\partial}{\partial z_0}\mathbf{k}$$

and let J denote the Jacobian

$$\frac{\partial(x, y, z)}{\partial(x_0, y_0, z_0)}.$$

Let **v** denote the velocity vector:

$$\mathbf{v} = \frac{\partial x}{\partial t}\mathbf{i} + \frac{\partial y}{\partial t}\mathbf{j} + \frac{\partial z}{\partial t}\mathbf{k} = \frac{\partial \phi}{\partial t}\mathbf{i} + \frac{\partial \psi}{\partial t}\mathbf{j} + \frac{\partial \chi}{\partial t}\mathbf{k}.$$

a) Show that $J = \nabla x \cdot \nabla y \times \nabla z$.

b) Show that

$$\frac{\partial x_0}{\partial x} = \mathbf{i} \cdot \frac{\nabla y \times \nabla z}{J}, \qquad \frac{\partial y_0}{\partial x} = \mathbf{j} \cdot \frac{\nabla y \times \nabla z}{J}, \qquad \frac{\partial z_0}{\partial x} = \mathbf{k} \cdot \frac{\nabla y \times \nabla z}{J},$$

and obtain similar expressions for

$$\frac{\partial x_0}{\partial y}, \quad \frac{\partial y_0}{\partial y}, \quad \frac{\partial z_0}{\partial y}, \quad \frac{\partial x_0}{\partial z}, \quad \frac{\partial y_0}{\partial z}, \quad \frac{\partial z_0}{\partial z}.$$

(Hint: See Problem 5 following Section 2.12).

c) Show that

$$\frac{\partial J}{\partial t} = \nabla v_x \cdot \nabla y \times \nabla z + \nabla v_y \cdot \nabla z \times \nabla x + \nabla v_z \cdot \nabla x \times \nabla y.$$

d) Show that

$$\operatorname{div} \mathbf{v} = \frac{1}{J}\frac{\partial J}{\partial t}.$$

[Hint: By the chain rule,

$$\frac{\partial v_x}{\partial x} = \frac{\partial v_x}{\partial x_0}\frac{\partial x_0}{\partial x} + \frac{\partial v_x}{\partial y_0}\frac{\partial y_0}{\partial x} + \frac{\partial v_x}{\partial z_0}\frac{\partial z_0}{\partial x}.$$

Use the result of (b) to show that

$$\frac{\partial v_x}{\partial x} = \frac{\nabla v_x \cdot \nabla y \times \nabla z}{J}$$

and obtain similar expressions for $\partial v_y/\partial y$, $\partial v_z/\partial z$. Add the results and use the result of (c).]

Remark. The Jacobian J can be interpreted as the ratio of the volume occupied by a small piece of the fluid at time t to the volume occupied by this piece when $t = 0$, as in Fig. 5.39. Hence by (d) the divergence of the velocity vector can be interpreted as measuring the percentage of change in this ratio per unit time or simply as the *rate of change of volume per unit volume* of the moving piece of fluid.

Figure 5.39   Problem 6.

8. Consider a piece of the fluid of Problem 6 (not necessarily a "small" piece) occupying a region $R = R(t)$ at time t and a region $R_0 = R(0)$ when $t = 0$. Let $F(x, y, z, t)$ be a function differentiable throughout the part of space concerned.

a) Show that

$$\iiint\limits_{R(t)} F(x, y, z, t)\, dx\, dy\, dz = \iiint\limits_{R_0} F[\phi(x_0, y_0, z_0, t), \dots] J\, dx_0\, dy_0\, dz_0.$$

[Hint: Use Eq. (5.109), noting that the degree δ must be 1 here.]

b) Show that

$$\frac{d}{dt} \iiint\limits_{R(t)} F(x, y, z, t)\, dx\, dy\, dz = \iiint\limits_{R(t)} \left[\frac{\partial F}{\partial t} + \text{div}\,(Fv)\right] dx\, dy\, dz.$$

[Hint: Use (a) and apply Leibnitz's Rule of Section 4.9 to differentiate the right-hand side. Use the result of part (d) of Problem 7 to simplify the result. Then return to the original variables by (a) again.]

9. Let $\rho = \rho(x, y, z, t)$ be the density of the fluid motion of Problems 7 and 8. The integral

$$\iiint\limits_{R(t)} \rho\, dx\, dy\, dz$$

represents the mass of the fluid filling $R(t)$. The conservation of mass implies that this integral is constant:

$$\frac{d}{dt} \iiint\limits_{R(t)} \rho\, dx\, dy\, dz = 0.$$

Use this result and that of Problem 8(b) to establish the *continuity equation*

$$\frac{\partial \rho}{\partial t} + \text{div}\,(\rho v) = 0.$$

[Hint: Cf. the derivation of the heat equation (5.144).]

*5.16 POTENTIAL THEORY IN THE PLANE

In this section and the following section we introduce potential theory in space of two or three dimensions. The 3-dimensional theory reduces to the 2-dimensional one if one considers mass or charge distributions in xyz-space that do not depend on z and modifies the potential. The process is illustrated in Problem 3 following Section 5.15. The Newtonian potential m/r in space turns into a logarithmic potential $m \log 1/r$ in the plane.

We consider a force field F in the xy-plane given by the equation

$$F = \frac{km}{r^2} r, \tag{5.157}$$

where $r = \overrightarrow{PQ}$, $r = |r|$, Q is fixed at (ξ, η), P is (x, y), a variable point, k and m are constants, $k > 0$. Thus F is directed from P to Q and hence is an attractive force, if $m > 0$. Its magnitude is

$$|F| = \frac{km}{r^2} r = \frac{km}{r}. \tag{5.158}$$

Thus the force varies as the inverse *first* power of the distance.

The force $\mathbf{F}$ is the gradient of a scalar $U = U(x, y)$:

$$\mathbf{F} = \text{grad } U, \qquad U = km \log \frac{1}{r} \tag{5.159}$$

For $r = [(x - \xi)^2 + (y - \eta)^2]^{1/2}$ and hence

$$\frac{\partial U}{\partial x} = -\frac{km}{r} \frac{\partial r}{\partial x} = \frac{km}{r^2}(\xi - x),$$

$$\frac{\partial U}{\partial y} = -\frac{km}{r} \frac{\partial r}{\partial y} = \frac{km}{r^2}(\eta - y),$$

so that, as asserted,

$$\text{grad } U = \frac{km}{r^2}[(\xi - x)\mathbf{i} + (\eta - y)\mathbf{j}] = \frac{km}{r^2}\mathbf{r} = \mathbf{F}.$$

One terms $U = U(x, y)$ the *logarithmic potential* due to the point mass m at Q and unit mass at P.

By proper choice of units, one can assume that $k = 1$, and we shall do so. Thus

$$U = m \log \frac{1}{r}, \qquad \mathbf{F} = \text{grad } U = \frac{m}{r^2}\mathbf{r}. \tag{5.160}$$

We could add an arbitrary constant to U and still have grad $U = \mathbf{F}$, but we shall not do so.

For a finite set of particles of masses m_i at Q_i $(i = 1, \ldots, n)$, one has an associated force field

$$\mathbf{F} = \sum_{i=1}^{n} \frac{m_i}{r_i^2}\mathbf{r}_i, \qquad \mathbf{r}_i = \overrightarrow{PQ_i}, \tag{5.161}$$

and as previously, $\mathbf{F} = \text{grad } U$,

$$U = \sum_{i=1}^{n} m_i \log \frac{1}{r_i}; \tag{5.162}$$

here U is the logarithmic potential due to the finite set of masses.

By passage to the limit, one obtains logarithmic potentials of other mass distributions. For mass spread over a region E with density $\mu(\xi, \eta)$ the potential is

$$U(x, y) = \int_E \int \mu \log \frac{1}{r} d\xi \, d\eta,$$

$$r = \sqrt{(\xi - x)^2 + (\eta - y)^2}. \tag{5.163}$$

The corresponding force is

$$\mathbf{F} = \text{grad } U = \int_E \int \nabla_{xy}\left[\mu \log \frac{1}{r}\right] d\xi \, d\eta$$

$$= \int_E \int \frac{\mu}{r^2}[(\xi - x)\mathbf{i} + (\eta - y)\mathbf{j}] \, d\xi \, d\eta. \tag{5.164}$$

Here $\nabla_{xy} = (\partial/\partial x)\mathbf{i} + (\partial/\partial y)\mathbf{j}$. The vector integral is interpreted, as in Section 4.5, as

$$\int_E\int \frac{\mu}{r^2}(\xi - x)\, d\xi\, d\eta\, \mathbf{i} + \int_E\int \frac{\mu}{r^2}(\eta - y)\, d\xi\, d\eta\, \mathbf{j}. \qquad (5.165)$$

In (5.163) we assume that the density $\mu(\xi, \eta)$ is continuous in E and E is a bounded closed region, as is usual for double integrals. If $P: (x, y)$ is outside E, then $1/r$ is continuous in (ξ, η) on E, so that $U(x, y)$ is well defined. If P is inside E or on the boundary of E, then the integral in (5.163) is improper. However, as in Section 4.8, the integral does exist. For example, if $\mu(\xi, \eta)$ does not change sign, then as in Section 4.8, one is led to consider a limit process at (x, y). We use polar coordinates R, θ with origin at (x, y), so that $R = r$. In these coordinates, one must consider

$$\lim_{h\to 0+} \int_0^{2\pi}\int_h^0 \mu(\xi, \eta) \log \frac{1}{R} R\, dR\, d\theta,$$

and, since $R \log 1/R \to 0$ as $R \to 0$, the limit does exist. The conclusion also holds for μ of variable sign (as occurs in the electrostatic case) (see Section 6.27).

Obtaining $\mathbf{F}$ from U requires differentiating under the integral sign. For P outside E, Eq. (5.164) follows from Leibnitz's Rule of Section 4.9. For P in E the result is correct, but a more careful analysis is needed. One can in fact show that grad U is given by (5.164) for all (x,y) and both U and grad U are continuous everywhere.

One can also, by a passage to the limit from (5.162), obtain the logarithmic potential of a distribution of mass on a curve C. One is led to a line integral

$$U = \int_C \mu(s) \log \frac{1}{r}\, ds. \qquad (5.166)$$

Here we find that if C is piecewise smooth and μ is continuous, then U is continuous everywhere. However, the corresponding force field grad U is defined only for P off of C.

Another limit process starts with two "parallel" curves C_1, C_2, close together with mass (or charge) densities equal but opposite in sign at adjacent points, as suggested in Fig. 5.40. If the two curves are brought to coincidence while the densities become infinite in the proper manner, one obtains as limit a new potential

$$U(x, y) = \int_C \gamma(s) \frac{\partial}{\partial n} \log \frac{1}{r}\, ds, \qquad (5.167)$$

called the logarithmic potential of a *double layer* on C. In electrostatics this corresponds to a dipole layer on C. The curve C is assumed to have a continuously varying unit normal vector $\mathbf{n}$, with respect to which the directional derivative $\partial/\partial n$ is formed; $\gamma(s)$ is assumed to be continuous on C. By contrast to (5.167) the potential (5.166) is called the potential of a single layer on C. For (5.167), one finds that U itself is discontinuous on C. However, for P off of C, U

is well defined and continuous, and Leibnitz's Rule can be applied to form the force field **F**, which is also continuous off of C.

Figure 5.40   Formation of double layer.

EXAMPLE   The logarithmic potential of a mass distribution of constant density b over a disk of radius a with center at $(0,0)$ is given by

$$U = b \int_0^a \int_0^{2\pi} \log \frac{1}{\sqrt{R^2 + \rho^2 - 2R\rho \cos(\theta - \alpha)}} \, \rho \, d\theta \, d\rho. \qquad (5.168)$$

Here $x = R\cos\alpha$, $y = R\sin\alpha$, and we used the law of cosines to express r in terms of R, ρ, θ, α, as in Fig. 5.41. With the help of advanced integration formulas, one finds:

$$U = \begin{cases} \dfrac{\pi}{2} b \left[a^2 - R^2 - 2a^2 \log a \right], & R \le a \\[2mm] \pi b a^2 \log \dfrac{1}{R}, & R > a \end{cases} \qquad (5.169)$$

(see Problem 1 following Section 5.18). At $R = a$ the two expressions give $U = \pi b a^2 \log 1/a$ so that U is continuous for all (x, y). Also

$$\mathbf{F} = \begin{cases} -\pi b (x\mathbf{i} + y\mathbf{j}), & R \le a \\[2mm] -\dfrac{\pi b a^2}{R^2} (x\mathbf{i} + y\mathbf{j}), & R > a, \end{cases} \qquad (5.170)$$

from which we see that **F** is continuous everywhere. The force is radial everywhere, directed toward the origin. The force magnitude depends only on R; this magnitude and U are graphed as functions of R in Fig. 5.42.   ∎

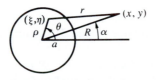

Figure 5.41   Distances and angles for Eq. (5.168).

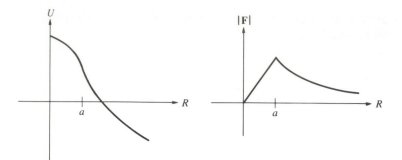

Figure 5.42   Potential force magnitude for Eq. (5.168).

U as a harmonic function.   For fixed (ξ, η) the function $\log 1/r$ is *harmonic* in (x, y) for (x, y) not at (ξ, η). For

$$\frac{\partial^2}{\partial x^2} \log \frac{1}{r} = \frac{\partial}{\partial x} \frac{\xi - x}{r^2} = \frac{(x - \xi)^2 - (y - \eta)^2}{r^4},$$

$$\frac{\partial^2}{\partial y^2} \log \frac{1}{r} = \frac{\partial}{\partial y} \frac{\eta - y}{r^2} = \frac{(y - \eta)^2 - (x - \xi)^2}{r^4},$$

and the sum is 0. Leibnitz's Rule permits a similar differentiation under the integral sign for the potentials (5.163), (5.166), and (5.167). Hence we conclude that *away from the masses $U(x,y)$ is harmonic.* Equivalently,

$$\operatorname{div} \mathbf{F} = \operatorname{div} \operatorname{grad} U = 0.$$

This is illustrated by the example of Eq. (5.168), for which U is a constant times $\log 1/R$ ($R^2 = x^2 + y^2$), away from the mass.

Behavior of U and F for large R.   The results to be presented hold for all types of mass distributions, but we formulate them only for the case of a continuous mass distribution of E, as in (5.163). We ask how U and $\mathbf{F}$ behave when P: (x,y) recedes to infinite distance, that is, when $R \to \infty$, $R = (x^2 + y^2)^{1/2}$. We let

$$M = \int_E \int \mu(\xi, \eta) \, d\xi \, d\eta; \tag{5.171}$$

we interpret M as the total mass.

For U itself we have the following assertion: For large R, $R \geq R_0$,

$$U(x, y) = M \log \frac{1}{R} + \frac{p(x, y)}{R}, \tag{5.172}$$

where $p(x, y)$ is bounded, $|p(x, y)| \leq$ const. Thus for large R the potential behaves as if the mass were concentrated at the origin, with a small error, which approaches 0 as $R \to \infty$.

To prove this, we let d be so large that E is included in the circular region $x^2 + y^2 \leq d^2$. Then by the triangle inequality ($a + b \geq c$ for a triangle of sides a, b, c),

$$R \leq r + d, \qquad r \leq R + d$$

Figure 5.43   Derivation of Eq. (5.172)

(see Fig. 5.43). Therefore for $R > d$,

$$\frac{R}{r} \le 1 + \frac{d}{r} \le 1 + \frac{d}{R - d},$$

$$\frac{R}{r} \ge 1 - \frac{d}{r} \ge 1 - \frac{d}{R - d},$$

so that

$$1 - \frac{d}{R - d} \le \frac{R}{r} \le 1 + \frac{d}{R - d} \quad \text{or} \quad \left|\frac{R}{r} - 1\right| \le \frac{d}{R - d}. \qquad (5.173)$$

In particular, $R/r \ge 1/2$ for $R \ge R_0 = 3d$. Next by (5.171),

$$U - M\log\frac{1}{R} = \int_E\int \mu(\xi, \eta)\left(\log\frac{1}{r} - \log\frac{1}{R}\right) d\xi \, d\eta$$

$$= \int_E\int \mu(\xi, \eta) \log\frac{R}{r} \, d\xi \, d\eta.$$

Now for $x > 0.5$, $|\log x| \le |2(x - 1)|$, as one sees from a graph of $y = \log x$ and $y = 2(x - 1)$ (Fig. 5.43). Therefore by (5.173), for $R \ge R_0$,

$$|p(x, y)| = R\left|U - M\log\frac{1}{R}\right| \le R\int_E\int |\mu(\xi, \eta)| \left|\log\frac{R}{r}\right| d\xi \, d\eta$$

$$\le R\frac{2d}{R - d}\int_E\int |\mu(\xi, \eta)| \, d\xi \, d\eta = \text{const}\,\frac{Rd}{R - d}.$$

The last expression is continuous for $R \ge R_0 = 3d$ and has limit const $\cdot d$ as $R \to \infty$. Therefore $p(x, y)$ is bounded, as was asserted.   □

Next we turn to $\mathbf{F} = \text{grad}\, U$. From the preceding result we expect that for large R, $\mathbf{F}$ is approximately equal to the gradient of a potential due to mass M at the origin; that is, for large R,

$$\mathbf{F} \sim \text{grad}\, M\log\frac{1}{R} = -\frac{M}{R^2}(x\mathbf{i} + y\mathbf{j}) = -\frac{M}{R^2}\overline{OP}.$$

Thus **F** should be approximately purely radial for large R with magnitude M/R. Now we introduce the unit vectors

$$\mathbf{u} = \cos\theta\,\mathbf{i} + \sin\theta\,\mathbf{j}, \qquad \mathbf{v} = \mathbf{k} \times \mathbf{u} = -\sin\theta\,\mathbf{i} + \cos\theta\,\mathbf{j},$$

which we term the *unit radial* and *tangential vectors*, respectively (Fig. 5.43). As the figure shows, we have corresponding components of **F**:

$$F_{\text{rad}} = \mathbf{F} \cdot \mathbf{u}, \qquad F_{\text{tan}} = \mathbf{F} \cdot \mathbf{v}.$$

We can now assert: For $R \geq R_0$,

$$F_{\text{rad}} = -\frac{M}{R} + \frac{q(x, y)}{R^2}, \tag{5.174}$$

$$F_{\text{tan}} = \frac{s(x, y)}{R^2}, \tag{5.175}$$

where $|q(x, y)| \leq \text{const}$ and $|s(x, y)| \leq \text{const}$. The proofs are left as exercises (Problem 5).

Mass relationship.   We know that $\text{div}\,\mathbf{F} = 0$ away from the masses. It follows from the divergence theorem that

$$\oint_C \mathbf{F} \cdot \mathbf{n}\, ds = 0$$

for every simple closed path C in a simply connected domain containing no mass. Here **n** is the exterior normal with respect to the region enclosed by C. If C does enclose mass, then this line integral will generally not be 0. For example, for the case of one particle of mass m at the origin,

$$\mathbf{F} = -\frac{m}{R^2}\mathbf{R} \qquad (\mathbf{R} = x\mathbf{i} + y\mathbf{j}),$$

and hence by Eq. (5.39),

$$\oint_C \mathbf{F} \cdot \mathbf{n}\, ds = -m \oint_C \frac{-y\, dx + x\, dy}{x^2 + y^2} = -2\pi m$$

for every path C enclosing the origin (Example 2 in Section 5.6).

As in Section 5.7 (see especially Problem 8 following that section), if C encloses a finite set of masses $m_1, \ldots, m_k$, then

$$\oint_C \mathbf{F} \cdot \mathbf{n}\, ds = -2\pi(m_1 + \cdots + m_k).$$

In general, we are led to assert: For each mass distribution the gradient **F** of the corresponding logarithmic potential has the property that

$$-\oint_C \mathbf{F} \cdot \mathbf{n}\, ds = \text{total mass enclosed by } C. \tag{5.176}$$

This conclusion is valid, provided that C does not pass through a point at which a mass particle is located or which lies on a curve on which there is a single or double layer.

For the case of a distribution over a region E we first take the case in which C completely encloses E. We then observe that the value of

$$\oint_C \mathbf{F} \cdot \mathbf{n} \, ds \tag{5.177}$$

is the same for all such C. For since the integral is 0 for all curves enclosing no mass, we are dealing with a line integral $\int P \, dx + Q \, dy$ for which $\partial Q / \partial x = \partial P / \partial y$ wherever there is no mass. Thus as in Section 5.7, the integral has the same value for all closed paths C enclosing the same "holes."

To evaluate the integral, we take C to be a circle $x^2 + y^2 = a^2$, where a is large. Then along $C, \mathbf{F} \cdot \mathbf{n}$ is simply the radial component of $\mathbf{F}$. By (5.174),

$$F_{\text{rad}} = -\frac{M}{a} + \frac{q(x, y)}{a^2},$$

where $|q(x, y)| \le K = \text{const}$ for a sufficiently large. Hence

$$\oint_C \mathbf{F} \cdot \mathbf{n} \, ds = \oint_C \left(-\frac{M}{a} + \frac{q(x, y)}{a^2} \right) ds$$

$$= -2\pi M + \frac{1}{a^2} \oint_C q(x, y) \, ds.$$

The last term is in absolute value at most

$$\frac{1}{a^2} K \cdot 2\pi a = \frac{2\pi K}{a},$$

and hence it can be made as small as desired by taking a sufficiently large. But the value of the integral (5.177) is independent of a. Thus the second term must be 0. This gives

$$\oint_C \mathbf{F} \cdot \mathbf{n} \, ds = -2\pi M = -2\pi \, (\text{total mass}).$$

If C encloses only some of the mass, as in Fig. 5.44, then we can write $\mathbf{F} = \mathbf{F}_{\text{int}} + \mathbf{F}_{\text{ext}}$, where $\mathbf{F}_{\text{int}}$ is the force field arising from the mass inside C and $\mathbf{F}_{\text{ext}}$ is that arising from the mass outside C. For $\mathbf{F}_{\text{ext}}$ there is now no mass inside C, so that

$$\oint_C \mathbf{F}_{\text{ext}} \cdot \mathbf{n} \, ds = 0.$$

Figure 5.44   Proof of mass relation.

For $\mathbf{F}_{\text{int}}$ we conclude as previously that

$$\oint_C \mathbf{F}_{\text{int}} \cdot \mathbf{n} \, ds = -2\pi \, (\text{mass inside } C).$$

Hence for $\mathbf{F} = \mathbf{F}_{\text{int}} + \mathbf{F}_{\text{ext}}$,

$$\oint_C \mathbf{F} \cdot \mathbf{n} \, ds = -2\pi \, (\text{mass inside } C).$$

Poisson equation.   For the mass distribution just considered, let us take a point (x_0, y_0) interior to E and take C to be a circle C_δ of radius δ about this point, bounding a circular region E_δ. Then we can write

$$\oint_{C_\delta} \mathbf{F} \cdot \mathbf{n} \, ds = -2\pi \int_{E_\delta}\!\!\int \mu(\xi, \eta) \, d\xi \, d\eta.$$

For the integral on the right-hand side gives the total mass inside C_δ. We divide both sides by $\pi\delta^2$, the area of E_δ:

$$\frac{1}{\pi\delta^2} \oint_C \mathbf{F} \cdot \mathbf{n} \, ds = -2\pi \frac{1}{\pi\delta^2} \int_{E_\delta}\!\!\int \mu(\xi, \eta) \, d\xi \, d\eta.$$

We now let $\delta \to 0$. As in Section 5.11, the left-hand side approaches $\operatorname{div} \mathbf{F}$, evaluated at (x_0, y_0); the right-hand side approaches $-2\pi\mu(x_0, y_0)$, as in Section 4.3 [Eq. (4.49)]. Therefore for every point (x, y) interior to E,

$$\operatorname{div} \mathbf{F} = -2\pi\mu(x, y). \tag{5.178}$$

Since $\mathbf{F} = \operatorname{grad} U$, we have also

$$\operatorname{div} \operatorname{grad} U = \nabla^2 U = -2\pi\mu(x, y) \tag{5.179}$$

at every such point. Equation (5.177) is called the *Poisson equation.*

Remark.   The replacement of div grad U by $\nabla^2 U$ in (5.179) assumes that the potential U has continuous second partial derivatives at (x, y). This can be shown to be valid if, for example, $\mu(x, y)$ has continuous first derivatives. (See Chapter 5 of the book by Sternberg and Smith listed at the end of the chapter.)

*5.17   GREEN'S THIRD IDENTITY

The first two identities of Green are given in Problems 11 and 12 following Section 5.7. The third identity is the equation

$$u(x, y) = \frac{1}{2\pi} \oint_C \left[\log \frac{1}{r} \frac{\partial u}{\partial n} - u \frac{\partial}{\partial n} \log \frac{1}{r} \right] ds. \tag{5.180}$$

Here u is assumed to have continuous first partial derivatives in a domain containing the closed region E bounded by the curve C and to be *harmonic* inside

C. The point (x, y) is interior to C, $\mathbf{n}$ is the exterior normal on C, and r is as above the distance from a general point (ξ, η) on C to (x, y).

We remark that the term

$$\frac{1}{2\pi} \oint_C \frac{\partial u}{\partial n} \log \frac{1}{r} \, ds$$

is a logarithmic potential of a single layer on C and the term

$$-\frac{1}{2\pi} \oint_C u \frac{\partial}{\partial n} \log \frac{1}{r} \, ds$$

is the logarithmic potential of a double layer on C. Hence the identity asserts that a harmonic function can be represented as the sum of two such logarithmic potentials.

To prove the identity, one starts with the second identity of Green:

$$\int_E \int (u\nabla^2 v - v\nabla^2 u) \, d\xi \, d\eta = \int_{B_E} \left(u \frac{\partial v}{\partial n} - v \frac{\partial u}{\partial n} \right) ds \qquad (5.181)$$

(Problem 12(a) following Section 5.7). We apply this first to our given region E, taking $v = \log 1/r$, where r is as above the distance from (ξ, η) to (x, y), but take (x, y) exterior to E. Then v is harmonic as a function of (ξ, η) in E, so that (5.181) becomes

$$\int_E \int \log \frac{1}{r} \nabla^2 u \, d\xi \, d\eta = \oint_C \left[\frac{\partial u}{\partial n} \log \frac{1}{r} - u \frac{\partial}{\partial n} \log \frac{1}{r} \right] ds. \qquad (5.182)$$

If, in addition, u is harmonic in E, then (5.182) becomes

$$\oint_C \left[\frac{\partial u}{\partial n} \log \frac{1}{r} - u \frac{\partial}{\partial n} \log \frac{1}{r} \right] ds = 0. \qquad (5.183)$$

This is valid for (x, y) outside of E.

Now for (x, y) interior to E we draw a small circle C_δ of radius δ, with center at (x, y), and let E_δ be the region obtained from E by deleting the interior of this circle from E (Fig. 5.45). We again take $v = \log 1/r$ and can apply Eq. (5.181) to

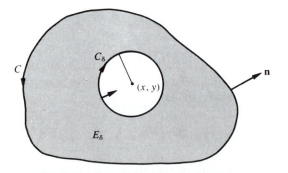

Figure 5.45   Proof of third Green identity.

the region E_δ, since (x, y) is exterior to this region. We obtain

$$\int_{E_\delta}\int \log\frac{1}{r}\nabla^2 u\, d\xi\, d\eta = \oint_C \left[\frac{\partial u}{\partial n}\log\frac{1}{r} - u\frac{\partial}{\partial n}\log\frac{1}{r}\right] ds$$

$$+ \oint_{C_\delta} \left[\frac{\partial u}{\partial n}\log\frac{1}{r} - u\frac{\partial}{\partial n}\log\frac{1}{r}\right] ds.$$

On C_δ the normal is exterior to E_δ, so that $\partial/\partial n = -\partial/\partial r$. Thus in polar coordinates r, θ, with center at (x, y), so that $ds = -\delta\, d\theta$ on C_δ, traced clockwise,

$$\oint_{C_\delta} \frac{\partial u}{\partial n}\log\frac{1}{r}\, ds = \delta\cdot\log\frac{1}{\delta}\int_{2\pi}^0 \frac{\partial u}{\partial r}(\delta, \theta)\, d\theta.$$

We assume that u has continuous partial derivatives in E, from which it follows that, as $\delta \to 0$, the last integral has a finite limit. But $\delta\log 1/\delta \to 0$ as $\delta \to 0$. Thus

$$\lim_{\delta\to 0}\oint_{C_\delta}\frac{\partial u}{\partial n}\log\frac{1}{r}\, ds = 0.$$

Next

$$\oint_{C_\delta} u\frac{\partial}{\partial n}\log\frac{1}{r}\, ds = -\int_{2\pi}^0 u(\delta, \theta)\frac{\partial}{\partial r}\log\frac{1}{r}\bigg|_{r=\delta}(-\delta)\, d\theta$$

$$= \int_0^{2\pi} u(\delta, \theta)\left(-\frac{1}{\delta}\right)(-\delta)\, d\theta = \int_0^{2\pi} u(\delta, \theta)\, d\theta.$$

As $\delta \to 0$, this has limit

$$\int_0^{2\pi} u(0, \theta)\, d\theta = 2\pi u \quad \text{at} \quad (x, y).$$

Accordingly, as $\delta \to 0$, the double integral over E_δ has a limit, which is just the integral over E, as is usual for logarithmic potentials. Thus

$$\int_E\int \log\frac{1}{r}\nabla^2 u\, d\xi\, d\eta = \oint_C \left(\frac{\partial u}{\partial n}\log\frac{1}{r} - u\frac{\partial}{\partial n}\log\frac{1}{r}\right) ds - 2\pi u(x, y).$$

$$(5.184)$$

If, finally, u is harmonic in the interior of E, then the left-hand side is 0, and the identity (5.180) is proved.

Remark.   The proof applies to a general region E bounded by several closed curves $C_1,\ldots,C_k$, so that as for the first and second identities, we can replace the integral over C in (5.180) by one over B_E, properly directed.

Green's function.   The identity (5.180) gives the harmonic function u in the interior of E in terms of the values of u and $\partial u/\partial n$ on the boundary. One can show that the interior values of u can be found from those of u alone on the boundary. There is a corresponding formula

$$u(x, y) = \int_{B_E} u(\xi, \eta) G(x, y; \xi, \eta)\, ds, \qquad (5.185)$$

using a *Green's function* G, a function of x, y, ξ, and η having special properties. For the case of a circle of radius a, Eq. (5.185) can be written as follows:

$$u(x, y) = \frac{1}{2\pi} \int_0^{2\pi} u(a\cos\phi, a\sin\phi) \frac{a^2 - R^2}{a^2 + R^2 - 2aR\cos(\theta - \phi)} d\phi.$$

$$(5.186)$$

Here $R = (x^2 + y^2)^{1/2}$. This is the *Poisson integral formula* for the circle.

For further details on logarithmic potentials, see the books by Kellogg and Sternberg and Smith listed at the end of the chapter.

*5.18   POTENTIAL THEORY IN SPACE

There is a potential theory for 3-dimensional space analogous to the theory of Sections 5.16 and 5.17 for the plane. In space the force law is that of gravitation. Hence for a single particle of mass m at Q, for appropriate choice of units,

$$\mathbf{F} = \frac{m}{r^3}\mathbf{r}, \qquad (5.187)$$

where as before $\mathbf{r} = \overrightarrow{PQ}$, Q is fixed at (ξ, η, ζ), P is at (x, y, z), $r = |\mathbf{r}|$. The force magnitude is, for $m > 0$,

$$|\mathbf{F}| = \frac{m}{r^3} r = \frac{m}{r^2}. \qquad (5.188)$$

We verify that

$$\mathbf{F} = \operatorname{grad} U(x, y, z),$$

$$U(x, y, z) = \frac{m}{r}. \qquad (5.189)$$

We call $U(x, y, z)$ the *Newtonian potential* due to the single mass m at Q.

We generalize as in Eqs. (5.161) and (5.162) to several particles and by passage to the limit to the distribution of mass over a region with density $\mu(\xi, \eta, \zeta)$:

$$U(x, y, z) = \int\int\int_E \frac{\mu}{r}\, d\xi\, d\eta\, d\zeta,$$

$$\mathbf{F} = \int\int\int_E \frac{\mu}{r^3}[(\xi - x)\mathbf{i} + (\eta - y)\mathbf{j} + (\zeta - z)\mathbf{k}]\, d\xi\, d\eta\, d\zeta.$$

$$(5.190)$$

One verifies as in Section 5.16 that the improper integrals exist and that U and $\mathbf{F}$ are continuous in all of 3-dimensional space (see Section 6.27).

One has also single- and double-layer potentials:

$$U = \int_S \int \frac{\mu}{r} \, d\sigma,$$

(5.191)

$$U = \int_S \int \gamma \frac{\partial}{\partial n} \left(\frac{1}{r} \right) d\sigma.$$

(5.192)

In (5.192), S must have a continuous unit normal vector field $\mathbf{n}$.

Again U is found to be harmonic away from the mass:

$$\operatorname{div} \mathbf{F} = \operatorname{div} \operatorname{grad} U = \nabla^2 U = 0.$$

(5.193)

For large $R = (x^2 + y^2 + z^2)^{1/2}$, again U and F behave as if all the mass M were concentrated at the origin $(0, 0, 0)$: For (5.190),

$$U = \frac{M}{R} + \frac{p(x, y, z)}{R^2},$$

(5.194)

$$\mathbf{F} = \operatorname{grad} \left(\frac{M}{R} \right) + \frac{1}{R^3} \mathbf{q}(x, y, z)$$

(5.195)

for R sufficiently large, where $|p(x, y, z)|$ and $|\mathbf{q}(x, y, z)|$ are bounded. Thus the force field is mainly radial for large R, and one finds as in the plane that

$$- \int_S \int \mathbf{F} \cdot \mathbf{n} \, d\sigma = 4\pi \cdot \text{mass enclosed by } S.$$

(5.196)

Here S is the boundary of a bounded closed region E, and $\mathbf{n}$ is the unit exterior normal on S. From (5.196), one concludes as in the plane that for a distribution over a solid region E with continuous density μ, one has

$$\operatorname{div} \mathbf{F} = \nabla^2 U = -4\pi\mu(\xi, \eta, \zeta)$$

(5.197)

at each interior point of E.

Green's third identity generalizes to space in the form

$$u(x, y, z) = \frac{1}{4\pi} \int_S \int \left(\frac{1}{r} \frac{\partial u}{\partial n} - u \frac{\partial}{\partial n} \frac{1}{r} \right) d\sigma$$

(5.198)

where S bounds E, u has continuous first partial derivatives in a domain containing E, (x, y, z) is interior to E, and $u(x, y, z)$ is harmonic on the interior.

There is also a formula analogous to (5.185) for space, using a Green's function. For the sphere $S: x^2 + y^2 + z^2 = a^2$ this becomes a *Poisson integral formula*:

$$u(x, y, z) = \frac{1}{4\pi a} \int_S \int u \frac{a^2 - R^2}{(a^2 + R^2 - 2aR \cos \alpha)^{3/2}} \, d\sigma.$$

(5.199)

On the right, $R^2 = x^2 + y^2 + z^2$, u is evaluated at points (ξ, η, ζ) on S, α is the angle between the vectors $x\mathbf{i} + y\mathbf{j} + z\mathbf{k}$, and $\xi\mathbf{i} + \eta\mathbf{j} + \zeta\mathbf{k}$, (x, y, z) is a point interior to S.

For details, one is referred to the books of Kellogg and Sternberg and Smith listed at the end of the chapter.

Helmholtz's theorem. Let **G** be a vector field whose components G_x, G_y, G_z are continuous over the bounded closed region E. Then

$$\mathbf{G} = \nabla \times \mathbf{H} + \nabla \phi, \tag{5.200}$$

for appropriate **H** and ϕ, on the interior of E.

To prove this, we let

$$\mathbf{U} = \iiint_E \frac{\mathbf{G}(\xi, \eta, \zeta)}{r} \, d\xi \, d\eta \, d\zeta, \tag{5.201}$$

so that the component of **U** are Newtonian potentials. By applying (5.197) to each component of **U** we conclude that

$$\nabla^2 \mathbf{U} = -4\pi \mathbf{G} \tag{5.202}$$

on the interior of E (see Eq. (3.39) in Section 3.6). By Eq. (3.38) in Section 3.6,

$$\nabla^2 \mathbf{U} = -\text{curl curl } \mathbf{U} + \text{grad div } \mathbf{U},$$

so that by (5.202),

$$\mathbf{G} = \frac{1}{4\pi} \text{curl curl } \mathbf{U} - \frac{1}{4\pi} \text{grad div } \mathbf{U},$$

and this gives (5.200), with $\mathbf{H} = (1/4\pi) \text{ curl } \mathbf{U}$, $\phi = -(1/4\pi) \text{ div } \mathbf{U}$. Thus the proof is complete.

One can extend the result to unbounded regions, provided that the corresponding improper integrals converge.

PROBLEMS

1. Use the result of Problem 6 following Section 4.9:

$$\int_0^{2\pi} \log (1 - c \cos \theta) \, d\theta = 2\pi \log \frac{1 + \sqrt{1 - c^2}}{2}, 0 \le c < 1,$$

to evaluate the logarithmic potential U of the example in Section 5.16.

2. Another way to evaluate the potential U of the example in Section 5.16 is to use the following information:

a) $\nabla^2 U = -2\pi b$ for $R < a$,

b) $\nabla^2 U = 0$ for $R > a$,

c) $U = M \log \frac{1}{R} + \frac{p(x, y)}{R}$ for large R, as in Eq. (5.172),

d) U and grad U are continuous for all (x, y),

e) U depends on R alone, by symmetry.

Since $\nabla^2 U = \partial^2 U / \partial R^2 + (1/R) \partial U / \partial R$ in polar coordinates, for a function depending only on R (Section 2.17), (a) and (b) give differential equations for U. By (c) and (d) the arbitrary constants in the solution can be determined.

Carry out the process suggested to find U.

3. Let U be the logarithmic potential of a distribution of mass with constant density b on the circle $x^2 + y^2 = a^2$.

a) Find U by integrating, with the aid of the formula given in Problem 1.

b) Find U as in Problem 2, omitting (a) and replacing (d) by the condition that U is continuous for all (x, y) and (b) by $\nabla^2 u = 0$ for $R \neq a$.

c) Find grad U and show that it is discontinuous for $R = a$.

4. Let U be the logarithmic potential of a double layer of constant moment γ on the circle $x^2 + y^2 = a^2$, with $\mathbf{n}$ the outer normal.

a) Find U by integration. [Hint: To find $(\partial/\partial n) \log 1/r$, write $r^2 = \rho^2 + R^2 - 2\rho R \cos(\theta - \alpha)$, where ρ, θ are the polar coordinates of (ξ, η) and R, α are those of (x, y). Then $\partial/\partial n = \partial/\partial \rho$, to be evaluated for $\rho = a$.]

b) Find grad U and show that the normal derivatives of U have a discontinuity on the circle.

5. a) Prove (5.174) with the conditions stated. [Hint: Use (5.164) to obtain $F_{rad} = \mathbf{F} \cdot \mathbf{u}$, where $\mathbf{u} = \cos\theta \mathbf{i} + \sin\theta \mathbf{j} = (x\mathbf{i} + y\mathbf{j})/R$. Show by (5.174) that q can be written as

$$\int_E \int \mu \left[\frac{R^2}{r^2} (\xi\mathbf{i} + \eta\mathbf{j}) \cdot \mathbf{u} + R\left(1 - \frac{R}{r}\right)\left(1 + \frac{R}{r}\right) \right] d\xi \, d\eta.$$

Show by (5.173) that each of the terms inside the brackets is bounded by a constant for $R \geq r_0 = 2d$.]

b) Prove (5.175) with the conditions stated. [Hint: Show by (5.175) that

$$s(x, y) = \int_E \int \mu \frac{R^2}{r^2} (\xi\mathbf{i} + \eta\mathbf{j}) \cdot \mathbf{v} \, d\xi \, d\eta$$

and use (5.173) to show that the integrand is bounded for $R \geq R_0 = 2d$.]

6. Find the Newtonian potential U of the solid sphere $x^2 + y^2 + z^2 \leq a^2$ with density $\mu = k\sqrt{x^2 + y^2 + z^2}$. Verify that $\nabla^2 U = -4\pi\mu$ inside the sphere.

7. Show that for a mass distribution of constant density on a spherical surface there is no net force on a particle inside the surface.

8. Justify Eq. (5.194) for the Newtonian potential (5.190), with continuous density μ, where $p(x,y,z)$ is bounded for R large.

*5.19   DIFFERENTIAL FORMS

The expressions occurring in our integrals: $f(x)\,dx, F(x,y)\,dx\,dy, P\,dx + Q\,dy, L\,dy\,dz + M\,dz\,dx + N\,dx\,dy, \ldots$ are termed *differential forms*. The mathematician Élie Cartan (1869–1951) developed procedures for working with such forms which yield our rules and many others. We here outline the principal ideas, giving references for a full treatment.

For this discussion it is convenient to generally regard the integrals as line or surface integrals or their generalizations to higher dimensions. Thus

$$\int_a^b f(x)\,dx \quad\quad \text{and} \quad\quad \int\int_S N(x,y)\,dx\,dy$$

are regarded, respectively, as a line integral of f on the path from a to b and as a surface integral of N over an oriented surface S in the xy-plane, as in Fig. 5.28

in Section 5.10. We even distinguish between $dx\,dy$ and $dy\,dx$. The order $dx\,dy$ is considered to be associated with the normal orientation of the xy-plane and, if S has that orientation, $\iint_S N\,dx\,dy$ has the value $\iint_S N\,dA$ as in Chapter 4. If S has the opposite orientation, this value is multiplied by -1. In either case

$$\iint_S N\,dy\,dx = -\iint_S N\,dx\,dy.$$

Similar conventions apply to triple integrals $\iiint f(x,y,z)\,dx\,dy\,dz$ and to multiple integrals in general. In E^n the coordinates $x^1,\ldots,x^n$ or the corresponding basis vectors $\mathbf{e}_1 = (1,0,\ldots,0),\ldots,\mathbf{e}_n = (0,\ldots,0,1)$ of V^n (Section 1.13) in that order will be considered to provide the positive orientation; any interchange of two coordinates will reverse the orientation, as will any odd number of interchanges (but not an even number of interchanges). Each region R of integration is assumed oriented, either positively or negatively, and

$$\int\cdots\int_R f(x^1,\ldots,x^n)\,dx^{i_1}\ldots dx^{i_n} = \pm\int\cdots\int_R f\,dV,$$

where the integral on the right is the usual integral with respect to n-dimensional volume over the unoriented region R; the $+$ sign holds if $i_1,\ldots,i_n$ is an even permutation of $1,\ldots,n$ and R is positively oriented; replacing "even" by "odd" or reversing the orientation of R multiply the value by -1. (See Section 1.3 and Problem 16 (d) following Section 1.4.)

From these conventions we are led to our first rule for differential forms:

$$f\,dx\,dy = -f\,dy\,dx$$

and in general

$$f\,dx^{i_1}\ldots dx^{i_m} = \pm f\,dx^{j_1}\ldots dx^{j_m}, \tag{5.203}$$

with $+$ or $-$ according as $(j_1,\ldots,j_m)$ is an even or odd permutation of $i_1,\ldots,i_m$. We clearly have no use for $dx\,dx$ and we write

$$f\,dx\,dx = 0$$

and, in general,

$$f\,dx^{i_1}\ldots dx^{i_m} = 0 \quad\text{if two of}\quad i_1,\ldots,i_m \quad\text{are equal.} \tag{5.204}$$

Here 0 is a *zero form,* which can be identified with $0\,dx^1\ldots dx^m$.

We now define a *differential form of degree m,* or an *m-form,* in n *variables* to be a sum of terms of form

$$A\,dx^{i_1}dx^{i_2}\ldots dx^{i_m}, \tag{5.205}$$

where A is a function of $x^1,\ldots,x^n$, $1 \le i_1 \le n,\ldots,1 \le i_m \le n$. We also allow $m = 0$ and consider as 0-form a function $A(x^1,\ldots,x^n)$. All functions will be assumed to be differentiable as required in an appropriate open region.

Two m-forms are considered to be *equal* if both can be reduced to the same form by applying the rules (5.203), (5.204) and combining terms of the same indices

$i_1, \ldots, i_m$. Thus each m-form can be written as a sum of terms (5.205) with no two indices equal and no two terms having the same set of indices—for example, each sum can be written in exactly one way as a sum

$$\sum_{1 \le i_1 < \ldots < i_m \le n} A_{i_1 \ldots i_m} dx^{i_1} \ldots dx^{i_m}.$$

We remark that as a special case, a form of degree n can be written in just one way as

$$B \, dx^1 dx^2 \ldots dx^n$$

and hence thought of as $B \, dV$, where V is n-dimensional volume. Further, a form of degree $m > n$ can be written uniquely as a zero form, as follows from (5.204), since for $m > n$, the $i_1, \ldots, i_m$ cannot be all distinct.

We can now perform the algebraic operations of addition and multiplication of m-forms, for given n, in the obvious way, using the rules (5.203) and (5.204) to simplify, combining terms where possible. For addition we combine only terms of the same degree m and obtain results of this degree. The multiplication, called *exterior product*, combines two forms of degrees m_1, m_2 to obtain one of degree $m_1 + m_2$ (reducing to a zero form if $m_1 + m_2 > n$).

One verifies that these definitions respect the definition of equality of forms given above: that is, replacement of α, β by equal forms α_1, β_1, replaces the sum or product γ by an equal form γ_1 (see Problem 2 following Section 5.22).

Remark.   The term (5.205) is sometimes written as $A_{i_1 \ldots i_m} dx^{i_1} \wedge dx^{i_2} \ldots \wedge dx^{i_m}$ and exterior products (also called *wedge products*) are also written with a $\wedge$.

EXAMPLE 1   With $n = 5$

$$(Adx^1 + Bdx^2)(Cdx^1 dx^3 + Ddx^3 dx^4)$$
$$= ACdx^1 dx^1 dx^3 + BCdx^2 dx^1 dx^3 + ADdx^1 dx^3 dx^4 + BDdx^2 dx^3 dx^4$$
$$= -BCdx^1 dx^2 dx^3 + ADdx^1 dx^3 dx^4 + BDdx^2 dx^3 dx^4. \quad \blacksquare$$

One now verifies algebraic laws:

$$\alpha + \beta = \beta + \alpha, \quad (\alpha + \beta) + \gamma = \alpha + (\beta + \gamma), \tag{5.206}$$

$$\alpha\beta = (-1)^{m_1 m_2} \beta\alpha, \quad \alpha(\beta\gamma) = (\alpha\beta)\gamma, \tag{5.207}$$

$$\alpha(\beta + \gamma) = \alpha\beta + \alpha\gamma, \quad (\beta + \gamma)\alpha = \beta\alpha + \gamma\alpha. \tag{5.208}$$

Here α, β and γ are forms of degrees m_1, m_2, m_3 respectively and in (5.206) $m_1 = m_2 = m_3$, in (5.208) $m_2 = m_3$.

EXAMPLE 2   We illustrate the first rule (5.207):

$$(Adx^1 dx^3 dx^5)(Bdx^2 dx^4) = ABdx^1 dx^3 dx^5 dx^2 dx^4,$$
$$(Bdx^2 dx^4)(Adx^1 dx^3 dx^5) = BAdx^2 dx^4 dx^1 dx^3 dx^5.$$

Here $AB = BA$ (scalar multiplication) and

$$
\begin{aligned}
dx^1 dx^3 dx^5 dx^2 dx^4 &= -dx^1 dx^3 dx^2 dx^5 dx^4 \\
&= dx^1 dx^2 dx^3 dx^5 dx^4 \\
&= -dx^2 dx^1 dx^3 dx^5 dx^4 \\
&= dx^2 dx^1 dx^3 dx^4 dx^5 \\
&= -dx^2 dx^1 dx^4 dx^3 dx^5 \\
&= dx^2 dx^4 dx^1 dx^3 dx^5;
\end{aligned}
$$

thus six interchanges were used to go from one order of the dx^i to the other and $m_1 m_2 = 3 \times 2 = 6$, so the rule is satisfied. For the general case, see Problem 3 following Section 5.22. ■

Now we introduce a derivative, the *exterior derivative*. If ω is an m-form (here usually $m < n$), the exterior derivative of ω is an $(m + 1)$-form, denoted by $d\omega$. If ω is a single term (5.205), then

$$
d\omega = \left(\frac{\partial A}{\partial x^1} dx^1 + \cdots + \frac{\partial A}{\partial x^n} dx^n \right) dx^{i_1} dx^{i_2} \ldots dx^{i_m}. \tag{5.209}
$$

For $m = 0$, $\omega = A$, a scalar function, and

$$
d\omega = \frac{\partial A}{\partial x^1} dx^1 + \cdots + \frac{\partial A}{\partial x^n} dx^n \tag{5.210}
$$

as usual. For ω a sum of terms, $d\omega$ is obtained by applying (5.209) to each term and adding. The results can be simplified by the rules given above.

From the definition just given one deduces further rules. For an m_1-form α and m_2-form β,

$$
d(\alpha + \beta) = d\alpha + d\beta \quad (\text{here} \quad m_1 = m_2), \tag{5.211}
$$

$$
d(\alpha\beta) = (d\alpha)\beta + (-1)^{m_1}\alpha(d\beta), \tag{5.212}
$$

$$
d(d\alpha) = 0. \tag{5.213}
$$

(See Problem 4 following Section 5.22.)

EXAMPLE 3   Let $n = 3$ and $\omega = A_1 dx^1 + A_2 dx^2 + A_3 dx^3$. Then

$$
\begin{aligned}
d\omega &= \left(\frac{\partial A_1}{\partial x^1} dx^1 + \cdots \right) dx^1 + \left(\frac{\partial A_2}{\partial x^1} dx^1 + \cdots \right) dx^2 + (\ldots) dx^3 \\
&= \left(\frac{\partial A_3}{\partial x^2} - \frac{\partial A_2}{\partial x^3} \right) dx^2 dx^3 + \left(\frac{\partial A_1}{\partial x^3} - \frac{\partial A_3}{\partial x^1} \right) dx^3 dx^1 + \left(\frac{\partial A_2}{\partial x^1} - \frac{\partial A_1}{\partial x^2} \right) dx^1 dx^2.
\end{aligned}
$$

In familiar notation, this is just curl $\mathbf{v} \cdot d\boldsymbol{\sigma}$, where $\mathbf{v} = A_1 \mathbf{i} + A_2 \mathbf{j} + A_3 \mathbf{k}$. ■

EXAMPLE 4   Let $n = 3$ and $\omega = A_1 dx^2 dx^3 + A_2 dx^3 dx^1 + A_3 dx^1 dx^2$. Then

$$d\omega = \left(\frac{\partial A_1}{\partial x^1} dx^1 + \cdots \right) dx^2 dx^3 + \left(\frac{\partial A_2}{\partial x^1} dx^1 + \cdots \right) dx^3 dx^1 + \cdots$$

$$= \left(\frac{\partial A_1}{\partial x^1} + \frac{\partial A_2}{\partial x^2} + \frac{\partial A_3}{\partial x^3} \right) dx^1 dx^2 dx^3.$$

In familiar notation, this is div **v** $dx\, dy\, dz$.   ∎

*5.20   CHANGE OF VARIABLES IN AN m-FORM AND GENERAL STOKES THEOREM

If one is given equations

$$x^i = x^i(y^1, \ldots, y^p), \qquad i = 1, \ldots, n,$$

describing a differentiable mapping from a region in E^p into our given region of Section 5.19 in E^n, then each m-form in $x^1, \ldots, x^n$ is automatically converted into an m-form in $y^1, \ldots, y^p$ by expressing each scalar function in terms of the new variables and applying the rules of Section 5.19.

EXAMPLE 1   The 2-form ω of Example 4 of the preceding section becomes, after the substitution

$$x^i = x^i(u, v), \quad i = 1, 2, 3, \tag{5.214}$$

$$A_1(x^1(u, v), \ldots) \left(\frac{\partial x^2}{\partial u} du + \frac{\partial x^2}{\partial v} dv \right) \left(\frac{\partial x^3}{\partial u} du + \frac{\partial x^3}{\partial v} dv \right) +$$

$$A_2(\ldots) \left(\frac{\partial x^3}{\partial u} du + \cdots \right) \left(\frac{\partial x^1}{\partial u} du + \cdots \right) + A_3(\ldots)(\ldots)(\ldots)$$

$$= \left[A_1(x^1(u, v), \ldots) \frac{\partial(x^2, x^3)}{\partial(u, v)} + A_2(\ldots) \frac{\partial(x^3, x^1)}{\partial(u, v)} + A_3(\ldots) \frac{\partial(x^1, x^2)}{\partial(u, v)} \right] du\, dv.$$

This is exactly the procedure used in evaluating the surface integral of our 2-form. Thus

$$\iint_S A_1 dx^2 dx^3 + A_2 dx^3 dx^1 + A_3\, dx^1 dx^2 =$$

$$\iint_{R_{uv}} \left[A_1(x^1(u, v), \ldots) \frac{\partial(x^2, x^3)}{\partial(u, v)} + \cdots \right] du\, dv$$

as in Eq. (5.81) in Section 5.10. Here we are assuming that the uv-plane has its usual orientation, that R_{uv} is positively oriented and that the orientation of S is determined by that of R_{uv} by the mapping equations (5.214)—that is, that the $+$ sign holds in Eq. (5.81).   ∎

A similar discussion applies to other substitutions.

EXAMPLE 2   A line integral (integral of a 1-form)

$$\int_C P\,dx + Q\,dy$$

in which C is given by $x = f(t), y = g(t)$, as in Section 5.2, becomes

$$\int_a^b \left\{ P[f(t), g(t)]\frac{dx}{dt} + Q[f(t), g(t)]\frac{dy}{dt} \right\} dt$$

as in Eqs. (5.4) and (5.5). Here we assume that C has the orientation (direction) given by increasing t.

EXAMPLE 3   A double integral

$$\iint_{R_{xy}} F(x,y)\,dx\,dy$$

can be considered as the integral of a 2-form $F(x,y)\,dx\,dy$ over the positively oriented region R_{xy}. The substitution $x = x(u,v), y = y(u,v)$, as in Example 1, gives the integral

$$\iint_{R_{uv}} F\,[x(u,v),y(u,v)]\frac{\partial(x,y)}{\partial(u,v)}\,du\,dv$$

over the oriented region R_{uv} in the uv-plane (in which $du\,dv$ is associated with the positive orientation). The two double integrals are equal if the mapping is one-to-one and preserves orientation—that is, if $\partial(x,y)/\partial(u,v) > 0$; they are also equal if $\partial(x,y)/\partial(u,v) < 0$, for then R_{uv} is negatively oriented and there are two changes of sign. Equality also holds under more general conditions, as in Section 5.14.   ∎

In general, one verifies in the same way that, under a differentiable mapping

$$x^i = x^i(\bar{x}^1, \ldots, \bar{x}^n), \quad i = 1, \ldots, n,$$

with differentiable inverse, from one open region in E^n to another, each line, surface or hypersurface or multiple integral of an m-form over an appropriate, curve, surface etc. is converted by the change of variables procedure into a new integral of an m-form and the value of the integral is unchanged.

General Stokes Theorem.   Green's theorem, the 2- and 3-dimensional divergence theorems and Stokes' theorem can be considered as special cases of one theorem, called the *general Stokes theorem*:

$$\int \cdots \int_S d\omega = \int \cdots \int_B \omega . \tag{5.215}$$

Here ω is a general m-form in n-dimensional space, $m < n$, and S is a hypersurface of dimension $m + 1$ (($m + 1$)-dimensional manifold) with boundary B, an m-dimensional hypersurface; S and B are consistently oriented (as suggested by our rule for Stokes' theorem in Section 5.12).

EXAMPLE 4   Let $m = 1$, $n = 2$, let B be the simple closed curve C bounding a positively oriented plane region R in the xy-plane, as in Section 5.5. Then (5.215) becomes

$$\iint_R d(P\,dx + Q\,dy) = \int_C P\,dx + Q\,dy.$$

Since $d(P\,dx + Q\,dy) = (\partial Q/\partial x - \partial P/\partial y)\,dx\,dy$ as in Example 3 in Section 5.19, we recover Green's theorem and hence, as in Section 5.5, the 2-dimensional divergence theorem.   ■

Examples 3 and 4 of Section 5.19 show how Eq. (5.215) leads to the familiar Stokes' theorem and Gauss theorem.

The proof of the general theorem is given in the books by Boothby (pages 256–266), Hodge (pages 75–78) and Flanders (pages 64–66), listed at the end of the chapter. These three books are valuable general references for differential forms.

*5.21   TENSOR ASPECTS OF DIFFERENTIAL FORMS

Let an m-form $(m \geq 1)$ be given in E^n. We can write it as

$$\frac{1}{m!} A_{i_1\,i_2\dots\,i_m} dx^{i_1} dx^{i_2}\dots dx^{i_m}, \tag{5.216}$$

where we follow the summation convention of Section 3.9, summing over the repeated indices $i_1, i_2, \dots$, each going from 1 to n. Thus a general 2-form in E^2 is

$$\frac{1}{2} A_{ij}(x^1, x^2)\,dx^i dx^j = \frac{1}{2} A_{11} dx^1 dx^1 + \frac{1}{2} A_{12}\,dx^1 dx^2 + \frac{1}{2} A_{21} dx^2 dx^1$$
$$+ \frac{1}{2} A_{22}\,dx^2 dx^2.$$

By the rules of Section 5.19, the m-form can be written in many different ways. In particular, we can write it (in precisely one way) as in (5.216), with

$$A_{i_1\dots i_m} = 0 \quad \text{if} \quad i_p = i_q \quad \text{for some} \quad p \neq q, \tag{5.217}$$

$$A_{j_1\dots j_m} = \pm A_{i_1\dots i_m} \tag{5.218}$$

if $(j_1, \dots, j_m)$ is a permutation of distinct integers $i_1, \dots, i_m$, with + (−) if the permutation is even (odd). (See Problem 7 following Section 5.22.) For the rest of this section we shall assume that all our forms are written in this way. We further assume as before that the coefficients are differentiable as required in an open region D of E^n.

At a chosen point of D, for given numerical choices of $dx^1, \dots, dx^n$, the form (5.216) has a numerical value. Hence, as a tensor, it should have order 0, or be an invariant. Thus, as in Section 3.9, to show that we have a tensor, we must show the values are unaffected by a change of coordinates

$$x^i = x^i(\bar{x}^1, \dots, \bar{x}^n), \quad i = 1, \dots, n. \tag{5.219}$$

The dx^i are the components of a contravariant vector since

$$dx^i = \frac{\partial x^i}{\partial \bar{x}^j} d\bar{x}^j.$$

From our discussion of the $A_{i_1 \ldots i_m}$, they appear to be the components of an *alternative covariant tensor* as in Section 3.11. To verify this, we make the substitution (5.219). As in Section 5.20 the given form becomes

$$\frac{1}{m!} A_{i_1 \ldots i_m}(x^1(\bar{x}^1, \ldots), \ldots) \frac{\partial x^{i_1}}{\partial \bar{x}^{j_1}} \cdots \frac{\partial x^{i_m}}{\partial \bar{x}^{j_m}} d\bar{x}^{j_1} \ldots d\bar{x}^{j_m}.$$

Thus

$$\bar{A}_{j_1 \ldots j_m} = A_{i_1 \ldots i_m} \frac{\partial x^{i_1}}{\partial \bar{x}^{j_1}} \cdots \frac{\partial x^{i_m}}{\partial \bar{x}^{j_m}}. \tag{5.220}$$

This shows that $A_{i_1 \ldots i_m}$ is indeed a covariant tensor of order m. As in Section 3.11, it is alternating in all allowed coordinate systems. [We note that the fact that the integral of an m-form does not change its value under a change of coordinates, as pointed out in Section 5.21, also implies that the m-form is an invariant.]

Thus an m-form is an invariant obtained by contraction of a mixed tensor

$$\frac{1}{m!} A_{i_1 \ldots i_m} dx^{j_1} \ldots dx^{j_m}$$

(setting $j_1 = i_1, \ldots, j_m = i_m$ and summing as in Section 3.9). The result is an invariant, just as $ds^2 = g_{ij} dx^i dx^j$ is an invariant obtained by contraction.

One verifies that the exterior product of alternating covariant tensors, as defined in Section 3.11, is equivalent to the exterior product of differential forms. (See Problem 8 following Section 5.22).

Finally formation of the exterior derivative is a valid tensor operation: that is, validity of the relation $\beta = d\alpha$ in one allowed coordinate system implies its validity in all allowed coordinate systems. (See Problem 10 following Section 5.22).

*5.22   TENSORS AND DIFFERENTIAL FORMS WITHOUT COORDINATES

The elaborate formalism of tensors and differential forms is due to their expressions in different coordinate systems. It is natural to seek ways of describing the objects and operations in so far as possible without coordinates. An example is provided by the interpretation of the gradient of a function F in Section 3.14 as a vector whose direction is that in which F increases most rapidly and whose length equals that rate of increase.

Throughout, all functions will be assumed differentiable as needed.

We begin by considering vectors in space, considered here as bound vectors as in a vector field. With each vector $\mathbf{v}$ at a point P we can associate the operator $\mathbf{v} \cdot \nabla$ (Problem 12 following Section 3.6). When this operator is applied to a function f, the result is a scalar $\mathbf{v} \cdot \operatorname{grad} f$. (If $\mathbf{v}$ is a unit vector, then this equals the directional derivative of f at P in the direction of $\mathbf{v}$, but we do *not* assume that $\mathbf{v}$ is a unit vector.) We write L for the differential operator $\mathbf{v} \cdot \nabla$ and verify that L has the two properties:

> I. $L(c_1f_1 + c_2f_2) = c_1Lf_1 + c_2Lf_2$ (linearity),
>
> II. $L(f_1f_2) = f_1Lf_2 + f_2Lf_1$ (product rule).

(In I, c_1, c_2 are arbitrary scalars.) We now *define* the vectors at P to be the set of all linear operators L satisfying the rules I. and II. at P for all functions f_1, f_2 differentiable in a neighborhood of P. One shows that each such operator L must be representable as $\mathbf{v} \cdot \nabla$ for a unique $\mathbf{v}$ (so that there is a one-to-one correspondence between the old vectors $\mathbf{v}$ and operators L).

The definition applies to n-dimensional space. The set of all vectors at a point P (by the new definition) is called the *tangent space* at P. The term is suggested by the fact that we can apply the same idea to vectors tangent to a curve, surface or hypersurfaces at a point P (cf. the "directional derivative along a curve" in Section 2.14). The operators L satisfying I. and II. at P for all functions defined and differentiable in a neighborhood of P are again representable as the operators $\mathbf{v} \cdot \text{grad}$ (where, for the curve or surface or hypersurface S, $\mathbf{v}$ is tangent to S).

The operators L at a point can be added and multiplied by scalars and form a vector space. The dimension of this vector space is n if we are considering vectors in n-dimensional space or a hypersurface of dimension n.

We next observe that, in an arbitrary vector space of dimension n, we can define a *linear function* as a real-valued function ω satisfying the rule:

> III. $\omega(a\mathbf{u} + b\mathbf{v}) = a\omega(\mathbf{u}) + b\omega(\mathbf{v})$

for all vectors $\mathbf{u}, \mathbf{v}$ in the space and all real scalars a, b. In 3-dimensional space V^3 an example of such a function is $\mathbf{v}_0 \cdot \mathbf{v}$ for fixed $\mathbf{v}_0$; in fact, every linear function on V^3 can be so represented, and an analogous statement holds for V^n.

Again the functions ω satisfying III. can be added and multiplied by scalars and themselves form a vector space (also of dimension n). This vector space is called the *dual space* of the vector space we started with and its elements ω are referred to as *covectors*.

We apply this idea to the tangent space at P introduced above. The space of covectors is called the *cotangent space* at P.

Finally we consider a function

$$\Phi(L_1, \ldots, L_p, \quad \omega_1, \ldots, \omega_q)$$

assigning a real number to each choice of p vectors $L_1, \ldots, L_p$ in the tangent space at P and of q covectors $\omega_1, \ldots, \omega_q$ in the cotangent space at P and satisfying a rule of *multilinearity*: Φ is to be linear in each variable. For example, for $p = 2, q = 1$,

$$\Phi(a_1L_1' + a_2L_1'', \quad b_1L_2' + b_2L_2'', \quad c_1\omega_1' + c_2\omega_1'')$$

$$= a_1\Phi(L_1', b_1L_2' + b_2L_2'', \dots,) + a_2\ \Phi(L_1'', b_1L_2' + \dots, \dots)$$

$$= \cdots = a_1b_1c_1\ \Phi(L_1', L_2', \omega_1') + a_1b_1c_2\ \Phi(L_1', L_2', \omega_1'')$$

$$+ \cdots + a_2b_2c_2\ \Phi(L_1'', L_2'', \omega_1'').$$

We define each such function Φ to be a *tensor at P*, covariant of order p and contravariant of order q. As usual we include as tensors with $p = 0, q = 0$ the scalar functions defined at P. One shows that, if coordinates are introduced with reference to appropriate bases in the tangent space and its dual, then Φ becomes a real function of $2n$ real variables $\alpha^1, \dots, \alpha^n, \beta_1, \dots, \beta_n$:

$$\Phi = A_{i_1\dots i_p}^{j_1\dots j_q}\alpha^{i_1}\cdots\alpha^{i_p}\beta_{j_1}\cdots\beta_{j_q} \tag{5.221}$$

(summation convention), where the coefficients $A_{\dots}^{\dots}$ define a tensor at P of the type indicated.

Thus we have achieved (in outline) our goal: a definition of tensors which does not rely on coordinates. Of course the coordinates are in the background and do motivate the definition.

We have emphasized tensors at a point. By allowing P to vary over an open region, we obtain vector fields, covector fields and tensor fields as in Section 3.9. For the tensor fields Φ becomes a function of P, in addition to $L_1, \dots, \omega_1, \dots,$ and one assumes appropriate differentiability with respect to P.

In the discussion at the end of Section 5.21 differential forms are related to alternating covariant tensors. One can in fact define a differential form of order m as a multilinear function Φ of tangent vectors $L_1, \dots, L_m$ (depending differentiably on P), which is alternating: that is, $\Phi = \Phi(P, L_1, \dots, L_m)$, where

$$\Phi(P, L_1, L_2, L_3, \dots, L_m) = -\Phi(P, L_2, L_1, L_3, \dots, L_m)$$

and, in general, Φ reverses sign when two of the L_i are interchanged. (for $m = 0$ or 1, the condition is satisfied automatically.) In a coordinate system, the tangent vectors can be represented in terms of components $dx^1, \dots, dx^n$ and, as in (5.216) (Section 5.21),

$$\Phi = \frac{1}{m!}A_{i_1i_2\dots i_m}dx^{i_1}dx^{i_2}\dots dx^{i_m}. \tag{5.222}$$

For a full discussion, one is referred to the book of Boothby cited at the end of the chapter.

PROBLEMS

1. In E^4 let $\alpha = x^1 x^4$ (0-form), $\beta = x^4 dx^1 - x^1 dx^4$, $\gamma = x^1 dx^2 dx^3 + x^2 dx^3 dx^1 + x^3 dx^1 dx^2$. Calculate and simplify:

   a) $\alpha\beta$ **b)** $\beta^2 - \gamma$ **c)** $\beta\gamma + \gamma\beta$ **d)** γ^2

   e) $d\alpha$ **f)** $d^2\beta$ **g)** $d\gamma$

2. Illustrate by examples of 1-forms and 2-forms in E^4 that $\alpha = \alpha_1$ and $\beta = \beta_1$, imply $\alpha + \beta = \alpha_1 + \beta_1$ and $\alpha\beta = \alpha_1\beta_1$.

3. Prove the rule $\alpha\beta = (-1)^{m_1 m_2}\beta\alpha$ of (5.207). [Hint: By the distributive laws (5.208) it suffices to prove it for

$$\alpha = A dx^{i_1}\ldots dx^{i_{m_1}}, \qquad \beta = B dx^{j_1}\ldots dx^{j_{m_2}}.$$

If $i_k = j_\ell$ for some k, ℓ, both sides are 0. Otherwise, as in Example 2 of Section 5.21, one has to verify that one can go from $(i_1, \ldots, i_{m_1}, j_1, \ldots, j_{m_2})$ to $(j_1, \ldots, j_{m_2}, i_1, \ldots, i_{m_1})$ by $m_1 m_2$ interchanges.]

4. a) Prove the rule (5.211).

   b) Prove the rule (5.212) for $m_1 = 1$, $m_2 = 1$, $n = 4$.

   c) Prove the rule (5.213) for $m_1 = 1$, $n = 4$.

5. For γ as in Problem 1, apply the given change of variables to obtain a form $\bar{\gamma}$ in $\bar{x}^1, \ldots, \bar{x}^4$ and simplify:

   a) $x^1 = \bar{x}^1 + \bar{x}^2, x^2 = \bar{x}^2 + 2\bar{x}^3, x^3 = \bar{x}^3 + 3\bar{x}^4, x^4 = 4\bar{x}^1 + \bar{x}^4$

   b) $x^1 = (\bar{x}^1)^2, x^2 = \bar{x}^1\bar{x}^2, x^3 = \bar{x}^1\bar{x}^3, x^4 = \bar{x}^1\bar{x}^4$.

6. Under the (invertible) change of coordinates in $E^3: x^1 = \bar{x}^1, x^2 = \bar{x}^1 + \bar{x}^2, x^3 = \bar{x}^1 + \bar{x}^3$, a smooth path $C: x^i = x^i(t)$, $a \le t \le b$ and a smooth surface $S: x^i = x^i(u, v)$, $i = 1, 2, 3$, (u, v) in R_{uv}, become $\bar{C}$ and $\bar{S}$ respectively. Let $\bar{X}_i$ denote $X_i(x^1, x^2, x^3)$ with x^1, x^2, x^3 expressed in terms of $\bar{x}^1, \bar{x}^2, \bar{x}^3$.

   a) Express the line integral $\int_C X_1 dx^1 + X_2 dx^2 + X_3 dx^3$ as an integral over $\bar{C}$.

   b) Express the surface integral $\iint_S X_1 dx^2 dx^3 + X_2 dx^3 dx^1 + X_3 dx^1 dx^2$ as a surface integral over $\bar{S}$.

7. In Section 5.21 it is asserted that each m-form in E^n can be written uniquely in the form (5.216) where $A_{i_1, \ldots, i_m}$ satisfies (5.217) and (5.218). With the help of the alternation operation $\mathcal{A}$ of Section 3.11 prove this

   a) for $m = 2$ **b)** for arbitrary $m \le n$.

8. Let A_i and B_{ij} be alternating covariant tensors in an open region of E^4. Show that the exterior product of $A_i dx^i$ and $\frac{1}{2} B_{ij} dx^i dx^j$ equals $\frac{1}{3!} C_{ijk} dx^i dx^j dx^k$, where C_{ijk} is the exterior product of the tensors A_i and B_{ij}.

9. State Stokes' theorem (5.215) explicitly in terms of coordinates in E^4.

   a) for line integrals ($m = 1$)

  b) for surface integrals ($m = 2$).

10. (Difficult). Let A_{ij} be a covariant alternating tensor in an open region of $E^n (n \geq 3)$ and
let

$$d\left(\frac{1}{2!}A_{ij}dx^idx^j\right) = \frac{1}{3!}B_{kij}\,dx^k dx^i dx^j.$$

Let an allowable change of coordinates (5.219) be made under which the A_{ij} become $\overline{A}_{ij}$
and B_{kij} become $\overline{B}_{kij}$ (as in (5.220)). Show that in the new coordinates $d\left(\frac{1}{2!}\overline{A}_{ij}\,d\overline{x}^i d\overline{x}^j\right) = \frac{1}{3!}\overline{B}_{kij}\,d\overline{x}^k d\overline{x}^i d\overline{x}^j$.

Suggested References

Boothby, William M., An Introduction to Differentiable Manifolds and Riemannian Geom-
    etry. New York: Academic Press, 1975.

Brand, Louis, Vector and Tensor Analysis. New York: John Wiley and Sons, Inc., 1947.

Buck, R. C., Advanced Calculus, 3rd ed. New York: McGraw-Hill, 1978.

Courant, Richard J., Differential and Integral Calculus, transl. by E. J. McShane, Vol. 2.
    New York: Interscience, 1947.

Flanders, Harley, Differential Forms, with Applications to the Physical Sciences. New York:
    Academic Press, 1963.

Franklin, Philip, A Treatise on Advanced Calculus. New York: John Wiley and Sons, Inc.,
    1940.

Gibbs, J. W., Vector Analysis. New Haven: Yale University Press, 1913.

Goursat, Édouard, A Course in Mathematical Analysis, transl. by E. R. Hendrick, Vol. 1.
    New York: Dover Publications, 1970.

Hodge, W. V. D., The Theory and Applications of Harmonic Integrals, 2nd ed. Cambridge:
    Cambridge University Press, 1952.

Kellogg, O. D., Foundation of Potential Theory. Berlin: Springer, 1929.

Lamb, H., Hydrodynamics, 6th ed. Cambridge: Cambridge University Press, 1932.

Sternberg, Wolfgang J., and Turner L. Smith, The Theory of Potential and Spherical Har-
    monics. Toronto: The University of Toronto Press, 1946.

6
Infinite Series

6.1 INTRODUCTION

An infinite series is an indicated sum of the form:

$$a_1 + a_2 + \cdots + a_n + \cdots,$$

going on to infinitely many terms. Such series are familiar even in the simplest operations with numbers. Thus one writes:

$$\frac{1}{3} = 0.33333\ldots;$$

this is the same as saying

$$\frac{1}{3} = \frac{3}{10} + \frac{3}{100} + \frac{3}{1000} + \frac{3}{10,000} + \frac{3}{100,000} + \cdots.$$

Of course, we do not interpret this as an ordinary addition problem, which would take forever to carry out. Instead we say, for example, $1/3$ equals 0.33333 to *very good accuracy*. We "round off" after a certain number of decimal places and use the resulting *rational number*

$$0.33333 = \frac{33,333}{100,000}$$

as a sufficiently good *approximation* to the number $\frac{1}{3}$.

The procedure just used applies to the general series $a_1 + a_2 + \cdots + a_n + \cdots$. To evaluate it, we round off after k terms and replace the series by the finite sum

$$a_1 + a_2 + \cdots + a_k.$$

However, the rounding-off procedure must be justified; we must be sure that taking more than k terms would not significantly affect the result. For the series

$$1 + 1 + \cdots + 1 + \cdots,$$

such a justification is impossible. For one term gives 1 as sum, two terms give 2, three terms give 3, and so on; rounding off is of no help here. This series is an example of a *divergent series*.

On the other hand, for the series

$$1 + \frac{1}{4} + \frac{1}{9} + \frac{1}{16} + \cdots + \frac{1}{n^2} + \cdots$$

it seems *safe* to round off. Thus one has as sums of the first k terms:

$$1, \quad 1 + \frac{1}{4} = \frac{5}{4}, \quad 1 + \frac{1}{4} + \frac{1}{9} = \frac{49}{36}, \quad 1 + \frac{1}{4} + \frac{1}{9} + \frac{1}{16} = \frac{205}{144}, \cdots.$$

The sums do not appear to change much, and one would hazard the guess that the sum of 50 terms would not differ from that of four terms by more than, say, $\frac{1}{4}$. While it will be seen below that such appearances can be misleading, in this case our instinct happens to be right. This series is an example of a *convergent series*.

It is the purpose of the present chapter to systematize the procedure indicated and to formulate tests that enable one to decide when rounding off is meaningful (convergent case) or meaningless (divergent case). As the example of the decimal expansion of $\frac{1}{3}$ shows, the notion of infinite series lies right at the heart of the concept of the real number system. Accordingly, a complete theory of series would require a profound analysis of the real number system. It is not our purpose to carry this out here, so some of the rules will be justified in an intuitive manner. For a complete treatment the reader is referred to the books by Hardy and K. Knopp listed at the end of the chapter. Sections 2.23, 4.10, and 4.11 cover real variable theory in more depth and provide proofs for some of the key results of this chapter.

Because of the large number of theorems appearing in this chapter, the theorems will be numbered serially from 1 to 59 throughout the chapter.

6.2 INFINITE SEQUENCES

If to each positive integer n there is assigned a number s_n, then the numbers s_n are said to form an *infinite sequence*. The numbers are thought of as arranged in order, according to subscript:

$$s_1, s_2, s_3, \ldots, s_n, s_{n+1}, \ldots.$$

Examples of such sequences are the following:

$$\frac{1}{2}, \frac{1}{4}, \cdots, \frac{1}{2^n}, \cdots; \tag{6.1}$$

$$2, \left(\frac{3}{2}\right)^2, \left(\frac{4}{3}\right)^3, \cdots, \left(\frac{n+1}{n}\right)^n, \cdots; \tag{6.2}$$

$$1, 1 + \frac{1}{2}, 1 + \frac{1}{2} + \frac{1}{3}, \cdots, 1 + \frac{1}{2} + \frac{1}{3} + \cdots + \frac{1}{n}, \cdots. \tag{6.3}$$

These are formed by the rules:

$$s_n = \frac{1}{2^n}, \qquad s_n = \left(\frac{n+1}{n}\right)^n, \qquad s_n = 1 + \frac{1}{2} + \frac{1}{3} + \cdots + \frac{1}{n}.$$

At times it is convenient to number the members of the sequence starting with 0, with 2, or with some other integer.

A sequence s_n is said to *converge* to the number s or to have the *limit s*:

$$\lim_{n \to \infty} s_n = s \tag{6.4}$$

if to each number $\epsilon > 0$ a value N can be found such that

$$|s_n - s| < \epsilon \quad \text{for} \quad n > N. \tag{6.5}$$

This is illustrated in Fig. 6.1(a). If s_n does not converge, it is said to *diverge*.

The limit s is clearly unique. For if s' is a limit different from s, we let $\epsilon = |s - s'|/2$. Then as in (6.5), $|s_n - s| < \epsilon$ for $n > N$, and similarly, $|s_n - s'| < \epsilon$ for $n > N'$. Therefore if N_0 is the larger of N and N', for $n > N_0$ we have both $|s_n - s| < \epsilon$ and $|s_n - s'| < \epsilon$, so that

$$2\epsilon = |s - s'| = |s - s_n + s_n - s'| \leq |s - s_n| + |s_n - s'| < \epsilon + \epsilon = 2\epsilon,$$

and that is impossible.

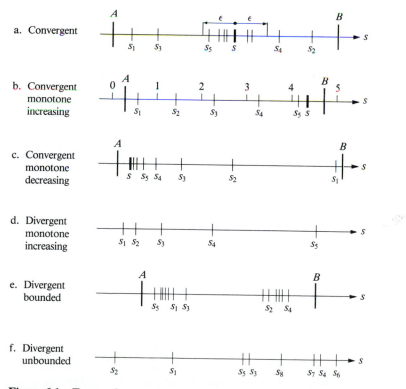

Figure 6.1 Types of sequences.

By a similar reasoning, other limits in the calculus are unique (for example, limits of functions, the definite integral as a limit).

One can regard a sequence s_n as a function $s(n)$ of the *integer* variable n. The limit definition (6.4), (6.5) is then formally the same as that for a function $f(x)$ of the real variable x.

The sequences (6.1) and (6.2) can be shown to converge:

$$\lim_{n \to \infty} \frac{1}{2^n} = 0, \qquad \lim_{n \to \infty} \left(\frac{n+1}{n} \right)^n = e,$$

while (6.3) diverges.

A sequence s_n is said to be *monotone increasing* if $s_1 \leq s_2 \leq \cdots \leq s_n \leq s_{n+1} \cdots$; this is illustrated in Figs. 6.1(b) and (d). Since the $=$ sign is permitted, it is perhaps clearer to call these sequences *monotone nondecreasing*. Similarly, a sequence s_n is called *monotone decreasing* (or *monotone nonincreasing*) if $s_n \geq s_{n+1}$ for all n; this is shown in Fig. 6.1(c).

A sequence s_n is said to be *bounded* if there are two numbers A and B such that $A \leq s_n \leq B$ for all n. All the sequences suggested in Fig. 6.1 are bounded except those of (d) and (f).

THEOREM 1   Every bounded monotone sequence converges.

Thus let s_n be monotone increasing as in Fig. 6.1(b). The figure suggests that since the numbers s_n move to the right as n increases but are not allowed to pass B, they must pile up on a number s to the left of B or at B. One can determine the decimal expansion of s as follows: Let k_1 be the largest integer such that $s_n \geq k_1$ for n sufficiently large; in Fig. 6.1(b), $k_1 = 4$. Let k_2 be the largest integer between 0 and 9 (inclusive) such that

$$s_n \geq k_1 + \frac{k_2}{10}$$

for n sufficiently large; in Fig. 6.1(b), $k_2 = 3$. Let k_3 be the largest integer between 0 and 9 such that

$$s_n \geq k_1 + \frac{k_2}{10} + \frac{k_3}{100}$$

for n sufficiently large. Proceeding thus, we obtain the complete decimal expansion of the limit s:

$$s = k_1 + \frac{k_2}{10} + \frac{k_3}{100} + \frac{k_4}{1000} + \cdots .$$

If $s = 4\frac{1}{3}$, one would find

$$s = 4 + \frac{3}{10} + \frac{3}{100} + \frac{3}{1000} + \cdots .$$

A careful perusal of the discussion just given shows that the number s has actually been precisely defined as a real number and that the difference $s - s_n$ can be made as small as desired by making n sufficiently large. Thus Theorem 1 is proved. (See Section 2.23 for a more detailed proof.)

If a sequence s_n is monotone increasing, there are then only two possibilities: The sequence is bounded and has a limit, or the sequence is unbounded. In the latter case, for n sufficiently large the sequence becomes larger than any given number K:

$$s_n > K \quad \text{for} \quad n > N; \tag{6.6}$$

this we also describe by writing:

$$\lim_{n \to \infty} s_n = \infty, \tag{6.7}$$

or "s_n diverges to ∞." This is illustrated by the sequence $s_n = n$; see Fig. 6.1(d). It can happen that $\lim s_n = \infty$ without s_n being a monotone sequence; for example, $s_n = n + (-1)^n$.

Similarly, one defines:

$$\lim_{n \to \infty} s_n = -\infty, \tag{6.8}$$

if for every number K an N can be found such that

$$s_n < K \quad \text{for} \quad n > N. \tag{6.9}$$

Every monotone decreasing sequence is either bounded or divergent to $-\infty$.

THEOREM 2 *Every unbounded sequence diverges.*

Proof. Suppose on the contrary that the unbounded sequence s_n converges to s; then

$$s - \epsilon < s_n < s + \epsilon$$

for $n > N$. Thus all but a finite number of s_n lie in the interval between $s - \epsilon$ and $s + \epsilon$, and the sequence is necessarily bounded, contrary to assumption. □

6.3   UPPER AND LOWER LIMITS

It remains to consider sequences s_n that are bounded but not monotone. These can converge, as in Fig. 6.1(a), but can equally well diverge, as the example

$$1, -1, 1, -1, \ldots, (-1)^{n+1}, \ldots$$

illustrates. This example also suggests the way in which a general bounded sequence can diverge, namely, by *oscillating* between various limiting values. In this example the sequence oscillates between 1 and -1. The sequence

$$1, 0, -1, 0, 1, 0, -1, 0, 1, \ldots$$

oscillates with three limit values $1, 0, -1$; the oscillation suggests trigonometric functions. In fact, the general term of this sequence can be written thus:

$$s_n = \sin\left(\tfrac{1}{2}n\pi\right).$$

In general the bounded sequence can wander back and forth in its interval, coming arbitrarily close to many (even all) values of the interval; it diverges not for lack of limiting values, but because it has too many, namely, more than one.

Of greatest interest for such bounded sequences are the largest and least limiting values, which are termed the *upper* and *lower limits* of the sequence s_n. The upper limit is defined as follows:

$$\overline{\lim_{n \to \infty}} \; s_n = k$$

if, for every positive ϵ,

$$|s_n - k| < \epsilon$$

for infinitely many values of n and if no number larger than k has this property. Similarly one defines the lower limit:

$$\underline{\lim_{n \to \infty}} \; s_n = h$$

if, for every positive ϵ,

$$|s_n - h| < \epsilon$$

for infinitely many values of n and if no number less than h has this property. It is a theorem that *every bounded sequence has an upper limit and a lower limit*; this is proved in much the same way as Theorem 1. (The proof of Theorem C in Section 2.23 shows the existence of a finite upper limit s_0 for a sequence that is bounded above.)

EXAMPLES

$$\overline{\lim_{n \to \infty}} \; (-1)^n = 1, \qquad \underline{\lim_{n \to \infty}} \; (-1)^n = -1,$$

$$\overline{\lim_{n \to \infty}} \; \sin\left(\tfrac{1}{3}n\pi\right) = \tfrac{1}{2}\sqrt{3}, \qquad \underline{\lim_{n \to \infty}} \; \sin\left(\tfrac{1}{3}n\pi\right) = -\tfrac{1}{2}\sqrt{3},$$

$$\overline{\lim_{n \to \infty}} \; (-1)^n(1 + 1/n) = 1, \qquad \underline{\lim_{n \to \infty}} \; (-1)^n(1 + 1/n) = -1,$$

$$\overline{\lim_{n \to \infty}} \; \sin n = 1, \qquad \underline{\lim_{n \to \infty}} \; \sin n = -1.$$

These should be verified by graphing the sequences as in Fig. 6.1. The proofs of the relations are not difficult, with the exception of the last two, which are quite subtle. ■

The definitions of upper and lower limits can be extended to unbounded sequences. One must then consider ∞ and $-\infty$ as possible "limiting values"; thus ∞ is a limiting value if for every number K, $s_n > K$ for infinitely many n; $-\infty$ is a limiting value if for every number K, $s_n < K$ for infinitely many n. The upper limit is then defined as the "largest limiting value" and the lower limit as the "smallest limiting value."

EXAMPLES

$$\overline{\lim_{n \to \infty}} \; (-1)^n n = \infty, \qquad \underline{\lim_{n \to \infty}} \; (-1)^n n = -\infty,$$

$$\overline{\lim_{n \to \infty}} \; n^2 \sin^2 \left(\tfrac{1}{2} n\pi \right) = \infty, \qquad \underline{\lim_{n \to \infty}} \; n^2 \sin^2 \left(\tfrac{1}{2} n\pi \right) = 0.$$

Every sequence will then possess upper and lower limits. It should be noted that when $\lim s_n = -\infty$, the upper and lower limits are both $-\infty$. In general, the lower limit cannot exceed the upper limit. ■

THEOREM 3   If the sequence s_n converges to s, then

$$\overline{\lim_{n \to \infty}} \; s_n = \underline{\lim_{n \to \infty}} \; s_n = s;$$

conversely, if upper and lower limits are equal and finite, the sequence converges.

The first part of the theorem follows from the definition of convergence; for the limit s satisfies all conditions for both upper and lower limits. If the upper and lower limits are both finite and equal to s, the sequence is necessarily bounded, for otherwise the upper limit would be $+\infty$ or the lower limit would be $-\infty$. If s_n does not converge to s, then for some ϵ there are infinitely many members of the sequence such that $|s_n - s| > \epsilon$. One can then proceed as in the proof of Theorem 1 to produce a limiting value k other than s. This contradicts the fact that s is both the largest and least limiting value.

6.4   FURTHER PROPERTIES OF SEQUENCES

We first point out connections between sequences and continuity of functions.

THEOREM 4   Let $y = f(x)$ be defined for $a \le x \le b$. If $f(x)$ is continuous at x_0, then

$$\lim_{n \to \infty} f(x_n) = f(x_0)$$

for every sequence x_n in the interval converging to x_0. Similarly, if $f(x, y)$ is defined in a domain D and is continuous at (x_0, y_0), then

$$\lim_{n \to \infty} f(x_n, y_n) = f(x_0, y_0)$$

for all sequences x_n, y_n such that (x_n, y_n) is in D and x_n converges to x_0, y_n converges to y_0.

We prove the statement for $f(x)$, the proof for a function of two (or more) variables being similar. Given $\epsilon > 0$, a δ can be found such that $|f(x) - f(x_0)| < \epsilon$

for $|x - x_0| < \delta$; this is simply the definition of continuity. One can then choose N so that $|x_n - x_0| < \delta$ for $n > N$, since x_n converges to x_0. Accordingly, $|f(x_n) - f(x_0)| < \epsilon$ for $n > N$; that is, $f(x_n)$ converges to $f(x_0)$.

We remark that there is a converse to Theorem 4: If $f(x)$ has the property that $f(x_n)$ converges to $f(x_0)$ for every sequence x_n that converges to x_0, then $f(x)$ must be continuous at x_0 (Theorem G in Section 2.23).

THEOREM 5   If

$$\lim_{n \to \infty} x_n = x, \qquad \lim_{n \to \infty} y_n = y,$$

then

$$\lim_{n \to \infty} (x_n \pm y_n) = x \pm y, \qquad \lim_{n \to \infty} (x_n \cdot y_n) = x \cdot y, \qquad \lim_{n \to \infty} \frac{x_n}{y_n} = \frac{x}{y},$$

provided, in the last case, that there is no division by 0.

Proof.   The function $f(x, y) = x + y$ is continuous for all x and y. Hence by Theorem 4

$$\lim_{n \to \infty} f(x_n, y_n) = \lim_{n \to \infty} (x_n + y_n) = x + y.$$

The other rules follow in the same way, with $f(x, y)$ chosen successively as $x - y$, $x \cdot y$, x/y.   □

THEOREM 6 (*Cauchy criterion*)   If the sequence s_n has the property that for every $\epsilon > 0$ an N can be found such that

$$|s_m - s_n| < \epsilon \quad \text{for} \quad n > N \quad \text{and} \quad m > N, \tag{6.10}$$

then s_n converges. Conversely, if s_n converges, then for every $\epsilon > 0$ an N can be found such that (6.10) holds.

Proof.   If (6.10) holds, then the sequence s_n is necessarily bounded. Indeed, if we choose a fixed ϵ and N according to (6.10), then for a fixed $m > N$,

$$s_m - \epsilon < s_n < s_m + \epsilon$$

for all n greater than N. Hence all but a finite number of the s_n lie in this interval, so that the sequence must be bounded. If the sequence fails to converge, then by Theorem 3 the upper limit $\bar{s}$ and lower limit $\underline{s}$ must be unequal. Let $\bar{s} - \underline{s} = a$. Then

$$|s_m - \bar{s}| < \frac{a}{3}, \qquad |s_n - \underline{s}| < \frac{a}{3}$$

for infinitely many m and n. Hence $|s_m - s_n| > a/3$ for infinitely many m and n, as Fig. 6.2 shows. If ϵ is chosen as $a/3$, condition (6.10) cannot be satisfied for any N. This is a contradiction; hence $\bar{s} = \underline{s}$, and the sequence converges.

Figure 6.2 Cauchy condition.

Conversely, if s_n converges to s, then, for a given $\epsilon > 0$, N can be chosen so that $|s_n - s| < \frac{1}{2}\epsilon$ for $n > N$. Hence for $m > N$ and $n > N$,

$$|s_m - s_n| = |s_m - s + s - s_n| \leq |s_m - s| + |s - s_n| < \tfrac{1}{2}\epsilon + \tfrac{1}{2}\epsilon = \epsilon.$$

Accordingly, condition (6.10) is satisfied. □

PROBLEMS

1. Show that the following sequences converge and find their limits:

   a) $\dfrac{(n^2 + 1)}{(n^3 + 1)}$,   b) $\dfrac{\log n}{n}$,   c) $\dfrac{n}{2^n}$,

   d) $n \log \left(1 + \dfrac{1}{n}\right)$,   e) $s_n = 1$ for $n = 1, 2, 3, \ldots$

2. Determine the upper and lower limits of the sequences:

   a) $\cos n\pi$,   b) $\sin \frac{1}{3}n\pi$,   c) $n \sin \frac{1}{2}n\pi$.

3. Construct sequences having the properties stated:

   a) $\overline{\lim\limits_{n \to \infty}} \, s_n = 2$, $\varliminf\limits_{n \to \infty} s_n = 0$;

   b) $\overline{\lim\limits_{n \to \infty}} \, s_n = 0$, $\varliminf\limits_{n \to \infty} s_n = -\infty$;

   c) $\overline{\lim\limits_{n \to \infty}} \, s_n = +\infty$, $\varliminf\limits_{n \to \infty} s_n = +\infty$.

4. Show that the Cauchy criterion is satisfied for the sequence $s_n = 1/n$.

5. Evaluate e to 2 decimal places from its definition as limit of the sequence $(1 + 1/n)^n$.

6. Evaluate $\overline{\lim\limits_{n \to \infty}} \, x^n$ and $\varliminf\limits_{n \to \infty} x^n$.

7. The number π can be defined as the limit, as $n \to \infty$, of the area of a regular polygon of 2^n sides inscribed in a circle of radius 1. Show that the sequence is monotone and use it to evaluate π approximately.

6.5 INFINITE SERIES

An infinite series is an indicated sum:

$$a_1 + a_2 + \cdots + a_n + \cdots \tag{6.11}$$

of the members of a sequence a_n. The series can be abbreviated by the Σ sign:

$$a_1 + a_2 + \cdots + a_n + \cdots = \sum_{n=1}^{\infty} a_n, \qquad (6.12)$$

and the Σ notation is to be preferred, except for very simple series.

Associated with each infinite series Σa_n is the *sequence of partial sums* S_n:

$$S_n = a_1 + a_2 + \cdots + a_n = \sum_{j=1}^{n} a_j. \qquad (6.13)$$

(Note that the index j in the sum here is a *dummy* index, so that the result does not depend on j. Thus $\sum_{j=1}^{n} a_j = \sum_{m=1}^{n} a_m$. This is like the case of the integration variable in a definite integral.) Accordingly,

$$S_1 = a_1, \qquad S_2 = a_1 + a_2, \qquad \ldots, \qquad S_n = a_1 + \cdots + a_n.$$

DEFINITION The infinite series $\sum_{n=1}^{\infty} a_n$ is *convergent* if the sequence of partial sums is convergent; the series is *divergent* if the sequence of partial sums is divergent. If the series is convergent and the sequence of partial sums S_n converges to S, then S is called the *sum* of the series, and one writes

$$\sum_{n=1}^{\infty} a_n = S. \qquad (6.14)$$

Thus by definition,

$$\sum_{n=1}^{\infty} a_n = \lim_{n \to \infty} \sum_{j=1}^{n} a_j$$

when the limit exists.

If $\lim S_n = +\infty$ or $\lim S_n = -\infty$, the series Σa_n is said to be *properly divergent*. From Theorem 1 of Section 6.2, one concludes:

THEOREM 7 If $a_n \geq 0$ for $n = 1, 2, \ldots$, then the series $\sum_{n=1}^{\infty} a_n$ is either convergent or properly divergent.

For, since $a_n \geq 0$,

$$S_1 = a_1 \leq S_2 = a_1 + a_2 \leq S_3 = a_1 + a_2 + a_3 \ldots;$$

that is, the S_n form a monotone increasing sequence. If this sequence is bounded, it converges by Theorem 1; if it is unbounded, then necessarily $\lim S_n = \infty$, so that the series is properly divergent.

EXAMPLE 1 The series

$$\frac{1}{2} + \frac{3}{4} + \frac{7}{8} + \cdots + \frac{2^n - 1}{2^n} + \cdots$$

is properly divergent. For each term is at least equal to $\frac{1}{2}$, and accordingly,

$$S_n \geqq \frac{1}{2} + \frac{1}{2} + \cdots + \frac{1}{2} = n \cdot \frac{1}{2}.$$

The sequence S_n is monotone but unbounded, and $\lim S_n = \infty$. ∎

It should be noted that the *terms* of this series form a convergent sequence; for

$$\lim_{n \to \infty} \left(\frac{2^n - 1}{2^n} \right) = \lim_{n \to \infty} \left(1 - \frac{1}{2^n} \right) = 1.$$

This does not imply convergence of the series. In general, one must carefully distinguish the notions of series, sequence of terms of a series, and sequence of partial sums of a series. The series is simply another way of describing the sequence of partial sums; the terms of the series describe the steps taken from each partial sum to the next. This is suggested in Fig. 6.3.

Figure 6.3 Series Σa_n versus sequence S_n of partial sums.

EXAMPLE 2 The series

$$\frac{1}{2} + \frac{1}{4} + \frac{1}{8} + \cdots + \frac{1}{2^n} + \cdots$$

is convergent. Here the *partial sums* form the sequence

$$\frac{1}{2}, \quad \frac{3}{4}, \quad \frac{7}{8}, \quad \cdots, \quad \frac{2^n - 1}{2^n}, \quad \cdots,$$

which converges, as before, to 1. Hence

$$\sum_{n=1}^{\infty} \frac{1}{2^n} = 1. \quad ∎$$

From Theorem 5 of Section 6.4, one deduces a rule for addition and subtraction of convergent series, as well as a rule for multiplication by a constant:

THEOREM 8 If $\displaystyle\sum_{n=1}^{\infty} a_n$ and $\displaystyle\sum_{n=1}^{\infty} b_n$ are convergent with sums A and B respectively and k is a constant, then

$$\sum_{n=1}^{\infty} (a_n + b_n) = A + B, \qquad \sum_{n=1}^{\infty} (a_n - b_n) = A - B,$$

$$\sum_{n=1}^{\infty} (ka_n) = k \sum_{n=1}^{\infty} a_n = kA. \tag{6.15}$$

For if one introduces the partial sums:

$$A_n = a_1 + \cdots + a_n, \qquad B_n = b_1 + \cdots + b_n,$$
$$S_n = (a_1 + b_1) + \cdots + (a_n + b_n),$$

then $S_n = A_n + B_n$. Since A_n converges to A and B_n to B, S_n converges to $A + B$. A similar reasoning holds for the series $\Sigma(a_n - b_n)$, Σka_n.

The Cauchy criterion (Theorem 6) can also be interpreted in terms of series:

THEOREM 9   The series $\displaystyle\sum_{n=1}^{\infty} a_n$ is convergent if and only if to each $\epsilon > 0$ an N can be found such that

$$|a_{n+1} + a_{n+2} + \cdots + a_m| < \epsilon \quad \text{for} \quad m > n > N. \qquad (6.16)$$

This is a simple rewriting of condition (6.10). For, when $m > n$,

$$S_m - S_n = (a_1 + a_2 + \cdots + a_m) - (a_1 + \cdots + a_n) = a_{n+1} + \cdots + a_m.$$

6.6   TESTS FOR CONVERGENCE AND DIVERGENCE

A topic of prime importance is the formulation of rules or "tests" that permit one to determine whether a particular series converges or diverges. The problem is similar to and, in fact, closely related to that of improper integrals. This connection will be made clear later.

A few preliminary remarks will be helpful. The convergence or divergence of a series is unaffected if one modifies a *finite* number of terms of the series; thus the first ten terms can be replaced by zeros without affecting the convergence, though it will of course affect the sum if the series converges. The terms of the series can all be multiplied by the *same* nonzero constant k without affecting convergence or divergence, for the partial sums S_n converge or diverge according to whether kS_n converges or diverges. In particular, all terms can be multiplied by -1 without affecting convergence; however, other changes in sign can change the character of the series completely.

The operations of *grouping terms* and of *rearrangement* of the terms of a series require considerable care. They will be discussed briefly later.

THEOREM 10 (*The nth term test*)   If it is not true that

$$\lim_{n \to \infty} a_n = 0,$$

then $\displaystyle\sum_{n=1}^{\infty} a_n$ diverges.

Proof.   If the series were to converge to S, then

$$\lim_{n \to \infty} a_n = \lim_{n \to \infty} (S_{n+1} - S_n) = \lim_{n \to \infty} S_{n+1} - \lim_{n \to \infty} S_n = S - S = 0.$$

Hence if a_n does not converge to 0, the series cannot converge. $\square$

It is to be emphasized that this test can be used only to prove divergence. If $\lim_{n \to \infty} a_n = 0$, the series Σa_n may converge or diverge. It should also be noted that to show that a_n does not converge to 0, one need not show that a_n converges to a number different from 0, for the same conclusion is reached if one can show that the sequence a_n diverges.

EXAMPLES

1. $\sum_{n=1}^{\infty} (-1)^n$. Here $a_n = \pm 1$, hence a_n diverges and the series diverges.

2. $\sum_{n=1}^{\infty} n$. $\lim_{n \to \infty} a_n = \lim_{n \to \infty} n = \infty$. The series diverges.

3. $\sum_{n=1}^{\infty} \dfrac{3n-1}{4n+5}$. $\lim_{n \to \infty} a_n = \dfrac{3}{4}$. The series diverges.

4. $\sum_{n=1}^{\infty} \dfrac{1}{n}$. $\lim_{n \to \infty} a_n = 0$. The test proves nothing. The series is the *harmonic* series and will be shown to diverge.

5. $\sum_{n=1}^{\infty} \dfrac{1}{2^n}$. $\lim_{n \to \infty} a_n = 0$. The test proves nothing. The series converges to 1, as shown in Section 6.5. ∎

If a series Σa_n is such that $\Sigma |a_n|$ is convergent, then the series Σa_n is called *absolutely convergent*.

THEOREM 11 (*Theorem on absolute convergence*) If $\sum_{n=1}^{\infty} |a_n|$ converges, then $\sum_{n=1}^{\infty} a_n$ converges; that is, every absolutely convergent series is convergent.

Proof. The theorem follows from the Cauchy criterion (Theorem 9 of Section 6.5). For

$$|a_{n+1} + \cdots + a_m| \leq |a_{n+1}| + |a_{n+2}| + \cdots + |a_m|.$$

If $\sum_{n=1}^{\infty} |a_n|$ converges, then the sum on the right is less than ϵ when $n > N$, for suitable choice of N; hence the sum on the left is less than ϵ for $n > N$, so that the series Σa_n converges. □

One can interpret this theorem as stating that introduction of minus signs before various terms of a positive term series tends to *help* convergence; if the original series diverges, introduction of enough minus signs may make the series converge; if the original series converges, introduction of the minus signs will make the series converge even more rapidly.

EXAMPLES

6. $\displaystyle\sum_{n=0}^{\infty} \frac{(-1)^n}{2^n} = 1 - \frac{1}{2} + \frac{1}{4} \cdots$. Since the series $\displaystyle\sum_{n=1}^{\infty} \frac{1}{2^n}$ converges, this series converges.

7. $\displaystyle\sum_{n=1}^{\infty} (-1)^n$. The series of absolute values is $1 + 1 + \cdots + 1 + \cdots$. This series diverges, so that Theorem 11 gives no help. However the nth term fails to converge to 0, so that the series diverges.

8. $\displaystyle\sum_{n=1}^{\infty} \frac{(-1)^n}{-n} = 1 - \frac{1}{2} + \frac{1}{3} - \frac{1}{4} + \cdots$. The series of absolute values is the harmonic series of Example 4. Although the harmonic series diverges, it will be seen that this series *converges*. ■

A series Σa_n that converges but is not absolutely convergent is called *conditionally convergent*. Such a series converges because the minus signs have been properly introduced. An example is the series of Example 8.

> **THEOREM 12** (*Comparison test for convergence*)   If $|a_n| \leqq b_n$ for $n = 1$, $2, \ldots$ and $\displaystyle\sum_{n=1}^{\infty} b_n$ converges, then $\displaystyle\sum_{n=1}^{\infty} a_n$ is absolutely convergent.

Proof.   By Theorem 7 the series $\Sigma |a_n|$ is either convergent or properly divergent. If it were properly divergent, then

$$\lim_{n \to \infty} \sum_{j=1}^{n} |a_j| = \infty;$$

since $b_n \geqq |a_n|$, one would then have

$$\lim_{n \to \infty} \sum_{j=1}^{n} b_j = \infty,$$

so that Σb_n would diverge, contrary to assumption. Hence $\Sigma |a_n|$ converges, and Σa_n is absolutely convergent.   □

EXAMPLE 9 $\displaystyle\sum_{n=1}^{\infty} \frac{1}{n2^n}$. Since

$$\left| \frac{1}{n2^n} \right| = \frac{1}{n2^n} \leqq \frac{1}{2^n}$$

and the series $\Sigma 1/2^n$ converges, the given series converges.   ■

THEOREM 13 (*Comparison test for divergence*)   If $a_n \geq b_n \geq 0$ for $n = 1, 2, \ldots$ and $\sum\limits_{n=1}^{\infty} b_n$ diverges, then $\sum\limits_{n=1}^{\infty} a_n$ diverges.

Proof.   If Σa_n were to converge, then $\Sigma b_n = \Sigma|b_n|$ would converge by Theorem 12. Hence Σa_n diverges.   □

It should be emphasized that the test of Theorem 13 applies only to positive term series.

EXAMPLE 10 $\sum\limits_{n=1}^{\infty} \dfrac{n-1}{n^2}$. It is apparent that the terms are close to those of the divergent harmonic series $\sum\limits_{n=1}^{\infty} \dfrac{1}{n}$. However, the inequality

$$\frac{n-1}{n^2} > \frac{1}{n}$$

is false since it is equivalent to the statement $n^2 - n > n^2$. On the other hand,

$$\frac{n-1}{n^2} > \frac{1}{2n}$$

for $n > 2$, since this is equivalent to the inequality: $2n^2 - 2n > n^2$ or to the inequality $n > 2$. The constant factor $\frac{1}{2}$ and the failure of the condition for $n < 3$ have no effect on convergence or divergence. Hence Theorem 13 applies, and one concludes that the series diverges.   ■

It should be stressed that one can conclude nothing about the series Σa_n from the inequality $|a_n| \leq |b_n|$, where Σb_n *diverges*; nor can one conclude anything from the inequality $a_n \geq b_n > 0$, when Σb_n *converges*.

THEOREM 14 (*Integral test*)   Let $y = f(x)$ satisfy the following conditions:

a) $f(x)$ is defined and continuous for $c \leq x < \infty$;
b) $f(x)$ decreases as x increases and $\lim\limits_{x \to \infty} f(x) = 0$;
c) $f(n) = a_n$.

Then the series $\sum\limits_{n=1}^{\infty} a_n$ converges or diverges according to whether the improper integral $\int_c^{\infty} f(x)\, dx$ converges or diverges.

Proof.   Let us suppose the improper integral converges. Assumptions (b) and (c) imply that $a_n > 0$ for n sufficiently large. Hence by Theorem 7 of Section 6.5 the series Σa_n must either converge or properly diverge. Let the integer m be chosen so that $m > c$. Then, since $f(x)$ is decreasing,

$$\int_n^{n+1} f(x)\, dx \geq f(n+1) = a_{n+1} \quad \text{for} \quad n \geq m,$$

as illustrated in Fig. 6.4. Hence $0 < a_{m+1} + \cdots + a_{m+p} \leq \int_m^{m+p} f(x)\, dx \leq \int_m^{\infty} f(x)\, dx$. It is thus impossible for the series Σa_n to diverge properly, and the series must be convergent. $\square$

Figure 6.4 Proof of integral test for convergence.

The proof in the case of divergence is left as an exercise (Problem 10 following Section 6.7).

THEOREM 15 The harmonic series of order p:

$$\sum_{n=1}^{\infty} \frac{1}{n^p} = 1 + \frac{1}{2^p} + \frac{1}{3^p} + \cdots$$

converges for $p > 1$ and diverges for $p \leq 1$.

Proof. The nth term fails to converge to 0 when $p \leq 0$, so that the series surely diverges when $p \leq 0$. For $p > 0$ the integral text can be applied, with $f(x) = 1/x^p$. Now for $p \neq 1$,

$$\int_1^{\infty} \frac{1}{x^p}\, dx = \lim_{b \to \infty} \int_1^b \frac{1}{x^p}\, dx = \lim_{b \to \infty} \left[\frac{1}{p-1} \left(1 - \frac{1}{b^{p-1}} \right) \right].$$

This limit exists and has the value $1/(p-1)$ for $p > 1$. For $p < 1$ the integral diverges, and for $p = 1$,

$$\int_1^{\infty} \frac{1}{x}\, dx = \lim_{b \to \infty} \log b = \infty,$$

so that again there is divergence. The theorem is now established. $\square$

When $p = 1$, one obtains the divergent series

$$1 + \frac{1}{2} + \cdots + \frac{1}{n} + \cdots,$$

commonly called the *harmonic series*; the connection with harmony (that is, music) will be pointed out in connection with the Fourier series in the next chapter. For general (even complex) values of p the series of Theorem 15 defines a function $\zeta(p)$, the *zeta-function of Riemann*.

THEOREM 16 The geometric series

$$a + ar + ar^2 + \cdots + ar^n + \cdots = \sum_{n=0}^{\infty} ar^n \qquad (a \neq 0)$$

converges for $-1 < r < 1$:

$$\sum_{n=0}^{\infty} ar^n = \frac{a}{1 - r}, \qquad -1 < r < 1, \tag{6.17}$$

and diverges for $|r| \geq 1$.

Proof. The algebraic identity

$$1 - r^n = (1 - r)(1 + r + \cdots + r^{n-1})$$

leads to the rule for geometric progressions:

$$a + ar + \cdots + ar^{n-1} = a(1 + r + \cdots + r^{n-1}) = a\frac{1 - r^n}{1 - r} \qquad (r \neq 1).$$

Therefore for the geometric series,

$$S_n = a\frac{1 - r^n}{1 - r}$$

for $r \neq 1$. For $-1 < r < 1$, r^n converges to 0; here S_n converges to $a/(1 - r)$. For $|r| \geq 1$ the nth term of the series does not converge to 0, so that the series diverges. $\square$

THEOREM 17 (*Ratio test*) If $a_n \neq 0$ for $n = 1, 2, \ldots$ and

$$\lim_{n \to \infty} \left| \frac{a_{n+1}}{a_n} \right| = L,$$

then

if $L < 1$, $\sum_{n=1}^{\infty} a_n$ is absolutely convergent,

if $L = 1$, the test fails,

if $L > 1$, $\sum_{n=1}^{\infty} a_n$ is divergent.

Proof. Let us assume first that $L < 1$. Let $r = L + (1 - L)/2 = (1 + L)/2$, so that $L < r < 1$. Let $b_n = |a_{n+1}/a_n|$. By assumption, b_n converges to L, so that $0 < b_n < r$ for $n > N$ and appropriate choice of N. Now one can write

$$|a_{N+1}| + |a_{N+2}| + \cdots + |a_{N+k}| + \cdots$$

$$= |a_{N+1}|\left(1 + \left|\frac{a_{N+2}}{a_{N+1}}\right| + \left|\frac{a_{N+2}}{a_{N+1}}\right|\left|\frac{a_{N+3}}{a_{N+2}}\right| + \cdots\right)$$

$$= |a_{N+1}|(1 + b_{N+1} + b_{N+1}b_{N+2} + b_{N+1}b_{N+2}b_{N+3} + \cdots).$$

Since $0 < b_n < r$ for $n > N$, each term of the series in parentheses is less than the corresponding term of the geometric series

$$|a_{N+1}|(1 + r + r^2 + r^3 + \cdots).$$

This series converges, by Theorem 16, since $r < 1$. Hence by Theorem 12 the series $\sum_{n=N+1}^{\infty} |a_n|$ converges, so that $\sum_{n=1}^{\infty} |a_n|$ converges and Σa_n is absolutely convergent.

If $L > 1$, then $|a_{n+1}/a_n| \geq 1$ for n sufficiently large, so that $|a_n| \leq |a_{n+1}| \leq |a_{n+2}| \ldots$. Since the terms are increasing in absolute value (and none is 0), it is impossible for a_n to converge to 0. Accordingly, the series diverges.

For $L = 1$ the series can converge or diverge. This is illustrated by the harmonic series of order p, for

$$\lim_{n \to \infty} \frac{n^p}{(n+1)^p} = \lim_{n \to \infty} \left(\frac{n}{n+1}\right)^p = 1.$$

The limit of the ratio is 1; however, the series converges when $p > 1$ and diverges when $p \leq 1$ by Theorem 15. $\square$

THEOREM 18 (*Alternating series test*) An alternating series

$$a_1 - a_2 + a_3 - a_4 + \cdots = \sum_{n=1}^{\infty} (-1)^{n+1} a_n, \qquad a_n > 0,$$

converges if the following two conditions are satisfied:

a) its terms are decreasing in absolute value:

$$a_{n+1} \leq a_n \quad \text{for} \quad n = 1, 2, \ldots,$$

b) $\lim_{n \to \infty} a_n = 0.$

Proof. Let $S_n = a_1 - a_2 + a_3 - a_4 + \cdots \pm a_n$. Then $S_1 = a_1$, $S_2 = a_1 - a_2 < S_1$, $S_3 = S_2 + a_3 > S_2$, $S_3 = S_1 - (a_2 - a_3) < S_1$, so that $S_2 < S_3 < S_1$. Reasoning in this way, we conclude that

$$S_1 > S_3 > S_5 > S_7 > \cdots > S_6 > S_4 > S_2,$$

as shown in Fig. 6.5. Thus the odd partial sums form a bounded monotone decreasing sequence, and the even partial sums form a bounded monotone increasing sequence. By Theorem 1 of Section 6.2, both sequences converge:

$$\lim_{n \to \infty} S_{2n+1} = S^*, \qquad \lim_{n \to \infty} S_{2n} = S^{**}.$$

Now

$$\lim_{n \to \infty} a_{2n+1} = \lim_{n \to \infty} (S_{2n+1} - S_{2n}) = S^* - S^{**}.$$

By assumption (b) this limit is 0, so that $S^* = S^{**}$. It follows that the series converges to S^*.   □

Figure 6.5   Alternating series.

THEOREM 19 (*Root test*)   Let a series $\displaystyle\sum_{n=1}^{\infty} a_n$ be given and let

$$\lim_{n \to \infty} \sqrt[n]{|a_n|} = R.$$

Then

> if $R < 1$, the series is absolutely convergent;
>
> if $R > 1$, the series diverges;
>
> if $R = 1$, the test fails.

Proof.   If $R < 1$, then, as in the proof for the ratio test, one can choose r and N such that $r < 1$ and

$$\sqrt[n]{|a_n|} < r \quad \text{for} \quad n > N.$$

Hence

$$|a_n| < r^n \quad \text{for} \quad n > N,$$

and the series $\sum |a_n|$ converges by comparison with the geometric series $\sum r^n$. If $R > 1$, $\sqrt[n]{|a_n|} > 1$ for n sufficiently large, so that $|a_n| > 1$, and the nth term cannot converge to zero. The failure for $R = 1$ is again shown by the harmonic series; thus for $a_n = 1/n^p$,

$$\lim_{n \to \infty} \sqrt[n]{\frac{1}{n^p}} = \lim_{n \to \infty} \frac{1}{n^{p/n}} = \lim_{n \to \infty} \frac{1}{e^{(p/n)\log n}} = \frac{1}{e^0} = 1,$$

since $(1/n)\log n$ converges to 0. These series converge for $p > 1$ and diverge for $p \leq 1$; thus the root test can yield no information when $R = 1$.   □

6.7   EXAMPLES OF APPLICATIONS OF TESTS FOR CONVERGENCE AND DIVERGENCE

The results of the preceding section provide the following tests:

a) *nth term test for divergence* (Theorem 10);
b) *comparison test for convergence* (Theorem 12);
c) *comparison test for divergence* (Theorem 13);
d) *integral test* (Theorem 14);
e) *ratio test* (Theorem 17);
f) *alternating series test* (Theorem 18);
g) *root test* (Theorem 19).

In addition, the theorem on absolute convergence relates the convergence of a series to that of the series of absolute values of the terms. It is simplest to remember this as a basic principle and then to regard the tests (b), (c), (d), (e), and (g) as tests for *positive* term series.

The general theorems on sequences and series in Sections 6.2 to 6.5 can also be of aid in testing convergence. In particular, one can in certain cases use the definition of convergence directly; that is, one can show that S_n converges to a definite number S. This was done earlier for the geometric series (Theorem 16). On occasion the Cauchy criterion (Theorem 9) can also be used. Theorem 8 suggests a method of proving convergence by representing the given series as the "sum" of two convergent series; this idea can also be used to prove divergence, as shown in Example 12 below. On the basis of these remarks we enlarge the preceding list to include the tests:

h) *partial sum test*;
 i) *Cauchy criterion* (Theorem 9);
 j) *addition of series* (Theorem 8).

In general, the problem of deciding convergence or divergence of a given series can require great ingenuity and resourcefulness. A vast number of particular series and classes of series have been studied; it is therefore important to be able to use the literature on the subject. The book of Knopp cited at the end of this chapter is suggested as a starting point for this. Some additional information is given in Section 6.8 and in Chapters 7 and 9, but this can only be regarded as a sampling of a huge field.

It is recommended that the nth term test (Theorem 10) be used first; if this yields no information, other tests can then be tried.

EXAMPLE 1 $\displaystyle\sum_{n=1}^{\infty} \frac{(-1)^{n+1}}{n} = 1 - \frac{1}{2} + \frac{1}{3} \cdots$. The series of absolute values is the harmonic series $\Sigma \frac{1}{n}$, which diverges by Theorem 15. The series is hence not absolutely convergent, so that tests (b), (d), (e) are of no use. Since the terms are of variable sign, (c) is of no use. The nth term does approach 0 as n becomes infinite, so that (a) yields no information. The alternating series test (f) is applicable, since $1 > \frac{1}{2} > \frac{1}{3}\ldots$, and the nth term approaches 0. Hence the series *converges*. Since it is not absolutely convergent, it is *conditionally convergent*. ∎

It is of interest to remark that the sum of this series is $\log 2 = 0.69315\ldots$. There exist methods of applying (h), that is, of showing that the nth partial sum does approach $\log 2$ as limit; however, these are too involved to describe here. The Cauchy criterion can also be applied. For one can show, without difficulty, that

$$\left| \frac{1}{n} - \frac{1}{n+1} + \frac{1}{n+2} - \cdots \pm \frac{1}{n+p} \right| < \frac{1}{n} \qquad (p > 0).$$

As a matter of fact, the Cauchy criterion could also be used to prove Theorem 18 on alternating series; it will always apply when Theorem 18 applies (cf. Problem 11).

The operations of addition and subtraction, at least in their simplest form, do not help here.

EXAMPLE 2 $\displaystyle\sum_{n=1}^{\infty} (-1)^{n+1} \frac{n+1}{n}$. This at first suggests the alternating series test. However, since $(n+1)/n$ converges to 1, the nth term does not converge to 0. The series diverges. ∎

EXAMPLE 3 $\displaystyle\sum_{n=1}^{\infty} \frac{n+1}{3n^2 + 5n + 2}$. The nth term converges to zero, so that (a) is of no help. For large values of n the general term is approximately $n/3n^2 = 1/3n$, since the terms of lower degree in the numerator and denominator become small in comparison to those of highest degree. This suggests comparing with $\Sigma(1/3n)$ for divergence. The inequality

$$\frac{n+1}{3n^2 + 5n + 2} > \frac{1}{3n}$$

is not correct. However, the inequality

$$\frac{n+1}{3n^2 + 5n + 2} > \frac{1}{4n}$$

is correct for $n > 2$. Since $\Sigma(1/n)$ diverges, $\Sigma(1/4n)$ diverges, and the given series diverges. ∎

EXAMPLE 4 $\displaystyle\sum_{n=2}^{\infty} \frac{1}{n \log n}$. The terms are smaller than those of the harmonic series, so that one might hope for convergence. However,

$$\int_2^{\infty} \frac{dx}{x \log x} = \lim_{b \to \infty} \log\log x \Big|_2^b = \infty;$$

since all the conditions of the integral test are satisfied, the series diverges. ∎

EXAMPLE 5 $\displaystyle\sum_{n=1}^{\infty} \frac{\log n}{n}$. Here there is no doubt concerning divergence, for

$$\frac{\log n}{n} > \frac{1}{n} \qquad (n \geq 3). \quad \blacksquare$$

EXAMPLE 6 $\displaystyle\sum_{n=1}^{\infty} \frac{\log n}{n^2}$. This is like Example 5, but the higher power of n makes a considerable difference. The function $\log n$ grows very slowly as n increases; in fact,

$$\lim_{n \to \infty} \frac{\log n}{n^p} = 0$$

for $p > 0$, as a consideration of indeterminate forms shows. Hence the inequality

$$\frac{\log n}{n^2} = \frac{\log n}{n^{\frac{1}{2}}} \cdot \frac{1}{n^{\frac{3}{2}}} < \frac{1}{n^{\frac{3}{2}}}$$

can be justified for n sufficiently large. Since a finite number of terms of the series can be disregarded in testing convergence, one concludes from Theorem 15 that the series converges. $\blacksquare$

EXAMPLE 7 $\displaystyle\sum_{n=1}^{\infty} \frac{2^n}{n!}$. Here the ratio test applies:

$$\lim_{n \to \infty} \frac{2^{n+1}}{(n+1)!} \cdot \frac{n!}{2^n} = \lim_{n \to \infty} \frac{2}{n+1} = 0,$$

since

$$\frac{n!}{(n+1)!} = \frac{1 \cdot 2 \cdot 3 \cdots n}{1 \cdot 2 \cdot 3 \cdots n(n+1)} = \frac{1}{n+1}.$$

Hence $L = 0$, and the series converges. $\blacksquare$

EXAMPLE 8 $\displaystyle\sum_{n=1}^{\infty} \frac{n^n}{n!}$. Again the ratio test applies:

$$\lim_{n \to \infty} \frac{(n+1)^{n+1}}{(n+1)!} \frac{n!}{n^n} = \lim_{n \to \infty} \frac{n+1}{n+1} \cdot \left(\frac{n+1}{n}\right)^n = \lim_{n \to \infty} \left(1 + \frac{1}{n}\right)^n.$$

The limit on the right is e; since $e > 1$, the series diverges. One can in fact conclude from this result that

$$\lim_{n \to \infty} \frac{n^n}{n!} = \infty;$$

that is, the nth term approaches ∞ as n becomes infinite. $\blacksquare$

EXAMPLE 9 $\displaystyle\sum_{n=1}^{\infty} \frac{2^n}{1 \cdot 3 \cdot 5 \cdots (2n + 1)}$. The ratio test applies:

$$\lim_{n \to \infty} \frac{2^{n+1}}{1 \cdot 3 \cdot 5 \cdots (2n + 1)(2n + 3)} \cdot \frac{1 \cdot 3 \cdot 5 \cdots (2n + 1)}{2^n}$$

$$= \lim_{n \to \infty} \frac{2}{2n + 3} = 0.$$

The series converges.   ∎

EXAMPLE 10 $\displaystyle\sum_{n=1}^{\infty} \log \frac{n}{n + 1}$. Here the simplest procedure is to consider the partial sum:

$$S_n = \log \frac{1}{2} + \log \frac{2}{3} + \cdots + \log \frac{n}{n + 1}$$

$$= \log \left(\frac{1}{2} \cdot \frac{2}{3} \cdot \frac{3}{4} \cdots \frac{n}{n + 1} \right) = \log \frac{1}{n + 1}.$$

Thus $\lim S_n = -\infty$, and the series diverges.   ∎

EXAMPLE 11 $\displaystyle\sum_{n=1}^{\infty} \frac{n^2 + 2^n}{2^n n^2}$. Here the principle of addition applies, for the general term is

$$\frac{n^2 + 2^n}{2^n n^2} = \frac{1}{2^n} + \frac{1}{n^2}.$$

Since $\Sigma(1/2^n)$ and $\Sigma(1/n^2)$ converge, the given series converges.   ∎

EXAMPLE 12 $\displaystyle\sum_{n=1}^{\infty} \left(\frac{1}{n} - \frac{1}{2^n} \right)$. Here the addition principle applies in reverse. If this series converges, then the sum

$$\sum_{n=1}^{\infty} \left(\frac{1}{n} - \frac{1}{2^n} \right) + \sum_{n=1}^{\infty} \frac{1}{2^n} = \sum_{n=1}^{\infty} \frac{1}{n}$$

would also have to converge. Hence the series diverges. In general, the "sum" of a convergent series and a divergent series must be divergent. However, the "sum" of two divergent series can be convergent; this is illustrated by the divergent series:

$$\sum_{n=1}^{\infty} \left(\frac{1}{n^2} + \frac{1}{n} \right), \qquad \sum_{n=1}^{\infty} \left(\frac{1}{n^2} - \frac{1}{n} \right),$$

which, when added, become the convergent series

$$\sum_{n=1}^{\infty} \frac{2}{n^2}.$$ ∎

EXAMPLE 13 $\displaystyle\sum_{n=2}^{\infty} \frac{1}{(\log n)^n}$. Here the root test is easily applicable:

$$\sqrt[n]{|a_n|} = \frac{1}{\log n},$$

so that the root approaches 0 as limit, and the series converges. ∎

EXAMPLE 14 $\displaystyle\sum_{n=2}^{\infty} \left(\frac{n}{1+n^2}\right)^n$. Again the root test proves convergence:

$$\lim_{n\to\infty} \sqrt[n]{|a_n|} = \lim_{n\to\infty} \frac{n}{1+n^2} = 0. \quad ∎$$

PROBLEMS

1. Prove divergence by the nth term test:

a) $\displaystyle\sum_{n=1}^{\infty} \sin\left(\frac{n^2\pi}{2}\right)$

b) $\displaystyle\sum_{n=1}^{\infty} \frac{2^n}{n^3}$

2. Prove convergence by the comparison test:

a) $\displaystyle\sum_{n=2}^{\infty} \frac{1}{n^3 - 1}$

b) $\displaystyle\sum_{n=1}^{\infty} \frac{\sin n}{n^2}$

3. Prove divergence by the comparison test:

a) $\displaystyle\sum_{n=1}^{\infty} \frac{n+5}{n^2 - 3n - 5}$

b) $\displaystyle\sum_{n=2}^{\infty} \frac{1}{\sqrt{n}\,\log n}$

4. Prove convergence by the integral test:

a) $\displaystyle\sum_{n=1}^{\infty} \frac{1}{n^2 + 1}$

b) $\displaystyle\sum_{n=2}^{\infty} \frac{1}{n\log^2 n}$

5. Prove divergence by the integral test:

a) $\displaystyle\sum_{n=1}^{\infty} \frac{n}{n^2 + 1}$

b) $\displaystyle\sum_{n=10}^{\infty} \frac{1}{n\log n \log\log n}$

6. Determine convergence or divergence by the ratio test:

a) $\displaystyle\sum_{n=1}^{\infty} \frac{(-1)^n}{n!}$

b) $\displaystyle\sum_{n=1}^{\infty} \frac{2^n + 1}{3^n + n}$

7. Prove convergence by the alternating series test:

a) $\displaystyle\sum_{n=2}^{\infty} \frac{(-1)^n}{\log n}$

b) $\displaystyle\sum_{n=2}^{\infty} \frac{(-1)^n \log n}{n}$

8. Prove convergence by the root test:

a) $\displaystyle\sum_{n=1}^{\infty} \frac{1}{n^n}$

b) $\displaystyle\sum_{n=1}^{\infty} \left(\frac{n}{n+1}\right)^{n^2}$

9. Prove convergence by showing that the nth partial sum converges:

a) $\displaystyle\sum_{n=1}^{\infty} \frac{1}{(n+2)(n+1)} = \sum_{n=1}^{\infty} \left(\frac{n+1}{n+2} - \frac{n}{n+1}\right)$

b) $\displaystyle\sum_{n=1}^{\infty} \frac{1-n}{2^{n+1}} = \sum_{n=1}^{\infty} \left(\frac{n+1}{2^{n+1}} - \frac{n}{2^n}\right)$

10. Prove the validity of the integral test (Theorem 14) for divergence.

11. Prove the validity of the alternating series test (Theorem 18) by applying the Cauchy criterion (Theorem 9).

12. Determine convergence or divergence:

a) $\displaystyle\sum_{n=1}^{\infty} \frac{n+4}{2n^3-1}$

b) $\displaystyle\sum_{n=1}^{\infty} \frac{3n-5}{n2^n}$

c) $\displaystyle\sum_{n=1}^{\infty} \frac{e^n}{n+1}$

d) $\displaystyle\sum_{n=1}^{\infty} \frac{n^2}{n!+1}$

e) $\displaystyle\sum_{n=1}^{\infty} \frac{n!}{3\cdot5\cdots(2n+3)}$

f) $\displaystyle\sum_{n=1}^{\infty} \frac{(-1)^n \log n}{2n+3}$

g) $\displaystyle\sum_{n=2}^{\infty} \frac{1+\log^2 n}{n\log^2 n}$

h) $\displaystyle\sum_{n=1}^{\infty} \frac{\cos n\pi}{n+2}$

i) $\displaystyle\sum_{n=1}^{\infty} \frac{\log n}{n+\log n}$

j) $\displaystyle\sum_{n=1}^{\infty} \left(\frac{n+1}{2n}\right)^n$

13. (*Ratio form of comparison test*) Let $a_n > 0$ and $b_n > 0$ for $n = 1,2,\ldots$ and let the sequence a_n/b_n have limit k, possibly infinite.

a) Show that if $0 < k < \infty$, then the series Σa_n and Σb_n either both converge or both diverge.

b) Show that if $k = 0$, then convergence of Σb_n implies convergence of Σa_n, but Σb_n may diverge while Σa_n converges.

c) Discuss the case $k = \infty$. [Hint: Use (b), reversing the roles of a_n and b_n.]

14. Apply the results of Problem 13 to test for convergence of the given series Σa_n.

a) $\displaystyle\sum_{n=1}^{\infty} \frac{2n+1}{3n^2+n+1}$ (take $b_n = 1/n$)

b) $\displaystyle\sum_{n=1}^{\infty} \frac{n^3-3n^2+5}{n^5+n+1}$ (take $b_n = 1/n^2$)

c) $\displaystyle\sum_{n=1}^{\infty} \sin \frac{1}{n}$

d) $\displaystyle\sum_{n=1}^{\infty} \left(1 - \cos \frac{1}{n}\right)$

*6.8 EXTENDED RATIO TEST AND ROOT TEST

It may happen that the test ratio $|a_{n+1}/a_n|$ has no limit as n becomes infinite. The behavior of this ratio may still give information as to the convergence or divergence of the series.

THEOREM 20   If $a_n \neq 0$ for $n = 1, 2, \ldots$ and there exist a number r such that $0 < r < 1$ and an integer N such that

$$\left| \frac{a_{n+1}}{a_n} \right| \leq r \quad \text{for} \quad n > N,$$

then the series $\sum\limits_{n=1}^{\infty} a_n$ is absolutely convergent. If, on the other hand,

$$\left| \frac{a_{n+1}}{a_n} \right| \geq 1 \quad \text{for} \quad n > N$$

for some integer N, then the series diverges.

The proof of Theorem 20 is the same as that of Theorem 17; for in that proof the existence of the limit L is not actually used. In the case of convergence, one uses only the existence of the number r such that

$$\left| \frac{a_{n+1}}{a_n} \right| \leq r$$

for $n > N$; in the case of divergence, one uses only the fact that the terms are increasing in absolute value and cannot approach 0.

EXAMPLE   In the series

$$1 + \frac{1}{3} + \frac{1}{4} + \frac{1}{12} + \frac{1}{16} + \frac{1}{48} + \cdots + \frac{5 - (-1)^n}{3 \cdot 2^n} + \cdots,$$

the odd terms and even terms each form a geometric series with ratio $\frac{1}{4}$. The ratios of successive terms are

$$\tfrac{1}{3} \div 1 = \tfrac{1}{3}, \qquad \tfrac{1}{4} \div \tfrac{1}{3} = \tfrac{3}{4}, \qquad \tfrac{1}{12} \div \tfrac{1}{4} = \tfrac{1}{3}, \qquad \tfrac{1}{16} \div \tfrac{1}{12} = \tfrac{3}{4}, \ldots;$$

in general the ratio of an even term to the preceding one is $\frac{1}{3}$ and the ratio of an odd term to the preceding one is $\frac{3}{4}$. The ratios approach no definite limit but are always less than or equal to $r = \frac{3}{4}$. Accordingly, the series converges. ∎

It is of interest to note that part of Theorem 20 can be stated in terms of upper limits (Section 6.3):

THEOREM 20(a)   If $a_n \neq 0$ for $n = 1, 2, \ldots$ and

$$\varlimsup_{n \to \infty} \left| \frac{a_{n+1}}{a_n} \right| < 1,$$

then the series Σa_n converges absolutely.

For if the upper limit is $k < 1$, then the ratio must remain below $k + \epsilon$, for n sufficiently large, for each given $\epsilon > 0$. One can choose ϵ so that $k + \epsilon = r < 1$, and Theorem 20 applies.

It should be remarked that no conclusions can be drawn from either of the relations:

$$\overline{\lim_{n \to \infty}} \left| \frac{a_{n+1}}{a_n} \right| = 1, \qquad \underline{\lim_{n \to \infty}} \left| \frac{a_{n+1}}{a_n} \right| = 1.$$

The condition

$$\lim_{n \to \infty} \left| \frac{a_{n+1}}{a_n} \right| > 1$$

implies divergence, but this is not as precise as Theorem 20.

The root test (Theorem 19) can be extended in the same way as the ratio test:

THEOREM 21   If there exist a number r such that $0 < r < 1$ and an integer N such that

$$\sqrt[n]{|a_n|} \leqq r \quad \text{for} \quad n > N,$$

then the series $\sum\limits_{n=1}^{\infty} a_n$ converges absolutely. If, on the other hand,

$$\sqrt[n]{|a_n|} \geqq 1$$

for infinitely many values of n, then the series diverges.

The proof in the case of convergence is the same as that of Theorem 19. If $\sqrt[n]{|a_n|} \geq 1$ for infinitely many values of n, then $|a_n| \geq 1$ for infinitely many values of n, so that the nth term cannot converge to 0, and the series diverges.

The theorem has a counterpart in terms of upper limits:

THEOREM 21(a)   If

$$\overline{\lim_{n \to \infty}} \sqrt[n]{|a_n|} < 1,$$

then the series $\sum\limits_{n=1}^{\infty} a_n$ is absolutely convergent. If

$$\overline{\lim_{n \to \infty}} \sqrt[n]{|a_n|} > 1,$$

then the series $\sum\limits_{n=1}^{\infty} a_n$ diverges.

No conclusion can be drawn from the relation

$$\overline{\lim_{n \to \infty}} \sqrt[n]{|a_n|} = 1,$$

as the harmonic series show.

Again the previous proof can be repeated.

*6.9   COMPUTATION WITH SERIES—ESTIMATE OF ERROR

Up to this point we have considered only the question of convergence or divergence of infinite series. If a series is known to be convergent, one still has the problem of evaluating the sum. In principle, the sum can always be found to desired accuracy by adding up enough terms of the series, that is, by computing the partial sum S_n for n sufficiently large. However, "sufficiently large" can mean very different things for different series. In order to make the procedure precise, one must have information, for each particular series, as to how large n must be to ensure the accuracy desired. This amounts to saying that one must determine, for each series, a function $N(\epsilon)$ such that

$$|S_n - S| < \epsilon \quad \text{for} \quad n \geq N;$$

in other words, the first N terms are sufficient to give the desired sum S, with an error less than ϵ.

One can phrase this in another manner. One can write

$$S = S_n + R_n,$$

where R_n is the "remainder." One then seeks $N(\epsilon)$ such that

$$|R_n| < \epsilon \quad \text{for} \quad n \geq N(\epsilon).$$

It will be seen that for certain convergent series, one can find a useful explicit *upper estimate* for R_n; more precisely, one can find a sequence T_n converging to 0 for which

$$|R_n| \leq T_n \quad (n \geq n_1).$$

If the sequence T_n is *monotone decreasing*, then we can choose $N(\epsilon)$ as the smallest integer n for which $T_n < \epsilon$. For, if $N(\epsilon)$ is so chosen and $n \geq N(\epsilon)$, then

$$|R_n| \leq T_n \leq T_{N(\epsilon)} < \epsilon.$$

It will now be shown how such a monotone decreasing sequence T_n can be found when the series converges by one of the following tests: comparison test, integral test, ratio test, root test, alternating series test.

THEOREM 22   If $|a_n| \leq b_n$ for $n \geq n_1$ and $\displaystyle\sum_{n=1}^{\infty} b_n$ converges, then

$$|R_n| \leq \sum_{m=n+1}^{\infty} b_m = T_n$$

for $n \geq n_1$; the sequence T_n is monotone decreasing and converges to 0.

Proof.   By definition

$$R_n = a_{n+1} + \cdots + a_{n+p} + \cdots = \lim_{p \to \infty} \sum_{m=n+1}^{n+p} a_m.$$

Since, for $n \geq n_1$,

$$|a_{n+1} + \cdots + a_{n+p}| \leq |a_{n+1}| + \cdots + |a_{n+p}|$$

$$\leq b_{n+1} + \cdots + b_{n+p} \leq \sum_{m=n+1}^{\infty} b_m,$$

one concludes that

$$|R_n| = \lim_{p \to \infty} \left| \sum_{m=n+1}^{n+p} a_m \right| \leq \sum_{m=n+1}^{\infty} b_m = T_n.$$

Since T_n is the remainder, after n terms, of a convergent series of positive terms, it is necessarily a monotone sequence converging to 0.   □

The theorem can be stated as follows: If a series converges by the comparison test, then the remainder is in absolute value at most equal to that of the comparison series.

EXAMPLE 1   The series $\displaystyle\sum_{n=1}^{\infty} \frac{1}{n2^n}$ can be compared with the geometric series $\Sigma(1/2^n)$. Hence the remainder after five terms is at most

$$T_5 = \frac{1}{2^6} + \frac{1}{2^7} + \cdots = \frac{1}{2^6}\left(1 + \frac{1}{2} + \cdots\right) = \frac{1}{32}.$$

In general, $T_n = 2^{-n-1} + 2^{-n-2} + \cdots = 2^{-n}$. The condition $T_n < \epsilon$ leads to the inequality $2^n > 1/\epsilon$, so that

$$n > -(\log \epsilon / \log 2);$$

we can thus choose $N(\epsilon)$ as the smallest integer n satisfying this inequality.   ■

THEOREM 23   If the series $\displaystyle\sum_{n=1}^{\infty} a_n$ converges by the integral test of Theorem 14, with the function $f(x)$ decreasing for $x \geq c$, then

$$|R_n| < \int_n^{\infty} f(x)\, dx = T_n$$

for $n \geq c$; the sequence T_n is monotone decreasing and converges to 0.

Proof.   This theorem follows from Theorem 22. For one can interpret the improper integral as a series:

$$\int_n^{\infty} f(x)\, dx = \sum_{m=n+1}^{\infty} b_m, \qquad b_m = \int_{m-1}^{m} f(x)\, dx;$$

one then has

$$\int_n^\infty f(x)\, dx = \lim_{p \to \infty} \int_n^{n+p} f(x)\, dx = \lim_{p \to \infty} \sum_{m=n+1}^{n+p} b_m .$$

As in the proof of Theorem 14,

$$|a_n| < \int_{n-1}^n f(x)\, dx = b_n ,$$

so that

$$|R_n| < \sum_{m=n+1}^\infty b_m = \int_n^\infty f(x)\, dx. \quad \square$$

EXAMPLE 2 For the harmonic series, with order $p > 1$, one finds

$$0 < R_n = \sum_{m=n+1}^\infty \frac{1}{m^p} < \int_n^\infty \frac{1}{x^p}\, dx = \frac{1}{(p-1)n^{p-1}} .$$

This result can now be used, by Theorem 22, for any series whose convergence is established by comparison with a harmonic series of order p.

If, for example, $p = 6$, then $T_n = 0.2n^{-5}$; it follows that we can choose $N(\epsilon)$ as the smallest integer n such that $n^5 > 0.2\epsilon^{-1}$. Accordingly, to evaluate the harmonic series of order 6 with an error less than 0.0001, five terms are sufficient:

$$\sum_{n=1}^\infty \frac{1}{n^6} \sim 1 + \frac{1}{2^6} + \frac{1}{3^6} + \frac{1}{4^6} + \frac{1}{5^6} = 1.0173. \quad \blacksquare$$

THEOREM 24 If

$$\left| \frac{a_{n+1}}{a_n} \right| \leqq r < 1$$

for $n > n_1$, so that the series Σa_n converges by the ratio test, then

$$|R_n| \leqq \frac{|a_{n+1}|}{1 - r} = T_n, \qquad n \geqq n_1; \tag{6.18}$$

the sequence T_n is monotone decreasing and converges to 0. If

$$\lim_{n \to \infty} \left| \frac{a_{n+1}}{a_n} \right| = L < 1,$$

then r will be at least equal to L. If

$$1 > \left| \frac{a_{n+2}}{a_{n+1}} \right| \geqq \left| \frac{a_{n+3}}{a_{n+2}} \right|$$

for $n \geqq n_1$, then

$$|R_n| \leqq \frac{|a_{n+1}^2|}{|a_{n+1}| - |a_{n+2}|} = T_n^*, \qquad n \geqq n_1; \tag{6.19}$$

the sequence T_n^* is monotone decreasing and converges to 0.

Proof.   The first statement follows from Theorem 22 and the fact that the ratio test is a comparison of the given series with a geometric series. Thus one has, as in the proof of Theorem 17,

$$|R_n| \le |a_{n+1}| + |a_{n+2}| + \cdots \le |a_{n+1}|(1 + r + \cdots) = \frac{|a_{n+1}|}{1 - r} = T_n.$$

The second statement emphasizes the fact that if the test ratio converges to L, then the ratio cannot remain less than or equal to a number r less than L. Thus $r \ge L$. One can use $r = L$ only when

$$\left| \frac{a_{n+1}}{a_n} \right| \le L$$

for $n > N$, so that the limit L is approached from *below*.

The third statement concerns the case in which the test ratios are steadily *decreasing* and hence approaching a limit L. One cannot use L in such a case; however, under the assumptions made, one can use $r = |a_{n+2}/a_{n+1}|$ for $n \ge n_1$, so that

$$R_n \le |a_{n+1}| \frac{1}{1 - \left| \dfrac{a_{n+2}}{a_{n+1}} \right|} = \frac{|a_{n+1}|^2}{|a_{n+1}| - |a_{n+2}|} = T_n^*.$$

This gives formula (6.19). $\square$

EXAMPLE 3 $\displaystyle\sum_{n=1}^{\infty} \frac{n + 1}{n \cdot 2^n}$. The test ratio is found to be

$$\frac{n^2 + 2n}{n^2 + 2n + 1} \cdot \frac{1}{2}.$$

This converges to $\frac{1}{2}$ but is always less than $\frac{1}{2}$. Hence we can use $r = \frac{1}{2}$ and, for example,

$$R_5 = \frac{7}{6 \cdot 2^6} + \frac{8}{7 \cdot 2^7} + \cdots < \frac{7}{6 \cdot 2^6}\left(1 + \frac{1}{2} + \cdots\right) = \frac{7}{192} = 0.037. \quad \blacksquare$$

EXAMPLE 4 $\displaystyle\sum_{n=1}^{\infty} \frac{n}{(n + 1)2^n}$. Here the test ratio is

$$\frac{n^2 + 2n + 1}{n^2 + 2n} \cdot \frac{1}{2}.$$

Again the limit is $\frac{1}{2}$, but the ratio is always greater than $\frac{1}{2}$. The successive ratios are decreasing since

$$\frac{n^2 + 2n + 1}{n^2 + 2n} \cdot \frac{1}{2} > \frac{(n + 1)^2 + 2(n + 1) + 1}{(n + 1)^2 + 2(n + 1)} \cdot \frac{1}{2},$$

as algebraic manipulation shows. Hence inequality (6.19) applies (for $n \ge 1$), and one finds, for example,

$$R_5 = \frac{6}{7 \cdot 2^6} + \cdots < \frac{\left(\dfrac{6}{7 \cdot 2^6}\right)^2}{\dfrac{6}{7 \cdot 2^6} - \dfrac{7}{8 \cdot 2^7}} = \frac{9}{329} = 0.027. \quad \blacksquare$$

THEOREM 25   If

$$\sqrt[n]{|a_n|} \leq r < 1$$

for $n > n_1$, so that the series Σa_n converges by the root test, then

$$|R_n| \leq \frac{r^{n+1}}{1 - r} = T_n, \qquad n \geq n_1. \tag{6.20}$$

If

$$\lim_{n \to \infty} \sqrt[n]{|a_n|} = R < 1,$$

then r will be at least equal to R. If

$$1 > |a_{n+1}|^{1/(n+1)} \geq |a_{n+2}|^{1/(n+2)}$$

for $n \geq n_1$, then

$$|R_n| \leq \frac{|a_{n+1}|}{1 - |a_{n+1}|^{1/(n+1)}} = T_n^*, \qquad n \geq n_1. \tag{6.21}$$

The sequences T_n and T_n^* are monotone decreasing and converge to 0.

The proofs parallel those for Theorem 24, since again the test is based on comparison with a geometric series.

EXAMPLE 5 $\displaystyle\sum_{n=2}^{\infty} \frac{1}{(\log n)^n}$. Here

$$\sqrt[n]{|a_n|} = \frac{1}{\log n}.$$

This is decreasing and less than 1 for $n = 3, 4, \ldots$. Thus, for example,

$$R_5 = \frac{1}{(\log 6)^6} + \frac{1}{(\log 7)^7} + \cdots \leq \frac{\dfrac{1}{(\log 6)^6}}{1 - \dfrac{1}{\log 6}} = 0.06 \ldots. \quad \blacksquare$$

THEOREM 26   If the series

$$a_1 - a_2 + a_3 - a_4 + \cdots = \sum_{n=1}^{\infty} (-1)^{n+1} a_n, \qquad a_n > 0, \tag{6.22}$$

converges by the alternating series test, then

$$0 < |R_n| < a_{n+1} = T_n.$$

Hence $N(\epsilon)$ can be chosen as the smallest integer such that $a_{n+1} < \epsilon$.

The theorem can be stated in words as follows: When a series converges by the alternating series test, the error made in stopping at n terms is in absolute value less than the first term neglected.

Proof of Theorem 26.   As was pointed out in the proof of the alternating series test (Theorem 18), one has

$$S_2 < S_4 < \cdots < S_{2n} < \cdots < S_{2n-1} \cdots < S_3 < S_1.$$

It follows that the sum S lies between each two successive partial sums (one odd, one even):

$$S_{2n} < S < S_{2n+1}, \qquad S_{2n} < S < S_{2n-1}.$$

Hence

$$0 < R_{2n} = S - S_{2n} < S_{2n+1} - S_{2n} = a_{2n+1},$$
$$0 > R_{2n-1} = S - S_{2n-1} > S_{2n} - S_{2n-1} = -a_{2n};$$

that is,

$$0 < R_{2n} < a_{2n+1}, \qquad -a_{2n} < R_{2n-1} < 0.$$

This shows that for every n,

$$0 < |R_n| < a_{n+1},$$

but actually tells more: R_n is positive if n is even; R_n is negative if n is odd. $\square$

EXAMPLE 6 $\displaystyle\sum_{n=1}^{\infty} \frac{(-1)^{n+1}}{n}$. Here the partial sums are $S_1 = 1$, $S_2 = 1 - \frac{1}{2} = \frac{1}{2}$, $S_3 = \frac{5}{6}$, $S_4 = \frac{7}{12},\ldots$. The preceding theorem asserts that $|R_1| < \frac{1}{2}$, $|R_2| < \frac{1}{3}$, $|R_3| < \frac{1}{4},\ldots$ or, more precisely, that $-\frac{1}{2} < R_1 < 0, 0 < R_2 < \frac{1}{3}, -\frac{1}{4} < R_3 < 0$. Thus in particular, if three terms are used, the sum is between $\frac{5}{6}$ and $\frac{5}{6} - \frac{1}{4} = \frac{7}{12}$. In order to compute the sum with an error of less than 0.01, one would need 100 terms, for the 101st term, $1/101$, is the first one less than 0.01.   ∎

PROBLEMS

1. Determine how many terms are sufficient to compute the sum with given allowed error ϵ and find the sum to this accuracy:

a) $\displaystyle\sum_{n=1}^{\infty} \frac{1}{n^2}$, $\epsilon = 1$

b) $\displaystyle\sum_{n=1}^{\infty} \frac{(-1)^{n+1}}{n^2}$, $\epsilon = 0.1$

c) $\displaystyle\sum_{n=1}^{\infty} \frac{n}{n^3 + 5}$, $\epsilon = 0.2$

d) $\displaystyle\sum_{n=1}^{\infty} \frac{1}{n^2 + 1}$, $\epsilon = 0.5$

e) $\displaystyle\sum_{n=1}^{\infty} \frac{1}{n^n}$, $\epsilon = 0.01$ f) $\displaystyle\sum_{n=1}^{\infty} \frac{1}{n!}$, $\epsilon = 0.01$

g) $\displaystyle\sum_{n=1}^{\infty} \frac{(-1)^{n+1}}{(2n-1)!}$, $\epsilon = 0.001$ h) $\displaystyle\sum_{n=2}^{\infty} \frac{(-1)^n}{n\log n}$, $\epsilon = 0.5$

i) $\displaystyle\sum_{n=2}^{\infty} \frac{1}{n^3 \log n}$, $\epsilon = 0.5$ j) $\displaystyle\sum_{n=1}^{\infty} \frac{2^n}{3^n + 1}$, $\epsilon = 0.1$

2. Let Σa_n be the geometric series $1 + r + r^2 + \cdots$.

   a) Determine how many terms are needed to compute the sum with error less than 0.01 when $r = \frac{1}{2}$, $r = 0.9$, $r = 0.99$.

   b) Show that for any positive ϵ, $|R_n| < \epsilon$ when

$$n > \frac{\log \epsilon(1-r)}{\log |r|}, \qquad -1 < r < 1,$$

   and that no smaller value of n will suffice.

   c) Show that as r approaches 1, the number of terms needed to compute the sum with error less than a fixed ϵ becomes infinite.

3. Show that for $p > 0$ the sum of the series $1 - 1/2^p + 1/3^p - \cdots$ is positive.

6.10   OPERATIONS ON SERIES

It has been pointed out (Theorem 8) that convergent series can be added and subtracted term by term:

$$\sum_{n=1}^{\infty} a_n + \sum_{n=1}^{\infty} b_n = \sum_{n=1}^{\infty} (a_n + b_n), \qquad \sum_{n=1}^{\infty} a_n - \sum_{n=1}^{\infty} b_n = \sum_{n=1}^{\infty} (a_n - b_n).$$

We also know that a convergent series can be multiplied by a constant:

$$k \sum_{n=1}^{\infty} a_n = \sum_{n=1}^{\infty} (ka_n).$$

Three other operations will be considered in the present section—multiplication, grouping, and rearrangement.

We consider first the operation of *grouping*, that is, of inserting parentheses in a series Σa_n. For example, the series Σa_n might be replaced by the series $(a_1 + a_2) + \cdots +(a_{2n-1} + a_{2n}) + \cdots$, that is, by Σb_n, where $b_n = a_{2n-1} + a_{2n}$.

> **THEOREM 27**   If the series $\displaystyle\sum_{n=1}^{\infty} a_n$ is convergent, then insertion of parentheses yields a new convergent series having the same sum as $\displaystyle\sum_{n=1}^{\infty} a_n$. If $\displaystyle\sum_{n=1}^{\infty} a_n$ is properly divergent, insertion of parentheses yields a properly divergent series.

Proof. The effect of insertion of parentheses is to cause one to *skip* certain partial sums. Thus the partial sums of the series

$$(a_1 + a_2) + (a_3 + a_4) + (a_5 + a_6) + \cdots + (a_{2n-1} + a_{2n}) + \cdots$$

are the partial sums $S_2, S_4, S_6, \ldots, S_{2n}, \ldots$ of the series Σa_n. If S_n converges to S, then the new sequence obtained by skipping must also converge to S; if $\lim S_n = +\infty$, then the sequence obtained by skipping must also diverge to $+\infty$. Thus the theorem follows. □

Since a positive term series is either convergent or properly divergent, one can insert parentheses freely without affecting convergence or sum in such a series. This can be used to prove divergence of the harmonic series, for one has

$$\sum_{n=1}^{\infty} \frac{1}{n} = 1 + \frac{1}{2} + \left(\frac{1}{3} + \frac{1}{4}\right) + \left(\frac{1}{5} + \frac{1}{6} + \frac{1}{7} + \frac{1}{8}\right)$$

$$+ \left(\frac{1}{9} + \cdots + \frac{1}{16}\right) + \cdots + \left(\frac{1}{2^{n-1}+1} + \cdots + \frac{1}{2^n}\right) + \cdots;$$

each set of parentheses contributes at least $\frac{1}{2}$, so that the general term of the grouped series does not converge to zero.

For a divergent series of variable signs, insertion of parentheses can on occasion produce a convergent series. Thus the series $\Sigma(-1)^{n+1}$ becomes

$$(1 - 1) + (1 - 1) + \cdots + (1 - 1) + \cdots = 0,$$

with the parentheses inserted as shown. Therefore in testing a series of variable signs for convergence, parentheses should not be inserted.

A series $\sum_{m=1}^{\infty} b_m$ is said to be a *rearrangement* of a series $\sum_{n=1}^{\infty} a_n$ if there exists a one-to-one correspondence between the indices n and m such that $a_n = b_m$ for corresponding indices. Thus the series

$$1 + \tfrac{1}{3} + \tfrac{1}{2} + \tfrac{1}{5} + \tfrac{1}{4} + \tfrac{1}{7} + \tfrac{1}{6} + \cdots,$$

$$1 + \tfrac{1}{2} + \tfrac{1}{3} + \tfrac{1}{4} + \tfrac{1}{5} + \tfrac{1}{6} + \tfrac{1}{7} + \cdots,$$

are rearrangements of each other. Also

$$1 + \tfrac{1}{2} + \tfrac{1}{4} + \tfrac{1}{3} + \tfrac{1}{6} + \tfrac{1}{8} + \tfrac{1}{5} + \tfrac{1}{10} + \tfrac{1}{12} + \tfrac{1}{7} + \cdots$$

is a rearrangement of the harmonic series.

THEOREM 28 If $\sum_{n=1}^{\infty} a_n$ is absolutely convergent and $\sum_{m=1}^{\infty} b_m$ is a rearrangement of $\sum_{n=1}^{\infty} a_n$, then $\sum_{m=1}^{\infty} b_m$ is absolutely convergent and has the same sum as $\sum_{n=1}^{\infty} a_n$.

Proof. One has clearly

$$\sum_{m=1}^{N} |b_m| \leq \sum_{n=1}^{\infty} |a_n|,$$

since every b_m equals an a_n for appropriate n and no two m's correspond to the same n. Hence the series $\sum_{m=1}^{\infty} |b_m|$ converges, so that Σb_m is absolutely convergent. Let S_n and S be the nth partial sum and sum for Σa_n and let S'_m and S' be the corresponding quantities for Σb_m. For given positive ϵ, let N be chosen so large that

$$|S_n - S| < \tfrac{1}{2}\epsilon \quad \text{and} \quad |a_{n+1}| + \cdots + |a_{n+p}| < \tfrac{1}{2}\epsilon$$

for $n > N$ and $p \geq 1$; such an N can be found, since Σa_n converges to S and since the Cauchy criterion holds for $\Sigma|a_n|$. For m sufficiently large, S'_m will be a sum of terms including all of $a_1 \ldots a_n$ and perhaps more:

$$S'_m = S_n + a_{k_1} + \cdots a_{k_s},$$

where $k_1, \ldots, k_s$ are all larger than n. Let $k_0 = n + p_0$ be the largest of these indices. Then

$$|S'_m - S_n| \leq |a_{k_1}| + \cdots + |a_{k_s}| \leq |a_{n+1}| + \cdots + |a_{n+p_0}| < \tfrac{1}{2}\epsilon.$$

Now

$$|S'_m - S| = |S'_m - S_n + S_n - S| \leq |S'_m - S_n| + |S_n - S| < \tfrac{1}{2}\epsilon + \tfrac{1}{2}\epsilon = \epsilon.$$

It follows that S'_m converges to S, so that $S' = S$ as asserted. $\square$

The preceding theorem is of importance in connection with the *multiplication* of series. If one multiplies two series as in algebra:

$$(a_1 + a_2 + a_3 + \cdots + a_n + \cdots)(b_1 + b_2 + \cdots + b_n + \cdots)$$
$$= a_1 b_1 + a_1 b_2 + \cdots + a_2 b_1 + a_2 b_2 + \cdots,$$

one obtains a collection of terms of form $a_n b_m$ without any special order. It will be seen that when Σa_n and Σb_m are absolutely convergent, the products $a_n b_m$ can be arranged in order to form an absolutely convergent infinite series. It follows from Theorem 28 that the same result is obtained no matter how the products are arranged and that the same sum is always obtained. Of the many possible arrangements the following one, known as the *Cauchy product*, is most commonly used:

$$a_1 b_1 + a_1 b_2 + a_2 b_1 + a_1 b_3 + a_2 b_2 + a_3 b_1 + \cdots .$$

This is illustrated in Fig. 6.6. If the series Σa_n and Σb_m are indexed beginning with $n = 0$ and $m = 0$, the Cauchy product of the series appears as

$$a_0 b_0 + a_0 b_1 + a_1 b_0 + a_0 b_2 + a_1 b_1 + a_2 b_0 + \cdots .$$

This series is suggested by the multiplication of two power series (to be considered later):

$$\left(a_0 + a_1x + a_2x^2 + \cdots a_nx^n + \cdots\right)\left(b_0 + b_1x + b_2x^2 + \cdots b_mx^m + \cdots\right)$$
$$= a_0b_0 + x(a_0b_1 + a_1b_0) + x^2(a_0b_2 + a_1b_1 + a_2b_0) + \cdots ,$$

the terms of same degree in x being collected together.

Figure 6.6   Cauchy product of series.

THEOREM 29   If the series $\sum\limits_{n=1}^{\infty} a_n$ and $\sum\limits_{m=1}^{\infty} b_m$ are absolutely convergent, then the products $a_n b_m$ can be arranged to form an absolutely convergent series $\sum\limits_{i=1}^{\infty} c_i$, and

$$\sum_{n=1}^{\infty} a_n \cdot \sum_{m=1}^{\infty} b_m = \sum_{i=1}^{\infty} c_i. \tag{6.23}$$

Proof.   We choose the series Σc_i as the series

$$a_1b_1 + a_1b_2 + a_2b_2 + a_2b_1 + a_1b_3 + a_2b_3 + a_3b_3 + a_3b_2 + a_3b_1 + \cdots ,$$

as suggested in Fig. 6.7. From this definition, one has in particular

$$c_1 = a_1b_1, \qquad c_1 + c_2 + c_3 + c_4 = (a_1 + a_2)(b_1 + b_2),\ldots$$

and in general

$$c_1 + (c_2 + c_3 + c_4) + \cdots + (\cdots + c_{n^2}) = (a_1 + \cdots + a_n)(b_1 + \cdots + b_n).$$

Similarly,

$$|c_1| + (|c_2| + |c_3| + |c_4|) + \cdots + (\cdots + |c_{n^2}|)$$
$$= (|a_1| + \cdots + |a_n|)(|b_1| + \cdots |b_n|).$$

Figure 6.7   Proof of Theorem 29.

Since the series $\Sigma|a_n|$ and $\Sigma|b_n|$ converge, the equation just written shows that the left-hand side approaches a limit as $n \to \infty$; dropping parentheses, on the basis of Theorem 27, we conclude that $\Sigma|c_n|$ converges, so that Σc_n is absolutely convergent. The preceding equation then shows that

$$\sum_{i=1}^{\infty} c_i = \lim_{n \to \infty} \left[c_1 + (\cdots) + \cdots + (\cdots + c_{n^2}) \right] = \sum_{n=1}^{\infty} a_n \cdot \sum_{m=1}^{\infty} b_m,$$

the insertion of parentheses being justified by Theorem 27. The theorem is thus proved. $\square$

On the basis of Theorem 28, one can now use any arrangement desired for the c's, in particular that of the Cauchy product. One can also *group* the terms as suggested by the preceding power series multiplication:

$$\left(\sum_{n=0}^{\infty} a_n \right) \left(\sum_{m=0}^{\infty} b_m \right) = a_0 b_0 + (a_0 b_1 + a_1 b_0) + (a_0 b_2 + a_1 b_1 + a_2 b_0)$$

$$+ \cdots + (a_0 b_n + a_1 b_{n-1} + \cdots + a_{n-1} b_1 + a_n b_0) + \cdots .$$

The insertion of parentheses is justified by Theorem 27.

If the series Σa_n and Σb_n converge but are not both absolutely convergent, one can nevertheless form the Cauchy product of the series as Σc_n, where $c_n = a_0 b_n + a_1 b_{n-1} + \cdots + a_{n-1} b_1 + a_n b_0$. It can be shown that if the series Σc_n converges, then always

$$\sum_{n=0}^{\infty} a_n \cdot \sum_{n=0}^{\infty} b_n = \sum_{n=0}^{\infty} c_n. \tag{6.24}$$

Furthermore, if one of the two series Σa_n, Σb_n converges absolutely, then Σc_n must converge. For proofs, refer to page 321 of the book by Knopp listed at the end of the chapter.

PROBLEMS

1. Assuming the relations (proved in the next chapter):

$$\sum_{n=1}^{\infty} \frac{1}{n^2} = \frac{\pi^2}{6}, \qquad \sum_{n=1}^{\infty} \frac{1}{n^4} = \frac{\pi^4}{90}, \qquad \sum_{n=1}^{\infty} \frac{1}{n^6} = \frac{\pi^6}{945},$$

evaluate the series:

a) $\displaystyle\sum_{n=1}^{\infty} \frac{6}{n^2}$ **b)** $\displaystyle\sum_{n=1}^{\infty} \frac{n^2+1}{n^4}$ **c)** $\displaystyle\sum_{n=1}^{\infty} \frac{2n^2-3}{n^4}$

d) $\displaystyle\sum_{n=1}^{\infty} \frac{9+3n^2+5n^4}{n^6}$ **e)** $\displaystyle\sum_{n=3}^{\infty} \frac{n^4-1}{n^6}$ **f)** $\displaystyle\sum_{n=2}^{\infty} \frac{n^2+1}{(n^2-1)^2}$

2. Verify the following relations:

a) $\displaystyle\sum_{n=1}^{\infty} \frac{1}{n^3} = \sum_{n=2}^{\infty} \frac{1}{(n-1)^3};$

b) $\displaystyle\sum_{n=1}^{\infty} [f(n+1) - f(n)] = \lim_{n \to \infty} f(n) - f(1)$, if the limit exists;

c) $\displaystyle\sum_{n=2}^{\infty} [f(n+1) - f(n-1)] = \lim_{n \to \infty} [f(n) + f(n+1)] - f(1) - f(2)$, if the limit exists.

3. Use $f(n) = 1/n$ and $f(n) = 1/n^2$ in 2(b) and 2(c) to prove

a) $\displaystyle\sum_{n=1}^{\infty} \frac{2n+1}{n^2(n+1)^2} = 1$

b) $\displaystyle\sum_{n=1}^{\infty} \frac{1}{n(n+1)} = 1$

c) $\displaystyle\sum_{n=2}^{\infty} \frac{1}{n^2-1} = \frac{3}{4}$

d) $\displaystyle\sum_{n=2}^{\infty} \frac{4n}{(n^2-1)^2} = \frac{5}{4}$

4. From the relation

$$\frac{1}{1-r} = 1 + r + \cdots + r^n + \cdots = \sum_{n=0}^{\infty} r^n, \qquad -1 < r < 1,$$

prove the following:

a) $\displaystyle\frac{1}{(1-r)^2} = 1 + 2r + 3r^2 + \cdots + (n+1)r^n + \cdots, \qquad -1 < r < 1$

b) $\displaystyle\frac{1}{(1-r)^3} = 1 + 3r + \cdots + \frac{(n+2)(n+1)r^n}{2} + \cdots, \qquad -1 < r < 1$

5. Prove the general binomial formula for negative exponents:

$$(1-r)^{-k} = 1 + kr + \frac{k(k+1)}{1 \cdot 2}r^2 + \cdots + \frac{k(k+1) \cdots (k+n-1)}{1 \cdot 2 \cdots n}r^n + \cdots$$

$$-1 < r < 1, \qquad k = 1, 2, \ldots.$$

6. Assuming (as will be proved later) that $\sin x$ and $\cos x$ are representable for all x by the absolutely convergent series:

$$\sin x = x - \frac{x^3}{3!} + \frac{x^5}{5!} - \cdots + (-1)^{n+1}\frac{x^{2n-1}}{(2n-1)!} + \cdots,$$

$$\cos x = 1 - \frac{x^2}{2!} + \frac{x^4}{4!} - \cdots + (-1)^n\frac{x^{2n}}{(2n)!} + \cdots,$$

verify by means of series the identity:

$$2 \sin x \cos x = \sin 2x.$$

7. Let all sums be from 1 to ∞. If $\sum b_n$ has been evaluated and $\{a_n\}$ is "close to $\{b_n\}$", it may be of advantage to write $\sum a_n = \sum b_n - \sum(b_n - a_n)$. For example, to evaluate $\sum(2^n + 1)^{-1} = \sum a_n$, one can choose $\sum b_n = \sum 2^{-n}$ and write

$$\sum(2^n + 1)^{-1} = \sum 2^{-n} - \sum(4^n + 2^n)^{-1}.$$

The last series converges much more rapidly than $\sum (2^n + 1)^{-1}$, since its test ratio has limit $\frac{1}{4}$, as opposed to $\frac{1}{2}$. One can repeat the process, using $a_n = (4^n + 2^n)^{-1}$ and $b_n = 4^{-n}$, and so on.

a) Follow the steps described through one stage of the process, to find $\sum (2^n + 1)^{-1} = 0.7645$.

b) Evaluate $\sum a_n = \sum (2^n + 9^{-1})^{-1}$, using $\sum b_n = \sum 2^{-n}$.

c) Evaluate $\sum a_n = \sum (n^2 + 1)^{-1}$ using $\sum b_n = \sum n^{-2}$, then using $\sum b_n = \sum n^{-4}$, finally using $\sum b_n = \sum n^{-6}$ (see Problem 1).

Remarks.   In general, for a_n a rational function of n, as in (c), one can take b_n to be the ratio of the leading terms in numerator and denominator. For a convergent series, the degree of the denominator must exceed that of the numerator by at least 2. By repeating the process several times one obtains

$$\sum a_n = k_2 \sum n^{-2} + k_3 \sum n^{-3} + \cdots + k_m \sum n^{-m} + \sum c_n,$$

where c_n is a rational function of n whose denominator has degree exceeding that of the numerator by at least $m + 1$. The coefficients $k_2, k_3, \ldots$ can also be obtained as coefficients of the power series $\sum k_i x^i$ of the rational function $f(x)$ such that $f(1/n) = a_n$ for all n. Thus for $a_n = (n^2 + 1)^{-1}, f(x) = (x^{-2} + 1)^{-1} = x^2 - x^4 + x^6 - \cdots$ as in Section 6.15 and 6.16. The sums $\sum n^{-p}$ are values of the zeta function $\zeta(p)$ (Section 6.16) and are tabulated (see for example the tables of Jahnke and Emde listed at the end of Chapter 7).

6.11   SEQUENCES AND SERIES OF FUNCTIONS

If for each positive integer n a function $f_n(x)$ is given, then the functions $f_n(x)$ are said to form a *sequence of functions*. It will usually be assumed that the functions are all defined over the same interval (perhaps infinite) on the x-axis.

For each fixed x of the interval the sequence $f_n(x)$ may converge or diverge. The first question of interest for a particular sequence is the determination of the values of x for which it converges. Thus the sequence $f_n(x) = x^n/2^n$ converges for $-2 < x \le 2$ and diverges otherwise.

Similar remarks hold for infinite series whose terms are functions:

$$u_1(x) + u_2(x) + \cdots + u_n(x) + \cdots = \sum_{n=1}^{\infty} u_n(x).$$

The nth partial sum of such a series is itself a function of x:

$$S_n = S_n(x) = u_1(x) + \cdots + u_n(x).$$

Convergence of the series is by definition equivalent to convergence of the sequence of partial sums; if $S_n(x)$ converges to $S(x)$ for certain values of x, then

$$\sum_{n=1}^{\infty} u_n(x) = S(x)$$

for this range of x.

Examples of this have been considered already: the series $\sum_{n=1}^{\infty} 1/n^p$, in which the terms are functions of p, and the geometric series $\sum_{n=0}^{\infty} r^n$, in which the terms are functions of r. The first converges for $p > 1$; the second for $|r| < 1$. Other examples are the following:

$$\sum_{n=0}^{\infty} \frac{x^n}{n!}, \quad \sum_{n=1}^{\infty} \frac{x^n}{n^2}, \quad \sum_{n=1}^{\infty} \frac{\cos nx}{n^2},$$

which can be shown to converge for all x, for $|x| \leq 1$, and for all x, respectively.

The tests for convergence of a series of functions are the same as those for a series of constants. For each particular choice of x (or whatever the independent variable may be) the series $\Sigma u_n(x)$ is nothing but a series of constants. Thus when x ranges over an interval, the problem is really that of discussing the convergence of many (in fact, infinitely many) series. One could, of course, test a variety of x-values separately; this is rarely advisable, although a particular application might make it unavoidable. It will be seen that for many series it is possible to determine by a single analysis all the values of x for which the series converges.

For power series, precise results follow. In many cases the ratio test alone gives the answer; for example, for the series

$$\sum_{n=0}^{\infty} \frac{x^n}{n!},$$

which represents e^x, one has

$$\lim_{n \to \infty} \left| \frac{a_{n+1}}{a_n} \right| = \lim_{n \to \infty} \left(\frac{|x^{n+1}|}{(n+1)!} \cdot \frac{n!}{|x^n|} \right) = \lim_{n \to \infty} \frac{|x|}{n+1}.$$

Since this limit is 0 for each x, the series converges for all x.

6.12   UNIFORM CONVERGENCE

Although it may thus be possible to prove that a series $\Sigma u_n(x)$ converges for all x of a certain range, another question remains to be answered: Is the convergence equally rapid for all x of the range? This question turns out to be important both for practical applications and for basic theory. The practical application is as follows. Suppose it is known that a series $\Sigma u_n(x)$ converges for $a < x < b$ to a function $f(x)$. If the function $f(x)$ is a complicated one, while the functions $u_n(x)$ are simple, it would be a considerable simplification if we could replace the function $f(x)$ by a partial sum $S_n(x) = u_1(x) + \cdots + u_n(x)$; for example, if the $u_n(x)$ are of the form $c_n x^n$, we would be replacing $f(x)$ by a *polynomial*. This amounts to "rounding off" the series of functions, just as we rounded off the decimal expansion of $\frac{1}{3}$ in Section 6.1. The procedure can be justified if the error made is below a prescribed value ϵ, that is, if $|f(x) - S_n(x)| < \epsilon$ for $a < x < b$. However, it can happen, as will be seen below, that no matter how large n is chosen, the error $|f(x) - S_n(x)|$ exceeds ϵ for some x of the interval $a < x < b$. This does not contradict the fact that the series converges for each x; it is simply due to the fact that the number of terms needed to reduce the error below ϵ varies from one x to another in such a manner that no one n will serve for the whole interval. When this difficulty arises, rounding off over the whole interval $a < x < b$ cannot be justified. On the other hand, it may happen that, however ϵ is chosen, an N can be found such that for each $n \geq N$, $|S_n(x) - f(x)| < \epsilon$ for all x of the interval $a < x < b$; if this is so, the series is termed *uniformly* convergent in the interval $a < x < b$. A uniformly convergent series can be rounded off to any accuracy desired:

DEFINITION   The series $\displaystyle\sum_{n=1}^{\infty} u_n(x)$ is uniformly convergent to $S(x)$ for a set E of values of x if for each $\epsilon > 0$ an integer N can be found such that

$$|S_n(x) - S(x)| < \epsilon$$

for $n \geq N$ and all x in E.

The essential idea is that N can be found as soon as ϵ (or the number of decimal places desired) is known, without considering which x of the set E is going to be used. The set E must, of course, be a set of values of x for which the series converges, but it need not consist of all such values. In many cases it can be shown that only by restricting attention to a part of the range for which convergence is known can it be ensured that the convergence is uniform. In the worst case, one could restrict attention to a finite number of values of x, for which the convergence *must* be uniform.

The definition of uniform convergence is given in terms of *series*; it can equally well be given in terms of *sequences*: The sequence $f_n(x)$ is uniformly convergent to $f(x)$ for a set E of values of x, if for each $\epsilon > 0$ an N can be found such that

$$|f_n(x) - f(x)| < \epsilon$$

for $n \geq N$ and all x in E. Thus the uniform convergence of the series $\Sigma u_n(x)$ is equivalent to the uniform convergence of its sequence of partial sums $S_n(x)$. [Actually, sequences and series are simply two different ways of describing the same limit process; every series can be interpreted in terms of its partial sums, and every sequence S_n can be interpreted as the partial sums of a series, namely, the series $S_1 + (S_2 - S_1) + \cdots + (S_n - S_{n-1}) + \cdots$.]

The most common case of uniform convergence is that in which the functions $u_n(x)$ are continuous in a closed interval $a \leq x \leq b$ and the series $\Sigma u_n(x)$ converges in this interval to a continuous function $f(x)$. [It is shown below, in Theorem 31 of Section 6.14, that uniform convergence of the series guarantees the continuity of $f(x)$.] In such a case the absolute error $|R_n(x)|$ in replacement of $f(x)$ by the partial sum $S_n(x)$:

$$|R_n(x)| = |f(x) - S_n(x)| \tag{6.25}$$

is continuous for $a \leq x \leq b$ and has a definite maximum $\bar{R}_n$ in this interval:

$$\bar{R}_n = \max_{a \leq x \leq b} |f(x) - S_n(x)|; \tag{6.26}$$

$\bar{R}_n$ is simply the worst absolute error occurring in the interval. Uniform convergence of the series for $a \leq x \leq b$ is then equivalent to the statement:

$$\lim_{n \to \infty} \bar{R}_n = 0; \tag{6.27}$$

that is, *the series is uniformly convergent for $a \leq x \leq b$ if and only if the maximum error approaches 0 as $n \to \infty$* (see Fig. 6.8). The new condition is simply a rewording of the preceding definition for the case considered, for convergence of $\bar{R}_n$ to 0 is the same as the condition $\bar{R}_n < \epsilon$ for $n \geq N(\epsilon)$ and hence as the condition $|f(x) - S_n(x)| < \epsilon$ for $n \geq N(\epsilon)$.

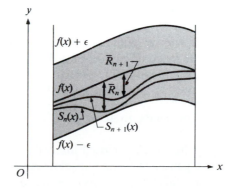

Figure 6.8   Uniform convergence.

A similar formulation can be given for the general case (when, for example, E need not be a closed interval). One simply defines $\bar{R}_n$ as the *least upper bound* of $|f(x) - S_n(x)|$ on the set E, that is, as the smallest number K such that $|f(x) - S_n(x)| \le K$ for all x in E; the existence of such a number is established in the same way as for the upper limit of a sequence (Section 6.3; see also Section 2.23). This reduces to the maximum of $|f(x) - S_n(x)|$ on the set E if this function has a maximum; however, the function, even when continuous, need not have a maximum (for example, when E is an open interval; cf. Section 2.19). Just as for upper limits, the least upper bound may be $+\infty$. Uniform convergence of $\Sigma u_n(x)$ to $f(x)$ in the set E is then equivalent to the statement: $\bar{R}_n \to 0$ as $n \to \infty$, that is, to the statement:

$$\lim_{n \to \infty} \text{l.u.b.} |f(x) - S_n(x)| = 0, \tag{6.27'}$$

where "l.u.b." stands for "least upper bound."

EXAMPLE 1   The geometric series $\displaystyle\sum_{n=0}^{\infty} x^n$. The series converges for $-1 < x < 1$. The nth partial sum is

$$S_n(x) = 1 + x + \cdots + x^{n-1} = \frac{1 - x^n}{1 - x},$$

and the sum is

$$S(x) = \frac{1}{1 - x};$$

these are plotted in Fig. 6.9. The remainder $R_n(x) = S(x) - S_n(x)$ satisfies the equation:

$$|R_n(x)| = \frac{|x^n|}{|1 - x|}.$$

We consider first a closed interval: $-\frac{1}{2} \le x \le \frac{1}{2}$. The convergence is uniform in this interval. We find

$$\bar{R}_n = \max_{-\frac{1}{2} \le x \le \frac{1}{2}} |R_n(x)| = \frac{\left(\frac{1}{2}\right)^n}{1 - \frac{1}{2}} = \frac{1}{2^{n-1}};$$

for when $x = \frac{1}{2}$, the numerator $|x^n|$ has its largest value and the denominator $|1 - x|$ has its smallest value in the interval. Accordingly,

$$\lim_{n \to \infty} \bar{R}_n = \lim_{n \to \infty} \frac{1}{2^{n-1}} = 0;$$

the maximum absolute error tends to 0 as $n \to \infty$, and the convergence is uniform.

A similar reasoning applies to each closed interval $-a \le x \le a \ (0 < a < 1)$ within the interval of convergence. The worst absolute error occurs for $x = a$, and its value is $\bar{R}_n = a^n/(1 - a)$; this tends to 0 as $n \to \infty$, so that the convergence is uniform in the interval.

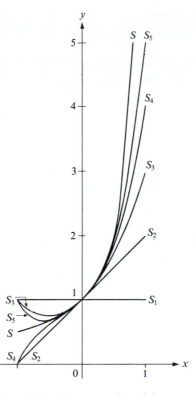

Figure 6.9 Sequence of partial sums of the geometric series.

One might expect from this that the convergence would be uniform over the entire *open* interval $-1 < x < 1$. This is not the case. Indeed, for each n the absolute error $|R_n(x)|$ is unbounded for $-1 < x < 1$, since

$$\lim_{x \to 1-} |R_n(x)| = \lim_{x \to 1-} \left| \frac{x^n}{1-x} \right| = \infty.$$

The least upper bound $\bar{R}_n$ is always $+\infty$! We can see the difficulty in detail by asking how many terms are needed to compute the sum with an absolute error less than $\epsilon = 0.01$, for example. For $x = 0$, one term is sufficient; for $x = 0.5$, one must have

$$\frac{(0.5)^n}{0.5} < 0.01 \quad \text{or} \quad 2^{n-1} > 100;$$

hence n must be at least 8. For $x = 0.9$, one must have

$$\frac{(0.9)^n}{0.1} < 0.01 \quad \text{or} \quad (0.9)^n < 0.001;$$

this requires $n > 65$, as a computation with logarithms shows. As x increases toward 1, more and more terms are needed; one can easily show that, as $x \to 1$, the number of terms required approaches ∞ (see Problem 2 following Section 6.9).

It appears from this discussion that the difficulty is due to the fact that the sum $S(x) = 1/(1 - x)$ becomes infinite as x approaches 1. However, the convergence is not uniform in the interval $-1 < x \leq 0$. For the absolute error $|R_n(x)|$ lies between 0 and $\frac{1}{2}$ in this interval; this is suggested graphically in Fig. 6.9 and can be proved strictly by calculus. Since $|R_n(x)| \to \frac{1}{2}$ as $x \to -1$, the least upper bound $\bar{R}_n$ is always $\frac{1}{2}$; this is not the maximum absolute error, since $x = -1$ is excluded from the interval under consideration. Accordingly, $\bar{R}_n$ cannot converge to 0 as $n \to \infty$, and the convergence is nonuniform for $-1 < x \leq 0$. Again we can verify that the number of terms needed to compute $S(x)$ with an error less than $\epsilon = 0.01$ approaches ∞ as $x \to -1$.

EXAMPLE 2 The *sequence* $f_n(x) = x^n$ converges to 0 for $-1 < x < 1$ and to 1 for $x = 1$; it diverges for all other values of x. The convergence for $-1 < x < 1$ is not uniform. For $0 \leq x < 1$ the error is precisely x^n; for this to be less than ϵ, one must have

$$x^n < \epsilon \quad \text{or} \quad n > \frac{\log \epsilon}{\log x}.$$

As x approaches 1, the value of n required approaches $+\infty$. For $x = 1$ the error is always 0. Successive members of the sequence $f_n(x)$ are plotted in Fig. 6.10. Since the worst errors occur near $x = \pm 1$, one can obtain uniform convergence by restricting x as in the preceding example: $-0.5 \leq x \leq 0.5$. ∎

6.13 WEIERSTRASS *M*-TEST FOR UNIFORM CONVERGENCE

The following test is adequate for determining the uniform convergence of a large number of familiar series.

THEOREM 30 (*Weierstrass M-test*) Let $\sum\limits_{n=1}^{\infty} u_n(x)$ be a series of functions all defined for a set E of values of x. If there is a convergent series of constants $\sum\limits_{n=1}^{\infty} M_n$, such that

$$|u_n(x)| \leq M_n \quad \text{for all } x \text{ in } E,$$

then the series $\sum\limits_{n=1}^{\infty} u_n(x)$ converges absolutely for each x in E and is uniformly convergent in E.

Proof. The Weierstrass *M*-test is first of all a *comparison* test. For each fixed x, each term of the series $\sum |u_n(x)|$ is less than or equal to the nth term M_n of the convergent series ΣM_n. Hence by the comparison test (Section 6.6, Theorem 12) the series $\Sigma u_n(x)$ is absolutely convergent.

But the comparison series ΣM_n is the *same* for all x of the range considered. It is from this fact that the uniform convergence follows. For if $R_n = S - S_n(x)$ is the remainder after n terms of the series $\Sigma u_n(x)$, then

$$|R_n(x)| = |u_{n+1}(x) + u_{n+2}(x) + \cdots | \leq |u_{n+1}(x)|$$
$$+ |u_{n+2}(x)| + \cdots \leq M_{n+1} + M_{n+2} + \cdots ,$$

as in Theorem 22 of Section 6.9. In other words, if T_n denotes the remainder after n terms of the convergent series ΣM_n:

$$T_n = M_{n+1} + M_{n+2} + \cdots ,$$

then

$$|R_n(x)| \leq T_n.$$

Since ΣM_n is a series of constants, for each given $\epsilon > 0$, N can be found such that $T_n < \epsilon$ for $n \geq N$; for this same N, one has

$$|R_n(x)| \leq T_n < \epsilon \quad \text{for} \quad n \geq N.$$

Since N does not depend on x, but only on ϵ, the uniform convergence has been established. At the same time it has been shown that T_n serves as an upper estimate for the error committed in using only n terms of the series $\Sigma u_n(x)$, regardless of the x. The error will, of course, vary from one x to another, but the upper estimate is the same for all x.   □

EXAMPLE 1 $\displaystyle\sum_{n=1}^{\infty} \frac{x^n}{n^2}$. Here the ratio test gives

$$\lim_{n \to \infty} \left| \frac{a_{n+1}}{a_n} \right| = \lim_{n \to \infty} \left| \frac{x^{n+1}}{(n+1)^2} \cdot \frac{n^2}{x^n} \right| = \lim_{n \to \infty} |x| \cdot \frac{n^2}{(n+1)^2} = |x|.$$

Hence the series converges for $|x| < 1$ and diverges for $|x| > 1$. For $x = \pm 1$ the series converges by comparison with the harmonic series of order 2:

$$\left| \frac{(\pm 1)^n}{n^2} \right| \leq \frac{1}{n^2}.$$

Hence the series converges for $-1 \leq x \leq 1$. The convergence is uniform for this range since the comparison

$$\left| \frac{x^n}{n^2} \right| \leq M_n = \frac{1}{n^2}.$$

holds for all x of the range and the series ΣM_n converges.   ■

EXAMPLE 2 $\displaystyle\sum_{n=1}^{\infty} \frac{\cos nx}{2^n}$. This series converges uniformly for all x, since

$$\left| \frac{\cos nx}{2^n} \right| \leq \frac{1}{2^n} = M_n$$

for all x, and the series $\Sigma(1/2^n)$ is convergent. ■

EXAMPLE 3 $\displaystyle\sum_{n=1}^{\infty} \frac{x^n}{n}$. The ratio test shows, as in Example 1, that the series
converges for $-1 < x < 1$ and diverges for $|x| > 1$. For $x = 1$ the series is the
divergent harmonic series; for $x = -1$ the series is a convergent alternating
series. One might try to prove uniform convergence for $0 \leq x < 1$ by using the
inequality

$$\left| \frac{x^n}{n} \right| \leq x^n, \qquad 0 \leq x < 1,$$

and the fact that the series $\displaystyle\sum_{n=1}^{\infty} x^n$ converges. However, this reasoning is incorrect,
since the comparison series Σx^n *depends on x* and is not a series of constants as
required in Theorem 30. As it happens, the given series is *not* uniformly convergent for $0 \leq x < 1$. ■

PROBLEMS

1. Determine the values of x for which each of the following series converges:

a) $\displaystyle\sum_{n=1}^{\infty} \frac{x^n}{2n^2 - n}$

b) $\displaystyle\sum_{n=1}^{\infty} \frac{nx^n}{2^n}$

c) $\displaystyle\sum_{n=1}^{\infty} \frac{1}{nx^{2n}}$

d) $\displaystyle\sum_{n=0}^{\infty} \frac{1}{2^{nx}}$

e) $\displaystyle\sum_{n=1}^{\infty} \frac{x^n}{(1-x)^n}$

f) $\displaystyle\sum_{n=1}^{\infty} \frac{2^n \sin^n x}{n^2}$

g) $\displaystyle\sum_{n=1}^{\infty} \frac{(x-1)^n}{n^2}$

h) $\displaystyle\sum_{n=1}^{\infty} \frac{1}{x^n \log(n+1)}$

i) $\displaystyle\sum_{n=1}^{\infty} \frac{(x-2)^{3n}}{n!}$

j) $\displaystyle\sum_{n=2}^{\infty} \frac{x^n}{(\log n)^n}$

2. Prove that each of the following series is uniformly convergent over the set of values
of x given:

a) $\displaystyle\sum_{n=1}^{\infty} \frac{x^n}{n^3}$, $-1 \leq x \leq 1$

b) $\displaystyle\sum_{n=1}^{\infty} \frac{(\tanh x)^n}{n!}$,   all x

c) $\displaystyle\sum_{n=1}^{\infty} \frac{\sin nx}{n^2 + 1}$,   all x

d) $\displaystyle\sum_{n=1}^{\infty} \frac{e^{nx}}{2^n}$, $x \leq \log \frac{3}{2}$

e) $\sum_{n=0}^{\infty} \frac{x^n}{n!}$, $-1 \le x \le 1$

f) $\sum_{n=1}^{\infty} nx^n$, $-\frac{1}{2} \le x \le \frac{1}{2}$

g) $\sum_{n=1}^{\infty} nx^n$, $-0.9 \le x \le 0.9$

h) $\sum_{n=1}^{\infty} nx^n$, $-a \le x \le a$, $a < 1$

3. Prove: If $\sum_{n=1}^{\infty} u_n(x)$ is uniformly convergent for $a \le x \le b$, then the series is uniformly convergent in each smaller interval contained in the interval $a \le x \le b$. More generally, if a series is uniformly convergent for a given set E of values of x, then it is uniformly convergent for any set E_1 that is part of E.

4. Prove: If $\sum_{n=1}^{\infty} v_n(x)$ is uniformly convergent for a set E of values of x and $|u_n(x)| \le v_n(x)$ for x in E, then $\sum_{n=1}^{\infty} u_n(x)$ is uniformly convergent for x in E.

5. Prove: If $0 < u_n(x) < 1/n$ and $u_{n+1}(x) \le u_n(x)$ for $a \le x \le b$, then the series $\sum_{n=1}^{\infty} (-1)^n u_n(x)$ is uniformly convergent for $a \le x \le b$.

6. Prove: If the series $\sum_{n=1}^{\infty} M_n$ of constants M_n is convergent and $|f_{n+1}(x) - f_n(x)| \le M_n$ for x in E, then the *sequence* $f_n(x)$ is uniformly convergent for x in E.

7. Prove that the following sequences are uniformly convergent for the range of x given (cf. Problem 6):

a) $\frac{n + x}{n}$, $0 \le x \le 1$

b) $\frac{x^n}{n!}$, $-1 \le x \le 1$

c) $\frac{\log(1 + nx)}{n}$, $1 \le x \le 2$

d) $\frac{n}{e^{nx^2}}$, $\frac{1}{2} \le x \le 1$

6.14   PROPERTIES OF UNIFORMLY CONVERGENT SERIES AND SEQUENCES

Let $\Sigma u_n(x)$ be a series of functions, each of which is defined for $a \le x \le b$. Let it be assumed further that this series converges to a sum $f(x)$ for $a \le x \le b$, so that one has

$$f(x) = \sum_{n=1}^{\infty} u_n(x), \qquad a \le x \le b.$$

One can then ask questions such as the following: If each function $u_n(x)$ is continuous, is the sum $f(x)$ continuous? If each $u_n(x)$ has a derivative, does $f(x)$ have a derivative? The following theorems answer such questions.

THEOREM 31   The sum of a uniformly convergent series of continuous functions is continuous; that is, if each $u_n(x)$ is continuous for $a \le x \le b$, then so is $f(x) = \sum_{n=1}^{\infty} u_n(x)$, provided that the series converges uniformly for $a \le x \le b$.

Proof. Let x_0 be given, $a \leq x_0 \leq b$, and let $\epsilon > 0$ be given. We then seek a δ such that

$$|f(x) - f(x_0)| < \epsilon \quad \text{when} \quad |x - x_0| < \delta$$

and x is in the given interval. We choose N so large that

$$|S_n(x) - f(x)| < \tfrac{1}{3}\epsilon, \qquad a \leq x \leq b, \qquad n \geq N, \tag{6.28}$$

where $S_n(x) = u_1(x) + \cdots + u_n(x)$; this is possible since the series is uniformly convergent. The function $S_N(x)$, as sum of a *finite* number of continuous functions, is itself continuous. One can hence choose a δ such that

$$|S_N(x) - S_N(x_0)| < \tfrac{1}{3}\epsilon \quad \text{for} \quad |x - x_0| < \delta. \tag{6.29}$$

By (6.28), one has

$$|S_N(x) - f(x)| < \tfrac{1}{3}\epsilon, \qquad |S_N(x_0) - f(x_0)| < \tfrac{1}{3}\epsilon. \tag{6.30}$$

Hence

$$
\begin{aligned}
|f(x) - f(x_0)| &= |f(x) - S_N(x) + S_N(x) - S_N(x_0) + S_N(x_0) - f(x_0)| \\
&\leq |f(x) - S_N(x)| + |S_N(x) - S_N(x_0)| + |S_N(x_0) - f(x_0)| \\
&< \tfrac{1}{3}\epsilon + \tfrac{1}{3}\epsilon + \tfrac{1}{3}\epsilon = \epsilon, \quad \text{for} \quad |x - x_0| < \delta,
\end{aligned}
$$

by (6.29) and (6.30). Thus continuity is proved. $\square$

Remark 1. The property of convergence alone, for a series of continuous functions, does not guarantee continuity of the sum. This is seen by the example:

$$f(x) = x + \sum_{n=2}^{\infty} (x^n - x^{n-1}), \qquad 0 \leq x \leq 1.$$

Here the nth partial sums form the sequence $S_n(x) = x^n$ plotted in Fig. 6.10. As

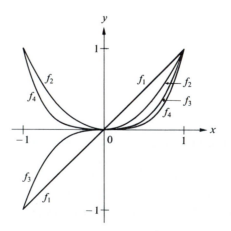

Figure 6.10   The sequence $f_n(x) = x^n$.

was pointed out in Section 6.12, this sequence does not converge uniformly. The sum of the series is 0 for $0 \leq x < 1$ and 1 for $x = 1$; there is a jump discontinuity at $x = 1$.

Remark 2.   Theorem 31 can be interpreted in terms of sequences as follows: If $S_n(x)$ is a sequence of functions all continuous for $a \leq x \leq b$ and this sequence converges uniformly to $f(x)$ for $a \leq x \leq b$, then $f(x)$ is continuous for $a \leq x \leq b$; furthermore, if $a \leq x_0 \leq b$, then

$$\lim_{x \to x_0} \left[\lim_{n \to \infty} S_n(x) \right] = \lim_{n \to \infty} \left[\lim_{x \to x_0} S_n(x) \right]. \tag{6.31}$$

The left-hand side is precisely $\lim\limits_{x \to x_0} f(x)$, and the right-hand side is $\lim\limits_{n \to \infty} S_n(x_0)$ $= f(x_0)$, by the continuity of $S_n(x)$; Eq. (6.31) is simply the assertion that

$$\lim_{x \to x_0} f(x) = f(x_0),$$

that is, that $f(x)$ is continuous at x_0. Accordingly, as Eq. (6.31) shows, *uniform convergence permits one to interchange two limit processes.*

   THEOREM 32   A uniformly convergent series of continuous functions can be integrated term by term; that is, if each $u_n(x)$ is continuous for $a \leq x \leq b$, then

$$\int_a^b f(x) \, dx = \int_a^b u_1(x) \, dx + \int_a^b u_2(x) \, dx + \cdots + \int_a^b u_n(x) \, dx + \cdots .$$

$$\tag{6.32}$$

Proof.   As in the preceding proof, let $S_n(x)$ be the nth partial sum of the series $\Sigma u_n(x)$. Then

$$\int_a^b S_n(x) \, dx = \int_a^b u_1(x) \, dx + \cdots + \int_a^b u_n(x) \, dx.$$

To prove (6.32), one must show that the sequence $\int_a^b S_n(x) \, dx$ converges to $\int_a^b f(x) \, dx$, that is, for each $\epsilon > 0$ an N can be found such that

$$\left| \int_a^b f(x) \, dx - \int_a^b S_n(x) \, dx \right| < \epsilon, \qquad n \geq N.$$

To establish this, we choose N so large that

$$|f(x) - S_n(x)| < \frac{\epsilon}{b - a}, \qquad n \geq N, \qquad a \leq x \leq b;$$

this is possible because of the uniform convergence. Hence

$$\left| \int_a^b f(x) \, dx - \int_a^b S_n(x) \, dx \right| = \left| \int_a^b [f(x) - S_n(x)] \, dx \right| \leq \frac{\epsilon}{b - a} \cdot (b - a) = \epsilon,$$

$$n \geq N,$$

by the rule (4.6) in Section 4.1. Theorem 32 is thus established.

Remark 3.   This theorem can be formulated in terms of sequences and interchange of limit processes as in the preceding Remark 2.

EXAMPLE   In Section 6.12 the series $\sum_{n=0}^{\infty} x^n$ was shown to converge uniformly to $1/(1 - x)$ in each interval $-a \leq x \leq a$, where $a < 1$. On integrating the equation

$$\frac{1}{1 - x} = 1 + x + \cdots + x^n + \cdots$$

from 0 to x_1, one finds

$$\int_0^{x_1} \frac{1}{1 - x} dx = \log \frac{1}{1 - x_1} = x_1 + \frac{1}{2} x_1^2 + \cdots + \frac{x_1^{n+1}}{n + 1} + \cdots.$$

This holds for every x_1 between -1 and 1 (the ends excluded), so that one can write

$$\log \frac{1}{1 - x} = x + \frac{1}{2} x^2 + \cdots + \frac{x^{n+1}}{n + 1} + \cdots, \qquad -1 < x < 1. \quad (6.33)$$

The same procedure could be phrased in terms of *indefinite* integrals. Thus

$$\int \frac{1}{1 - x} dx = \int (1 + x + x^2 + \cdots + x^n + \cdots) \, dx + c, \qquad -1 < x < 1;$$

$$\log \frac{1}{1 - x} = x + \frac{1}{2} x^2 + \frac{1}{3} x^3 + \cdots + \frac{x^{n+1}}{n + 1} + \cdots + c.$$

Of course, c is no longer arbitrary; by choosing $x = 0$ we recognize that $c = 0$, so that (6.33) is again obtained.   ■

THEOREM 33   A convergent series can be differentiated term by term, provided that the functions of the series have continuous derivatives and that the series of derivatives is uniformly convergent; that is, if $u'_n(x) = du_n/dx$ is continuous for $a \leq x \leq b$, if the series $\sum_{n=1}^{\infty} u_n(x)$ converges for $a \leq x \leq b$ to $f(x)$, and if the series $\sum_{n=1}^{\infty} u'_n(x)$ converges uniformly for $a \leq x \leq b$, then

$$f'(x) = \sum_{n=1}^{\infty} u'_n(x), \qquad a \leq x \leq b. \quad (6.34)$$

The derivatives at a and b are understood as right- and left-sided derivatives, respectively.

Proof of Theorem 33.   Let $g(x)$ be the sum of the series of derivatives:

$$g(x) = \sum_{n=1}^{\infty} u'_n(x), \qquad a \leq x \leq b.$$

Then $g(x)$ is continuous by Theorem 31, and by Theorem 32,

$$\int_a^{x_1} g(x)\,dx = \sum_{n=1}^{\infty} \int_a^{x_1} u_n'(x)\,dx, \qquad a \le x_1 \le b.$$

Hence

$$\int_a^{x_1} g(x)\,dx = \sum_{n=1}^{\infty} [u_n(x_1) - u_n(a)]$$

$$= \sum_{n=1}^{\infty} u_n(x_1) - \sum_{n=1}^{\infty} u_n(a).$$

Accordingly,

$$\int_a^{x_1} g(x)\,dx = f(x_1) - f(a).$$

If both sides are differentiated with respect to x_1, the fundamental theorem of calculus (Section 4.1) gives

$$g(x_1) = f'(x_1), \qquad a \le x_1 \le b.$$

In other words,

$$f'(x) = g(x) = \sum_{n=1}^{\infty} u_n'(x), \qquad a \le x \le b. \quad \square$$

THEOREM 34   If $\displaystyle\sum_{n=1}^{\infty} u_n(x)$ and $\displaystyle\sum_{n=1}^{\infty} v_n(x)$ are uniformly convergent for $a \le x \le b$ and $h(x)$ is continuous for $a \le x \le b$, then the series

$$\sum_{n=1}^{\infty} [u_n(x) + v_n(x)], \quad \sum_{n=1}^{\infty} [u_n(x) - v_n(x)], \quad \sum_{n=1}^{\infty} [h(x)u_n(x)]$$

are uniformly convergent for $a \le x \le b$.

Proof.   Let $f(x)$ and $g(x)$ be the sums of $\Sigma u_n(x)$ and $\Sigma v_n(x)$, respectively; let $S_n(x)$ and $Q_n(x)$ be the corresponding partial sums. The nth partial sum of $\Sigma(u_n + v_n)$ is $S_n + Q_n$. Let N be chosen, for given ϵ, so that

$$|S_n(x) - f(x)| < \tfrac{1}{2}\epsilon, \qquad |Q_n(x) - g(x)| < \tfrac{1}{2}\epsilon,$$

for $n \ge N$ and $a \le x \le b$. Then

$$|\{S_n(x) + Q_n(x)\} - \{f(x) + g(x)\}|$$
$$\le |S_n(x) - f(x)| + |Q_n(x) - g(x)| < \tfrac{1}{2}\epsilon + \tfrac{1}{2}\epsilon = \epsilon.$$

Thus $\Sigma(u_n + v_n)$ converges uniformly to $f(x) + g(x)$. A similar proof applies to the difference.

Since $h(x)$ is continuous for $a \le x \le b$, it is necessarily bounded: $|h(x)| \le M$ for $a \le x \le b$. Hence

$$|h(x)S_n(x) - h(x)f(x)| = |h(x)||S_n(x) - f(x)| < M \cdot \tfrac{1}{2}\epsilon < M\epsilon$$

for $n \geq N$ as before. This shows that the series $\Sigma h(x) u_n$ converges uniformly to $h(x)f(x)$. It should be noted that actually only the *boundedness* of $h(x)$ was required. □

6.15   POWER SERIES

By a *power series* in powers of x is meant a series of the form

$$\sum_{n=0}^{\infty} c_n x^n = c_0 + c_1 x + \cdots + c_n x^n + \cdots, \tag{6.35}$$

where $c_0, c_1, \ldots, c_n, \ldots$ are constants. By a power series in powers of $(x - a)$ is meant a series:

$$\sum_{n=0}^{\infty} c_n (x - a)^n = c_0 + c_1 (x - a) + \cdots + c_n (x - a)^n + \cdots. \tag{6.36}$$

By a power series in *negative* powers of x is meant a series:

$$\sum_{n=0}^{\infty} \frac{c_n}{x^n} = c_0 + \frac{c_1}{x} + \cdots + \frac{c_n}{x^n} + \cdots. \tag{6.37}$$

The term *power series* alone usually refers to (6.36), of which (6.35) is a special case ($a = 0$). The substitutions $t = x - a$ or $t = 1/x$ can be used to reduce series of form (6.36) or (6.37) to the form (6.35).

The power series (6.36) converges when $x = a$. It can happen that this is the only value of x for which the series converges. If there are other values of x for which the series converges, then it will be seen that these form an interval, the "convergence interval," having midpoint $x = a$. This is pictured in Fig. 6.11. The interval can be infinite. These properties are summarized in the following basic theorem.

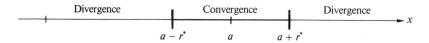

Figure 6.11   Interval of convergence of a power series.

THEOREM 35 Every power series

$$c_0 + c_1 (x - a) + \cdots + c_n (x - a)^n + \cdots$$

has a "radius of convergence" r^* such that the series converges absolutely when $|x - a| < r^*$ and diverges when $|x - a| > r^*$.

The number r^* can be 0 (in which case the series converges only for $x = a$), a positive number, or ∞ (in which case the series converges for all x).

If r^* is not zero and r_1 is such that $0 < r_1 < r^*$, then the series converges uniformly for $|x - a| \le r_1$. The number r^* can be evaluated as follows:

$$r^* = \lim_{n \to \infty} \left| \frac{c_n}{c_{n+1}} \right|, \quad \text{if the limit exists,} \tag{6.38}$$

$$r^* = \lim_{n \to \infty} \frac{1}{\sqrt[n]{|c_n|}}, \quad \text{if the limit exists,} \tag{6.39}$$

and in any case by the formula:

$$r^* = \frac{1}{\overline{\lim_{n \to \infty}} \sqrt[n]{|c_n|}}. \tag{6.40}$$

Proof.   We begin by considering the case in which the limit (6.38) exists, as this at once suggests the general situation. The ratio test, applied to the given series, proceeds as follows:

$$\lim_{n \to \infty} \left| \frac{a_{n+1}}{a_n} \right| = \lim_{n \to \infty} \left| \frac{c_{n+1}(x - a)^{n+1}}{c_n(x - a)^n} \right| = \lim_{n \to \infty} \left| \frac{c_{n+1}}{c_n} \right| \cdot |x - a|$$

$$= \frac{|x - a|}{\lim_{n \to \infty} \left| \frac{c_n}{c_{n+1}} \right|} = \frac{|x - a|}{r^*}.$$

The series converges absolutely when

$$\frac{|x - a|}{r^*} < 1, \quad \text{that is,} \quad |x - a| < r^*$$

and diverges when

$$\frac{|x - a|}{r^*} > 1, \quad \text{that is,} \quad |x - a| > r^*.$$

If $r^* = 0$, the test shows divergence except for $x = a$. If $r^* = \infty$, so that

$$\lim_{n \to \infty} \left| \frac{c_{n+1}}{c_n} \right| = 0,$$

the test shows that the series converges absolutely for all x.

The case in which the limit (6.39) exists is treated in the same way by the root test. This case is included in the final formula (6.40), which follows at once from Theorem 21(a) of Section 6.8, for

$$\overline{\lim_{n \to \infty}} \sqrt[n]{|a_n|} = \overline{\lim_{n \to \infty}} \left(\sqrt[n]{|c_n|} \cdot |x - a| \right) = \left(\overline{\lim_{n \to \infty}} \sqrt[n]{|c_n|} \right) \cdot |x - a|.$$

The series converges absolutely when the upper limit is less than 1, that is, when

$$|x - a| < \frac{1}{\overline{\lim_{n \to \infty}} \sqrt[n]{|c_n|}} = r^*,$$

and diverges when $|x - a| > r^*$.

Thus however it may be determined, there is a number r^*, $0 \leq r^* \leq \infty$, with the properties described. It remains to prove that if $0 < r_1 < r^*$, then the series converges uniformly for $|x - a| \leq r_1$. This follows from the M-test, with $M_n = |c_n| r_1^n$, for

$$\sum_{n=0}^{\infty} M_n = \sum_{n=0}^{\infty} |c_n| |x_1 - a|^n, \qquad x_1 = a + r_1,$$

and this converges, since $|x_1 - a| = r_1 < r^*$. For $|x - a| \leq r_1$,

$$|c_n| |x - a|^n \leq |c_n| r_1^n = M_n,$$

so that the convergence is uniform. The main idea of this proof is emphasized in Section 6.12—the slowest convergence of a power series occurs toward the *ends* of the interval of convergence. $\square$

It should be emphasized that when r^* is a finite positive number, the series may converge or diverge for each of the end-values $x = a + r^*$, $x = a - r^*$. These must be investigated separately for each series.

EXAMPLE 1 $\displaystyle\sum_{n=1}^{\infty} \frac{x^n}{n^2}$. Here $c_n = 1/n^2$, and (6.38) gives

$$r^* = \lim_{n \to \infty} \frac{(n + 1)^2}{n^2} = 1;$$

similarly, (6.39) gives

$$r^* = \lim_{n \to \infty} \sqrt[n]{n^2} = \lim_{n \to \infty} n^{2/n} = \lim_{n \to \infty} e^{(2/n) \log n} = e^{\lim(2/n) \log n} = e^0 = 1,$$

by Theorem 4 of Section 6.4. Hence the series converges absolutely for $-1 < x < 1$ and diverges for $|x| > 1$. For $x = \pm 1$ the series converges absolutely, since

$$\left| \frac{(\pm 1)^n}{n^2} \right| = \frac{1}{n^2}.$$

In fact, the series converges absolutely and uniformly for $-1 \leq x \leq 1$, since

$$\left| \frac{x^n}{n^2} \right| \leq \frac{1}{n^2} = M_n$$

in this range. ∎

EXAMPLE 2 $\displaystyle\sum_{n=1}^{\infty} \frac{x^n}{n}$. The ratio formula and root formula apply as in Example 1 to give $r^* = 1$. Hence the series converges absolutely for $|x| < 1$ and diverges for $|x| > 1$. For $x = 1$ the series is the harmonic series and diverges; for $x = -1$ the series converges conditionally by the alternating series test. Hence the series converges for $-1 \leq x < 1$ and diverges otherwise. ∎

It should be noted that Example 2 is obtained from Example 1 essentially by term-by-term differentiation:

$$\sum_{n=1}^{\infty} \frac{d}{dx}\left(\frac{x^n}{n^2}\right) = \sum_{n=1}^{\infty} \frac{nx^{n-1}}{n^2} = \sum_{n=1}^{\infty} \frac{x^{n-1}}{n};$$

an extra factor of x is needed. The effect of term-by-term differentiation is thus to multiply each term by n (while lowering the degree of x^n by 1). This cannot affect the radius of convergence, since

$$\lim_{n\to\infty} \sqrt[n]{n} = \lim_{n\to\infty} n^{1/n} = \lim_{n\to\infty} e^{(1/n)\log n} = 1.$$

However, it does slow down convergence at the ends of the interval, as the second example shows. If one differentiates the series of Example 2, one obtains the geometric series:

$$\sum_{n=1}^{\infty} \frac{nx^{n-1}}{n} = \sum_{n=1}^{\infty} x^{n-1} = \sum_{n=0}^{\infty} x^n = 1 + x + x^2 + \cdots,$$

which converges for $-1 < x < 1$ and *diverges* at the ends of the interval.

THEOREM 36   A power series represents a continuous function within the interval of convergence; that is, if r^* is the radius, then

$$f(x) = \sum_{n=0}^{\infty} c_n(x - a)^n$$

is continuous for $a - r^* < x < a + r^*$.

Proof.   By Theorem 35 the series converges uniformly for $a - r_1 \le x \le a + r_1$, $r_1 < r^*$. Each term $c_n(x - a)^n$ of the series is continuous for all x; hence by Theorem 31, $f(x)$ is continuous for $a - r_1 \le x \le a + r_1$. This holds for every r_1 between 0 and r^*. Accordingly, $f(x)$ is continuous for $a - r^* < x < a + r^*$.   □

Remark.   If the series converges at either end of the interval, then $f(x)$ is continuous at the end concerned. For a proof refer to page 177 of the book by Knopp listed at the end of the chapter.

THEOREM 37   A power series can be integrated term by term within the interval of convergence; that is, if

$$f(x) = \sum_{n=0}^{\infty} c_n(x - a)^n, \qquad a - r^* < x < a + r^*,$$

then for $a - r^* < x_1 < x_2 < a + r^*$,

$$\int_{x_1}^{x_2} f(x)\,dx = \sum_{n=0}^{\infty} c_n \int_{x_1}^{x_2} (x - a)^n\,dx = \sum_{n=0}^{\infty} c_n \frac{(x_2 - a)^{n+1} - (x_1 - a)^{n+1}}{n + 1},$$

or in terms of indefinite integrals,

$$\int f(x)\, dx = \sum_{n=0}^{\infty} c_n \frac{(x-a)^{n+1}}{n+1} + C, \qquad a - r^* < x < a + r^*.$$

Proof. By Theorem 35 the series converges uniformly in each interval $a - r_1 \le x \le a + r_1$. It is therefore uniformly convergent for each interval $x_1 \le x \le x_2$, for the latter interval will be included in the former for appropriate choice of r_1. One can now apply Theorem 32 to justify term-by-term integration. The indefinite integral can be regarded as a special case of this, for if $x_1 = a$, then

$$\int_a^{x_2} f(x)\, dx = \sum_{n=0}^{\infty} c_n \frac{(x_2-a)^{n+1}}{n+1}.$$

This defines a function $F(x_2)$:

$$F(x_2) = \int_a^{x_2} f(x)\, dx$$

and

$$\frac{dF(x_2)}{dx_2} = f(x_2)$$

by the fundamental theorem of calculus; that is,

$$\frac{dF(x)}{dx} = f(x), \qquad F(x) = \sum_{n=0}^{\infty} c_n \frac{(x-a)^{n+1}}{n+1},$$

or

$$\int f(x)\, dx = \sum_{n=0}^{\infty} c_n \frac{(x-a)^{n+1}}{n+1} + C.$$

The number x_2 can be chosen anywhere in the interval $a - r^* < x < a + r^*$, and accordingly, the last equation is valid for all x of this interval. $\square$

THEOREM 38 A power series can be differentiated term by term within the interval of convergence; that is if

$$f(x) = \sum_{n=0}^{\infty} c_n (x-a)^n, \qquad a - r^* < x < a + r^*,$$

then

$$f'(x) = \sum_{n=1}^{\infty} n c_n (x-a)^{n-1}, \qquad a - r^* < x < a + r^*.$$

Proof. The general coefficient of the differentiated series is $n c_n$ [or, more precisely, $(n+1)c_{n+1}$]. The extra factor of n has no effect on the upper limit of the nth root, as was pointed out in connection with Example 2. Hence the *differentiated series has the same radius of convergence r^**; this series is then uniformly convergent in each interval $a - r_1 \leq x \leq a + r_1, r_1 < r^*$. Hence by Theorem 33,

$$f'(x) = \sum_{n=1}^{\infty} nc_n(x - a)^{n-1}$$

for each x of such an interval. Every x such that $a - r^* < x < a + r^*$ lies in such an interval, so that the result is justified for all x within the interval of convergence. □

Application.   From the relation

$$\frac{1}{1 - x} = 1 + x + \cdots + x^n + \cdots = \sum_{n=0}^{\infty} x^n, \qquad -1 < x < 1,$$

one now obtains by successive differentiation the relations

$$\frac{1}{(1 - x)^2} = 1 + 2x + \cdots + nx^{n-1} + \cdots = \sum_{n=0}^{\infty} (n + 1)x^n, \qquad -1 < x < 1,$$

$$\frac{2}{(1 - x)^3} = 2 + 6x + \cdots + n(n - 1)x^{n-2} + \cdots = \sum_{n=0}^{\infty} (n + 2)(n + 1)x^n,$$

$$-1 < x < 1.$$

The general case can be written thus:

$$\frac{1}{(1 - x)^k} = (1 - x)^{-k}$$

$$= 1 + \frac{kx}{1} + \frac{k(k + 1)}{1 \cdot 2}x^2$$

$$+ \cdots + \frac{k(k + 1) \cdots (k + n - 1)}{1 \cdot 2 \cdots n}x^n + \cdots,$$

$$-1 < x < 1, \qquad k = 1, 2, 3, \ldots . \qquad (6.41)$$

This is termed the *binomial theorem for negative integral exponents.* Equation (6.41) is actually valid for every *real number k*, as will be seen.

6.16 TAYLOR AND MACLAURIN SERIES

Let $f(x)$ be the sum of a power series with convergence interval $a - r^* < x < a + r^*$ $(r^* > 0)$:

$$f(x) = \sum_{n=0}^{\infty} c_n(x - a)^n, \qquad a - r^* < x < a + r^*. \qquad (6.42)$$

This series is called the *Taylor series of f(x) at x = a* if the coefficients c_n are given by the rule:

$$c_0 = f(a), \quad c_1 = \frac{f'(a)}{1!}, \quad c_2 = \frac{f''(a)}{2!}, \ldots, c_n = \frac{f^{(n)}(a)}{n!}, \ldots,$$

so that

$$f(x) = f(a) + \frac{f'(a)}{1!}(x - a) + \cdots + \frac{f^{(n)}(a)}{n!}(x - a)^n + \cdots. \quad (6.43)$$

THEOREM 39 Every power series with nonzero convergence radius is the Taylor series of its sum.

Proof. Let $f(x)$ be given by (6.42). Then by repeated differentiation, on the basis of Theorem 38, one finds

$$f(x) = c_0 + c_1(x - a) + \cdots + c_n(x - a)^n + \cdots,$$

$$f'(x) = c_1 + 2c_2(x - a) + \cdots + n \cdot c_n(x - a)^{n-1} + \cdots,$$

$$f''(x) = 2c_2 + 6c_3(x - a) + \cdots + n(n - 1) \cdot c_n(x - a)^{n-2} + \cdots,$$

$$\cdots,$$

$$f^{(n)}(x) = n(n - 1)(n - 2) \cdots 2 \cdot 1 \cdot c_n$$
$$+ (n + 1)n(n - 1) \cdots 2 \cdot c_{n+1}(x - a) + \cdots,$$

$$\cdots.$$

Here all series converge for $a - r^* < x < a + r^*$. If one now sets $x = a$, one finds

$$f(a) = c_0, \quad f'(a) = c_1, \quad f''(a) = 2c_2, \ldots, f^{(n)}(a) = n!c_n, \ldots.$$

This gives $c_0 = f(a)$ and

$$c_n = \frac{f^{(n)}(a)}{n!}, \quad n = 1, 2, \ldots,$$

as asserted. □

In the case in which $a = 0$ the expression (6.43) for the Taylor series of $f(x)$ becomes

$$f(x) = f(0) + \frac{f'(0)}{1!}x + \frac{f''(0)}{2!}x^2 + \cdots + \frac{f^{(n)}(0)x^n}{n!} + \cdots. \quad (6.44)$$

This is called the *Maclaurin series* of $f(x)$. For many purposes it is easier to employ. The substitution $t = x - a$ reduces the general Taylor series to the form of Maclaurin series.

THEOREM 40 If two power series

$$\sum_{n=0}^{\infty} c_n(x - a)^n, \quad \sum_{n=0}^{\infty} C_n(x - a)^n$$

have nonzero convergence radii and have equal sums wherever both series converge, then the series are identical; that is,

$$c_n = C_n, \quad n = 0, 1, 2, \ldots.$$

Proof.   Let the convergence radii be r^* and R^*, respectively, where $0 < r^* \le R^*$. Then, by assumption,

$$\sum_{n=0}^{\infty} c_n(x-a)^n = \sum_{n=0}^{\infty} C_n(x-a)^n = f(x),$$

$$a - r^* < x < a + r^*.$$

By the preceding theorem, one must then have $c_0 = C_0 = f(a)$, and

$$c_n = C_n = \frac{f^{(n)}(a)}{n!}, \qquad n = 1, 2, \dots .$$

Thus the coefficients are equal for every value of n.   □

COROLLARY   If a power series has a nonzero convergence radius and has a sum that is identically zero, then every coefficient of the series is zero.

PROBLEMS

1. a) Obtain the Maclaurin series

$$\log \frac{1}{1-x} = x + \frac{1}{2}x^2 + \cdots + \frac{x^n}{n} + \cdots, \qquad -1 < x < 1,$$

by integration of the series for $1/(1-x)$. Verify that (6.43) holds.

b) Show that the series converges for $x = -1$ and hence prove (on the basis of the remark preceding Theorem 37) that

$$\log 2 = \sum_{n=1}^{\infty} \frac{(-1)^{n+1}}{n}.$$

2. Prove (6.41) by induction.

3. a) Expand $1/x$ in a Taylor series about $x = 1$. [Hint: Write $1/[1 - (1 - x)] = 1/(1 - r)$ and use the geometric series.]

b) Expand $1/(x + 2)$ in a Maclaurin series. [Hint: Write $1/(x + 2) = 1/\{2[1 - (-\frac{1}{2}x)]\} = \frac{1}{2}\{1/(1 - r)\}$.]

c) Expand $1/(3x + 5)$ in a Maclaurin series.

d) Expand $1/(3x + 5)$ in a Taylor series about $x = 1$. [Hint: Write $1/(3x + 5) = 1/[3(x - 1) + 8] = \frac{1}{8}/[1 + \frac{3}{8}(x - 1)]$.]

e) Expand $1/(ax + b)$ in a Taylor series about $x = c$, where $ac + b \ne 0$, $a \ne 0$.

f) Expand $1/(1 - x^2)$ in a Maclaurin series.

g) Expand $1/[(x - 2)(x - 3)]$ in a Maclaurin series. [Hint: Write $1/[(x - 2)(x - 3)] = A/(x - 2) + B/(x - 3)$.]

h) Expand $1/x^2$ in a Taylor series about $x = 1$. [Hint: Write $1/x^2 = 1/[1 - (1 - x)]^2 = 1/(1 - r)^2$ and use (6.41).]

i) Expand $1/(3x + 5)^2$ in a Taylor series about $x = 1$.

j) Expand $1/(ax + b)^k$ in a Taylor series about $x = c$, where $ac + b \neq 0$, $a \neq 0$, and $k = 1, 2, \ldots$.

4. Let $f(x) = \displaystyle\sum_{n=1}^{\infty} \frac{x^n}{n^n}$.

   a) Show that $f(x)$ is defined for all x.

   b) Evaluate (approximately where necessary) $f(0)$, $f(1)$, $f'(0)$, $f'(1)$, $f''(0)$.

   c) Obtain Maclaurin series for $f'(x)$, $f''(x)$.

5. Let $y = f(x)$ be a function (if there is one) such that $f(x)$ is defined for all x, $f(x)$ has a Maclaurin series valid for all x, $f(0) = 1$ and $dy/dx = y$ for all x. Show that necessarily

$$f(x) = 1 + x + x^2/2! + \cdots + x^n/n! + \cdots$$

and that this function satisfies all requirements. [It will be shown later that $f(x) = e^x$.]

6.17 TAYLOR'S FORMULA WITH REMAINDER

The preceding discussion concentrated on the power series, rather than on the functions that they represent. The opposite point of view is also of fundamental importance, the first question being the following: Given a function $f(x)$, $a < x < b$, can $f(x)$ be represented by a power series in this interval? If $f(x)$ can be so represented, $f(x)$ is termed *analytic* in the given interval. More generally, $f(x)$ is termed analytic for $a < x < b$ if for each x_0 of this interval, $f(x)$ can be represented by a power series in some interval $x_0 - \delta < x < x_0 + \delta$. Most of the familiar functions—polynomials, rational functions, e^x, $\sin x$, $\cos x$, $\log x$, $\sqrt{x}$—and those constructed from them by operations of algebra and by substitutions are analytic in every interval in which the function considered is continuous. The exceptions are not too difficult to recognize. Thus $\sqrt{x^2} = |x|$ is continuous for all x but has a discontinuous derivative for $x = 0$. Hence the function cannot be analytic in an interval containing this value. The function $f(x) = e^{-1/x^2}$ is defined and continuous for all x other than 0. If one defines $f(0)$ to be 0, the function becomes continuous for all x and can in fact be shown to have derivatives of all orders for all x (Problem 6 following Section 6.18). However, the function fails to be analytic in any interval containing $x = 0$. For example, one finds that for this function, $f(0) = 0$, $f'(0) = 0, \ldots, f^{(n)}(0) = 0, \ldots$, so that the Maclaurin series would reduce identically to zero. The series converges but does not represent the function.

A satisfactory theory of analytic functions is most easily achieved with the aid of complex variables. The reader is referred to Chapter 9 for this topic. However, the following theorem does prove useful in establishing analyticity without recourse to complex numbers.

THEOREM 41 (*Taylor's formula with remainder*)   Let $f(x)$ be defined and continuous and have continuous derivatives up to the $(n + 1)$-st order for

$a - r_0 < x < a + r_0$. Then for each x of this interval except $x = a$,

$$f(x) = f(a) + \frac{f'(a)}{1}(x - a) \cdots + \frac{f^{(n)}(a)}{n!}(x - a)^n$$

$$+ \frac{f^{(n+1)}(x_1)}{(n + 1)!}(x - a)^{n+1}$$

for some x_1 such that $a < x_1 < x$ or (if $x < a$) $x < x_1 < a$.

It should be noted that for $n = 0$ the theorem reduces to the law of the mean: $f(x) = f(a) + f'(x_1)(x - a)$. For general n it gives an expansion identical with the Taylor series up to the term in $(x - a)^n$, with the rest of the series replaced by a single term.

We prove the theorem for $n = 1$, leaving the general case as a problem (Problem 3 following Section 6.18). Let x_2 be a fixed number in the given interval, $x_2 \neq a$ and let

$$F(x) = f(x_2) - f(x) - (x_2 - x)f'(x)$$

$$- \left(\frac{x_2 - x}{x_2 - a}\right)^2 [f(x_2) - f(a) - (x_2 - a)f'(a)].$$

Then F is defined and continuous for x in the given interval, and $F(a) = 0$, $F(x_2) = 0$. Hence by the law of the mean, $F'(x_1) = 0$ for some x_1 between a and x_2. But a calculation shows that

$$F'(x) = \frac{2(x_2 - x)}{(x_2 - a)^2} \left\{ f(x_2) - f(a) - (x_2 - a)f'(a) - \tfrac{1}{2}f''(x)(x_2 - a)^2 \right\}.$$

The equation $F'(x_1) = 0$ thus becomes the equation

$$f(x_2) = f(a) + (x_2 - a)f'(a) + \tfrac{1}{2}f''(x_1)(x_2 - a)^2.$$

If x_2 is now replaced by a variable x, one has the desired result:

$$f(x) = f(a) + (x - a)f'(a) + \tfrac{1}{2}f''(x_1)(x - a)^2.$$

Now let $f(x)$ be a function having derivatives of all orders in the given interval so that one can form all terms of the (hypothetical) Taylor series of f about $x = a$. Although this series may fail to converge, except for $x = a$, and, even if convergent, may fail to have $f(x)$ as sum, one can nevertheless write for each n:

$$f(x) = f(a) + \frac{f'(a)}{1!}(x - a) + \cdots + \frac{f^{(n)}(a)}{n!}(x - a)^n + R_n,$$

where R_n is the *remainder* term:

$$R_n = \frac{f^{(n+1)}(x_1)}{(n + 1)!}(x - a)^{n+1}. \tag{6.45}$$

Since x_1 is not explicitly given, the remainder is not explicitly known. However, Eq. (6.45) can often be used to obtain an *upper estimate* for $|R_n|$. From this estimate, one may then be able to show that

$$\lim_{n \to \infty} R_n = 0$$

for all x of the chosen interval. If this has been demonstrated, then one concludes that

$$f(x) = f(a) + \frac{f'(a)}{1}(x - a) + \cdots + \frac{f^{(n)}(a)}{n!}(x - a)^n + \cdots,$$

that is, that $f(x)$ is represented by a Taylor series over the given interval and is analytic; at the same time, one has proved convergence of the series.

EXAMPLE   Let $f(x) = e^x$. Then, for $a = 0$ and $x > 0$,

$$R_n = \frac{e^{x_1} x^{n+1}}{(n+1)!}, \qquad 0 < x_1 < x.$$

Hence

$$0 < R_n < \frac{e^x x^{n+1}}{(n+1)!}.$$

This implies that R_n is less than the nth term of the series

$$\sum_{n=1}^{\infty} \frac{e^x x^{n+1}}{(n+1)!},$$

which converges, by the ratio test, for all x. Thus one has

$$\lim_{n \to \infty} \frac{e^x x^{n+1}}{(n+1)!} = 0,$$

and accordingly, $\lim R_n = 0$. A similar discussion applies for $x < 0$. Accordingly, e^x can be represented by a Taylor series:

$$e^x = 1 + \frac{x}{1!} + \frac{x^2}{2!} + \cdots + \frac{x^n}{n!} + \cdots = \sum_{n=0}^{\infty} \frac{x^n}{n!}, \qquad \text{all } x. \qquad (6.46)$$

In a similar manner, one can prove that the following expansions are valid:

$$\sin x = \frac{x}{1!} - \frac{x^3}{3!} + \frac{x^5}{5!} + \cdots + \frac{(-1)^{n+1} x^{2n-1}}{(2n-1)!} + \cdots, \qquad \text{all } x;$$

$$(6.47)$$

$$\cos x = 1 - \frac{x^2}{2!} + \frac{x^4}{4!} + \cdots + \frac{(-1)^n x^{2n}}{(2n)!} + \cdots, \qquad \text{all } x; \qquad (6.48)$$

$$(1 + x)^m = 1 + \frac{m}{1!}x + \frac{m(m-1)}{2!}x^2 + \cdots$$

$$+ \frac{m(m-1)\cdots(m-n+1)}{n!}x^n + \cdots,$$

$$-1 < x < 1, \qquad m \text{ any real number.} \qquad (6.49)$$

[For the derivation of (6.49), see CLA, Section 6–20.]   ■

A variety of other expansions can be obtained from these by appropriate substitutions and combinations. The series in (6.49) becomes the geometric series when $m = -1$ and x is replaced by $-x$. This series can be used, as indicated in Problem 3 following Section 6.16, to obtain expansions of other rational functions. Then by differentiation and integration, further results can be obtained.

EXAMPLE 1   With $m = -1$ and x replaced by x^2, (6.49) becomes

$$\frac{1}{1 + x^2} = 1 - x^2 + \cdots + (-1)^n x^{2n} + \cdots, \qquad -1 < x < 1. \quad (6.50)$$

If this is integrated, one finds

$$\arctan x = x - \tfrac{1}{3}x^3 + \cdots + \frac{(-1)^n x^{2n+1}}{2n + 1} + \cdots, \qquad -1 < x < 1.$$

$$(6.51) \quad \blacksquare$$

EXAMPLE 2   Since $\cosh x = \tfrac{1}{2}(e^x + e^{-x})$, one finds

$$\cosh x = \frac{1}{2}\left[\left(1 + \frac{x}{1!} + \frac{x^2}{2!} + \cdots + \frac{x^n}{n!} + \cdots\right)\right.$$

$$\left. + \left(1 - \frac{x}{1!} + \frac{x^2}{2!} + \cdots + \frac{(-1)^n x^n}{n!} + \cdots\right)\right] \qquad (6.52)$$

$$= 1 + \frac{x^2}{2!} + \cdots + \frac{x^{2n}}{(2n)!} + \cdots, \qquad -\infty < x < \infty. \quad \blacksquare$$

EXAMPLE 3   Since $\sin x \cos x = \tfrac{1}{2}\sin 2x$, one finds

$$\sin x \cos x = \frac{1}{2}\left\{\frac{2x}{1!} - \frac{2^3 x^3}{3!} + \cdots + \frac{(-1)^{n-1} 2^{2n-1} x^{2n-1}}{(2n - 1)!} + \cdots\right\},$$

$$-\infty < x < \infty. \quad (6.53) \quad \blacksquare$$

Remark.   Since the remainder formula does provide a method for estimating R_n, it enables one to estimate the error committed in computing the sum of a power series. Thus if $f(x)$ is known to be analytic in a given interval and it is known that in this interval,

$$|f^{(n+1)}(x)| \leqq M_{n+1}$$

for a certain constant M_{n+1}, then

$$|R_n| \leqq \frac{M_{n+1}|x - a|^{n+1}}{(n + 1)!}. \qquad (6.54)$$

This formula can be added to those developed in Section 6.9.

As Theorem 41 shows, one does not require analyticity of the function $f(x)$ in order to be able to apply the remainder formula; $f(x)$ need only have continuous derivatives through the $(n + 1)$-st order. Actually, the $(n + 1)$-st derivative need only *exist* between a and x, continuity not being required. Thus in principle, one can use the formula as a method for evaluating a nonanalytic $f(x)$ by a *finite* series, with remainder estimated by (6.54).

6.18 FURTHER OPERATIONS ON POWER SERIES

Four other operations by which new Taylor series can be obtained are described by the following theorems.

THEOREM 42 Convergent power series can be multiplied; that is, if

$$f(x) = \sum_{n=0}^{\infty} c_n(x - a)^n, \qquad F(x) = \sum_{n=0}^{\infty} C_n(x - a)^n$$

are power series with convergence radii r_0^* and r_1^*, respectively, $0 < r_0^* \le r_1^*$, then

$$f(x)F(x) = \sum_{n=0}^{\infty} k_n(x - a)^n, \qquad a - r_0^* < x < a + r_0^*,$$

where

$$k_n = c_0 C_n + c_1 C_{n-1} + c_2 C_{n-2} + \cdots + c_{n-1}C_1 + c_n C_0.$$

This is simply an application of the Cauchy product rule (Section 6.10) to the absolutely convergent series for $f(x)$ and $F(x)$. □

THEOREM 43 Convergent power series can be divided, provided that there is no division by zero; that is, if $f(x)$ and $F(x)$ are given as in Theorem 42 and $F(a) = C_0 \ne 0$, then

$$\frac{f(x)}{F(x)} = \sum_{n=0}^{\infty} p_n(x - a)^n, \qquad a - r_2^* < x < a + r_2^*$$

for some positive number r_2^*, where the p_n satisfy the equations

$$c_n = p_0 C_n + p_1 C_{n-1} + \cdots + p_{n-1}C_1 + p_n C_0. \tag{6.55}$$

The rule (6.55) expresses the fact that the series $\sum p_n(x - a)^n$ multiplied by the series $\sum C_n(x - a)^n$ gives the series $\sum c_n(x - a)^n$. A proof of this theorem and a more precise determination of r_2^* require complex variables; this is taken up in Chapter 9. It should be noted that the equations (6.55) are implicit equations for the coefficients p_n:

$$c_0 = p_0 C_0, \qquad c_1 = p_0 C_1 + p_1 C_0, \ldots.$$

These can be solved in turn to obtain as many coefficients as are desired:

$$p_0 = \frac{c_0}{C_0}, \qquad p_1 = \frac{c_1 C_0 - c_0 C_1}{C_0^2}, \ldots;$$

it will usually be difficult to obtain a formula for the general coefficient p_n.

THEOREM 44 A Taylor series with constant term a can be substituted for the variable x in a Taylor series about $x = a$; that is, if

$$f(x) = \sum_{n=0}^{\infty} c_n (x - a)^n, \qquad g(x) = a + \sum_{n=1}^{\infty} d_n (x - b)^n,$$

have nonzero convergence radii r_0^* and r_1^*, respectively, and $|g(x) - a| < r_0^*$ for $|x - b| < r_2$, where $r_2 \leqq r_1^*$, then

$$f[g(x)] = \sum_{n=0}^{\infty} c_n \left\{ \sum_{m=1}^{\infty} d_m (x - b)^m \right\}^n = \sum_{n=0}^{\infty} q_n (x - b)^n, \qquad |x - b| < r_2,$$

where the coefficients q_n are obtained by collecting terms of same degree.

An example will make clear the rule of formation of coefficients:

$$e^{\sin x} = \sum_{n=0}^{\infty} \frac{(\sin x)^n}{n!} = \sum_{n=0}^{\infty} \frac{1}{n!} \left(x - \frac{x^3}{3!} + \frac{x^5}{5!} \cdots \right)^n$$

$$= 1 + \left(x - \frac{x^3}{3!} + \cdots \right) + \frac{1}{2!} \left(x - \frac{x^3}{3!} \cdots \right)^2$$

$$+ \frac{1}{3!} \left(x - \frac{x^3}{3!} + \cdots \right)^2 + \cdots$$

$$= 1 + x + \frac{x^2}{2!} + x^3 \left(-\frac{1}{3!} + \frac{1}{3!} \right) + \cdots.$$

In this case, $r_2 = \infty$.

The proof of the theorem is simplest in terms of complex variables and will not be considered here. The theorem is a special case of a theorem of Weierstrass on double series, for which the reader is referred to the book by Knopp given at the end of the chapter.

THEOREM 45 A power series can be inverted, provided that the first-degree term is not zero; that is, if

$$y = f(x) = \sum_{n=0}^{\infty} c_n (x - a)^n, \qquad |x - a| < r_0,$$

and $c_1 \neq 0$, then there is an inverse function

$$x = g(y) = a + \sum_{n=1}^{\infty} b_n (y - c_0)^n, \qquad |y - c_0| < r_1, \qquad r_1 > 0.$$

The coefficients b_n are determined from the identity

$$x - a \equiv \sum_{n=1}^{\infty} b_n \left[\sum_{m=1}^{\infty} c_m (x - a)^m \right]^n.$$

Again an example will clarify the method of determining coefficients. From

the series (6.51):
$$\arctan x = x - \tfrac{1}{3}x^3 + \cdots,$$
one can seek to determine a series for $x = \tan y$:

$$x \equiv \sum_{n=1}^{\infty} b_n y^n = \sum_{n=1}^{\infty} b_n \left(x - \tfrac{1}{3}x^3 + \cdots \right)^n,$$

$$x \equiv b_1 \left(x - \tfrac{1}{3}x^3 + \cdots \right) + b_2 \left(x - \tfrac{1}{3}x^3 + \cdots \right)^2 + b_3 \left(x - \tfrac{1}{3}x^3 + \cdots \right)^3 + \cdots$$

$$\equiv b_1 x + b_2 x^2 + x^3 \left(-\tfrac{1}{3}b_1 + b_3 \right) + \cdots .$$

Hence
$$b_1 = 1, \qquad b_2 = 0, \qquad b_3 - \tfrac{1}{3}b_1 = 0, \ldots,$$

so that
$$x = \tan y = y + \tfrac{1}{3}y^3 + \cdots .$$

For a proof of the theorem, one is referred to page 184 of the book of Knopp referred to earlier.

This last theorem and the one on division suggest a principle that has a broad field of applications: In order to determine a function that is to satisfy a given condition, one can postulate that the function is expressible by a power series and then try to determine the coefficients of such a series in order to satisfy the given condition. If such a series can be found, one can then investigate the convergence of the series and determine whether it actually defines a function satisfying the given condition.

Remark.   The operations on power series can be reduced to calculations with $n \times n$ matrices, with n chosen sufficiently large. For details one is referred to pp. 14–51 of the book by Henrici given at the end of the chapter.

PROBLEMS

1. Obtain the following Taylor series expansions:

a) $\sinh x = \displaystyle\sum_{n=1}^{\infty} \frac{x^{2n-1}}{(2n-1)!}$,   all x

b) $\cos^2 x = 1 + \displaystyle\sum_{n=1}^{\infty} \frac{(-1)^n 2^{2n-1} x^{2n}}{(2n)!}$,   all x

c) $\sin^2 x = \displaystyle\sum_{n=1}^{\infty} \frac{(-1)^{n+1} 2^{2n-1} x^{2n}}{(2n)!}$,   all x

d) $\log x = \displaystyle\sum_{n=1}^{\infty} \frac{(-1)^{n+1}(x-1)^n}{n}$, $|x-1| < 1$

e) $\sqrt{1-x} = 1 - \dfrac{x}{2} - \dfrac{1}{2^2 2!}x^2 - \dfrac{1\cdot 3}{2^3 3!}x^3 - \cdots,$ $|x| < 1$

f) $\dfrac{1}{\sqrt{1-x^2}} = 1 + \dfrac{x^2}{2} + \dfrac{1\cdot 3}{2^2 2!}x^4 + \dfrac{1\cdot 3\cdot 5}{2^3 3!}x^6 + \cdots,$ $|x| < 1$

g) $\arcsin x = x + \dfrac{x^3}{2\cdot 3} + \dfrac{1\cdot 3}{2^2 \cdot 2!}\dfrac{x^5}{5} + \dfrac{1\cdot 3\cdot 5}{2^3 \cdot 3!}\dfrac{x^7}{7} + \cdots,$ $|x| < 1.$

2. Find the first three nonzero terms of the following Taylor series:

   a) $e^x \sin x$ about $x = 0$ b) $\tan x$ about $x = 0$

   c) $\log^2(1 + x)$ about $x = 0$ d) $\log(1 - x^2)$ about $x = 0$

   e) $x^3 + 3x + 1$ about $x = 2$ f) $e^{\tan x}$ about $x = 0$

   g) $y = \sinh^{-1} x$ about $x = 0$ h) $y = \tanh x$ about $x = 0$

   i) $y = \tanh^{-1} x$ about $x = 0$ j) $y = \log \sec x$ about $x = 0$

3. Prove Taylor's remainder formula (Theorem 41) for general n. [Hint: Replace the function $F(x)$ by the function $G(x) - [(x_2 - x)/(x_2 - a)]^n G(a)$, where

$$G(x) = f(x_2) - f(x) - (x_2 - x)f'(x) - \cdots - \frac{(x_2 - x)^{n-1}}{(n-1)!} f^{(n-1)}(x).$$

4. Show that the remainder term R_n of Taylor's formula can be written in the *integral form*:

$$R_n = \int_a^x \frac{(x - t)^n}{n!} f^{(n+1)}(t) \, dt.$$

[Hint: Use induction. For $n = 0$ the formula becomes the known rule

$$f(x) - f(a) = \int_a^x f'(t) \, dt.$$

Suppose the formula is true with n replaced by m and prove it is true for $m + 1$; for this step, integrate the remainder term by parts taking $u = f^{(m+1)}(t)$, $dv = (x - t)^m \, dt/m!$.]

5. Evaluate to three decimal places:

   a) $\displaystyle\int_0^1 e^{-x^2} \, dx$ b) $\displaystyle\int_0^{0.5} \frac{dx}{\sqrt{1 + x^4}}$

6. Let $f(x) = e^{-1/x^2}$ for $x \neq 0$ and let $f(0) = 0$.

   a) Prove that $f(x)$ is continuous for all x.

   b) Prove that $f'(x)$ is continuous for $x \neq 0$ and that

   $$\lim_{x \to 0} f'(x) = f'(0) = 0,$$

   so that $f'(x)$ is continuous for all x.

   c) Prove that $f^{(n)}(x)$ is continuous for all x and $f^{(n)}(0) = 0$.

   d) Graph the function $f(x)$.

7. Use the remainder formula to estimate the error in the following computations:

   a) $e = 1 + 1 + \dfrac{1}{2!} + \dfrac{1}{3!} + \dfrac{1}{4!} + \dfrac{1}{5!}$ [assume that $e < 3$ is known]

   b) $\sin 1 = 1 - \dfrac{1}{3!} + \dfrac{1}{5!}$

   c) $\log\dfrac{3}{2} = \dfrac{1}{2} - \dfrac{1}{2 \cdot 2^2} + \dfrac{1}{3 \cdot 2^3}$

8. Let $f(x)$ satisfy the conditions stated in Theorem 41. Let $f'(a) = f''(a) = \cdots = f^{(n)}(a) = 0$, but $f^{(n+1)}(a) \neq 0$. Show that $f(x)$ has a maximum, minimum, or

horizontal inflection point at $x = a$ according to whether the function $f^{(n+1)}(a)(x - a)^{n+1}$ has a maximum, minimum, or inflection point for $x = a$. (This gives another proof of the rule deduced in Section 2.19.)

9. (*Derivatives and differences*) In numerical analysis, one uses approximations for derivatives. For the first derivative of $f(x)$ at x, one chooses a small positive h and uses

$$g_1(x, h) = \frac{f(x + h) - f(x)}{h} \quad \text{and} \quad g_2(x, h) = \frac{f(x + h) - f(x - h)}{2h}.$$

For the second derivative, one uses

$$g_3(x, h) = \frac{f(x + h) - 2f(x) + f(x - h)}{h^2}.$$

With the aid of Taylor's formula with remainder (assuming sufficient differentiability), establish the following:

a) $g_1(x, h) - f'(x)$ has limit 0 as $h \to 0$ and $\dfrac{g_1(x, h) - f'(x)}{h}$ has a finite limit as $h \to 0$.

b) $g_2(x, h) - f'(x)$ has limit 0 as $h \to 0$ and $\dfrac{g_2(x, h) - f'(x)}{h^2}$ has a finite limit as $h \to 0$.

c) $g_3(x, h) - f''(x)$ has limit 0 as $h \to 0$ and $\dfrac{g_3(x, h) - f''(x)}{h}$ has a finite limit as $h \to 0$.

Remark   We describe these results by saying that the error in approximating $f'(x)$ by $g_1(x, h)$ or $f''(x)$ by $g_3(x, h)$ is of the order h, while the error in approximating $f'(x)$ by $g_2(x, h)$ is of the order h^2. (The higher the power of h, the better the approximation for small h.)

*6.19   SEQUENCES AND SERIES OF COMPLEX NUMBERS

We assume familiarity with complex numbers, written typically as $z = x + iy$, where $i^2 = -1$. The *absolute value* of z is defined as

$$|z| = \sqrt{x^2 + y^2},$$

and hence $|z|$ equals the magnitude of the vector (x, y) of V^2 or the distance from z to the origin in the xy-plane. Since addition and subtraction of complex numbers is the same as for vectors in V^2, $|z_1 - z_2|$ can be interpreted as the distance from z_1 to z_2. These concepts are illustrated in Fig. 6.12.

Sequences of complex numbers are defined as for real numbers. The following are examples:

$$z_n = i^n, \quad z_n = \frac{n(1 - i)}{1 + n^2}, \quad z_n = \left(1 + \frac{i}{n}\right)^n.$$

A sequence z_n is said to *converge to z_0*:

$$\lim_{n \to \infty} z_n = z_0$$

if, given $\epsilon > 0$, an integer N can be found such that

$$|z_n - z_0| < \epsilon \quad \text{for} \quad n > N.$$

This resembles the definition for real numbers. The inequality states that z_n is within distance ϵ of z_0 or that z_n is within the circle of radius ϵ about z_0; this is illustrated in Fig. 6.13. If a sequence z_n does not converge, it is said to *diverge*.

Figure 6.12   Complex numbers.

Figure 6.13   Sequence converging to z_0.

THEOREM 46   Let $z_n = x_n + iy_n$ $(n = 1, 2, \ldots)$ be a sequence of complex numbers. If this sequence converges to $z_0 = x_0 + iy_0$, then

$$\lim_{n \to \infty} x_n = x_0, \qquad \lim_{n \to \infty} y_n = y_0.$$

Conversely, if x_n converges to x_0 and y_n converges to y_0, then

$$\lim_{n \to \infty} z_n = \lim_{n \to \infty} (x_n + iy_n) = x_0 + iy_0.$$

This theorem shows that the convergence of sequences of complex numbers can be referred back to that of real numbers simply by studying real and imaginary parts.

To prove the theorem, we remark that, if $|z_n - z_0| < \epsilon$, then (x_n, y_n) is within the circle of radius ϵ about (x_0, y_0), so that necessarily

$$|x_n - x_0| < \epsilon, \qquad |y_n - y_0| < \epsilon.$$

Thus convergence of z_n to z_0 implies convergence of x_n to x_0 and y_n to y_0. Conversely, if x_n converges to x_0 and y_n converges to y_0, then for given ϵ, N can be chosen so large that

$$|x_n - x_0| < \tfrac{1}{2}\epsilon, \qquad |y_n - y_0| < \tfrac{1}{2}\epsilon \quad \text{for} \quad n > N.$$

These inequalities force (x_n, y_n) to lie within a *square* with center (x_0, y_0) and side ϵ; hence (x_n, y_n) must lie within the *circle* of radius ϵ about (x_0, y_0), so that $|z_n - z_0| < \epsilon$ for $n > N$. Accordingly, z_n must converge to z_0.

THEOREM 47 (*Cauchy criterion*) A sequence z_n of complex numbers converges if and only if for each $\epsilon > 0$ an N can be found such that

$$|z_m - z_n| < \epsilon \quad \text{for} \quad m > N, \quad n > N.$$

This is proved by referring the convergence of z_n back to that of the two real sequences x_n, y_n and then applying the Cauchy criterion (Theorem 6) to the real sequences.

An *infinite series* of complex numbers is defined as for real numbers:

$$\sum_{n=1}^{\infty} z_n = z_1 + z_2 + \cdots + z_n + \cdots .$$

The series is said to converge or diverge according to whether the nth partial sums:

$$S_n = z_1 + \cdots + z_n$$

form a convergent or divergent sequence. The *sum* of the series is then the limit

$$S = \lim_{n \to \infty} S_n$$

$$= \lim_{n \to \infty} \sum_{m=1}^{n} z_m ,$$

when the limit exists.

On the basis of Theorem 46, one can at once assert the following theorem:

THEOREM 48 If $z_n = x_n + iy_n$, then the series

$$\sum_{n=1}^{\infty} z_n = \sum_{n=1}^{\infty} (x_n + iy_n)$$

converges and has sum $S = A + Bi$ if and only if

$$\sum_{n=1}^{\infty} x_n = A, \qquad \sum_{n=1}^{\infty} y_n = B.$$

Thus the convergence of series of complex numbers is also referred back to that for real numbers. This can also be done in a second way.

THEOREM 49 If $\displaystyle\sum_{n=1}^{\infty} |z_n|$ converges, then $\displaystyle\sum_{n=1}^{\infty} z_n$ converges.

In words: If a complex series is *absolutely convergent*, then it is convergent. The proof is the same as that of the corresponding theorem for real numbers (Theorem 11 of Section 6.6).

On the basis of Theorems 48 and 49, one can now obtain tests for convergence and divergence of series of complex numbers. In particular, the following rules hold:

the nth term test (Theorem 10);
the Cauchy criterion (Theorem 9);
the comparison test for convergence (Theorem 12);
the ratio test (Theorems 17 and 20);
the root test (Theorems 19, 21, 21(a)).

The rule for addition or subtraction of convergent series (Theorem 8) and the product rule (Theorem 29) also hold for complex series. The estimates of remainders (Theorems 22, 23, 24, 25) can also be used for complex series.

EXAMPLE 1 The series

$$1 + \frac{i}{2!} + \cdots + \frac{i^n}{n!} + \cdots = \sum_{n=0}^{\infty} \frac{i^n}{n!}$$

is absolutely convergent since the series of absolute values is the real series

$$\sum_{n=0}^{\infty} \frac{1}{n!},$$

which converges (for example, by the ratio test). ∎

EXAMPLE 2 The series

$$\frac{i}{1} + \frac{i^2}{2} + \frac{i^3}{3} + \cdots + \frac{i^n}{n} + \cdots = \sum_{n=1}^{\infty} \frac{i^n}{n}$$

is not absolutely convergent since $\Sigma 1/n$ diverges. However, the series of real parts is

$$0 - \tfrac{1}{2} + 0 + \tfrac{1}{4} + 0 - \tfrac{1}{6} + \cdots,$$

and the series of imaginary parts is

$$1 + 0 - \tfrac{1}{3} + 0 + \tfrac{1}{5} + \cdots.$$

If the zeros are disregarded, these are convergent alternating series. Hence $\Sigma i^n/n$ converges. ∎

The general theory of functions of a complex variable is developed in Chapter 9. Here we consider briefly the functions z^n, where n is a positive integer or 0, and the corresponding *power series*:

$$\sum_{n=0}^{\infty} c_n z^n = c_0 + c_1 z + c_2 z^2 + \cdots + c_n z^n + \cdots. \tag{6.56}$$

Since each term is defined for all z, the series may converge for some or all z. The basic theorem on power series (Theorem 35) can be restated and proved just as for real numbers, and one concludes that each power series (6.56) has a *radius of convergence* r^* such that the series converges when $|z| < r^*$ and diverges when $|z| > r^*$. The use of the term *radius of convergence* now receives its justification; for the power series (6.56) converges within the circle with center O and radius r^*. This is illustrated in Fig. 6.14.

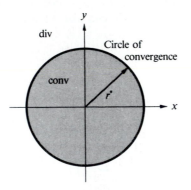

Figure 6.14 Radius of convergence of power series.

It is of special interest that the power series (6.46), (6.47), (6.48) for e^x, $\sin x$, and $\cos x$ continue to converge when x is replaced by an arbitrary complex number z. One can then use the equations

$$e^z = 1 + z + \cdots + \frac{z^n}{n!} + \cdots, \tag{6.57}$$

$$\sin z = z - \frac{z^3}{3!} + \cdots + (-1)^{n-1} \frac{z^{2n-1}}{(2n-1)!} + \cdots, \tag{6.58}$$

$$\cos z = 1 - \frac{z^2}{2!} + \cdots + (-1)^n \frac{z^{2n}}{(2n)!} + \cdots \tag{6.59}$$

to *define* these functions for complex z. From these series, one derives the *Euler identity* (Problem 4 below):

$$e^{iy} = \cos y + i \sin y, \tag{6.60}$$

or the more general relation:

$$e^{x+iy} = e^x(\cos y + i \sin y). \tag{6.61}$$

This last equation can also be used as a *definition* of e^z for complex z, the series expression being then a consequence of this definition. This is shown in Chapter 9.

PROBLEMS

1. Evaluate the limits:

a) $\displaystyle\lim_{n\to\infty} \frac{i^n}{n}$

b) $\displaystyle\lim_{n\to\infty} \frac{(1+i)n^3 - 2in + 3}{in^3 - 1}$

2. Test for absolute convergence and for convergence:

a) $\left(\dfrac{1+i}{2}\right) + \left(\dfrac{1+i}{2}\right)^2 + \cdots + \left(\dfrac{1+i}{2}\right)^n + \cdots$

b) $\displaystyle\sum_{n=1}^{\infty} ni^n$

c) $\displaystyle\sum_{n=1}^{\infty} \frac{ni^n}{n^2 + 1}$

d) $\displaystyle\sum_{n=1}^{\infty} \frac{1}{(n+i)^2}$

3. Prove that the series (6.57), (6.58), and (6.59) converge for all z.

4. a) Prove the Euler identity (6.60) from the series definition of e^z, $\cos z$, $\sin z$.

b) Prove (6.61) from the series definitions of e^z, $\cos z$, $\sin z$.

5. Use the series expressions (6.57), (6.58), (6.59) to prove the identities

a) $\cos z = \dfrac{e^{iz} + e^{-iz}}{2}$

b) $\sin z = \dfrac{e^{iz} - e^{-iz}}{2i}$

c) $e^{z_1 + z_2} = e^{z_1} \cdot e^{z_2}$

d) $\sin(-z) = -\sin z$

e) $\cos(-z) = \cos z$

f) $\sin^2 z + \cos^2 z = 1$

g) $\cos 2z = \cos^2 z - \sin^2 z$

h) $\sin 2z = 2 \sin z \cos z$

6. Show that the following series converge for $|z| < 1$ and diverge for $|z| > 1$:

a) $z + \dfrac{z^2}{2} + \cdots + \dfrac{z^n}{n} + \cdots$

b) $1 + z + z^2 + \cdots + z^n + \cdots$

*6.20   SEQUENCES AND SERIES OF FUNCTIONS OF SEVERAL VARIABLES

The notions of sequence and series of functions extend at once to functions of several variables. Thus

$$\sum_{n=1}^{\infty} (xy)^n = xy + x^2y^2 + \cdots + x^ny^n + \cdots \tag{6.62}$$

is a series of functions of the two variables x and y. The notion of uniform convergence also extends at once, as well as the M-test and the properties described in Section 6.14.

One can in particular consider *power series* in several variables. For two variables x, y, such a power series is a series

$$\sum_{n=0}^{\infty} f_n(x, y) = f_0(x, y) + f_1(x, y) + \cdots + f_n(x, y) + \cdots, \tag{6.63}$$

where

$$f_n(x, y) = c_{n,0}x^n + c_{n,1}x^{n-1}y + \cdots + c_{n,n-1}xy^{n-1} + c_{n,n}y^n, \tag{6.64}$$

the c's being constants. Thus f_n is a *homogeneous polynomial of degree n in x and y*. The series (6.62) is an example of this, with $f_{2n} = x^ny^n$ and $f_0 = 0, f_1 = f_3 = \cdots = 0$. This series also illustrates the fact that the values (x, y) for which a power series in x and y converges form a more complicated set than the convergence interval for series in x alone. For (6.62) the series converges when $|xy| < 1$; this region is pictured in Fig. 6.15. In general the convergence region can be quite complicated.

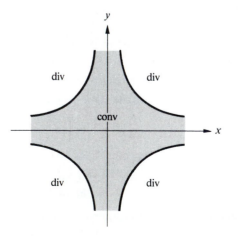

Figure 6.15   Convergence set for (6.62).

If a function $F(x, y)$ can be represented by such a power series in a neighborhood of the origin,

$$F(x, y) = c_{0,0} + (c_{1,0}x + c_{1,1}y) + (c_{2,0}x^2 + c_{2,1}xy + c_{2,2}y^2) + \cdots,$$

then a term-by-term differentiation (which can be justified) shows that

$$F(0,0) = c_{0,0}, \qquad \frac{\partial F}{\partial x}(0,0) = c_{1,0},$$

$$\frac{\partial F}{\partial y}(0,0) = c_{1,1}, \qquad \frac{1}{2!}\frac{\partial^2 F}{\partial x^2} = c_{2,0},$$

$$\frac{2}{2!}\frac{\partial^2 F}{\partial x \, \partial y} = c_{2,1}, \qquad \frac{1}{2!}\frac{\partial^2 F}{\partial y^2} = c_{2,2}.$$

In general one finds

$$f_n(x, y) = \frac{1}{n!}\left(\frac{\partial^n F}{\partial x^n}x^n + n\frac{\partial^n F}{\partial x^{n-1}\, \partial y}x^{n-1}y \right.$$

$$+ \frac{n(n-1)}{2!}\frac{\partial^n F}{\partial x^{n-2}\, \partial y^2}x^{n-2}y^2 + \cdots$$

$$+ \frac{n(n-1)\cdots(n-k+1)}{k!}\frac{\partial^n F}{\partial x^{n-k}\, \partial y^k}x^{n-k}y^k$$

$$\left. + \cdots + \frac{\partial^n F}{\partial y^n}y^n \right), \tag{6.65}$$

all derivatives being evaluated at $(0,0)$. A series $\Sigma f_n(x, y)$, in which the f_n are given by (6.65), is known as a Taylor series in x and y, about $(0,0)$, and the function $F(x, y)$ that it represents is termed *analytic* in the corresponding region. The expansion about a general point (x_1, y_1) is obtained by a translation of origin:

$$F(x, y) = F(x_1, y_1) + \left[\frac{\partial F}{\partial x}(x - x_1) + \frac{\partial F}{\partial y}(y - y_1) \right]$$

$$+ \frac{1}{2!}\left[\frac{\partial^2 F}{\partial x^2}(x - x_1)^2 + 2\frac{\partial^2 F}{\partial x \, \partial y}(x - x_1)(y - y_1) + \frac{\partial^2 F}{\partial y^2}(y - y_1)^2 \right]$$

$$+ \cdots + \frac{1}{n!}\left[\frac{\partial^n F}{\partial x^n}(x - x_1)^n + \cdots \right] + \cdots, \tag{6.66}$$

all derivatives being evaluated at (x_1, y_1).

The general term of the series (6.66) can be interpreted in terms of an *nth differential* $d^n F$ of the function $F(x, y)$:

$$d^n F = \frac{\partial^n F}{\partial x^n}(x - x_1)^n + \cdots$$

$$= \sum_{r=0}^{n} \binom{n}{r}\frac{\partial^n F}{\partial x^r \, \partial y^{n-r}}(x_1, y_1)(x - x_1)^r(y - y_1)^{n-r},$$

where the $\binom{n}{r}$ are the binomial coefficients.* To indicate the dependence of d^nF on x_1, y_1 and the differences $x - x_1, y - y_1$, we write:

$$d^nF = d^nF(x_1, y_1; x - x_1, y - y_1).$$

When $n = 1$ and $x - x_1 = dx$, $y - y_1 = dy$, one finds:

$$d^1F(x_1, y_1; dx, dy) = \frac{\partial F}{\partial x} dx + \frac{\partial F}{\partial y} dy = dF;$$

this is the familiar expression for the first differential. The series (6.66) can now be written more concisely:

$$F(x, y) = F(x_1, y_1) + dF(x_1, y_1; x - x_1, y - y_1)$$
$$+ \frac{1}{2!} d^2F(x_1, y_1; x - x_1, y - y_1) + \cdots$$
$$+ \frac{1}{n!} d^nF(x_1, y_1; x - x_1, y - y_1) + \cdots . \qquad (6.66')$$

The theory of analytic functions of several variables is again best studied with the aid of complex numbers. As with one variable, the familiar functions are in general analytic. Thus

$$e^x \sin y = y + xy + \frac{3x^2y - y^3}{6} + \cdots ,$$

the series converging for all x and y.

The subject of analytic functions of several variables has not been studied intensively until recent times, and most books on the subject are very advanced. In O. D. Kellogg's *Foundations of Potential Theory* (Berlin: Springer, 1929) a brief treatment is given on pages 135–140. An introduction to the topic is also given in the last chapter of the book *Introduction to Analytic Functions* by W. Kaplan (Reading, Mass.: Addison-Wesley, 1966).

*6.21   TAYLOR'S FORMULA FOR FUNCTIONS OF SEVERAL VARIABLES

There is a Taylor formula with remainder for functions of several variables:

$$F(x, y) = F(x_1, y_1) + dF(x_1, y_1; x - x_1, y - y_1)$$
$$+ \cdots + \frac{1}{n!} d^nF(x_1, y_1; x - x_1, y - y_1)$$
$$+ \frac{1}{(n + 1)!} d^{n+1}F(x^*, y^*; x - x_1, y - y_1);$$
$$x^* = x_1 + t^*(x - x_1), \qquad y^* = y_1 + t^*(y - y_1), \qquad 0 < t^* < 1.$$
$$(6.67)$$

$$*\binom{n}{r} = \frac{n!}{r!(n - r)!} \qquad \text{for} \quad r = 0, 1, \ldots, n.$$

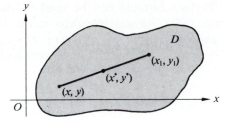

Figure 6.16   Taylor's formula for $F(x, y)$.

The point (x^*, y^*) lies between (x_1, y_1) and (x, y) on the line segment joining these points, as in Fig. 6.16. For $n = 1$ the formula becomes

$$F(x, y) = F(x_1, y_1) + (x - x_1)F_x(x^*, y^*) + (y - y_1)F_y(x^*, y^*). \quad (6.68)$$

This is known as the *law of the mean* for functions of two variables.

To prove (6.68), one writes:

$$\phi(t) = F\big[x_1 + t(x - x_1), \, y_1 + t(y - y_1)\big], \qquad 0 \le t \le 1.$$

Thus x and y are considered as fixed, and ϕ depends only on t. By the law of the mean for ϕ,

$$\phi(1) = \phi(0) + \phi'(t^*), \qquad 0 < t^* < 1.$$

But $\phi(1) = F(x, y)$, $\phi(0) = F(x_1, y_1)$, and

$$\phi'(t) = (x - x_1)F_x\big[x_1 + t(x - x_1), \, y_1 + t(y - y_1)\big]$$
$$+ (y - y_1)F_y\big[x + t(x - x_1), \, y_1 + t(y - y_1)\big].$$

If t is replaced by t^*, one obtains (6.68). The general formula (6.67) is proved in the same way, on the basis of Taylor's formula for ϕ:

$$\phi(1) = \phi(0) + \phi'(0) + \cdots + \frac{\phi^{(n)}(0)}{n!} + \frac{\phi^{(n+1)}(t^*)}{(n+1)!},$$

where $0 < t^* < 1$. For one finds by induction that

$$\phi^{(n)}(t) = d^n F\big[x_1 + t(x - x_1), \, y_1 + t(y - y_1); \, x - x_1, \, y - y_1\big]. \quad (6.69)$$

The validity of (6.67) is ensured if $F(x, y)$ has continuous derivatives through the $(n + 1)$-st order in a domain D containing the line segment joining (x, y) to (x_1, y_1).

Taylor's series or Taylor's formula can be used to study the nature of a function near a particular point. As was remarked earlier, the linear terms give dF, the best "linear approximation" to $F(x, y) - F(x_1, y_1)$. If $dF = 0$, the quadratic terms $d^2F/2!$ become of importance. In particular, if the quadratic expression

$$d^2F = A(x - x_1)^2 + 2B(x - x_1)(y - y_1) + C(y - y_1)^2$$

is positive, except for $x = x_1, y = y_1$, then $F(x,y)$ has a minimum at (x_1, y_1). Pursuing this further, we rediscover the criteria for maxima and minima developed in Section 2.19.

Remark.   The first differential dF of $F(x,y)$ can be considered as an "invariant differential form." By this we mean that, under a differentiable substitution $x = x(u,v), y = y(u,v)$, the differential of the composite function $F(x(u,v),y(u,v))$ equals that of F at corresponding points:

$$dF(x(u,v),\, y(u,v)) = \frac{\partial F}{\partial x}\, dx + \frac{\partial F}{\partial y}\, dy, \tag{6.70}$$

where on the right $x = x(u,v), y = y(u,v), dx = (\partial x/\partial u)\, du + (\partial x/\partial v)\, dv, dy = (\partial y/\partial u)\, du + (\partial y/\partial v)\, dv$. The assertion follows at once by the chain rule (as in the theorem concerning Eq. (2.36) in Section 2.8). For the second and higher order differentials, the analogous invariance fails in general. However, at a point (x_1, y_1) at which $dF = 0, d^2F = 0, \ldots, d^{n-1}F = 0$, the differential d^nF is invariant, so that under the appropriately differentiable substitution $x = x(u,v), y = y(u,v)$, $d^nF(x(u,v),y(u,v))$ is calculated from $d^nF(x,y)$ by expressing x, y, dx, dy in terms of u, v, du, dv and expanding by algebra. For example, from $d^2F = F_{xx}(x,y)\, dx^2 + 2F_{xy}(x,y)\, dx\, dy + F_{yy}(x,y)\, dy^2$ one obtains

$$d^2F(x(u,v),\, y(u,v)) = F_{xx}(x(u,v),\, y(u,v)) \left(\frac{\partial x}{\partial u}\, du + \frac{\partial x}{\partial v}\, dv\right)^2 + \cdots \tag{6.71}$$

at each point (u_1, v_1) at which $x = x_1, y = y_1$ and $F_x(x_1, y_1) = 0, F_y(x_1, y_1) = 0$. (See Problem 6 below.)

The invariance of d^2F at a critical point explains its importance for maxima and minima as in the preceding paragraph. The concept also occurs in discussion of the *second fundamental form* on a surface (Problem 23(d) following Section 3.11). See also Sections 5.19 and 5.20.

PROBLEMS

1. Expand in power series, stating the region of convergence:

a) $e^{x^2 - y^2}$

b) $\sin(xy)$

c) $\dfrac{1}{1 - x - y}$

d) $\dfrac{1}{1 - x - y - z}$

2. Prove Eq. (6.69) by induction.

3. Prove that if a power series

$$c_{0,0} + (c_{1,0}x + c_{1,1}y) + (c_{2,0}x^2 + c_{2,1}xy + c_{2,2}y^2) + \cdots$$

converges at (x_0, y_0), then it converges at every point $(\lambda x_0, \lambda y_0)$, for $|\lambda| < 1$.

4. Evaluate $\int_0^1 \int_0^1 \sin(xy)\, dx\, dy$ with the aid of power series.

5. As in Problem 9 following Section 6.18, use Taylor series or the remainder formula to analyze the following difference approximations to partial derivative expressions:

a) $g_1(x, h) = \dfrac{f(x + h, y + h) + f(x + h, y - h)}{h^2}$

$$+ \dfrac{f(x - h, y + h) + f(x - h, y - h) - 4f(x, y)}{h^2}$$

as approximation to $\nabla^2 f$ at (x, y);

b) $g_2(x, h) =$

$$\dfrac{f(x + h, y + h) - f(x + h, y - h) - f(x - h, y + h) + f(x - h, y - h)}{4h^2}$$

as approximation to $\partial^2 f / \partial x\, \partial y$ at (x, y).

6. Prove the validity of Eq. (6.71) at (u_1, v_1) under the hypotheses stated. [Hint: Use the rules of Section 2.16.]

*6.22   IMPROPER INTEGRALS VERSUS INFINITE SERIES

In Section 4.1, improper integrals are defined. In this section we show that there is a close relationship between improper integrals and infinite series. The relation-ship is suggested by the integral test of Section 6.6. It leads us here to some valuable tests for convergence of improper integrals. Attention will be confined to integrals from a to ∞, but the methods are applicable to other forms of improper integrals.

Let us consider first, as an example, the integral

$$\int_1^\infty \frac{dx}{x}.$$

This would be termed *convergent* if the limit

$$\lim_{b \to \infty} \int_1^b \frac{dx}{x}$$

exists. However, since the integrand $1/x$ is *positive*, the integral from 1 to b must *increase* as b increases. From this it follows that either the above limit is $+\infty$ (that is, there is no limit) and the integral diverges or else the limit is a finite number I. Now, in order to determine which of the two cases holds it is clearly sufficient to let b approach ∞ through integral values n, that is, to consider the limit of a sequence:

$$\lim_{n \to \infty} \int_1^n \frac{dx}{x} \qquad (n = 1, 2, \dots).$$

But one can write

$$\int_1^n \frac{dx}{x} = \int_1^2 \frac{dx}{x} + \int_2^3 \frac{dx}{x} + \cdots + \int_{n-1}^n \frac{dx}{x}.$$

In other words, the integral exists precisely when the series

$$\sum_{n=1}^\infty \int_n^{n+1} \frac{dx}{x} = \sum_{n=1}^\infty a_n$$

converges and the two have the same value. This is suggested in Fig. 6.17.

In this particular case,

$$a_n = \int_n^{n+1} \frac{dx}{x}$$

$$= \log \frac{n+1}{n},$$

Figure 6.17   Improper integral as a series.

and the series is

$$\sum_{n=1}^{\infty} \log \frac{n+1}{n}.$$

One easily verifies that both series and integral diverge.

We now formulate the relationships in general terms.

THEOREM 50   Let $f(x)$ be continuous and let $f(x) \geq 0$ for $a \leq x < \infty$. Then the integral

$$\int_a^{\infty} f(x)\, dx$$

converges and has value I if and only if the series

$$\sum_{n=1}^{\infty} a_n, \qquad a_n = \int_{a+n-1}^{a+n} f(x)\, dx$$

converges and has sum I.

If $f(x)$ changes sign infinitely often, then convergence of the integral will certainly imply that of the series; however, the converse need not hold, as the example

$$\int_0^{\infty} \sin 2\pi x\, dx \tag{6.72}$$

shows (Problem 2 following Section 6.25). The proper connection between series and integrals in this case is given by the following theorem.

THEOREM 51   Let $f(x)$ be continuous for $a \leq x < \infty$. Let $f(x) \geq 0$ for $a = b_0 \leq x \leq b_1$, $f(x) \leq 0$ for $b_1 \leq x \leq b_2$ and, in general, $(-1)^n f(x) \geq 0$ for $b_n \leq x \leq b_{n+1}$, where b_n is a monotone sequence such that

$$\lim_{n \to \infty} b_n = \infty.$$

Then the integral

$$\int_a^{\infty} f(x)\, dx$$

converges and has value I if and only if the alternating series

$$\sum_{n=1}^{\infty} a_n, \qquad a_n = \int_{b_{n-1}}^{b_n} f(x)\, dx$$

converges and has sum I.

Proof.   If the integral converges to I, then as before,

$$\lim_{n \to \infty} \int_a^{b_n} f(x)\, dx = \lim_{n \to \infty} (a_1 + \cdots + a_n) = I,$$

so that the series converges and has sum I.

Conversely, let the series converge to I. Let $\epsilon > 0$ be given and choose N so large that

$$|a_1 + \cdots + a_n - I| < \tfrac{1}{2}\epsilon \quad \text{and} \quad |a_n| < \tfrac{1}{2}\epsilon \quad \text{for} \quad n \geq N.$$

The latter condition can be satisfied by the nth term test. If $x_1 > b_N$, then $b_n \leq x_1 \leq b_{n+1}$ for some $n \geq N$. Hence

$$\int_a^{x_1} f(x)\, dx = \int_a^{b_n} f(x)\, dx + \int_{b_n}^{x_1} f(x)\, dx = a_1 + \cdots + a_n + \int_{b_n}^{x_1} f(x)\, dx$$

and

$$\left| \int_a^{x_1} f(x)\, dx - I \right| = \left| a_1 + \cdots + a_n - I + \int_{b_n}^{x_1} f(x)\, dx \right|$$

$$\leq |a_1 + \cdots + a_n - I| + \left| \int_{b_n}^{x_1} f(x)\, dx \right|$$

$$\leq |a_1 + \cdots + a_n - I| + \left| \int_{b_n}^{b_{n+1}} f(x)\, dx \right| ;$$

the last step is justified, since $f(x)$ does not change sign between b_n and b_{n+1}. Hence

$$\left| \int_a^{x_1} f(x)\, dx - I \right| \leq |a_1 + \cdots + a_n - I| + |a_{n+1}| < \tfrac{1}{2}\epsilon + \tfrac{1}{2}\epsilon = \epsilon.$$

It follows that

$$\lim_{x_1 \to \infty} \int_a^{x_1} f(x)\, dx = I. \quad \square$$

COROLLARY Let $f(x)$ be continuous for $a \leq x < \infty$, let $f(x)$ decrease as x increases, and let $\lim_{x \to \infty} f(x) = 0$. Then the integrals

$$\int_a^\infty f(x) \sin x\, dx, \qquad \int_a^\infty f(x) \cos x\, dx$$

converge.

Proof. We consider the sine integral, the cosine integral being similar. Under the assumptions made, the alternating series of Theorem 51 is, except perhaps for the first term, of the form

$$\sum_{n=k}^\infty a_n, \qquad a_n = \int_{n\pi}^{(n+1)\pi} f(x) \sin x\, dx.$$

Since $f(x)$ decreases as x increases, $|a_n|$ is decreasing; since $f(x)$ has limit 0 as $x \to \infty$, a_n converges to 0. The alternating series test (Theorem 18) then guarantees convergence. $\square$

EXAMPLES   The integrals

$$\int_1^\infty \frac{\sin x}{x}\, dx, \quad \int_0^\infty \frac{\cos x}{1+x}\, dx, \quad \int_1^\infty \frac{\sin x}{\sqrt{x}}\, dx$$

all exist, by virtue of the preceding corollary.   ■

THEOREM 52 (*Cauchy criterion*)   Let $f(x)$ be continuous for $a \leqq x < \infty$. The integral

$$\int_a^\infty f(x)\, dx$$

exists if and only if for each $\epsilon > 0$ a number B can be found such that

$$\left| \int_p^q f(x)\, dx \right| < \epsilon \quad \text{for} \quad B < p < q.$$

Proof.   If the integral converges to I, then for given $\epsilon > 0$ a B can be found such that

$$\left| \int_a^b f(x)\, dx - I \right| < \tfrac{1}{2}\epsilon \quad \text{for} \quad b > B.$$

Hence for $B < p < q$,

$$\left| \int_a^p f\, dx - I \right| < \tfrac{1}{2}\epsilon, \quad \left| \int_a^q f\, dx - I \right| < \tfrac{1}{2}\epsilon$$

and

$$\left| \int_p^q f\, dx \right| = \left| \int_a^q f\, dx - \int_a^p f\, dx \right| = \left| \int_a^q f\, dx - I + I - \int_a^p f\, dx \right|$$

$$\leqq \left| \int_a^q f\, dx - I \right| + \left| \int_a^p f\, dx - I \right| < \tfrac{1}{2}\epsilon + \tfrac{1}{2}\epsilon = \epsilon.$$

Conversely, let the condition hold. Then the Cauchy criterion of Theorem 6 applies to the sequence

$$S_n = \int_a^n f(x)\, dx \qquad (n > a);$$

hence this sequence converges to a number I. Let $\epsilon > 0$ be given and let a corresponding number B be chosen as in the theorem; let N be an integer larger than B and such that $|S_n - I| < \epsilon$ for $n \geqq N$. For $x_1 > N$,

$$\left| \int_a^{x_1} f(x)\, dx - I \right| = \left| \int_a^N f(x)\, dx - I + \int_N^{x_1} f(x)\, dx \right|$$

$$\leqq |S_N - I| + \left| \int_N^{x_1} f(x)\, dx \right|$$

$$< \epsilon + \epsilon = 2\epsilon.$$

It follows that

$$\lim_{x_1 \to \infty} \int_a^{x_1} f(x)\, dx = I. \quad \square$$

Remark. It follows from Theorem 52 that for a convergent integral $\int_a^\infty f(x)\, dx$,

$$\lim_{b \to \infty} \int_b^{b+k} f(x)\, dx = 0$$

for every fixed k. However, it is not necessary that $f(x)$ itself approach 0. This is illustrated by the integral (Problem 1 following Section 6.25):

$$\int_1^\infty \sin x^2\, dx.$$

THEOREM 53   Let $f(x)$ be continuous for $a \leq x < \infty$. If $\int_a^\infty |f(x)|\, dx$ converges, then

$$\int_a^\infty f(x)\, dx$$

converges, and

$$\left| \int_a^\infty f(x)\, dx \right| \leq \int_a^\infty |f(x)|\, dx. \tag{6.73}$$

In words: An *absolutely convergent* improper integral is convergent.

Proof. Since $|f|$ is a continuous function of f, $|f(x)|$ is a continuous function of a continuous function and must hence be continuous. Now

$$\left| \int_p^q f(x)\, dx \right| \leq \int_p^q |f(x)|\, dx.$$

Hence if the Cauchy criterion holds for the integral of $|f|$, it must also hold for the integral of f itself; thus the convergence of $\int |f|\, dx$ implies that of $\int f\, dx$. Furthermore,

$$\left| \int_a^b f(x)\, dx \right| \leq \int_a^b |f(x)|\, dx \leq \int_a^\infty |f(x)|\, dx.$$

This is true for every b; hence in the limit,

$$\left| \int_a^\infty f(x)\, dx \right| \leq \int_a^\infty |f(x)|\, dx. \quad \square$$

THEOREM 54 (*Comparison test*)   Let $f(x)$ and $g(x)$ be continuous for $a \leq x < \infty$. If $0 \leq |f(x)| \leq g(x)$ and $\int_a^\infty g(x)\, dx$ converges, then

$$\int_a^\infty f(x)\, dx$$

is absolutely convergent, and

$$\left| \int_a^\infty f(x) \, dx \right| \leq \int_a^\infty g(x) \, dx.$$

If $0 \leq g(x) \leq f(x)$ and

$$\int_a^\infty g(x) \, dx$$

diverges, then

$$\int_a^\infty f(x) \, dx$$

diverges.

The proof is just like that for series and need not be repeated.

The preceding discussion has been confined to integrals from a to ∞. A similar one holds for integrals from $-\infty$ to b and for integrals from a to b, which are improper at an endpoint. Thus the convergence of integrals

$$\int_0^1 f(x) \, dx,$$

where f is unbounded near 0 and continuous for $0 < x \leq 1$, is related to the convergence of series

$$\sum_{n=1}^\infty a_n, \qquad a_n = \int_{b_{n+1}}^{b_n} f(x) \, dx,$$

where b_n is a monotone sequence converging to 0.

*6.23   IMPROPER INTEGRALS DEPENDING ON A PARAMETER—UNIFORM CONVERGENCE

The analogue of a series of functions

$$\sum_{n=1}^\infty f_n(x)$$

is an improper integral

$$\int_a^\infty f(t, x) \, dt;$$

thus the variable t replaces the index n. Both of these, when convergent, define a function $F(x)$. Because of the close relationship between series and improper integrals demonstrated in the preceding section, we can expect the discussion of functions defined by integrals to parallel that of Sections 6.11 to 6.14 for functions defined by series.

One defines the improper integral to be *uniformly convergent* to $F(x)$ for a given range of x, if, given $\epsilon > 0$, a number B can be found such that

$$\left| \int_a^b f(t, x)\, dt - F(x) \right| < \epsilon \quad \text{for} \quad b > B,$$

where B is independent of x.

THEOREM 55 (*M-test for integrals*) Let $M(t)$ be continuous for $a \le t < \infty$; let $f(t, x)$ be continuous in t for $a \le t < \infty$ for each x of a set E. If

$$|f(t, x)| \le M(t)$$

for x in E and

$$\int_a^\infty M(t)\, dt$$

converges, then

$$\int_a^\infty f(t, x)\, dt$$

is uniformly and absolutely convergent for x in E.

THEOREM 56 If $f(t, x)$ is continuous in t and x for $a \le t < \infty, c \le x \le d$, and

$$\int_a^\infty f(t, x)\, dt$$

is uniformly convergent for $c \le x \le d$, then the function $F(x)$ defined by this integral is continuous for $c \le x \le d$.

THEOREM 57 If $f(t, x)$ is continuous in t and x for $a \le t < \infty, c \le x \le d$, and

$$\int_a^\infty f(t, x)\, dt$$

is uniformly convergent to $F(x)$ for $c \le x \le d$, then

$$\int_c^d F(x)\, dx = \int_a^\infty \int_c^d f(t, x)\, dx\, dt.$$

THEOREM 58 If $f(t, x)$ is continuous in t and x and has a derivative $\partial f/\partial x$, which is continuous in t and x for $a \le t < \infty$ and $c \le x \le d$, and the integrals

$$\int_a^\infty f(t, x)\, dt, \qquad \int_a^\infty \frac{\partial f}{\partial x}(t, x)\, dt$$

converge, the second one uniformly, for $c \le x \le d$, then

$$F(x) = \int_a^\infty f(t, x)\, dt$$

has a continuous derivative for $c \le x \le d$, and

$$F'(x) = \int_a^\infty \frac{\partial f}{\partial x}(t, x)\, dt.$$

These theorems are proved exactly as for series.

EXAMPLE   The integral

$$\int_0^\infty e^{-xt^2}\, dt \tag{6.74}$$

is uniformly convergent for $x \ge 1$, since

$$0 \le e^{-xt^2} \le e^{-t^2} = M(t) \quad \text{for} \quad x \ge 1,$$

and the integral

$$\int_0^\infty e^{-t^2}\, dt$$

exists, as a comparison with

$$\int_0^\infty e^{-t}\, dt$$

reveals. Hence (6.74) defines a function $F(x)$. Theorem 56 shows that $F(x)$ is continuous for $x \ge 1$. The integral

$$\int_0^\infty \frac{\partial}{\partial x}\left(e^{-xt^2}\right) dt = -\int_0^\infty t^2 e^{-xt^2}\, dt$$

is also uniformly convergent, since

$$0 \le t^2 e^{-xt^2} \le t^2 e^{-t^2} < e^{-t}$$

for $x \ge 1$ and t sufficiently large. Hence

$$\frac{d}{dx}\int_0^\infty e^{-xt^2}\, dt = -\int_0^\infty t^2 e^{-xt^2}\, dt, \qquad x \ge 1. \quad \blacksquare$$

As in the preceding section, the theory extends to improper integrals over a finite interval without essential change.

*6.24   PRINCIPAL VALUE OF IMPROPER INTEGRALS

An integral from $-\infty$ to ∞ would normally be decomposed into integrals from $-\infty$ to 0 and from 0 to ∞, as in Section 4.1. If the last two integrals converge, the given integral converges; otherwise, it diverges. When this process leads to divergence, one may be able to salvage the integral by the following procedure, which has important applications to Laplace and Fourier transforms.

Let f be continuous for $-\infty < x < \infty$ (or, more generally, let f be such that f has an integral over each finite interval). Then the *Cauchy principal value* (briefly, *principal value*) of the integral of f from $-\infty$ to ∞ is the limit

$$\lim_{a \to \infty} \int_{-a}^{a} f(x)\, dx,$$

if this limit exists. When it does, one denotes the value by

$$(P) \int_{-\infty}^{\infty} f(x)\, dx.$$

One sees at once that if $\displaystyle\int_{-\infty}^{\infty} f(x)\, dx$ exists in the usual sense, then

$$\lim_{a \to \infty} \int_{0}^{a} f(x)\, dx \quad \text{and} \quad \lim_{a \to \infty} \int_{-a}^{0} f(x)\, dx$$

both exist, and hence the principal value exists and equals the usual improper integral. However, the principal value may exist even though the usual value does not. For example,

$$(P) \int_{-\infty}^{\infty} \frac{x-1}{1+(x-1)^2}\, dx = \lim_{a \to \infty} \int_{-a}^{a} \frac{x-1}{1+(x-1)^2}\, dx$$

$$= \lim_{a \to \infty} \tfrac{1}{2} \log \frac{1+(a-1)^2}{1+(a+1)^2} = 0.$$

Here the integral from 0 to ∞ is $+\infty$, from $-\infty$ to 0 is $-\infty$. There is in effect a cancellation of the two infinities. This is related to a certain symmetry of the graph of the function.

The concept of principal value can be extended to the integral of a function f from a to b, where f is discontinuous only at c, with $a < c < b$. Here one defines:

$$(P) \int_{a}^{b} f(x)\, dx = \lim_{\epsilon \to 0+} \left[\int_{a}^{c-\epsilon} f(x)\, dx + \int_{c+\epsilon}^{b} f(x)\, dx \right].$$

For example,

$$(P) \int_{0}^{3} \frac{1}{(x-1)^3}\, dx = \lim_{\epsilon \to 0+} \left[\int_{0}^{1-\epsilon} \frac{1}{(x-1)^3}\, dx + \int_{1+\epsilon}^{3} \frac{1}{(x-1)^3}\, dx \right]$$

$$= \lim_{\epsilon \to 0+} \left[\left(-\frac{1}{2\epsilon^2} + \frac{1}{2} \right) + \left(-\frac{1}{8} + \frac{1}{2\epsilon^2} \right) \right] = \frac{3}{8}.$$

Here the integrals from 0 to 1 and from 1 to 3 are $+\infty$ and $-\infty$, respectively, and again there is a cancellation of the two infinities, made clear in the evaluation of the last limit. It is easily seen, as previously, that whenever the integral of f from a to b exists, as usual, as an improper integral, the principal value exists and equals the previous value; the example just given shows that the principal value can exist even when the usual improper integral does not exist. Of course, the principal value can also fail to exist, as the following example shows:

$$\int_{-1}^{1}\frac{1}{x^2}\,dx = \lim_{\epsilon \to 0+}\left[\int_{-1}^{-\epsilon}\frac{1}{x^2}\,dx + \int_{\epsilon}^{1}\frac{1}{x^2}\,dx\right]$$

$$= \lim_{\epsilon \to 0+}\left[\left(\frac{1}{\epsilon}+1\right)+\left(-1+\frac{1}{\epsilon}\right)\right] = \infty.$$

Here there is no cancellation because the function is always positive.

If f is continuous for $-\infty < x < \infty$ except at c, then one can use a principal value both for c and for large x as follows:

$$(P)\int_{-\infty}^{\infty} f(x)\,dx = \lim_{a \to \infty}\lim_{\epsilon \to 0+}\left[\int_{-a}^{c-\epsilon}f(x)\,dx + \int_{c+\epsilon}^{a}f(x)\,dx\right].$$

This procedure can be adapted to the case of several discontinuities $c_1, \ldots, c_m$.

*6.25   LAPLACE TRANSFORMATION ▪ Γ-FUNCTION AND B-FUNCTION

In view of the analogy between infinite series and improper integrals it is natural to seek an improper integral corresponding to a power series. If we write the power series as

$$\sum_{n=0}^{\infty} f(n)x^n, \tag{6.75}$$

then a natural analogue is the improper integral

$$\int_{0}^{\infty} f(t)x^t\,dt. \tag{6.76}$$

Except for a minor change in notation, this is the *Laplace transform* of the function $f(t)$. More precisely, the Laplace transform of $f(t)$ is the integral

$$\int_{0}^{\infty} f(t)e^{-st}\,dt; \tag{6.77}$$

(6.77) is obtained from (6.76) by replacing x by e^{-s}. Just as (6.75) defines a function $G(x)$ within the interval of convergence, the interval (6.77) defines a function $F(s)$, for those values of s for which the integral converges:

$$F(s) = \int_{0}^{\infty} f(t)e^{-st}\,dt. \tag{6.78}$$

In fact, the integral (6.76) can be shown to have a "radius of convergence" r^* such that (6.76) converges for $0 \le x < r^*$; it is necessary to restrict to positive x, since x^t would be imaginary for x negative and $t = \frac{1}{2}, \frac{1}{4}$, etc. Accordingly, the integral (6.78) converges and defines a function $F(s)$ for $0 < e^{-s} < r^*$, that is, for $s > \log(1/r^*)$.

Integrals of form (6.78) have proved to be exceedingly useful in the theory of ordinary and partial differential equations. The integral can be regarded as an "operator" $\mathscr{L}$ transforming the function $f(t)$ into the function $F(s)$, and one writes

$$\mathscr{L}[f] = F.$$

This explains the word *transform* and also the use of the term *operational method* in connection with the applications to differential equations. (See Section 7.19 and the book by Kaplan listed at the end of Chapter 7.)

An important particular Laplace transform is:

$$\int_0^\infty t^k e^{-st}\, dt, \tag{6.79}$$

in which $f(t) = t^k$; the parameter k must be greater than -1, to avoid divergence of the integral at $t = 0$. With k so restricted, the integral converges for $s > 0$ (Problem 1). When k is 0 or a positive integer, the integral is easily evaluated:

$$\int_0^\infty e^{-st}\, dt = \frac{1}{s}, \qquad \int_0^\infty t e^{-st}\, dt = \frac{1}{s^2}, \ldots (s > 0).$$

In general, an integration by parts (Problem 10 below) shows that

$$\int_0^\infty t^k e^{-st}\, dt = \frac{k}{s} \int_0^\infty t^{k-1} e^{-st}\, dt, \qquad s > 0. \tag{6.80}$$

Accordingly, we can apply induction to conclude that

$$\mathscr{L}[t^k] = \int_0^\infty t^k e^{-st}\, dt = \frac{k}{s} \frac{k-1}{s} \cdots \frac{1}{s} \frac{1}{s} = \frac{k!}{s^{k+1}}, \qquad s > 0. \tag{6.81}$$

For $s = 1$, (6.81) gives

$$k! = \int_0^\infty t^k e^{-t}\, dt. \tag{6.82}$$

This suggests a method of generalizing the factorial; that is, we could use Eq. (6.82) to define $k!$ for k an arbitrary real number greater than -1. It is customary to denote this generalized factorial by $\Gamma(k + 1)$; the Gamma function $\Gamma(k)$ is then defined by the equation

$$\Gamma(k) = \int_0^\infty t^{k-1} e^{-t}\, dt, \qquad k > 0; \tag{6.83}$$

when k is a positive integer or 0,

$$\Gamma(k + 1) = k! = \begin{cases} 1 \cdot 2 \cdots k, & k > 0, \\ 1, & k = 0. \end{cases} \tag{6.84}$$

The general integral (6.79) is expressible in terms of the Gamma function (Problem 14 below):

$$\mathscr{L}[t^k] = \int_0^\infty t^k e^{-st}\, dt = \frac{\Gamma(k + 1)}{s^{k+1}} \qquad (s > 0). \tag{6.85}$$

Equation (6.80) then states that

$$\frac{\Gamma(k+1)}{s^{k+1}} = \frac{k}{s}\frac{\Gamma(k)}{s^k};$$

that is,

$$\Gamma(k+1) = k\Gamma(k). \tag{6.86}$$

This is the *functional equation of the Gamma function*. The functional equation can be used to define $\Gamma(k)$ for negative k; thus we write

$$\Gamma\left(\frac{1}{2}\right) = \left(-\frac{1}{2}\right)\Gamma\left(-\frac{1}{2}\right),$$

$$\Gamma\left(-\frac{1}{2}\right) = \left(-\frac{3}{2}\right)\Gamma\left(-\frac{3}{2}\right),\dots$$

in order to define $\Gamma(-\frac{1}{2})$, $\Gamma(-\frac{3}{2}),\dots$ in terms of the known value of $\Gamma(\frac{1}{2})$. This procedure fails only for $k = -1, -2,\dots$. In fact, we can show (Problem 13) that

$$\lim_{k\to 0+} \Gamma(k) = +\infty,$$

and if the Gamma function is extended to negative nonintegral k as before,

$$\lim_{k\to -n} |\Gamma(k)| = +\infty$$

for every negative integer $-n$. These properties are indicated in Fig. 6.18. The Gamma function has other important properties:

$$\frac{1}{\Gamma(k)} = ke^{\gamma k} \lim_{n\to\infty} \left[(1+k)\left(1+\frac{k}{2}\right)\cdots\left(1+\frac{k}{n}\right)e^{-k-\frac{k}{2}-\cdots-\frac{k}{n}}\right]; \tag{6.87}$$

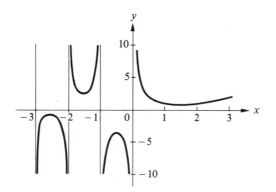

Figure 6.18 $y = \Gamma(x)$.

here γ is the Euler-Mascheroni constant (Problem 15 below):

$$\gamma = 1 + \sum_{n=2}^{\infty} \left(\frac{1}{n} + \log \frac{n-1}{n} \right) = 0.5772\ldots; \tag{6.88}$$

$$\Gamma(k) = k^{k-\frac{1}{2}} e^{-k} \sqrt{2\pi} \, e^{\theta(k)/12k}, \qquad k > 0, \tag{6.89}$$

where $\theta(k)$ denotes a function of k such that $0 < \theta(k) < 1$. Proof of (6.87) and (6.89) and other properties of the Gamma function are given in Chapter 12 of the book of Whittaker and Watson listed at the end of this chapter. From (6.89), one can prove (Problem 16) that

$$\lim_{k \to \infty} \frac{\Gamma(k+1)}{k^{k+\frac{1}{2}} \sqrt{2\pi} \, e^{-k}} = 1. \tag{6.90}$$

When k is an integer, this gives the *Stirling approximation to $k!$*:

$$k! \sim k^{k+\frac{1}{2}} \sqrt{2\pi} \, e^{-k}. \tag{6.91}$$

For $k = 10$ the left-hand side is 3.629×10^6 whereas the right-hand side is 3.599×10^6.

The Beta function $B(p, q)$ is defined by the equation

$$B(p, q) = \int_0^1 x^{p-1}(1-x)^{q-1} \, dx \qquad (p > 0, \quad q > 0). \tag{6.92}$$

This can be expressed in terms of the Gamma function (Problem 17):

$$B(p, q) = \frac{\Gamma(p)\Gamma(q)}{\Gamma(p+q)}. \tag{6.93}$$

PROBLEMS

1. Determine convergence or divergence of the following improper integrals:

   a) $\displaystyle \int_1^\infty \frac{e^{\sin x}}{x} \, dx$

   b) $\displaystyle \int_1^\infty \frac{dx}{(\log x)^x}$

   c) $\displaystyle \int_1^\infty \frac{dx}{x^x}$

   d) $\displaystyle \int_0^\infty t^k e^{-st} \, dt, \quad k > -1$

   e) $\displaystyle \int_1^\infty \sin x^2 \, dx$ (Hint: Let $u = x^2$.)

2. Prove that if $f(x) = \sin 2\pi x$, then the series

   $$\sum_{n=1}^{\infty} \int_{n-1}^n f(x) \, dx$$

   converges, but $\displaystyle \int_0^\infty f(x) \, dx$ diverges.

3. Prove the *ratio test for integrals*: If $f(x)$ is continuous for $a \leq x < \infty$ and

$$\lim_{x \to \infty} \left| \frac{f(x+1)}{f(x)} \right| = k < 1,$$

then $\int_a^\infty f(x)\, dx$ is absolutely convergent.

4. Apply the ratio test of Problem 3 to prove convergence of the integrals:

a) $\int_1^\infty \dfrac{x^2}{e^x}\, dx$

b) $\int_1^\infty \dfrac{1}{x^x}\, dx$

5. Prove the *root test for integrals*: If $f(x)$ is continuous for $a \leq x < \infty$ and

$$\lim_{x \to \infty} |f(x)|^{1/x} = k < 1,$$

then $\int_a^\infty f(x)\, dx$ is absolutely convergent.

6. Apply the root test of Problem 5 to prove convergence of the integrals:

a) $\int_a^\infty e^{-x^2}\, dx$

b) $\int_2^\infty \dfrac{dx}{(\log x)^x}$

7. Prove that the following integrals are uniformly convergent for $0 \leq x \leq 1$:

a) $\int_1^\infty \dfrac{dt}{(x^2 + t^2)^{\frac{5}{2}}}$

b) $\int_1^\infty \dfrac{\sin t}{x^2 + t^2}\, dt$

8. a) Prove that for every choice of $x_1 > 0$ the integral

$$\int_0^\infty t^n e^{-xt^2}\, dt, \qquad n > 0,$$

is uniformly convergent for $x \geq x_1$.

b) Use the known result that

$$\int_0^\infty e^{-x^2}\, dx = \tfrac{1}{2}\sqrt{\pi}$$

(Problem 1 following Section 4.8) to prove that

$$\int_0^\infty e^{-xt^2}\, dt = \frac{1}{2}\sqrt{\frac{\pi}{x}}, \qquad x > 0.$$

c) Use the results of parts (a) and (b) and Theorem 58 to show that for $n = 1, 2, \ldots$, and $x > 0$,

$$\int_0^\infty t^{2n} e^{-xt^2}\, dt = \tfrac{1}{2} \int_0^\infty t^{n-\frac{1}{2}} e^{-xt}\, dt = \tfrac{1}{2}\Gamma(n + \tfrac{1}{2}) x^{-n-\frac{1}{2}} = \frac{\sqrt{\pi}}{2} \frac{1 \cdot 3 \cdot 5 \cdots (2n-1)}{2^n x^{n+\frac{1}{2}}}.$$

9. a) Prove that if $n > 0$, the integrals

$$\int_0^\infty t^n e^{-t^2} \cos(tx)\, dt, \qquad \int_0^\infty t^n e^{-t^2} \sin(tx)\, dt$$

are uniformly convergent for all x.

b) Let

$$F(x) = \int_0^\infty e^{-t^2} \cos(tx)\, dt.$$

Use integration by parts (cf. Problem 10) to show that $F'(x) = -\frac{1}{4}xF(x)$. From this, deduce that $d\log F(x) = -\frac{1}{2}x\,dx$ and hence that $F(x) = ce^{-\frac{1}{4}x^2}$. Let $x = 0$ and use Problem 8 to find c and thus prove that

$$F(x) = \frac{1}{2}\sqrt{\pi}\,e^{-\frac{1}{4}x^2}.$$

10. a) Prove that if $u(x)$, $u'(x)$, $v(x)$, $v'(x)$ are continuous for $a \le x < \infty$ and $\lim_{x \to \infty} [u(x)v(x)]$ exists, then

$$\int_a^\infty u(x)v'(x)\,dx = \lim_{x\to\infty} [u(x)v(x)] - u(a)v(a) - \int_a^\infty u'(x)v(x)\,dx;$$

that is, if one of the two improper integrals converges, then the other converges and the equation holds.

b) Use the result of part (a) to justify the derivation of Eq. (6.80).

c) Prove Abel's formula:

$$\sum_{k=1}^{n} u_k(v_{k+1} - v_k) = u_n v_{n+1} - u_1 v_1 - \sum_{k=1}^{n-1} v_{k+1}(u_{k+1} - u_k)$$

and hence obtain the analogue of part (a) for infinite series:

$$\sum_{k=1}^{\infty} u_k(v_{k+1} - v_k) = \lim_{n\to\infty} (u_n v_{n+1}) - u_1 v_1 - \sum_{k=1}^{\infty} v_{k+1}(u_{k+1} - u_k);$$

that is, if $\lim_{n\to\infty} (u_n v_{n+1})$ exists, then convergence of one of the two series implies convergence of the other and validity of the equation. This "integration by parts" for series provides valuable tests for convergence; cf. Chapter X of the book of Knopp listed at the end of the chapter.

11. Prove the following:

a) if f is continuous for $-\infty < x < \infty$ and f is odd, then (P)$\int_{-\infty}^\infty f(x)\,dx = 0$;

b) if f is continuous for $-a \le x \le a$, except at $x = 0$, and f is odd, then (P)$\int_{-a}^a f(x)\,dx = 0$.

12. For each of the following integrals, obtain the principal value, if it exists; if it does, determine whether the usual value also exists [Hint: Note the result of Problem 11.]:

a) $\int_{-\infty}^\infty \dfrac{x^3}{x^4 + 1}\,dx$

b) $\int_{-\infty}^\infty \dfrac{x}{x^4 + 1}\,dx$

c) $\int_{-\infty}^\infty \sin x\,dx$

d) $\int_{-\infty}^\infty \dfrac{x^3}{x^2 + 1}\,dx$

e) $\int_{-1}^1 \dfrac{1}{x}\,dx$

f) $\int_{-2}^4 \dfrac{1}{(x+1)^{1/3}}\,dx$

g) $\int_{-\infty}^\infty \dfrac{1}{x}\,dx$

h) $\int_{-\infty}^\infty \dfrac{x}{(x-1)^2}\,dx$

13. a) Prove from Eq. (6.83) that $\Gamma(k)$ is continuous and positive for $k > 0$.

b) Prove from Eq. (6.86) that

$$\lim_{k\to 0+} \Gamma(k) = +\infty;$$

c) Prove that if Eq. (6.86) is used to define $\Gamma(k)$ for k negative but nonintegral, then

$$\lim_{k \to -n} |\Gamma(k)| = +\infty \quad \text{for} \quad n = 0, 1, 2 \cdots .$$

14. Prove (6.85)

15. a) Prove by the integral test that the series (6.88) defining γ converges.
    b) Use Theorem 23 to prove that $\gamma = 0.6$ to one significant figure.

16. Prove (6.90) from (6.89).

17. Prove the identity (6.93). [Hint: Show that

$$\Gamma(p) = 2 \int_0^\infty x^{2p-1} e^{-x^2} \, dx, \qquad \Gamma(q) = 2 \int_0^\infty y^{2q-1} e^{-y^2} \, dy$$

and hence, as in Problem 1 following Section 4.8, that

$$\Gamma(p)\Gamma(q) = 4 \int_0^{\frac{1}{2}\pi} \int_0^\infty \sin^{2p-1} \theta \cos^{2q-1} \theta \, r^{2p+2q-1} e^{-r^2} \, dr \, d\theta$$

$$= \left(2 \int_0^\infty r^{2p+2q-1} e^{-r^2} \, dr \right) \left(2 \int_0^{\frac{1}{2}\pi} \sin^{2p-1} \theta \cos^{2q-1} \theta \, d\theta \right).$$

The first factor on the right is $\Gamma(p+q)$; the second reduces to $B(p, q)$ if one sets $x = \sin^2 \theta$.]

18. Verify that the following functions $F(s)$ are Laplace transforms of the functions $f(t)$ given:

a) $F(s) = \dfrac{1}{s-k}$, $s > k$; $f(t) = e^{kt}$;

b) $F(s) = \dfrac{k}{s^2 + k^2}$, $s > 0$; $f(t) = \sin kt$

c) $F(s) = \dfrac{1}{s^k}$, $s > 0$, $k > 0$; $f(t) = \dfrac{t^{k-1}}{\Gamma(k)}$

d) $F(s) = \displaystyle\sum_{k=1}^\infty \dfrac{b_k}{s^k}$, $s > s_1$; $f(t) = \displaystyle\sum_{k=0}^\infty \dfrac{1}{k!} b_{k+1} t^k$.

*6.26   CONVERGENCE OF IMPROPER MULTIPLE INTEGRALS

For simplicity we confine attention to double integrals and consider the case of $f(x, y)$ continuous but unbounded on a set R consisting of a bounded closed region minus one point (x_0, y_0) interior to the region. We can further take R to be a circular region with center (x_0, y_0) (excluded) and radius a since we can always consider the integral over R to be an improper one over such a circular region plus a "proper" integral over the rest of R.

The improper integral of f over R is said to exist and have value c (Section 4.8) if for every $\epsilon > 0$ we can find a bounded closed region R_1 contained in R such that for every bounded closed region R_2 contained in R and containing R_1 we have

$$\left| \int\!\!\int_{R_2} f(x, y) \, dx \, dy - c \right| < \epsilon.$$

Our purpose is to show that in certain cases this limit process can be simplified by considering only

$$\lim_{h \to 0+} \int_{E_h} \int f(x, y) \, dx \, dy, \tag{6.94}$$

where E_h is the part of R outside a circular neighborhood of (x_0, y_0) of radius h. By analogy with the definitions of Section 6.24 we call (6.94) the *principal value* of the double integral of f over R.

We point out that since R is a circular region, we can write the limit (6.94) as

$$\lim_{h \to 0+} \int_h^a \int_0^{2\pi} f r \, dr \, d\theta. \tag{6.95}$$

Here we have introduced polar coordinates at (x_0, y_0): $x = x_0 + r \cos \theta$, $y = y_0 + r \sin \theta$, so that f is continuous in r, θ for all θ and $0 < r \le a$. By Theorem M of Section 4.11 the inner integral defines a function

$$g(r) = \int_0^{2\pi} f(x_0 + r \cos \theta, \, y_0 + r \sin \theta) \, d\theta$$

continuous for $0 < r \le a$, and we are really considering an improper integral

$$\int_0^a r g(r) \, dr = \lim_{h \to 0+} \int_h^a r g(r) \, dr. \tag{6.96}$$

We can parallel the discussion of Section 6.22 for such integrals. For example, if $a > 1$ and $g(x) \ge 0$ for $0 \le x \le a$, then the integral exists if and only if the monotone increasing sequence

$$\int_{1/n}^a r g(r) \, dr \qquad (n = 1, 2, 3, \ldots)$$

is bounded. Also, in general, the integral exists if and only if there is a $\delta > 0$ such that

$$\left| \int_{h_2}^{h_1} r g(r) \, dr \right| < \epsilon \quad \text{for} \quad 0 < h_2 < h_1 < \delta \tag{6.97}$$

(Cauchy criterion). The theorem on absolute convergence and the comparison test also have their counterparts.

THEOREM 59   Let R be the region $0 < (x - x_0)^2 + (y - y_0)^2 \le a^2$, where a is a positive constant. Let f be continuous and unbounded on R and let the principal value of the double integral of f over R be absolutely convergent, so that in the notation given earlier,

$$\lim_{h \to 0} \int_h^a \int_0^{2\pi} |f| r \, d\theta \, dr \tag{6.98}$$

exists. Then the improper double integral of f over R exists.

Proof.   We introduce $g(r)$ as before for f and also $g_1(r)$ for $|f|$, so that

$$g_1(r) = \int_0^{2\pi} |f(x_0 + \ldots, \, y_0 + \cdots)| \, d\theta. \tag{6.99}$$

The existence of the limit (6.98) is equivalent to existence of the improper integral

$$\int_0^a r g_1(r) \, dr. \tag{6.100}$$

Hence by the Cauchy condition we can choose δ so small that for given $\epsilon > 0$,

$$\int_{h_2}^{h_1} r g_1(r) \, dr < \frac{\epsilon}{2} \quad \text{for} \quad 0 < h_2 < h_1 < \delta. \tag{6.101}$$

Now

$$|g(r)| \le \int_0^{2\pi} |f(x_0 + \cdots, y_0 + \cdots)| \, d\theta = g_1(r).$$

Hence existence of the integral (6.100) implies, by comparison, that the integral (6.96) exists, so that the principal value (6.94) for f exists. We let c be this value.

Given $\epsilon > 0$, we now choose δ as before but also so small that

$$\left| \int_{E_h} \int f(x, y) \, dx \, dy - c \right| < \frac{\epsilon}{2}$$

for $0 < h < \delta$. We choose h_0, h_1 so that $0 < h_1 < h_0 < \delta$ and let R_i be the annular region E_{h_i} for $i = 0, 1$. Then

$$\left| \int_{R_0} \int f(x, y) \, dA - c \right| < \frac{\epsilon}{2}.$$

Now let R_2 be any bounded closed region containing R_1 and contained in R. Then R_2 can be considered as the union of the annular region R_0 and a region R_3 contained in the circular region $(x - x_0)^2 + (y - y_0)^2 \le h_0^2$ but not containing (x_0, y_0) (Fig. 6.19). Therefore R_3 lies in some annulus $h_2 \le r \le h_0$ with $h_2 > 0$. Hence

$$\left| \int_{R_3} \int f(x, y) \, dA \right| \le \int_{R_3} \int |f(x, y)| \, dx \, dy$$

$$\le \int_{h_2}^{h_0} r g_1(r) \, dr < \frac{\epsilon}{2}.$$

Thus

$$\left| \int_{R_2} \int f(x, y) \, dA - c \right| = \left| \int_{R_0} \int f(x, y) \, dA - c + \int_{R_3} \int f(x, y) \, dA \right|$$

$$< \frac{\epsilon}{2} + \frac{\epsilon}{2} = \epsilon,$$

as was to be proved. $\square$

COROLLARY   In Theorem 59 let $f(x, y) = p(x, y) \log(1/r)$ or $p(x, y) r^{-p}$ for $0 < p < 2$, where $p(x, y)$ is bounded in R. Then the improper double integral of f over R exists.

The proof is left as an exercise (Problem 2 below).

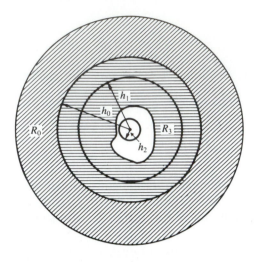

Figure 6.19 Proof of Theorem 59.

PROBLEMS

1. Let $f(x, y) = (x + y)/r^p$, where $r = (x^2 + y^2)^{1/2}$ and $p = $ const > 0. Show that the principal value of the double integral of f over R: $0 < r \le 1$ exists, but the double integral itself exists only for $p < 3$.

2. Prove the corollary to Theorem 59.

3. **a)** Let R be the solid sphere of radius a and center (x_0, y_0, z_0) with the center deleted. Let $f(x, y, z)$ be continuous and unbounded on R. Define the principal value of the triple integral of f over R by analogy with (6.94) and state and prove the analogue of Theorem 59 for $f(x,y,z)$.

   b) On the basis of the theorem of part (a) deduce that if $g(x, y, z)$ is bounded and continuous over R, then the improper integral

   $$\int\int_R\int \frac{g(x, y, z)}{r^p}\, dx\, dy\, dz, \qquad r = \left[(x - x_0)^2 + (y - y_0)^2 + (z - z_0)^2\right]^{1/2},$$

   exists if $0 < p < 3$.

Suggested References

Courant, Richard J., Differential and Integral Calculus, transl. by E. J. McShane, 2 vols. New York: Interscience, 1947.

Green, J. A., Sequences and Series, Glencoe, Ill.: The Free Press, 1958.

Hardy, G. H., Pure Mathematics, 9th ed. Cambridge: Cambridge University Press, 1947.

Henrici, P., Applied and Computational Complex Analysis, Vol. I. New York: John Wiley & Sons, 1974.

Knopp, K., Theory and Application of Infinite Series, transl. by Miss R. C. Young. London and Glasgow: Blackie and Son, 1928.

Rudin, W., Principles of Mathematical Analysis, 2nd ed. New York: McGraw-Hill, 1964.

Whittaker, E. T., and G. N. Watson, Modern Analysis, 4th ed. Cambridge: Cambridge University Press, 1940.

Widder, D. V., The Laplace Transform. Princeton: Princeton University Press, 1946.

7

Fourier Series and Orthogonal Functions

7.1 TRIGONOMETRIC SERIES

A trigonometric series is a series of form

$$\tfrac{1}{2}a_0 + a_1 \cos x + b_1 \sin x + \cdots + a_n \cos nx + b_n \sin nx + \cdots, \qquad (7.1)$$

where the coefficients a_n and b_n are constants. If these constants satisfy certain conditions, to be specified in Section 7.2, then the series is called a Fourier series. Almost all trigonometric series encountered in physical problems are Fourier series.

Each term in (7.1) has the property of repeating itself in intervals of 2π:

$$\cos(x + 2\pi) = \cos x, \qquad \sin(x + 2\pi) = \sin x, \ldots,$$

$$\cos[n(x + 2\pi)] = \cos(nx + 2n\pi) = \cos nx, \ldots.$$

It follows that if (7.1) converges for all x, then its sum $f(x)$ must also have this property:

$$f(x + 2\pi) = f(x). \qquad (7.2)$$

We say: $f(x)$ has period 2π. In general, a function $f(x)$ such that

$$f(x + p) = f(x) \qquad (p \neq 0) \qquad (7.3)$$

for all x is said to be *periodic* and have *period* p. It should be noted that $\cos 2x$ has, in addition to the period 2π, the period π and, in general, $\cos nx$ and $\sin nx$ have the periods $2\pi/n$. However, 2π is the smallest period shared by all terms of the series.

If $f(x)$ has period p, then the substitution:

$$x = p\frac{t}{2\pi} \tag{7.4}$$

converts $f(x)$ into a function of t having period 2π; for when t increases by 2π, x increases by p.

A function $f(x)$ having period 2π is illustrated in Fig. 7.1. Such periodic functions appear in a great variety of physical problems: the vibrations of a spring; the motion of the planets about the sun; the rotation of the earth about its axis; the motion of a pendulum; the tides and wave motion in general; vibrations of a violin string, of an air column (for example, in a flute); and musical sounds in general. The modern theory of light is based on "wave mechanics," with periodic vibrations a characteristic feature; the spectrum of a molecule is simply a picture of the different vibrations taking place simultaneously within it. Electric circuits involve many periodically varying variables, for example, the alternating current. The fact that a journey around the globe involves a total change in longitude of 360° is an expression of the fact that the rectangular coordinates of position on the globe are periodic functions of longitude, with period 360°; many other examples of such periodic functions of angular coordinates can be given.

Figure 7.1   Function with period 2π.

Now it can be shown that *every* periodic function of x satisfying certain very general conditions can be represented in the form (7.1), that is, as a trigonometric series (see Section 7.3). This mathematical theorem is a reflection of a physical experience most vividly illustrated in the case of *sound*, for example, that of a violin string. The term $\frac{1}{2}a_0$ represents the neutral position, the terms $a_1 \cos x + b_1 \sin x$ the fundamental tone, the terms $a_2 \cos 2x + b_2 \sin 2x$ the first overtone (octave); the other terms represent higher overtones. The variable x must here be thought of as *time* and the function $f(x)$ as the displacement of an instrument, such as a phonograph needle, which is recording the sound, or of a point on the string. Thus the musical tone heard is a combination of simple harmonic vibrations—the terms $(a_n \cos nx + b_n \sin nx)$. Each such pair can be written in the form

$$A_n \sin(nx + \alpha),$$

where

$$A_n = \sqrt{a_n^2 + b_n^2}, \qquad a_n = A_n \sin \alpha, \qquad b_n = A_n \cos \alpha.$$

The "amplitude" A_{n+1} is a measure of the importance of the nth overtone in the whole sound. The differences in the tones of different musical instruments can be ascribed mainly to the differences in the weights A_n of the overtones.

7.2 FOURIER SERIES

Let us suppose now that a periodic function $f(x)$ is the sum of a trigonometric series (7.1), that is, that

$$f(x) = \frac{a_0}{2} + \sum_{n=1}^{\infty} (a_n \cos nx + b_n \sin nx). \tag{7.5}$$

What is the relation between the coefficients a_n and b_n and the function $f(x)$? To answer this, we multiply $f(x)$ by $\cos mx$ and integrate from $-\pi$ to π:

$$\int_{-\pi}^{\pi} f(x) \cos mx \, dx$$

$$= \int_{-\pi}^{\pi} \left[\frac{a_0}{2} \cos mx + \sum_{n=1}^{\infty} (a_n \cos nx \cos mx + b_n \sin nx \cos mx) \right] dx.$$

If term-by-term integration of the series is allowed, then we find

$$\int_{-\pi}^{\pi} f(x) \cos mx \, dx = \frac{a_0}{2} \int_{-\pi}^{\pi} \cos mx \, dx$$

$$+ \sum_{n=1}^{\infty} \left\{ a_n \int_{-\pi}^{\pi} \cos nx \cos mx \, dx + b_n \int_{-\pi}^{\pi} \sin nx \cos mx \, dx \right\}. \tag{7.6}$$

The integrals on the right-hand side are easily evaluated with the help of the identities

$$\cos x \cos y = \tfrac{1}{2} [\cos (x + y) + \cos (x - y)],$$

$$\sin x \cos y = \tfrac{1}{2} [\sin (x + y) + \sin (x - y)], \tag{7.7}$$

$$\sin x \sin y = -\tfrac{1}{2} [\cos (x + y) - \cos (x - y)].$$

They give

$$\int_{-\pi}^{\pi} \cos nx \cos mx \, dx = \begin{cases} 0, & n \neq m, \\ \pi, & n = m \neq 0, \end{cases}$$

$$\int_{-\pi}^{\pi} \sin nx \cos mx \, dx = 0.$$

If $m = 0$, then all terms on the right-hand side of (7.6) are 0 except the first one, and one finds

$$\int_{-\pi}^{\pi} f(x)\, dx = \pi a_0. \tag{7.8'}$$

For any positive integer m, only the term in a_m gives a result different from 0. Thus

$$\int_{-\pi}^{\pi} f(x) \cos mx\, dx = \pi a_m \qquad (m = 1, 2, \ldots). \tag{7.8''}$$

Multiplying $f(x)$ by $\sin mx$ and proceeding in the same way, we find

$$\int_{-\pi}^{\pi} f(x) \sin mx\, dx = \pi b_m \qquad (m = 1, 2, \ldots). \tag{7.8'''}$$

From the last three formulas, one now concludes that

$$a_n = \frac{1}{\pi} \int_{-\pi}^{\pi} f(x) \cos nx\, dx \qquad (n = 0, 1, 2, \ldots),$$

$$b_n = \frac{1}{\pi} \int_{-\pi}^{\pi} f(x) \sin nx\, dx \qquad (n = 1, 2, \ldots). \tag{7.9}$$

This is the fundamental rule for coefficients in a Fourier series. Without concerning ourselves with the validity of the steps leading to (7.9) we *define* a Fourier series to be any trigonometric series

$$\tfrac{1}{2} a_0 + a_1 \cos x + b_1 \sin x + \cdots + a_n \cos nx + b_n \sin nx + \cdots \tag{7.10}$$

in which the coefficients a_n, b_n are computed from a function $f(x)$ by (7.9); the series is then called *the Fourier series of* $f(x)$. Concerning $f(x)$ we assume only that the integrals in (7.9) exist; for this it is sufficient that $f(x)$ be continuous except for a finite number of jumps between $-\pi$ and π.

No parentheses are used in the general definition (7.10). It is common practice to group the terms as in (7.5). However, the series will always be understood in the ungrouped form (7.10). We recall that *insertion* of parentheses in a *convergent* series is always permissible (Theorem 27 of Section 6.10).

THEOREM 1 Every uniformly convergent trigonometric series is a Fourier series. More precisely, if the series (7.10) converges uniformly for all x to $f(x)$, then $f(x)$ is continuous for all x, $f(x)$ has period 2π, and the series (7.10) is the Fourier series of $f(x)$.

Proof. Since the series converges uniformly for all x, its sum $f(x)$ is continuous for all x (Theorem 31 of Section 6.14). The series remains uniformly convergent if all terms are multiplied by $\cos mx$ or by $\sin mx$ (Theorem 34 of Section 6.14). Accordingly, the term-by-term integration of Eq. (7.6) is justified (Theorem 32 of Section 6.14); (7.9) now follows as previously so that the series is the Fourier series of $f(x)$. The periodicity of $f(x)$ is a consequence of the periodicity of the terms of the series, as remarked in Section 7.1. □

COROLLARY If two trigonometric series converge uniformly for all x and have the same sum for all x:

$$\tfrac{1}{2}a_0 + \sum_{n=1}^{\infty} (a_n \cos nx + b_n \sin nx) \equiv \tfrac{1}{2}a_0' + \sum_{n=1}^{\infty} (a_n' \cos nx + b_n' \sin nx),$$

then the series are identical: $a_0 = a_0'$, $a_n = a_n'$, $b_n = b_n'$ for $n = 1, 2, \ldots$. In particular, if a trigonometric series converges uniformly to 0 for all x, then all coefficients are 0.

Proof. Let $f(x)$ denote the sum of both series. Then by Theorem 1,

$$a_n = a_n' = \frac{1}{\pi} \int_{-\pi}^{\pi} f(x) \cos nx \, dx \qquad (n = 0, 1, 2, \ldots),$$

and similarly $b_n = b_n'$ for all n. If $f(x) \equiv 0$, then all coefficients are 0. $\square$

7.3 CONVERGENCE OF FOURIER SERIES

Although the Fourier series of $f(x)$ is well defined when $f(x)$ is merely "piecewise continuous," it is too much to expect that the series will converge to $f(x)$ under such general conditions. However, it turns out that very little more is required to ensure convergence to $f(x)$. In particular, if f is periodic with period 2π and has continuous first and second derivatives for all x, then the Fourier series of $f(x)$ will converge uniformly to $f(x)$ for all x. The result is in itself remarkable when one considers the fact that expansion of f in a convergent power series requires continuous derivatives of all orders—plus the condition that the remainder R_n of Taylor's formula converges to 0. One can even go further and guarantee uniform convergence of the Fourier series of $f(x)$ to $f(x)$ when $f(x)$ has "corners," that is, points at which $f'(x)$ has a jump discontinuity, while f has continuous first and second derivatives between the corners; this is illustrated in Fig. 7.2. Indeed, one can enlarge the concept of corner to include jump discontinuities of $f(x)$, as illustrated in Fig. 7.3; one can hardly expect convergence of the series to $f(x)$ at the discontinuity points, where $f(x)$ may even be ambiguously defined. But the

Figure 7.2 Piecewise very smooth continuous function.

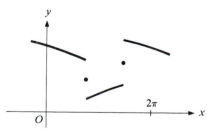

Figure 7.3 Piecewise very smooth function with jump discontinuities.

Fourier series makes up our minds for us, in a most reasonable way: It converges to the *average of the left- and right-hand limits*, that is, to the number

$$\tfrac{1}{2}\left[\lim_{x\to x_1-} f(x) + \lim_{x\to x_1+} f(x)\right]$$

at the discontinuity x_1. One cannot expect the series to converge uniformly near the discontinuity (cf. Section 6.14), but it will converge uniformly in each closed interval containing no discontinuity.

Although we have up to this point considered only periodic functions $f(x)$ (with period 2π), it must be remarked that the basic coefficient formulas (7.9) use *only the values of $f(x)$ between $-\pi$ and π*. Thus if $f(x)$ is given only in this interval and is, for example, continuous, then the corresponding Fourier series can be formed, and we continue to call the series the Fourier series of $f(x)$. If the series converges to $f(x)$ between $-\pi$ and π, then it will converge outside this interval to a function $F(x)$, which is the "periodic extension of $f(x)$"; this is illustrated in Fig. 7.4. It should be noted that unless $f(\pi) = f(-\pi)$, the process of extension will introduce jump discontinuities at $x = \pi$ and $x = -\pi$. At these points the series will converge to the number midway between the two "values" of $F(x)$.

Figure 7.4   Periodic extension of a function defined between $-\pi$ and π.

We term a function $f(x)$, defined for $a \le x \le b$, *piecewise continuous* in this interval if the interval can be subdivided into a finite number of subintervals, inside each of which $f(x)$ is continuous and has finite limits at the left and right ends of the interval. Accordingly, inside the ith subinterval the function $f(x)$ coincides with a function $f_i(x)$ that is continuous in the *closed* subinterval; if, in addition, the functions $f_i(x)$ have continuous first derivatives, we term $f(x)$ *piecewise smooth*; if, in addition, the functions $f_i(x)$ have continuous second derivatives, we term $f(x)$ *piecewise very smooth*.

FUNDAMENTAL THEOREM   Let $f(x)$ be piecewise very smooth in the interval $-\pi \le x \le \pi$. Then the Fourier series of $f(x)$:

$$\frac{a_0}{2} + \sum_{n=1}^{\infty} (a_n \cos nx + b_n \sin nx),$$

$$a_n = \frac{1}{\pi}\int_{-\pi}^{\pi} f(x)\cos nx\, dx, \qquad b_n = \frac{1}{\pi}\int_{-\pi}^{\pi} f(x)\sin nx\, dx,$$

converges to $f(x)$ wherever $f(x)$ is continuous inside the interval. The series converges to

$$\frac{1}{2}\left[\lim_{x \to x_1-} f(x) + \lim_{x \to x_1+} f(x)\right]$$

at each point of discontinuity x_1 inside the interval, and to

$$\frac{1}{2}\left[\lim_{x \to \pi-} f(x) + \lim_{x \to -\pi+} f(x)\right]$$

at $x = \pm\pi$. The convergence is uniform in each closed interval containing no discontinuity.

This theorem is adequate for most applications: The hypotheses can be weakened, "very smooth" being replaced by "smooth"; in fact, $f(x)$ need only be expressible as the difference of two functions, both of which are steadily increasing as x increases. For extensions to these cases the reader is referred to the books of Jackson and Zygmund listed at the end of the chapter. The proof of the fundamental theorem will be given in Section 7.9.

7.4   EXAMPLES—MINIMIZING OF SQUARE ERROR

We now proceed to consider several examples that will bring out more clearly the relation between $f(x)$ and its Fourier series.

EXAMPLE 1   Let $f(x)$ have the value -1 for $-\pi \le x < 0$ and the value $+1$ for $0 \le x \le \pi$. The periodic extension of $f(x)$ then gives the "square wave" of Fig. 7.5. One finds

$$a_n = -\frac{1}{\pi}\int_{-\pi}^{0} \cos nx\, dx + \frac{1}{\pi}\int_{0}^{\pi} \cos nx\, dx = 0, \qquad n = 0,1,2,\ldots,$$

$$b_n = -\frac{1}{\pi}\int_{-\pi}^{0} \sin nx\, dx + \frac{1}{\pi}\int_{0}^{\pi} \sin nx\, dx = \begin{cases} 0, & n = 2,4,\ldots, \\ \dfrac{4}{n\pi}, & n = 1,3,5,\ldots. \end{cases}$$

Hence for $0 < |x| < \pi$,

$$f(x) = \frac{4}{\pi}\sin x + \frac{4}{3\pi}\sin 3x + \cdots = \frac{4}{\pi}\sum_{n=1}^{\infty}\frac{\sin(2n-1)x}{2n-1}.$$

The figure shows the first three partial sums

$$S_1 = \frac{4}{\pi}\sin x, \qquad S_2 = S_1 + \frac{4}{3\pi}\sin 3x, \qquad S_3 = S_2 + \frac{4}{5\pi}\sin 5x$$

of this series. If the graphs are studied carefully, then it becomes clear that $f(x)$ is being approached as limit. For $x = 0$, each partial sum equals 0, so that the series does converge to the average value at the jump; there is a similar situation at

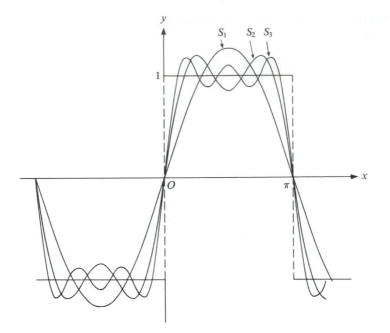

Figure 7.5   Representation of square wave by Fourier series.

$x = \pm\pi$. However, it should be noted that the approximation to $f(x)$ by each partial sum is poorest immediately to the left and right of the jump points. ■

EXAMPLE 2   Let $f(x) = \frac{1}{2}\pi + x$ for $-\pi \le x \le 0$ and $f(x) = \frac{1}{2}\pi - x$ for $0 \le x \le \pi$. The periodic extension of $f(x)$ is the "triangular wave" of Fig. 7.6. In this example the extended function is continuous for all x. One finds

$$a_n = \frac{1}{\pi}\int_{-\pi}^{0}\left(\frac{\pi}{2} + x\right)\cos nx\, dx + \frac{1}{\pi}\int_{0}^{\pi}\left(\frac{\pi}{2} - x\right)\cos nx\, dx$$

$$= \frac{1}{\pi}\left[\left(\frac{\pi}{2} + x\right)\frac{\sin nx}{n}\Big|_{-\pi}^{0} - \frac{1}{n}\int_{-\pi}^{0}\sin nx\, dx + \cdots\right]$$

$$= \frac{1}{\pi}\left[\frac{1}{n^2}(1 - \cos n\pi) + \frac{1}{n^2}(1 - \cos n\pi)\right] = \frac{2}{n^2\pi}(1 - \cos n\pi)$$

for $n = 1, 2, \ldots$. For $n = 0$ a separate computation is needed:

$$a_0 = \frac{1}{\pi}\int_{-\pi}^{0}\left(\frac{\pi}{2} + x\right)dx + \frac{1}{\pi}\int_{0}^{\pi}\left(\frac{\pi}{2} - x\right)dx = 0.$$

The computation of the b_n's is like that of the a_n's, and one finds $b_n = 0$ for $n = 1, 2, \ldots$. Hence

$$f(x) = \frac{4}{\pi}\cos x + \frac{4}{9\pi}\cos 3x + \cdots = \frac{4}{\pi}\sum_{n=1}^{\infty}\frac{\cos(2n - 1)x}{(2n - 1)^2}.$$

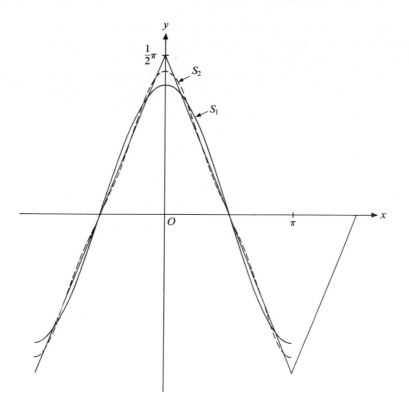

Figure 7.6   Representation of triangular wave by Fourier series.

The first two partial sums are plotted in Fig. 7.6. Since there are no jumps, one must expect convergence everywhere. It should, however, be noted that at the corners (where $f'(x)$ has a jump), the convergence is poorer than elsewhere. ∎

Thus far we have proceeded formally, evaluating coefficients and verifying graphically that the series converges to the function. We now proceed to examine the steps more carefully.

The constant term $a_0/2$ of the series is given by the formula

$$\frac{a_0}{2} = \frac{1}{2\pi} \int_{-\pi}^{\pi} f(x) \, dx.$$

The right-hand member is simply the *average* or *arithmetic mean* of $f(x)$ over the interval $-\pi \leq x \leq \pi$ (Section 4.1). One can also write

$$\int_{-\pi}^{\pi} \left[f(x) - \frac{a_0}{2} \right] dx = 0.$$

In words: The line $y = \frac{1}{2}a_0$ must be such that the *area between the line and the curve $y = f(x)$ lying above the line equals the area between the line and the curve $y = f(x)$ lying below the line.* Thus the line $y = \frac{1}{2}a_0$ is a sort of symmetry line for the graph of $y = f(x)$.

From either of these points of view it is clear that in the two examples considered, one must have $\frac{1}{2}a_0 = 0$; the average of $f(x)$ is 0, and there is as much area above the x-axis as below.

Still another point of view yields the same formula for $\frac{1}{2}a_0$. We define the total *square error* of a function $g(x)$ relative to $f(x)$ as the integral

$$E = \int_{-\pi}^{\pi} [f(x) - g(x)]^2 \, dx. \tag{7.11}$$

This error is 0 when $g = f$ (or when $g = f$ except for a finite number of points) is otherwise positive. We now seek a constant function $y = g_0$ such that this error is as small as possible. In other words, we try to approximate $y = f(x)$ as accurately as possible, in terms of least square error, by a constant g_0. The error is now

$$E(g_0) = \int_{-\pi}^{\pi} [f(x) - g_0]^2 \, dx = \int_{-\pi}^{\pi} [f(x)]^2 \, dx - 2g_0 \int_{-\pi}^{\pi} f(x) \, dx + g_0^2 \cdot 2\pi$$

$$= A - 2Bg_0 + 2\pi g_0^2,$$

where A and B are constants. Thus $E(g_0)$ is a quadratic function of g_0, having a minimum when $dE/dg_0 = 0$:

$$-2B + 4\pi g_0 = 0.$$

Hence the error is minimized when

$$g_0 = \frac{B}{2\pi} = \frac{1}{2\pi} \int_{-\pi}^{\pi} f(x) \, dx = \frac{a_0}{2}.$$

Thus the constant function $y = \frac{1}{2}a_0$ is the best constant approximation, in the sense of least square error, to the function $f(x)$.

This last point of view holds for the coefficients of the general partial sum:

THEOREM 2   Let $f(x)$ be piecewise continuous for $-\pi \leq x \leq \pi$. The coefficients of the partial sum

$$\tfrac{1}{2}a_0 + a_1 \cos x + b_1 \sin x + \cdots + a_n \cos nx + b_n \sin nx$$

of the Fourier series of $f(x)$ are precisely those among all coefficients of the function

$$g_n(x) = p_0 + p_1 \cos x + q_1 \sin x + \cdots + p_n \cos nx + q_n \sin nx$$

that render the square error

$$\int_{-\pi}^{\pi} [f(x) - g_n(x)]^2 \, dx$$

a minimum. Furthermore, the minimum square error E_n satisfies the equation:

$$E_n = \int_{-\pi}^{\pi} [f(x)]^2 \, dx - \pi \left[\tfrac{1}{2}a_0^2 + \sum_{k=1}^{n} (a_k^2 + b_k^2) \right]. \tag{7.12}$$

The proof is left to Problem 7 below.

COROLLARY   If $f(x)$ is piecewise continuous for $-\pi \leqq x \leqq \pi$ and $a_0, a_1, \ldots, b_1, b_2, \ldots$ are the Fourier coefficients of $f(x)$, then

$$\frac{1}{2}a_0^2 + \sum_{k=1}^{n} \left(a_k^2 + b_k^2\right) \leqq \frac{1}{\pi} \int_{-\pi}^{\pi} [f(x)]^2 \, dx, \qquad (7.13)$$

so that the series $\sum_{n=1}^{\infty}(a_n^2 + b_n^2)$ converges. Furthermore,

$$\lim_{n \to \infty} a_n = 0, \qquad \lim_{n \to \infty} b_n = 0. \qquad (7.14)$$

Proof.   Since the square error $\int (f - g)^2 \, dx$ is always positive or 0, the minimum square error E_n is always positive or 0. Accordingly, (7.13) follow from (7.12). By Theorem 7 of Section 6.5 the series $\Sigma(a_n^2 + b_n^2)$ must converge or diverge properly; because of (7.13), the series cannot be properly divergent. Therefore the series converges; (7.14) then follows from the fact that the nth term of the series converges to 0. $\square$

It is shown in Section 7.12 that E_n can be made as small as desired by choosing n sufficiently large; that is, the sequence E_n converges to 0. The relation (7.13) is *Bessel's inequality*.

Theorem 2 can be made the basis of a *graphical* estimation of Fourier coefficients. Thus for the function of Example 1, one first chooses the best-fitting constant term $\frac{1}{2}a_0$; this is clearly 0. One then tries to add a function $p_1 \cos x + p_2 \sin x$ to make the square error as small as possible. It is apparent that the best approximation is achieved by taking a sine term alone. The function $\sin x$ itself fits fairly well, though the errors are large near $\pm\pi$ and 0. To reduce these, we *overshoot* at $\frac{1}{2}\pi$, taking, say, $1.3 \sin x$. New errors are introduced near $\frac{1}{2}\pi$, but the total square error will be less. If we subtract $1.3 \sin x$ from $f(x)$ graphically, we find the function of Fig. 7.7. To eliminate this error, it is clear that a function $p_6 \sin 3x$ is called for. Again we overshoot and estimate $p_6 = 0.4$. We subtract $0.4 \sin 3x$ from the function graphed in Fig. 7.7 and so on. We thus obtain the

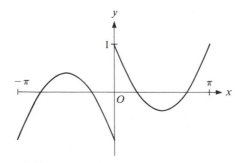

Figure 7.7   Graphically determined error.

expression

$$f(x) = 1.3 \sin x + 0.4 \sin 3x.$$

This agrees, up to the number of significant figures carried, with the first two terms of the Fourier series

$$\frac{4}{\pi} \sin x + \frac{4}{3\pi} \sin 3x + \cdots$$

obtained above.

It is recommended that the following problems be solved first roughly by this graphical procedure. In this way, considerable insight into the structure of the series can be gained. In constructing the series in this way it is best to think of the pair $a_n \cos nx + b_n \sin nx$ in the *amplitude-phase* form:

$$a_n \cos nx + b_n \sin nx = A_n \sin (nx + \alpha) \tag{7.15}$$

pointed out at the end of Section 7.1. The phase angle α effectively determines how much the curve $y = \sin nx$ has to be shifted to the left or right to match the given oscillation; the amplitude A_n merely adjusts the vertical scale.

It should be remarked that the process of decomposing a periodic phenomenon into its component simple harmonic parts is used in a great variety of common experiences. The rattling of an old-fashioned vehicle corresponds to a high-frequency (large n) component with large amplitude; we automatically separate this from a low-frequency vibration, or *swaying*. In a less precise sense, a large day-to-day fluctuation in weather conditions also corresponds to a high-frequency component of large amplitude; the seasonal changes are of low frequency and are much less disturbing.

PROBLEMS

1. Find the Fourier series for each of the following functions:

   a) $f(x) = 0, \quad -\pi \le x < 0; \quad f(x) = 1, \quad 0 \le x \le \pi$

   b) $f(x) = 0, \quad -\pi \le x < 0; \quad f(x) = x, \quad 0 \le x \le \pi$

   c) $f(x) = -x, \quad -\pi \le x \le 0; \quad f(x) = x, \quad 0 \le x \le \pi$

   d) $f(x) = x^2, \quad -\pi \le x \le \pi$

   e) $F(x) = -\dfrac{1}{2} - \dfrac{x}{2\pi}, \quad -\pi \le x < 0; \quad F(x) = \dfrac{1}{2} - \dfrac{x}{2\pi}, \quad 0 < x \le \pi; \quad F(0) = 0$

   f) $G(x) = \dfrac{\pi}{2} - \dfrac{x}{2} - \dfrac{x^2}{4\pi}, \quad -\pi \le x \le 0; \quad G(x) = \dfrac{\pi}{2} + \dfrac{x}{2} - \dfrac{x^2}{4\pi}, \quad 0 \le x \le \pi$

   (It is suggested that the first few partial sums be graphed and compared with the function in each case.)

2. It follows from the fundamental theorem of Section 7.3 that if $f(x)$ is defined between 0 and 2π and is piecewise very smooth in that interval, then $f(x)$ can be represented by a series of form (7.1) in that interval.

a) Show that the coefficients a_n and b_n are given by the formulas:

$$a_n = \frac{1}{\pi}\int_0^{2\pi} f(x)\cos nx\, dx, \qquad b_n = \frac{1}{\pi}\int_0^{2\pi} f(x)\sin nx\, dx.$$

b) Extend this result to a function defined from $x = c$ to $x = c + 2\pi$, where c is any constant.

3. Using the results of Problem 2(a), find Fourier series for the following functions:

    a) $f(x) = x,\quad 0 \le x \le 2\pi$ **b)** $f(x) = |\cos x|,\quad 0 \le x \le 2\pi$

4. Determine which of the following functions are periodic and find the smallest period of those that are periodic:

    a) $\sin 5x$ **b)** $\cos \dfrac{x}{3}$ **c)** $\sin \pi x$ **d)** $x \sin x$

    e) $\sin 3x + \sin 5x$ **f)** $\sin \dfrac{x}{3} + \sin \dfrac{x}{5}$ **g)** $\sin x + \sin \pi x$

5. The result of Problem 1(a) implies that for $0 < x < \pi$,

$$1 = \frac{1}{2} + \frac{2}{\pi}\left(\sin x + \frac{\sin 3x}{3} + \cdots\right).$$

Use $x = \frac{1}{2}\pi$ in this equation to show that

$$\frac{\pi}{4} = 1 - \frac{1}{3} + \frac{1}{5} - \frac{1}{7} + \cdots.$$

6. Use the method of Problem 5 to obtain, with the aid of the series developed in Problems 1 and 3, the relations

    a) $\dfrac{\pi}{\sqrt{8}} = 1 + \dfrac{1}{3} - \dfrac{1}{5} - \dfrac{1}{7} + \dfrac{1}{9} + \dfrac{1}{11} - \dfrac{1}{13} - \dfrac{1}{15} + \cdots$

    b) $\dfrac{\pi^2}{8} = \dfrac{1}{1^2} + \dfrac{1}{3^2} + \dfrac{1}{5^2} + \cdots + \dfrac{1}{(2n-1)^2} + \cdots$

    c) $\dfrac{\pi^2}{12} = 1 - \dfrac{1}{2^2} + \dfrac{1}{3^2} - \dfrac{1}{4^2} + \cdots$

    d) $\dfrac{\pi^2}{6} = \dfrac{1}{1^2} + \dfrac{1}{2^2} + \dfrac{1}{3^2} + \dfrac{1}{4^2} + \cdots$

7. Prove Theorem 2. [Hint: Show that

$$\int (f-g)^2\, dx = \int f^2\, dx + \left(2\pi p_0^2 - 2p_0\int f\, dx\right)$$

$$+ \left(\pi p_1^2 - 2p_1\int f\cos x\, dx\right) + \cdots + \left(\pi q_n^2 - 2q_n\int f\sin nx\, dx\right);$$

accordingly, if $p_0, p_1, \ldots, q_n$ are chosen to give each term on the right its smallest value, the error will be minimized.]

8. a) Prove the trigonometric identities:

$$\sin^3 x = \tfrac{3}{4}\sin x - \tfrac{1}{4}\sin 3x, \qquad \cos^3 x = \tfrac{3}{4}\cos x + \tfrac{1}{4}\cos 3x.$$

    b) Obtain analogous expressions for $\sin^n x$ and $\cos^n x$. [Hint: Use the identities $\sin x = \frac{1}{2}(e^{ix} - e^{-ix})/i$, $\cos x = \frac{1}{2}(e^{ix} + e^{-ix})$.]

    c) Show that the identities of parts (a) and (b) can be interpreted as Fourier series expansions.

7.5   GENERALIZATIONS ▪ FOURIER COSINE SERIES ▪ FOURIER SINE SERIES

The Fourier series up to this point have been considered only for functions of period 2π or, more restrictedly, for functions defined between $-\pi$ and π. We now proceed to enlarge the scope of the theory.

If $f(x)$ is a function of period 2π, one can use as basic interval any interval $c \leq x \leq c + 2\pi$, that is, any interval of length 2π. For such an interval the same reasoning as previously leads to a Fourier series

$$\frac{a_0}{2} + \sum_{n=1}^{\infty} (a_n \cos nx + b_n \sin nx),$$

where

$$a_n = \frac{1}{\pi} \int_c^{c+2\pi} f(x) \cos nx \, dx,$$

$$b_n = \frac{1}{\pi} \int_c^{c+2\pi} f(x) \sin nx \, dx. \tag{7.16}$$

If $f(x)$ is given for all x, with period 2π, this is merely another way of computing the coefficients a_n, b_n. If $f(x)$ is given only for $c \leq x \leq c + 2\pi$, the series can be used to represent f in this interval; it will then (if convergent) represent the periodic extension of f outside this interval.

The interval $-\pi \leq x \leq \pi$ has certain advantages for utilization of symmetry properties. Let $f(x)$ be defined in this interval and let

$$f(-x) = f(x), \qquad -\pi \leq x \leq \pi. \tag{7.17}$$

Then f is called an *even* function of x (in the given interval). If, on the other hand,

$$f(-x) = -f(x), \qquad -\pi \leq x \leq \pi, \tag{7.18}$$

then f is called an *odd* function of x. We note that the product of two even functions or of two odd functions is even whereas the product of an odd function and an even function is odd. Furthermore,

$$\int_{-a}^a f(x) \, dx = \begin{cases} 0, & f \text{ odd} \\ 2\int_0^a f(x) \, dx, & f \text{ even}. \end{cases} \tag{7.19}$$

Let f now be even in the interval $-\pi \leq x \leq \pi$. Then $f(x)\cos nx$ is even (product of two even functions) whereas $f(x)\sin nx$ is odd (product of odd function and even function). Hence by (7.19),

$$a_n = \frac{2}{\pi} \int_0^\pi f(x) \cos nx \, dx \qquad (n = 0, 1, 2, \dots),$$

$$b_n = 0 \qquad (n = 1, 2, \dots). \tag{7.20}$$

Similarly, if f is odd,

$$a_n = 0, \qquad b_n = \frac{2}{\pi} \int_0^\pi f(x) \sin nx \, dx. \tag{7.21}$$

We have thus the expansions (for a function piecewise very smooth):

$$f(x) = \frac{a_0}{2} + \sum_{n=1}^{\infty} a_n \cos nx \qquad (f\ \text{even}),$$

$$a_n = \frac{2}{\pi} \int_0^\pi f(x) \cos nx \, dx \qquad\qquad (7.22)$$

and

$$f(x) = \sum_{n=1}^{\infty} b_n \sin nx \qquad (f\ \text{odd}),$$

$$b_n = \frac{2}{\pi} \int_0^\pi f(x) \sin nx \, dx. \qquad\qquad (7.23)$$

Now (7.22) uses only the values of $f(x)$ between $x = 0$ and $x = \pi$. Hence for any function $f(x)$ given only over this interval, one can form the series (7.22). This is called the *Fourier cosine series* of $f(x)$. It follows from the fundamental theorem that the series will converge to $f(x)$ for $0 \le x \le \pi$ and outside this interval to the even periodic function that coincides with $f(x)$ for $0 \le x \le \pi$. This is illustrated in Fig. 7.8.

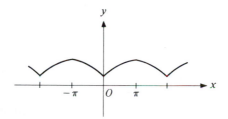

Figure 7.8   Even periodic extension of function defined between 0 and π.

In the same way, (7.23) defines the *Fourier sine series* of a function $f(x)$ defined only between 0 and π. The series represents an odd periodic function that coincides with $f(x)$ for $0 < x < \pi$, as illustrated in Fig. 7.9.

Figure 7.9   Odd periodic extension of function defined between 0 and π.

EXAMPLE Let $f(x) = \pi - x$. Then one can represent $f(x)$ by a Fourier series over the interval $-\pi < x < \pi$. The formulas (7.9) give

$$a_0 = \frac{1}{\pi} \int_{-\pi}^{\pi} (\pi - x)\, dx = 2\pi, \qquad a_1 = a_2 = \cdots = 0,$$

$$b_n = \frac{1}{\pi} \int_{-\pi}^{\pi} (\pi - x)\sin nx\, dx = \frac{2(-1)^n}{n}.$$

Hence one has

$$\pi - x = \pi + 2 \sum_{n=1}^{\infty} \frac{(-1)^n \sin nx}{n}, \qquad -\pi < x < \pi. \qquad \text{(a)}$$

The same function, $\pi - x$, can be represented by a Fourier cosine series over the interval $0 \le x \le \pi$. The formulas (7.22) give

$$a_0 = \frac{2}{\pi} \int_0^{\pi} (\pi - x)\, dx = \pi,$$

$$a_n = \frac{2}{\pi} \int_0^{\pi} (\pi - x)\cos nx\, dx = \frac{2}{\pi n^2}(1 - (-1)^n), \qquad n = 1, 2, \dots .$$

Hence one has

$$\pi - x = \frac{\pi}{2} + \frac{2}{\pi} \sum_{n=1}^{\infty} \frac{1 - (-1)^n}{n^2} \cos nx$$

$$= \frac{\pi}{2} + \frac{2}{\pi}\left(2\cos x + \frac{2\cos 3x}{3^2} + \frac{2\cos 5x}{5^2} + \cdots\right), \qquad 0 \le x \le \pi. \quad \text{(b)}$$

Finally, the same function, $\pi - x$, can be represented by a Fourier sine series over the interval $0 < x < \pi$. The formulas (7.23) give

$$\pi - x = 2 \sum_{n=1}^{\infty} \frac{\sin nx}{n}, \qquad 0 < x < \pi. \qquad \text{(c)}$$

Figure 7.10 shows the graphs of the three functions represented by the series (a), (b), (c). ■

Change of period. If $f(x)$ has period p:

$$f(x + p) = f(x) \qquad (p \ne 0),$$

then the substitution

$$x = \frac{p}{2\pi} t$$

transforms $f(x)$ into a function $g(t)$:

$$g(t) = f\left(\frac{p}{2\pi} t\right),$$

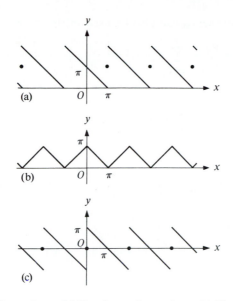

Figure 7.10   (a) Fourier series.   (b) Fourier cosine series.   (c) Fourier sine series.

and $g(t)$ has period 2π, since

$$g(t + 2\pi) = f\left[\frac{p}{2\pi}(t + 2\pi)\right]$$

$$= f\left(\frac{pt}{2\pi} + p\right) = f\left(\frac{pt}{2\pi}\right) = g(t).$$

The change from x to t is simply a change of scale. Since g has period 2π, one has a Fourier series for g (assumed piecewise very smooth):

$$g(t) = \frac{a_0}{2} + \sum_{n=1}^{\infty} (a_n \cos nt + b_n \sin nt),$$

where, for example,

$$a_n = \frac{1}{\pi}\int_{-\pi}^{\pi} g(t) \cos nt \, dt, \qquad b_n = \frac{1}{\pi}\int_{-\pi}^{\pi} g(t) \sin nt \, dt.$$

If now t is replaced by $(2\pi/p)x$, one finds a Fourier series for $f(x)$:

$$f(x) = \frac{a_0}{2} + \sum_{n=1}^{\infty}\left[a_n \cos\left(n \cdot \frac{2\pi}{p}x\right) + b_n \sin\left(n \cdot \frac{2\pi}{p}x\right)\right]. \qquad (7.24)$$

The coefficients a_n, b_n can be expressed directly in terms of $f(x)$. Thus:

$$a_n = \frac{1}{L}\int_{-L}^{L} f(x) \cos\left(n \cdot \frac{2\pi}{p}x\right) dx,$$

$$b_n = \frac{1}{L}\int_{-L}^{L} f(x) \sin\left(n \cdot \frac{2\pi}{p}x\right) dx, \qquad (7.25)$$

where $p = 2L$.

The Fourier cosine series can also be used in this case. One finds

$$f(x) = \frac{a_0}{2} + \sum_{n=1}^{\infty} a_n \cos\left(n \cdot \frac{2\pi}{p} x\right), \qquad 0 \le x \le L, \tag{7.26}$$

where

$$a_n = \frac{2}{L} \int_0^L f(x) \cos\left(n \cdot \frac{2\pi}{p} x\right) dx. \tag{7.27}$$

Similarly, $f(x)$ has a Fourier sine series:

$$f(x) = \sum_{n=1}^{\infty} b_n \sin\left(n \cdot \frac{2\pi}{p} x\right), \qquad 0 < x < L, \tag{7.28}$$

where

$$b_n = \frac{2}{L} \int_0^L f(x) \sin\left(n \cdot \frac{2\pi}{p} x\right) dx. \tag{7.29}$$

EXAMPLE Let $f(x) = 2x + 1$. Then $f(x)$ can be represented by a Fourier series over the interval $0 < x < 2$. Here $p = 2$ and $t = \pi x$, so that

$$f(x) = 2x + 1 = \frac{2t}{\pi} + 1 = g(t),$$

where $g(t)$ is defined for $0 \le t \le 2\pi$. Hence $g(t)$ can be represented by a Fourier series:

$$g(t) = \frac{a_0}{2} + \sum_{n=1}^{\infty} (a_n \cos nt + b_n \sin nt),$$

where

$$a_n = \frac{1}{\pi} \int_0^{2\pi} \left(\frac{2t}{\pi} + 1\right) \cos nt \, dt, \qquad b_n = \frac{1}{\pi} \int_0^{2\pi} \left(\frac{2t}{\pi} + 1\right) \sin nt \, dt.$$

One finds

$$a_0 = 6, \qquad a_1 = a_2 = \cdots = 0, \qquad b_n = -\frac{4}{n\pi},$$

so that

$$g(t) = 3 - \frac{4}{\pi} \sum_{n=1}^{\infty} \frac{\sin nt}{n}$$

and hence

$$f(x) = g(\pi x) = 3 - \frac{4}{\pi} \sum_{n=1}^{\infty} \frac{\sin n\pi x}{n}.$$

One could use the formulas (7.25) directly, but the change to t simplifies the calculations. ∎

PROBLEMS

1. Let $f(x) = 2x + 1$.

   a) Expand $f(x)$ in a Fourier series for $-\pi < x < \pi$.

   b) Expand $f(x)$ in a Fourier series for $0 < x < 2\pi$.

   c) Expand $f(x)$ in a Fourier cosine series for $0 \leq x \leq \pi$.

   d) Expand $f(x)$ in a Fourier sine series for $0 < x < \pi$.

   e) Expand $f(x)$ in a Fourier series for $0 < x < \pi$.

   f) Graph the functions represented by the series of parts (a), (b), (c), (d), and (e).

2. Let $f(x) = x^2$.

   a) Expand $f(x)$ in a Fourier series for $\pi < x < 3\pi$.

   b) Expand $f(x)$ in a Fourier series for $1 < x < 2$.

   c) Graph the functions represented by the series of parts (a) and (b).

3. Let $f(x) = \sin x$.

   a) Expand $f(x)$ in a Fourier series for $0 \leq x \leq 2\pi$.

   b) Expand $f(x)$ in a Fourier series for $0 \leq x \leq \pi$.

   c) Expand $f(x)$ in a Fourier cosine series for $0 \leq x \leq \pi$.

4. Let $f(x) = x$.

   a) Expand $f(x)$ in a Fourier cosine series for $0 \leq x \leq \pi$.

   b) Expand $f(x)$ in a Fourier sine series for $0 \leq x < \pi$.

   c) Expand $f(x)$ in a Fourier cosine series for $0 \leq x \leq 1$.

   d) Expand $f(x)$ in a Fourier sine series for $0 \leq x < 1$.

5. Let $f(x) = \dfrac{a_0}{2} + \sum\limits_{n=1}^{\infty} (a_n \cos nx + b_n \sin nx)$, where the series converges uniformly for all x. State what conclusions can be drawn concerning the coefficients a_n, b_n from each of the following properties of $f(x)$:

   a) $f(-x) = f(x)$

   b) $f(-x) = -f(x)$

   c) $f(\pi - x) = f(x)$

   d) $f\left(\dfrac{\pi}{2} - x\right) = f(x)$

   e) $f(-x) = f(x) = f\left(\dfrac{\pi}{2} - x\right)$

   f) $f(\pi - x) = -f(x)$

   g) $f(\pi + x) = f(x)$

   h) $f\left(\dfrac{\pi}{2} + x\right) = f(x)$

   i) $f\left(\dfrac{\pi}{3} + x\right) = f(x)$

   j) $f(x) = f(2x)$

   [Hint: Use the corollary to Theorem 1.]

7.6 REMARKS ON APPLICATIONS OF FOURIER SERIES

As was indicated in Section 7.1, the natural field of application of Fourier series is to periodic phenomena. The fact that a periodic function $f(t)$ can be resolved into its simple harmonic components $A_n \sin(nt + \alpha_n)$ is of fundamental physical significance. For all "linear" problems this resolution permits one to reduce the problem to the simpler one of a single simple harmonic vibration and then to build up the general case by addition (superposition) of the simple ones.

The concrete application of Fourier series to such problems takes two main forms: A periodic function $f(t)$ may be given in graphical or tabulated form; an understanding of the physical mechanism leading to such a function requires a "harmonic analysis" of $f(t)$, or representation of $f(t)$ as a Fourier series. Second, the function $f(t)$ is known to be periodic and is known to satisfy some implicit relation, such as a differential equation; it is desired to determine $f(t)$ as a Fourier series on the basis of this information.

The first problem is thus one of *interpretation* of experimental data; the second is one of *prediction* of the result of an experiment, on the basis of a mathematical theory.

Although the application of Fourier series to periodic phenomena is basic, there is a much wider field of application. As has been shown, an "arbitrary" function $f(x)$, given for $a \le x \le b$, has a representation as a Fourier series over that interval. Thus in any problem concerning a function over an interval it may be advantageous to represent the function by the corresponding series. This permits a tremendous variety of applications. As before, the applications usually take the form of interpreting given data or of predicting functions satisfying given conditions.

In this book, particular applications will be considered in Chapter 10 on partial differential equations. In the following problems, several illustrations of the form of solutions to differential equations are provided.

PROBLEMS

1. Show that the linear differential equation

$$\frac{d^2y}{dt^2} + 4\frac{dy}{dt} + y = p\cos\omega t + q\sin\omega t$$

is satisfied by

$$y = A\cos\omega t + B\sin\omega t,$$

where

$$A = \frac{p(1-\omega^2) - 4\omega q}{(1-\omega^2)^2 + 16\omega^2}, \qquad B = \frac{4\omega p + q(1-\omega^2)}{(1-\omega^2)^2 + 16\omega^2},$$

and determine a solution, in series form, of the differential equation

$$\frac{d^2y}{dt^2} + 4\frac{dy}{dt} + y = f(t) = \sum_{n=1}^{\infty}(a_n\cos nt + b_n\sin nt).$$

2. Show that, granting the correctness of the necessary term-by-term differentiations of

series, the function

$$f(x,t) = \sum_{n=1}^{\infty} [A_n \cos nct + B_n \sin nct] \sin nx,$$

where the A_n and B_n are constants, satisfies the partial differential equation

$$\frac{\partial^2 f}{\partial t^2} = c^2 \frac{\partial^2 f}{\partial x^2}.$$

The differential equation is that of the vibrating string, and the series represents the general solution when the ends are fixed, π units apart.

3. Show that, granting the correctness of the necessary term-by-term differentiations of series, the function

$$f(r,\theta) = A_0 + \sum_{n=1}^{\infty} (A_n r^n \cos n\theta + B_n r^n \sin n\theta)$$

satisfies the Laplace equation in polar coordinates:

$$\frac{\partial^2 f}{\partial r^2} + \frac{1}{r^2}\frac{\partial^2 f}{\partial \theta^2} + \frac{1}{r}\frac{\partial f}{\partial r} = 0.$$

As was shown in Section 5.15, this equation describes equilibrium temperature distributions, electrostatic potentials, and velocity potentials. Every function f that is harmonic in a circular domain $r < R$ can be represented by such a series, as will be shown in Section 9.11.

7.7   UNIQUENESS THEOREM

If two functions $f(x)$ and $f_1(x)$ have the same set of Fourier coefficients:

$$\int_{-\pi}^{\pi} f(x) \cos nx \, dx = \int_{-\pi}^{\pi} f_1(x) \cos nx \, dx, \qquad (n = 0,1,2,\dots),$$

$$\int_{-\pi}^{\pi} f(x) \sin nx \, dx = \int_{-\pi}^{\pi} f_1(x) \sin nx \, dx, \qquad (n = 1,2,\dots), \qquad (7.30)$$

are the functions necessarily identical? In other words, is a function *uniquely* *determined* by its Fourier coefficients? The answer is in the affirmative:

THEOREM 3 (*Uniqueness theorem*)   Let $f(x)$ and $f_1(x)$ be piecewise continuous in the interval $-\pi \le x \le \pi$ and satisfy (7.30), so that the two functions have the same Fourier coefficients. Then $f(x) = f_1(x)$ except perhaps at points of discontinuity.

Proof.   Let $h(x) = f(x) - f_1(x)$. Then $h(x)$ is piecewise continuous, and from (7.30) it at once follows that all Fourier coefficients of $h(x)$ are 0. We then show that $h(x) = 0$ except perhaps at discontinuity points.

Let us suppose $h(x_0) \ne 0$ at a point of continuity x_0, for example, $h(x_0) = 2c > 0$. Then, by continuity, $h(x) > c$ for $|x - x_0| < \delta$ and δ sufficiently small (cf. Problem 7 following Section 2.4). We can assume $-\pi < x_0 < \pi$.

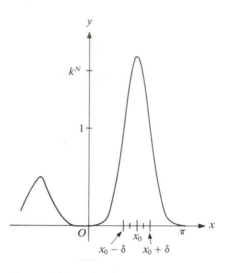

Figure 7.11   Pulse function.

We now achieve a contradiction by showing that there exists a "trigonometric polynomial"

$$P(x) = p_0 + p_1 \cos x + p_2 \sin x + \cdots + p_{2k-1} \cos kx + p_{2k} \sin kx$$

that represents a "pulse" at x_0 of arbitrarily large amplitude and arbitrarily small width. This is pictured in Fig. 7.11. If such a pulse can be constructed, then one has a contradiction. On one hand,

$$\int_{-\pi}^{\pi} h(x)P(x)\, dx = p_0 \int_{-\pi}^{\pi} h(x)\, dx + p_1 \int_{-\pi}^{\pi} h(x) \cos x\, dx + \cdots = 0.$$

On the other hand, the major portion of the integral $\int h(x)P(x)\, dx$ is concentrated in the interval in which the pulse occurs; here $h(x)$ is positive, and $P(x)$ is large and positive. Hence the integral is positive and cannot be 0.

To make this precise, we take

$$P(x) = [\psi(x)]^N, \qquad \psi(x) = 1 + \cos(x - x_0) - \cos \delta$$

for an appropriate positive integer N. Since the functions $\sin^n x$ and $\cos^n x$ are expressible as trigonometric polynomials (Problem 8 following Section 7.4), the function $P(x)$ is a trigonometric polynomial. Let

$$k = \psi\left(x_0 + \frac{\delta}{2}\right) = 1 + \cos\frac{\delta}{2} - \cos \delta.$$

Then $k > 1$ and $P \geq k^N$ for $|x - x_0| \leq \frac{1}{2}\delta$. Since ψ is positive (greater than 1) for $\frac{1}{2}\delta \leq |x - x_0| < \delta$, P is positive in this range. On the other hand, $|\psi(x)| < 1$ for $-\pi \leq x < x_0 - \delta$ and for $x_0 + \delta < x \leq \pi$, so that $|P| < 1$ in this range.

Now the function $h(x)$, being piecewise continuous, is bounded by a constant M for $-\pi \leq x \leq \pi$: $|h(x)| \leq M$. It follows from the properties of $P(x)$ just listed that $P(x)h(x) > -M$ for $-\pi \leq x \leq x_0 - \frac{1}{2}\delta$ and for $x_0 + \frac{1}{2}\delta \leq x \leq \pi$, while $P(x)h(x) \geq ck^N$ for $x_0 - \frac{1}{2}\delta \leq x \leq x_0 + \frac{1}{2}\delta$. Accordingly, by rule (4.17) of Section 4.1,

$$\int_{-\pi}^{\pi} P(x)h(x)\,dx = \int_{-\pi}^{x_0-\frac{1}{2}\delta} P(x)h(x)\,dx + \int_{x_0+\frac{1}{2}\delta}^{\pi} P(x)h(x)\,dx$$

$$+ \int_{x_0-\frac{1}{2}\delta}^{x_0+\frac{1}{2}\delta} P(x)h(x)\,dx > -M(2\pi - \delta) + ck^N\delta.$$

Since $k^N \to +\infty$ as $N \to \infty$, the right-hand member of the inequality is surely positive when N is sufficiently large. Accordingly, the left-hand member is positive for appropriate choice of N. This contradicts the fact that the left-hand member is 0. Accordingly, $h(x) = f(x) - f_1(x) = 0$ wherever $f(x)$ and $f_1(x)$ are continuous. $\qquad\square$

Remarks. The uniqueness theorem can be looked at in another way, namely, as asserting that the system of functions

$$1, \quad \cos x, \quad \sin x, \ldots, \cos nx, \quad \sin nx, \ldots$$

is "large enough," that is, that there are *enough* functions in this system to construct series for all the periodic functions envisaged. It should be noted that omission of any *one* function of the system would destroy this property. Thus if $\cos x$ were omitted, one could still form a series

$$\tfrac{1}{2}a_0 + b_1 \sin x + a_2 \cos 2x + b_2 \sin 2x + \cdots$$

as before. But there are very smooth periodic functions whose Fourier series in this deficient form could never converge to the function, namely, all functions $A\cos x$ for $A = \text{const.} \neq 0$. For each such function would have all coefficients 0:

$$\int_{-\pi}^{\pi} A \cos x\,dx = 0, \qquad \int_{-\pi}^{\pi} A \cos x \sin x\,dx = 0, \qquad \ldots .$$

The series reduces to 0 and cannot represent the function. The essence of the proof of the uniqueness theorem is the demonstration that there are *enough* functions in the system of sines and cosines to construct a pulse function $P(x)$.

THEOREM 4 Let the function $f(x)$ be continuous for $-\pi \leq x \leq \pi$ and let the Fourier series of $f(x)$ converge uniformly in this interval. Then the series converges to $f(x)$ for $-\pi \leq x \leq \pi$.

Proof. Let the sum of the Fourier series of $f(x)$ be denoted by $f_1(x)$:

$$f_1(x) = \tfrac{1}{2}a_0 + \sum_{n=1}^{\infty} (a_n \cos nx + b_n \sin nx).$$

Since the series converges uniformly, it follows from Theorem 1 that $f_1(x)$ is continuous and that a_n, b_n are the Fourier coefficients of $f_1(x)$. But the series was given as the Fourier series of $f(x)$. Hence $f(x)$ and $f_1(x)$ have the same Fourier coefficients, and by Theorem 3, $f(x) \equiv f_1(x)$; that is, $f(x)$ is the sum of its Fourier series for $-\pi \leq x \leq \pi$. ☐

7.8 PROOF OF FUNDAMENTAL THEOREM FOR CONTINUOUS, PERIODIC, AND PIECEWISE VERY SMOOTH FUNCTIONS

THEOREM 5   Let $f(x)$ be continuous and piecewise very smooth for all x and let $f(x)$ have period 2π. Then the Fourier series of $f(x)$ converges uniformly to $f(x)$ for all x.

Proof.   Let us first assume that $f(x)$ has continuous first and second derivatives for all x. One has (for $n \neq 0$)

$$a_n = \frac{1}{\pi} \int_{-\pi}^{\pi} f(x) \cos nx \, dx = \frac{f(x) \sin nx}{n\pi} \Big|_{-\pi}^{\pi} - \frac{1}{n\pi} \int_{-\pi}^{\pi} f'(x) \sin nx \, dx$$

by integration by parts. The first term on the right is zero. A second integration by parts gives

$$a_n = \frac{f'(x) \cos nx}{n^2\pi} \Big|_{-\pi}^{\pi} - \frac{1}{n^2\pi} \int_{-\pi}^{\pi} f''(x) \cos nx \, dx = -\frac{1}{n^2\pi} \int_{-\pi}^{\pi} f''(x) \cos nx \, dx,$$

the first term being zero because of the periodicity of $f'(x)$. The function $f''(x)$ is continuous in the interval $-\pi \leq x \leq \pi$, and hence $|f''(x)| \leq M$ for an appropriate constant M. One concludes that

$$|a_n| = \left| \frac{1}{n^2\pi} \int_{-\pi}^{\pi} f''(x) \cos nx \, dx \right| \leq \frac{2M}{n^2} \qquad (n = 1, 2, \ldots).$$

In exactly the same way we prove that $|b_n| \leq 2M/n^2$ for all n. Hence each term of the Fourier series of $f(x)$ is in absolute value at most equal to the corresponding term of the convergent series

$$\tfrac{1}{2}|a_0| + \frac{2M}{1} + \frac{2M}{1} + \frac{2M}{2^2} + \frac{2M}{2^2} + \cdots.$$

Application of the Weierstrass M-test (Section 6.13) now establishes that the Fourier series converges uniformly for all x. By Theorem 4 the sum is $f(x)$.

We now consider the case of a function $f(x)$ that is periodic, continuous, and piecewise very smooth. The only step that requires reexamination is the proof that $|a_n| \leq 2Mn^{-2}$ and that $|b_n| \leq 2Mn^{-2}$. The integration by parts must now be carried out separately over each interval within which $f''(x)$ is continuous. If the results are added, one obtains, for example, for b_n:

$$b_n = \left[\frac{-f(x) \cos nx}{n\pi} \Big|_{-\pi}^{x_1} + \frac{-f(x) \cos nx}{n\pi} \Big|_{x_1}^{x_2} + \cdots \right]$$

$$+ \left[\frac{f'(x) \sin nx}{n^2\pi} \Big|_{-\pi}^{x_1-} + \cdots \Big|_{x_1+}^{x_2} + \cdots \right] - \frac{1}{n^2\pi} \int_{-\pi}^{\pi} f''(x) \sin nx \, dx.$$

The integrals are technically improper but exist as such. Since f is continuous and periodic, the terms in the first bracket add up to

$$\left| \frac{-f(x)\cos nx}{n\pi} \right|_{-\pi}^{\pi} = 0.$$

The functions $f'(x)$ and $f''(x)$ are bounded in each subinterval. Hence one constant M can be chosen so that $|f'(x)| \leq M$ and $|f''(x)| \leq M$ throughout. If there are k subintervals, we conclude that

$$|b_n| \leqq \frac{2kM}{n^2\pi} + \frac{2M}{n^2} = \frac{M_1}{n^2}.$$

A similar result holds for a_n. Hence *the Fourier series of a periodic, continuous, piecewise very smooth function converges uniformly to the function for all x.*   □

7.9   PROOF OF FUNDAMENTAL THEOREM

Before proceeding to the general case of a function with jump discontinuities we consider an example illustrating the result just obtained. Let

$$G(x) = \frac{\pi}{2} - \frac{x}{2} - \frac{x^2}{4\pi}, \qquad -\pi \leqq x \leqq 0,$$

$$G(x) = \frac{\pi}{2} + \frac{x}{2} - \frac{x^2}{4\pi}, \qquad 0 \leqq x \leqq \pi,$$

and let G be repeated periodically outside this interval, as shown in Fig. 7.12. The resulting function $G(x)$ is continuous for all x and is piecewise smooth. Its Fourier series is found in Problem 1(f) following Section 7.4 to be the series

$$\frac{2\pi}{3} - \frac{1}{\pi} \sum_{n=1}^{\infty} \frac{\cos nx}{n^2}.$$

Hence $|a_n| \leq Mn^{-2}$ as asserted, with $M = 1/\pi$; the b_n happen to be 0. By Theorem 5 this series converges uniformly to $G(x)$.

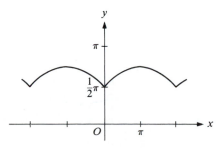

Figure 7.12   Auxiliary function $G(x)$.

We now ask: Is term-by-term differentiation of the series permissible? In other words, is

$$G'(x) = \frac{1}{\pi} \sum_{n=1}^{\infty} \frac{\sin nx}{n} \qquad (7.31)$$

wherever $G'(x)$ is defined? By Theorem 33 of Section 6.14 this is correct if x lies within an interval within which the differentiated series converges uniformly. It is shown in Problem 4 below that the series $\Sigma(\sin nx/n)$ converges uniformly for $a \le |x| \le \pi$, provided that $a > 0$. Accordingly, (7.31) *is correct for* $-\pi \le x \le \pi$, *except for* $x = 0$. Now let $F(x)$ be the periodic function of period 2π such that $F(0) = 0$ and

$$F(x) = G'(x) = \begin{cases} -\tfrac{1}{2} - (x/2\pi), & -\pi \le x < 0, \\ \tfrac{1}{2} - (x/2\pi), & 0 < x \le \pi. \end{cases}$$

The function $F(x)$ is shown in Fig. 7.13. We have now proved that

$$F(x) = \frac{1}{\pi} \sum_{n=1}^{\infty} \frac{\sin nx}{n}$$

for all x, the convergence being uniform for $0 < a \le |x| \le \pi$. The series on the right-hand side was computed as the Fourier series of $F(x)$ in Problem 1(e) following Section 7.4. Accordingly, we have shown that $F(x)$ is represented by its Fourier series for all x. The remarkable feature of this result is that $F(x)$ has a

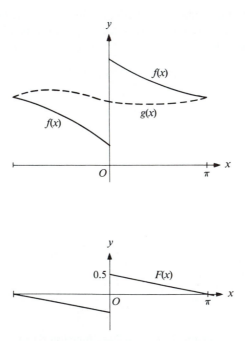

Figure 7.13   Removal of jump discontinuity.

jump, from $-\frac{1}{2}$ to $\frac{1}{2}$ at $x = 0$. The series converges to the average value $F(0) = 0$. We have therefore verified a special case of the following theorem:

THEOREM 6   Let $f(x)$ be defined and piecewise very smooth for $-\pi \le x \le \pi$ and let $f(x)$ be defined outside this interval in such a manner that $f(x)$ has period 2π. Then the Fourier series of $f(x)$ converges uniformly to $f(x)$ in each closed interval containing no discontinuity of $f(x)$. At each discontinuity x_0 the series converges to

$$\frac{1}{2}\left[\lim_{x \to x_0+} f(x) + \lim_{x \to x_0-} f(x) \right].$$

Proof. For convenience we redefine $f(x)$ at each discontinuity x_0 as the average of left and right limit values. Let us suppose, for example, that the only discontinuity is at $x = 0$ (and the points $2k\pi$, $k = \pm 1, \pm 2, \ldots$) as in Fig. 7.13. Let

$$\lim_{x \to 0+} f(x) - \lim_{x \to 0-} f(x) = s,$$

so that s is precisely the "jump." We then proceed to eliminate the discontinuity by subtracting from $f(x)$ the function $sF(x)$, where $F(x)$ is the function defined earlier. Since $sF(x)$ has also the jump s at $x = 0$ (and $x = 2k\pi$), $g(x) = f(x) - sF(x)$ has jump 0 at $x = 0$ and is continuous for all x; for

$$\lim_{x \to 0-} g(x) = \lim_{x \to 0-} f(x) - s \lim_{x \to 0-} F(x)$$

$$= [f(0) - \tfrac{1}{2}s] + \tfrac{1}{2}s = f(0) = g(0),$$

and a similar statement applies to the right-hand limit. Since $F(x)$ is piecewise linear, $g(x)$ is continuous and piecewise very smooth for all x and has period 2π. Hence by Theorem 5, $g(x)$ is representable by a uniformly convergent Fourier series for all x:

$$g(x) = \tfrac{1}{2}A_0 + \sum_{n=1}^{\infty} (A_n \cos nx + B_n \sin nx).$$

Accordingly,

$$f(x) = g(x) + sF(x) = \frac{1}{2}A_0 + \sum_{n=1}^{\infty} (A_n \cos nx + B_n \sin nx) + \frac{s}{\pi} \sum_{n=1}^{\infty} \frac{\sin nx}{n}$$

$$= \frac{1}{2}A_0 + \sum_{n=1}^{\infty} \left[A_n \cos nx + \left(B_n + \frac{s}{n\pi} \right) \sin nx \right],$$

so that $f(x)$ is represented by a trigonometric series for all x. The series is the Fourier series of $f(x)$, for

$$b_n = \frac{1}{\pi} \int_{-\pi}^{\pi} f(x) \sin nx \, dx = \frac{1}{\pi} \int_{-\pi}^{\pi} [g(x) + sF(x)] \sin nx \, dx$$

$$= \frac{1}{\pi} \int_{-\pi}^{\pi} g(x) \sin nx \, dx + \frac{s}{\pi} \int_{-\pi}^{\pi} F(x) \sin nx \, dx = B_n + \frac{s}{n\pi},$$

and similarly, $a_n = A_n$. Therefore the Fourier series of $f(x)$ converges to $f(x)$ for all x. At $x = 0$ the series converges to $f(0)$, which was defined to be the average of left and right limits at $x = 0$. Since the series for $g(x)$ is uniformly convergent for all x, while the series for $F(x)$ converges uniformly in each closed interval not containing $x = 0$ (or $x = 2k\pi$), the Fourier series of $f(x)$ converges uniformly in each such closed interval.

The theorem has now been proved for the case of just one jump discontinuity. If there are several jumps, at points $x_1, x_2, \ldots$, we simply remove them by subtracting from $f(x)$ the function

$$s_1 F(x - x_1) + s_2 F(x - x_2) + \cdots.$$

The resulting function $g(x)$ is again continuous and piecewise very smooth, so that the same conclusion holds. Thus the theorem is established in all generality. $\square$

Remarks. The proof just given uses the *principle of superposition*: The Fourier series of a linear combination of two functions is the same linear combination of the corresponding two series. This can be put to very good use to systematize the computation of Fourier series. Illustrations are given in Problem 1 which follows.

The idea of subtracting off the series corresponding to a jump discontinuity also has a practical significance. If a function $f(x)$ is defined by its Fourier series and is not otherwise explicitly known, one can, of course, use the series to tabulate the function. If $f(x)$ has a jump discontinuity, the convergence will be poor near the discontinuity; this will reveal itself in the presence of terms having coefficients approaching 0 like $1/n$. If the discontinuity x_1 and jump s_1 are known, as is often the case, one can subtract the corresponding function $s_1 F(x - x_1)$ as before; the new series will converge much more rapidly.

The same idea can be applied to functions $f(x)$ that are continuous but for which $f'(x)$ has a jump discontinuity s_1 at x_1. One now subtracts from f the function $s_1 G(x - x_1)$, for this continuous function has as derivative precisely the function $s_1 F(x - x_1)$, with jump s_1 at x_1. By integrating $G(x) - 2\pi/3$, one obtains a periodic continuous function having a jump in its second derivative at $x = 0$. Continuing in this way, one provides a jump function for each derivative; each such function can be used to remove corresponding slowly converging terms from the Fourier series.

PROBLEMS

1. Let $f_1(x)$ and $f_2(x)$ be defined by the equations:

$$f_1(x) = 0, \quad -\pi \leq x < 0; \quad f_1(x) = 1, \quad 0 \leq x \leq \pi;$$
$$f_2(x) = 0, \quad -\pi \leq x < 0; \quad f_2(x) = x, \quad 0 \leq x < \pi.$$

Then $f_1(x)$ and $f_2(x)$ can be represented by Fourier series:

$$f_1(x) = \frac{1}{2} + \frac{2}{\pi} \sum_{n=1}^{\infty} \frac{\sin (2n - 1) x}{2n - 1}, \qquad 0 < |x| < \pi;$$

$$f_2(x) = \frac{\pi}{4} - \frac{2}{\pi} \sum_{n=1}^{\infty} \frac{\cos (2n - 1) x}{(2n - 1)^2} - \sum_{n=1}^{\infty} (-1)^n \frac{\sin nx}{n}, \qquad -\pi < x < \pi.$$

Without further integration, find the Fourier series for the following functions:

a) $f_3(x) = 1, \quad -\pi \le x < 0; \quad f_3(x) = 0, \quad 0 \le x \le \pi;$

b) $f_4(x) = x, \quad -\pi \le x \le 0; \quad f_4(x) = 0, \quad 0 \le x \le \pi;$

c) $f_5(x) = 1, \quad -\pi \le x < 0; \quad f_5(x) = x, \quad 0 \le x \le \pi;$

d) $f_6(x) = 2, \quad -\pi \le x < 0; \quad f_6(x) = 0, \quad 0 \le x \le \pi;$

e) $f_7(x) = 2, \quad -\pi \le x < 0; \quad f_7(x) = 3, \quad 0 \le x \le \pi;$

f) $f_8(x) = 1, \quad -\pi \le x \le 0; \quad f_8(x) = 1 + 2x, \quad 0 \le x \le \pi;$

g) $f_9(x) = a + bx, \quad -\pi \le x < 0; \quad f_9(x) = c + dx, \quad 0 \le x \le \pi.$

2. Let $f(x) = \dfrac{a_0}{2} + \displaystyle\sum_{n=1}^{\infty} (a_n \cos nx + b_n \sin nx), \ -\pi \le x \le \pi$. By squaring this series and integrating (assuming the operations are permitted) show that

$$\frac{1}{\pi} \int_{-\pi}^{\pi} [f(x)]^2 \, dx = \frac{a_0^2}{2} + \sum_{n=1}^{\infty} (a_n^2 + b_n^2).$$

[This relation, known as *Parseval's equation*, can be justified for the most general function $f(x)$ considered before. See Sections 7.11 and 7.12.]

3. Use the result of Problem 2 and the Fourier series found in previous problems to establish the formulas:

a) $\dfrac{\pi^2}{8} = \dfrac{1}{1^2} + \dfrac{1}{3^2} + \cdots + \dfrac{1}{(2n - 1)^2} + \cdots,$

b) $\dfrac{\pi^2}{6} = \dfrac{1}{1^2} + \dfrac{1}{2^2} + \cdots + \dfrac{1}{n^2} + \cdots,$

c) $\dfrac{\pi^4}{90} = \dfrac{1}{1^4} + \dfrac{1}{2^4} + \cdots + \dfrac{1}{n^4} + \cdots,$

d) $\dfrac{\pi^6}{945} = \dfrac{1}{1^6} + \dfrac{1}{2^6} + \cdots + \dfrac{1}{n^6} + \cdots.$

4. Prove that the series

$$\sum_{n=1}^{\infty} \frac{\sin nx}{n} = \sin x + \frac{\sin 2x}{2} \cdots + \frac{\sin nx}{n} + \cdots$$

converges uniformly in each interval $-\pi \le x \le a, \ a \le x \le \pi$, provided that $a > 0$. This can be established by the following procedure:

a) Let $p_n(x) = \sin x + \cdots + \sin nx$. Prove the identity:

$$p_n(x) = \frac{\cos \frac{1}{2}x - \cos \left(n + \frac{1}{2}\right) x}{2 \sin \frac{1}{2}x},$$

$(x \neq 0, \ x \neq \pm 2\pi,\dots)$. [Hint: Multiply $p_n(x)$ by $\sin \frac{1}{2}x$ and apply the identity (7.7) for $\sin x \sin y$ to each term of the result.]

b) Show that if $a > 0$ and $a \leq |x| \leq \pi$, then $|p_n(x)| \leq 1/\sin \frac{1}{2}a$.

c) Show that the nth partial sum $S_n(x)$ of the series

$$\sin x + \frac{\sin 2x}{2} + \cdots + \frac{\sin nx}{n} + \cdots$$

can be written as follows

$$S_n(x) = \frac{p_1(x)}{1 \cdot 2} + \frac{p_2(x)}{2 \cdot 3} + \cdots + \frac{p_{n-1}(x)}{n(n-1)} + \frac{p_n(x)}{n}.$$

[Hint: Write $\sin x = p_1$, $\sin 2x = p_2 - p_1$, and so on.]

d) Show that the series $\displaystyle\sum_{n=1}^{\infty} \frac{\sin nx}{n}$ is uniformly convergent for $a \leq |x| \leq \pi$, where $a > 0$. [Hint: By (c),

$$S_n(x) = S_n^*(x) + \frac{p_n(x)}{n}.$$

Hence uniform convergence of the sequences S_n^* and p_n/n implies uniform convergence of $S_n(x)$. The sequence S_n converges uniformly, since it is the nth partial sum of the series

$$\sum_{n=1}^{\infty} \frac{p_n(x)}{n(n+1)},$$

which converges uniformly, because of (b), by the M-test. The sequence p_n/n converges uniformly to 0 by (b).]

7.10 ORTHOGONAL FUNCTIONS

If one reviews the theory of expansion of a function in a Fourier series, as presented in the preceding sections of this chapter, it is natural to ask why the trigonometric functions $\sin nx$ and $\cos nx$ play such a special part and whether these functions can be replaced by other functions. If one is interested only in periodic functions, there is indeed no natural alternative. However, if one is concerned with representation of a function over a given interval, it will be seen that a great variety of other series is available, in particular, series of Legendre polynomials, Bessel functions, Laguerre polynomials, Jacobi polynomials, Hermite polynomials, and general Sturm-Liouville series.

Let $f(x)$ be given in an interval $a \leq x \leq b$; this interval will be fixed throughout the following discussion. Let $\phi_1(x), \phi_2(x),\dots,\phi_n(x),\dots$ be functions all piecewise continuous in this interval; this sequence is to replace the system of sines and cosines. We then postulate a development,

$$f(x) = \sum_{n=1}^{\infty} c_n \phi_n(x), \tag{7.32}$$

just as in the case of Fourier series. Our next step, for Fourier series, was to multiply both sides by $\cos mx$ or $\sin mx$ and integrate from $-\pi$ to π; when we did this, all terms dropped out except the one in a_m or b_m, respectively, because of the relations

$$\int_{-\pi}^{\pi} \cos mx \cos nx \, dx = 0, \qquad m \neq n, \dots.$$

By analogy we multiply both sides of (7.32) by $\phi_m(x)$ and integrate term by term:

$$\int_a^b f(x)\phi_m(x) \, dx = \sum_{n=1}^{\infty} c_n \int_a^b \phi_m(x)\phi_n(x) \, dx. \tag{7.33}$$

In order to achieve a result analogous to that for Fourier series, we must postulate that

$$\int_a^b \phi_m(x)\phi_n(x) \, dx = 0, \qquad m \neq n. \tag{7.34}$$

The series on the right of (7.33) then reduces to one term:

$$\int_a^b f(x)\phi_m(x) \, dx = c_m \int_a^b [\phi_m(x)]^2 \, dx. \tag{7.35}$$

The integral on the right is a certain constant:

$$\int_a^b [\phi_m(x)]^2 \, dx = B_m. \tag{7.36}$$

The constant B_m will be positive unless $\phi_m(x) \equiv 0$ (except at a finite number of points); to avoid this trivial case, we assume that no B_m is 0. Then

$$c_m = \frac{1}{B_m} \int_a^b f(x)\phi_m(x) \, dx. \tag{7.37}$$

Thus under the simple conditions (7.34) and (7.36) we have a rule for formation of a series just like that for Fourier series and can hope that analogous convergence theorems can also be proved.

We now summarize the assumptions in formal definitions.

Definitions.   Two functions $p(x)$, $q(x)$, which are piecewise continuous for $a \leq x \leq b$, are *orthogonal* in this interval if

$$\int_a^b p(x)q(x) \, dx = 0. \tag{7.38}$$

A system of functions $\{\phi_n(x)\}$ ($n = 1, 2, \dots$) is termed an *orthogonal system* in the interval $a \leq x \leq b$ if ϕ_n and ϕ_m are orthogonal for each pair of distinct indices m, n:

$$\int_a^b \phi_m(x)\phi_n(x) \, dx = 0 \qquad (m \neq n), \tag{7.39}$$

and no $\phi_n(x)$ is identically 0 except at a finite number of points.

EXAMPLE   The trigonometric system in the interval $-\pi \leq x \leq \pi$:

$$1, \quad \cos x, \quad \sin x, \dots, \quad \cos nx, \quad \sin nx, \dots.$$

The function ϕ_1 is the constant 1; ϕ_2 is the function $\cos x, \dots$. ■

If $f(x)$ is piecewise continuous in the interval $a \leq x \leq b$ and $\{\phi_n(x)\}$ is an orthogonal system in this interval, then the series

$$\sum_{n=1}^{\infty} c_n \phi_n(x), \tag{7.40}$$

where

$$c_n = \frac{1}{B_n} \int_a^b f(x)\phi_n(x)\, dx, \qquad B_n = \int_a^b [\phi_n(x)]^2\, dx, \tag{7.41}$$

is called the *Fourier series of f with respect to the system* $\{\phi_n(x)\}$. The numbers $c_1, c_2, \ldots$ are called the Fourier coefficients of $f(x)$ with respect to the system $\{\phi_n(x)\}$.

The preceding formulas can be simplified if one assumes that the constant B_n is always 1, that is, that

$$\int_a^b [\phi_n(x)]^2 = 1 \qquad (n = 1, 2, \ldots).$$

This can always be achieved by dividing the original $\phi_n(x)$ by appropriate constants. When the condition $B_n = 1$ is satisfied for all n, the system of functions $\phi_n(x)$ is called *normalized*. A system that is both normalized and orthogonal is called *orthonormal*. This is illustrated by the functions:

$$\frac{1}{\sqrt{2\pi}}, \quad \frac{\cos x}{\sqrt{\pi}}, \quad \frac{\sin x}{\sqrt{\pi}}, \ldots, \quad \frac{\cos nx}{\sqrt{\pi}}, \quad \frac{\sin nx}{\sqrt{\pi}}, \ldots.$$

Although the general theory is simpler for normalized systems, the advantages for applications are slight, and we shall not use normalization in what follows.

The operations with orthogonal systems are strikingly similar to those with vectors. In fact, we can consider the piecewise continuous functions for $a \leq x \leq b$ as a sort of *vector space*, as in Section 1.17. The sum or difference of two such functions $f(x)$, $g(x)$ is again piecewise continuous, as is the product cf of $f(x)$ by a constant or *scalar* c. Equation (7.38) suggests a definition of *inner product* (or scalar product):

$$(f, g) = \int_a^b f(x)g(x)\, dx. \tag{7.42}$$

One can then define a *norm* (or absolute value):

$$\|f\| = \sqrt{(f, f)} = \left\{ \int_a^b [f(x)]^2\, dx \right\}^{\frac{1}{2}}. \tag{7.43}$$

The zero function 0^* is a function that is 0 except at a finite number of points; in general, in this vector theory of functions we consider two functions that differ only at a finite number of points to be the same function. It is now a straightforward exercise to verify that all the axioms for a Euclidean vector space, except the one concerning dimension, are satisfied. For convenience we restate the axioms here [cf. (1.102) in Section 1.13].

> I. $f + g = g + f$,
> II. $(f + g) + h = f + (g + h)$,
> III. $c(f + g) = cf + cg$,
> IV. $(c_1 + c_2)f = c_1 f + c_2 f$,
> V. $(c_1 c_2)f = c_1(c_2 f)$,
> VI. $1 \cdot f = f$,
> VII. $0 \cdot f = 0*$,
> VIII. $(f, g) = (g, f)$,
> IX. $(f + g, h) = (f, h) + (g, h)$,
> X. $(cf, g) = c(f, g)$,
> XI. $(f, f) \geq 0$,
> XII. $(f, f) = 0$ if and only if $f = 0*$.   (7.44)

The proof is left as an exercise (Problem 3 below).

We define k functions $f_1, f_2, \ldots, f_k$ to be *linearly independent* if the only scalars $c_1, \ldots, c_n$ for which

$$c_1 f_1 + \cdots + c_k f_k = 0* \qquad (7.45)$$

are the numbers $0, \ldots, 0$. In place of Rule XIII of Section 1.14 we have now the theorem that for every integer n there exist n linearly independent functions. For example, the functions $\sin x, \sin 2x, \ldots, \sin nx$ are linearly independent for $-\pi \leq x \leq \pi$. (See Problem 9 following Section 1-14.)

The theorem of Section 1.13 is proved as a consequence of the axioms (7.44) alone. Hence this theorem holds for functions.

THEOREM 7   Let $f(x)$ and $g(x)$ be piecewise continuous for $a \leq x \leq b$. Then

$$|(f, g)| \leq \|f\| \cdot \|g\|. \qquad (7.46)$$

The equality holds precisely when f and g are linearly dependent. Furthermore,

$$\|f + g\| \leq \|f\| + \|g\|; \qquad (7.47)$$

the equal sign holds precisely when $f = cg$ or $g = cf$, where c is a positive constant or 0.

The relation (7.46) is the *Schwarz inequality*. If we use the definitions (7.42) and (7.43), it can be written as follows:

$$\left[\int_a^b f(x) g(x)\, dx \right]^2 \leq \int_a^b [f(x)]^2\, dx \cdot \int_a^b [g(x)]^2\, dx. \qquad (7.48)$$

The relation (7.47) is the *Minkowski inequality*. In explicit form it reads:

$$\left\{ \int_a^b [f(x) + g(x)]^2\, dx \right\}^{\frac{1}{2}} \leq \left\{ \int_a^b [f(x)]^2\, dx \right\}^{\frac{1}{2}} + \left\{ \int_a^b [g(x)]^2\, dx \right\}^{\frac{1}{2}}. \quad (7.49)$$

*7.11   FOURIER SERIES OF ORTHOGONAL FUNCTIONS ▪ COMPLETENESS

The definition of orthogonal functions can be restated in terms of inner products: $f(x)$ is orthogonal to $g(x)$ if $(f, g) = 0$. An orthogonal system is a system $\{\phi_n(x)\}$ $(n = 1, 2, \ldots)$ of functions all piecewise continuous for $a \leq x \leq b$ and such that

$$(\phi_m, \phi_n) = 0, \quad m \neq n; \quad (\phi_n, \phi_n) = \|\phi_n\|^2 = B_n > 0. \tag{7.50}$$

The system is *orthonormal* if (7.50) holds, and $\|\phi_n\| = 1$ for $n = 1, 2, \ldots$, so that the ϕ_n are "unit vectors."

The Fourier series of $f(x)$ with respect to the orthogonal system $\{\phi_n(x)\}$ is the series

$$\sum_{n=1}^{\infty} c_n \phi_n(x), \quad c_n = (f, \phi_n)/\|\phi_n\|^2. \tag{7.51}$$

If the system is orthonormal, the series becomes

$$\sum_{n=1}^{\infty} c_n \phi_n(x), \quad c_n = (f, \phi_n), \tag{7.52}$$

which is analogous to the expressions

$$\mathbf{v} = v_x \mathbf{i} + v_y \mathbf{j} + v_z \mathbf{k}, \quad v_x = \mathbf{v} \cdot \mathbf{i}, \ldots,$$

$$\mathbf{v} = v_1 \mathbf{e}_1 + \cdots + v_n \mathbf{e}_n, \quad v_j = \mathbf{v} \cdot \mathbf{e}_j$$

for a vector in terms of base vectors in 3-dimensional space and n-dimensional space. However, since (7.51) is an infinite series, complications arise through convergence questions.

> **THEOREM 8**   Let $\{\phi_n(x)\}$ be an orthogonal system of continuous functions for the interval $a \leq x \leq b$. If the series $\sum_{n=1}^{\infty} c_n \phi_n(x)$ converges uniformly to $f(x)$ for $a \leq x \leq b$, then
>
> $$c_n = (f, \phi_n)/\|\phi_n\|^2, \tag{7.53}$$
>
> so that the series is the Fourier series of $f(x)$ with respect to $\{\phi_n(x)\}$. If the system is orthonormal, then $c_n = (f, \phi_n)$.

Proof.   Just as in the proof of Theorem 1 in Section 7.2, we reason that $f(x)$ is continuous. Then

$$(f, \phi_m) = (c_1 \phi_1 + \cdots + c_n \phi_n + \cdots, \phi_m)$$
$$= (c_1 \phi_1, \phi_m) + \cdots + (c_n \phi_n, \phi_m) + \cdots$$
$$= c_1 (\phi_1, \phi_m) + \cdots + c_n (\phi_n, \phi_m) + \cdots = c_m (\phi_m, \phi_m) = c_m \|\phi_m\|^2,$$

so that (7.53) holds. The operations on the series are justified by the uniform convergence. If the system is orthonormal, then $\|\phi_n\| = 1$, so that $c_n = (f, \phi_n)$.   □

COROLLARY If under the hypotheses of Theorem 8,

$$\sum_{n=1}^{\infty} c_n \phi_n(x) \equiv \sum_{n=1}^{\infty} c'_n \phi_n(x), \qquad a \leq x \leq b,$$

and both series converge uniformly over the interval, then $c_n = c'_n$ for $n = 1, 2, \ldots$.

THEOREM 9 Let $\{\phi_n(x)\}$ be an orthogonal system for the interval $a \leq x \leq b$ and let $f(x)$ be piecewise continuous for $a \leq x \leq b$. For each n the coefficients $c_1, \ldots, c_n$ of the Fourier series of f with respect to $\{\phi_n(x)\}$ are those constants that give the square error $\| f - g \|^2$ its smallest value, when g ranges over all linear combinations $p_1 \phi_1(x) + \cdots + p_n \phi_n(x)$. The minimum value of the error is

$$E_n = \| f - (c_1 \phi_1 + \cdots + c_n \phi_n) \|^2 = \| f \|^2 - c_1^2 \| \phi_1 \|^2 - \cdots - c_n^2 \| \phi_n \|^2. \tag{7.54}$$

COROLLARY Under the hypotheses of Theorem 9, one has

$$c_1^2 \| \phi_1 \|^2 + \cdots + c_n^2 \| \phi_n \|^2 \leq \| f \|^2, \tag{7.55}$$

so that the series $\Sigma c_k^2 \| \phi_k \|^2$ converges. Furthermore,

$$\lim_{k \to \infty} c_k \| \phi_k \| = 0. \tag{7.56}$$

The proofs are left as exercises (Problems 5, 6, 7 following Section 7.13); (7.55) is *Bessel's inequality*.

A crucial question is whether one can assert that E_n converges to 0, as $n \to \infty$. This is equivalent to asking whether $f(x)$ can be approximated, in the sense of least square error, as closely as desired by a linear combination of a finite number of the functions $\phi_n(x)$. If this is the case, the system $\{\phi_n(x)\}$ is termed complete:

DEFINITION An orthogonal system $\{\phi_n(x)\}$ for the interval $a \leq x \leq b$ is termed *complete* if for every piecewise continuous function $f(x)$ in the interval $a \leq x \leq b$ the minimum square error $E_n = \| f - (c_1 \phi_1 + \cdots + c_n \phi_n) \|^2$ converges to zero as n becomes infinite.

If the system is complete, then Theorem 9 shows that

$$\| f \|^2 = c_1^2 \| \phi_1 \|^2 + \cdots + c_n^2 \| \phi_n \|^2 + \cdots . \tag{7.57}$$

This is *Parseval's equation*. Conversely, if Parseval's equation holds for every piecewise continuous f, then E_n must converge to 0, and the system $\{\phi_n(x)\}$ is complete. Therefore *the validity of Parseval's equation is equivalent to completeness*. If the system $\{\phi_n(x)\}$ is orthonormal, Parseval's equation becomes

$$\| f \|^2 = c_1^2 + \cdots + c_n^2 + \cdots .$$

This is analogous to the vector relations

$$|\mathbf{v}|^2 = v_x^2 + v_y^2 + v_z^2, \qquad |\mathbf{v}|^2 = v_1^2 + \cdots + v_n^2.$$

If the orthogonal system $\{\phi_n(x)\}$ is complete, then the square error E_n converges to 0. This does not imply convergence of the Fourier series $\Sigma c_n\phi_n(x)$ to $f(x)$, although the successive partial sums do approach $f(x)$ in the sense of square error. We describe the situation by saying that the series *converges in the mean* to $f(x)$, and we write

$$\text{L.i.m.}_{n \to \infty} [c_1\phi_1(x) + \cdots + c_n\phi_n(x)] = f(x),$$

where "L.i.m." stands for "limit in the mean." In general, if functions $f(x)$ and $f_n(x)$ $(n = 1, 2, \ldots)$ are all piecewise continuous for $a \leq x \leq b$, we write

$$\text{L.i.m.}_{n \to \infty} f_n(x) = f(x) \tag{7.58}$$

if the sequence $\|f_n(x) - f(x)\|$ converges to 0, that is, if

$$\lim_{n \to \infty} \int_a^b [f(x) - f_n(x)]^2 \, dx = 0.$$

Remark. The following assertions can be proved: (a) if $f_n(x)$ converges uniformly to $f(x)$ for $a \leq x \leq b$, then $f_n(x)$ also converges in the mean to $f(x)$, provided that all functions are piecewise continuous; (b) if $f_n(x)$ converges in the mean to $f(x)$, then $f_n(x)$ need not converge uniformly to $f(x)$; in fact, the sequence $f_n(x)$ need not converge for $a \leq x \leq b$; (c) if $f_n(x)$ converges to $f(x)$ for $a \leq x \leq b$, but not uniformly, then $f_n(x)$ need not converge in the mean to $f(x)$. For proofs, refer to Problem 9 following Section 7.13.

DEFINITION   An orthogonal system $\{\phi_n(x)\}$ for the interval $a \leq x \leq b$ has the *uniqueness property* if every piecewise continuous function $f(x)$ for $a \leq x \leq b$ is uniquely determined by its Fourier coefficients with respect to $\{\phi_n(x)\}$; that is, if $f(x)$ and $g(x)$ are piecewise continuous for $a \leq x \leq b$ and $(f, \phi_n) = (g, \phi_n)$ for all n, then $f(x) - g(x) = 0^*$. This is equivalent to the statement that $h(x) = 0^*$ is the only piecewise continuous function orthogonal to all the functions $\phi_n(x)$; the system of orthogonal functions can therefore not be enlarged.

THEOREM 10   Let $\{\phi_n(x)\}$ be a complete system of orthogonal functions for the interval $a \leq x \leq b$. Then $\{\phi_n(x)\}$ has the uniqueness property.

The proof is left as an exercise (Problem 8 below). One might expect the converse to hold, that is, that the uniqueness property implies completeness. However, examples can be given to show that this is not the case. Theorem 12 gives more information on this point.

THEOREM 11   Let $\{\phi_n(x)\}$ be an orthogonal system of continuous functions for the interval $a \leq x \leq b$ and let $\{\phi_n(x)\}$ have the uniqueness property. Let $f(x)$ be continuous for $a \leq x \leq b$ and let the Fourier series of $f(x)$

with respect to $\{\phi_n(x)\}$ converge uniformly for $a \leq x \leq b$. Then the Fourier series converges to $f(x)$.

The proof is left as an exercise (Problem 10 following Section 7.13).

*7.12 SUFFICIENT CONDITIONS FOR COMPLETENESS

THEOREM 12 Let $\{\phi_n(x)\}$ be an orthogonal system of continuous functions for the interval $a \leq x \leq b$. Let the following two properties hold: (a) $\{\phi_n(x)\}$ has the uniqueness property; (b) for some positive integer k, the Fourier series of $g(x)$ with respect to $\{\phi_n(x)\}$ is uniformly convergent for every $g(x)$ having continuous derivatives through the kth order for $a \leq x \leq b$ and such that $g(a) = g'(a) = \cdots = g^{(k)}(a) = 0$, $g(b) = g'(b) = \cdots = g^{(k)}(b) = 0$. Then the system $\{\phi_n(x)\}$ is complete.

Proof. Let $f(x)$ be piecewise continuous for $a \leq x \leq b$. We must then show that, given $\epsilon > 0$, a linear combination $c_1\phi_1(x) + \cdots + c_n\phi_n(x) = \psi(x)$ can be found such that $\|f - \psi\| < \epsilon$.

For simplicity we assume the integer k of assumption (b) to be 2, as this case is typical.

The construction of the function $\psi(x)$ proceeds in several stages. We first determine a continuous function $F(x)$ such that $\|F - f\| < \frac{1}{4}\epsilon$ and $F(x) \equiv 0$ when $a \leq x \leq a + \delta$, $b - \delta \leq x \leq b$ for a proper choice of $\delta > 0$. This is suggested graphically in Fig. 7.14. We denote by $f_1(x)$ the piecewise continuous function that coincides with $f(x)$ except for $a \leq x \leq a + 2\delta$ and for $b - 2\delta \leq x \leq b$, where $f_1(x)$ is identically zero. We now bridge the jumps in $f_1(x)$ by line segments. At each jump the line joins $[x_0 - \delta, f_1(x_0 - \delta)]$ to $[x_0 + \delta, f_1(x_0 + \delta)]$. The function $F(x)$ coincides with $f_1(x)$ except between $x_0 - \delta$ and $x_0 + \delta$, where it follows the straight line. Accordingly, $F(x)$ is a continuous function, and $\|F - f\|^2$ is a sum of a finite number K of integrals of form

$$\int_{x_0 - \delta}^{x_0 + \delta} [F(x) - f(x)]^2 \, dx$$

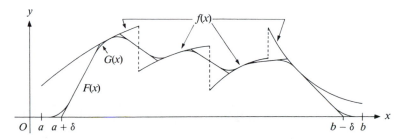

Figure 7.14 Approximation of the piecewise continuous function $f(x)$ by a continuous function $F(x)$ and by a smooth function $G(x)$.

plus two integrals of $[f(x)]^2$ from a to $a + \delta$ and from $b - \delta$ to b, where $F \equiv 0$. Since $f(x)$ is piecewise continuous, $|f(x)| \leq M$ for some constant M. By its construction, $|F(x)| \leq M$ also. Hence $|F(x) - f(x)| \leq 2M$ for every x, and

$$\int_{x_0 - \delta}^{x_0 + \delta} [F(x) - f(x)]^2 \, dx \leq 4M^2 \cdot 2\delta.$$

Adding the expressions for all K jumps and for the two ends, we find

$$\|F - f\|^2 \leq K \cdot 4M^2 \cdot 2\delta + 2M^2 \cdot \delta.$$

As $\delta \to 0$, the expression on the right approaches 0. Accordingly, for δ sufficiently small, $\|F - f\| < \frac{1}{4}\epsilon$.

We next choose a function $G(x)$ having a continuous first derivative for $a \leq x \leq b$, such that $\|G - F\| < \frac{1}{4}\epsilon$ and $G(x) \equiv 0$ for $a \leq x \leq a + \frac{1}{2}\delta$ and for $b - \frac{1}{2}\delta \leq x \leq b$. We define $F(x)$ to be identically 0 outside the interval $a \leq x \leq b$ and then set

$$G(x_1) = \frac{1}{h} \int_{x_1 - h}^{x_1 + h} F(x) \, dx, \qquad 0 < h < \frac{1}{2}\delta.$$

The constant h will be specified later. The function $G(x)$ is then defined for all x and $G(x) \equiv 0$ for $x \leq a + \frac{1}{2}\delta$, and for $x \geq b - \frac{1}{2}\delta$, as required. By the fundamental theorem of calculus [Sections 4.1 and 4.9],

$$G'(x_1) = \frac{1}{h} [F(x_1 + h) - F(x_1 - h)].$$

Accordingly, $G(x)$ has a continuous derivative for all x. Now

$$G(x_1) - F(x_1) = \frac{1}{2h} \int_{x_1 - h}^{x_1 + h} F(x) \, dx - F(x_1) = \frac{1}{2h} \int_{x_1 - h}^{x_1 + h} [F(x) - F(x_1)] \, dx,$$

and therefore

$$[G(x_1) - F(x_1)]^2 = \frac{1}{4h^2} \left\{ \int_{x_1 - h}^{x_1 + h} [F(x) - F(x_1)] \, dx \right\}^2.$$

We apply the Schwarz inequality (7.48) to the integral on the right-hand side, with $f(x)$ replaced by $F(x) - F(x_1)$ and $g(x)$ by 1. Accordingly,

$$[G(x_1) - F(x_1)]^2 \leq \frac{2h}{4h^2} \int_{x_1 - h}^{x_1 + h} [F(x) - F(x_1)]^2 \, dx,$$

and

$$\|G - F\|^2 = \int_a^b [G(x_1) - F(x_1)]^2 \, dx_1 \leq \frac{1}{2h} \int_a^b \int_{x_1 - h}^{x_1 + h} [F(x) - F(x_1)]^2 \, dx \, dx_1.$$

If we set $x_1 = u$, $x = u - v$ in the double integral on the right-hand side, it becomes (cf. Section 4.6)

$$\frac{1}{2h}\int_{-h}^{h}\int_{a}^{b}[F(u-v)-F(u)]^2\,du\,dv,$$

since the region of integration in the uv-plane is the rectangle $a \le u \le b$, $-h \le v \le h$. If we write $H(v)$ for the inner integral, then $H(v)$ is a continuous function of v (Section 4.11), and $H(0) = 0$. Now

$$\|G - F\|^2 \le \frac{1}{2h}\int_{-h}^{h} H(v)\,dv = H(v^*)\cdot\frac{2h}{2h} = H(v^*),$$

where $-h < v^* < h$, by the law of the mean. As $h \to 0$, $H(v^*)$ must approach 0; hence $\|G - F\| < \frac{1}{4}\epsilon$ for h sufficiently small.

Next we construct a function $g(x)$ such that $\|g - G\| < \frac{1}{4}\epsilon$, while g has continuous first and second derivatives and $g(x) \equiv 0$ for $a \le x \le a + \frac{1}{4}\delta$ and for $b - \frac{1}{4}\delta \le x \le b$. We need only repeat the averaging process of the preceding paragraph:

$$g(x_1) = \frac{1}{2p}\int_{x_1-p}^{x_1+p} G(x)\,dx, \qquad 0 < p < \tfrac{1}{4}\delta.$$

The other steps can be repeated, and the desired inequality is obtained by appropriate choice of p. Since

$$g'(x_1) = \frac{1}{2p}[G(x_1 + p) - G(x_1 - p)]$$

and G has a continuous derivative, $g(x)$ has continuous first and second derivatives for all x.

Finally, we construct a linear combination $\psi(x) = c_1\phi_1(x) + \cdots + c_n\phi_n(x)$ such that $\|g - \psi\| < \frac{1}{4}\epsilon$. Since $g(x)$ satisfies all the conditions of assumption (b) in the theorem, the Fourier series of $g(x)$ is uniformly convergent. By (a) and Theorem 11 the series converges to $g(x)$. By the remark preceding Theorem 10 the partial sums of this series also converge in the mean to $g(x)$. Accordingly, a partial sum $S_n(x) = \psi(x)$ can be chosen such that $\|g - \psi\| < \frac{1}{4}\epsilon$.

The function $\psi(x)$ is precisely the linear combination sought, for by (7.47),

$$\|f - \psi\| = \|f - F + F - G + G - g + g - \psi\|$$
$$\le \|f - F\| + \|F - G\| + \|G - g\| + \|g - \psi\|$$
$$< \tfrac{1}{4}\epsilon + \tfrac{1}{4}\epsilon + \tfrac{1}{4}\epsilon + \tfrac{1}{4}\epsilon = \epsilon. \quad \square$$

Remark.   The preceding proof shows that the assumption (a) can be omitted if one replaces "uniformly convergent" in (b) by "convergent in the mean to $g(x)$."

THEOREM 13   The trigonometric system 1, $\cos x$, $\sin x$,..., $\cos nx$, $\sin nx$,... is complete in the interval $-\pi \le x \le \pi$.

Proof.   The hypotheses (a) and (b) of Theorem 12 are satisfied by virtue of the preceding Theorems 3 and 5. Accordingly, the system is complete. $\quad\square$

*7.13   INTEGRATION AND DIFFERENTIATION OF FOURIER SERIES

THEOREM 14   An orthogonal system $\{\phi_n(x)\}$, for the interval $a \leq x \leq b$, is complete if and only if for every two functions $f(x)$ and $g(x)$ that are piecewise continuous for $a \leq x \leq b$, one has

$$(f, g) = \int_a^b f(x)g(x)\, dx = \sum_{n=1}^{\infty} c_n c_n' \|\phi_n\|^2, \tag{7.59}$$

where c_n, c_n' are the Fourier coefficients of $f(x)$ and $g(x)$, respectively, with respect to $\{\phi_n(x)\}$.

Remark.   Equation (7.59) is termed the *second form of Parseval's equation*. When the system is orthonormal, it reduces to the equation

$$(f, g) = c_1 c_1' + \cdots + c_n c_n' + \cdots, \qquad c_n = (f, \phi_n), \qquad c_n' = (g, \phi_n),$$

which is analogous to the equations

$$\mathbf{u} \cdot \mathbf{v} = u_x v_x + u_y v_y + u_z v_z, \qquad \mathbf{u} \cdot \mathbf{v} = u_1 v_1 + \cdots + u_n v_n$$

for vectors.

Proof of Theorem 14.   If (7.59) holds, then by taking $g = f$, one concludes that Parseval's equation (7.57) holds; hence the system is complete. Conversely, let $\{\phi_n(x)\}$ be complete. Then the algebraic identity $AB = \frac{1}{4}\{(A + B)^2 - (A - B)^2\}$ gives the equation

$$(f, g) = \frac{1}{4} \int_a^b (f + g)^2\, dx - \frac{1}{4} \int_a^b (f - g)^2\, dx.$$

Parseval's equation for $f + g$ and $f - g$ now gives

$$(f, g) = \frac{1}{4} \sum_{n=1}^{\infty} (c_n + c_n')^2 \|\phi_n\|^2 - \frac{1}{4} \sum_{n=1}^{\infty} (c_n - c_n')^2 \|\phi_n\|^2.$$

On adding the two series we obtain (7.59). $\square$

THEOREM 15   Let $\{\phi_n(x)\}$ be a complete orthonormal system for the interval $a \leq x \leq b$. Let $f(x)$ be piecewise continuous for $a \leq x \leq b$ and let $g(x)$ be piecewise continuous for $x_1 \leq x \leq x_2$, where $a \leq x_1 < x_2 \leq b$. Let $\Sigma c_n \phi_n(x)$ be the Fourier series of $f(x)$ with respect to $\{\phi_n(x)\}$. Then

$$\int_{x_1}^{x_2} f(x)g(x)\, dx = \sum_{n=1}^{\infty} c_n \int_{x_1}^{x_2} g(x)\phi_n(x)\, dx. \tag{7.60}$$

Remark.   The theorem asserts that the integral on the left-hand side can be computed by term-by-term integration of the series $\Sigma c_n g(x)\phi_n(x)$. This is remarkable since there is no assumption of convergence—let alone uniform convergence—of the series before integration.

Proof of Theorem 15. We extend the definition of $g(x)$ to the whole of the interval $a \leq x \leq b$ by setting $g(x) \equiv 0$ for $a \leq x \leq x_1$ and for $x_2 \leq x \leq b$. Then Parseval's equation (7.59) gives

$$\int_a^b f(x)g(x)\, dx = \int_{x_1}^{x_2} f(x)g(x)\, dx = \sum_{n=1}^{\infty} c_n c_n' \|\phi_n\|^2,$$

$$\|\phi_n\|^2 c_n' = (g, \phi_n) = \int_a^b g(x)\phi_n(x)\, dx = \int_{x_1}^{x_2} g(x)\phi_n(x)\, dx,$$

so that Eq. (7.60) is valid. $\square$

If we choose $g(x) \equiv 1$ for $x_1 \leq x \leq x_2$, we obtain the following result:

COROLLARY Under the hypotheses of Theorem 15,

$$\int_{x_1}^{x_2} f(x)\, dx = \sum_{n=1}^{\infty} c_n \int_{x_1}^{x_2} \phi_n(x)\, dx; \tag{7.61}$$

that is, term-by-term integration is permissible for every Fourier series with respect to a complete orthogonal system $\{\phi_n(x)\}$.

Although term-by-term integration causes no difficulties, term-by-term differentiation calls for great caution. For example, the series $\sin x + \cdots + (\sin nx/n) + \cdots$ converges for all x, but the differentiated series $\cos x + \cdots + \cos nx + \cdots$ diverges for all x. Differentiation *multiplies* the nth term by n, which interferes with convergence; integration *divides* the nth term by n, which aids convergence. The safest rule to follow is that of Theorem 33 of Section 6.14: Term-by-term differentiation of a convergent Fourier series is allowed, *provided that the differentiated series converges uniformly* in the interval considered.

PROBLEMS

1. Let $\phi_n(x) = \sin nx \ (n = 1, 2, \ldots)$.

   a) Show that the functions $\{\phi_n(x)\}$ form an orthogonal system in the interval $0 \leq x \leq \pi$.

   b) Show that the functions $\phi_n(x)$ have the uniqueness property. [Hint: Let $f(x)$ be a function orthogonal to all the ϕ_n. Let $F(x)$ be the odd function coinciding with $f(x)$ for $0 < x \leq \pi$. Show that all Fourier coefficients of $F(x)$ are 0.]

   c) Show that the Fourier sine series of $f(x)$ is uniformly convergent for every $f(x)$ having continuous first and second derivatives for $0 \leq x \leq \pi$ and such that $f(0) = f(\pi) = 0$.

   d) Show that $\{\phi_n(x)\}$ is a complete system for the interval $0 \leq x \leq \pi$.

2. Carry out the steps (a), (b), (c), and (d) of Problem 1 for the functions $\phi_n(x) = \cos nx$ $(n = 0, 1, 2, \ldots)$. Show that the condition $f(0) = f(\pi) = 0$ is not needed for (c).

3. Prove the validity of (7.44) for functions $f(x)$, $g(x)$ that are piecewise continuous for $a \leq x \leq b$.

4. Verify the correctness of the Schwarz inequality (7.48) and the Minkowski inequality (7.49) for $f(x) = x$ and $g(x) = e^x$ in the interval $0 \leq x \leq 1$.

5. Prove the corollary to Theorem 8 [see the proof of the corollary to Theorem 1, Section 7.2].

6. Prove Theorem 9 [cf. Problem 7 following Section 7.4].

7. Prove the corollary to Theorem 9 [cf. the proof of the corollary to Theorem 2, Section 7.4].

8. Prove Theorem 10. [Hint: Show by Parseval's equation (7.57) that if $(h, \phi_n) = 0$ for all n, then $\|h\| = 0$.]

9. a) Let the functions $f_n(x)$ be continuous for $a \leq x \leq b$ and let the sequence $f_n(x)$ converge uniformly to $f(x)$ for $a \leq x \leq b$. Prove that $\underset{n \to \infty}{\text{L.i.m.}} f_n(x) = f(x)$.

   b) Prove that the sequence $\cos^n x$ converges to 0 in the mean for $0 \leq x \leq \pi$ but does not converge for $x = \pi$. Prove that the sequence converges for $0 \leq x \leq \frac{1}{2}\pi$ but not uniformly.

   c) Let $f_n(x) = 0$ for $0 \leq x \leq 1/n$, $= n$ for $1/n \leq x \leq 2/n$, $= 0$ for $2/n \leq x \leq 1$. Show that the sequence converges to 0 for $0 \leq x \leq 1$ but not uniformly and that the sequence does not converge in the mean to 0.

10. Prove Theorem 11 [cf. the proof of Theorem 4 in Section 7.7].

11. a) Show that the functions $\phi_n(x) = \sin(2n + 1)x$, $n = 0, 1, 2, \ldots$ form a complete orthogonal system on the interval $I : 0 \leq x \leq \pi/2$. [Hint: Verify orthogonality directly. The functions all satisfy the symmetry conditions:

   (*) $$f(-x) = -f(x), \quad f(\pi - x) = f(x)$$

   and one verifies that they are the only members of the sequence $1, \cos x, \sin x, \cos 2x,$ $\sin 2x, \ldots$ which satisfy (*). Show that if $g(x)$ is defined on I, $g''(x)$ is continuous on I and $g(0) = g(\pi/2) = g'(0) = g'(\pi/2) = 0$, then $g(x)$ has an extension $f(x)$ of period 2π which satisfies (*) and is piecewise very smooth, so that its Fourier series converges uniformly to f for all x. As in Problem 5 following Section 7.5, the Fourier series of f contains only terms in $\sin x, \sin 3x, \ldots$ Now apply Theorem 12.]

   b) Show that the functions $\{\cos(2n + 1)x\}, n = 0, 1, 2, \ldots$ form a complete orthogonal system on the interval $0 \leq x \leq \pi/2$.

12. By a linear substitution $x = \alpha t + \beta$, $\alpha \neq 0$, an orthogonal system $\{\phi_n(x)\}$, $a \leq x \leq b$, becomes a system $\{\psi_n(t)\}$ on a new interval with endpoints $(a - \beta)/\alpha$, $(b - \beta)/\alpha$. Show that $\{\psi_n(t)\}$ is orthogonal and is complete, if $\{\phi_n(x)\}$ is complete.

13. Use the indicated substitution as in Problem 12 to obtain a new complete orthogonal system from the given system $\{\phi_n(x)\}$.

   a) $\phi_n(x) = \cos nx, n = 0, 1, 2, \ldots, 0 \leq x \leq \pi, x = 2t$

   b) ϕ_n as in (a), $x = (\pi/2) - t$

   c) $\phi_n(x) = \sin nx, n = 1, 2, \ldots, 0 \leq x \leq \pi, x = t + (\pi/2)$

14. In Problem 12 let the substitution be nonlinear: $x = g(t)$, where g and g' are continuous for $h \leq t \leq k$, $g'(t) \neq 0$, with $g(h) = a$ and $g(k) = b$ or $g(h) = b$ and $g(k) = a$. Show that the new system $\psi_n(t) = \phi_n(g(t)) |g'(t)|^{1/2}$ is orthogonal on the interval with endpoints h, k and is complete if $\{\phi_n(x)\}$ is complete [One says here that the functions

$\phi_n\big(g(t)\big)$ are orthogonal with respect to the weight function $w(t) = |g'(t)|$ (see Section 7.15).]

15. The choice $g(t) = \cos^{-1} t$ (principal value) for $0 \le t \le 1$ or for $-1 \le t \le 1$ does not meet the conditions of Problem 14 because of the discontinuities of $g'(t)$ at ± 1. However, the resulting improper integrals converge (Section 4.1) and the conclusions remain valid. Apply this substitution to each of the following cases and give the new orthogonal system:

a) $\cos nx$, $0 \le x \le \pi$, $n = 0, 1, 2, \ldots$, $-1 \le t \le 1$

b) $\sin nx$, $0 \le x \le \pi$, $n = 1, 2, \ldots$, $-1 \le t \le 1$

c) $\sin(2n + 1)x$, $0 \le x \le \pi/2$, $n = 0, 1, 2, \ldots$, $0 \le t \le 1$

d) $\cos(2n + 1)x$, $0 \le x \le \pi/2$, $n = 0, 1, 2, \ldots$, $0 \le t \le 1$

Remark 1.   For (c) and (d) see Problem 11.

Remark 2.   By trigonometry, for $n = 0, 1, 2, \ldots$ $\cos(n \cos^{-1} t)$ is a polynomial in t: $\cos(2 \cos^{-1} t) = 2t^2 - 1$, for example. Hence from (a) one obtains a system of polynomials $T_n(t)$ which are orthogonal for $-1 \le t \le 1$ with respect to a weight function. A similar statement applies to (b), (c) and (d), since $\sin(n \cos^{-1} t)$ equals a polynomial $U_n(t)$ times $\sin x = (1 - t^2)^{1/2}$. The polynomials $T_n(t)$ and $U_n(t)$ are the *Chebyshev polynomials* of first and second kinds, respectively. They are special cases of the Jacobi polynomials $P_n^{(\alpha, \beta)}(x)$ defined in Section 7.15: $\alpha = \beta = -\frac{1}{2}$ for the first kind, $\alpha = \beta = \frac{1}{2}$ for the second kind.

*7.14   FOURIER-LEGENDRE SERIES

In Theorem 13 we exhibited a key example of a complete orthogonal system: the trigonometric system of sines and cosines. In Problems 1, 2 and 11 to 15 following Section 7.13 we provided further examples, all related to the system of sines and cosines. In this section we present an example of a different kind, that of the Legendre polynomials; in the following section, other examples are considered.

The Legendre polynomials $P_n(x)$ $(n = 0, 1, 2, \ldots)$ can be defined by the formula of Rodrigues:

$$P_0(x) = 1, \qquad P_n(x) = \frac{1}{2^n n!} \frac{d^n}{dx^n} (x^2 - 1)^n \qquad (n = 1, 2, \ldots). \quad (7.62)$$

Thus

$$P_1(x) = \frac{1}{2} \frac{d}{dx} (x^2 - 1) = x,$$

$$P_2(x) = \frac{1}{8} \frac{d^2}{dx^2} (x^4 - 2x^2 + 1) = \frac{3}{2} x^2 - \frac{1}{2},$$

$$P_3(x) = \frac{5}{2} x^3 - \frac{3}{2} x, \qquad P_4(x) = \frac{35}{8} x^4 - \frac{15}{4} x^2 + \frac{3}{8}, \ldots.$$

THEOREM 16 $P_n(x)$ is a polynomial of degree n. $P_n(x)$ is an odd function or an even function according to whether n is odd or even. The following identities hold for $n = 1, 2, \ldots$:

a) $P_n'(x) = xP_{n-1}'(x) + nP_{n-1}(x)$,

b) $P_n(x) = xP_{n-1}(x) + \dfrac{x^2 - 1}{n}P_{n-1}'(x)$.　　　　　　　　　　　　(7.63)

Proof. The statements concerning the degree and the odd or even character can be verified from the definition. We recall the Leibnitz rule: $(u \cdot v)^{(n)} = uv^{(n)} + nu'v^{(n-1)} + \cdots + \binom{n}{r}u^{(r)}v^{(n-r)} + \cdots$, where $\binom{n}{r}$ is the general binomial coefficient $n!/[r!(n-r)!]$. To prove identity (a), we let $u = x^2 - 1$. Then

$$P_n'(x) = \frac{1}{2^n n!} \frac{d^{n+1}}{dx^{n+1}} u^n = \frac{1}{2^n n!} \frac{d^n}{dx^n}(2nxu^{n-1})$$

$$= \frac{1}{2^{n-1}(n-1)!}\left(x\frac{d^n}{dx^n} u^{n-1} + n\frac{d^{n-1}}{dx^{n-1}} u^{n-1}\right) = xP_{n-1}' + nP_{n-1},$$

by the Leibnitz Rule. To prove identity (b), we write $P_n(x)$ in two different ways:

$$P_n = \frac{1}{2^n n!} \frac{d^n}{dx^n}(u \cdot u^{n-1})$$

$$= \frac{1}{2^n n!}\left[u\frac{d^n}{dx^n} u^{n-1} + 2xn\frac{d^{n-1}}{dx^{n-1}} u^{n-1} + n(n-1)\frac{d^{n-2}}{dx^{n-2}} u^{n-1}\right];$$

$$P_n = \frac{1}{2^n n!} \frac{d^{n-1}}{dx^{n-1}}(2nxu^{n-1})$$

$$= \frac{1}{2^{n-1}(n-1)!}\left[x\frac{d^{n-1}}{dx^{n-1}} u^{n-1} + (n-1)\frac{d^{n-2}}{dx^{n-2}} u^{n-1}\right].$$

If the second equation is subtracted from twice the first, (b) is obtained.　□

It will be seen that the identities (a) and (b), plus the fact that $P_0(x) = 1$, are sufficient to establish all other properties needed, without reference to the definition (7.62).

THEOREM 17 The Legendre polynomials satisfy the following identities and relations:

c) $P_{n+1}'(x) - P_{n-1}'(x) = (2n+1)P_n(x)$　　　$(n \geq 1)$,

d) $\dfrac{d}{dx}[(1 - x^2)P_n'(x)] + n(n+1)P_n(x) = 0$,

e) $P_{n+1}(x) = \dfrac{(2n+1)xP_n(x) - nP_{n-1}(x)}{n+1}$　　　$(n \geq 1)$,

f) $P_n(1) = 1$, $P_n(-1) = (-1)^n$,

g) $\dfrac{1 - x^2}{n^2}P_n'^2 + P_n^2 = \dfrac{1 - x^2}{n^2}P_{n-1}'^2 + P_{n-1}^2$　　　$(n \geq 1)$,

h) $\dfrac{1 - x^2}{n^2}P_n'^2 + P_n^2 \leq 1$　　　$(n \geq 1, |x| \leq 1)$,

i) $|P_n(x)| \leq 1 \qquad (|x| \leq 1)$,

j) $\int_{-1}^{1} P_n(x)P_m(x)\, dx = 0 \qquad (n \neq m)$,

k) $\int_{-1}^{1} [P_n(x)]^2\, dx = \dfrac{2}{2n+1}$,

l) x^n can be expressed as a linear combination of $P_0(x), \dots, P_n(x)$. $\qquad$ (7.64)

The proofs are left to the exercises (Problems 2 to 9 following Section 7.15). Identity (d) is the *differential equation* satisfied by each $P_n(x)$; its importance will be seen in Chapter 10. Identity (e) is known as the *recursion formula* for Legendre polynomials. It expresses each polynomial in terms of the two preceding members of the sequence; hence knowing that $P_0 = 1$ and $P_1 = x$, one can successively determine $P_2, P_3, \dots$ from (e) alone. By virtue of (j) and (k) we can state:

THEOREM 18 The Legendre polynomials $P_n(x)$ $(n = 0, 1, 2, \dots)$ form an orthogonal system for the interval $-1 \leq x \leq 1$, and

$$\|P_n(x)\|^2 = \frac{2}{2n+1}. \qquad (7.65)$$

Accordingly, we can form the Fourier series of an arbitrary function $f(x)$ piecewise continuous for $-1 \leq x \leq 1$, with respect to the system $\{P_n(x)\}$:

$$\sum_{n=0}^{\infty} c_n P_n(x), \qquad c_n = \frac{(f, P_n)}{\|P_n\|^2} = \frac{2n+1}{2}\int_{-1}^{1} f(x)P_n(x)\, dx. \qquad (7.66)$$

It will be seen that this Fourier-Legendre series behaves in essentially the same way as the trigonometric Fourier series.

THEOREM 19 The system of Legendre polynomials in the interval $-1 \leq x \leq 1$ has the uniqueness property.

The proof of Theorem 3 (Section 7.7) can be repeated with slight modifications. By part (l) of Theorem 17, every polynomial in x is expressible as a finite linear combination of Legendre polynomials; hence it is sufficient to construct a polynomial pulse-function $P(x)$. One immediately verifies that the polynomial

$$P(x) = \left[1 - \tfrac{1}{2}(x - x_0)^2 + \tfrac{1}{2}\delta^2\right]^N$$

has the same properties as the function $P(x)$ used in Section 7.7, so that the proof can be completed in the same way. $\square$

THEOREM 20 If $f(x)$ is very smooth for $-1 \leq x \leq 1$, then the Fourier-Legendre series of $f(x)$ converges uniformly for $-1 \leq x \leq 1$.

Proof. By use of (c) and integration by parts we find for $n \geq 2$

$$c_n = \frac{2n+1}{2} \int_{-1}^{1} f(x) P_n(x)\, dx$$

$$= \frac{1}{2} \int_{-1}^{1} f(x)(P'_{n+1} - P'_{n-1})\, dx$$

$$= \frac{1}{2} [f(x)\{P_{n+1}(x) - P_{n-1}(x)\}]\Big|_{-1}^{1} - \frac{1}{2} \int_{-1}^{1} f'(x)(P_{n+1} - P_{n-1})\, dx.$$

Because of (f), the first term is 0. Integrating by parts again, we find

$$c_n = \frac{1}{2} \int_{-1}^{1} f''(x) \left[\frac{P_{n+2} - P_n}{2n+3} - \frac{P_n - P_{n-2}}{2n-1} \right] dx$$

$$= \frac{1}{4n+6}(f'', P_{n+2}) - \frac{1}{4n+6}(f'', P_n) - \frac{1}{4n-2}(f'', P_n)$$

$$+ \frac{1}{4n-2}(f'', P_{n-2}).$$

Accordingly, since $4n - 2$ is the smallest denominator,

$$|c_n| \leq \frac{1}{4n-2}[|(f'', P_{n+2})| + 2|(f'', P_n)| + |(f'', P_{n-2})|]$$

$$\leq \frac{1}{4n-2}(\|f''\| \cdot \|P_{n+2}\| + 2\|f''\| \cdot \|P_n\| + \|f''\| \cdot \|P_{n-2}\|),$$

by the Schwarz inequality (7.46). By (7.65),

$$|c_n| \leq \frac{\|f''\|}{4n-2}\left(\sqrt{\frac{2}{2n+5}} + 2\sqrt{\frac{2}{2n+1}} + \sqrt{\frac{2}{2n-3}} \right)$$

$$\leq \frac{\|f''\|}{4n-2} \frac{4\sqrt{2}}{\sqrt{2n-3}} \leq \frac{4\sqrt{2}\,\|f''\|}{(2n-3)^{\frac{3}{2}}} = M_n.$$

By part (i) of Theorem 17 the Weierstrass M-test is applicable to the Fourier-Legendre series of $f(x)$: for $n \geq 2$,

$$|c_n P_n(x)| \leq M_n = \text{const} \cdot (2n-3)^{-\frac{3}{2}}.$$

The series ΣM_n converges by the integral test. Therefore the Fourier-Legendre series converges uniformly for $-1 \leq x \leq 1$. $\square$

COROLLARY If $f(x)$ is very smooth for $-1 \leq x \leq 1$, then the Fourier-Legendre series of $f(x)$ converges uniformly to $f(x)$ for $-1 \leq x \leq 1$.

This is a consequence of Theorems 11, 19, and 20.

Remark. No condition of periodicity or other condition at $x = \pm 1$ is imposed on $f(x)$, as in the analogous theorem (Theorem 5) for trigonometric series. This is because of a symmetry of the Legendre polynomials, somewhat like that of the functions cos nx (Problem 2 following Section 7.13).

THEOREM 21   The Legendre polynomials form a complete orthogonal system for the interval $-1 \leq x \leq 1$.

This follows at once from the two preceding theorems, by virtue of the general Theorem 12 of Section 7.12.

Remark.   Given a sequence $f_n(x)$ of functions continuous for $a \leq x \leq b$, no finite number of which are linearly dependent, one can construct linear combinations $\phi_1 = f_1$, $\phi_2 = a_1 f_1 + a_2 f_2, \ldots$ such that the functions $\{\phi_n(x)\}$ form an orthogonal system in the interval $a \leq x \leq b$. This is carried out by the *Gram-Schmidt orthogonalization process*, described in Section 1.13. If we choose the sequence $f_n(x)$ to be $1, x, x^2, \ldots$ and the interval to be $-1 \leq x \leq 1$, the functions $\phi_n(x)$ turn out to be constants times the Legendre polynomials.

The case of a function $f(x)$ that is piecewise very smooth is covered as for Fourier series by studying a particular jump function $s(x)$. We do not carry through the details here but merely state the result: *Just as does the Fourier series, the Fourier-Legendre series converges uniformly to the function in each closed interval containing no discontinuity and converges at each jump discontinuity to the halfway value*

$$\tfrac{1}{2}\left[\lim_{x \to x_1-} f(x) + \lim_{x \to x_1+} f(x) \right].$$

It is to be emphasized that there is no peculiar behavior at the points $x = \pm 1$; here the series converges to the function (provided that the function remains continuous).

For a more complete treatment the reader is referred to the book by Jackson listed at the end of the chapter.

*7.15   FOURIER-BESSEL SERIES

We consider first a special case of the Bessel functions.

The Bessel function of first kind of order 0 is defined by the equation

$$J_0(x) = 1 - \frac{x^2}{2^2} + \frac{x^4}{2^4 (2!)^2} + \cdots + (-1)^n \frac{x^{2n}}{2^{2n}(n!)^2} + \cdots. \qquad (7.67)$$

The ratio test shows at once that this series converges for all values of x. The series is somewhat suggestive of that for cos x, and as the graph of Fig. 7.15 shows, $J_0(x)$ does resemble a trigonometric function. *In particular, $J_0(x)$ has infinitely many roots.* The positive roots will be denoted, in increasing order, by $\lambda_1, \lambda_2, \ldots, \lambda_k, \ldots$. It can be shown that as k increases, these roots approach more and more closely the spacing at intervals of π of the roots of sin x or cos x.

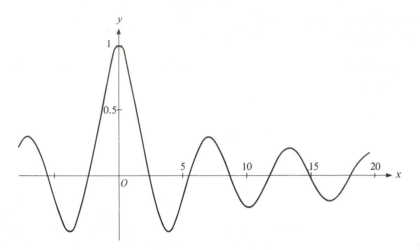

Figure 7.15   The Bessel function $J_0(x)$.

The function $J_0(x)$ deviates from the pattern of the trigonometric functions in that

$$\lim_{x \to \infty} J_0(x) = 0.$$

The oscillation represented by $J_0(x)$ is "damped." The rate of damping is shown by the following "asymptotic formula":

$$J_0(x) = \sqrt{\frac{2}{\pi x}} \sin\left(x + \frac{\pi}{4}\right) + \frac{r(x)}{x^{\frac{3}{2}}}, \qquad x > 0, \tag{7.68}$$

where $r(x)$ is bounded:

$$|r(x)| < K. \tag{7.69}$$

In other words, the function

$$\sqrt{\frac{\pi x}{2}} J_0(x)$$

differs from the function $\sin(x + \frac{1}{4}\pi)$ by a quantity $\text{const} \cdot r(x)/x$, which approaches zero as x becomes infinite.

Because of the orthogonality of the trigonometric functions, it is not surprising that there is a corresponding orthogonality relation for the functions $\sqrt{x}\, J_0(\lambda_n x)$:

$$\int_0^1 \left[\sqrt{x}\, J_0(\lambda_n x) \cdot \sqrt{x}\, J_0(\lambda_m x)\right] dx = 0, \qquad n \neq m. \tag{7.70}$$

The condition (7.70) is often described by stating that the functions $J_0(\lambda_n x)$ are *orthogonal with respect to the weight function* $\rho(x) = x$:

$$\int_0^1 J_0(\lambda_n x) J_0(\lambda_m x) \rho(x)\, dx = 0, \qquad n \neq m, \qquad \rho(x) = x.$$

The functions $\sqrt{x}\,J_0(\lambda_n x)$ can now be made the basis of a Fourier-Bessel series. For simplicity we write the function being expanded as $\sqrt{x}\,f(x)$, so that the series for $f(x)$ is

$$\sum_{n=1}^{\infty} c_n J_0(\lambda_n x).$$

Since it can be shown that

$$B_n = \int_0^1 x\left[J_0(\lambda_n x)\right]^2 dx = \tfrac{1}{2}\left[J_0'(\lambda_n)\right]^2,$$

the series for $f(x)$ is given by

$$\sum_{n=1}^{\infty} c_n J_0(\lambda_n x), \qquad c_n = \frac{2}{\left[J_0'(\lambda_n)\right]^2} \int_0^1 xf(x) J_0(\lambda_n x)\, dx. \qquad (7.71)$$

This series is usually referred to as the Fourier-Bessel series of order 0 of $f(x)$.

Because of the close relationship between the functions $\sqrt{x}\,J_0(\lambda_n x)$ and trigonometric functions, it is natural to expect convergence theorems analogous to those for Fourier series. It can indeed be shown that the system $\{\sqrt{x}\,J_0(\lambda_n x)\}$ is *complete* for the interval $0 \le x \le 1$. Hence by Theorem 11 of Section 7.11 the Fourier-Bessel series of $f(x)$ converges to $f(x)$ whenever the series is uniformly convergent. The series can be shown to be uniformly convergent if $f(x)$ has, for example, a continuous second derivative and $f(1) = 0$. The peculiar requirement at $x = 1$ is due to the fact that when $x = 1$, every function $J_0(\lambda_n x)$ is 0. If $f(x)$ is piecewise very smooth, an analysis of jump functions again shows that the series is uniformly convergent to $f(x)$ except near discontinuity points, where the series converges to the average of left- and right-hand limits.

For details on the theory of such expansions, the reader is referred to Chapter XVIII of the treatise of Watson listed at the end of the chapter.

The general *Bessel function of first kind and order m* is denoted by $J_m(x)$ and is defined by the series:

$$J_m(x) = \sum_{n=0}^{\infty} \frac{(-1)^n x^{m+2n}}{2^{m+2n} n!\, \Gamma(m+n+1)}. \qquad (7.72)$$

The function $\Gamma(x)$ was defined for all x except $0, -1, -2, -3, \dots$ in Section 6.25. The definition (7.72) therefore fails if m is a negative integer. However, it can be used for these values if we interpret $1/\Gamma(k)$ to be 0 for $k = 0, -1, -2, \dots$. If x^m is factored out of the series in (7.72), the resulting power series can be shown to converge for all x. However, if m is not an integer, the factor x^m gives difficulty for negative x and possibly for $x = 0$. Hence we shall concentrate on the functions $J_m(x)$ for $x \ge 0$. These functions are continuous for $x > 0$ and, if $m \ge 0$, for $x \ge 0$.

We can formulate a theorem analogous to Theorem 17 for these Bessel functions:

THEOREM 22   The Bessel functions of first kind satisfy the following identities and relations for $x > 0$:

a) $2J_m'(x) = J_{m-1}(x) - J_{m+1}(x)$.

b) $\dfrac{2m}{x} J_m(x) = J_{m-1}(x) + J_{m+1}(x)$.

c) $J_{m-1}(x) = J_m'(x) + \dfrac{m}{x} J_m(x)$,   or $x^m J_{m-1}(x) = \dfrac{d}{dx}(x^m J_m(x))$.

d) $J_{m+1}(x) = \dfrac{m}{x} J_m(x) - J_m'(x)$,   or $-x^{-m} J_{m+1}(x) = \dfrac{d}{dx}(x^{-m} J_m(x))$.

e) $x^2 J_m''(x) + x J_m'(x) + (x^2 - m^2) J_m(x) = 0$ (differential equation of the Bessel functions).

f) $\dfrac{d^2}{dx^2}\left[\sqrt{x}\, J_m(cx)\right] + \left(c^2 + \dfrac{1 - 4m^2}{4x^2}\right)\sqrt{x}\, J_m(cx) = 0$,
$c = \text{const} > 0$.

g) $(\alpha^2 - \beta^2)\displaystyle\int_0^1 x J_m(\alpha x) J_m(\beta x)\, dx = \beta J_m(\alpha) J_m'(\beta) - \alpha J_m(\beta) J_m'(\alpha)$,   for $m \geq 0$, $\alpha > 0$, $\beta > 0$.

h) $(\alpha^2 - \beta^2)\displaystyle\int_0^1 x J_m(\alpha x) J_m(\beta x)\, dx = \alpha J_m(\beta) J_{m+1}(\alpha) - \beta J_m(\alpha) J_{m+1}(\beta)$, for $m \geq 0$, $\alpha > 0$, $\beta > 0$.

i) $2\alpha \displaystyle\int_0^1 x J_m^2(\alpha x)\, dx = J_m(\alpha) J_{m+1}(\alpha) + \alpha[J_m(\alpha) J_{m+1}'(\alpha) - J_{m+1}(\alpha) J_m'(\alpha)]$, for $m \geq 0$, $\alpha > 0$.

j) The equation $J_m(x) = 0$ has infinitely many positive roots $\lambda_{m1}, \lambda_{m2}, \ldots,$ where $\lambda_{m1} < \lambda_{m2} < \cdots < \lambda_{mn} < \cdots,$ $\lambda_{mn} \to \infty$ as $n \to \infty$, and each root is simple (that is, $J_m'(x) \neq 0$ at each root).

k) The equations $J_m(x) = 0$ and $J_{m+1}(x) = 0$ have no common positive roots.

l) Between two successive positive roots of $J_m(x) = 0$ there is one and only one root of $J_{m-1}(x) = 0$ and one and only one root of $J_{m+1}(x) = 0$.

m) For each fixed $m \geq 0$ the functions $\sqrt{x}\, J_m(\lambda_{mn} x)$, $n = 1, 2, 3, \ldots,$ form an orthogonal system on the interval $0 \leq x \leq 1$.

n) $\displaystyle\int_0^1 x J_m^2(\lambda_{mn} x)\, dx = \tfrac{1}{2}[J_{m+1}(\lambda_{mn})]^2$.

o) For $m = 0, 1, 2, \ldots,$ $J_m(x) = \dfrac{1}{\pi}\displaystyle\int_0^\pi \cos(x \sin \theta - m\theta)\, d\theta$.

p) For $m = 0, 1, 2, \ldots,$ $|J_m(x)| \leq 1$ for $x \geq 0$.

The proofs are left to Problem 10 below. It can be shown that the orthogonal system in part (m) is complete, so that there is a corresponding Fourier-Bessel series of order m. This series is uniformly convergent for every function $\sqrt{x}\, f(x)$ such that $f(x)$ has a continuous second derivative in the interval and $f(1) = 0$.

For proofs, see Chapter XVIII of the treatise by Watson. On page 406 of the same book it is shown that assertion (p) can be extended to all $m \geq 0$.

The Bessel functions of the *second kind* are the functions $Y_m(x)$ ($m = 0, 1, 2, \ldots$) defined as follows:

$$Y_m(x) = \frac{1}{\pi} \sum_{k=0}^{\infty} \frac{(-1)^k \left(\frac{x}{2}\right)^{m+2k}}{k!(m+k)!} \left\{ 2 \log \frac{x}{2} + 2\gamma - b_{m+k} - b_k \right\}$$

$$- \frac{1}{\pi} \sum_{k=0}^{m-1} \left(\frac{x}{2}\right)^{-m+2k} \frac{(m-k-1)!}{k!}, \tag{7.73}$$

where $b_k = 1 + 2^{-1} + \cdots + k^{-1}$, $b_0 = 0$, and γ is defined in Eq. (6.88). These functions satisfy the same differential equations (e) as the functions of first kind and have other analogous properties. The complex-valued functions

$$H_m^{(1)}(x) = J_m(x) + iY_m(x), \qquad H_m^{(2)}(x) = J_m(x) - iY_m(x) \tag{7.74}$$

are called Bessel functions of the *third kind* or *Hankel functions*. For details, see the book by Watson.

Tables and graphs of both Legendre polynomials and Bessel functions are available. The book by Jahnke and Emde listed at the end of the chapter provides such data and also a convenient summary of properties of the functions.

We mention, without further discussion, several other important systems of orthogonal functions:

The Jacobi polynomials.   For each $\alpha > -1$, $\beta > -1$, the system of polynomials $\{P_n^{(\alpha, \beta)}(x)\}$ ($n = 0, 1, 2, \ldots$) is defined by the equations

$$P_n^{(\alpha, \beta)}(x) = \frac{(-1)^n}{2^n n!} (1-x)^{-\alpha} (1+x)^{-\beta} \frac{d^n}{dx^n} \left[(1-x)^{\alpha+n} (1+x)^{\beta+n} \right].$$

$$\tag{7.75}$$

The functions $\phi_n(x) = (1-x)^{\frac{1}{2}\alpha} (1+x)^{\frac{1}{2}\beta} P_n^{(\alpha, \beta)}(x)$ form a complete orthogonal system in the interval $-1 \leq x \leq 1$. When $\alpha = \beta = 0$, the functions $\phi_n(x)$ reduce to the Legendre polynomials.

The Hermite polynomials.   The system of polynomials $\{H_n(x)\}$ is defined by the equations:

$$H_n(x) = (-1)^n e^{\frac{1}{2}x^2} \frac{d^n}{dx^n} e^{-\frac{1}{2}x^2} \qquad (n = 0, 1, 2, \ldots). \tag{7.76}$$

They are orthogonal over the *infinite interval* $-\infty < x < \infty$ with respect to the weight function $e^{-\frac{1}{2}x^2}$.

The Laguerre polynomials.   For each $\alpha > -1$ the system of polynomials $\{L_n^{(\alpha)}(x)\}$ is defined by the equations:

$$L_n^{(\alpha)}(x) = (-1)^n x^{-\alpha} e^x \frac{d^n}{dx^n} (x^{\alpha+n} e^{-x}) \qquad (n = 0, 1, \ldots). \tag{7.77}$$

They are orthogonal over the *infinite interval* $0 \le x < \infty$ with respect to the weight function $x^\alpha e^{-x}$.

For the significance of the infinite interval and further information on these functions we refer to the book of Jackson listed at the end of the chapter.

All of these functions arise in a natural way in physical problems. This is illustrated in Chapter 10, where it will be shown that each of a large class of physical problems automatically provides a complete system of orthogonal functions.

PROBLEMS

1. Graph $P_0(x), P_1(x), \ldots, P_4(x)$.

2. Prove the following parts of Theorem 17:

   (i) part (c),   (ii) part (d),   (iii) part (e). Each of these can be deduced from (7.63) alone.

3. Prove part (f) of Theorem 17 by induction, with the aid of the recursion formula (e).

4. Prove part (g) of Theorem 17 by squaring (a) and (b) of (7.63) and eliminating the terms in $P_{n-1}P'_{n-1}$ from the equations obtained.

5. Prove part (h) of Theorem 17. [Hint: Show by induction that (g) gives the chain of inequalities:

$$\frac{1-x^2}{n^2}P_n'^2 + P_n^2 \le \frac{1-x^2}{(n-1)^2}P_{n-1}'^2 + P_{n-1}^2 \le \cdots \le 1, \qquad |x| \le 1.]$$

6. Prove part (i) of Theorem 17 as a consequence of part (h).

7. Prove the orthogonality condition (j). [Hint: For every very smooth function $f(x)$ for $-1 \le x \le 1$, let $f^*(x) = [(1-x^2)f'(x)]'$. Prove by integration by parts that $(f^*, g) - (f, g^*) = 0$. Take $f = P_n(x)$, $g = P_m(x)$ and use the differential equation (d) to replace f^* by $-n(n+1)f$ and g^* by $-m(m+1)g$, and conclude that $(f, g)[m(m+1) - n(n+1)] = 0$.]

8. Prove part (k) of Theorem 17. [Hint: Use the recursion formula (e) to express P_n in terms of P_{n-1} and P_{n-2} and use the orthogonality condition to show that $(P_n, P_n) = [(2n-1)/n](xP_n, P_{n-1})$. Apply the recursion formula to xP_n and use orthogonality to show that $(P_n, P_n) = (P_{n-1}, P_{n-1})[(2n-1)/(2n+1)]$. Now prove by induction that $(P_n, P_n) = 2/(2n+1)$.]

9. Prove part (l) of Theorem 17 by induction, using the fact that $P_n(x)$ is a polynomial of nth degree.

10. (a)...(p). Prove the corresponding parts of Theorem 22 with the aid of the following hints: For (a) and (b), use (7.72). For (c) and (d), use (a) and (b). For (e), use (a) through (d). For (f), use (e). For (g), let $u(x) = \sqrt{x}J_m(\alpha x)$, $v(x) = \sqrt{x}J_m(\beta x)$. Take $c = \alpha$ and then $c = \beta$ in (f) to obtain equations $u''(x) + \cdots = 0$, $v''(x) + \cdots = 0$. Multiply the first equation by $-v$, multiply the second by u, and add. Then integrate the result from b to 1, noting that $uv'' - vu'' = (uv' - vu')'$. Replace u and v by their expressions in terms of J_m and finally let $b \to 0 +$. For (h), use (g) and (d). For (i), divide both sides of (h) by $\alpha - \beta$ and then, with α fixed, let $\beta \to \alpha$. For (j), note that

if $J_m(\alpha) = 0$ and $J'_m(\alpha) = 0$, then by (g) $\int_0^1 x J_m(\alpha x) J_m(\beta x)\, dx = 0$ for all $\beta \neq \alpha$. Let $\beta \to \alpha$ to conclude that $J_m(\alpha x) \equiv 0$ for $0 \leq x \leq 1$, and this is impossible. Hence each root is simple. Deduce from this that there are only finitely many roots in each finite interval, so that the roots can be numbered in increasing order as stated and, if the sequence is infinite, it must diverge to ∞. If the sequence were to stop, then for example, $J_m(x) > 0$ for $x > x_0$. Take $c = 1$ in (f) to conclude that with $g(x) = \sqrt{x}\, J_m(x)$, $g(x) > 0$ and $g''(x) < 0$ for $x > x_1$, for some x_1. If $g(x) \to \infty$ as $x \to \infty$, then by (f), $g''(x) \to -\infty$ as $x \to \infty$; show that this is impossible, so that $g(x)$ is bounded and that $g(x)$ must have a limit k as $x \to \infty$. Show by (f) that either $k > 0$ or $k = 0$ is impossible. Thus the sequence of roots must be infinite. For (k), use (h). For (l), use (c), (d), and Rolle's theorem. For (m), use (g). For (n), use (i) and (d). For (o), use induction. For $m = 0$, replace $\cos(x \sin \theta)$ by $1 - (x^2 \sin^2 \theta / 2!) + \cdots$ and use the formula $\int_0^\pi \sin^{2m} \theta\, d\theta = \pi \cdot 1 \cdot 3 \cdots (2m-1)/[2 \cdot 4 \cdots 2m]$. Show, with the aid of integration by parts, that $g_m(x) = \pi^{-1} \int_0^\pi \cos(x \sin \theta - m\theta)\, d\theta$ satisfies $g_{m+1}(x) = (m/x) g_m(x) - g'_m(x)$, which by (d) is the same recursion formula as is satisfied by $J_m(x)$. For (p), use (o).

11. Prove that for $m = 1, 2, 3, \ldots, J_{-m}(x) = (-1)^m J_m(x)$.

12. Prove:

a) $J_{\frac{1}{2}}(x) = \sqrt{\dfrac{2}{\pi x}} \sin x$

b) $J_{-\frac{1}{2}}(x) = \sqrt{\dfrac{2}{\pi x}} \cos x$

c) $J_{\frac{3}{2}}(x) = \sqrt{\dfrac{2}{\pi x}} \left(\dfrac{\sin x}{x} - \cos x \right)$

*7.16 ORTHOGONAL SYSTEMS OF FUNCTIONS OF SEVERAL VARIABLES

The theory of Sections 7.10 to 7.12 can be generalized with minor changes to functions of several variables; one need only replace the interval by a bounded closed region R and the definite integral by a multiple integral over R. For example, in two dimensions the inner product and norm are defined as follows:

$$(f, g) = \int_R \int f(x, y) g(x, y)\, dx\, dy,$$

$$\|f\| = (f, f)^{\frac{1}{2}} = \left\{ \int_R \int [f(x, y)]^2\, dx\, dy \right\}^{\frac{1}{2}}. \tag{7.78}$$

Orthogonal systems are defined as before, and one can consider the corresponding Fourier series. Discussion of discontinuities becomes more involved, and it is sufficient for most purposes to restrict attention to continuous functions. The analogues of Theorems 7 to 11 can then be proved essentially without alteration. The generalization of Theorem 12 can be carried out, but this requires more care.

Corresponding to the trigonometric system, one has the following system in two dimensions:

$$1, \quad \sin x, \quad \cos x, \quad \sin y, \quad \cos y, \quad \sin x \cos y, \ldots, \sin px \sin qy,$$

$$\cos px \sin qy, \quad \sin px \cos qy, \quad \cos px \cos qy, \ldots . \quad (7.79)$$

This system can be arranged in a definite order to form a system $\{\phi_n(x, y)\}$, which is orthogonal and complete for the rectangle $R: -\pi \le x \le \pi, -\pi \le y \le \pi$ (see Problem 1 following Section 7.19). One can also write the Fourier series as a "double Fourier series":

$$\sum_{q=0}^{\infty} \sum_{p=0}^{\infty} \{a_{pq} \sin px \sin qy + b_{pq} \cos px \sin qy$$

$$+ c_{pq} \sin px \cos qy + d_{pq} \cos px \cos qy\}. \quad (7.80)$$

This amounts to a special sort of rearrangement and grouping of the series; if the series is absolutely convergent, then the reasoning of Section 6.10 shows that the double series has the same sum as the single series in any order. If $f(x, y)$ has continuous first and second derivatives for all (x, y) and is periodic in both variables:

$$f(x, y) = f(x + 2\pi, y) = f(x, y + 2\pi),$$

then an argument similar to that of Section 7.8 shows that the series is absolutely and uniformly convergent to $f(x, y)$.

*7.17   COMPLEX FORM OF FOURIER SERIES

From the identity

$$e^{ix} = \cos x + i \sin x \quad (i = \sqrt{-1})$$

of Section 6.19, one derives the relations

$$\cos x = \frac{e^{ix} + e^{-ix}}{2}, \quad \sin x = \frac{e^{ix} - e^{-ix}}{2i}. \quad (7.81)$$

A Fourier series

$$\frac{a_0}{2} + \sum_{n=1}^{\infty} (a_n \cos nx + b_n \sin nx)$$

can hence be written in the form

$$\frac{a_0}{2} + \sum_{n=1}^{\infty} (c_n e^{inx} + d_n e^{-inx}) = \sum_{n=-\infty}^{\infty} c_n e^{inx}, \quad (7.82)$$

$$c_0 = \frac{a_0}{2}, \quad c_n = \frac{a_n - ib_n}{2}, \quad d_n = \frac{a_n + ib_n}{2} = c_{-n} \quad (n = 1, 2, \ldots).$$

The summation from $-\infty$ to ∞ is understood to mean an addition of two series:

$$\sum_{n=-\infty}^{\infty} c_n e^{inx} = \sum_{n=0}^{\infty} c_n e^{inx} + \sum_{n=1}^{\infty} c_{-n} e^{-inx}.$$

If both series converge, the result is clearly the same as the single series on the left-hand side of (7.82).

The form (7.82) has various advantages. The coefficients c_n can be defined directly in terms of $f(x)$:

$$c_n = \frac{1}{2\pi} \int_{-\pi}^{\pi} f(x) e^{-inx}\, dx \qquad (n = 0, \pm 1, \pm 2, \ldots), \qquad (7.83)$$

for the integral on the right is interpreted as

$$\frac{1}{2\pi} \int_{-\pi}^{\pi} f(x)(\cos nx - i \sin nx)\, dx$$

$$= \frac{1}{2\pi} \int_{-\pi}^{\pi} f(x) \cos nx\, dx - \frac{i}{2\pi} \int_{-\pi}^{\pi} f(x) \sin nx\, dx.$$

See Section 9.2 for the theory of such integrals of complex functions. When $n = 0$, this gives $\frac{1}{2} a_0$; when $n > 0$, the integral equals $\frac{1}{2}(a_n - ib_n)$; and when $n < 0$, it equals $\frac{1}{2}(a_{-n} + ib_{-n})$. One has thus the concise statement,

$$f(x) = \sum_{n=-\infty}^{\infty} c_n e^{inx}, \qquad c_n = \frac{1}{2\pi} \int_{-\pi}^{\pi} f(x) e^{-inx}\, dx, \qquad (7.84)$$

whenever the series converges to $f(x)$.

For formal work with Fourier series, and even for computation of coefficients, the series (7.84) provides a considerable simplification.

*7.18   FOURIER INTEGRAL

By a suitable limiting process the equations (7.84) lead to the relations

$$f(x) = \frac{1}{\sqrt{2\pi}} \int_{-\infty}^{\infty} g(t) e^{ixt}\, dt, \qquad g(t) = \frac{1}{\sqrt{2\pi}} \int_{-\infty}^{\infty} f(x) e^{-ixt}\, dx. \quad (7.85)$$

Thus under appropriate hypotheses a function $f(x)$ defined for $-\infty < x < \infty$ can be represented as a "continuous sum" of sines and cosines ($e^{ixt} = \cos xt + i \sin xt$). The integral representing $f(x)$ is termed the *Fourier integral* of $f(x)$. The Fourier coefficients of $f(x)$ in this integral representation are the numbers $g(t)$, which form a new function. Equations (7.85) show that the relation between f and g is nearly symmetrical.

Equations (7.85) can also be written in real form as follows:

$$f(x) = \int_0^{\infty} \alpha(t) \cos xt\, dt + \int_0^{\infty} \beta(t) \sin xt\, dt,$$

$$\alpha(t) = \frac{1}{\pi} \int_{-\infty}^{\infty} f(x) \cos xt\, dx, \qquad \beta(t) = \frac{1}{\pi} \int_{-\infty}^{\infty} f(x) \sin xt\, dx. \quad (7.86)$$

This is directly analogous with the real form for Fourier series.

The validity of the formulas (7.85) or the equivalent formulas (7.86) can be established if $f(x)$ is piecewise smooth in each finite interval and the integral

$$\int_{-\infty}^{\infty} |f(x)| \, dx$$

converges. The Fourier integral must in general be taken as a *principal value* (Section 6.24). It converges to $f(x)$ wherever f is continuous and to the average of left and right limits at jump discontinuities. For a proof the reader is referred to Chapter 5 of the book by Kaplan listed at the end of the chapter.

The equations (7.85) are often rephrased in the language of transforms. If

$$F(t) = \int_{-\infty}^{\infty} f(x)e^{-ixt} \, dx, \qquad -\infty < t < \infty, \tag{7.87}$$

then one calls the function F the *Fourier transform* of the function f and writes

$$F = \Phi[f]. \tag{7.88}$$

Equations (7.85) then state that

$$f(x) = \frac{1}{2\pi} \int_{-\infty}^{\infty} F(t)e^{ixt} \, dt = \frac{1}{2\pi} \int_{-\infty}^{\infty} F(-t)e^{-ixt} \, dt. \tag{7.89}$$

Thus by interchanging the roles of x and t we see that f is itself the Fourier transform of $F(-t)/(2\pi)$.

The Fourier transform has a number of properties that make it exceptionally useful in applications to differential equations. We state several such properties formally here and refer to the books on Fourier theory listed at the end of the chapter for details. First of all, Φ is a *linear operator*:

$$\Phi[c_1 f_1 + c_2 f_2] = c_1 \Phi[f_1] + c_2 \Phi[f_2]. \tag{7.90}$$

Next

$$\Phi[f^{(k)}(x)] = (it)^k \Phi[f], \qquad k = 1, 2, \ldots . \tag{7.91}$$

This property is crucial for differential equations. Here it is assumed, in particular, that f has continuous derivatives through order k. The result must be modified when discontinuities occur. (See (7.101) for such a rule for Laplace transforms.) If c is a real constant, then

$$\Phi[f(x - c)] = e^{-ict} \Phi[f], \tag{7.92}$$

$$\Phi[e^{icx}f(x)] = F(t - c), \qquad \text{where } F = \Phi[f]. \tag{7.93}$$

From two functions $f(x)$ and $g(x)$ (under appropriate hypotheses), one forms their *convolution* $f * g = h$ as the function $h(x)$ such that

$$h(x) = \int_{-\infty}^{\infty} f(u)g(x - u) \, du. \tag{7.94}$$

Then

$$\Phi[f * g] = \Phi[f]\Phi[g]. \tag{7.95}$$

Examples of Fourier transforms.   Let $f_1(x) = e^{-x}$ for $x > 0$ and $f_1(x) = 0$ for $x \le 0$. Then

$$\Phi[f_1] = \int_0^\infty e^{-x} e^{-ixt}\, dx = \frac{1}{1+it}.$$

Similarly, if $f_2(x) = e^x$ for $x \le 0$ and $f_2(x) = 0$ for $x > 0$, then

$$\Phi[f_2] = \frac{1}{1-it}.$$

Now $f_1(x) + f_2(x) = e^{-|x|} = f(x)$, and hence by linearity,

$$\Phi[f] = \frac{1}{1+it} + \frac{1}{1-it} = \frac{2}{1+t^2} = F(t).$$

Since f is the Fourier transform of $(2\pi)^{-1}F(-t) = [\pi(1+t^2)]^{-1}$, we conclude (after interchanging x and t) that

$$\Phi[g] = e^{-|t|}, \quad \text{where} \quad g(x) = \frac{1}{\pi(1+x^2)},$$

or by linearity,

$$\Phi[g_1] = \pi e^{-|t|}, \quad \text{where} \quad g_1(x) = \frac{1}{1+x^2}.$$

*7.19   THE LAPLACE TRANSFORM AS A SPECIAL CASE OF THE FOURIER TRANSFORM

The Laplace transform of a function $f(x)$ is defined (Section 6.25) as $\mathscr{L}[f] = F$, where

$$F(s) = \int_0^\infty f(x) e^{-sx}\, dx. \tag{7.96}$$

Here f need only be defined for $x \ge 0$. However, it will be convenient here to define $f(x)$ to be 0 for negative x. The Laplace transform F is a function of s, defined wherever the integral in (7.96) exists. We now allow s to be complex and write $s = \sigma + it$, where σ and t are real. Since $f(x) = 0$ for $x < 0$, we can write

$$F(s) = \int_{-\infty}^\infty f(x) e^{-(\sigma+it)x}\, dx = \int_{-\infty}^\infty f(x) e^{-\sigma x} e^{-itx}\, dx.$$

Thus we see that for each fixed σ, $F(s)$ is the Fourier transform of $f(x)e^{-\sigma x}$:

$$\mathscr{L}[f] = \Phi[f(x) e^{-\sigma x}]. \tag{7.97}$$

It can be shown that if f is piecewise continuous on each finite interval and $|f(x)| < ke^{bx}$, $x \ge 0$, for some constants k and b, then for each $\sigma > b$, the Laplace transform $F(s) = F(\sigma + it)$ is defined for $-\infty < t < \infty$.

EXAMPLE   Let $f(x) = 1$ for $x = 0$ (and $f(x) = 0$ for $x < 0$). Then

$$\mathscr{L}[f] = \Phi[f(x) e^{-\sigma x}] = \int_0^\infty e^{-\sigma x} e^{-itx}\, dx = \frac{1}{\sigma + it} = \frac{1}{s}.$$

Here $|f(x)| < 2e^{0x}$ for $x \geq 0$, so that we can take $b = 0$, and $\mathscr{L}[f] = F$, where $F(s) = 1/s = 1/(\sigma + it)$ for $\sigma > 0$. ∎

Since the Laplace transform is related to the Fourier transform by (7.97), we can apply the theory of the Fourier transform to Laplace transforms. In particular, if $\mathscr{L}[f] = F(s) = F(\sigma + it)$, then Eq. (7.89) allows us to write, under appropriate conditions on f,

$$f(x)e^{-\sigma x} = \frac{1}{2\pi} \int_{-\infty}^{\infty} F(\sigma + it)e^{ixt}\,dt$$

(where, as for (7.89), the integral is a principal value). We can rewrite this result as follows:

$$f(x) = \frac{1}{2\pi} \int_{-\infty}^{\infty} F(\sigma + it)e^{(\sigma+it)x}\,dt = \frac{1}{2\pi} \int_{-\infty}^{\infty} F(s)e^{sx}\,dt. \qquad (7.98)$$

This is the *inversion formula* for Laplace transforms. It permits us to recover a function $f(x)$ from its Laplace transform $F(s)$. The existence of such a formula, as of the analogous formula for Fourier transforms, shows that we are dealing with *one-to-one* mappings.

For $f(x)$ as in the example, we found $F(s) = 1/s$ for $\sigma > 0$. Hence (with principal values understood)

$$\frac{1}{2\pi} \int_{-\infty}^{\infty} \frac{1}{s}e^{sx}\,dt = \frac{1}{2\pi} \int_{-\infty}^{\infty} \frac{e^{(\sigma+it)x}}{\sigma + it}\,dt = \begin{cases} 1, & x > 0 \\ 0, & x < 0. \end{cases} \qquad (7.99)$$

At $x = 0$, f has a jump discontinuity, and the integral converges to the expected average value, namely, $\frac{1}{2}$.

We can list formally other properties of Laplace transforms, all deducible from the corresponding properties of Fourier transforms: The Laplace transform is a *linear operator*. Corresponding to the formula

$$\Phi[f^{(k)}] = (it)^k \Phi[f],$$

one has

$$\mathscr{L}[f^{(k)}(x)] = s^k \mathscr{L}[f]. \qquad (7.100)$$

However, this is valid only if f (regarded as 0 for $x < 0$) and the derivatives occurring have no discontinuities on the infinite x-axis. If f has continuous derivatives through order n, when regarded as a function on the interval $x \geq 0$, then the formula becomes

$$\mathscr{L}[f^{(k)}(x)] = s^k \mathscr{L}[f] - [f^{(k-1)}(0) + sf^{(k-2)}(0) + \cdots + s^{k-1}f(0)], \qquad (7.101)$$

where all derivatives are regarded as *right-hand* derivatives.

The formulas (7.92) and (7.93) become

$$\mathscr{L}[f(x - c)] = e^{-cs}\mathscr{L}[f], \qquad c > 0, \qquad (7.102)$$

$$\mathscr{L}[e^{bx}f(x)] = F(s - b), \quad \text{where} \quad F(s) = \mathscr{L}[f]. \qquad (7.103)$$

The formula (7.95) becomes

$$\mathcal{L}[f*g] = \mathcal{L}[f]\mathcal{L}[g], \tag{7.104}$$

where, since $f = 0$ and $g = 0$ for $x < 0$, the convolution (7.94) can be simplified to $f * g = h$, where

$$h(x) = \int_0^x f(u)g(x-u)\, du. \tag{7.105}$$

This automatically makes $h(x) = 0$ for $x < 0$.

For further information on Laplace transforms, see the books by Churchill, Kaplan, and Widder listed at the end of the chapter.

PROBLEMS

1. **a)** Prove that the functions (7.79) form an orthogonal system in the rectangle R: $-\pi \leq x \leq \pi$, $-\pi \leq y \leq \pi$.

   b) Expand $f(x, y) = x^2y^2$ in a double Fourier series in R.

2. Represent the following functions as Fourier integrals:

   a) $f(x) = 0$, $x < 0$; $f(x) = e^{-x}$, $x \geq 0$

   b) $f(x) = 0$, $x < 0$; $f(x) = 1$, $0 \leq x \leq 1$; $f(x) = 0$, $x > 1$

3. Show that if f is even, then its Fourier integral reduces to

   $$\int_0^\infty \alpha(t) \cos xt \, dt, \qquad \alpha(t) = \frac{2}{\pi} \int_0^\infty f(x) \cos xt \, dx$$

   and that if f is odd, its Fourier integral reduces to

   $$\int_0^\infty \beta(t) \sin xt \, dt, \qquad \beta(t) = \frac{2}{\pi} \int_0^\infty f(x) \sin xt \, dx.$$

4. Find the Fourier transform of each function:

   a) $f(x) = 1$, $-1 \leq x \leq 1$; $f(x) = 0$ otherwise

   b) $f(x) = xe^{-x}$, $x \geq 0$; $f(x) = 0$ otherwise

   c) $f(x) = x^2$, $-1 \leq x \leq 1$; $f(x) = 0$ otherwise

5. **a)** Prove the rule (7.92).   **b)** Prove the rule (7.93).

6. It is shown in Section 6.25 that for $k = 0, 1, 2, \ldots$, $\mathcal{L}[x^k] = k!/s^{k+1}$ for real positive s.

   a) Show that the result is valid for complex $s = \sigma + it$, provided that $\sigma > 0$.

   b) Write out the corresponding inversion formula (7.98) for $f(x) = x^k$.

7. For each of the following choices of f and g, evaluate $h = f * g$ and verify that (7.104) holds. (Throughout, $f = 0$ and $g = 0$ for $x < 0$.)

   a) $f(x) = g(x) = 1$ for $x \geq 0$.

   b) $f(x) = 1$ and $g(x) = x$ for $x \geq 0$.

   c) $f(x) = 1$ and $g(x) = e^x$ for $x \geq 0$.

*7.20   GENERALIZED FUNCTIONS

A number of physical problems lead one to attempt to extend the concept of "function" to include new mathematical objects, called *generalized functions* or *distributions*, differing markedly from traditional functions. We give here an intuitive introduction to the subject and give references for a general discussion.

A simple example is that of density, say for mass distributed along a line, the x-axis. If $\rho(x)$ is the density (in units of mass per unit length), then $\int_a^b \rho(x)\, dx$ should give the total mass in the interval $a \leq x \leq b$. Now suppose that all our mass is concentrated in a single particle of mass 1, located at the origin. It is tempting to assign a density $\rho(x)$ such that $\int_a^b \rho(x)\, dx = 0$ if the interval $a \leq x \leq b$ does not contain the origin and such that $\int_a^b \rho(x)\, dx = 1$ if the interval does contain the origin. However, these conditions would make $\rho(x) = 0$ except "near the origin," where $\rho(x)$ must suddenly become infinite so rapidly that the integral of $\rho(x)$ over each small interval containing the origin is 1. This particular density "function" is the *Dirac delta function* $\delta(x)$. Thus we require:

$$\delta(x) = 0, \qquad x \neq 0; \qquad \delta(0) = \infty; \qquad \int_a^b \delta(x)\, dx = 1 \quad \text{if} \quad a < 0 < b.$$

$$(7.106)$$

The apparently contradictory nature of these properties led to their rejection as meaningless by mathematicians until recently. Then a number of mathematicians, notably L. Schwartz, developed systematic theories of generalized functions, which satisfy all the desired properties in a rigorous manner. The function $\delta(x)$ is such a generalized function.

We can arrive at the delta function in another way, which suggests a limit process for obtaining this and other generalized functions. Let $g(b)$ denote the total mass on the interval $-\infty < x \leq b$, for a distribution of mass (with finite total mass) on the x-axis. Thus for a continuous density $\rho(x)$, $g(b) = \int_{-\infty}^b \rho(x)\, dx$ and $g'(b) = \rho(b)$. Now for the case of a single particle of mass 1 at the origin we have difficulty in defining $\rho(x)$ but can easily find $g(x)$ for each x. For $g(x)$ is simply the total mass to the left of x; hence for the case of the single particle,

$$g(x) = 0 \quad \text{for} \quad x < 0, \qquad g(x) = 1 \quad \text{for} \quad x \geq 0. \qquad (7.107)$$

Thus $g(x)$ has a jump discontinuity at $x = 0$. We call the function $g(x)$ in (7.107) the *Heaviside unit function*.

Now from this $g(x)$ we can again try to obtain the density $\rho(x)$ as the *derivative* of $g(x)$. It is clear that $g'(x) = 0$ for $x \neq 0$, and it is reasonable to write $g'(0) = \infty$, because of the jump. Thus we have

$$g'(x) = 0 = \delta(x) \quad \text{for} \quad x \neq 0, \qquad g'(0) = \infty = \delta(0). \qquad (7.108)$$

Accordingly, we interpret the "function" $\delta(x)$ as the *derivative of the Heaviside unit function*. Now the unit function really has no derivative at $x = 0$, so we have not completely clarified the meaning of $\delta(x)$. To go further, we *approximate $g(x)$* by a smooth function $g_\epsilon(x)$ having a derivative for all x; we choose an $\epsilon > 0$ and set

$$g_\epsilon(x) = 1 - \tfrac{1}{2}e^{-x/\epsilon} \quad \text{for} \quad x \geq 0, \quad g_\epsilon(x) = \tfrac{1}{2}e^{x/\epsilon} \quad \text{for} \quad x < 0 \quad (7.109)$$

(see Fig. 7.16). The function $g_\epsilon(x)$ has been chosen so that as $\epsilon \to 0 +$, $g_\epsilon(x) \to g(x)$ (except at $x = 0$, where $g_\epsilon(x) = \tfrac{1}{2}$ for all ϵ). Furthermore, $g_\epsilon(x)$ has a derivative

$$g_\epsilon'(x) = \frac{1}{2\epsilon}e^{-x/\epsilon} \quad \text{for} \quad x \geq 0, \qquad g_\epsilon'(x) = \frac{1}{2\epsilon}e^{x/\epsilon} \quad \text{for} \quad x < 0,$$

$$(7.110)$$

and since $g_\epsilon(x) \to g(x)$ as $\epsilon \to 0 +$, it is reasonable to consider $\delta(x)$ as the limit of $g_\epsilon'(x)$ as $\delta \to 0 +$. Thus $\delta(x)$ is considered to be the limit of the "pulse" graphed in Fig. 7.17 (a smooth approximation to $g'(x)$), as $\epsilon \to 0 +$. However, if we do indeed pass to the limit, we again obtain the value 0 for $x \neq 0$, the "value" ∞ for $x = 0$. It is thus not clear what we have gained.

We make headway by, in effect, stopping just short of the limit: We regard $\delta(x)$ as a function *very closely approximated by* the pulse of Fig. 7.17; the smaller the ϵ, the better the approximation. Thus to evaluate the integral of $\delta(x)$ we write

$$\int_a^b \delta(x)\, dx = \lim_{\epsilon \to 0+} \int_a^b g_\epsilon'(x)\, dx. \qquad (7.111)$$

From (7.110) we easily find, say for $a < 0 < b$,

$$\lim_{\epsilon \to 0+} \int_a^b g_\epsilon'(x)\, dx = \lim_{\epsilon \to 0+} \left[1 - \tfrac{1}{2}(e^{a/\epsilon} + e^{-b/\epsilon})\right] = 1,$$

and hence the integral of $\delta(x)$ from a to b is 1. Similarly, if $a < b < 0$ or $0 < a < b$, the integral is found to be 0. (The secret here is in the interchange of two limit processes, as a closer examination shows.)

The procedure followed here for $\delta(x)$ can be extended to other generalized functions. In general, it provides a method for assigning derivatives to a large class of functions (including all piecewise continuous functions) not having

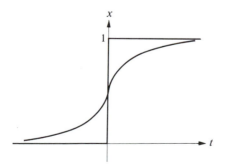

Figure 7.16 Smooth approximation of Heaviside unit function.

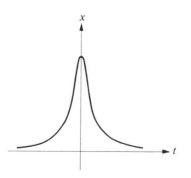

Figure 7.17 Function approximating the Dirac delta function.

derivatives in the ordinary sense. A systematic development based on the ideas suggested here is given in the book by Lighthill listed at the end of the chapter. Here we simply list some of the results obtained and confine attention to generalized functions related to $\delta(x)$.

We find that generalized functions can be differentiated arbitrarily often and hence obtain $\delta'(x), \delta''(x), \ldots$. The first derivative $\delta'(x)$ can be approximated by the derivative of the pulse of Fig. 7.17; hence it appears as in Fig. 7.18. Similar approximations are obtained for higher derivatives.

A simple translation along the x-axis provides us with generalized functions of form $\delta(x - c), \delta'(x - c), \ldots$ for constant c.

An ordinary function, say piecewise continuous, can also be regarded as a generalized function. We confine attention here to ordinary functions that are formed of a finite number of pieces having derivatives of all orders—for example,

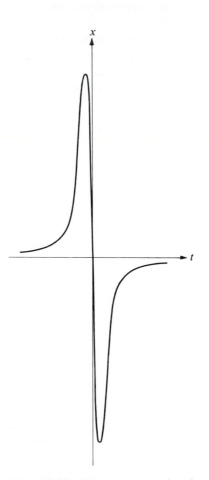

Figure 7.18   Function approximating the derivative $\delta'(x)$.

the function $f(x) = 0$ for $x < 0$, $= e^{-x}$ for $x \geq 0$. If we approximate f by a smooth function $f_{\epsilon}(x)$, as we did for $g(x)$ above, we are led to the conclusion that $f(x)$ has a derivative $f'(x) = \delta(x) + h(x)$, where $h(x) = 0$ for $x < 0$ and $h(x) = -e^{-x}$ for $x > 0$ (the value at 0 being immaterial). Similarly, $f''(x)$ is found to be $\delta'(x) - \delta(x) + f(x)$.

In general, we are led in this way to generalized functions of form

$$f(x) + a_1\delta(x - c_1) + a_2\delta(x - c_2) + \cdots + b_1\delta'(x - d_1) + \cdots$$
$$+ e_1\delta''(x - p_1) + \cdots, \quad (7.112)$$

that is, of the form ordinary function (of the type described) plus a linear combination of translated delta functions and translated derivatives up to a certain order of the delta function. Terms with 0 coefficient are to be suppressed. We now confine our attention to generalized function of this type.

Operations on generalized functions.   We *add* or *subtract* generalized functions of the type considered in the obvious way. For example,

$$(e^x + \delta(x)) + (\sin x + \delta(x) + \delta'(x - 1)) = e^x + \sin x + 2\delta(x) + \delta'(x - 1).$$

We can also *multiply* such a function by a scalar constant, by multiplying each term by the scalar. Multiplication of arbitrary generalized functions is not defined; in particular, $\delta(x) \cdot \delta(x)$ is not defined. However, if one of the two factors is an ordinary function $f(x)$ satisfying certain continuity conditions, then one can multiply f by each term of the other factor. In particular, one defines $f(x)\delta(x)$ to equal $f(0)\delta(x)$, provided that $f(x)$ is continuous at $x = 0$; one defines $f(x)\delta'(x)$ to equal $f(0)\delta'(x) - f'(0)\delta(x)$, provided that f and f' are continuous at $x = 0$. These rules can be justified by the limit process described before; the latter rule can also be justified by the rule for differentiating products. In general, one is led to the definition:

$$f(x)\delta^{(k)}(x - c) = \delta^{(k)}(x - c)f(x) = \sum_{r=0}^{k} (-1)^r \binom{k}{r} f^{(r)}(c)\delta^{(k-r)}(x - c),$$

$$(7.113)$$

provided that $f, \ldots, f^{(k)}$ are continuous at $x = c$. For example,

$$f(x)\delta''(x - c) = f(c)\delta''(x - c) - 2f'(c)\delta'(x - c) + f''(c)\delta(x - c).$$

One can differentiate a generalized function of the type considered, by differentiating each term: The derivative of $a\delta^{(k)}(x - c)$ is $a\delta^{(k+1)}(x - c)$; the derivative of $f(x)$ is the ordinary derivative plus contributions from the jump discontinuities. If c is such a discontinuity, then we add a term

$$\left[\lim_{x \to c+} f(x) - \lim_{x \to c-} f(x) \right] \delta(x - c).$$

Thus if $f(x) = 0$ for $x < 0$, $f(x) = \cos x$ for $0 < x < \pi/2$, and $f(x) = -1$ for $x > \pi/2$, then $f'(x) = g(x) + \delta(x) - \delta(x - (\pi/2))$, where $g(x) = 0$ for $x < 0$, $g(x) = -\sin x$ for $0 < x < \pi/2$, $g(x) = 0$ for $x > \pi/2$. One can verify that the usual rules for derivatives of sums and products hold.

Generalized functions can also be integrated. We consider here only definite integrals from a to b (where a may be $-\infty$ and b may be ∞) and exclude the case where terms of form $A \, \delta^{(k)}(x - c)$ occur, with $c = a$ or b. Then we integrate each term separately, with

$$\int_a^b \delta(x - c) \, dx = \begin{cases} 1 & \text{if } a < c < b, \\ 0 & \text{otherwise;} \end{cases}$$

$$\int_a^b \delta^{(k)}(x - c) \, dx = 0, \qquad k = 1, 2, \dots . \tag{7.114}$$

From these definitions we can now integrate certain products of the form $g(x) \, \delta^{(k)} \, (x - c)$: If $a < c < b$, then

$$\int_a^b g(x) \, \delta^{(k)}(x - c) \, dx = (-1)^k g^{(k)}(c), \tag{7.115}$$

provided that $g^{(k)}(x)$ is continuous at c. This rule follows from the previous rules, especially the rule (7.113) for multiplication. The rule (7.115) is a crucial one and is itself often made the starting point for the theory of generalized functions. It shows that $\delta^{(k)}(x - c)$ acts as a "linear functional," assigning real numbers to sufficiently smooth ordinary functions $g(x)$.

Fourier and Laplace transforms of generalized functions.   From the rule (7.115) we deduce that $\delta^{(k)}(x - c)$ has a Fourier transform:

$$\int_{-\infty}^{\infty} e^{-ixt} \, \delta^{(k)}(x - c) \, dx = (it)^k e^{-ict}. \tag{7.116}$$

In particular, $\Phi[\delta(x)] = 1$ (constant function). If we could write $\delta(x)$ in terms of its Fourier integral, we would have

$$\delta(x) = \frac{1}{2\pi} \int_{-\infty}^{\infty} e^{ixt} \, dt. \tag{7.117}$$

This equation states that the constant function 1 has a Fourier transform $2\pi\delta(t)$. One normally *defines* the Fourier transform of 1 to be $2\pi\delta(t)$, even though the reasoning leading to the definition is purely formal. The definition has been shown to lead to no inconsistencies and is a useful one. In a similar manner, from (7.116), one is led to assign to $x^k e^{-icx}$ the Fourier transform $2\pi i^k \, \delta^{(k)}(t + c)$.

The Fourier transform can be regarded as a one-to-one linear mapping of a certain vector space of functions onto a second vector space of functions. The process we are following can be regarded as an extension of the linear mapping to *larger* vector spaces. The extension is chosen to preserve linearity and one-to-one-ness.

There is a similar discussion of the Laplace transform, which we know to be essentially a special case of the Fourier transform. We find

$$\mathscr{L}\big[\delta^{(k)}(x - c)\big] = s^k e^{-cs}, \qquad c \geq 0. \tag{7.118}$$

It is a remarkable fact that for both Fourier and Laplace transforms, generalized functions permit one to circumvent the awkward rules for transforms of derivatives. One finds:

$$\Phi\big[f^{(k)}(x)\big] = (it)^k \Phi[f], \tag{7.119}$$

$$\mathscr{L}\big[f^{(k)}(x)\big] = s^k \mathscr{L}[f], \tag{7.120}$$

where f is as in the previous rules, except for a finite number of jump discontinuities, as allowed for generalized functions. In (7.119) and (7.120), $f^{(k)}(x)$ is treated as a *generalized* function so that it will contain terms in delta functions arising from jumps in f or its derivatives.

Convolutions can also be defined for generalized functions. If the rule "transform of convolution of two functions equals product of two transforms" is to hold, then we must require:

$$g(x) * \delta^{(k)}(x) = \delta^{(k)}(x) * g(x) = g^{(k)}(x), \tag{7.121}$$

where $g^{(k)}(x)$ is a generalized function. For in the case of Laplace transforms, (7.121) gives

$$\mathscr{L}\big[g(x) * \delta^{(k)}(x)\big] = s^k \mathscr{L}[g],$$

in agreement with (7.120). More generally, we are led to

$$g(x) * \delta^{(k)}(x - c) = \delta^{(k)}(x - c) * g(x) = g^{(k)}(x - c), \tag{7.122}$$

and then, by term-by-term operations (as for multiplication), we can obtain the convolution of two arbitrary generalized functions.

The fact that convolution becomes multiplication in "transform language" has also been used as a starting point for the theory of generalized functions. For details, see the book by Mikusiński listed at the end of the chapter.

PROBLEMS

1. **a)** Verify the graph of $g_\epsilon(x)$, shown in Fig. 7.16.

   b) Graph $g_\epsilon'(x)$, approximating $\delta(x)$ [cf. Fig. 7.17].

   c) Graph the derivative of the pulse of Fig. 7.17, to approximate $\delta'(x)$ [cf. Fig. 7.18].

   d) Graph the derivative of the function of Fig. 7.18 to approximate $\delta''(x)$.

2. Another smooth approximation of the Heaviside unit function is the function $h_\epsilon(x) = \frac{1}{2}(1 + \sin(x/\epsilon))$ for $-\pi/(2\epsilon) \leq x \leq \pi/(2\epsilon)$, $= 1$ for $x > \pi/(2\epsilon)$, $= 0$ for $x < -\pi/(2\epsilon)$.

   a) Graph $h_\epsilon(x)$.

   b) Graph $h_\epsilon'(x)$ and consider this as an approximation to $\delta(x)$ for small positive ϵ.

   c) Show that $\int_{-1}^{1} h'_\epsilon(x)\, dx$ has limit 1 as $\epsilon \to 0 +$.

3. Let $f(x) = 0$ for $x < 0$, $f(x) = 3 - x$ for $x > 0$. Let $g(x) = \cos x$ for $-\pi \leqq x \leqq \pi$, $g(x) = 0$ otherwise. Evaluate the following:

   a) $5f'(x) + 2g'(x)$ **b)** $f''(x)$

   c) $f(x)\delta(x - 1)$ **d)** $g(x)\delta'(x)$

   e) $f(x)g''(x)$ **f)** $\int_{-1}^{2} f(x)\delta'(x - 1)\, dx$

   g) $\int_{-1}^{\pi/2} g(x)\delta''(x)\, dx$ **h)** $\mathscr{L}\,[f''(x)]$

   i) $\Phi[g''(x)]$ **j)** $f(x) * \delta'(x)$

   k) $f(x) * f''(x)$

Suggested References

Churchill, Ruel V., and Brown, James W., Fourier Series and Boundary Value Problems, 4th ed. New York: McGraw-Hill, 1987.

Churchill, Ruel V., Operational Mathematics, 3rd ed. New York: McGraw-Hill, 1972.

Franklin, Philip, A Treatise on Advanced Calculus. New York: John Wiley and Sons, Inc., 1940.

Franklin, Philip, Fourier Methods. New York: McGraw-Hill, 1949.

Jackson, Dunham, Fourier Series and Orthogonal Polynomials (Carus Mathematical Monographs, No. 6). Menasha, Wisconsin: Mathematical Association of America, 1941.

Jahnke, E., and F. Emde, Tables of Higher Functions, 6th ed. revised by F. Losch. New York: McGraw-Hill, 1960.

Kaplan, Wilfred, Operational Methods for Linear Systems. Reading: Mass., Addison-Wesley, 1962.

Lighthill, M. J., An Introduction to Fourier Analysis and Generalized Functions. Cambridge: Cambridge University Press, 1958.

Mikusiński, Jan, Operational Calculus, 2nd ed. New York: Pergamon Press, 1983.

Rogosinski, Werner, Fourier Series (transl. by H. Cohn and F. Steinhardt). New York: Chelsea Publishing Co., 1950.

Szegö, Gabor, Orthogonal Polynomials (American Mathematical Society Colloquium Publications, Vol. 23), 4th ed. New York: American Mathematical Society, 1975.

Titchmarsh, E. C., Eigenfunction Expansions, 2nd ed. Oxford: Oxford University Press, 1962.

Titchmarsh, E. C., Theory of Fourier Integrals. Oxford: Oxford University Press, 1937.

Watson, G. N., Theory of Bessel Functions, 2nd ed. New York: Macmillan, 1944.

Whittaker, E. T., and G. N. Watson, Modern Analysis, 4th ed. Cambridge: Cambridge University Press, 1940.

Widder, D. V., The Laplace Transform. Princeton: Princeton University Press, 1946.

Wiener, Norbert, The Fourier Integral. Cambridge: Cambridge University Press, 1933.

Zygmund, A., Trigonometric Series. 2 Vols. London: Cambridge University Press, 1959.

8

Functions of A
Complex Variable

In this chapter we introduce the theory of analytic functions of a complex variable. The principal topics are series expansions, integrals, and residues. For a discussion of conformal mapping and its applications the reader is referred to Chapter 9.

8.1 COMPLEX FUNCTIONS

We assume familiarity with the complex number system. Figure 8.1 reviews standard notations for the complex z-plane, where $z = x + iy$. We shall denote real and imaginary parts as follows:

$$\text{for} \quad z = x + iy, \qquad x = \operatorname{Re} z, \qquad y = \operatorname{Im} z.$$

We shall make steady use of all these notations, as well as the analogous notations in a complex w-plane, where $w = u + iv$.

For addition and multiplication by real scalars, complex numbers can be regarded as vectors in the plane. In particular, $|z_1|, |z_2|, |z_1 + z_2|$ are the sides of a triangle, as in Fig. 8.1, so that one has the familiar *triangle inequality*

$$|z_1 + z_2| \le |z_1| + |z_2|. \tag{8.1}$$

Complex-valued functions of z. If to each value of the complex number $z = x + iy$, with certain exceptions, there is assigned a value of the complex number $w = u + iv$, then w is given as a *complex-valued function* of z, and we write $w = f(z)$. For

557

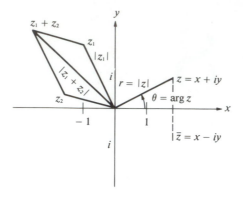

Figure 8.1   The complex z-plane.

example,

$$w = z^2, \qquad w = z^3 + 5z + 7, \qquad w = \frac{z+1}{z-2} \qquad (z \neq 2)$$

are such functions. Important functions of this type are

polynomials: $w = a_0 z^n + \cdots + a_{n-1} z + a_n,$

rational functions: $w = \dfrac{a_0 z^n + \cdots + a_n}{b_0 z^m + \cdots + b_m},$

exponential function: $\exp z = e^z = e^{x+iy} = e^x(\cos y + i \sin y),$

trigonometric functions: $\sin z = \dfrac{e^{iz} - e^{-iz}}{2i}, \qquad \cos z = \dfrac{e^{iz} + e^{-iz}}{2},$

hyperbolic functions: $\sinh z = \dfrac{e^z - e^{-z}}{2}, \qquad \cosh z = \dfrac{e^z + e^{-z}}{2}.$

The definition of the exponential function is motivated by interpreting e^z for complex z as the sum of its power series $\Sigma z^n/n!$; see Section 6.19. From the definition it follows that for real y,

$$e^{iy} = \cos y + i \sin y \qquad \text{and} \qquad e^{-iy} = \cos y - i \sin y,$$

so that

$$\sin y = \frac{e^{iy} - e^{-iy}}{2i}, \qquad \cos y = \frac{e^{iy} + e^{-iy}}{2}.$$

These equations suggest the definitions given above for $\sin z$ and $\cos z$. The definitions of the hyperbolic functions are based on the usual definitions for real variables. By the reasoning described, it follows that when z is real ($z = x + i0$), each of the five functions reduces to the familiar real function. For example, $e^{x+i0} = e^x$.

The other trigonometric and hyperbolic functions are defined in terms of the sine, cosine, sinh, and cosh in the usual way.

8.2 COMPLEX-VALUED FUNCTIONS OF A REAL VARIABLE

It will be convenient to represent paths for line integrals in the complex plane by equations of form

$$z = F(t), \qquad a \le t \le b. \tag{8.2}$$

Here t is a real variable and F is a function whose values are complex, so that we are dealing with a complex-valued function of a real variable. Examples are the following functions:

$$z = e^{it}, \qquad 0 \le t \le 2\pi; \qquad z = t + it^2, \qquad 0 \le t \le 1.$$

In (8.2) we can write $z = x + iy$ and $F(t) = f(t) + ig(t)$, where f and g are real-valued. Then (8.2) is equivalent to the *pair* of equations

$$x = f(t), \qquad y = g(t), \qquad a \le t \le b. \tag{8.2'}$$

We can also consider (8.2) as an alternative to the familiar vector function representation of a path.

The path $z = e^{it}$ given previously is equivalent to the path $x = \cos t$, $y = \sin t$, and hence its graph is a circle.

The calculus can be developed for complex-valued functions of t in strict analogy with the development for real-valued functions. We write

$$\lim_{t \to t_0} F(t) = c = x_0 + iy_0 \tag{8.3}$$

if, given $\epsilon > 0$, we can choose $\delta > 0$ so that $|F(t) - c| < \epsilon$ when $0 < |t - t_0| < \delta$. It is assumed here that $F(t)$ is defined for t sufficiently close to t_0, but not necessarily for $t = t_0$; if $F(t)$ is defined only in an interval $t_0 < t < \beta$, the limit is interpreted as a limit *to the right* and is written

$$\lim_{t \to t_0+} F(t).$$

Limits to the left are defined similarly. Limits as $t \to \infty$ or $t \to -\infty$ are defined as for real functions. [However,

$$\lim_{t \to t_0} F(t) = \infty$$

is defined to mean

$$\lim_{t \to t_0} |F(t)| = \infty;$$

there is no concept of $+\infty$ or $-\infty$ for complex numbers; see Section 8.14.] If $F(t)$ is defined for $\alpha < t < \beta$ and t_0 lies in this interval, then $F(t)$ is said to be *continuous* at t_0 if

$$\lim_{t \to t_0} F(t) = F(t_0).$$

If $F(t)$ is also defined at $t = \alpha$ and

$$\lim_{t \to \alpha+} F(t) = F(\alpha),$$

then $F(t)$ is said to be continuous to the right at $t = \alpha$. Continuity to the left and continuity in an interval are defined as for real functions.

If $F(t) = f(t) + ig(t)$, then Eq. (8.3) is equivalent to the two equations

$$\lim_{t \to t_0} f(t) = x_0, \qquad \lim_{t \to t_0} g(t) = y_0. \tag{8.4}$$

For Eq. (8.3) signifies that $F(t)$ is as close to c as desired, for t sufficiently close to t_0; by a geometric argument we see that this is equivalent to the requirement that $f = \operatorname{Re} F$ be as close to x_0 as desired and $g = \operatorname{Im} F$ be as close to y_0 as desired, for t sufficiently close to t_0. Similarly, continuity of $F(t)$ at t_0 is equivalent to continuity of $f(t)$ and $g(t)$ at t_0. Accordingly, such functions as $t^2 + it^3$ and e^{it} are continuous for all t. Furthermore, the rules for limits of sums, products, and quotients and the analogous continuity theorems must carry over to the complex case. For example, if $F_1(t) = f_1(t) + ig_1(t)$ and $F_2(t) = f_2(t) + ig_2(t)$ are continuous in an interval, then so also is

$$F_1(t) \cdot F_2(t) = \left[f_1(t) + ig_1(t) \right] \cdot \left[f_2(t) + ig_2(t) \right]$$
$$= f_1(t)f_2(t) - g_1(t)g_2(t) + i\left[f_1(t)g_2(t) + f_2(t)g_1(t) \right].$$

For $f_1(t), g_1(t), f_2(t), g_2(t)$ must be continuous, so that the real and imaginary parts of $F_1(t) \cdot F_2(t)$ are continuous, and hence $F_1(t) \cdot F_2(t)$ is continuous.

For the discussion thus far we are to some extent repeating the theory of vector functions. However, we remark that complex multiplication and division have no analogue for vectors.

Derivatives and integrals.   The derivative of $F(t)$ can be defined as for real functions:

$$F'(t_0) = \lim_{\Delta t \to 0} \frac{F(t_0 + \Delta t) - F(t_0)}{\Delta t}. \tag{8.5}$$

By taking real and imaginary parts and applying (8.4) we conclude that

$$F'(t_0) = f'(t_0) + ig'(t_0); \tag{8.6}$$

that is, $F(t)$ has a derivative at t_0 precisely when $f(t), g(t)$ have derivatives at t_0, and the derivatives are related by Eq. (8.6). Derivatives to the left or right are defined by requiring that $\Delta t < 0$ or $\Delta t > 0$, respectively, in Eq. (8.5); Eq. (8.6) also applies to these derivatives.

The rules for derivative of sum, product, quotient, constant, and constant times function all carry over, and the proofs for real functions can be repeated. We also note the rules

$$\frac{d}{dt}[F(t)]^n = n[F(t)]^{n-1}F'(t) \qquad (n = 1, 2, \ldots), \tag{8.7}$$

$$\frac{d}{dt} e^{(a+bi)t} = (a + bi)e^{(a+bi)t} \qquad (a, b \text{ real}). \tag{8.8}$$

Furthermore, if $F'(t) \equiv 0$ for $\alpha < t < \beta$, then $F(t)$ is identically constant for $\alpha < t < \beta$. The proofs are left as exercises (Problems 9 to 11 below).

Higher derivatives are obtained by repeated differentiation:

$$F''(t) = [F'(t)]' = D^2F, \qquad F'''(t) = [F''(t)]' = D^3F,\dots.$$

The first derivative of $F(t)$ can be thought of as the *velocity vector* of the point (x, y) as it moves on the path $x = f(t), y = g(t)$, with t as *time*. The second derivative can be interpreted as *acceleration*.

The *definite integral* of $F(t)$ over an interval $\alpha \le t \le \beta$ is defined as a limit of a sum $\Sigma F(t_k^*) \Delta_k t$ as for real functions. However, again the limit theorem permits us to take real and imaginary parts:

$$\int_\alpha^\beta F(t)\, dt = \int_\alpha^\beta f(t)\, dt + i \int_\alpha^\beta g(t)\, dt. \tag{8.9}$$

If $F(t)$ is continuous over the interval, then $f(t)$ and $g(t)$ are continuous, so the integral exists. More generally, the integral exists if $f(t)$ and $g(t)$ are *piecewise continuous* for $\alpha \le t \le \beta$. When f and g are piecewise continuous, we term $F = f + ig$ piecewise continuous.

An *indefinite integral* of $F(t)$ is defined as a function $G(t)$ whose derivative is $F(t)$. As in ordinary calculus, we find that if $G(t)$ is one indefinite integral, then $G(t) + c$ provides all indefinite integrals:

$$\int F(t)\, dt = G(t) + c,$$

c being an arbitrary complex constant. If an indefinite integral G of F is known, then it can be used to evaluate definite integrals of F as in calculus:

$$\int_\alpha^\beta F(t)\, dt = \int_\alpha^\beta G'(t)\, dt = G(\beta) - G(\alpha). \tag{8.10}$$

The proof is left as an exercise (Problem 12).

From Eq. (8.9) we can verify the familiar rules for the integral of a sum, the integral of constant times function, the combination of integrals from α to β and from β to γ, and integration by parts. We also have the basic inequality

$$\left| \int_\alpha^\beta F(t)\, dt \right| \le \int_\alpha^\beta |F(t)|\, dt \le M(\beta - \alpha); \tag{8.11}$$

this is valid if $\alpha < \beta$, if $F(t)$ is, for example, piecewise continuous for $\alpha \le t \le \beta$, and if $|F(t)| \le M$ on this interval. The inequality is most easily obtained from the definition of the integral as limit of a sum; for we have, by repeated application of the triangle inequality (8.1),

$$\left| \sum_{k=1}^n F(t_k^*) \Delta_k t \right| \le \sum_{k=1}^n |F(t_k^*)| \Delta_k t \le M(\beta - \alpha),$$

and passage to the limit gives (8.11). Definite integrals and indefinite integrals are related by the rule

$$\frac{d}{dt} \int_\alpha^t F(u)\, du = F(t) \tag{8.12}$$

(see Problem 12).

EXAMPLE 1 $\displaystyle\int_1^2 (t + it^2)\, dt = \left(\frac{t^2}{2} + i\frac{t^3}{3}\right)\Bigg|_1^2 = \frac{3}{2} + \frac{7}{3}i.$ ∎

EXAMPLE 2 $\displaystyle\int_0^1 e^{(a+bi)t}\, dt = \frac{e^{(a+bi)t}}{a + bi}\Bigg|_0^1 = \frac{e^{a+bi} - 1}{a + bi}$ $(a + bi \neq 0).$ ∎

EXAMPLE 3

$$\int p(t) e^{-at}\, dt = -e^{-at}\left[\frac{p(t)}{a} + \frac{p'(t)}{a^2} + \cdots + \frac{p^{(n)}(t)}{a^{n+1}}\right] + C,$$

where $p(t)$ is a polynomial of degree n and a is a complex constant, not 0. The equation is established by integration by parts [Problem 8(f)]. ∎

PROBLEMS

1. **a)** Let $z_1 = r_1(\cos \theta_1 + i \sin \theta_1)$, $z_2 = r_2(\cos \theta_2 + i \sin \theta_2)$. Show that
$$z_1 \cdot z_2 = r_1 r_2 [\cos (\theta_1 + \theta_2) + i \sin (\theta_1 + \theta_2)].$$
    b) Show that $e^{i(\theta_1 + \theta_2)} = e^{i\theta_1} \cdot e^{i\theta_2}$ [see part (a)].

2. Prove the following identities:
    a) $e^{z_1} \cdot e^{z_2} = e^{z_1 + z_2}$ [see Problem 1(b)]
    b) $(e^z)^n = e^{nz}$ $(n = 0, \pm 1, \pm 2, \dots)$
    c) $\sin^2 z + \cos^2 z = 1$
    d) $\sin (z_1 + z_2) = \sin z_1 \cos z_2 + \cos z_1 \sin z_2$
    e) $\mathrm{Re}\,(\sin z) = \sin x \cosh y$, $\mathrm{Im}\,(\sin z) = \cos x \sinh y$
    f) $\sin iz = i \sinh z$, $\cos iz = \cosh z$
    g) $\overline{e^z} = e^{\bar{z}}$, $\overline{\sin z} = \sin \bar{z}$, $\overline{\cos z} = \cos \bar{z}$

3. **a)** Prove that $e^z \neq 0$ for all z.
    b) Prove that $\sin z$ and $\cos z$ are 0 only for appropriate real values of z.

4. Represent the following functions graphically:
    a) $w = (1 + t) + i(1 - t)$ **b)** $w = t^4 + i(t^2 + 1)$
    c) $w = e^{3it}$ **d)** $w = 2e^{(-1+2i)t}$
    e) $w = te^{(-1+2i)t}$ **f)** $w = e^{-t} - ie^{it}$

5. Find the derivatives of the functions of Problem 4.

6. Graph $w = 3e^{2it}$ and indicate the first and second derivatives graphically for $t = 0$, $t = \pi/2$, $t = \pi$.

7. Integrate the functions of Problem 4 from 0 to 1.

8. Use integration by parts to evaluate each of the following:

    a) $\displaystyle\int (1 + it)^2 \sin t\, dt$

    b) $\displaystyle\int t^n e^{-at}\, dt$ $(n = 1, 2, \dots)$

c) $\int t^n \sin bt \, dt = \dfrac{1}{2i} \int t^n (e^{bit} - e^{-bit}) \, dt \quad (n = 1, 2, \cdots)$

d) $\int t^n \cos at \, dt = \text{Re} \int t^n e^{iat} \, dt \quad (a \text{ real}, n = 1, 2, \ldots)$

e) $\int t^n \cos at \cos bt \cos ct \, dt \quad (n = 1, 2, \ldots)$

f) $\int p(t) e^{-at} \, dt$, where $p(t)$ is a polynomial of degree n (Example 3 in the text)

9. Prove Eq. (8.7) by induction (repeated application of rule for differentiation of a product).
10. Prove (8.8) with the aid of (8.6).
11. Prove that if $F'(t) \equiv 0$, $\alpha < t < \beta$, then $F(t) \equiv$ const for $\alpha < t < \beta$.
12. **a)** Prove (8.12) by taking real and imaginary parts.
    b) Prove (8.10) either directly or as a consequence of (8.12).

8.3 COMPLEX-VALUED FUNCTIONS OF A COMPLEX VARIABLE · LIMITS AND CONTINUITY

We return to the general complex-valued function of a complex variable. These functions will be our principal concern for the remainder of this chapter. We write

$$w = f(z),$$

where $z = x + iy$, $w = u + iv$, to indicate such a function. An example is the function

$$w = z^2 \qquad (\text{all } z).$$

Here we can also write:

$$u + iv = (x + iy)^2 = x^2 - y^2 + 2ixy,$$

so that (on taking real and imaginary parts)

$$u = x^2 - y^2, \qquad v = 2xy.$$

In a similar manner, *every* complex function $w = f(z)$ is equivalent to a pair of real functions:

$$u = u(x, y) = \text{Re}\,[f(z)], \qquad v = v(x, y) = \text{Im}\,[f(z)],$$

of the two real variables x, y. Also from such a pair of real functions, defined on the same set, we obtain a complex function of z. For example,

$$u = x^2 + xy + y^2, \qquad v = xy^3$$

is equivalent to the complex function

$$w = f(z) = x^2 + xy + y^2 + xy^3 i,$$

for which $f(1 + 2i) = 1 + 2 + 4 + 8i = 7 + 8i$.

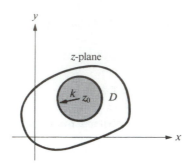

Figure 8.2   Domain and neighborhood.

The functions e^z, $\sin z$, $\cos z$, $\sinh z$, and $\cosh z$ were defined in Section 8.1. For these the corresponding pairs of real functions are as follows:

$$w = e^z: \qquad u = e^x \cos y, \qquad v = e^x \sin y,$$

$$w = \sin z: \qquad u = \sin x \cosh y, \qquad v = \cos x \sinh y,$$

$$w = \cos z: \qquad u = \cos x \cosh y, \qquad v = -\sin x \sinh y,$$

$$w = \sinh z: \qquad u = \sinh x \cos y, \qquad v = \cosh x \sin y,$$

$$w = \cosh z: \qquad u = \cosh x \cos y, \qquad v = \sinh x \sin y. \qquad (8.13)$$

The proofs are left as exercises (Problem 1 following Section 8.4). In (8.13), each function is defined for all z, that is, for all (x,y).

In general, we assume $w = f(z)$ to be defined in a *domain* (open region) D in the z-plane, as suggested in Fig. 8.2. If z_0 is a point of D, we can then find a circular *neighborhood* $|z - z_0| < k$ about z_0 in D. If $f(z)$ is defined in such a neighborhood, except perhaps at z_0, then we write

$$\lim_{z \to z_0} f(z) = w_0 \qquad (8.14)$$

if for every $\epsilon > 0$ we can choose $\delta > 0$, so that

$$|f(z) - w_0| < \epsilon \quad \text{for} \quad 0 < |z - z_0| < \delta. \qquad (8.15)$$

If $f(z_0)$ is defined and equals w_0 and (8.14) holds, then we call $f(z)$ *continuous at z_0.*

THEOREM 1   The function $w = f(z)$ is continuous at $z_0 = x_0 + iy_0$ if and only if $u(x, y) = \text{Re}[f(z)]$ and $v(x, y) = \text{Im}[f(z)]$ are continuous at (x_0, y_0).

Thus $w = z^2 = x^2 - y^2 + 2ixy$ is continuous for all z, since $u = x^2 - y^2$ and $v = 2xy$ are continuous for all (x, y). The proof of Theorem 1 is left as an exercise (Problem 5 following Section 8.4).

THEOREM 2   The sum, product, and quotient of continuous functions of z are continuous, except for division by zero; a continuous function of a continuous function is continuous. Similarly, if the limits exist,

$$\lim_{z \to z_0} [f(z) + g(z)] = \lim_{z \to z_0} f(z) + \lim_{z \to z_0} g(z),\dots . \tag{8.16}$$

These properties are proved as for real variables. (It is assumed in Theorem 2 that the functions are defined in appropriate domains.)

It follows from Theorem 2 that polynomials in z are continuous for all z and that each rational function is continuous except where the denominator is zero. From Theorem 1 it follows that

$$e^z = e^x \cos y + i e^x \sin y$$

is continuous for all z. Hence, by Theorem 2, so also are the functions

$$\sin z = \frac{e^{iz} - e^{-iz}}{2i}, \qquad \cos z = \frac{e^{iz} + e^{-iz}}{2}.$$

We write

$$\lim_{z \to z_0} f(z) = \infty \quad \text{if} \quad \lim_{z \to z_0} |f(z)| = +\infty;$$

that is, if for each real number K there is a positive δ such that $|f(z)| > K$ for $0 < |z - z_0| < \delta$. Similarly, if $f(z)$ is defined for $|z| > R$, for some R, then $\lim_{z \to \infty} f(z) = c$ if for each $\epsilon > 0$ we can choose a number R_0 such that $|f(z) - c| < \epsilon$ for $|z| > R_0$. All these definitions emphasize that there is but *one* complex number ∞ and that "approaching ∞" is equivalent to receding from the origin.

8.4   DERIVATIVES AND DIFFERENTIALS

Let $w = f(z)$ be given in D and let z_0 be a point of D. Then w is said to have a derivative $f'(z_0)$ if

$$\lim_{\Delta z \to 0} \frac{f(z_0 + \Delta z) - f(z_0)}{\Delta z} = f'(z_0).$$

In appearance this definition is the same as that for functions of a real variable, and it will be seen that the derivative does have the usual properties. However, it will also be shown that if $w = f(z)$ has a continuous derivative in a domain D, then $f(z)$ has a number of additional properties; in particular, the second derivative $f''(z)$, third derivative $f'''(z), \dots$, must also exist in D.

The reason for the remarkable consequences of possession of a derivative lies in the fact that the increment Δz is allowed to approach zero in any manner. If we restricted Δz so that $z_0 + \Delta z$ approached z_0 along a particular line, then we would obtain a "directional derivative." But here the limit obtained is required to be the *same for all directions*, so that the "directional derivative" has the same value in all directions. Moreover, $z_0 + \Delta z$ may approach z_0 in a quite arbitrary manner—for example, along a spiral path. The limit of the ratio $\Delta w / \Delta z$ must be the same for all manners of approach.

We say that $f(z)$ has a *differential* $dw = c\,\Delta z$ at z_0 if $f(z_0 + \Delta z) - f(z_0) = c\,\Delta z + \epsilon\,\Delta z$, where ϵ depends on Δz and is continuous at $\Delta z = 0$, with value zero when $\Delta z = 0$.

THEOREM 3 If $w = f(z)$ has a differential $dw = c\,\Delta z$ at z_0, then w has a derivative $f'(z_0) = c$. Conversely, if w has a derivative at z_0, then w has a differential at z_0: $dw = f'(z_0)\,\Delta z$.

This is proved just as for real functions. We also write $\Delta z = dz$, as for real variables, so that

$$dw = f'(z)\,dz, \qquad \frac{dw}{dz} = f'(z).$$

From Theorem 3 we see that existence of the derivative $f'(z_0)$ implies continuity of f at z_0, for

$$f(z_0 + \Delta z) - f(z_0) = c\,\Delta z + \epsilon\,\Delta z \to 0$$

as $\Delta z \to 0$.

THEOREM 4 If w_1 and w_2 are functions of z that have differentials in D, then

$$d(w_1 + w_2) = dw_1 + dw_2,$$
$$d(w_1 w_2) = w_1\,dw_2 + w_2\,dw_1, \tag{8.17}$$
$$d\frac{w_1}{w_2} = \frac{w_2\,dw_1 - w_1\,dw_2}{w_2^2} \qquad (w_2 \neq 0).$$

If w_2 is a differentiable function of w_1, and w_1 is a differentiable function of z, then wherever $w_2[w_1(z)]$ is defined,

$$\frac{dw_2}{dz} = \frac{dw_2}{dw_1} \cdot \frac{dw_1}{dz}. \tag{8.18}$$

If w is a differentiable function of z, and z is a differentiable function of the real variable t, then wherever $w[z(t)]$ is defined,

$$\frac{dw}{dt} = \frac{dw}{dz} \cdot \frac{dz}{dt}. \tag{8.19}$$

These rules are proved as in elementary calculus. We can now prove as usual the basic rule:

$$\frac{d}{dz}z^n = nz^{n-1} \qquad (n = 1, 2, \ldots). \tag{8.20}$$

Furthermore, the derivative of a constant is zero.

PROBLEMS

1. For each of the following, write the given function as two real functions of x and y and determine where the given function is continuous:

a) $w = (1 + i)z^2$

b) $w = \dfrac{z}{z + i}$

c) $w = \tan z = \dfrac{\sin z}{\cos z}$

d) $w = \dfrac{e^{-z}}{z + 1}$

e) $w = e^z$

f) $w = \sin z$

g) $w = \cos z$

h) $w = \sinh z$

i) $w = \cosh z$

j) $w = e^z \cos z$

2. Evaluate each of the following limits:

a) $\displaystyle\lim_{z \to \pi i} \dfrac{\sin z + z}{e^z + 2}$

b) $\displaystyle\lim_{z \to 0} \dfrac{z^2 - z}{2z}$

c) $\displaystyle\lim_{z \to 0} \dfrac{\cos z}{z}$

d) $\displaystyle\lim_{z \to \infty} \dfrac{z}{z^2 + 1}$

3. Differentiate each of the following complex functions:

a) $w = z^3 + 5z + 1$

b) $w = \dfrac{1}{z - 1}$

c) $w = [1 + (z^2 + 1)^3]^7$

d) $w = \dfrac{z^2}{(z + 1)^3}$

4. Prove the rule (8.20)

5. Prove Theorem 1.

8.5 INTEGRALS

The complex integral $\int f(z)\, dz$ is defined as a line integral, and its properties are closely related to those of the integral $\int P\, dx + Q\, dy$ (see Chapter 5).

Let C be a path from A to B in the complex plane: $x = x(t)$, $y = y(t)$, $a \le t \le b$. We assume C to have a direction, usually that of increasing t. We subdivide the interval $a \le t \le b$ into n parts by $t_0 = a, t_1, \ldots, t_n = b$. We let $z_j = x(t_j) + iy(t_j)$ and $\Delta_j z = z_j - z_{j-1}$, $\Delta_j t = t_j - t_{j-1}$. We choose an arbitrary value t_j^* in the interval $t_{j-1} \le t \le t_j$ and set $z_j^* = x(t_j^*) + iy(t_j^*)$. These quantities are all shown in Fig. 8.3. We then write

$$\int_C f(z)\, dz = \int_{C}^{B} {}_{A} f(z)\, dz = \lim_{\substack{n \to \infty \\ \max \Delta_j t \to 0}} \sum_{j=1}^{n} f(z_j^*)\, \Delta_j z. \qquad (8.21)$$

If we take real and imaginary parts in (8.21), we find

$$\int_C f(z)\, dz = \lim \sum (u + iv)(\Delta x + i\, \Delta y)$$

$$= \lim \left\{ \sum (u\, \Delta x - v\, \Delta y) + i \sum (v\, \Delta x + u\, \Delta y) \right\};$$

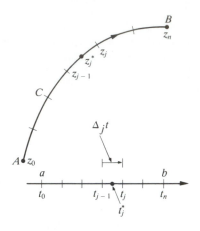

Figure 8.3 Complex line integral.

that is,

$$\int_C f(z) \, dz = \int_C (u + iv)(dx + i \, dy)$$

$$= \int_C (u \, dx - v \, dy) + i \int_C (v \, dx + u \, dy). \qquad (8.22)$$

The complex line integral is thus simply a combination of two real line integrals. Hence we can apply all the theory of real line integrals. In the following, each path is assumed to be *piecewise smooth*; that is, $x(t)$ and $y(t)$ are to be continuous with piecewise continuous derivatives. As in Section 5.2, we have:

THEOREM 5 If $f(z)$ is continuous in domain D, then the integral (8.21) exists, and

$$\int_C f(z) \, dz = \int_a^b \left(u \frac{dx}{dt} - v \frac{dy}{dt} \right) dt + i \int_a^b \left(v \frac{dx}{dt} + u \frac{dy}{dt} \right) dt. \qquad (8.23)$$

We now write our path as $z = z(t)$ as in Section 8.2. If we introduce the derivative

$$\frac{dz}{dt} = \frac{dx}{dt} + i \frac{dy}{dt}$$

of z with respect to the real variable t and also use the theory of integrals of such functions (Section 8.2), we can write (8.23) more consisely:

$$\int_C f(z) \, dz = \int_a^b f[z(t)] \frac{dz}{dt} \, dt. \qquad (8.24)$$

EXAMPLE 1 Let C be the path $x = 2t$, $y = 3t$, $1 \leq t \leq 2$. Let $f(z) = z^2$. Then

$$\int_C z^2 \, dz = \int_1^2 (2t + 3it)^2 (2 + 3i) \, dt = (2 + 3i)^3 \int_1^2 t^2 \, dt$$

$$= -107\tfrac{1}{3} + 21i. \quad \blacksquare$$

EXAMPLE 2 Let C be the circular path $x = \cos t$, $y = \sin t$, $0 \leq t \leq 2\pi$. This can be written more concisely thus: $z = e^{it}$, $0 \leq t \leq 2\pi$. Since $dz/dt = ie^{it}$,

$$\int_C \frac{1}{z} \, dz = \int_0^{2\pi} e^{-it} ie^{it} \, dt = i \int_0^{2\pi} dt = 2\pi i. \quad \blacksquare$$

Further properties of complex integrals follow from those of real integrals:

THEOREM 6 Let $f(z)$ and $g(z)$ be continuous in a domain D. Let C be a piecewise smooth path in D. Then

$$\int_C [f(z) + g(z)] \, dz = \int_C f(z) \, dz + \int_C g(z) \, dz,$$

$$\int_C kf(z) \, dz = k \int_C f(z) \, dz \qquad (k = \text{const}),$$

$$\int_C f(z) \, dz = \int_{C_1} f(z) \, dz + \int_{C_2} f(z) \, dz,$$

where C is composed of a path C_1 from z_0 to z_1 and a path C_2 from z_1 to z_2, and

$$\int_C f(z) \, dz = - \int_{C'} f(z) \, dz,$$

where C' is obtained from C by reversing direction on C.

Upper estimates for the absolute value of a complex integral are obtained by the following theorem.

THEOREM 7 Let $f(z)$ be continuous on C, let $|f(z)| \leq M$ on C, and let

$$L = \int_C ds = \int_a^b \sqrt{(dx/dt)^2 + (dy/dt)^2} \, dt$$

be the length of C. Then

$$\left| \int_C f(z) \, dz \right| \leq \int_C |f(z)| \, ds \leq M \cdot L. \tag{8.25}$$

Proof. The line integral $\int_C |f(z)| \, ds$ is defined as a limit:

$$\int_C |f(z)| \, ds = \lim \sum |f(z_j^*)| \, \Delta_j s,$$

where $\Delta_j s$ is the length of the jth arc of C. Now

$$|f(z_j^*)\,\Delta_j z| = |f(z_j^*)| \cdot |\Delta_j z| \leq |f(z_j^*)| \cdot \Delta_j s,$$

for $|\Delta_j z|$ represents the *chord* of the arc $\Delta_j s$. Hence

$$\left|\sum f(z_j^*)\,\Delta_j z\right| \leq \sum |f(z_j^*)\,\Delta_j z| \leq \sum |f(z_j^*)|\,\Delta_j s$$

by repeated application of the triangle inequality (8.1). Passing to the limit, we conclude that

$$\left|\int_C f(z)\,dz\right| \leq \int_C |f(z)|\,ds. \tag{8.26}$$

Also, if $|f| \leq M = \text{const}$,

$$\sum |f(z_j^*)|\,\Delta_j s \leq \sum M\Delta_j s = M \cdot L.$$

Hence

$$\int_C |f(z)|\,ds \leq M \cdot L. \tag{8.27}$$

Inequalities (8.25) follow from (8.26) and (8.27). $\square$

PROBLEMS

1. Evaluate the following integrals:

a) $\int_0^{1+i} (x^2 - iy^2)\,dz$ on the straight line from 0 to $1 + i$.

b) $\int_0^\pi z\,dz$ on the curve $y = \sin x$.

c) $\int_1^{1+i} \dfrac{dz}{z}$ on the line $x = 1$.

2. Write each of the following integrals in the form $\int u\,dx - v\,dy + i\int v\,dx + u\,dy$; then show that each of the two real integrals is independent of path in the xy-plane.

a) $\int(z + 1)\,dz$ **b)** $\int e^z\,dz$

c) $\int z^4\,dz$ **d)** $\int \sin z\,dz$

3. a) Evaluate

$$\oint \frac{1}{z}\,dz$$

on the circle $|z| = R$.

b) Show that

$$\oint \frac{1}{z}\,dz = 0$$

on every simple closed path not meeting or enclosing the origin.

c) Show that

$$\oint \frac{1}{z^2}\, dz = 0$$

on every simple closed path not passing through the origin.

8.6 ANALYTIC FUNCTIONS · CAUCHY-RIEMANN EQUATIONS

A function $w = f(z)$, defined in a domain D, is said to be an *analytic function* in D if w has a continuous derivative in D. Almost the entire theory of functions of a complex variable is confined to the study of such functions. Furthermore, almost all functions used in the applications of mathematics to physical problems are analytic functions or are derived from such.

It will be seen that possession of a continuous derivative implies possession of a continuous second derivative, third derivative,..., and, in fact, convergence of the Taylor series

$$f(z_0) + f'(z_0)\frac{(z - z_0)}{1!} + f''(z_0)\frac{(z - z_0)^2}{2!} + \cdots$$

in a neighborhood of each z_0 of D. Thus one could define an analytic function as one so representable by Taylor series, and this definition is often used. The two definitions are equivalent, for convergence of the Taylor series in a neighborhood of each z_0 implies continuity of the derivatives of all orders.

Although it is possible to construct continuous functions of z that are not analytic (examples will be given), it is impossible to construct a function $f(z)$ possessing a derivative, but not a continuous one, in D. In other words, if $f(z)$ has a derivative in D, the derivative is necessarily continuous, so that $f(z)$ is analytic. Therefore we could define an analytic function as one merely possessing a derivative in domain D, and this definition is also often used. For a proof that existence of the derivative implies its continuity, refer to Vol. I of the book by Knopp listed at the end of the chapter.

THEOREM 8 If $w = u + iv = f(z)$ is analytic in D, then u and v have continuous first partial derivatives in D and satisfy the Cauchy-Riemann equations

$$\frac{\partial u}{\partial x} = \frac{\partial v}{\partial y}, \qquad \frac{\partial u}{\partial y} = -\frac{\partial v}{\partial x} \tag{8.28}$$

in D. Furthermore,

$$\frac{dw}{dz} = \frac{\partial u}{\partial x} + i\frac{\partial v}{\partial x} = \frac{\partial v}{\partial y} + i\frac{\partial v}{\partial x} = \frac{\partial u}{\partial x} - i\frac{\partial u}{\partial y} = \frac{\partial v}{\partial y} - i\frac{\partial u}{\partial y}. \tag{8.29}$$

Conversely, if $u(x,y)$ and $v(x,y)$ have continuous first partial derivatives in D and satisfy the Cauchy-Riemann equations (8.28), then $w = u + iv = f(z)$ is analytic in D.

Proof.   Let z_0 be a fixed point of D and let

$$\Delta w = \Delta u + i\,\Delta v = f(z_0 + \Delta z) - f(z_0), \qquad \Delta z = \Delta x + i\,\Delta y,$$

as in Fig. 8.4. We consider several equivalent formulations of the condition that $f'(z_0)$ exists. Throughout, $\epsilon, \epsilon_1, \epsilon_2, \epsilon_3, \epsilon_4$ denote functions of $\Delta z = \Delta x + i\,\Delta y$, continuous and equal to zero at $\Delta z = 0$. By Theorem 3, existence of $f'(z_0)$ is equivalent to the statement

$$\Delta w = c \cdot \Delta z + \epsilon \cdot \Delta z, \qquad c = f'(z_0), \qquad c = a + ib; \tag{8.30}$$

this is equivalent to

$$\Delta w = c\,\Delta z + \epsilon\,\Delta x + i\epsilon\,\Delta y \tag{8.30'}$$

and also to

$$\Delta w = c\,\Delta z + \epsilon_1\,\Delta x + \epsilon_2\,\Delta y + i(\epsilon_3\,\Delta x + \epsilon_4\,\Delta y), \tag{8.30''}$$

where $\epsilon_1, \epsilon_2, \epsilon_3, \epsilon_4$ are real. For if (8.30') holds, then (8.30'') holds with $\epsilon_1 = \mathrm{Re}(\epsilon)$, $\epsilon_2 = -\mathrm{Im}(\epsilon)$, $\epsilon_3 = \mathrm{Im}(\epsilon)$, $\epsilon_4 = \mathrm{Re}(\epsilon)$. Conversely , if (8.30'') holds, then (8.30') holds with $\epsilon = 0$ for $\Delta z = 0$ and

$$\epsilon = (\epsilon_1 + i\epsilon_3)\frac{\Delta x}{\Delta z} + (\epsilon_2 + i\epsilon_4)\frac{\Delta y}{\Delta z} \qquad (\Delta z \neq 0). \tag{8.31}$$

As Fig. 8.4 shows,

$$\left|\frac{\Delta x}{\Delta z}\right| \leq 1, \qquad \left|\frac{\Delta y}{\Delta z}\right| \leq 1,$$

so that we deduce from (8.31) that $\epsilon \to 0$ as $\Delta z \to 0$. Thus (8.30), (8.30'), and (8.30'') are all equivalent to existence of $f'(z_0) = c = a + ib$. By taking real and imaginary parts in (8.30'') we obtain one more equivalent condition:

$$\Delta u = a\,\Delta x - b\,\Delta y + \epsilon_1\,\Delta x + \epsilon_2\,\Delta y,$$
$$\Delta v = b\,\Delta x + a\,\Delta y + \epsilon_3\,\Delta x + \epsilon_4\,\Delta y; \tag{8.30'''}$$

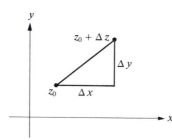

Figure 8.4   Complex derivative.

these equations state that u, v have differentials $du = a\,dx - b\,dy$, $dv = b\,dx + a\,dy$ at (x_0, y_0), and hence at this point

$$\frac{\partial u}{\partial x} = a = \frac{\partial v}{\partial y}, \qquad \frac{\partial u}{\partial y} = -b = -\frac{\partial v}{\partial x}.$$

Thus differentiability of $f'(z)$ at any z is equivalent to differentiability of u, v along with validity of the Cauchy-Riemann equations. Furthermore, $f'(z)$ and $\partial u/\partial x, \ldots$ are related by (8.29). By Theorem 1 these equations show that continuity of $f'(z)$ in D is equivalent to continuity of $\partial u/\partial x, \ldots$. Thus the theorem is proved. ☐

The theorem provides a perfect test for analyticity: If $f(z)$ is analytic, then the Cauchy-Riemann equations hold; if the equations hold (and the derivatives concerned are continuous), then $f(z)$ is analytic.

EXAMPLE 1 $w = z^2 = x^2 - y^2 + i \cdot 2xy$. Here $u = x^2 - y^2$, $v = 2xy$. Thus

$$\frac{\partial u}{\partial x} = 2x = \frac{\partial v}{\partial y}, \qquad \frac{\partial u}{\partial y} = -2y = -\frac{\partial v}{\partial x},$$

and w is analytic for all z. ■

EXAMPLE 2 $w = \dfrac{x}{x^2 + y^2} - \dfrac{iy}{x^2 + y^2}$. Here

$$\frac{\partial u}{\partial x} = \frac{y^2 - x^2}{\left(x^2 + y^2\right)^2} = \frac{\partial v}{\partial y}, \qquad \frac{\partial u}{\partial y} = \frac{-2xy}{\left(x^2 + y^2\right)^2} = -\frac{\partial v}{\partial x}.$$

Hence w is analytic except for $x^2 + y^2 = 0$, that is, for $z = 0$. ■

EXAMPLE 3 $w = x - iy = \bar{z}$. Here $u = x$, $v = -y$, and

$$\frac{\partial u}{\partial x} = 1, \qquad \frac{\partial v}{\partial y} = -1, \qquad \frac{\partial u}{\partial y} = 0 = \frac{\partial v}{\partial x}.$$

Thus w is not analytic in any domain. ■

EXAMPLE 4 $w = x^2y^2 + 2x^2y^2i$. Here

$$\frac{\partial u}{\partial x} = 2xy^2, \qquad \frac{\partial v}{\partial y} = 4x^2y, \qquad \frac{\partial u}{\partial y} = 2x^2y, \qquad \frac{\partial v}{\partial x} = 4xy^2.$$

The Cauchy-Riemann equations give $2xy^2 = 4x^2y$, $2x^2y = -4xy^2$. These equations are satisfied only along the lines $x = 0$, $y = 0$. There is *no domain* in which the Cauchy-Riemann equations hold, hence no domain in which $f(z)$ is analytic.

One does not consider functions analytic only at certain points unless these points form a domain. ∎

The terms *analytic at a point* or *analytic along a curve* are used, apparently in contradiction to the remark just made. However, we say that $f(z)$ is *analytic at the point* z_0 only if there is a domain containing z_0 within which $f(z)$ is analytic. Similarly, $f(z)$ is *analytic along a curve* C only if $f(z)$ is analytic in a domain containing C.

THEOREM 9   The sum, product, and quotient of analytic functions are analytic (provided that in the last case the denominator is not equal to zero at any point of the domain under consideration). All polynomials are analytic for all z. Every rational function is analytic in each domain containing no root of the denominator. An analytic function of an analytic function is analytic.

This follows from Theorem 4.

We readily verify (Problem 1 below) that the Cauchy-Riemann equations are satisfied for $u = \mathrm{Re}(e^z)$, $v = \mathrm{Im}(e^z)$. Hence e^z is analytic for all z. It then follows from Theorem 9 that $\sin z$, $\cos z$, $\sinh z$, and $\cosh z$ are analytic for all z, while $\tan z$, $\sec z$, and $\csc z$ are analytic except for certain points (Problem 6 below). Furthermore, the usual formula for derivatives hold:

$$\frac{d}{dz}e^z = e^z, \qquad \frac{d}{dz}\sin z = \cos z, \qquad \dots \tag{8.32}$$

(Problem 3).

Two basic theorems of more advanced theory are useful at this point. Proofs are given in Chapter IV of the book by Goursat listed at the end of the chapter.

THEOREM 10   Given a function $f(x)$ of the real variable x, $a \le x \le b$, there is at most one analytic function $f(z)$ that reduces to $f(x)$ when z is real.

THEOREM 11   If $f(z)$, $g(z)$,... are functions that are all analytic in a domain D that includes part of the real axis, and $f(z)$, $g(z)$,... satisfy an algebraic identity when z is real, then these functions satisfy the same identity for all z in D.

Theorem 10 implies that our definitions of e^z, $\sin z$,... are the only ones that yield analytic functions and agree with the definitions for real variables.

Because of Theorem 11, we can be sure that all familiar identities of trigonometry, namely,

$$\sin^2 z + \cos^2 z = 1, \qquad \sin\left(\frac{\pi}{2} - z\right) = \cos z, \dots$$

continue to hold for complex z. A general algebraic identity is formed by replacing the variables $w_1, \ldots, w_n$ in an algebraic equation by functions $f_1(z), \ldots, f_n(z)$. Thus in the two examples given, one has

$$w_1^2 + w_2^2 - 1 = 0 \qquad (w_1 = \sin z, \ w_2 = \cos z),$$

$$w_1 - w_2 = 0 \qquad \left[w_1 = \sin\left(\frac{\pi}{2} - z \right), \ w_2 = \cos z \right].$$

To prove identities such as

$$e^{z_1} \cdot e^{z_2} = e^{z_1 + z_2}, \tag{8.33}$$

it may be necessary to apply Theorem 11 several times. (See Problems 4 and 5 below.)

It should be remarked that although e^z is written as a power of e, it is best not to think of it as such. Thus $e^{1/2}$ has only one value, not two, as would a usual complex root. To avoid confusion with the general power function, to be defined later, we often write $e^z = \exp z$ and refer to e^z as the *exponential function of z*.

To obtain the real and imaginary parts of $\sin z$, we use the identity

$$\sin(z_1 + z_2) = \sin z_1 \cos z_2 + \cos z_1 \sin z_2,$$

which holds, by the reasoning described above, for all complex z_1 and z_2. Hence $\sin(x + iy) = \sin x \cos iy + \cos x \sin iy$. Now from the definitions (Section 8.1),

$$\sinh y = -i \sin iy,$$

$$\cosh y = \cos iy. \tag{8.34}$$

Hence

$$\sin z = \sin x \cosh y + i \cos x \sinh y. \tag{8.35}$$

Similarly, we prove, as in (8.13),

$$\cos z = \cos x \cosh y - i \sin x \sinh y,$$

$$\sinh z = \sinh x \cos y + i \cosh x \sin y, \tag{8.36}$$

$$\cosh z = \cosh x \cos y + i \sinh x \sin y.$$

Conformal mapping.   A complex function $w = f(z)$ can be considered as a *mapping* from the xy-plane to the uv-plane as in Section 2.7. In the case of an analytic function $f(z)$ this mapping has a special property: It is a *conformal mapping*. By this we mean that two curves in the xy-plane, meeting at (x_0, y_0) at angle α, correspond to two curves meeting at the corresponding point (u_0, v_0) at the *same angle* α (in value and in sense—positive or negative). This means that a small triangle in the xy-plane corresponds to a *similar* small (curvilinear) triangle in the uv-plane. (The properties described fail at the exceptional points where $f'(z) = 0$.) Furthermore, every conformal mapping from the xy-plane to the uv-plane is given by an analytic function. For a discussion of conformal mapping and its applications, see the next chapter.

PROBLEMS

1. Verify that the following are analytic functions of z:

   a) $2x^3 - 3x^2y - 6xy^2 + y^3 + i(x^3 + 6x^2y - 3xy^2 - 2y^3)$

   b) $w = e^z = e^x \cos y + i e^x \sin y$

   c) $w = \sin z = \sin x \cosh y + i \cos x \sinh y$

2. Test each of the following for analyticity:

   a) $x^3 + y^3 + i(3x^2y + 3xy^2)$

   b) $\sin x \cos y + i \cos x \sin y$

   c) $3x + 5y + i(3y - 5x)$

3. Prove the following properties directly from the definitions of the functions:

   a) $\dfrac{d}{dz} e^z = e^z$

   b) $\dfrac{d}{dz} \sin z = \cos z$, $\dfrac{d}{dz} \cos z = -\sin z$

   c) $\sin(z + \pi) = -\sin z$

   d) $\sin(-z) = -\sin z$, $\cos(-z) = \cos z$

4. Prove the identity $e^{z_1 + z_2} = e^{z_1} \cdot e^{z_2}$ by application of Theorem 11. [Hint: Let $z_2 = b$, a fixed real number, and $z_1 = z$, a variable complex number. Then $e^{z+b} = e^z \cdot e^b$ is an identity connecting analytic functions which is known to be true for z real. Hence it is true for all complex z. Now proceed similarly with the identity $e^{z_1 + z} = e^{z_1} \cdot e^z$.]

5. Prove the following identities by application of Theorem 11 (see Problem 4):

   a) $\cos(z_1 + z_2) = \cos z_1 \cos z_2 - \sin z_1 \sin z_2$

   b) $e^{iz} = \cos z + i \sin z$

   c) $(e^z)^n = e^{nz}$ $(n = 0, 1, 2, \dots)$

6. Determine where the following functions are analytic (see Problem 3 following Section 8.2):

   a) $\tan z = \dfrac{\sin z}{\cos z}$

   b) $\cot z = \dfrac{\cos z}{\sin z}$

   c) $\tanh z = \dfrac{\sinh z}{\cosh z}$

   d) $\dfrac{\sin z}{z}$

   e) $\dfrac{e^z}{z \cos z}$

   f) $\dfrac{e^z}{\sin z + \cos z}$

8.7 THE FUNCTIONS $\log z,\, a^z,\, z^a,\, \sin^{-1} z,\, \cos^{-1} z$

The function $w = \log z$ is defined as the inverse of the exponential function $z = e^w$. We write $z = re^{i\theta}$, in terms of polar coordinates r, θ, and $w = u + iv$, so that

$$re^{i\theta} = e^{u+iv} = e^u e^{iv},$$

$$e^u = r, \qquad v = \theta + 2k\pi \qquad (k = 0, \pm 1, \dots).$$

Accordingly,

$$w = \log z = \log r + i(\theta + 2k\pi) = \log |z| + i \arg z, \tag{8.37}$$

where $\log r$ is the real logarithm of r. Thus $\log z$ is a multiple-valued function of z, with infinitely many values except for $z = 0$. We can select one value of θ for each z and obtain a single-valued function, $\log z = \log r + i\theta$; however, θ cannot be chosen to depend continuously on z for all $z \neq 0$, since θ will increase by 2π each time one encircles the origin in the positive direction.

If we concentrate on an appropriate portion of the z-plane, we can choose θ to vary continuously within the domain. For example, the inequalities

$$-\pi < \theta < \pi, \qquad r > 0$$

together describe a domain (Fig. 8.5) and also tell how to assign the values of θ within the domain. With θ so restricted, $\log r + i\theta$ then defines a *branch* of $\log z$ in the domain chosen; this particular branch is called the *principal value* of $\log z$ and is denoted by $\mathrm{Log}\, z$. The points on the negative real axis are excluded from the domain, but we usually assign the values $\mathrm{Log}\, z = \log |x| + i\pi$ on this line. Within the domain of Fig. 8.5, $\mathrm{Log}\, z$ is *an analytic function of z* (Problem 4 below). Other branches of $\log z$ are obtained by varying the choice of θ or of the domain. For example, in the domain of Fig. 8.5 we might choose θ so that $\pi < \theta < 3\pi$ or so that $-3\pi < \theta < -\pi$. The inequalities $0 < \theta < 2\pi$, $\pi/2 < \theta < 5\pi/2, \ldots$ also suggest other domains and choices of θ. We can verify that as long as θ varies continuously in the domain, $\log z = \log r + i\theta$ is analytic there. The most general domain possible here is an arbitrary simply connected domain not containing the origin.

As a result of this discussion it appears that $\log z$ is formed of many branches, each analytic in some domain not containing the origin. The branches fit together in a simple way; in general, we can get from one branch to another by moving around the origin a sufficient number of times while varying the choice of $\log z$ continuously. We say that the branches form "analytic continuations" of each other.

We can further verify that for each branch of $\log z$ the rule

$$\frac{d}{dz} \log z = \frac{1}{z} \tag{8.38}$$

remains valid. The familiar identities are also satisfied (Problems 4 and 5 below).

Figure 8.5   Domain for $\log z$.

Before turning to exponential functions we recall the rule of algebra for the nth roots of a complex number $z = r(\cos \theta + i \sin \theta)$:

$$z^{1/n} = [r(\cos \theta + i \sin \theta)]^{1/n}$$

$$= r^{1/n} \left[\cos \left(\frac{\theta}{n} + k \frac{2\pi}{n} \right) + i \sin \left(\frac{\theta}{n} + k \frac{2\pi}{n} \right) \right], \qquad k = 0, 1, \ldots, n - 1.$$

$$(8.39)$$

The rule follows from the Demoivre formula:

$$[r(\cos \theta + i \sin \theta)]^n = r^n (\cos n\theta + i \sin n\theta).$$

The *general exponential function* a^z is defined, for $a \neq 0$, by the equation

$$a^z = e^{z \log a} = \exp(z \log a). \qquad (8.40)$$

Thus for $z = 0$, $a^0 = 1$. Otherwise, $\log a = \log |a| + i \arg a$, and we obtain many values: $a^z = \exp[z(\log |a| + i(\alpha + 2n\pi))]$, $(n = 0, \pm 1, \pm 2, \ldots)$, where α denotes one choice of arg a. For example,

$$(1 + i)^i = \exp \left[i \left\{ \log \sqrt{2} + i \left(\frac{\pi}{4} + 2n\pi \right) \right\} \right]$$

$$= e^{-(\pi/4) - 2n\pi} (\cos \log \sqrt{2} + i \sin \log \sqrt{2}).$$

If z is a positive integer m, a^z reduces to a^m and has only one value. The same holds for $z = -m$, and we have

$$a^{-m} = \frac{1}{a^m}. \qquad (8.41)$$

If z is a fraction p/q (in lowest terms), we find that a^z has q distinct values, which are the qth roots of a^p. (See Eq. 8.39).)

If a fixed choice of $\log a$ is made in (8.40), then a^z is simply e^{cz}, $c = \log a$, and is hence an analytic function of z for all z. Each choice of $\log a$ determines such a function.

If a and z are interchanged in (8.40), we obtain the *general power function*,

$$z^a = e^{a \log z}. \qquad (8.42)$$

If an analytic branch of $\log z$ is chosen as above, then this function becomes an analytic function of an analytic function and is hence analytic in the domain chosen. In particular, the *principal value* of z^a is defined as the analytic function $z^a = e^{a \operatorname{Log} z}$, in terms of the principal value of $\log z$.

For example, if $a = \frac{1}{2}$, we have

$$z^{1/2} = e^{(1/2) \log z} = e^{(1/2)(\log r + i\theta)} = e^{(1/2) \log r} e^{(1/2)i\theta}$$

$$= \sqrt{r} \left(\cos \frac{\theta}{2} + i \sin \frac{\theta}{2} \right),$$

as in Eq. (8.39) with $k = 0$. If $\operatorname{Log} z$ is used, then $\sqrt{z} = f_1(z)$ becomes analytic in the domain of Fig. 8.5. A second analytic branch $f_2(z)$ in the same domain is obtained by requiring that $\pi < \theta < 3\pi$. These are the only two analytic branches that can be obtained in this domain. It should be remarked that these two branches are related by the equation $f_2(z) = -f_1(z)$. For f_2 is obtained from f_1 by increasing θ by 2π, which replace $e^{(1/2)i\theta}$ by

$$e^{(1/2)i(\theta + 2\pi)} = e^{\pi i} e^{(1/2)i\theta} = -e^{(1/2)i\theta}.$$

The functions $\sin^{-1} z$ and $\cos^{-1} z$ are defined as the inverse of $\sin z$ and $\cos z$. We then find

$$\sin^{-1} z = \frac{1}{i} \log \left[iz \pm \sqrt{1 - z^2} \right],$$

$$\cos^{-1} z = \frac{1}{i} \log \left[z \pm i\sqrt{1 - z^2} \right]. \tag{8.43}$$

The proofs are left to Problem 2 below. It can be shown that analytic branches of both these functions can be defined in each simply connected domain not containing the points ± 1. For each z other than ± 1, one has two choices of $\sqrt{1 - z^2}$ and then an infinite sequence of choices of the logarithm, differing by multiples of $2\pi i$.

PROBLEMS

1. Obtain all values of each of the following:

   a) $\log 2$ b) $\log i$ c) $\log (1 - i)$ d) i^i

   e) $(1 + i)^{2/3}$ f) $i^{\sqrt{2}}$ g) $\sin^{-1} 1$ h) $\cos^{-1} 2$

2. Prove the formulas (8.43). [Hint: If $w = \sin^{-1} z$, then $2iz = e^{iw} - e^{-iw}$; multiply by e^{iw} and solve the resulting equation as a quadratic for e^{iw}.]

3. a) Evaluate $\sin^{-1} 0$, $\cos^{-1} 0$.

   b) Find all roots of $\sin z$ and $\cos z$ [compare part (a)].

4. Show that each branch of $\log z$ is analytic in each domain in which θ varies continuously and that

$$(d/dz) \log z = 1/z.$$

   [Hint: Show from the equations $x = r \cos \theta$, $y = r \sin \theta$ that $\partial \theta / \partial x = -y/r^2$, $\partial \theta / \partial y = x/r^2$. Show that the Cauchy-Riemann equations hold for $u = \log r$, $v = \theta$.]

5. Prove the following identities in the sense that for proper selection of values of the multiple-valued functions concerned the equation is correct for each allowed choice of the variables:

   a) $\log (z_1 \cdot z_2) = \log z_1 + \log z_2$ $(z_1 \neq 0, z_2 \neq 0)$ b) $e^{\log z} = z$ $(z \neq 0)$

   c) $\log e^z = z$ d) $\log z_1 z_2 = z_2 \log z_1$ $(z_1 \neq 0)$

6. For each of the following, determine all analytic branches of the multiple-valued function in the domain given:

   a) $\log z$, $x < 0$ b) $\sqrt[3]{z}$, $x > 0$

7. Prove that for the analytic function z^a (principal value),

$$(d/dz) z^a = (az^a)/z = az^{a-1}.$$

8. Plot the functions $u = \text{Re}(\sqrt{z})$ and $v = \text{Im}(\sqrt{z})$ as functions of x and y and show the two branches described in the text.

9. Let $w = u + iv = f(z)$, $z = x + iy$, be analytic in domain D.

   a) Show that $\partial(u, v)/\partial(x, y) = |f'(z)|^2$.

   b) Apply the inverse mapping theorem of Section 2.12 to conclude that if $f'(z_0) \neq 0$, where z_0 is in D, and $f(z_0) = w_0$, then there is an inverse $z = g(w)$ of f defined in a neighborhood D_0 of w_0, mapping D_0 onto a neighborhood of z_0.

   c) Show that $g(w)$ is analytic in D_0. [Hint: Find the corresponding Jacobian matrix and show that the Cauchy-Riemann equations in u, v are satisfied by $x(u, v) = \text{Re } g(w)$, $y(u, v) = \text{Im } g(w)$.]

   d) Show that $g'(w) = 1/f'(z)$ at $w = f(z)$ and that $g'(w) \neq 0$ in D_0.

8.8 INTEGRALS OF ANALYTIC FUNCTIONS ▪ CAUCHY INTEGRAL THEOREM

All paths in the integrals concerned here, as elsewhere in the chapter, are assumed to be piecewise smooth.

The following theorem is fundamental for the theory of analytic functions:

THEOREM 12 (*Cauchy integral theorem*)   If $f(z)$ is analytic in a simply connected domain D, then

$$\oint_C f(z)\, dz = 0$$

on every simple closed path C in D (Fig. 8.6).

Figure 8.6   Cauchy integral theorem.

Proof.   We have, by (8.22),

$$\oint_C f(z)\, dz = \oint_C u\, dx - v\, dy + i \oint_C v\, dx + u\, dy.$$

The two real integrals are equal to zero (see Section 5.6) provided that u and v

have continuous derivatives in D and

$$\frac{\partial u}{\partial y} = -\frac{\partial v}{\partial x}, \qquad \frac{\partial v}{\partial y} = \frac{\partial u}{\partial x}.$$

These are just the Cauchy-Riemann equations. Hence

$$\oint_C f(z)\,dz = 0 + i \cdot 0 = 0. \quad \square$$

This theorem can be stated in an equivalent form:

THEOREM 12′   If $f(z)$ is analytic in the simply connected domain D, then $\int f(z)\,dz$ is independent of the path in D.

For independence of path and equaling zero on closed paths are equivalent properties of line integrals. If C is a path from z_1 to z_2, we can now write

$$\int_C f(z)\,dz = \int_{z_1}^{z_2} f(z)\,dz,$$

the integral being the same for all paths C from z_1 to z_2.

THEOREM 13   Let $f(z) = u + iv$ be defined in domain D and let u and v have continuous partial derivatives in D. If

$$\oint_C f(z)\,dz = 0 \tag{8.44}$$

on every simple closed path C in D, then $f(z)$ is analytic in D.

Proof.   The condition (8.44) implies that

$$\oint_C u\,dx - v\,dy = 0, \qquad \oint_C v\,dx + u\,dy = 0$$

on all simple closed paths C; that is, the two real line integrals are independent of path in D. Therefore by Theorem III in Section 5.6,

$$\frac{\partial u}{\partial y} = -\frac{\partial v}{\partial x}, \qquad \frac{\partial v}{\partial y} = \frac{\partial u}{\partial x};$$

since the Cauchy-Riemann equations hold, f is analytic. $\square$

This theorem can be proved with the assumption that u and v have continuous derivatives in D replaced by the assumption that f is continuous in D; it is then known as *Morera's* theorem. For a proof, see Chapter 5 of Vol. I of the book by Knopp listed at the end of the chapter.

THEOREM 14   If $f(z)$ is analytic in D, then

$$\int_{z_1}^{z_2} f'(z)\,dz = f(z)\Big|_{z_1}^{z_2} = f(z_2) - f(z_1) \tag{8.45}$$

on every path in D from z_1 to z_2. In particular,

$$\oint f'(z)\, dz = 0$$

on every closed path in D.

Proof. By (8.29),

$$\int_{z_1}^{z_2} f'(z)\, dz = \int_{z_1}^{z_2}\left(\frac{\partial u}{\partial x} + i\frac{\partial v}{\partial x}\right)(dx + i\, dy)$$

$$= \int_{z_1}^{z_2}\frac{\partial u}{\partial x}\, dx + \frac{\partial u}{\partial y}\, dy + i\int_{z_1}^{z_2}\frac{\partial v}{\partial x}\, dx + \frac{\partial v}{\partial y}\, dy$$

$$= \int_{z_1}^{z_2} du + i\, dv = (u + iv)\big|_{z_1}^{z_2} = f(z_2) - f(z_1). \quad \square$$

This rule is the basis for evaluation of simple integrals, just as in elementary calculus. Thus we have

$$\int_i^{1+i} z^2\, dz = \frac{z^3}{3}\bigg|_i^{1+i} = \frac{(1+i)^3 - i^3}{3} = -\tfrac{2}{3} + i,$$

$$\int_i^{-i}\frac{1}{z^2}\, dz = -\frac{1}{z}\bigg|_i^{-i} = -i - i = -2i.$$

In the first of these, any path can be used; in the second, any path not through the origin.

THEOREM 15 If $f(z)$ is analytic in D and D is simply connected, then

$$F(z) = \int_{z_1}^{z} f(z)\, dz \qquad (z_1 \text{ fixed in } D) \tag{8.46}$$

is an indefinite integral of $f(z)$; that is, $F'(z) = f(z)$. Thus $F(z)$ is itself analytic.

Proof. Since $f(z)$ is analytic in D and D is simply connected, $\int_{z_1}^{z} f(z)\, dz$ is independent of path and defines a function F that depends only on the upper limit z. We have, further, $F = U + iV$, where

$$U = \int_{z_1}^{z} u\, dx - v\, dy, \qquad V = \int_{z_1}^{z} v\, dx + u\, dy,$$

and both integrals are independent of path. Hence $dU = u\, dx - v\, dy$, $dV = v\, dx + u\, dy$. Thus U and V satisfy the Cauchy-Riemann equations, so that $F = U + iV$ is analytic and

$$F'(z) = \frac{\partial U}{\partial x} + i\frac{\partial V}{\partial x} = u + iv = f(z). \quad \square$$

Figure 8.7 Cauchy theorem for doubly connected domain.

Figure 8.8 Cauchy theorem for triply connected domain.

Cauchy's theorem for multiply connected domains. If $f(z)$ is analytic in a multiply connected domain D, then we cannot conclude that

$$\oint f(z)\,dz = 0$$

on every simple closed path C in D. Thus if D is the doubly connected domain of Fig. 8.7 and C is the curve C_1 shown, then the integral around C need not be zero. However, by introducing cuts we can reason that

$$\oint_{C_1} f(z)\,dz = \oint_{C_2} f(z)\,dz; \tag{8.47}$$

that is, the integral has the same value on all paths that go around the inner "hole" once in the positive direction. For a triply connected domain, as in Fig. 8.8, we obtain the equation

$$\oint_{C_1} f(z)\,dz = \oint_{C_2} f(z)\,dz + \oint_{C_3} f(z)\,dz. \tag{8.48}$$

This can be written in the form

$$\oint_{C_1} f(z)\,dz + \oint_{C_2} f(z)\,dz + \oint_{C_3} f(z)\,dz = 0; \tag{8.49}$$

Eq. (8.49) states that the integral around the complete boundary of a certain region in D is equal to zero. More generally, we have the following theorem:

THEOREM 16 (*Cauchy's theorem for multiply connected domains*) Let $f(z)$ be analytic in a domain D and let $C_1,\ldots,C_n$ be n simple closed curves in D that together form the boundary B of a region R contained in D. Then

$$\int_B f(z)\,dz = 0,$$

where the direction of integration on B is such that the outer normal is $90°$ behind the tangent vector in the direction of integration.

8.9 CAUCHY'S INTEGRAL FORMULA

Now let D be a simply connected domain and let z_0 be a fixed point of D. If $f(z)$ is analytic in D, the function $f(z)/(z - z_0)$ will fail to be analytic at z_0. Hence

$$\oint \frac{f(z)}{z - z_0} \, dz$$

will in general not be zero on a path C enclosing z_0. However, as above, this integral will have the same value on all paths C about z_0. To determine this value, we reason that if C is a very small circle of radius R about z_0, then $f(z_0)$ has, by continuity, approximately the constant value $f(z_0)$ on the path. This suggests that

$$\oint_C \frac{f(z)}{z - z_0} \, dz = f(z_0) \cdot \oint_{|z-z_0|=R} \frac{dz}{z - z_0} = f(z_0) \cdot 2\pi i,$$

since we find

$$\oint_{|z-z_0|=R} \frac{dz}{z - z_0} = \int_0^{2\pi} \frac{Rie^{i\theta}}{Re^{i\theta}} \, d\theta = i \int_0^{2\pi} d\theta = 2\pi i,$$

with the aid of the substitution: $z - z_0 = Re^{i\theta}$. The correctness of the conclusion reached is the content of the following fundamental result:

> **THEOREM 17** (*Cauchy integral formula*)   Let $f(z)$ be analytic in a domain D. Let C be a simple closed curve in D, within which $f(z)$ is analytic and let z_0 be inside C. Then
>
> $$f(z_0) = \frac{1}{2\pi i} \oint_C \frac{f(z)}{z - z_0} \, dz. \tag{8.50}$$

Proof.   The domain D is not required to be simply connected, but since f is analytic within C, the theorem concerns only a simply connected part of D, as shown in Fig. 8.9. We reason as above to conclude that

$$\oint_C \frac{f(z)}{z - z_0} \, dz = \oint_{|z-z_0|=R} \frac{f(z)}{z - z_0} \, dz.$$

It remains to show that the integral on the right is indeed $f(z_0) \cdot 2\pi i$. Now, since $f(z_0) = \text{const}$,

$$\oint \frac{f(z_0)}{z - z_0} \, dz = f(z_0) \oint \frac{dz}{z - z_0}$$

$$= f(z_0) \cdot 2\pi i,$$

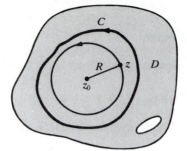

Figure 8.9 Cauchy integral formula.

where we integrate always on the circle $|z - z_0| = R$. Hence on the same path,

$$\oint \frac{f(z)}{z - z_0} \, dz - f(z_0) \cdot 2\pi i = \oint \frac{f(z) - f(z_0)}{z - z_0} \, dz. \qquad (8.51)$$

Now $|z - z_0| = R$ on the path, and since $f(z)$ is continuous at z_0, $|f(z) - f(z_0)| < \epsilon$ for $R < \delta$, for each preassigned $\epsilon > 0$. Hence by Theorem 7,

$$\left| \oint \frac{f(z) - f(z_0)}{z - z_0} \, dz \right| < \frac{\epsilon}{R} \cdot 2\pi R = 2\pi\epsilon.$$

Thus the absolute value of the integral can be made as small as desired by choosing R sufficiently small. But the integral has the same value for all choices of R. This is possible only if the integral is zero for all R. Hence the left-hand side of (8.51) is zero, and (8.50) follows. $\square$

The integral formula (8.50) is remarkable in that expresses the values of the function $f(z)$ at points z_0 inside the curve C in terms of the values along C alone. If C is taken as a circle $z = z_0 + Re^{i\theta}$, then (8.50) reduces to the following:

$$f(z_0) = \frac{1}{2\pi} \int_0^{2\pi} f(z_0 + Re^{i\theta}) \, d\theta. \qquad (8.52)$$

Thus the *value of an analytic function at the center of a circle equals the average* (*arithmetic mean*) *of the values on the circumference.*

Just as with the Cauchy integral theorem, the Cauchy integral formula can be extended to multiply connected domains. Under the hypotheses of Theorem 16,

$$f(z_0) = \frac{1}{2\pi i} \int_B \frac{f(z)}{z - z_0} \, dz = \frac{1}{2\pi i} \left(\oint_{C_1} \frac{f(z)}{z - z_0} \, dz + \oint_{C_2} \frac{f(z)}{z - z_0} \, dz + \cdots \right),$$

$$(8.53)$$

where z_0 is any point inside the region R bounded by C_1 (the outer boundary), $C_2, \ldots, C_n$. The proof is left as an exercise (Problem 6 below).

PROBLEMS

1. Evaluate the following integrals:

   a) $\oint z^2 \sin z \, dz$ on the ellipse $x^2 + 2y^2 = 1$

   b) $\oint \dfrac{z^2}{z+1} \, dz$ on the circle $|z - 2| = 1$

   c) $\displaystyle\int_1^{2i} z e^z \, dz$ on the line segment joining the endpoints

   d) $\displaystyle\int_{1+i}^{1-i} \dfrac{1}{z^2} \, dz$ on the parabola $2y^2 = x + 1$

2. a) Evaluate $\displaystyle\int_{-i}^{i} (dz/z)$ on the path $z = e^{it}$, $-\pi/2 \leq t \leq \pi/2$, with the aid of the relation $(\log z)' = 1/z$, for an appropriate branch of $\log z$.

   b) Evaluate $\displaystyle\int_i^{-i} (dz/z)$ on the path $z = e^{it}$, $\pi/2 \leq t \leq 3\pi/2$, as in part (a).

   c) Why does the relation $(\log z)' = 1/z$ not imply that the sum of the two integrals of parts (a) and (b) is zero?

3. A certain function $f(z)$ is known to be analytic except for $z = 1$, $z = 2$, $z = 3$, and it is known that

$$\oint_{C_k} f(z) \, dz = a_k \qquad (k = 1, 2, 3),$$

   where C_k is a circle of radius $\frac{1}{2}$ with center at $z = k$. Evaluate

$$\oint f(z) \, dz$$

   on each of the following paths:

   a) $|z| = 4$ b) $|z| = 2.5$ c) $|z + 2.5| = 1$

4. A certain function $f(z)$ is analytic except for $z = 0$, and it is known that

$$\lim_{z \to \infty} z f(z) = 0.$$

   Show that

$$\oint f(z) \, dz = 0$$

   on every simple closed path not passing through the origin. [Hint: Show that the value of the integral on a path $|z| = R$ can be made as small as desired by making R sufficiently large.]

5. Evaluate each of the following with the aid of the Cauchy integral formula:

   a) $\oint \dfrac{z}{z-3} \, dz$ on $|z| = 5$ b) $\oint \dfrac{e^z}{z^2 - 3z} \, dz$ on $|z| = 1$

   c) $\oint \dfrac{z+2}{z^2 - 1} \, dz$ on $|z| = 2$ d) $\oint \dfrac{\sin z}{z^2 + 1} \, dz$ on $|z| = 2$

   [Hint for (c) and (d): Expand the rational function in partial fractions.]

6. Prove (8.53) under the hypotheses stated.

7. Prove that if $f(z)$ is analytic in domain D and $f'(z) \equiv 0$, then $f(z) \equiv$ constant. [Hint: Apply Theorem 14.]

8.10   POWER SERIES AS ANALYTIC FUNCTIONS

We now proceed to enlarge the class of specific analytic functions still further by showing that every power series

$$\sum_{n=0}^{\infty} c_n(z - z_0)^n = c_0 + c_1(z - z_0) + \cdots + c_n(z - z_0)^n + \cdots$$

converging for some values of z other than $z = z_0$ represents an analytic function.

For the theory of series of complex numbers, see Section 6.19. The following fundamental theorem for complex power series is proved just as for real series (Theorem 35 in Section 6.15).

THEOREM 18   Every power series $\displaystyle\sum_{n=0}^{\infty} c_n(z - z_0)^n$ has a radius of convergence r^* such that the series converges absolutely when $|z - z_0| < r^*$ and diverges when $|z - z_0| > r^*$. The series converges uniformly for $|z - z_0| \leq r_1$, provided that $r_1 < r^*$.

The number r^* can be zero, in which case the series converges only for $z = z_0$, a positive number, or ∞, in which case the series converges for all z. The number r^* can be evaluated as follows:

$$r^* = \lim_{n \to \infty} \left| \frac{c_n}{c_{n+1}} \right|, \qquad \text{if the limit exists,}$$

$$r^* = \lim_{n \to \infty} \frac{1}{\sqrt[n]{|c_n|}}, \qquad \text{if the limit exists,} \qquad (8.54)$$

and in any case by the formula

$$r^* = \frac{1}{\displaystyle\varlimsup_{n \to \infty} \sqrt[n]{|c_n|}}. \qquad (8.55)$$

As for real variables, no general statement can be made about convergence on the boundary of the domain of convergence. This boundary (when $r^* \neq 0$, $r^* \neq \infty$) is a circle $|z - z_0| = r^*$, termed the *circle of convergence* (Fig. 8.10). The series may converge at some points, all points, or no points of this circle.

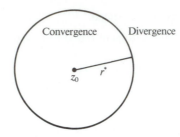

Figure 8.10   Circle of convergence
of a power series.

EXAMPLE 1 $\sum_{n=1}^{\infty} (z^n/n^2)$. The first formula (8.54) gives

$$r^* = \lim_{n \to \infty} \frac{(n+1)^2}{n^2} = 1.$$

The series converges absolutely on the circle of convergence, for when $|z| = 1$, the series of absolute value is the convergent series $\Sigma(1/n^2)$. ∎

EXAMPLE 2 $\sum_{n=0}^{\infty} z^n$. This complex geometric series converges for $|z| < 1$, as (8.54) shows. We have further

$$\sum_{n=0}^{\infty} z^n = \frac{1}{1-z} \qquad (|z| < 1),$$

as for real variables. On the circle of convergence the series diverges everywhere, since the nth term fails to converge to zero. ∎

The following theorems are proved as for real variables.

THEOREM 19   A power series with nonzero convergence radius represents a continuous function within the circle of convergence.

THEOREM 20   A power series can be integrated term by term within the circle of convergence; that is, if $r^* \neq 0$ and

$$f(z) = \sum_{n=0}^{\infty} c_n(z - z_0)^n \qquad (|z - z_0| < r^*),$$

then, for every path C inside the circle of convergence,

$$\int_{C}^{z_2} f(z)\, dz = \sum_{n=0}^{\infty} c_n \int_{z_1}^{z_2} (z - z_0)^n\, dz = \sum_{n=0}^{\infty} c_n \frac{(z - z_0)^{n+1}}{n+1} \Bigg|_{z_1}^{z_2},$$

or in terms of indefinite integrals,

$$\int f(z)\,dz = \sum_{n=0}^{\infty} c_n \frac{(z-z_0)^{n+1}}{n+1} + \text{const} \qquad (|z-z_0| < r^*).$$

THEOREM 21   A power series can be differentiated term by term; that is, if $r^* \neq 0$ and

$$f(z) = \sum_{n=0}^{\infty} c_n(z-z_0)^n \qquad (|z-z_0| < r^*),$$

then

$$f'(z) = \sum_{n=1}^{\infty} n c_n(z-z_0)^{n-1} \qquad (|z-z_0| < r^*),$$

$$f''(z) = \sum_{n=2}^{\infty} n(n-1) c_n(z-z_0)^{n-2} \qquad (|z-z_0| < r^*),$$

$$\vdots$$

Hence every power series with nonzero convergence radius defines an analytic function $f(z)$ within the circle of convergence, and the power series is the Taylor series of $f(z)$:

$$c_n = \frac{f^{(n)}(z_0)}{n!}.$$

THEOREM 22   If two power series $\sum_{n=0}^{\infty} c_n(z-z_0)^n$, $\sum_{n=0}^{\infty} C_n(z-z_0)^n$ have nonzero convergence radii and have equal sums wherever both series converge, then the series are identical; that is,

$$c_n = C_n \qquad (n=0,1,2,\ldots).$$

8.11   POWER SERIES EXPANSION OF GENERAL ANALYTIC FUNCTION

In Section 8.10 it was shown that every power series with nonzero convergence radius represents an analytic function. We now proceed to show that all analytic functions are obtainable in this way. If a function $f(z)$ is analytic in a domain D of general shape, we cannot expect to represent $f(z)$ by one power series, for the power series converges only in a circular domain. However, we can show that for each circular domain D_0 in D there is a power series converging in D_0 whose sum is $f(z)$. Thus several (perhaps infinitely many) power series are needed to represent $f(z)$ throughout all of D.

THEOREM 23 Let $f(z)$ be analytic in the domain D. Let z_0 be in D and let R be the radius of the largest circle with center at z_0 and having its interior in D. Then there is a power series

$$\sum_{n=0}^{\infty} c_n(z - z_0)^n$$

that converges to $f(z)$ for $|z - z_0| < R$. Furthermore,

$$c_n = \frac{f^{(n)}(z_0)}{n!} = \frac{1}{2\pi i} \oint_C \frac{f(z)}{(z - z_0)^{n+1}} \, dz, \tag{8.56}$$

where C is a simple closed path in D enclosing z_0 and within which $f(z)$ is analytic.

Proof. For simplicity we take $z_0 = 0$. The general case can then be obtained by the substitution $z' = z - z_0$. Let the circle $|z| = R$ be the largest circle with center at z_0 and having its interior within D; the radius R is then positive or $+\infty$ (in which case D is the whole z-plane). Let z_1 be a point within this circle, so that $|z_1| < R$. Choose R_2 so that $|z_1| < R_2 < R$ (see Fig. 8.11). Then $f(z)$ is analytic in a domain including the circle C_2: $|z| = R_2$ plus interior. Hence by the Cauchy integral formula,

$$f(z_1) = \frac{1}{2\pi i} \oint_{C_2} \frac{f(z)}{z - z_1} \, dz.$$

Now the factor $1/(z - z_1)$ can be expanded in a geometric series:

$$\frac{1}{z - z_1} = \frac{1}{z\left(1 - \dfrac{z_1}{z}\right)} = \frac{1}{z}\left(1 + \frac{z_1}{z} + \cdots + \frac{z_1^n}{z^n} + \cdots\right).$$

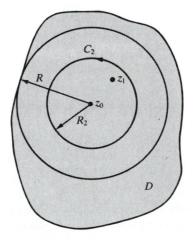

Figure 8.11 Taylor series of an analytic function.

The series can be considered as a power series in powers of $1/z$, for fixed z_1. It converges for $|z_1/z| < 1$ and converges uniformly for $|z_1/z| \le |z_1|/R_2 < 1$.

If we multiply by $f(z)$, we find

$$\frac{f(z)}{z - z_1} = \frac{f(z)}{z} + z_1 \frac{f(z)}{z^2} + \cdots + z_1^n \frac{f(z)}{z^{n+1}} + \cdots;$$

since $f(z)$ is continuous for $|z| = R_2$, the series remains uniformly convergent on C_2. Hence we can integrate term by term on C_2:

$$\frac{1}{2\pi i} \oint_{C_2} \frac{f(z)}{z - z_1} dz = \frac{1}{2\pi i} \oint_{C_2} \frac{f(z)}{z} dz$$

$$+ \frac{z_1}{2\pi i} \oint_{C_2} \frac{f(z)}{z^2} dz + \cdots + \frac{z_1^n}{2\pi i} \oint_{C_2} \frac{f(z)}{z^{n+1}} dz + \cdots.$$

The left-hand side is precisely $f(z_1)$, by the integral formula. Hence

$$f(z_1) = \sum_{n=0}^{\infty} c_n z_1^n, \qquad c_n = \frac{1}{2\pi i} \oint_{C_2} \frac{f(z)}{z^{n+1}} dz.$$

The path C_2 can be replaced by any path C as described in the theorem, since $f(z)/z^{n+1}$ is analytic in D except for $z = z_0 = 0$.

By Theorem 21 the series obtained is the Taylor series of f, so that

$$c_n = \frac{f^{(n)}(z_0)}{n!} \qquad (z_0 = 0).$$

The theorem is now completely proved.   ◻

The consequences of this theorem are far-reaching. First of all, not only does it guarantee that every analytic function is representable by power series, but it also ensures that the Taylor series converges to the function within each circular domain within the domain in which the function is given. Thus *without further analysis* we at once conclude that

$$e^z = 1 + z + \frac{z^2}{2!} + \cdots + \frac{z^n}{n!} + \cdots,$$

$$\sin z = z - \frac{z^3}{3!} + \frac{z^5}{5!} + \cdots + (-1)^n \frac{z^{2n+1}}{(2n+1)!} + \cdots,$$

$$\cos z = 1 - \frac{z^2}{2!} + \cdots + (-1)^n \frac{z^{2n}}{(2n)!} + \cdots$$

for all z. A variety of other familiar expansions can be obtained in the same way.

It should be recalled that a function $f(z)$ is defined to be analytic in a domain D if $f(z)$ has a continuous derivative $f'(z)$ in D (Section 8.6). By Theorem 23, $f(z)$ must have derivatives of all orders at every point of D. In particular, the derivative of an analytic function is itself analytic:

THEOREM 24   If $f(z)$ is analytic in domain D, then $f'(z), f''(z),\ldots,$
$f^{(n)}(z),\ldots$ exist and are analytic in D. Furthermore, for each n,

$$f^{(n)}(z_0) = \frac{n!}{2\pi i} \oint_C \frac{f(z)}{(z - z_0)^{n+1}}\, dz, \qquad (8.57)$$

where C is any simple closed path in D enclosing z_0 and within which $f(z)$ is
analytic.

Equation (8.57) is a restatement of (8.56).

Circle of convergence of the Taylor series.   Theorem 23 guarantees convergence of
the Taylor series of $f(z)$ about each z_0 in D in the largest circular domain
$|z - z_0| < R$ in D, as shown in Fig. 8.11. However, this does not mean that
R is the radius of convergence r^* of the series, for r^* can be larger than R, as
suggested in Fig. 8.12. When this happens, the function $f(z)$ can be prolonged
into a larger domain, while retaining analyticity. For example, if $f(z) = \text{Log } z$
$(0 < \theta < \pi)$ is expanded in a Taylor series about the point $z = -1 + i$, the series
has convergence radius $\sqrt{2}$, whereas $R = 1$ [Problem 4(c) below].

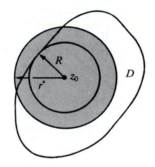

Figure 8.12   Analytic continuation.

The process of prolonging the function suggested here is called *analytic
continuation*.

Harmonic functions.   Let $w = f(x)$ be analytic in D and let
$$u(x, y) = \text{Re}\,[f(z)], \qquad v(x, y) = \text{Im}\,[f(z)].$$
By Theorem 8 (Section 8.6), $f'(z) = u_x + iv_x = v_y - iu_y = \cdots$, and by Theorem 24,
$f'(z)$ is analytic in D, so that
$$f''(z) = u_{xx} + iv_{xx} = v_{yx} - iu_{yx} = \cdots.$$
Thus $u_{xx}, u_{xy}, v_{xx},\ldots$ are also continuous in D. In general, Theorem 24 shows
that u and v have partial derivatives of all orders in D. Furthermore, by taking
real and imaginary parts in the Taylor series of $f(z)$, as given in Theorem 23, we

obtain convergent power series for $u(x, y)$ and $v(x, y)$. For example, if $f(z) = \Sigma c_n z^n$ for $|z| < R$, with $c_n = a_n + ib_n$, then

$$u(x, y) = a_0 + (a_1 x - b_1 y) + (a_2 x^2 - 2b_2 xy - a_2 y^2) + \cdots, \quad x^2 + y^2 < R^2.$$

The same series can be written in terms of polar coordinates r, θ by writing $c_n = \rho_n \exp(i\gamma_n) (\rho_n \geq 0)$, $z^n = r^n \exp(in\theta)$, so that

$$c_n z^n = \rho_n r^n \exp[i(\gamma_n + n\theta)].$$

Hence we obtain for $0 \leq r < R$, θ unrestricted,

$$u(x, y) = \mathrm{Re}\left(\sum c_n z^n\right) = \sum_{n=0}^{\infty} r^n (\alpha_n \cos n\theta + \beta_n \sin n\theta),$$

with $\alpha_n = \rho_n \cos \gamma_n$, $\beta_n = -\rho_n \sin \gamma_n$. (For fixed r, the series is a Fourier series in θ.) Also, by the Cauchy-Riemann equations, $u_x = v_y, u_y = -v_x$, so that

$$\frac{\partial^2 u}{\partial x^2} + \frac{\partial^2 u}{\partial y^2} = \frac{\partial^2 v}{\partial x \, \partial y} - \frac{\partial^2 v}{\partial y \, \partial x} = 0;$$

that is, $u(x, y)$ is *harmonic* in D. Similarly $v(x, y)$ is harmonic in D. Conversely, it can be shown that, in a simply connected domain D, each harmonic function $\phi(x, y)$ can be interpreted as the real or imaginary part of an analytic function in D. (See Problem 7 below.)

PROBLEMS

1. Determine the radius of convergence of each of the following series:

   a) $\displaystyle\sum_{n=0}^{\infty} \frac{z^n}{n^2 + 1}$
   b) $\displaystyle\sum_{n=0}^{\infty} \frac{(z-1)^n}{3^n}$
   c) $\displaystyle\sum_{n=0}^{\infty} n! z^n$
   d) $\displaystyle\sum_{n=0}^{\infty} \frac{z^n}{n!}$

2. Given the series $\displaystyle\sum_{n=1}^{\infty} (z^n/n)$, show that

   a) the series has radius of convergence 1; b) the series diverges for $z = 1$;
   c) the series converges for $z = i$ and for $z = -1$. It can be shown that the series converges for $|z| = 1$, except for $z = 1$.

3. By means of (8.57), evaluate each of the following:

   a) $\displaystyle\oint \frac{ze^z}{(z-1)^4} \, dz$ on $|z| = 2$
   b) $\displaystyle\oint \frac{\sin z}{z^4} \, dz$ on $|z| = 1$

   c) $\displaystyle\oint \frac{dz}{z^3(z+4)}$ on $|z| = 2$

4. Expand in a Taylor series about the point indicated, and determine the radius of convergence r^* and the radius R of the largest circle within which the series converges to the function:

   a) $\sin z$ about $z = 0$ b) $1/(z-1)$ about $z = 2$
   c) $\mathrm{Log}\, z$ $(0 < \theta < \pi)$ about $z = -1 + i$

5. (*Cauchy's inequalities*) Let $f(z)$ be analytic in a domain including the circle C:

$|z - z_0| = R$ and interior and let $|f(z)| \leq M = \text{const}$ on C. Prove that

$$|f^{(n)}(z_0)| \leq \frac{Mn!}{R^n} \qquad (n = 0, 1, 2, \dots).$$

[Hint: Apply (8.57).]

6. A function $f(z)$ that is analytic in the whole z-plane is termed an *entire* function or an *integral* function. Examples are polynomials, e^z, $\sin z$, $\cos z$. Prove *Liouville's theorem*: If $f(z)$ is an entire function and $|f(z)| \leq M$ for all z, where M is constant, then $f(z)$ reduces to a constant. [Hint: Take $n = 1$ in the Cauchy inequalities of Problem 5 to show that $f'(z_0) = 0$ for every z_0.]

7. Let $\phi(x, y)$ be harmonic in the simply connected domain D. Let

$$\psi(x, y) = \int_{(x_0, y_0)}^{(x, y)} -\phi_y \, dx + \phi_x \, dy, \quad (x, y) \text{ in } D,$$

where (x_0, y_0) is a fixed point of D and the line integral is taken on a path in D.

a) Show that the line integral is independent of path, so that $\psi(x, y)$ is well defined.

b) Show that $\phi_x = \psi_y$, $\phi_y = -\psi_x$, so that $f(z) = \phi + i\psi$ is analytic in D.

8.12   POWER SERIES IN POSITIVE AND NEGATIVE POWERS ▪ LAURENT EXPANSION

We have shown that every power series $\Sigma a_n (z - z_0)^n$ with nonzero convergence radius represents an analytic function and that every analytic function can be built up out of such series. It thus appears unnecessary to seek other explicit expressions for analytic functions. However, the power series represent functions only in circular domains and are hence awkward for representing a function in a more complicated type of domain. It is therefore worthwhile to consider other types of representations. A series of form

$$\sum_{n=1}^{\infty} \frac{b_n}{(z - z_0)^n} = \frac{b_1}{z - z_0} + \cdots + \frac{b_n}{(z - z_0)^n} + \cdots \tag{8.58}$$

will also represent an analytic function in a domain in which the series converges. For the substitution $z_1 = 1/(z - z_0)$ reduces the series to an ordinary power series,

$$\sum_{n=1}^{\infty} b_n z_1^n.$$

If this series converges for $|z_1| < r_1^*$, then its sum is an analytic function $F(z_1)$: hence the series (8.58) converges for

$$|z - z_0| > \frac{1}{r_1^*} = r_0^* \tag{8.59}$$

to the analytic function $g(z) = F(1/(z - z_0))$. The value $z_1 = 0$ corresponds to $z = \infty$, in a limiting sense; accordingly, we can also say that $g(z)$ is analytic at ∞ and $g(\infty) = 0$. This will be justified more fully in Section 8.14.

The domain of convergence of the series (8.58) is the region (8.59), which is the *exterior* of a circle. It can happen that $r_1^* = \infty$, in which case the series converges for all z except z_0; if $r_1^* = 0$, the series diverges for all z (except $z = \infty$, as above).

If we add to a series (8.58) a usual power series

$$\sum_{n=0}^{\infty} a_n (z - z_0)^n = a_0 + a_1(z - z_0) + \cdots,$$

converging for $|z - z_0| < r_2^*$, we obtain a sum

$$\sum_{n=1}^{\infty} \frac{b_n}{(z - z_0)^n} + \sum_{n=0}^{\infty} a_n (z - z_0)^n. \tag{8.60}$$

If $r_0^* < r_2^*$, the sum converges and represents an analytic function $f(z)$ in the *annular domain*: $r_0^* < |z - z_0| < r_2^*$, for each series has an analytic sum in this domain, so that the sum of the two series is analytic there. We can write this sum in the more compact form (after some relabeling)

$$f(z) = \sum_{n=-\infty}^{\infty} a_n (z - z_0)^n, \tag{8.61}$$

though this should be interpreted as the sum of two series, as in (8.60).

In this way we build up a new class of analytic functions, each defined in a ring-shaped domain. Every function analytic in such a domain can be obtained in this way:

THEOREM 25 (*Laurent's theorem*)   Let $f(z)$ be analytic in the ring:

$$R_1 < |z - z_0| < R_2.$$

Then

$$f(z) = \sum_{n=-\infty}^{\infty} a_n (z - z_0)^n = \left[a_0 + a_1(z - z_0) + \cdots \right]$$

$$+ \left[\frac{a_{-1}}{z - z_0} + \frac{a_{-2}}{(z - z_0)^2} + \cdots \right],$$

Figure 8.13   Laurent's theorem.

where

$$a_n = \frac{1}{2\pi i} \oint_C \frac{f(z)}{(z - z_0)^{n+1}} dz \qquad (8.62)$$

and C is any simple closed curve separating $|z| = R_1$ from $|z| = R_2$. The series converges uniformly for $R_1 < k_1 \le |z - z_0| \le k_2 < R_2$.

Proof. For simplicity we take $z_0 = 0$. Let z_1 be any point of the ring and choose r_1, r_2 so that $R_1 < r_1 < |z_1| < r_2 < R_2$, as in Fig. 8.13. We then apply the Cauchy integral formula in general form [Eq. (8.53)] to the region bounded by $C_1 : |z| = r_1$ and $C_2 : |z| = r_2$. Hence

$$f(z_1) = \frac{1}{2\pi i} \oint_{C_2} \frac{f(z)}{z - z_1} dz - \frac{1}{2\pi i} \oint_{C_1} \frac{f(z)}{z - z_1} dz.$$

The first term can be replaced by a power series

$$\sum_{n=0}^{\infty} a_n z_1^n, \qquad a_n = \frac{1}{2\pi i} \oint_{C_2} \frac{f(z)}{z^{n+1}} dz$$

as in the proof of Theorem 23 (Section 8.11). For the second term the series expansion

$$\frac{1}{z - z_1} = -\frac{1}{z_1}\left(\frac{1}{1 - z/z_1} \right) = -\frac{1}{z_1} - \frac{z}{z_1^2} - \frac{z^2}{z_1^3} - \cdots ,$$

valid for $|z_1| > |z| = r_1$, leads similarly to the series

$$\sum_{n=1}^{\infty} \frac{b_n}{z_1^n} = \sum_{n=-\infty}^{-1} a_n z_1^n, \qquad a_n = \frac{1}{2\pi i} \oint_{C_1} \frac{f(z)}{z^{n+1}} dz.$$

Hence

$$f(z_1) = \sum_{n=-\infty}^{\infty} a_n z_1^n, \qquad a_n = \frac{1}{2\pi i} \oint_{C_1} \frac{f(z)}{z^{n+1}} dz ; \cdot$$

the path C_2 or C_1 can be replaced by any path C separating $|z| = R_1$ from $|z| = R_2$, since the function integrated is analytic throughout the annulus. The uniform convergence follows as for ordinary power series (Theorem 18). The theorem is now established. $\square$

Laurent's theorem continues to hold when $R_1 = 0$ or $R_2 = \infty$ or both. In the case $R_1 = 0$ the Laurent expansion represents a function $f(z)$ analytic in a *deleted neighborhood* of z_0, that is, in the circular domain $|z - z_0| < R_2$ minus its center z_0. If $R_2 = \infty$, we can say similarly that the series represents $f(z)$ in a *deleted neighborhood* of $z = \infty$.

8.13 ISOLATED SINGULARITIES OF AN ANALYTIC FUNCTION ▪ ZEROS AND POLES

Let $f(z)$ be defined and analytic in domain D. We say that $f(z)$ has an *isolated singularity* at the point z_0 if $f(z)$ is analytic throughout a neighborhood of z_0 except at z_0 itself; that is, to use the term mentioned at the end of the preceding section, $f(z)$ is analytic in a deleted neighborhood of z_0 but not at z_0. The point z_0 is then a boundary point of D and would be called an *isolated boundary point* (see Fig. 8.14).

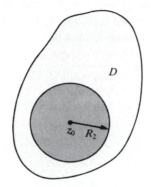

Figure 8.14   Isolated singularity.

A deleted neighborhood $0 < |z - z_0| < R_2$ forms a special case of the annular domain for which Laurent's theorem is applicable. Hence in this deleted neighborhood, $f(z)$ has a representation as a Laurent series:

$$f(z) = \sum_{n=-\infty}^{\infty} a_n(z - z_0)^n.$$

The form of this series leads to a classification of isolated singularities into three fundamental types:

Case I. No terms in negative powers of $z - z_0$ appear.   In this case the series is a Taylor series and represents a function analytic in a neighborhood of z_0. Thus the singularity can be removed by setting $f(z_0) = a_0$. We call this a *removable singularity* of $f(z)$. It is illustrated by

$$\frac{\sin z}{z} = 1 - \frac{z^2}{3!} + \frac{z^4}{5!} - \cdots$$

at $z = 0$. In practice we automatically remove the singularity by defining the function properly.

Case II.   A finite number of negative powers of $z - z_0$ appear.   Thus we have

$$f(z) = \frac{a_{-N}}{(z - z_0)^N} + \cdots + \frac{a_{-1}}{z - z_0} + a_0 + \cdots$$

$$+ a_n(z - z_0)^n + \cdots, \qquad (N \geq 1, a_{-N} \neq 0). \qquad (8.63)$$

Here $f(z)$ is said to have a *pole of order N* at z_0. We can write

$$f(z) = \frac{1}{(z - z_0)^N} g(z), \qquad g(z) = a_{-N} + a_{-N+1}(z - z_0) + \cdots, \qquad (8.64)$$

so that $g(z)$ is analytic for $|z - z_0| < R_2$ and $g(z_0) \neq 0$. Conversely, every function $f(z)$ representable in the form (8.64) has a pole of order N at z_0. Poles are illustrated by rational functions of z, such as

$$f(z) = \frac{z - 2}{(z^2 + 1)(z - 1)^3}, \qquad (8.65)$$

which has poles of order 1 at $\pm i$ and of order 3 at $z = 1$.

The rational function

$$\frac{a_{-N}}{(z - z_0)^N} + \cdots + \frac{a_{-1}}{z - z_0} = p(z) \qquad (8.66)$$

is called the *principal part* of $f(z)$ at the pole z_0. Thus $f(z) - p(z)$ is analytic at z_0.

EXAMPLE 1

$$f(z) = \frac{e^z \cos z}{z^3} \quad \text{at} \quad z = 0.$$

To obtain the Laurent series, we expand the numerator in a Taylor series:

$$e^z \cos z = \left(1 + z + \frac{z^2}{2!} + \cdots\right)\left(1 - \frac{z^2}{2!} + \cdots\right) = 1 + z - \frac{z^3}{3} + \cdots.$$

Hence

$$\frac{e^z \cos z}{z^3} = \frac{1}{z^3} + \frac{1}{z^2} - \frac{1}{3} + \cdots.$$

Here the first two terms form the principal part; the pole is of order 3. ∎

EXAMPLE 2

$$f(z) = \frac{z}{(z + 1)^2(z^3 + 2)} \quad \text{at} \quad z = -1.$$

We expand $z/(z^3 + 2)$ in a Taylor series about $z = -1$:

$$w = \frac{z}{z^3 + 2}, \qquad w' = \frac{2 - 2z^3}{(z^3 + 2)^2}, \qquad w'' = \frac{6(z^5 - 4z^2)}{(z^3 + 2)^3}, \dots,$$

$$\frac{z}{z^3 + 2} = -1 + 4(z + 1) - 15(z + 1)^2 + \cdots,$$

$$\frac{z}{(z + 1)^2(z^3 + 2)} = \frac{-1}{(z + 1)^2} + \frac{4}{z + 1} - 15 + \cdots.$$

The first two terms form the principal part; the pole is of order 2.   ■

Case III. Infinitely many negative powers of $z - z_0$ appear.   In this case, $f(z)$ is said to have an *essential singularity* at z_0. This is illustrated by the function

$$f(z) = e^{1/z}$$

$$= 1 + \frac{1}{z} + \frac{1}{2!}\frac{1}{z^2} + \frac{1}{3!}\frac{1}{z^3} + \cdots,$$

which has an essential singularity at $z = 0$.

In Case I, $f(z)$ has a finite limit at z_0, and accordingly, $|f(z)|$ is bounded near z_0; that is, there is a real constant M such that $|f(z)| < M$ for z sufficiently close to z_0.

In Case II, $\lim_{z \to z_0} f(z) = \infty$, and it is customary to assign the value ∞ (complex) to $f(z)$ at a pole. At an essential singularity, $f(z)$ has a very complicated discontinuity. In fact, for every complex number c we can find a sequence z_n converging to z_0 such that $\lim_{n \to \infty} f(z_n) = c$ (see Problem 8 following Section 8.14). It follows from this that if $|f(z)|$ is bounded near z_0, then z_0 must be a removable singularity, and if $\lim f(z) = \infty$ at z_0, then z_0 must be a pole.

If $f(z)$ is analytic at a point z_0, and $f(z_0) = 0$, then z_0 is termed a *root* or *zero* of $f(z)$. Thus the zeros of $\sin z$ are the numbers $n\pi$ ($n = 0, \pm1, \pm2,\ldots$). The Taylor series about z_0 has the form

$$f(z) = a_N(z - z_0)^N + a_{N+1}(z - z_0)^{N+1} + \cdots,$$

where $N \geq 1$ and $a_N \neq 0$, or else $f(z) \equiv 0$ in a neighborhood of z_0. It will be seen that the latter case can occur only if $f(z) \equiv 0$ throughout the domain in which it is given. If now $f(z)$ is not identically zero, then

$$f(z) = (z - z_0)^N \phi(z),$$

$$\phi(z) = a_N + a_{N+1}(z - z_0) + \cdots,$$

$$\phi(z_0) = \frac{f^{(N)}(z_0)}{N!} = a_N \neq 0.$$

We say that $f(z)$ has a zero of *order* N or *multiplicity* N at z_0. For example, $1 - \cos z$ has a zero of order 2 at $z = 0$, since

$$1 - \cos z = \frac{z^2}{2} - \frac{z^4}{24} + \cdots .$$

If $f(z)$ has a zero of order N at z_0, then $F(z) = 1/f(z)$ has a pole of order N at z_0, and the converse is also true. For if f has a zero of order N, then

$$f(z) = (z - z_0)^N \phi(z)$$

as above, with $\phi(z_0) \neq 0$. It follows from continuity that $\phi(z) \neq 0$ in a sufficiently small neighborhood of z_0. Hence $g(z) = 1/\phi(z)$ is analytic in the neighborhood, and $g(z_0) \neq 0$. Now in this neighborhood, except for z_0,

$$F(z) = \frac{1}{f(z)} = \frac{1}{(z - z_0)^N \phi(z)} = \frac{g(z)}{(z - z_0)^N},$$

so that F has a pole at z_0. The converse is proved in the same way.

It remains to consider the case when $f \equiv 0$ in a neighborhood of z_0. This is covered by the following theorem.

> **THEOREM 26**   The zeros of an analytic function are isolated, unless the function is identically zero; that is, if $f(z)$ is analytic in domain D and $f(z)$ is not identically zero, then for each zero z_0 of $f(z)$ there is a deleted neighborhood of z_0 in which $f(z) \neq 0$.

The proof is given in Section 3.2 of the book by Ahlfors listed at the end of this chapter.

8.14 THE COMPLEX NUMBER ∞

The complex number ∞ has been introduced several times in connection with limiting processes, for example, in the discussion of poles in the preceding section. In each case, ∞ has appeared in a natural way as the limiting position of a point receding indefinitely from the origin. We can incorporate this number into the complex number system with special algebraic rules:

$$\frac{z}{\infty} = 0 \; (z \neq \infty), \qquad z \pm \infty = \infty \;\; (z \neq \infty), \qquad \frac{z}{0} = \infty \;\; (z \neq 0),$$

$$(8.67)$$

$$z \cdot \infty = \infty \;\; (z \neq 0), \qquad \frac{\infty}{z} = \infty \;\; (z \neq \infty).$$

Expressions such as $\infty + \infty$, $\infty - \infty$, and ∞/∞ are not defined.

A function $f(z)$ is said to be analytic in a deleted neighborhood of ∞ if $f(z)$ is analytic for $|z| > R_1$ for some R_1. In this case the Laurent expansion with $R_2 = \infty$ and $z_0 = 0$ is available, and we have

$$f(z) = \sum_{n=-\infty}^{\infty} a_n z^n \qquad (|z| > R_1).$$

If there are no *positive* powers of z here, $f(z)$ is said to have a *removable singularity* at ∞, and we make f *analytic at* ∞ by defining $f(\infty) = a_0$:

$$f(z) = a_0 + \frac{a_{-1}}{z} + \cdots + \frac{a_{-n}}{z^n} + \cdots \qquad (|z| > R_1). \qquad (8.68)$$

This is clearly equivalent to the statement that if we set $z_1 = 1/z$, then $f(z)$ becomes a function of z_1 with removable singularity at $z_1 = 0$.

If a finite number of positive powers occurs, we have, with $N \geq 1$,

$$f(z) = a_N z^N + \cdots + a_1 z + a_0 + \frac{a_{-1}}{z} + \cdots$$

$$= z^N \phi(z),\qquad\qquad (8.69)$$

$$\phi(z) = a_N + \frac{a_{N-1}}{z} + \cdots,$$

where $\phi(z)$ is analytic at ∞ and $\phi(\infty) = a_N \neq 0$. In this case, $f(z)$ is said to have a *pole of order* N at ∞. The same holds for $f(1/z_1)$ at $z_1 = 0$. Furthermore,

$$\lim_{z \to \infty} f(z) = \infty.\qquad\qquad (8.70)$$

If infinitely many positive powers appear, $f(z)$ is said to have an *essential singularity* at $z = \infty$.

If $f(z)$ is analytic at ∞ as in (8.68) and $f(\infty) = a_0 = 0$, then $f(z)$ is said to have a *zero* at $z = \infty$. If f is not identically zero, then necessarily some $a_{-N} \neq 0$, and

$$f(z) = \frac{a_{-N}}{z^N} + \frac{a_{-N-1}}{z^{N+1}} + \cdots \qquad (|z| > R_1)$$

$$= \frac{1}{z^N} g(z),\qquad\qquad (8.71)$$

$$g(z) = a_{-N} + \frac{a_{-N-1}}{z} + \cdots.$$

Thus $g(z)$ is analytic at ∞, and $g(\infty) = a_{-N} \neq 0$. We say that $f(z)$ has a zero of order (or multiplicity) N at ∞. We can show that if $f(z)$ has a zero of order N at ∞, then $1/f(z)$ has a pole of order N at ∞, and the converse is also true.

The significance of the complex number ∞ can be shown geometrically by the device of *stereographic projection*, that is, a projection of the plane onto a sphere tangent to the plane at $z = 0$, as shown in Fig. 8.15. The sphere is given in

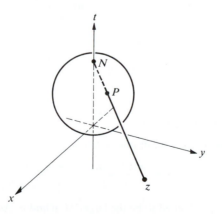

Figure 8.15   Stereographic projection.

xyt-space by the equation

$$x^2 + y^2 + \left(t - \tfrac{1}{2}\right)^2 = \tfrac{1}{4}, \tag{8.72}$$

so that the radius is $\tfrac{1}{2}$. The letter N denotes the "north pole" of the sphere, the point $(0, 0, 1)$. If N is joined to an arbitrary point z in the xy-plane, the line segment Nz will meet the sphere at one other point P, which is the projection of z on the sphere. For example, the points of the circle $|z| = 1$ project on the "equator" of the sphere, that is, the great circle $t = \tfrac{1}{2}$. As z recedes to infinite distance from the origin, P approaches N as limiting position. Thus N *corresponds to the complex number* ∞.

We refer to the z-plane plus the number ∞ as the *extended z-plane*. To emphasize that ∞ is *not* included, we refer to the *finite z-plane*.

PROBLEMS

1. For each of the following, expand in a Laurent series at the isolated singularity given and state the type of singularity:

   a) $\dfrac{e^z - 1}{z}$ at $z = 0$

   b) $\dfrac{1}{z^2(z - 3)}$ at $z = 0$

   c) $\dfrac{z - \cos z}{z}$ at $z = 0$

   d) $\csc z$ at $z = 0$

   [Hint for (d): Write

   $$\csc z = \frac{1}{\sin z} = \frac{1}{z - (z^3/3!) + \cdots} = \frac{a_{-1}}{z} + a_0 + \cdots$$

   and determine the coefficients $a_{-1}, a_0, a_1, \ldots$ so that

   $$1 = \left(z - z^3/3! + \cdots\right) \cdot \left(a_{-1}z^{-1} + a_0 + a_1 z + \cdots\right).]$$

2. For each of the following, find the principal part at the pole given:

   a) $\dfrac{z^2 + 3z + 1}{z^4}$ $(z = 0)$

   b) $\dfrac{z^2 - 2}{z(z + 1)}$ $(z = 0)$

   c) $\dfrac{e^z \sin z}{(z - 1)^2}$ $(z = 1)$

   d) $\dfrac{1}{z^2(z^3 + z + 1)}$ $(z = 0)$

3. For each of the following, expand in a Laurent series at $z = \infty$ and state the type of singularity:

   a) $\dfrac{1}{1 - z} = -\dfrac{1}{z}\dfrac{1}{1 - (1/z)}$

   b) $\dfrac{z^2}{z + 2}$

   c) $e^z + e^{1/z}$

4. Let $f(z)$ be a rational function in lowest terms:

   $$f(z) = \frac{a_0 z^n + a_1 z^{n-1} + \cdots + a_n}{b_0 z^m + b_1 z^{m-1} + \cdots + b_m}.$$

   The *degree d* of $f(z)$ is defined to be the larger of m and n. Assuming the fundamental theorem of algebra, show that $f(z)$ has precisely d zeros and d poles in the extended z-plane, a pole or zero of order N being counted as N poles or zeros.

5. For each of the following, locate all zeros and poles in the extended plane (compare Problem 4):

   a) $\dfrac{z}{z-1}$

   b) $\dfrac{z-1}{z^2+3z+2}$

   c) $\dfrac{z^3+3z^2+3z+1}{z}$

6. Let $A(z)$ and $B(z)$ be analytic at $z = z_0$; let $A(z_0) \neq 0$ and let $B(z)$ have a zero of order N at z_0, so that

$$f(z) = \frac{A(z)}{B(z)} = \frac{a_0 + a_1(z-z_0) + \cdots}{b_N(z-z_0)^N + b_{N+1}(z-z_0)^{N+1} + \cdots}$$

has a pole of order N at z_0. Show that the principal part of $f(z)$ at z_0 is

$$\frac{a_0}{b_N}\frac{1}{(z-z_0)^N} + \frac{a_1 b_N - a_0 b_{N+1}}{b_N^2}\frac{1}{(z-z_0)^{N-1}} + \cdots$$

and obtain the next term explicitly. [Hint: Set

$$\frac{a_0 + a_1(z-z_0) + \cdots}{b_N(z-z_0)^N + b_{N+1}(z-z_0)^{N+1} + \cdots} = \frac{C_{-N}}{(z-z_0)^N} + \frac{C_{-N+1}}{(z-z_0)^{N-1}} + \cdots.$$

Multiply across and solve for $C_{-N}, C_{-N+1}, \ldots$.]

7. Prove Riemann's theorem: If $|f(z)|$ is bounded in a deleted neighborhood of an isolated singularity z_0, then z_0 is a removable singularity of $f(z)$. [Hint: Proceed as in Problem 6 following Section 8.11 with the aid of (8.62).]

8. Prove the Theorem of Weierstrass and Casorati: If z_0 is an essential singularity of $f(z)$, c is an arbitrary complex number, and $\epsilon > 0$, then $|f(z) - c| < \epsilon$ for some z in every neighborhood of z_0. [Hint: If the property fails, then $1/[f(z) - c]$ is analytic and bounded in absolute value in a deleted neighborhood of z_0. Now apply Problem 7 and conclude that $f(z)$ has a pole or removable singularity at z_0.]

8.15   RESIDUES

Let $f(z)$ be analytic throughout a domain D except for an isolated singularity at a certain point z_0 of D. The integral

$$\oint f(z)\, dz$$

will not in general be zero on a simple closed path in D. However, the integral will have the same value on all curves C which enclose z_0 and no other singularity of f. This value, divided by $2\pi i$, is known as the *residue* of $f(z)$ at z_0 and is denoted by $\text{Res}[f(z), z_0]$. Thus

$$\text{Res}\left[f(z), z_0\right] = \frac{1}{2\pi i}\oint_C f(z)\, dz, \tag{8.73}$$

where the integral is taken over any path C within which $f(z)$ is analytic except at z_0 (Fig. 8.16).

THEOREM 27   The residue of $f(z)$ at z_0 is given by the equation

$$\text{Res}\left[f(z), z_0\right] = a_{-1}, \tag{8.74}$$

Figure 8.16 Residue.

where

$$f(z) = \cdots + \frac{a_{-N}}{(z - z_0)^N} + \cdots + \frac{a_{-1}}{z - z_0} + a_0 + a_1(z - z_0) + \cdots$$

(8.75)

is the Laurent expansion of $f(z)$ at z_0.

Proof. By (8.62),

$$a_{-1} = \frac{1}{2\pi i} \oint_C f(z) \, dz,$$

where C is chosen as in the definition of residue. Hence (8.74) follows at once. □

If C is a simple closed path in D, within which $f(z)$ is analytic except for isolated singularities at $z_1, \ldots, z_k$, then by Theorem 16,

$$\oint_C f(z) \, dz = \oint_{C_1} f(z) \, dz + \cdots + \oint_{C_k} f(z) \, dz,$$

where C_1 encloses only the singularity at z_1, C_2 encloses only $z_2, \ldots$, as in Fig. 8.17. We thus obtain the following basic theorem:

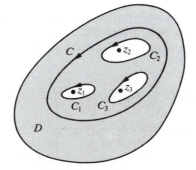

Figure 8.17 Cauchy residue theorem.

THEOREM 28 (*Cauchy's residue theorem*)   If $f(z)$ is analytic in a domain D and C is a simple closed curve in D within which $f(z)$ is analytic except for isolated singularities at $z_1, \ldots, z_k$, then

$$\oint_C f(z)\, dz = 2\pi i \{ \operatorname{Res}[f(z), z_1] + \cdots + \operatorname{Res}[f(z), z_k] \}. \quad (8.76)$$

This theorem permits rapid evaluation of integrals on closed paths, whenever it is possible to compute the coefficient a_{-1} of the Laurent expansion at each singularity inside the path. Various techniques for obtaining the Laurent expansion are illustrated in the problems preceding this section. However, if we wish only the term in $(z - z_0)^{-1}$ of the expansion, various simplifications are possible. We give several rules here:

RULE I.   At a simple pole z_0 (that is, a pole of first order),

$$\operatorname{Res}[f(z), z_0] = \lim_{z \to z_0} (z - z_0) f(z).$$

RULE II.   At a pole z_0 of order N ($N = 2, 3, \ldots$),

$$\operatorname{Res}[f(z), z_0] = \lim_{z \to z_0} \frac{g^{(N-1)}(z)}{(N-1)!},$$

where $g(z) = (z - z_0)^N f(z)$.

RULE III.   If $A(z)$ and $B(z)$ are analytic in a neighborhood of z_0, $A(z_0) \neq 0$, and $B(z)$ has a zero at z_0 of order 1, then

$$f(z) = \frac{A(z)}{B(z)}$$

has a pole of first order at z_0, and

$$\operatorname{Res}[f(z), z_0] = \frac{A(z_0)}{B'(z_0)}.$$

RULE IV.   If $A(z)$ and $B(z)$ are as in Rule III, but $B(z)$ has a zero of second order at z_0, so that $f(z)$ has a pole of second order at z_0, then

$$\operatorname{Res}[f(z), z_0] = \frac{6A'B'' - 2AB'''}{3B''^2}, \quad (8.77)$$

where A and the derivatives A', B'', B''' are evaluated at z_0.

Proofs of rules.   Let $f(z)$ have a pole of order N:

$$f(z) = \frac{1}{(z - z_0)^N} [a_{-N} + a_{-N+1}(z - z_0) + \cdots] = \frac{1}{(z - z_0)^N} g(z),$$

where

$$g(z) = (z - z_0)^N f(z), \qquad g(z_0) = a_{-N}$$

and g is analytic at z_0. The coefficient of $(z - z_0)^{-1}$ in the Laurent series for $f(z)$ is the coefficient of $(z - z_0)^{N-1}$ in the Taylor series for $g(z)$. This coefficient, which is the residue sought, is

$$\frac{g^{(N-1)}(z_0)}{(N-1)!} = \lim_{z \to z_0} \frac{g^{(N-1)}(z)}{(N-1)!}.$$

For $N = 1$ this gives Rule I; for $N = 2$ or higher we obtain Rule II. Rules III and IV follow from the identity of Problem 6 following Section 8.14:

$$\frac{A(z)}{B(z)} = \frac{a_0 + a_1(z - z_0) + \cdots}{b_N(z - z_0)^N + b_{N+1}(z - z_0)^{N+1} + \cdots}$$

$$= \frac{a_0}{b_N} \frac{1}{(z - z_0)^N} + \frac{a_1 b_N - a_0 b_{N+1}}{b_N^2} \frac{1}{(z - z_0)^{N-1}} + \cdots.$$

For a first-order pole, $N = 1$, and the residue is

$$\frac{a_0}{b_1} = \frac{A(z_0)}{B'(z_0)}.$$

For a second-order pole, $N = 2$, and the residue is $[(a_1 b_2 - a_0 b_3)/b_2^2]$. Since

$$a_0 = A(z_0), \qquad a_1 = A'(z_0),$$

$$b_2 = \frac{B''(z_0)}{2!}, \qquad b_3 = \frac{B'''(z_0)}{3!},$$

this reduces to the expression (8.77). □

EXAMPLE 1

$$\oint_{|z|=2} \frac{ze^z}{z^2 - 1} \, dz = 2\pi i \{ \text{Res}\,[\, f(z), 1] + \text{Res}\,[\, f(z), -1] \}.$$

Since $f(z)$ has first-order poles at ± 1, we find, by Rule I,

$$\text{Res}\,[\, f(z), 1] = \lim_{z \to 1} (z - 1) \cdot \frac{ze^z}{z^2 - 1} = \lim_{z \to 1} \frac{ze^z}{z + 1} = \frac{e}{2},$$

$$\text{Res}\,[\, f(z), -1] = \lim_{z \to -1} (z + 1) \cdot \frac{ze^z}{z^2 - 1} = \lim_{z \to -1} \frac{ze^z}{z - 1} = \frac{-e^{-1}}{-2}.$$

Accordingly,

$$\oint_{|z|=2} \frac{ze^z}{z^2 - 1} \, dz = 2\pi i \left(\frac{e}{2} + \frac{e^{-1}}{2} \right) = 2\pi i \cosh 1.$$

Rule III could also have been used:

$$\text{Res}\,[\, f(z), 1] = \frac{ze^z}{2z} \bigg|_{z=1} = \frac{e}{2}, \qquad \text{Res}\,[\, f(z), -1] = \frac{ze^z}{2z} \bigg|_{z=-1} = \frac{-e^{-1}}{-2}.$$

This is simpler than Rule I, since the expression $A(z)/B'(z)$, once computed, serves for all poles of the prescribed type.

EXAMPLE 2

$$\oint_{|z|=2} \frac{z}{z^4 - 1}\, dz = 2\pi i \{\operatorname{Res}\left[f(z),1\right] + \operatorname{Res}\left[f(z),-1\right]$$

$$+ \operatorname{Res}\left[f(z), i\right] + \operatorname{Res}\left[f(z), -i\right]\}.$$

All poles are of first order. Rule III gives $A(z)/B'(z) = z/(4z^3) = 1/(4z^2)$ as the expression for the residue at any one of the four points. Moreover, $z^4 = 1$ at each pole, so that

$$\frac{1}{4z^2} = \frac{z^2}{4z^4} = \frac{z^2}{4}.$$

Hence

$$\oint_{|z|=2} \frac{z}{z^4 - 1}\, dz = \frac{2\pi i}{4}(1 + 1 - 1 - 1) = 0. \quad\blacksquare$$

EXAMPLE 3

$$\oint_{|z|=2} \frac{e^z}{z(z-1)^2}\, dz = 2\pi i \{\operatorname{Res}\left[f(z),0\right] + \operatorname{Res}\left[f(z),1\right]\}.$$

At the first-order pole $z = 0$, application of Rule I gives the residue 1. At the second-order pole $z = 1$, Rule II gives

$$\operatorname{Res}\left[f(z),1\right] = \frac{d}{dz}\left(\frac{e^z}{z}\right)\bigg|_{z=1} = \frac{e^z(z-1)}{z^2}\bigg|_{z=1} = 0.$$

Rule IV could also be used, with $A = e^z$, $B = z^3 - 2z^2 + z$:

$$\operatorname{Res}\left[f(z),1\right] = \frac{6e^z(6z-4) - 2e^z \cdot 6}{3(6z-4)^2}\bigg|_{z=1} = 0.$$

Accordingly,

$$\oint_{|z|=2} \frac{e^z}{z(z-1)^2}\, dz = 2\pi i(1 + 0) = 2\pi i. \quad\blacksquare$$

8.16   RESIDUE AT INFINITY

Let $f(z)$ be analytic for $|z| > R$. The *residue of* $f(z)$ *at* ∞ is defined as follows:

$$\operatorname{Res}\left[f(z),\infty\right] = \frac{1}{2\pi i}\oint_C f(z)\, dz,$$

where the integral is taken in the *negative* direction on a simple closed path C, in the domain of analyticity of $f(z)$, and *outside* of which $f(z)$ has no singularity other than ∞. This is suggested in Fig. 8.18. Theorem 27 has an immediate extension to this case:

THEOREM 29   The residue of $f(z)$ at ∞ is given by the equation

$$\text{Res}\,[\,f(z),\infty] = -a_{-1}, \tag{8.78}$$

where a_{-1} is the coefficient of z^{-1} in the Laurent expansion of $f(z)$ at ∞:

$$f(z) = \cdots + \frac{a_{-n}}{z^n} + \cdots + \frac{a_{-1}}{z} + a_0 + a_1 z + \cdots . \tag{8.79}$$

Figure 8.18   Residue at infinity.

The proof is the same as for Theorem 27. It should be stressed that the presence of a nonzero residue at ∞ is not related to presence of a pole or essential singularity at ∞. That is, $f(z)$ can have a nonzero residue whether or not there is a pole or essential singularity, for the pole or essential singularity at ∞ is due to the *positive powers* of z, not to negative powers (Section 8.14). Thus the function $e^{1/z} = 1 + z^{-1} + (2!z^2)^{-1} + \cdots$ is analytic at ∞ but has the residue -1 there.

Cauchy's residue theorem has also an extension to include ∞:

THEOREM 30   Let $f(z)$ be analytic in a domain D that includes a deleted neighborhood of ∞. Let C be a simple closed path in D outside of which $f(z)$ is analytic except for isolated singularities at $z_1,\ldots,z_k$. Then

$$\oint_C f(z)\,dz = 2\pi i\,\{\text{Res}\,[\,f(z),z_1] + \cdots + \text{Res}\,[\,f(z),z_k] + \text{Res}\,[\,f(z),\infty]\}$$

$$\tag{8.80}$$

The proof, which is like that of Theorem 28, is left as an exercise (Problem 4 following Section 8.18). It is to be emphasized that the integral on C is taken in the *negative* direction (see Fig. 8.19) and that the *residue at ∞ must be included* on the right.

Figure 8.19   Residue theorem for exterior.

For a particular integral

$$\oint_C f(z)\, dz$$

on a simple closed path C we have now two modes of evaluation: The integral equals $2\pi i$ times the sum of the residues inside the path (provided that there are only a finite number of singularities there), and it also equals *minus* $2\pi i$ times the sum of the residues outside the path plus that at ∞ (provided there are only a finite number of singularities in the exterior domain). We can evaluate the integral both ways to check results. The principle involved here is summarized in the following theorem:

THEOREM 31   If $f(z)$ is analytic in the extended z-plane except for a finite number of singularities, then the sum of all residues of $f(z)$ (including ∞) is zero.

To evaluate the residues at ∞, we can formulate a set of rules like the preceding ones. However, the following two rules are adequate for most purposes:

RULE V.   If $f(z)$ has a zero of first order at ∞, then

$$\mathrm{Res}\left[f(z), \infty \right] = -\lim_{z \to \infty} zf(z).$$

If $f(z)$ has a zero of second or higher order at ∞, the residue at ∞ is zero.

RULE VI.

$$\mathrm{Res}\left[f(z), \infty \right] = -\mathrm{Res}\left[\frac{1}{z^2} f(1/z), 0 \right].$$

The proof of Rule V is left as an exercise (Problem 8 following Section 8.18). To prove Rule VI, we write

$$f(z) = \cdots + a_n z^n + \cdots + a_1 z + a_0 + \frac{a_{-1}}{z} + \frac{a_{-2}}{z^2} + \cdots \qquad (|z| > R).$$

Then for $0 < |z| < R^{-1}$,

$$f\left(\frac{1}{z}\right) = \cdots + \frac{a_n}{z^n} + \cdots + \frac{a_1}{z} + a_0 + a_{-1}z + a_{-2}z^2 + \cdots,$$

$$\frac{1}{z^2} f\left(\frac{1}{z}\right) = \cdots + \frac{a_0}{z^2} + \frac{a_{-1}}{z} + a_{-2} + \cdots.$$

Hence

$$\operatorname{Res}\left[\frac{1}{z^2} f\left(\frac{1}{z}\right), 0\right] = a_{-1},$$

and the rule follows. This result reduces the problem to evaluation of a residue at zero, to which Rules I through IV are applicable. □

EXAMPLE 1   We consider the integral

$$\oint_{|z|=2} \frac{z}{z^4 - 1}\, dz$$

of Example 2 in the preceding section. There is no singularity outside the path other than ∞, and at ∞ the function has a zero of order 3; hence the integral is zero. ∎

EXAMPLE 2

$$\oint_{|z|=2} \frac{1}{(z + 1)^4 (z^2 - 9)(z - 4)}\, dz.$$

Here there is a fourth-order pole inside the path, at which evaluation of the residue is tedious. Outside the path there are first-order poles at ± 3 and 4 and a zero of order 7 at ∞. Hence by Rule I the integral equals

$$-2\pi i\left(\frac{1}{4^4 6(-1)} + \frac{1}{(-2)^4(-6)(-7)} + \frac{1}{5^4 \cdot 7}\right). ∎$$

8.17 LOGARITHMIC RESIDUES ▪ ARGUMENT PRINCIPLE

Let $f(z)$ be analytic in a domain D. Then $f'(z)/f(z)$ is analytic in D except at the zeros of $f(z)$. If an analytic branch of $\log f(z)$ is chosen in part of D [necessarily excluding the zeros of $f(z)$], then

$$\frac{d}{dz} \log f(z) = \frac{f'(z)}{f(z)}. \tag{8.81}$$

For this reason the expression f'/f is termed the *logarithmic derivative* of $f(z)$. Its value is demonstrated by the following theorem:

THEOREM 32   Let $f(z)$ be analytic in domain D. Let C be a simple closed path in D within which $f(z)$ is analytic except for a finite number of poles and let $f(z) \neq 0$ on C. Then

$$\frac{1}{2\pi i} \oint_C \frac{f'(z)}{f(z)} \, dz = N_0 - N_p,$$

where N_0 is the total number of zeros of f inside C and N_p is the total number of poles of f inside C, zeros and poles being counted according to multiplicities.

Proof.   The logarithmic derivative f'/f has isolated singularities precisely at the zeros and poles of f. At a zero z_0,

$$f(z) = (z - z_0)^N g(z) \qquad [g(z_0) \neq 0],$$

$$f'(z) = (z - z_0)^N g'(z) + N(z - z_0)^{N-1} g(z),$$

$$\frac{f'(z)}{f(z)} = \frac{(z - z_0)^N g'(z) + N(z - z_0)^{N-1} g(z)}{(z - z_0)^N g(z)} = \frac{g'(z)}{g(z)} + \frac{N}{z - z_0}.$$

Hence the logarithmic derivative has a pole of first order, with residue N equal to the multiplicity of the zero. A similar analysis applies to each pole of f, with N replaced by $-N$. The theorem then follows from the Cauchy residue theorem (Theorem 28), provided that we show that there are only a finite number of singularities. The poles of f are finite in number by assumption, and it is easily shown that there can be only a finite number of zeros. (See Section 6.6 of the book by Kaplan listed at the end of this chapter.) Thus the theorem is proved.   □

Remarks.   Since

$$\frac{f'(z)}{f(z)} \, dz = d \log f(z) = d \log w = d(\log |w| + i \arg w),$$

we have

$$\frac{1}{2\pi i} \int_C \frac{f'(z)}{f(z)} \, dz = \frac{1}{2\pi i} \int_C d \log |w| + \frac{1}{2\pi} \int_C d \arg w.$$

Since $w \neq 0$ on C, $\log |w|$ is continuous on C, and the first integral on the right is zero. The second measures the total charge in $\arg w$, divided by 2π, as w traces the path C_w, the image of C, in the w-plane. Hence it also measures the number of times C_w winds about the origin of the w-plane; in Fig. 8.20 the number is $+2$. (For more on this *winding number* see Problem 5 following Section 5.7). The statement

$$\frac{1}{2\pi} [\text{increases in } \arg f(z) \text{ on path}] = N_0 - N_p \tag{8.82}$$

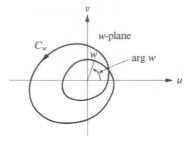

Figure 8.20 Argument principle.

is known as the *argument principle*. This is of great value in finding zeros and poles of analytic functions.

THEOREM 33 Let $f(z)$ be analytic in domain D. Let C_z be a piecewise smooth simple closed path in D, within which $f(z)$ is analytic. If the function $f(z)$ maps the curve C_z in a one-to-one fashion onto a simple closed path C_w in the w-plane and $f'(z) \neq 0$ inside C_z, then $f(z)$ maps the interior of C_z in a one-to-one fashion onto the entire interior of C_w.

Proof. We first remark that the Jacobian of the transformation from the xy-plane to the uv-plane is

$$J = \frac{\partial(u,v)}{\partial(x,y)} = \begin{vmatrix} \dfrac{\partial u}{\partial x} & \dfrac{\partial u}{\partial y} \\[2mm] \dfrac{\partial v}{\partial x} & \dfrac{\partial v}{\partial y} \end{vmatrix} = \left(\frac{\partial u}{\partial x}\right)^2 + \left(\frac{\partial v}{\partial x}\right)^2 = |f'(z)|^2.$$

Hence $J > 0$ inside C_z. It then follows (see Section 5.14) that, as z traces C_z in the positive direction, $w = f(z)$ traces C_w in the positive direction. Hence, if w_0 is a point interior to C_w (Fig. 8.21),

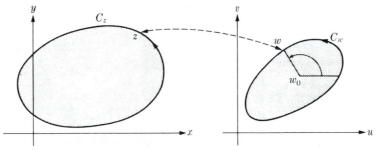

Figure 8.21 One-to-one mapping.

$\arg(w - w_0) = \arg[f(z) - w_0]$ increases by precisely 2π. We apply the argument principle to the function $f(z) - w_0$ and, since this function has no poles inside C_z, conclude that

$$N_0 = \frac{1}{2\pi} \cdot 2\pi = 1;$$

that is $f(z) - w_0$ has precisely one zero z_0 inside C_z.

Every z_0 inside C_z occurs as such a zero, for let $f(z_0) = w_0$. Then w_0 cannot be outside C_w, for then $\arg(w - w_0)$ increases by 0 as C_w is traced, so that $f(z) - w_0$ would have *no zeros* inside C_z; also w_0 cannot be on C_w, for then by the inverse mapping theorem (see Problem 9 following Section 8.7), f would map a neighborhood of z_0 (contained in the interior of C_z) onto a neighborhood of w_0 and hence $f(z_1) = w_1$ would be outside C_w for some z_1 inside C_z. Thus f establishes a one-to-one correspondence between the interiors of C_z and C_w.

COROLLARY. Let $f(z)$ be analytic in domain D_z and map D_z in a one-to-one fashion into the w plane. Then $f'(z) \neq 0$ in D_z.

Proof. If $f'(z_0) = 0$, then choose a circle C_z about z_0 lying with its interior in D_z. The function $f(z)$ maps C_z onto C_w as in the above theorem and, as in the above proof, $N_0 = 1$ for the function $f(z) - w_0$. However, since $f'(z_0) = 0$, $f(z) - w_0$ has a multiple root at z_0, so that $N_0 \geq 2$. This is a contradiction, hence $f'(z) \neq 0$ in D_z.

Remarks. As in Problem 9 following Section 8.7, the inverse $z = g(w)$ of $w = f(z)$ in Theorem 33 is analytic inside C_w and $g'(w) \neq 0$. Thus f and its inverse provide one-to-one conformal mappings between the interiors of C_z and C_w.

The condition $f'(z) \neq 0$ in D_z is not sufficient to guarantee a one-to-one transformation throughout all of D_z. For example, $w = z^2$ satisfies this condition for $z \neq 0$, but the transformation is not one-to-one, since the inverse is the two-valued function $z = \sqrt{w}$. Sufficient conditions for a one-to-one mapping, supplementing Theorem 33, are given in Section 9.1.

8.18 PARTIAL FRACTION EXPANSION OF RATIONAL FUNCTIONS

The theory of analytic functions provides a simple proof for the familiar rules for partial fraction expansions.

Let

$$f(z) = \frac{a_0 z^n + \cdots + a_n}{b_0 z^m + \cdots + a_m} \qquad (a_0 \neq 0, \quad b_0 \neq 0) \tag{8.83}$$

be given. We assume that $n < m$, so that f is a proper fraction. We also assume that the numerator and denominator have no common zeros. Let $z_1, z_2, \ldots, z_N$ be the *distinct* zeros of the *denominator* (no repetitions); these are the poles of $f(z)$. At z_1, $f(z)$ has a Laurent expansion:

$$f(z) = p_1(z) + g_1(z),$$

$$p_1(z) = \frac{A_{-k_1}}{(z - z_1)^{k_1}} + \frac{A_{-k_1+1}}{(z - z_1)^{k_1-1}} + \cdots + \frac{A_{-1}}{z - z_1}. \tag{8.84}$$

Here $p_1(z)$ is the principal part of $f(z)$ at z_1 and $g_1(z)$ is analytic at z_1; k_1 is the order of the pole at z_1. Similar expressions hold at the other poles.

The partial fraction expansion of $f(z)$ is now simply the identity

$$f(z) = p_1(z) + p_2(z) + \cdots + p_N(z). \tag{8.85}$$

To justify this, we let

$$F(z) = f(z) - \left[p_1(z) + p_2(z) + \cdots + p_N(z) \right].$$

Now $f(z) - p_1(z)$ has a removable singularity at z_1, while $p_2(z),\ldots,p_N(z)$ are analytic at z_1. Hence $F(z)$ has a removable singularity at z_1. In general, $F(z)$ has only removable singularities for finite z. At ∞, $f(z)$, $p_1(z),\ldots,p_N(z)$ all have zeros; hence $F(z)$ has a zero at ∞. But $F(z)$ is a rational function with no poles. Hence $F(z)$ must be a polynomial. Thus $F(z)$ has a pole at ∞ unless F is constant; since we know that $F = 0$ at ∞, F must be a constant, namely, zero. This proves (8.85).

If $f(z)$ has only simple poles, the principal part at each pole z_j is simply $A_j/(z - z_j)$, where A_j is the residue of f at z_j, and hence in this case

$$f(z) = \frac{A_1}{z - z_1} + \cdots + \frac{A_m}{z - z_m} \qquad \left(A_j = \operatorname{Res}\left[f, z_j \right] \right) \tag{8.86}$$

is the partial fraction expansion. If we write

$$f(z) = \frac{A(z)}{B(z)},$$

then Rule III can be applied:

$$A_j = \frac{A(z_j)}{B'(z_j)}, \qquad f(z) = \sum_{j=1}^{m} \frac{A(z_j)}{B'(z_j)} \frac{1}{z - z_j}. \tag{8.87}$$

At a multiple pole z_j of order k we can write

$$f(z) = \frac{1}{(z - z_j)^k} \phi(z).$$

The principal part $p_j(z)$ is then

$$p_j(z) = \frac{\phi(z_j)}{(z - z_j)^k} + \frac{\phi'(z_j)}{1!(z - z_j)^{k-1}} + \cdots + \frac{\phi^{(k-1)}(z_j)}{(k-1)!(z - z_j)}. \tag{8.88}$$

Hence $p_j(z)$ can be found without knowledge of the other poles.

EXAMPLE 1

$$f(z) = \frac{z^2 + 1}{z^3 + 4z^2 + 3z} = \frac{z^2 + 1}{z(z + 1)(z + 3)}.$$

There are simple poles at $0, -1, -3$. By (8.87),

$$f(z) = \frac{A(0)}{B'(0)} \frac{1}{z} + \frac{A(-1)}{B'(-1)} \frac{1}{z + 1} + \frac{A(-3)}{B'(-3)} \frac{1}{z + 3};$$

with $A = z^2 + 1$, $B = z^3 + 4z^2 + 3z$, $B' = 3z^2 + 8z + 3$ we find

$$f(z) = \frac{1}{3} \frac{1}{z} + \frac{2}{-2} \frac{1}{z + 1} + \frac{10}{6} \frac{1}{z + 3}. \quad \blacksquare$$

EXAMPLE 2

$$f(z) = \frac{z}{(z - 1)^2(z^3 + z + 1)}.$$

At the pole $z = 1$ we write

$$f = \frac{1}{(z-1)^2}\phi(z) \qquad \left(\phi = \frac{z}{z^3 + z + 1}\right).$$

Since $\phi(1) = \frac{1}{3}$, $\phi'(1) = -\frac{1}{9}$, the principal part at 1 is

$$\frac{1}{3}\frac{1}{(z-1)^2} - \frac{1}{9}\frac{1}{z-1}.$$

The cubic $z^3 + z + 1$ has one real root z_1 and two complex roots z_2, z_3. These are all simple. Hence we can write

$$f(z) = \frac{A(z_1)}{B'(z_1)}\frac{1}{z - z_1} + \frac{A(z_2)}{B'(z_2)}\frac{1}{z - z_2}$$

$$+ \frac{A(z_3)}{B'(z_3)}\frac{1}{z - z_3} + \frac{1}{3}\frac{1}{(z-1)^2} - \frac{1}{9}\frac{1}{z-1},$$

where $A(z) = z$, $B(z) = (z-1)^2(z^3 + z + 1)$. At the poles z_1, z_2, z_3, $B'(z)$ reduces to $z^2 + z + 7$, by the relation $z^3 + z + 1 = 0$. ∎

PROBLEMS

1. Evaluate the following integrals on the paths given:

a) $\displaystyle\oint \frac{z\,dz}{(z-1)(z-3)}$ $(|z| = 2)$ **b)** $\displaystyle\oint \frac{e^{3z}\,dz}{z^2 + 4}$ $(|z| = 3)$

c) $\displaystyle\oint \frac{\sin z}{z^4}\,dz$ $(|z| = 1)$ **d)** $\displaystyle\oint \frac{z\,dz}{z^3 + z + 1}$ $(|z| = 4)$

e) $\displaystyle\oint \frac{dz}{(z+1)^4(z+3)}$ $(|z| = 2)$ **f)** $\displaystyle\oint \frac{dz}{(z+1)^5(z+3)}$ $(|z| = 4)$

g) $\displaystyle\oint \frac{2z + 2}{z^2 + 2z + 2}\,dz$ $(|z| = 2)$ **h)** $\displaystyle\oint \frac{3z^2 - 6z + 1}{z^3 - 3z^2 + z - 3}\,dz$ $(|z| = 2)$

2. Expand each of the following in partial fractions:

a) $\dfrac{1}{z^2 - 4}$ **b)** $\dfrac{z + 1}{(z-1)(z-2)(z-3)}$

c) $\dfrac{z^2}{z^5 + 1}$ **d)** $\dfrac{1}{z^n - 1}$

e) $\dfrac{z}{(z-1)^2(z+1)^2}$ **f)** $\dfrac{\phi(z)}{z^n(z+1)}$ $(\phi = c_0 + c_1 z + \cdots + c_n z^n)$

g) $\dfrac{1}{(z-1)^3(z^4 + z + 1)}$ **h)** $\dfrac{1}{(z^2 + 1)(z^2 + 2z + 2)}$

3. Prove the fundamental theorem of algebra: Every polynomial of degree at least 1 has a zero. [Hint: Show that $\mathrm{Res}\,[f'(z)/f(z), \infty]$ is not 0, but is in fact minus the degree n of the polynomial $f(z)$. Then use Theorem 32 to show that f has n zeros.]

4. Prove Theorem 30.

5. Formulate and prove Theorem 32 for integration around the boundary B of a region R in D, bounded by simple closed curves $C_1, \ldots, C_k$.

6. Prove that under the hypotheses of Theorem 32, if $g(z)$ is analytic in D and within C, then

$$\frac{1}{2\pi i} \oint_C \frac{g(z)f'(z)}{f(z)} \, dz = \sum_{k=1}^{n} g(z_k') - \sum_{l=1}^{m} g(z_l''),$$

where $z_1', \ldots, z_n'$ are the zeros of f, and $z_1'', \ldots, z_m''$ are the poles of f inside C, repeated according to multiplicity.

7. Extend Rule IV of Section 8.15 to the case in which $A(z)$ has a first-order zero at z_0 and $B(z)$ has a second-order zero.

8. Prove Rule V of Section 8.16.

8.19 APPLICATION OF RESIDUES TO EVALUATION OF REAL INTEGRALS

A variety of real definite integrals between special limits can be evaluated with the aid of residues.

For example, an integral

$$\int_0^{2\pi} R(\sin\theta, \cos\theta) \, d\theta,$$

where R is a rational function of $\sin\theta$ and $\cos\theta$, is converted to a complex line integral by the substitution:

$$z = e^{i\theta}, \qquad dz = ie^{i\theta} \, d\theta = iz \, d\theta,$$

$$\cos\theta = \frac{e^{i\theta} + e^{-i\theta}}{2} = \frac{1}{2}\left(z + \frac{1}{z}\right),$$

$$\sin z = \frac{e^{i\theta} - e^{-i\theta}}{2i} = \frac{1}{2i}\left(z - \frac{1}{z}\right);$$

the path of integration is the circle $|z| = 1$.

EXAMPLE 1 $\int_0^{2\pi} \dfrac{1}{\cos\theta + 2} \, d\theta$. The substitution reduces this to

$$\oint_{|z|=1} \frac{-2i}{z^2 + 4z + 1} \, dz = 4\pi \operatorname{Res}\left[z^2 + 4z + 1, -2 + \sqrt{3}\,\right],$$

since $-2 + \sqrt{3}$ is the only root of the denominator inside the circle. Accordingly,

$$\int_0^{2\pi} \frac{1}{\cos\theta + 2} \, d\theta = \frac{2\pi}{\sqrt{3}}.$$

The substitution can be summarized in the one rule:

$$\int_0^{2\pi} R(\sin\theta, \cos\theta) \, d\theta = \oint_{|z|=1} R\left(\frac{z^2 - 1}{2iz}, \frac{z^2 + 1}{2z}\right) \frac{dz}{iz}. \tag{8.89}$$

The complex integral can be evaluated by residues, provided that R has no poles on the circle $|z| = 1$.   ∎

A second example is provided by integrals of the type

$$\int_{-\infty}^{\infty} f(x)\, dx.$$

We illustrate the procedure with an example and formulate a general principle below.

EXAMPLE 2 $\displaystyle\int_{-\infty}^{\infty} \frac{dx}{x^4 + 1}$. This integral can be regarded as a line integral of $f(z) = 1/(z^4 + 1)$ along the real axis. The path is not closed (unless one adjoins ∞), but we show that it acts like a closed path "enclosing" the upper half-plane, so that the integral along the path equals the sum of the residues in the upper half-plane.

To establish this, we consider the integral of $f(z)$ on the semicircular path C_R shown in Fig. 8.21. When R is sufficiently large, the path encloses the two poles: $z_1 = \exp(\frac{1}{4} i\pi)$, $z_2 = \exp(\frac{3}{4} i\pi)$ of $f(z)$. Hence

$$\oint_{C_R} f(z)\, dz = 2\pi i \{\operatorname{Res}[f(z), z_1] + \operatorname{Res}[f(z), z_2]\}.$$

As R increases, the integral on C_R cannot change, since it always equals the sum of the residues times 2π. Hence

$$\oint_{C_R} f(z)\, dz = \lim_{R \to \infty} \oint_{C_R} f(z)\, dz = \lim_{R \to \infty} \int_{-R}^{R} \frac{dx}{x^4 + 1} + \lim_{R \to \infty} \int_{D_R} \frac{1}{z^4 + 1}\, dz,$$

where D_R is the semicircle $z = Re^{i\theta}$, $0 \le \theta \le \pi$. The limit of the first term is the integral desired (since the limits at $+\infty$ and $-\infty$ exist separately, as required; cf. Section 4.1). The limit of the second term is 0, since $|z|^4 = |z^4| = |z^4 + 1 - 1| \le |z^4 + 1| + 1$, so that on D_R,

$$\left| \frac{1}{z^4 + 1} \right| \le \frac{1}{|z|^4 - 1} = \frac{1}{R^4 - 1},$$

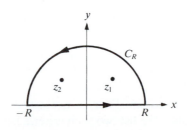

Figure 8.22   Evaluation of $\int_{-\infty}^{\infty} f(x)\, dx$ by residues.

and

$$\left| \int_{D_R} \frac{1}{z^4 + 1} \, dz \right| \leqq \frac{\pi R}{R^4 - 1}.$$

Accordingly,

$$\int_{C_R} f(z) \, dz = \int_{-\infty}^{\infty} \frac{dx}{x^4 + 1} = 2\pi i \{ \operatorname{Res} [f(z), z_1] + \operatorname{Res} [f(z), z_2] \}.$$

By Rule III the sum of the residues is

$$\frac{1}{4z_1^3} + \frac{1}{4z_2^3} = -\frac{1}{4}(z_1 + z_2) = -\frac{\sqrt{2}}{4} i,$$

and hence

$$\int_{-\infty}^{\infty} \frac{dx}{x^4 + 1} = \frac{\pi\sqrt{2}}{2}. \quad \blacksquare$$

We now formulate the general principle:

THEOREM 33   Let $f(z)$ be analytic in a domain D that includes the real axis and all of the half-plane $y > 0$ except for a finite number of points. If

$$\lim_{R \to \infty} \int_0^\pi f(Re^{i\theta}) Re^{i\theta} \, d\theta = 0 \tag{8.90}$$

and

$$\int_{-\infty}^{\infty} f(x) \, dx \tag{8.91}$$

exists, then

$$\int_{-\infty}^{\infty} f(x) \, dx = 2\pi i \{ \text{sum of residues of } f(z) \text{ in the upper half-plane} \}.$$

$$\tag{8.92}$$

The proof is a repetition of the reasoning used in the above example. It is of interest to note that even when the integral (8.91) fails to exist as improper integral, condition (8.90) implies that the principal value

$$(P) \int_{-\infty}^{\infty} f(x) \, dx = \lim_{R \to \infty} \int_{-R}^{R} f(x) \, dx \tag{8.93}$$

exists. (See Section 6.24.)

In order to apply Theorem 33 it is necessary to have simple criteria to guarantee that (8.90) holds. We list two such criteria here:

I. If $f(z)$ is rational and has a zero of order greater than 1 at ∞, then (8.90) holds.

For when $|z|$ is sufficiently large,

$$f(z) = \frac{a_{-N}}{z^N} + \frac{a_{-N+1}}{z^{N-1}} + \cdots, \quad N > 1,$$

so that $zf(z)$ has a zero at infinity. Now, by (8.11),

$$\left| \int_0^\pi f(Re^{i\theta}) Re^{i\theta} \, d\theta \right| \le \int_0^\pi |zf(z)| \, d\theta,$$

so that the integral must converge to 0 as $R \to \infty$.

II. If $g(z)$ is rational and has a zero of order 1 or greater at ∞, then (8.90) holds for $f(z) = e^{miz} g(z)$, $m > 0$.

For a proof and further criteria we refer to page 115 of the treatise of Whittaker and Watson listed at the end of the chapter. Rule II makes possible the evaluation of the *Fourier integral* (Section 7.18),

$$\int_{-\infty}^\infty g(x) e^{mix} \, dx = \int_{-\infty}^\infty g(x) \cos mx \, dx + i \int_{-\infty}^\infty g(x) \sin mx \, dx,$$

provided that both real integrals exist.

EXAMPLE 3   The integrals

$$\int_{-\infty}^\infty \frac{x \cos x}{x^2 + 1} \, dx \quad \text{and} \quad \int_{-\infty}^\infty \frac{x \sin x}{x^2 + 1} \, dx$$

both exist by the corollary to Theorem 51 of Section 6.22. Hence

$$\int_{-\infty}^\infty \frac{x e^{ix}}{x^2 + 1} \, dx = 2\pi i \operatorname{Res} \left[\frac{z e^{iz}}{z^2 + 1}, i \right] = \frac{\pi i}{e}.$$

Taking real and imaginary parts, we find

$$\int_{-\infty}^\infty \frac{x \cos x}{x^2 + 1} \, dx = 0, \qquad \int_{-\infty}^\infty \frac{x \sin x}{x^2 + 1} \, dx = \frac{\pi}{e}.$$

Since the first integral is an integral of an *odd* function, the value of 0 could be predicted. ∎

PROBLEMS

1. Evaluate the following integrals:

a) $\displaystyle\int_0^{2\pi} \frac{1}{5 + 3 \sin \theta} \, d\theta$

b) $\displaystyle\int_0^{2\pi} \frac{1}{5 - 4 \cos \theta} \, d\theta$

c) $\displaystyle\int_0^{2\pi} \frac{1}{(\cos \theta + 2)^2} \, d\theta$

d) $\displaystyle\int_0^{2\pi} \frac{1}{(3 + \cos^2 \theta)^2} \, d\theta$

2. Evaluate the following integrals:

a) $\displaystyle\int_{-\infty}^\infty \frac{1}{x^2 + x + 1} \, dx$

b) $\displaystyle\int_{-\infty}^\infty \frac{1}{(x^2 + 1)(x^2 + 4)} \, dx$

c) $\displaystyle\int_{-\infty}^{\infty} \frac{1}{x^6 + 1}\, dx$

d) $\displaystyle\int_{0}^{\infty} \frac{1}{(x^2 + 1)^2}\, dx$

3. Evaluate the following integrals:

a) $\displaystyle\int_{-\infty}^{\infty} \frac{\cos x}{x^2 + 4}\, dx$

b) $\displaystyle\int_{-\infty}^{\infty} \frac{\sin 2x}{x^2 + x + 1}\, dx$

c) $\displaystyle\int_{0}^{\infty} \frac{x^3 \sin x}{x^4 + 1}\, dx$

d) $\displaystyle\int_{0}^{\infty} \frac{x^2 \cos 3x}{(x^2 + 1)^2}\, dx$

4. Prove that $\displaystyle\int_{0}^{\infty} \frac{\sin x}{x}\, dx = \tfrac{1}{2}\pi$. [Hint: Let C be a path formed of the semicircular paths D_r: $|z| = r$ and D_R: $|z| = R$, where $0 \le \theta \le \pi$ and $0 < r < R$, plus the intervals $-R \le x \le -r,\ r \le x \le R$ on the real axis. Show that

$$\lim_{r \to 0} \int_{-r}^{r} \frac{e^{iz} - 1}{z}\, dz = 0.$$
$$\scriptstyle D_r$$

Use this result and Rule II to conclude that

$$\lim_{r \to 0} \int_{-r}^{r} \frac{e^{iz}}{z}\, dz = -\pi i, \qquad \lim_{R \to \infty} \int_{R}^{-R} \frac{e^{iz}}{z}\, dz = 0.$$
$$\scriptstyle D_r \qquad\qquad\qquad\qquad D_R$$

Accordingly,

$$\lim_{R \to \infty} \left\{ \lim_{r \to 0} \oint_{C} \frac{e^{iz}}{z}\, dz \right\} = \lim_{R \to \infty} \left\{ \lim_{r \to 0} \left[\int_{-R}^{-r} \frac{e^{ix}}{x}\, dx + \int_{r}^{R} \frac{e^{ix}}{x}\, dx \right] \right\} - \pi i.$$

The left-hand side is zero by the Cauchy integral theorem. Show that the imaginary part of the right-hand side has the value $2\displaystyle\int_{0}^{\infty} \frac{\sin x}{x}\, dx - \pi$.]

Suggested References

Ahlfors, Lars V., Complex Analysis, 3rd ed. New York: McGraw-Hill, 1969.

Betz, Albert, Konforme Abbildung, 2nd ed. Berlin: Springer, 1964.

Bieberbach, Ludwig, Lehrbuch der Funktionentheorie (2 vols.), 4th ed. Leipzig: B. G. Teubner, 1934.

Churchill, Ruel V., and Brown, James W., Complex Variables and Applications, 5th ed. New York: McGraw-Hill, 1990.

Courant, R., Dirichlet's Principle, Conformal Mapping and Minimal Surfaces. New York: Springer-Verlag, 1977.

Frank, P., and R. von Mises, Die Differentialgleichungen und Integralgleichungen der Mechanik and Physik, (2 vols.), 2nd ed. New York: Dover Publications, 1953.

Goursat, Édouard, A Course in Mathematical Analysis, Vol. II, Part 1 (transl. by E. R. Hedrick and O. Dunkel). New York: Dover Publications, 1970.

Hurwitz, A. and R. Courant, Funktionentheorie, 4th ed. Berlin: Springer, 1964.

Kaplan, Wilfred, Introduction to Analytic Functions. Reading, Mass.: Addison-Wesley, 1966.

Knopp, Konrad, Theory of Functions (2 vols.), transl. by F. Bagemihl. New York: Dover Publications, 1953.

Osgood, W. F., Lehrbuch der Funktionentheorie (2 vols.). New York: Chelsea Pub. Co., 1965.

Picard, Emile, Traité d'Analyse, 3rd ed., Vol. II. Paris: Gauthier-Villars, 1922.

Titchmarsh, E. C., The Theory of Functions, 2nd ed. Oxford: Oxford University Press, 1953.

Whittaker, E. T., and G. N. Watson, A Course of Modern Analysis, 4th ed. Cambridge: Cambridge University Press, 1950.

9

Conformal Mapping

9.1 DEFINITION OF CONFORMAL MAPPING

As pointed out in Section 8.6, a function $w = f(z)$ can be interpreted as a *transformation* or *mapping* from the z plane to the w plane. The term "mapping" is really justified only when the correspondence between z values and w values is one-to-one, i.e., to each z of a domain D_z corresponds just one $w = f(z)$ of a domain D_w and conversely. The domain D_w is then a distorted picture or "image" of the domain D_z; circles in D_z correspond to closed curves in D_w, triangles in D_z to curvilinear triangles in D_w, as illustrated in Fig. 9.1.

Figure 9.1 Conformal mapping.

If $f(z)$ is analytic, such a mapping has an additional property, that of being *conformal*. A mapping from D_z to D_w is termed *conformal* if to each pair of curves

623

in D_z intersecting at angle α there corresponds a pair of curves in D_w intersecting at angle α. The mapping is termed *conformal* and *sense-preserving* if the angles are equal and have the same sign, as illustrated in Fig. 9.1.

> **THEOREM**   Let $w = f(z)$ be analytic in the domain D_z, and map D_z in a one-to-one fashion on a domain D_w. If $f'(z) \neq 0$ in D_z, then $f(z)$ is conformal and sense-preserving.

Proof.   Let $z(t) = x(t) + iy(t)$ be parametric equations of a smooth curve through z_0 in D_z. By proper choice of parameter (e.g., by using arc length) we can ensure that the tangent vector

$$\frac{dz}{dt} = \frac{dx}{dt} + i\frac{dy}{dt}$$

is not 0 at z_0. The given curve corresponds to a curve $w = w(t)$ in the w plane, with tangent vector

$$\frac{dw}{dt} = \frac{dw}{dz}\frac{dz}{dt}$$

as in Section 8.2. Hence

$$\arg\frac{dw}{dt} = \arg\frac{dw}{dz} + \arg\frac{dz}{dt}.$$

This equation asserts that, at $w_0 = f(z_0)$, the argument of the tangent vector differs from that of dz/dt by the angle $\arg f'(z_0)$, which is *independent of the particular curve chosen through z_0*. Accordingly, as the direction of the curve through z_0 is varied, the direction of the corresponding curve through w_0 must vary through the same angle (in magnitude and sign). The theorem is thus established.   □

Conversely, it can be shown that all conformal and sense-preserving maps $w = f(z)$ are given by analytic functions; more explicitly, if u and v have continuous first partial derivatives in D_z and $J = \partial(u,v)/\partial(x,y) \neq 0$ in D_z, then the fact that the mapping by $w = u + iv = f(z)$ is conformal and sense-preserving implies that $u_x = v_y$, $u_y = -v_x$, so that w is analytic (see Section 2.3 of the book by Ahlfors listed at the end of the chapter). From this geometric characterization it is also clear that the inverse of a one-to-one conformal, sense-preserving mapping has the same property and *is itself analytic*. (See Problem 9 following Section 8.7.)

If $f'(z_0) = 0$ at a point z_0 of D_z, then $\arg f'(z_0)$ has no meaning and the argument above breaks down. One can in fact show that conformality breaks down at z_0; indeed, the transformation is not one-to-one in any neighborhood of z_0 by the Corollary to Theorem 33 in Section 8.17; see also Problem 14 below.

In practice, the term "conformal" is used loosely to mean "conformal and sense-preserving"; that will be done here. It should be noted that a reflection, such as the mapping $w = \bar{z}$, is conformal but sense-reversing.

Tests for one-to-one-ness.   For the applications of conformal mapping, it is crucial that the mapping be one-to-one in the domain chosen. In most examples, the mapping will also be defined and continuous on the boundary of the domain; failure of one-to-one-ness on the boundary is less serious.

By the Corollary to Theorem 33 in Section 8.17, if $f'(z) = 0$ at some point of the domain, then the mapping cannot be one-to-one. One should therefore verify that $f'(z) \neq 0$ in the domain as a first step. Even if this is satisfied, the mapping need not be one-to-one, and one must apply additional tests. The following are useful tests for one-to-one-ness:

I. *Explicit formula for the inverse function.* If an explicit formula for the inverse function $z = z(w)$ is available and one can show that, by this formula, there is at most one z in D_z for each w, then the mapping must be one-to-one. For example, $w = z^2$ is one-to-one in the first quadrant of the z plane, for to each w there is at most one square root $\sqrt{w}$ in the quadrant. As z varies over D_z, w varies over D_w: the upper half-plane $v > 0$, as will be seen below.

II. *Analysis of level curves of u and v.* One can verify one-to-one-ness and at the same time obtain a very clear picture of the mapping by plotting the level curves: $u(x,y) = c_1$, $v(x,y) = c_2$. If for arbitrary choice of c_1, c_2 the loci $u = c_1$, $v = c_2$ intersect at most once in D_z, then the mapping from D_z to the uv plane is one-to-one.

III. *One-to-one-ness on the boundary.* This is the test formulated in Theorem 33 in Section 8.17: if $f(z)$ is analytic in C plus interior D_z and one-to-one on C, then $f(z)$ is one-to-one on D_z. As one can show, it is sufficient that f be continuous on C plus interior, and analytic on the interior; in various important examples this is the case, analyticity being violated at one or more points on the boundary. One can also reason thus: if one can show that, for every simple closed curve C' approximating the boundary sufficiently closely, f is one-to-one on C', then f must be one-to-one in D_z; for if f fails to be one-to-one in D_z, then $f(z_1) = f(z_2)$ for two points z_1, z_2 of D_z. A curve C' sufficiently close to the boundary would enclose both z_1 and z_2, so that the failure of one-to-one-ness would have to reveal itself on C'. Other natural extensions of the principle will be pointed out in the examples to follow.

IV. *Real or imaginary part maxima and minima on the boundary.* Let $f(z)$ be analytic in domain D_z, bounded by the simple closed curve C. Let $u = \text{Re}[f(z)]$ be continuous in D_z plus C. *If u has precisely one relative maximum and one relative minimum on C, then w* $= f(z)$ *is one-to-one in* D_z. A similar criterion holds for $v = \text{Im}[f(z)]$. Furthermore, the same conclusion holds for mappings: $u = u(x,y)$, $v = v(x,y)$ by *nonanalytic* functions, provided the Jacobian $J = \partial(u,v)/\partial(x,y)$ is always positive (or always negative) in D_z. For a proof see the paper by Kaplan listed at the end of the chapter.

V. Let $f(z) = u + iv$ be analytic in D_z and let D_z be *convex;* that is, for each pair of points z_1 and z_2 of D_z, let the line segment from z_1 to z_2 lie in D_z. If real constants a, b can be found such that

$$a\frac{\partial u}{\partial x} + b\frac{\partial u}{\partial y} > 0 \qquad \text{in } D_z,$$

then f is one-to-one in D_z (see Problem 15 below, following Section 9.2).

VI.   Let $f(z) = u + iv$ be analytic in D_z and let D_z be the half-plane: $y > 0$. If a complex constant c and a real number a can be found, such that

$$\text{Re}[c(z - a)f'(z)] > 0$$

in D_z, then f is one-to-one in D_z (see Problem 16 below, following Section 9.2).

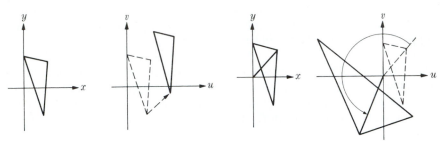

Figure 9.2   Translation.   **Figure 9.3**   Rotation-stretching.

9.2   EXAMPLES OF CONFORMAL MAPPING

EXAMPLE 1   *Translations.*   The general form is

$$w = z + a + bi \qquad (a, b \text{ real constants}).$$

Each point z is displaced through the vector $a + bi$, as in Fig. 9.2.   ∎

EXAMPLE 2   *Rotation-stretchings.*   The general form is

$$w = Ae^{i\alpha}z \qquad (A, \alpha \text{ real constants}, A > 0).$$

If one writes $z = re^{i\theta}$, $w = \rho e^{i\phi}$, so that ρ, ϕ are polar coordinates in the w plane, then one has

$$\rho = Ar, \qquad \phi = \theta + \alpha.$$

Thus distances from the origin are stretched in the ratio A to 1, while all figures are rotated about the origin through the angle α (Fig. 9.3).   ∎

EXAMPLE 3   *The general linear integral transformation*

$$w = az + b \qquad (a, b \text{ complex constants}). \tag{9.1}$$

This is equivalent to a rotation-stretching, as in Example 2, with $a = Ae^{i\alpha}$, followed by a translation through the vector b.   ∎

EXAMPLE 4   *The reciprocal transformation*

$$w = \frac{1}{z}.$$

In polar coordinates one has

$$\rho = \frac{1}{r}, \qquad \phi = -\theta.$$

Thus this transformation involves both a reflection in the real axis and "inversion" in the circle of radius 1 about the origin (Fig. 9.4). Figures outside the circle correspond to smaller ones inside. It can be shown that circles (including straight lines, as "circles through ∞") correspond to circles (Problem 5 below). Thus the line $x = 1$ becomes the circle

$$(u - \frac{1}{2})^2 + v^2 = \frac{1}{4}. \quad \blacksquare$$

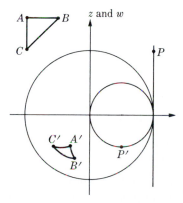

Figure 9.4   The transformation $w = \frac{1}{z}$.

EXAMPLE 5   *The general linear fractional transformation*

$$w = \frac{az + b}{cz + d} \qquad (a, b, c, d \text{ complex constants}),$$

$$\begin{vmatrix} a & b \\ c & d \end{vmatrix} \neq 0. \tag{9.2}$$

If $ad - bc$ were equal to 0, w would reduce to a constant; hence this is ruled out. Examples 1, 2, 3, and 4 are special cases of (9.2). Moreover the general transformation (9.2) is equivalent to a succession of transformations,

$$z_1 = cz + d, \qquad z_2 = \frac{1}{z_1}, \qquad w = \frac{a}{c} + \frac{bc - ad}{c} z_2 \tag{9.3}$$

of the types of Examples 3 and 4; if $c = 0$, (9.2) is already of type (9.1).

The transformation (9.2) is analytic except for $z = -d/c$. It is one-to-one, for (9.2) can be solved for z to give the inverse:

$$z = \frac{-dw + b}{cw - a},$$ (9.4)

which is single-valued; w has a pole at $z = -d/c$ and z has a pole at $w = a/c$; in other words, $z = -d/c$ corresponds to $w = \infty$ and $z = \infty$ corresponds to $w = a/c$. If one includes these values, then the transformation (9.2) is a *one-to-one transformation of the extended plane on itself.* When $c = 0$, $z = \infty$ corresponds to $w = \infty$. ∎

Since the transformations (9.3) all have the property of mapping circles (including straight lines) on circles, the general transformation (9.2) has also this property. By considering special domains bounded by circles and lines, one obtains a variety of interesting one-to-one mappings. The following three are important cases:

EXAMPLE 6   *Unit circle on unit circle.*   The transformations

$$w = e^{i\alpha} \frac{z - z_0}{1 - \bar{z}_0 z} \qquad (\alpha \text{ real, } |z_0| < 1)$$ (9.5)

all map $|z| \leq 1$ on $|w| \leq 1$, and every linear fractional (or even one-to-one conformal) transformation of $|z| \leq 1$ on $|w| \leq 1$ has this form. (See Problem 11 below). ∎

EXAMPLE 7   *Half-plane on half-plane.*   The transformations

$$w = \frac{az + b}{cz + d} \qquad (a, b, c, d \text{ real and } ad - bc > 0)$$ (9.6)

all map $\text{Im}(z) \geq 0$ on $\text{Im}(w) \geq 0$, and every linear fractional transformation of $\text{Im}(z) \geq 0$ on $\text{Im}(w) \geq 0$ has this form (see Problem 12 below). ∎

EXAMPLE 8   *Half-plane on unit circle.*   The transformations

$$w = e^{i\alpha} \frac{z - z_0}{z - \bar{z}_0} \qquad (\alpha \text{ real, } \text{Im}(z_0) > 0)$$ (9.7)

all map $\text{Im}(z) \geq 0$ on $|w| \leq 1$, and every linear fractional transformation of $\text{Im}(z) \geq 0$ on $|w| \leq 1$ has this form (see Problem 13 below). ∎

EXAMPLE 9   *The transformation*

$$w = z^2.$$

Here the transformation is not one-to-one in the whole z plane, for the inverse is $z = \sqrt{w}$ and has two values for each w. In polar coordinates one has

$$\rho = r^2, \qquad \phi = 2\theta.$$

Thus each sector in the z plane with vertex at the origin and angle α is mapped onto a sector in the w plane with vertex at $w = 0$ and angle 2α. In particular the half-plane $D_z: \text{Im}(z) > 0$ is mapped on the w plane minus the positive real axis, as shown in Fig. 9.5. Since the two square roots of w are negatives of each other, each w has at most one square root in D_z, so that the mapping is one-to-one. The structure of the level curves

$$u = x^2 - y^2 = \text{const}, \qquad v = 2xy = \text{const},$$

shown in Fig. 9.5, also reveals the one-to-one-ness, in accordance with test II of Section 9.1. Test V gives the inequality: $ax - by > 0$. Such an inequality is satisfied in each half-plane bounded by a line through the origin; hence $w = z^2$ is one-to-one in every such half-plane.

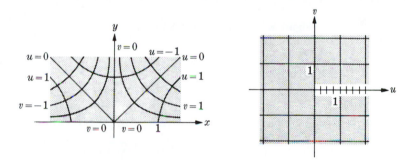

Figure 9.5   The transformation $w = z^2$.

It should be noted that $dw/dz = 0$ for $z = 0$ and that the mapping is not conformal at this point; the angles between curves are *doubled*. ∎

EXAMPLE 10   *The transformations*

$$w = z^n \qquad (n = 2, 3, 4, \ldots).$$

Here the inverse $z = w^{1/n}$ is n-valued. In polar coordinates

$$\rho = r^n, \qquad \phi = n\theta,$$

so that each sector is expanded by a factor of n. One obtains a one-to-one mapping by restricting to a sector: $0 < \arg z < 2\pi/n$, as illustrated in Fig. 9.6. The one-to-one-ness follows from the explicit formula:

$$r = \sqrt[n]{\rho}, \qquad \theta = \frac{\phi}{n}, \qquad 0 < \phi < 2\pi$$

for the inverse function, or by one of the other tests. Actually n need not be an integer; for n any positive real number, similar results hold. ∎

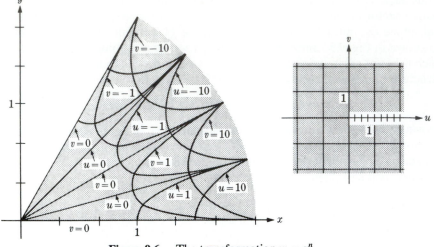

Figure 9.6 The transformation $w = z^n$.

EXAMPLE 11 *The exponential transformation*

$$w = e^z.$$

Here the inverse $z = \text{Log}\, w$ has infinitely many values. In polar coordinates:

$$\rho = e^x, \qquad \phi = y \pm 2n\pi \quad (n = 0, 1, 2, 3, \dots).$$

This shows that lines $x = $ const in the z plane become circles $\rho = $ const in the w plane, while lines $y = $ const become rays: $\phi = $ const. One obtains a one-to-one mapping by restricting z to a strip: $-\pi < y < \pi$, which is then mapped onto the w plane minus the negative real axis, as shown in Fig. 9.7. ∎

EXAMPLE 12 *The transformation*

$$w = \sin z.$$

Here the inverse

$$z = \sin^{-1} w = \frac{1}{i} \text{Log}\, (iw + \sqrt{1 - w^2})$$

is infinitely multiple-valued. One obtains a one-to-one mapping by restricting z to the strip $-\frac{1}{2}\pi < x < \frac{1}{2}\pi$, as the structure of the level curves of Fig. 9.8 reveals. One can also verify that $v = \cos x \sinh y$ has precisely one maximum and one minimum on each rectangle with vertices $(\pm\frac{1}{2}\pi, \pm k)$; hence by test IV we conclude that the mapping is one-to-one inside each such rectangle, and accordingly in the whole strip. Test V is also applicable, with $a = 1$ and $b = 0$. ∎

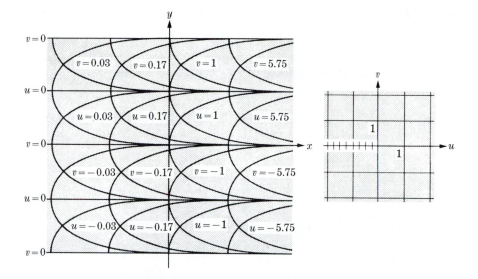

Figure 9.7   The transformation $w = e^z$.

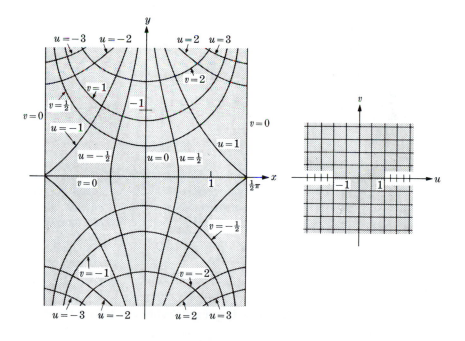

Figure 9.8   The transformation $w = \sin z$.

EXAMPLE 13   *The transformation*

$$w = z + \frac{1}{z}.$$

The inverse is the two-valued solution of the quadratic equation

$$z^2 - zw + 1 = 0.$$

The mapping is one-to-one in the domain: $|z| > 1$, as Fig. 9.9 reveals (see also Problem 3 following Section 9.7). ■

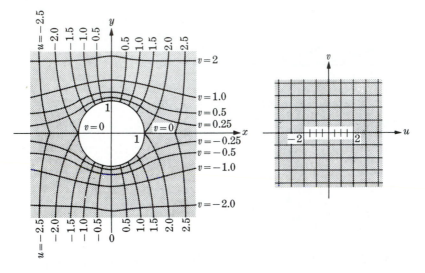

Figure 9.9   The transformation $w = z + (1/z)$.

A variety of other mappings can be obtained by applying first one of the above mappings and then a second one, since the composite of two conformal mappings is conformal. Also the inverse of a conformal transformation is analytic and one-to-one in the appropriate domain (see Section 9.1). Section 9.8 discusses general classes of mappings.

PROBLEMS

1. Determine the images of the circle $|z - 1| = 1$ and the line $y = 1$ under the following transformations:

a) $w = 2z$ **b)** $w = z + 3i - 1$ **c)** $w = 2iz$

d) $w = \dfrac{1}{z}$ **e)** $w = \dfrac{z + i}{z - i}$ **f)** $w = z^2$.

2. For each of the following transformations verify that the transformation is one-to-one in the given domain, determine the corresponding domain in the w-plane, and plot the level curves of u and v:

a) $w = \sqrt{z} = \sqrt{r}e^{i\theta/2}$, $\quad -\pi < \theta < \pi$.

b) $w = \dfrac{z-i}{z+i}$, $\quad |z| < 1$.   c) $w = \dfrac{1}{z}$, $\quad 1 < x < 2$.

d) $w = \text{Log}\, z$, $\quad \text{Im}(z) > 0$.   e) $w = \text{Log}\, \dfrac{z-1}{z+1}$, $\quad \text{Im}(z) > 0$.

                                                 [Hint: Use test VI, with $c = i$ and $a = 0$.]

f) $w = z + \dfrac{1}{z}$, $\quad \text{Im}(z) > 0$.

     [Hint: Use test VI, with $c = -i$ and $a = 0$.]

g) $w = z - \dfrac{1}{z}$, $\quad |z| > 1$

h) $w = \text{Log}\, \dfrac{z}{z+1} - 2\text{Log}\, z$, $\quad \text{Im}(z) > 0$

     [Hint: Use test V, with $a = 0$ and $b = -1$.]

i) $w = e^z + z$, $\quad -\pi < y < \pi$.

3. Verify the mapping onto the domain shown and the level curves of u and v for the transformations of

a) Fig. 9.5   b) Fig. 9.6   c) Fig. 9.7   d) Fig. 9.8   e) Fig. 9.9.

4. By combining particular transformations given above determine a one-to-one conformal mapping of

a) the quadrant $x > 0, y > 0$ onto the domain $|w| < 1$

b) the sector $0 < \theta < \frac{1}{3}\pi$ on the quadrant $u > 0, v > 0$

c) the half-plane $y > 0$ on the strip $0 < v < \pi$

d) the half-strip $-\frac{1}{2}\pi < x < \frac{1}{2}\pi, y > 0$ on the quadrant $u > 0, v > 0$

e) the domain $r > 1, 0 < \theta < \pi$ on the strip $0 < v < \pi$

f) the strip $1 < x + y < 2$ on the half-plane $v > 0$

g) the half-plane $x + y + 1 > 0$ on the quadrant $u > 0, v > 0$.

5. a) Show that the equation of an arbitrary circle or straight line can be written in the form

$$a z\bar{z} + bz + \bar{b}\bar{z} + c = 0 \qquad (a, c \text{ real}).$$

b) Using the result of a), show that circles or lines become circles or lines under the transformation $w = 1/z$.

6. If $w = f(z)$ is a linear fractional function (9.2), show that $f'(z) \neq 0$.

7. A point where $f'(z) = 0$ is called a *critical point* of the analytic function $f(z)$. Locate the critical points of the following functions and show that none lies within the domain of the corresponding mapping in Examples 11, 12, 13:

a) $w = e^z$

b) $w = \sin z$

c) $w = z + \dfrac{1}{z}$.

8. a) Show that the determinant of coefficients of the inverse (9.4) of the transformation (9.2) is not zero. Thus the *inverse of a linear fractional transformation is linear fractional.*

b) Show that, if

$$w_1 = \frac{az + b}{cz + d}, \qquad w_2 = \frac{a_1 w_1 + b_1}{c_1 w_1 + d_1}$$

are linear fractional transformations: $w_1 = f(z)$, $w_2 = g(w_1)$, then $w_2 = g[f(z)]$ is linear fractional. Thus a *linear fractional transformation followed by a linear fractional transformation gives rise to a linear fractional transformation.*

9. The *cross-ratio* of four complex numbers z_1, z_2, z_3, z_4, denoted by $[z_1, z_2, z_3, z_4]$, is defined as follows:

$$[z_1, z_2, z_3, z_4] = \frac{z_1 - z_3}{z_1 - z_4} \div \frac{z_2 - z_3}{z_2 - z_4}.$$

This is meaningful, provided at least three of the four numbers are distinct; if $z_1 = z_4$ or $z_2 = z_3$ it has the value ∞; the expression remains meaningful if one, but only one, of the numbers is ∞.

a) Prove that, if z_1, z_2, z_3 are distinct, then

$$w = [z, z_1, z_2, z_3]$$

defines a linear fractional function of z.

b) Prove that, if $w = f(z)$ is a linear fractional function and $w_1 = f(z_1)$, $w_2 = f(z_2)$, $w_3 = f(z_3)$, $w_4 = f(z_4)$, then

$$[z_1, z_2, z_3, z_4] = [w_1, w_2, w_3, w_4],$$

provided at least three of z_1, z_2, z_3, z_4 are distinct. Thus the cross-ratio is invariant under a linear fractional transformation.

c) Let z_1, z_2, z_3 be distinct and let w_1, w_2, w_3 be distinct. Prove that there is one and only one linear fractional transformation $w = f(z)$ such that $f(z_1) = w_1$, $f(z_2) = w_2$, $f(z_3) = w_3$, namely the transformation defined by the equation

$$[z, z_1, z_2, z_3] = [w, w_1, w_2, w_3].$$

d) Using the result of **c)** find linear fractional transformations taking each of the following triples of z values into the corresponding w's:

(i) $z = 1, i, 0$; $w = 0, i, 1$;

(ii) $z = 0, \infty, 1$; $w = \infty, 0, i$.

10. Let C be a circle of radius a and center Q. Points P, P' are said to be inverses of each other with respect to C if P is on the segment QP' or P' is on QP and $\overline{QP} \cdot \overline{QP'} = a^2$. Q and ∞ are also considered as a pair of inverse points. If C is a straight line, points P, P' symmetric with respect to C are called inverses of each other.

a) Prove: P, P' are inverses with respect to C, if and only if every circle through P, P' cuts C at right angles.

b) Prove: if P, P' are inverses with respect to C and a linear fractional transformation is applied, taking P to P_1, P' to P_1', C to C_1, then P_1, P_1' are inverses with respect to C_1. [Hint: Use **a**), noting that circles become circles and right angles become right angles under a linear fractional transformation.]

11. **a)** Let $w = f(z)$ be a linear fractional transformation mapping $|z| \leq 1$ on $|w| \leq 1$. Prove that $f(z)$ has the form (9.5). [Hint: Let z_0 map on $w = 0$. By Problem 10 b) $1/\bar{z}_0$ must map on $w = \infty$. The point $z = 1$ must map on a point $w = e^{i\beta}$. Take $z_1 = 1$, $z_2 = z_0, z_3 = 1/\bar{z}_0$ in 9 c).]

   b) Prove that every transformation (9.5) maps $|z| \leq 1$ on $|w| \leq 1$.

12. Prove that every linear fractional transformation of $\text{Im}(z) \geq 0$ onto $\text{Im}(w) \geq 0$ has the form (9.6) and that every transformation (9.6) maps $\text{Im}(z) \geq 0$ onto $\text{Im}(w) \geq 0$.

13. Prove that every linear fractional transformation of $\text{Im}(z) \geq 0$ on $|w| \leq 1$ has form (9.7) and that every transformation (9.7) maps $\text{Im}(z) \geq 0$ onto $|w| \leq 1$.

14. Let $w = f(z)$ be analytic at z_0 and let $f'(z_0) = 0, \ldots, f^{(n)}(z_0) = 0, f^{(n+1)}(z_0) \neq 0$. Prove that angles between curves intersecting at z_0 are multiplied by $n + 1$ under the transformation $w = f(z)$. [Hint: Near $z_0, w = w_0 + c_{n+1}(z-z_0)^{n+1} + c_{n+2}(z-z_0)^{n+2} + \cdots$, where $c_{n+1} \neq 0$. A curve starting at z_0 at $t = 0$ with tangent vector (complex number) $a \neq 0$ can be represented as $z = z_0 + at + \epsilon t, 0 \leq t \leq t_1$, where $\epsilon = \epsilon(t) \to 0$ as $t \to 0+$. Show that the corresponding curve in the w plane can be represented as $w = w_0 + c_{n+1}a^{n+1}t^{n+1} + \epsilon_1(t)t^{n+1}$. Show that $(w - w_0)/t^{n+1}$ has as limit, for $t \to 0+$, a tangent vector to the curve at w_0, and that this tangent vector has argument $\arg c_{n+1} + (n+1)\arg a$. From this conclude that two curves in the z plane forming angle α at z_0 become two curves in the w plane forming angle $(n+1)\alpha$ at w_0.]

15. Prove the validity of test V. [Hint: Let $c = a + ib$. Let z_1 and $z_2 = z_1 + re^{i\alpha}$ be distinct points of D. Then

$$c[f(z_2) - f(z_1)] = c\int_{z_1}^{z_2} f'(z)\,dz = e^{i\alpha}\int_0^r cf'(z)\,dt,$$

where $z = z_1 + te^{i\alpha}$ on the straight-line path of integration. Show that the real part of the last integral is positive, so that $f(z_2) \neq f(z_1)$. Cf. F. Herzog and G. Piranian, *Proceedings of the American Mathematical Society*, Vol. 2 (1951), pp. 625–633.]

16. Prove the validity of test VI. [Hint: Set $z_1 = \text{Log}(z - a)$ and show that the function $w_1 = f[z(z_1)] = g(z_1)$ is analytic and satisfies the inequality: $\text{Re}[cg'(z_1)] > 0$ in the domain: $0 < \text{Im}(z_1) < \pi$. Now apply test V.]

9.3   APPLICATIONS OF CONFORMAL MAPPING · THE DIRICHLET PROBLEM

The following problem, known as the Dirichlet problem, arises in a variety of situations in fluid dynamics, electric field theory, heat conduction, and elasticity: *given a domain D, to find a function $u(x,y)$ harmonic in D and having given values on the boundary of D.*

The statement is somewhat loose as to the boundary values. It will be seen immediately how it can be made more precise.

Let D be a simply connected domain bounded by a simple closed curve C, as in Fig. 9.10. Then assignment of boundary values is achieved by giving a function $h(z)$

for z on C and requiring that $u = h$ on C. If h is continuous, the natural formulation of the problem is to require that $u(x,y)$ be harmonic in D, continuous in D plus C, and equal to h on C. If h is piecewise continuous, it is natural to require that u be harmonic in D, continuous in D plus C, except where h is discontinuous, and equal to h except at the points of discontinuity.

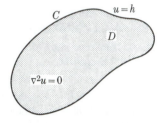

Figure 9.10   Dirichlet problem.

If C is a circle, $|z| = R$, the problem is solved by the Poisson integral formula of Section 5.17. For example, if $R = 1$, h can be written as $h(\theta)$, a function of period 2π, and

$$u(r_0, \theta_0) = \frac{1}{2\pi} \int_0^{2\pi} \frac{(1 - r_0^2)h(\theta)}{1 + r_0^2 - 2r_0 \cos(\theta_0 - \theta)}\, d\theta \tag{9.8}$$

defines a function harmonic in $|z| < 1$. If $h(\theta)$ is continuous and we define $u(1, \theta)$ to be $h(\theta)$, then u can be shown to be continuous for $|z| \le 1$ and therefore satisfies all conditions. If $h(\theta)$ is discontinuous, the same procedure is successful and (9.8) again provides a solution.

While the formula (9.8) does provide a solution of the problem, we must also ask: is this the only solution? The answer is yes when h is continuous for $|z| = 1$; the answer is no if h has discontinuities. In the latter case (9.8) provides a solution $u(x,y)$ which is bounded for $|z| < 1$, and it can be shown that this is the only bounded solution. With the supplementary requirement that the solution be bounded, (9.8) provides the one and only solution.

Now let D be a domain bounded by C as in Fig. 9.10 and let us suppose that a one-to-one conformal mapping: $z_1 = f(z)$ of D onto the circular domain $|z_1| < 1$ has been found and that this mapping is also continuous and one-to-one on D plus C, taking C to $|z_1| = 1$. By assigning to each point z_1 on $|z_1| = 1$ the value of h at the corresponding point on C, we obtain boundary values $h_1(z_1)$ on $|z_1| = 1$. Let $u(z_1)$ solve the Dirichlet problem for $|z_1| < 1$ with these boundary values. *Then $u[f(z)]$ is harmonic in D and solves the given Dirichlet problem in D.* For $u(z_1)$ can be written as $\mathrm{Re}[F(z_1)]$ where F is analytic and $u[f(z)] = \mathrm{Re}\{F[f(z)]\}$; i.e., $u[f(z)]$ is the real part of an analytic function in D. Hence u is harmonic in D. Since continuous functions of continuous functions are continuous, $u[f(z)]$ will have the proper behavior on the boundary C and hence does solve the problem.

Accordingly, conformal mapping appears as a powerful tool for solution of the Dirichlet problem. For any domain which can be mapped conformally and one-to-one on the circle $|z| < 1$, the problem is explicitly solved by (9.8). It can be shown that *every simply connected domain D can be mapped in a one-to-one conformal manner on the circular domain:* $|z| < 1$, *provided D does not consist of the entire z-plane. Furthermore, if D is bounded by a simple closed curve C, the mapping can always be defined on C so as to remain continuous and one-to-one.* Proofs of these theorems appear in the following texts listed at the end of the chapter: Ahlfors (pp. 229–232), Bieberbach (pp. 67–71), and Heins (pp. 130–134). For proofs of the stated properties of the Poisson integral formula, see Ahlfors (pp. 166–168) and Nevanlinna (pp. 22–24).

There now remains the question of how to map a particular simply connected domain D onto a circular domain. Further information on this is given in Section 9.8 below. We shall not consider extension of the theory to multiply connected domains.

9.4   DIRICHLET PROBLEM FOR THE HALF-PLANE

Since the transformation

$$iz = \frac{z_1 - i}{z_1 + i} \tag{9.9}$$

maps the domain $|z_1| < 1$ on the half-plane $\text{Im}(z) > 0$, the Dirichlet problem for the half-plane is reducible to that for the circle as above. However, it is simpler to treat the half-plane by itself. We shall develop the equivalent of (9.8) for the half-plane. Accordingly, if a domain D can be mapped on the half-plane, the Dirichlet problem for D will be immediately solved.

Figure 9.11   Example 1.

We consider first several examples.

EXAMPLE 1 $u(x,y)$ harmonic for $y > 0$; $u = \pi$ for $y = 0, x < 0$; $u = 0$ for $y = 0$, $x > 0$ as shown in Fig. 9.11. The function

$$u = \arg z = \theta \qquad 0 \leq \theta \leq \pi$$

clearly satisfies all conditions. It is harmonic in the upper half-plane, since

$$\arg z = \theta = \text{Im}(\text{Log}\, z);$$

it is continuous on the boundary $y = 0$, except for $x = 0$, and has the correct boundary values. Furthermore $0 < u < \pi$ in the upper half-plane, so that the solution is bounded. ■

Figure 9.12   Example 3.

Figure 9.13   Example 4.

EXAMPLE 2 $u(x,y)$ harmonic for $y > 0$; $u = h = $ const, for $y = 0, x > 0$; $u = 0$ for $y = 0, x < 0$. The solution is obtained as in Example 1:

$$u = h\left(1 - \frac{\arg z}{\pi}\right), \qquad 0 \leqq \arg z \leqq \pi.$$

This function is again harmonic and has the proper boundary values. ■

EXAMPLE 3 $u(x,y)$ harmonic for $y > 0$; $u = h = $ const, for $y = 0, x > x_1$; $u = 0$ for $y = 0, x < x_1$; see Fig. 9.12. A translation reduces this to Example 2:

$$u = h\left(1 - \frac{\arg(z - z_1)}{\pi}\right), \qquad 0 \leqq \arg(z - z_1) \leqq \pi; \qquad z_1 = x_1 + 0i.$$

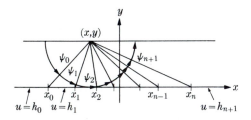

Figure 9.14   Example 5.

EXAMPLE 4 $u(x,y)$ harmonic for $y > 0$; $u = h = $ const, for $y = 0, x_1 < x < x_2$; $u = 0$ for $y = 0, x < x_1$ and $x > x_2$; see Fig. 9.13. The solution is obtained by subtracting two solutions of the type of Example 3:

$$u = h\left(1 - \frac{\arg(z - z_1)}{\pi}\right) - h\left(1 - \frac{\arg(z - z_2)}{\pi}\right)$$

$$= \frac{h}{\pi}[\arg(z-z_2) - \arg(z-z_1)] = \frac{h}{\pi}\arg\frac{z-z_2}{z-z_1}. \tag{9.10}$$

The result has an interesting geometric interpretation, for

$$\arg\frac{z-z_2}{z-z_1} = \psi, \qquad 0 < \psi < \pi,$$

where ψ is the angle shown in Fig. 9.13. Accordingly,

$$u = \frac{h\psi}{\pi}.$$

One can directly verify the boundary values on the figure. ■

Remark. Examples 1, 2, and 3 can be regarded as limiting cases of Example 4. In Example 1, $x_1 = -\infty, x_2 = 0$, and $h = \pi$, so that ψ becomes the angle θ, as shown in Fig. 9.11, and $u = \psi = \theta$. In Example 3, $x_2 = +\infty$; the angle ψ is shown in Fig. 9.12.

EXAMPLE 5 $u(x,y)$ harmonic for $y > 0$; $u = h_0 =$ const, for $y = 0, x < x_0$; $u = h_1$ for $y = 0, x_0 < x < x_1$; $u = h_2$ for $y = 0, x_1 < x < x_2; \ldots; u = h_n$ for $y = 0$, $x_{n-1} < x < x_n$; $u = h_{n+1}$ for $y = 0, x > x_n$ as in Fig. 9.14. The solution is obtained by addition of solutions of problems like those in the previous examples:

$$u = \frac{1}{\pi}\left[h_0 \arg(z - z_0) + h_1 \arg\frac{z - z_1}{z - z_0} + \cdots \right.$$
$$\left. + h_n \arg\frac{z - z_n}{z - z_{n-1}} + h_{n+1}\{\pi - \arg(z - z_n)\}\right]; \tag{9.11}$$

$$u = \frac{1}{\pi}[h_0\psi_0 + h_1\psi_1 + \cdots + h_{n+1}\psi_{n+1}], \qquad 0 \leq \psi_0 \leq \pi, \tag{9.12}$$
$$0 \leq \psi_1 \leq \pi, \ldots$$

The angles $\psi_0, \ldots, \psi_{n+1}$ are shown in Fig. 9.14. The sum of these angles is π; therefore u is the weighted mean of the numbers $h_0, h_1, \ldots, h_{n+1}$ and hence

$$h' \leq u \leq h'',$$

where h' is the smallest of these numbers and h'' is the largest.

Now let $h_0 = 0, h_{n+1} = 0$. Let z be fixed in the upper half-plane and let $t + 0i$ be a variable point on the x axis; write

$$g(t) = \arg[z - (t + 0i)],$$

where the angle is always taken between 0 and π. Then (9.11) can be written thus:

$$u = \frac{1}{\pi}\{h_1[g(t_1) - g(t_0)] + h_2[g(t_2) - g(t_0)] + \cdots$$

$$+ h_n[g(t_n) - g(t_{n-1})]\}.$$

Figure 9.15   Example 6.

This formula suggests passage to the limit: $n \to \infty$; the expression suggests an integral. In fact, the expression can be interpreted (with the aid of the law of the mean) as a sum

$$u = \frac{1}{\pi} \left[h(t_1)g'(t_1^*)\Delta_1 t + \cdots + h(t_n)g'(t_n^*)\Delta_n t \right]$$

which (under appropriate assumptions) converges as $n \to \infty$ to an integral:

$$u = \frac{1}{\pi} \int_\alpha^\beta h(t)g'(t)\,dt.$$

Now

$$g(t) = \arg[z - (t + 0i)] = \arctan \frac{y}{x - t},$$

so that

$$g'(t) = \frac{y}{(x - t)^2 + y^2}.$$

We are thus led to the formula:

$$u(x,y) = \frac{1}{\pi} \int_\alpha^\beta \frac{h(t)y}{(x - t)^2 + y^2}\,dt \tag{9.14}$$

as the expression for a harmonic function in the upper half-plane with boundary values $h(t)$ for $z = t + 0i$, $\alpha < t < \beta$, and $u = 0$ otherwise on the boundary. We get complete generality by letting $\alpha = -\infty$, $\beta = +\infty$:

$$u(x,y) = \frac{1}{\pi} \int_{-\infty}^\infty \frac{h(t)y}{(x - t)^2 + y^2}\,dt. \tag{9.15}$$

This integral is easily shown to converge, provided h is piecewise continuous and bounded. *The formula (9.15) is precisely the Poisson integral formula for the half-plane.* In fact a change of variables (Problem 6 below) transforms (9.8) into (9.15).

Just as important as the general formula (9.15) is the formula (9.11), which can be considered as the rectangular sum for approximate evaluation of the integral (9.15). ∎

EXAMPLE 6 $u(x,y)$ harmonic in the half-strip of Fig. 9.15, with the boundary values shown. One seeks the mapping of this half-strip onto a half-plane and finds it to be given by $z_1 = \sin z$, as in Example 12 of Section 9.2. Under this mapping, $z = -\frac{1}{2}\pi$ maps on $z_1 = -1$, $z = \frac{1}{2}\pi$ on $z_1 = 1$, and the whole boundary onto the real axis of the z_1 plane. The new problem in the z_1 plane requires a function u harmonic in the upper half-plane, with boundary values: $u = 2$ for $-1 < x_1 < 1$, $u = 0$ for $x_1 > 1$ and for $x_1 < 0$. This is solved as in Example 4 above by (9.10). Hence

$$u = \frac{2}{\pi} \arg \frac{z_1 - 1}{z_1 + 1}$$

and, in the z plane,

$$u = \frac{2}{\pi} \arg \frac{\sin z - 1}{\sin z + 1}.$$

To apply this formula, the simplest procedure is to use the diagram of Fig. 9.8, showing the mapping from the z plane to the z_1 plane. For each z, we locate the corresponding $z_1 = \sin z$, measure the angle

$$\psi = \arg \frac{z_1 - 1}{z_1 + 1}$$

as in Figure 9.13, and divide by $\frac{1}{2}\pi$. The answer can also be written explicitly in real form:

$$u = \frac{2}{\pi} \arctan \frac{\text{Im}\left\{\dfrac{\sin z - 1}{\sin z + 1}\right\}}{\text{Re}\left\{\dfrac{\sin z - 1}{\sin z + 1}\right\}}$$

$$= \frac{2}{\pi} \arctan \frac{2\cos x \sinh y}{\sin^2 x \cosh^2 y + \cos^2 x \sinh^2 y - 1}. \quad\blacksquare$$

EXAMPLE 7 $u(x,y)$ harmonic for $|z| < 1$, $u = 1$ for $r = 1$, $\alpha < \theta < \beta$ as in Fig. 9.16; $u = 0$ on the rest of the boundary. The function

$$z_1 = e^{(1/2)(\alpha-\beta)i} \frac{z - e^{i\beta}}{z - e^{i\alpha}}$$

maps $|z| < 1$ on the half-plane $\text{Im}(z_1) > 0$. The point $e^{i\alpha}$ has as image $z_1 = \infty$; the point $e^{i\beta}$ has as image $z_1 = 0$; the arc $\alpha < \theta < \beta$ corresponds to the negative real axis in the z_1 plane. Thus the solution is given by

$$u = \frac{1}{\pi} \arg z_1 = \frac{1}{\pi} \arg \left[e^{(1/2)(\alpha-\beta)i} \frac{z - e^{i\beta}}{z - e^{i\alpha}} \right]. \tag{9.16}$$

This rather clumsy formula is the formula for the circle corresponding to (9.10) for the half-plane. By use of geometry, this can be written in the much simpler form:

$$u = \frac{s}{2\pi}, \tag{9.17}$$

where s is the arc length of the arc on $|z| = 1$ determined by the chords through $e^{i\alpha}$ and z and through $e^{i\beta}$ and z; see Fig. 9.16.

This problem could also have been solved directly by the Poisson integral formula (9.8):

$$u(r_0, \theta_0) = \frac{1}{2\pi} \int_\alpha^\beta \frac{1 - r_0^2}{1 + r_0^2 - 2r_0 \cos(\theta_0 - \theta)} \, d\theta. \tag{9.18}$$

The integration, though awkward, can be carried out with elementary functions.

Figure 9.16   Example 7.

PROBLEMS

1. Solve the following boundary value problems:

a) u harmonic and bounded in the first quadrant,

$$\lim_{y \to 0+} u(x,y) = 1 \text{ for } 0 < x < 1,$$

$$\lim_{y \to 0+} u(x,y) = 0 \text{ for } x > 1,$$

$$\lim_{x \to 0+} u(x,y) = 1 \text{ for } 0 < y < 1,$$

$$\lim_{x \to 0+} u(x,y) = 0 \text{ for } y > 1.$$

See Fig. 9.17.

b) u harmonic and bounded in the sector: $0 < \theta < \frac{1}{4}\pi$, with boundary values 1 on the x axis and 0 on the line $y = x$, as in Fig. 9.18.

c) u harmonic and bounded for $0 < \theta < 2\pi$, with limiting value of 1 as z approaches the positive real axis from the upper half-plane and -1 as z approaches the positive real axis from the lower half-plane.

d) u harmonic and bounded in the strip: $0 < y < 1$ with boundary values 0 for $y = 0$, $x < 0$ and for $y = 1$, $x < 0$ and boundary values 2 for $y = 0$, $x > 0$ and for

$y = 1$, $x > 0$.

e) u harmonic and bounded in the domain: $0 < \theta < \pi$, $r > 1$ with boundary values of 1 on the real axis and -1 on the circle.

Figure 9.17   Figure 9.18

2. a) Verify that the transformation

$$w = 2\text{Log}\,z - z^2$$

maps the half-plane $\text{Im}(z) > 0$ in a one-to-one conformal manner on the w plane minus the lines $v = 2\pi$, $u < -1$ and $v = 0$, $u < -1$.

b) Find the electrostatic potential U between two condenser plates, which are idealized as two half-planes perpendicular to the uv plane along the lines $v = a$, $u < 0$ and $v = 0$, $u < 0$, if the potential difference between the plates is U_0; that is, solve the boundary value problem: $U(u,a) = U_0$ for $u < 0$ and $U(u,0) = 0$ for $u < 0$, $U(u,v)$ harmonic in the remaining portion of the uv plane.

3. a) Show that the transformation $w = z^2$ maps the domain: $\text{Im}(z) > 1$ in a one-to-one conformal manner on the parabolic domain

$$u < \frac{v^2}{4} - 1.$$

b) Let a solid be idealized as an infinite cylinder perpendicular to the uv plane, whose cross section in the uv plane is the region $u \leq v^2$. Let this solid be in temperature equilibrium, with temperatures T_1 maintained on the part of the boundary surface where $v > 0$ and T_2 where $v < 0$. Find the temperature distribution inside the solid; that is, solve the boundary value problem: $T(u,v)$ harmonic for $u < v^2$ with boundary values $T(u,v) = T_1$ for $u = v^2$, $v > 0$, and $T(u,v) = T_2$ for $u = v^2$, $v < 0$.

4. a) Verify that the functions (in polar coordinates)

$$r^n \cos n\theta, \qquad r^n \sin n\theta \qquad (n = 0, 1, 2, \ldots)$$

are harmonic for $r < 1$ and have the boundary values $\cos n\theta$, $\sin n\theta$ for $r = 1$.

b) Let $h(\theta)$ be continuous and have period 2π and let its Fourier series converge to $h(\theta)$

for all θ:

$$h(\theta) = \frac{1}{2}a_0 + \sum_{n=1}^{\infty}(a_n \cos n\theta + b_n \sin n\theta),$$

$$a_n = \frac{1}{\pi}\int_{-\pi}^{\pi} h(\theta)\cos n\theta\, d\theta, \qquad b_n = \frac{1}{\pi}\int_{-\pi}^{\pi} h(\theta)\sin n\theta\, d\theta.$$

Solve the boundary value problem: $u(r,\theta)$ harmonic for $r < 1$, $\lim u(r,\theta) = h(\theta)$ as $r \to 1$. [Hint: Use the result of **a)** to construct the function

$$u = \frac{1}{2}a_0 + \sum_{n=1}^{\infty}(a_n r^n \cos n\theta + b_n r^n \sin n\theta).$$

Verify convergence to $h(\theta)$ with the aid of the remarks following Theorem 36 in Section 6.15.]

5. **a)** Verify that the function (9.11) has the limiting value $\frac{1}{2}(h_j + h_{j+1})$ if the point (x,y) approaches the point $(x_j, 0)$ along the line $x = x_j$, $y > 0$.

   b) What is the limiting value if the approach to $(x_j, 0)$ is along a line making an angle $\alpha > 0$ with the positive x direction?

   c) Discuss the limiting values of the function (9.15) as $(x,y) \to (t_1, 0)$ at a point of discontinuity t_1 of $h(t)$.

Remark.   Problem 5 has an analogue in terms of the Poisson integral formula for the circle. As Problem 4 suggests, the topic is closely related to Fourier series. Actually, the properties of the Poisson integral stated in Section 9.3 above imply that, if $h(\theta)$ is piecewise continuous, then for the series obtained in Problem 4 **b)**,

$$\lim_{\substack{r \to 1 \\ \theta \to \theta_0}} u(r,\theta) = h(\theta_0)$$

at every point of continuity of $h(\theta)$ and

$$\lim_{r \to 1} u(r,\theta_0) = \frac{h(\theta_0+) + h(\theta_0-)}{2}$$

at each point of discontinuity of $h(\theta)$. The Fourier series for $h(\theta)$ may fail to converge to $h(\theta)$ for all θ under these assumptions, but the function $h(\theta)$ can be recovered from the series by multiplying the terms by r^n and letting $r \to 1$. This process is called Poisson (or Abel) summation of the series. (See Volume 1, Chapter 3, Section 6 of the book by Zygmund listed at the end of the chapter.)

6. Show that the Poisson integral formula (9.8) is transformed into the formula (9.15) for the half-plane, if the circle $|z| < 1$ is mapped on the half-plane $\mathrm{Im}(z') > 0$ by the equation

$$z' = i\frac{1+z}{1-z}.$$

[Hint: Write (9.8) in the form

$$u = \mathrm{Re}[f(z_0)], \qquad f(z_0) = \frac{1}{2\pi}\int_0^{2\pi} \frac{z+z_0}{z-z_0} h(z)\, d\theta,$$

where $z = e^{i\theta}$. The change of variable must be made both for z_0 and for z:

$$z_0' = i\frac{1 + z_0}{1 - z_0}, \qquad z' = i\frac{1 + z}{1 - z} = t + 0i.$$

Show that $f(z_0)$ becomes

$$\frac{1}{\pi i}\int_{-\infty}^{\infty}\left\{\frac{1}{t - z_0'} - \frac{t}{1 + t^2}\right\}h_1(t)\,dt,$$

where $h_1(t) = h[z(t)]$.]

9.5   CONFORMAL MAPPING IN HYDRODYNAMICS

As pointed out in Problem 2 following Section 5.15, the conditions

$$\operatorname{div} \mathbf{V} = 0, \qquad \operatorname{curl} \mathbf{V} = \mathbf{0}$$

for a two-dimensional vector field $\mathbf{V} = u\mathbf{i} - v\mathbf{j}$ are the same as the Cauchy-Riemann equations for u and v, so that the complex function $u + iv = f(z)$ is analytic. These equations therefore describe an incompressible, irrotational, steady 2-dimensional fluid motion with velocity vector $\mathbf{V}$.

If we restrict attention to a simply connected domain, then $f(z)$ has an indefinite integral $f(z)$ (determined up to a constant). If we write

$$f = \phi(x,y) + i\psi(x,y),$$

then

$$f'(z) = \frac{\partial\phi}{\partial x} - i\frac{\partial\phi}{\partial y} = u + iv.$$

Hence

$$\frac{\partial\phi}{\partial x} = u, \qquad \frac{\partial\phi}{\partial y} = -v$$

or

$$\operatorname{grad}\phi = u\mathbf{i} - v\mathbf{j} = \mathbf{V}.$$

The function ϕ is termed the velocity potential and $f(z)$ is termed the complex velocity potential. We can write

$$\overline{F'}(z) = \frac{\partial\phi}{\partial x} + i\frac{\partial\phi}{\partial y} = u - iv.$$

Hence *the conjugate of the derivative of the complex velocity potential is the velocity vector.*

The lines $\phi(x,y) = \text{const}$ are called *equipotential lines;* they are orthogonal to the velocity vector $\operatorname{grad}\phi$ at each point. The lines $\psi(x,y) = \text{const}$ are called *stream lines;* the velocity vector is tangent to such a line at each point, so that these stream lines can be considered as paths of actual fluid particles.

Conformal mapping can be applied to hydrodynamics problems in several ways. First of all, particular problems can be formulated as boundary value problems and solved with the aid of conformal mapping as in the preceding section. Second, starting with a known flow pattern, one can obtain a variety of others in an empirical way by simply applying different conformal transformations.

We treat here briefly only one example of the first type. We consider the problem of flow past an obstacle, as suggested in Fig. 9.19. The domain D of the flow is the exterior of a (piecewise smooth) simple closed curve C. Since the domain is not simply connected, we cannot be sure that there is a (single-valued) complex potential $f(z)$. It will turn out that, if we make an appropriate assumption about the flow "at ∞," then $f(z)$ does exist. The natural assumption is that the flow approaches a uniform flow at constant velocity at ∞. Hence $f(z) = F'(z)$ is analytic at ∞:

$$f(z) = a_0 + \frac{a_{-1}}{z} + \frac{a_{-2}}{z^2} + \cdots, \qquad |z| > R,$$

$$F(z) = \text{const} + a_0 z + a_{-1}\text{Log}\, z - \frac{a_{-2}}{z} + \cdots, \qquad |z| > R.$$

If $f(z)$ is to be single-valued, the term in $\text{Log}\, z$ must not be present; hence we also assume the flow to be such that $a_{-1} = 0$, so that

$$f(z) = a_0 + \frac{a_{-2}}{z^2} + \cdots,$$

$$F(z) = \text{const} + a_0 z - a_{-2}z^{-1} + \cdots. \tag{9.19}$$

The constant a_0 is precisely the value of f at ∞, so that $\bar{a}_0$ is the velocity of the limiting uniform flow.

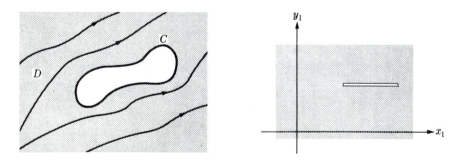

Figure 9.19   Flow past an obstacle.   **Figure 9.20**   Slit-domain.

The stream function $\psi = \text{Im}[F(z)]$ is constant along C, since (in the absence of viscosity) the velocity vector must be tangent to C. It would be natural to formulate a boundary value problem for ψ, but it is simpler to remark that the assumption (9.19) about the behavior of $f(z)$ at ∞ and the condition: $\text{Im}[F(z)] = \text{const}$ on C imply that $z_1 = F(z)$ *maps D in a one-to-one conformal manner onto a domain* D_1 *of the* z_1 *plane.* This is established by the argument principle (Problem 3 below). Since $\text{Im}(z_1) = y_1$ is constant on C, the image of C must be a line segment $y_1 = \text{const}$ in the $x_1 y_1$ plane; thus D_1 *consists of the entire* z_1 *plane minus a slit,* as in

Fig. 9.20. Conversely, if $f(z)$ is analytic in D and maps D in one-to-one fashion on such a slit-domain D_1, then f must have a pole of first order at ∞ and necessarily $\psi = \text{Im}[F(z)]$ is constant on C; thus every mapping of D onto a slit-domain D_1 provides an appropriate complex velocity potential $f(z)$.

EXAMPLE   Let C be the circle: $|z| = 1$. Then

$$z_1 = a_0 \left(z + \frac{1}{z} \right) = F(z) \qquad (a_0 \text{ real})$$

maps D onto a slit-domain D_1, the slit lying on the real axis (cf. Example 13 of Section 9.2). Hence this is an appropriate potential for the flow past the circle. The stream function is

$$\psi = a_0 y - a_0 \frac{y}{x^2 + y^2};$$

the stream lines are shown in Fig. 9.9 (Section 9.2).

It can be shown that, given C and the velocity $\bar{a}_0$ at ∞, a mapping function $f(z)$ having the expansion (9.19) at ∞ exists and is uniquely determined up to an additive constant [which has no effect on the velocity vector $\overline{F'(z)}$]. A similar theorem holds for flow past several obstacles, bounded by curves $C_1, \ldots, C_n$; the complex velocity potential maps the domain of the flow onto the z_1 plane minus n slits. For proofs one is referred to the book of Courant listed at the end of the chapter. ■

9.6   APPLICATIONS OF CONFORMAL MAPPING IN THE THEORY OF ELASTICITY

Two-dimensional problems in the theory of elasticity are reducible to solution of the biharmonic equation

$$\frac{\partial^4 U}{\partial x^4} + 2 \frac{\partial^4 U}{\partial x^2 \partial y^2} + \frac{\partial^4 U}{\partial y^4} = 0 \tag{9.20}$$

or, in terms of the operator $\nabla^2 = (\partial^2/\partial x^2) + (\partial^2/\partial y^2)$,

$$\nabla^4 U \equiv \nabla^2(\nabla^2 U) = 0.$$

The function U is *Airy's stress function;* its second derivatives

$$\frac{\partial^2 U}{\partial x^2}, \qquad \frac{\partial^2 U}{\partial x \partial y}, \qquad \frac{\partial^2 U}{\partial y^2}$$

give components of the *stress tensor,* which describes the forces acting on an arbitrary plane cross section of the solid being studied.

Solution of (9.20) in a domain D is equivalent to solution of two equations:

$$\nabla^2 U = P, \qquad \nabla^2 P = 0.$$

The solutions P of the second equation are harmonic functions. Furthermore, if

U_1, U_2 satisfy the first equation for given P, then

$$\nabla^2(U_1 - U_2) = P - P = 0;$$

hence the solutions of the first equation are of form

$$U_1 + W,$$

where U_1 is a particular solution and W is harmonic. Now let harmonic functions u and v be chosen, if possible, so that

$$\frac{\partial u}{\partial x} = P = \frac{\partial v}{\partial y}.$$

Then

$$\nabla^2(xu + yv) = x\nabla^2 u + 2\frac{\partial u}{\partial x} + y\nabla^2 v + 2\frac{\partial v}{\partial y} = 4P.$$

Hence

$$U_1 = \frac{1}{4}(xu + yv)$$

is the desired particular solution, provided u and v can be found. If D is simply connected, one can choose Q so that $P + iQ = F(z)$ is analytic in D; then

$$u + iv = f(z) = \int F(z)\,dz$$

defines (up to additive constants) harmonic functions u and v such that $\partial u/\partial x = \partial v/\partial y = P$ and U_1 can be written as follows:

$$U_1 = \frac{1}{4}(xu + yv) = \frac{1}{4}\,\mathrm{Re}[\bar{z}f(z)].$$

Finally, the solutions U of (9.20) can be written thus:

$$U = U_1 + W = \mathrm{Re}\left[\frac{\bar{z}f(z)}{4} + g(z)\right].$$

The factor of $\frac{1}{4}$ can be absorbed in $f(z)$, and we have the conclusion: *if $f(z)$ and $g(z)$ are analytic in domain D, then*

$$U = \mathrm{Re}[\bar{z}f(z) + g(z)] \tag{9.21}$$

is biharmonic in D; if D is simply connected, then all biharmonic functions in D can be represented in this form.

The boundary value problems for the stress function U can be formulated in terms of the analytic functions f and g. If D is mapped conformally on a second domain D_1, the problem is transformed into a boundary value problem for D_1. By mapping onto a simple domain D_1, such as the half-plane or unit circle, we reduce the problem to a simpler one. Hence, just as for the Dirichlet problem, conformal mapping is a powerful aid. The importance of (9.21) is that it expresses U in terms of analytic functions, which *remain analytic* if a conformal change of variables is made. A biharmonic function does not in general remain biharmonic under such a transformation.

For further details on the applications to elasticity one is referred to Chapter 5 of the book by Sokolnikoff listed at the end of the chapter.

9.7   FURTHER APPLICATIONS OF CONFORMAL MAPPING

In general, conformal mapping can be helpful in solving all boundary value problems associated with the Laplace equation or the more general *Poisson equation*

$$\frac{\partial^2 u}{\partial x^2} + \frac{\partial^2 u}{\partial y^2} = g(x,y) \tag{9.22}$$

in the plane. The crucial fact is that harmonic functions remain harmonic under conformal mapping. While the boundary values may be transformed in a complicated manner, this disadvantage is usually offset by the possibility of simplifying the domain by an appropriate mapping.

We mention here one example of the possibilities and refer to the books by Kellogg and by Frank and von Mises listed at the end of the chapter for further information. Let it be required to find a function $u(x,y)$ harmonic in a given domain D bounded by a smooth simple closed curve C and satisfying the boundary conditions: u has given values $h(x,y)$ on an arc of C; $\partial u/\partial n = 0$ on the rest of C, where $\mathbf{n}$ is the normal vector to C. The condition $\partial u/\partial n = 0$ is equivalent to the condition that the level curves of u are orthogonal to C; hence this condition is invariant under conformal mapping.

To solve the problem, we map D, if possible, on the *quarter-plane:* $x_1 > 0$, $y_1 > 0$, so that the arc on which $\partial u/\partial n = 0$ becomes the y_1 axis. The function $h(x,y)$ becomes a function $h_1(x_1)$, giving the values of u for $y_1 = 0$. We now solve the boundary value problem: $u(x_1,y_1)$ harmonic in the upper half-plane; $u(x_1,0) = h_1(x_1)$ for $x_1 > 0$ and $u(x_1,0) = h_1(-x_1)$ for $x_1 < 0$. In other words, we *reflect* the boundary values in the line $x_1 = 0$. The function u obtained is harmonic in the quarter-plane and has the correct boundary values for $x_1 > 0$. Furthermore, (9.15) shows that $u(x_1,y_1) = u(-x_1,y_1)$, that is, u shows the same symmetry as the boundary values. This implies that $\partial u/\partial x_1 = 0$ for $x_1 = 0$, that is, $\partial u/\partial n = 0$ on the boundary of the quarter-plane. Accordingly, u satisfies all conditions. If we return to the xy plane, u becomes a function of x and y which solves the given boundary value problem. This can be shown to be the only bounded solution, provided h is piecewise continuous.

PROBLEMS

1. Prove that

$$f(z) = a_0 \left(ze^{i\alpha} + \frac{1}{ze^{i\alpha}} \right) + \text{const} \qquad (\alpha \text{ real})$$

maps $D: |z| > 1$ in a one-to-one conformal manner on a slit domain D_1 (this is the most general such map). Interpret f as a complex velocity potential.

2. Show that the vector

$$\mathbf{V} = \left(1 + \frac{y}{x^2 + y^2} - \frac{x^2 - y^2}{(x^2 + y^2)^2}\right)\mathbf{i} + \left(\frac{-x}{x^2 + y^2} - \frac{2xy}{(x^2 + y^2)^2}\right)\mathbf{j}$$

can be interpreted as the velocity of an irrotational, incompressible flow past the obstacle bounded by the circle $x^2 + y^2 = 1$. Find the complex velocity potential and the stream function, and plot several stream lines.

3. Let $w = F(z)$ be analytic in the domain D outside the simple closed curve C and have a pole of first order at ∞. Let $F(z)$ be continuous in D plus C and let $\text{Im}[F(z)] = \text{const}$ on C. Show that $F(z)$ maps D in a one-to-one fashion on a slit-domain. [Hint: Show that the argument principle of Section 8.17 here takes the form: the increase in $\arg F(z)$ as z traces C in the negative direction is 2π times $(N_0 - N_\infty)$, where N_0 and N_∞ are the numbers of zeros and poles of $F(z)$ in D plus the point $z = \infty$. Show that the increase in $\arg[F(z) - w_0]$ is 0 for w_0 not on the image of C. Hence $N_0 - N_\infty = 0$; but $N_\infty = 1$, so that $N_0 = 1$.]

4. Let $U(x,y)$ be biharmonic for $x^2 + y^2 < 1$. Show that U can be expanded in a Taylor series in this domain, so that U is *analytic* in x and y.

5. Find the equilibrium temperature distribution T in the half-strip: $0 < x < 1$, $y > 0$ if the edge $x = 0$ is maintained at a temperature T_0, the edge $x = 1$ is maintained at a temperature T_1, while the edge $y = 0$ is insulated $(\partial T / \partial n = 0)$.

9.8 GENERAL FORMULAS FOR ONE-TO-ONE MAPPING ▪ SCHWARZ-CHRISTOFFEL TRANSFORMATION

The examples above have indicated the importance of conformal mapping for applications. There still remains the problem of exhibiting a reasonably large class of explicit mappings to meet the needs of applications. Here we give several formulas that help in this direction.

Mapping onto infinite strip with slits.   Let real constants $h_1, h_2, \ldots, h_n, h_{n+1}$ be chosen such that for some m,

$$h_1 < h_2 < \cdots < h_m; \qquad h_m > h_{m+1} > \cdots > h_{n+1}. \tag{9.23}$$

As extreme cases one can choose $m = 1$ or $m = n + 1$. Let points $x_1, \ldots, x_n$ be chosen on the real axis such that $x_1 < x_2 < \cdots < x_n$. A bounded harmonic function v in the half-plane $y > 0$, with the following boundary values for $y = 0$:

$$v = h_1 \text{ for } x < x_1,$$
$$v = h_2 \text{ for } x_1 < x < x_2, \ldots,$$
$$v = h_n \text{ for } x_{n-1} < x < x_n,$$
$$v = h_{n+1} \text{ for } x > x_n,$$

is provided by (9.11):

$$v = \frac{1}{\pi}\left[h_1 \arg(z - z_1) + h_2 \arg\frac{z - z_2}{z - z_1} + \cdots\right.$$

$$\left. + h_n \arg\frac{z - z_n}{z - z_{n-1}} + h_{n+1}\{\pi - \arg(z - z_n)\}\right];$$

here $z_1 = x_1 + 0i$, $z_2 = x_2 + 0i$, $\ldots$. A corresponding $f(z) = u + iv = w$ analytic in the upper half-plane is

$$f(z) = \frac{1}{\pi}\left[h_1 \mathrm{Log}\,(z - z_1) + h_2 \mathrm{Log}\,\frac{z - z_2}{z - z_1} + \cdots\right.$$

$$\left. + h_n \mathrm{Log}\,\frac{z - z_n}{z - z_{n-1}} + h_{n+1}\{i\pi - \mathrm{Log}\,(z - z_n)\}\right] + a, \qquad (9.24)$$

where a is a real constant. *When the constants $h_1, \ldots, h_{n+1}$ satisfy (9.23), this function gives a one-to-one conformal mapping of the half-plane;* for a proof, see Problem 5 below. The image domain D_1 is bounded by lines $v = $ const.

EXAMPLE

$$f(z) = \frac{1}{\pi}\left[\mathrm{Log}\,(z + 1) + 3\mathrm{Log}\,\frac{z}{z + 1} + 2(i\pi - \mathrm{Log}\,z)\right]$$

$$= \frac{1}{\pi}\left[2\pi i + \mathrm{Log}\,z - 2\mathrm{Log}\,(z + 1)\right]$$

maps the upper half-plane on the domain D_1 shown in Fig. 9.21. ■

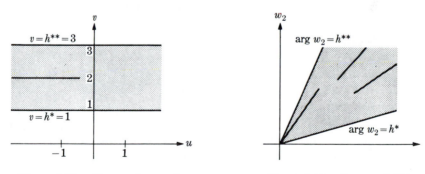

Figure 9.21   Range of example.   Figure 9.22   Range of $f(z)$.

In general the image domain D_1 lies between the lines $v = h^*$, $v = h^{**}$, where h^* and h^{**} are the smallest and largest of $h_1, h_2, \ldots, h_{n+1}$; the boundary of D_1 consists of these lines plus rays $v = $ const corresponding to the remaining h's. Every such image domain D_1 can be obtained by appropriate choice of the numbers $x_1, \ldots, x_n$, a. For the general mapping theorem (Section 9.3) ensures existence of the map; since v is bounded and has the given boundary values, the function v must have the form given above, so that f must have form (9.24).

Mapping onto sector with radial slits.   If $h^{**} - h^* \leq 2\pi$, so that the domain D_1 has width at most 2π, then the function e^w takes each value at most once in D_1. Hence $f(z) = \exp[f(z)]$ also gives a one-to-one mapping, provided (9.23) holds. The function $f(z)$ can be written as follows:

$$w_2 = F(z) = \exp[f(z)]$$

$$= \frac{\exp(a + ih_{n+1})}{(z - z_1)^{k_1}(z - z_2)^{k_2} \cdots (z - z_n)^{k_n}},\tag{9.25}$$

$$k_1 = \frac{h_2 - h_1}{\pi}, \cdots, \qquad k_n = \frac{h_{n+1} - h_n}{\pi};$$

the powers of $(z - z_1), \cdots$ are taken as *principal values*: $(z - z_1)^{k_1} = \exp[k_1 \mathrm{Log}\,(z - z_1)]$. The function e^w maps D_1 onto a domain D_2 lying in the sector: $h^* < \arg w_2 < h^{**}$ and bounded by the edges of the sector and other lines and rays as suggested in Fig. 9.22; hence $w_2 = F(z)$ maps the upper half-plane on the domain D_2. When $h^{**} - h^* = 2\pi$, the sector has opening 2π and the boundary lines coincide.

Schwarz-Christoffel mappings.   Another important class of mappings is obtainable from $f(z)$, namely, the mappings

$$w = G(z) = \int_{z_0}^{z} F(z)\,dz + \text{const}$$

$$= \int_{z_0}^{z} e^{f(z)}\,dz + \text{const}, \qquad \mathrm{Im}(z_0) > 0,$$

or, more explicitly,

$$w = G(z) = A \int_{z_0}^{z} \frac{dz}{(z - z_1)^{k_1}(z - z_2)^{k_2} \cdots (z - z_n)^{k_n}} + B,\tag{9.26}$$

where A and B are complex constants. The function $w = G(z)$ defines the *Schwarz-Christoffel transformation.* Every one-to-one conformal mapping of the upper half-plane onto a simply connected domain bounded by straight lines and line segments is representable in the form (9.26). This includes every mapping onto the interior of a polygon.

The function $G(z)$ satisfies the boundary conditions:

$$\arg G'(z) = h_1 \text{ for } y = 0, \qquad x < x_1,$$
$$\arg G'(z) = h_2 \text{ for } y = 0, \qquad x_1 < x < x_2, \ldots,$$

for

$$G'(z) = e^{f(z)} = e^{u+iv}, \qquad \arg G'(z) = v = \mathrm{Im}[f(z)].$$

The quantity $\arg G'(z)$ measures the amount by which directions are rotated in going from the z plane to the w plane (Section 9.1). Hence along the boundary the amount of rotation is piecewise constant; accordingly, each interval $x < x_1$, $x_1 < x < x_2$, ... must be mapped on a *straight line* by $w = G(z)$. The numbers $h_1, h_2, \ldots$ give the angles from the direction of the positive real axis to these lines, as suggested in Fig. 9.23 for the case of a convex polygonal domain; the numbers

$$k_1\pi = h_2 - h_1,$$
$$k_2\pi = h_3 - h_2, \ldots ,$$
$$k_n\pi = h_{n+1} - h_n$$

are successive *exterior angles* of the polygon; the $(n + 1)$-th exterior angle is $h_1 + 2\pi - h_{n+1}$. It can happen that $h_{n+1} = h_1 + 2\pi$, in which case the polygon has only n vertices; otherwise, for a proper convex polygon $h_1 + \pi < h_{n+1} < h_1 + 2\pi$. Since

$$k_1\pi + \cdots + k_n\pi = (h_2 - h_1) + \cdots + (h_{n+1} - h_n) = h_{n+1} - h_1,$$

the two cases are as follows:

$$k_1 + k_2 + \cdots + k_n = 2 \qquad (n \text{ vertices}),$$
$$1 < k_1 + \cdots + k_n < 2 \qquad (n + 1 \text{ vertices}).$$

The vertices $w_1, \ldots, w_n$ of the polygon are the images of $z_1, \ldots, z_n$. The $(n+1)$-th vertex, if present, is the image of $z = \infty$. We note that, for a convex polygon, $0 < k_1\pi < \pi$, $0 < k_2\pi < \pi$, ..., so that the k's are between 0 and 1. Given any set of k's such that

$$1 < k_1 + k_2 + \cdots + k_n \leqq 2,$$
$$0 < k_1 < 1, \ldots , 0 < k_n < 1,$$

the corresponding function $w = G(z)$ determines a one-to-one mapping of the upper half-plane onto the interior of a convex polygon with exterior angles $k_1\pi$, $k_2\pi$, ... Conversely, given a polygon (convex or not) with vertices $w_1, \ldots, w_{n+1}$, there exists a one-to-one mapping $w = G(z)$ of the upper half-plane onto the interior of the polygon. Determination of the points $z_1, \ldots, z_n$ and the constants A and B is in general a difficult problem in implicit equations; for special cases, as illustrated in the problems below, the constants can all be determined. For more details, see pages 253–258 of the book by Ahlfors listed at the end of the chapter.

As mentioned above, the image domain can also be bounded by infinite lines. Such a case is suggested in Fig. 9.24, on which the values of the h's are shown. The boundary of the domain can be regarded as a polygon, some of whose vertices are at ∞. In the example shown, there are 10 vertices in all. As the example shows, some of the sides can overlap, so that $\arg G'(z)$ has values differing by $\pm\pi$ on opposite "edges" of the side.

Remark.   Let a simply connected domain D of the w plane be given and let it be required to map the half-plane $y > 0$ of the z plane in a one-to-one conformal manner on D. It follows from the general mapping theorem of Section 9.3 that such

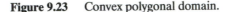

Figure 9.23   Convex polygonal domain.

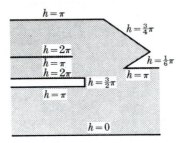

Figure 9.24   Possible range of $G(z)$.

a mapping can be found, provided D is not the finite w plane. In fact, infinitely many such mappings can be provided. Let $w = f(z)$ be one such mapping and let $z = g(z_1)$ be a linear fractional transformation [cf. (9.6)] of the half-plane $\mathrm{Im}(z_1) > 0$ on the half-plane $\mathrm{Im}(z) > 0$. Then $w = f[g(z_1)] = h(z_1)$ maps $\mathrm{Im}(z_1) > 0$ on D; that is, $w = h(z)$ maps $\mathrm{Im}(z) > 0$ on D. For every choice of the mapping g of half-plane on half-plane, we get a new mapping h of the half-plane on D. It can be shown that *all* mappings of the half-plane on the given domain D are obtainable from a single mapping f by compounding in this way with a linear fractional transformation g of half-plane on half-plane.

One can single out a particular mapping f by imposing additional conditions. For example, one can require that three specified points z_1, z_2, z_3 on the real axis correspond to three specified points w_1, w_2, w_3 on the boundary of D, provided the cyclic order of the triples matches the corresponding positive directions on the boundaries. The justification for this rule is given by the fact that one can map the half-plane onto itself to take three given points on the real axis to three given points on the real axis, provided the cyclic orders match (cf. Problem 9 following Section 9.2).

The points $z_1, \ldots, z_n$, ∞ play special roles in the transformations (9.24), (9.25), (9.26). In seeking a transformation of one of these types onto a given domain, one can assume three of the points at convenient positions, provided the cyclic order chosen agrees with the sense-preserving property of the transformation. It will be found most convenient to always make ∞ a special point, so that $h_{n+1} \neq h_1$ and, for (9.26), $h_{n+1} \neq h_1 + 2\pi$. The other two special points can be chosen, for example, as 0 and 1.

Maps onto circle with teeth. As a final example of a class of explicit one-to-one conformal mappings, we mention functions $w = H(z)$ having the following form:

$$H(z) = z + \sum_{s=1}^{n} k_s z_s \left[1 - \left(1 - \frac{z}{z_s} \right)^{\alpha_s} \right]. \tag{9.27}$$

Here the numbers k_s and α_s are real and $k_s > 0$, $0 < \alpha_s < 1$ for $s = 1, \ldots, n$; the numbers $z_1, \ldots, z_n$ represent distinct points on the circle $|z| = 1$; the principal value of the α_s power is used. The function $H(z)$ is then analytic for $|z| < 1$ and is moreover one-to-one in this domain (Problem 6 below). The image domain is approximately the region $|w| < 1$ plus n sharp "teeth" projecting from it; the points of the teeth are the images of the points $z_1, \ldots, z_n$. The smaller each α_s, the sharper the corresponding tooth. Further properties of these functions and other mapping classes are described in a paper by P. Erdös, F. Herzog, and G. Piranian, *Pacific Coast Journal of Mathematics,* Vol. 1 (1951), pp. 75–82.

PROBLEMS

1. Verify that the following functions define one-to-one conformal mappings of the upper half-plane and determine the image domains:

   a) $w = 2\text{Log}\,(z + 1) - \text{Log}\,z$

   b) $w = \text{Log}\,\frac{z-1}{z} + 2\text{Log}\,\frac{z-2}{z-1} + 3\text{Log}\,\frac{z-3}{z-2} - 2\text{Log}\,(z - 3)$

   c) $w = \sqrt{\frac{(z-2)(z-3)}{z(z-1)}}$ (principal value).

2. Determine a one-to-one conformal transformation of the half-plane $\text{Im}(z) > 0$ onto each of the following domains:

   a) the domain bounded by the lines $v = 0$, $v = 2$ and the ray $v = 1$, $0 \leq u < \infty$. [Hint: Seek a transformation of form (9.24), with $z_1 = 0$, $z_2 = 1$, $h_1 = 0$, $h_2 = 1$, $h_3 = 2$.]

   b) the domain bounded by the lines $v = 0$, $v = 2$ and the rays: $v = 1$, $-\infty < u \leq -1$; $v = 1$, $1 \leq u < \infty$. [Hint: Use (9.24) with $z_1 = -1$, $z_2 = 0$, $z_3 = p > 0$, $h_1 = 0$, $h_2 = 1$, $h_3 = 2$, $h_4 = 1$ and determine p so that the mapping is correct.]

   c) the first quadrant of the w plane minus the segment from

   $$w = 0 \text{ to } w = \exp\left(i\frac{\pi}{4}\right).$$

   [Hint: First apply the transformation $w_1 = \text{Log}\,w$ and then map the half-plane on the resulting domain in the w_1 plane.]

3. Show that each of the following transformations of the upper half-plane can be regarded

as a special case of the Schwarz-Christoffel transformation:

a) $w = \sqrt{z}$

b) $w = \sin^{-1} z$, $-\frac{1}{2}\pi < \text{Re}(w) < \frac{1}{2}\pi$

c) the transformations of form (9.24)

d) the transformations of form (9.25).

4. With the aid of the Schwarz-Christoffel transformation, determine a one-to-one conformal transformation of the half-plane $\text{Im}(z) > 0$ on the w plane minus the lines $v = 2\pi$, $u < -1$ and $v = 0$, $u < -1$ [cf. Problem 2a) following Section 9.4].

5. Prove that, if (9.23) holds, then (9.24) defines a one-to-one conformal transformation of the half-plane $\text{Im}(z) > 0$. [Hint: Prove that each nonconstant term g of $f = -\sum k_i \text{Log}\,(z - z_j) + \text{const}$ satisfies the inequality: $\text{Re}[i(z - z_m)g'(z)] > 0$ in the half-plane. Then apply test VI of Section 9.1.]

6. Let $w = u + iv = H(z)$ be defined by (9.27), under the conditions described above. Show that $H(z)$ is analytic and one-to-one for $|z| < 1$. [Hint: Show that each term h of $H(z)$ satisfies the condition: $\text{Re}[h'(z)] > 0$ for $|z| < 1$. Now apply test V of Section 9.1.]

Suggested References

Ahlfors, Lars V., Complex Analysis, 3rd ed. New York: McGraw-Hill, 1979.

Betz, Albert, Konforme Abbildung, 2nd ed. Berlin: Springer, 1964.

Bieberbach, Ludwig, Einführung in die konforme Abbildung. Berlin: Walter de Gruyter & Co., 1967.

Courant, R., Dirichlet's Principle, Conformal Mapping and Minimal Surfaces, New York: Springer-Verlag, 1977.

Frank, P., and R. von Mises, Die Differentialgleichungen und Integralgleichnungen der Mechanik und Physik, 2nd ed. New York: Dover Publications, 1953.

Heins, Maurice, Selected Topics in the Classical Theory of Functions of a Complex Variable. New York: Holt, Rinehart and Winston, 1962.

Kaplan, Wilfred, Close-to-convex Schlicht functions. *Michigan Mathematical Journal*, vol. 1, pp. 169–185, 1952.

Kellogg, O. D., Foundations of Potential Theory. New York: Springer, 1967.

Nehari, Zeev, Conformal Mapping. New York: McGraw-Hill, 1952.

Nevanlinna, Rolf, Analytic Functions, transl. by P. Emig. New York: Springer-Verlag, 1970.

Sokolnikoff, I. S., Mathematical Theory of Elasticity, 2nd ed. New York: McGraw-Hill, 1956.

Zygmund, A., Trigonometric Series, 2 Vols, London: Cambridge University Press, 1959.

10

Partial Differential Equations

10.1 INTRODUCTION

A partial differential equation is an equation expressing a relationship between an unknown function of several variables and its derivatives with respect to these variables. For example,

$$\frac{\partial u}{\partial t} - \frac{\partial^2 u}{\partial x^2} = 0, \tag{10.1}$$

$$\frac{\partial^2 u}{\partial x^2} + \frac{\partial^2 u}{\partial y^2} = 0 \tag{10.2}$$

are partial differential equations.

By a *solution* of a partial differential equation is meant a particular function that satisfies the equation identically in a domain of the independent variables. For example,

$$u = e^{-t}\sin x, \qquad u = x^2 - y^2$$

are solutions of (10.1) and (10.2), respectively.

Although partial differential equations were not studied as such in the earlier chapters, they occurred in a number of important connections. For example, the solutions of (10.2) are precisely the *harmonic* functions of x and y; such functions were studied in Chapter 2. See also Sections 5.16 and 8.11. The Cauchy-Riemann equations $u_x = v_y$, $u_y = -v_x$ form a system of partial differential equations; they were studied in Chapter 8. Other systems encountered earlier in this book are the following:

$$F_x = P(x, y), \qquad F_y = Q(x, y);$$

$$F_x = X(x, y, z), \qquad F_y = Y(x, y, z), \qquad F_z = Z(x, y, z);$$

$$Z_y - Y_z = L(x, y, z), \qquad X_z - Z_x = M(x, y, z), \qquad Y_x - X_y = N(x, y, z).$$

The first and second sets arose in consideration of line integrals in the plane and in space (Sections 5.6 and 5.13); the third set occurred in the study of solenoidal vector fields (Section 5.13). The results of Section 2.22 also concern partial differential equations, formed from Jacobians.

In this chapter we shall not attempt to consider general methods of determining the solutions of partial differential equations, but shall rather confine ourselves mainly to one class of *linear* partial differential equations—namely, equations of the form

$$\rho \frac{\partial^2 u}{\partial t^2} + H \frac{\partial u}{\partial t} - K^2 \nabla^2 u = F(t, x, \ldots), \tag{10.3}$$

where u is a function of t and one, two, or three space variables $x, \ldots; \rho, H$, and K^2 depend on the space variables, and F depends on t and the space variables. This equation is a natural generalization to continuous media of the equation

$$m \frac{d^2 x}{dt^2} + h \frac{dx}{dt} + k^2 x = F(t) \tag{10.4}$$

for forced vibrations of a spring. It will be seen that the parallel between (10.3) and (10.4) is far-reaching.

In order to make clear the relationship between (10.3) and (10.4) we shall first study the case of the motion of *two* particles attached by springs. The results obtained will then be generalized to the case of N particles. It will be seen that the equations governing the motion form a system of form

$$m_\sigma \frac{d^2 u_\sigma}{dt^2} + h_\sigma \frac{du_\sigma}{dt} + [\cdots] = F_\sigma(t, u_1, \ldots, u_n) \qquad (\sigma = 1, \ldots, n); \tag{10.5}$$

the quantities $u_1, \ldots, u_n$ are coordinates measuring the displacements of the various particles from their *equilibrium* positions. The expression in brackets $[\cdots]$ depends on the u_σ and, in particular, on their *differences*: $u_2 - u_1, u_3 - u_2, \ldots$. We shall show that such general systems are capable of "harmonic motion," damped vibrations, exponential decay, and forced motion, just as is the single mass governed by (10.4). If we let N become infinite, then n becomes infinite in (10.5), and we obtain as a "limiting case" precisely the partial differential equation (10.3). The *differences* in the expression $[\cdots]$ turn into the partial *derivatives* out of which $\nabla^2 u$ is built. Finally, it will be seen that the partial differential equation obtained as limit continues to exhibit all the properties observed for the systems of $1, 2, \ldots, N$ particles—harmonic motion, exponential decay, and so on.

The one basic difference between the various cases is that whereas for one particle there is only *one* frequency of oscillation, for two particles moving on a line there are it two frequencies, for N particles moving on a line there are N frequencies, and for infinitely many particles there are *infinitely* many frequencies.

The reader is assumed to be familiar with ordinary differential equations, especially linear equations and linear systems with constant coefficients. The first and second order equations are reviewed in the next section. For homogenous linear systems we make use of the basic tool of exponential substitution: seeking solutions of the form $x_1 = k_1 e^{\lambda t}, x_2 = k_2 e^{\lambda t}, \ldots$ This tool is applied to systems of the form

$$m_i \frac{d^2 x_i}{dt^2} + \sum_{j=1}^{n} a_{ij} x_j = 0, \qquad i = 1, \ldots, n, \tag{10.6}$$

where all $m_i > 0$ and (a_{ij}) is a positive definite symmetric $n \times n$ matrix. Such equations arise in physics in the study of "small vibrations," as will be illustrated below. (The physics first leads one to equations of the form

$$m_i \frac{d^2 x_i}{dt^2} + \frac{\partial V}{\partial x_i} = 0,$$

where $V = V(x_1, \ldots, x_n)$, the potential energy, has a minimum at $(0, \ldots, 0)$. Replacement of V by the terms through degree 2 in its power series leads to (10.6).)

For (10.6) one verifies that $x_i = k_i e^{\lambda t}$ is a solution only for *pure imaginary* λ and hence one is led to seek solutions of the form $x_i = k_i \sin(\omega t + \epsilon)$. Each such solution describes a *normal mode* of oscillation of the system. One finds that the general solution can be written in the form

$$x_i = \sum_{j=1}^{n} e_{ij} A_j \sin(\omega_j t + \epsilon_j), \tag{10.7}$$

where the A_j and ϵ_j are arbitrary real constants and (e_{ij}) is a nonsingular matrix. One can thus regard the general solution as a linear combination of normal modes. (See Problem 10 below, following Section 10.3.)

For the homogenous first order system with constant coefficients:

$$\frac{dx_i}{dt} = \sum_{j=1}^{n} a_{ij} x_j \qquad \text{or} \qquad \frac{d\mathbf{x}}{dt} = A\mathbf{x}, \tag{10.8}$$

the exponential substitution is used to obtain solutions of form $\mathbf{x} = e^{\lambda t} \mathbf{v}$, where λ is an eigenvalue of A (root of the characteristic equation $\det(A - \lambda I) = 0$) and $\mathbf{v}$ is a corresponding eigenvector. For most of the cases arising here, A has distinct real eigenvalues $\lambda_1, \ldots, \lambda_n$ and hence the general solution has the form

$$\mathbf{x} = \sum_{j=1}^{n} c_j e^{\lambda_j t} \mathbf{v}_j.$$

10.2   REVIEW OF EQUATION FOR FORCED VIBRATIONS OF A SPRING

We recall briefly some basic facts concerning Eq. (10.4). Throughout we assume $m = 0, h = 0, k > 0$.

a) *Simple harmonic motion.* Here $h = 0$, $F(t) \equiv 0$, and $m > 0$. The equation becomes

$$m\frac{d^2x}{dt^2} + k^2 x = 0. \tag{10.9}$$

The solutions are sinusoidal oscillations:

$$x = A \sin(\lambda t + \epsilon), \qquad \lambda = \frac{k}{\sqrt{m}}.$$

b) *Damped vibrations.* Here $m > 0$, $F(t) \equiv 0$, and $h > 0$, but h is small: $h^2 < 4mk^2$. The equation becomes

$$m\frac{d^2x}{dt^2} + h\frac{dx}{dt} + k^2 x = 0. \tag{10.10}$$

The solutions are oscillations with decreasing amplitude:

$$x = A e^{-at} \sin(\beta t + \epsilon),$$

where $a = h/2m$ and $\beta = (4mk^2 - h^2)^{\frac{1}{2}}/2m$.

c) *Exponential decay.* Here $m = 0$, $h > 0$, $F(t) \equiv 0$. The equation reads

$$h\frac{dx}{dt} + k^2 x = 0. \tag{10.11}$$

The solutions are decaying exponential functions:

$$x = c e^{-at}, \qquad a = \frac{k^2}{h}.$$

A similar result is obtained if we consider an equation (10.10) in which h is large in comparison to m: $h^2 > 4mk^2$; in fact, (10.11) can be considered as the limiting case: $m \to 0$ of (10.10).

d) *Equilibrium.* We assume $F(t) = F_0$, a constant. The equation becomes

$$m\frac{d^2x}{dt^2} + h\frac{dx}{dt} + k^2 x = F_0. \tag{10.12}$$

The equilibrium value of x is that for which x remains constant; hence

$dx/dt = 0$, $d^2x/dt^2 = 0$. Accordingly, at equilibrium,

$$x = x^* = \frac{F_0}{k^2}.$$ (10.13)

e) *Approach to equilibrium.* To study the approach to equilibrium, we let

$$u = x - x^*,$$ (10.14)

so that u measures the difference between x and the equilibrium value. We find that

$$m\frac{d^2u}{dt^2} + h\frac{du}{dt} + k^2u = 0,$$ (10.15)

so that u has the form (10.7), (10.9), or (10.11). For example, when $m = 0$ and $h > 0$ [case (c)],

$$u = ce^{-at}, \qquad a = \frac{k^2}{h},$$ (10.16)

$$x = u + x^*$$

$$= ce^{-at} + x^*;$$ (10.17)

x approaches its equilibrium value exponentially.

f) *Forced motion.* We allow F to be a general function of t. If, for example, $m = 0$, $h > 0$, then

$$h\frac{dx}{dt} + k^2x = F(t).$$ (10.18)

Each solution consists of a transient plus a particular solution:

$$x = ce^{-at} + x^*(t),$$ (10.19)

$$x^*(t) = \frac{1}{h}e^{-at}\int e^{at}F(t)\,dt.$$ (10.20)

The solutions attempt to follow the "input" $F(t)/k^2$, are hindered by the friction.

Remark. The equations for which $m = 0$ can be realized by other physical models—for example, the cooling of a hot mass or an electric circuit containing resistance and capacitance.

10.3   CASE OF TWO PARTICLES

We consider the model illustrated in Fig. 10.1. Two particles of masses m_1, m_2 are attached to each other and to "walls" by springs. The particles move on an x-axis and have coordinates x_1, x_2; the walls are at x_0, x_3. For simplicity we assume that all springs have the same natural length l and the same spring constant k^2. We assume the particles to be subject to resistances $-h_1(dx_1/dt)$, $-h_2(dx_2/dt)$ and

Figure 10.1 Linear system of two masses.

to outside forces $F_1(t)$, $F_2(t)$. The differential equations have the form

$$m_1 \frac{d^2 x_1}{dt^2} = -k^2(x_1 - x_0 - l) + k^2(x_2 - x_1 - l) - h_1 \frac{dx_1}{dt} + F_1(t),$$

$$m_2 \frac{d^2 x_2}{dt^2} = -k^2(x_2 - x_1 - l) + k^2(x_3 - x_2 - l) - h_2 \frac{dx_2}{dt} + F_2(t).$$

They simplify to the following:

$$m_1 \frac{d^2 x_1}{dt^2} + h_1 \frac{dx_1}{dt} - k^2(x_2 - 2x_1 + x_0) = F_1(t),$$

$$m_2 \frac{d^2 x_2}{dt^2} + h_2 \frac{dx_2}{dt} - k^2(x_3 - 2x_2 + x_1) = F_2(t). \tag{10.21}$$

We have not explicitly stated that the walls are fixed at x_0, x_3, and the differential equations remain correct even if the walls are moved in a manner controlled from outside the system. However, if x_0 and x_3 are constants x_0^*, x_3^* and $F_1(t)$, $F_2(t)$ are 0, the system has precisely one equilibrium state, namely, the solution of the equations

$$x_2 - 2x_1 + x_0^* = 0, \qquad x_3^* - 2x_2 + x_1 = 0. \tag{10.22}$$

The solution is at once found to be

$$x_1 = x_1^* = x_0^* + \tfrac{1}{3}(x_3^* - x_0^*), \qquad x_2 = x_2^* = x_0^* + \tfrac{2}{3}(x_3^* - x_0^*); \tag{10.23}$$

at equilibrium the particles are equally spaced between the walls.

We now refer each particle to its equilibrium position by introducing new variables:

$$u_0 = x_0 - x_0^*, \qquad u_1 = x_1 - x_1^*, \qquad u_2 = x_2 - x_2^*, \qquad u_3 = x_3 - x_3^*, \tag{10.24}$$

as suggested in Fig. 10.2. The differential equations (10.21) then have the

Figure 10.2 Displacements from equilibrium positions.

appearance

$$m_1 \frac{d^2 u_1}{dt^2} + h_1 \frac{du_1}{dt} - k^2 (u_2 - 2u_1 + u_0) = F_1(t),$$

$$m_2 \frac{d^2 u_2}{dt^2} + h_2 \frac{du_2}{dt} - k^2 (u_3 - 2u_2 + u_1) = F_2(t). \qquad (10.25)$$

When the walls are fixed, $x_0 = x_0^*$, $x_3 = x_3^*$, u_0 and u_3 are 0; however, (10.25) allows for a general motion of the walls and, in particular, for constant nonzero values of u_0 and u_2, which would signify a shift of the equilibrium position of the walls.

We now consider several special cases of (10.25), paralleling the cases (a), (c), (d), (e), and (f) of the preceding section. The discussion of the analogue of (b) is left to Problem 9 below.

a) *Harmonic motion.* We assume $m_1 > 0$, $m_2 > 0$, $h_1 = h_2 = 0$, $u_0 = u_3 = 0$, and $F_1(t) = F_2(t) = 0$, so that the differential equations (10.25) become

$$m_1 \frac{d^2 u_1}{dt^2} - k^2 (u_2 - 2u_1) = 0, \qquad m_2 \frac{d^2 u_2}{dt^2} - k^2 (-2u_2 + u_1) = 0.$$

$$(10.26)$$

These are equivalent to four first-order equations:

$$\frac{du_1}{dt} = w_1, \qquad \frac{dw_1}{dt} = \frac{k^2}{m_1} (u_2 - 2u_1),$$

$$\frac{du_2}{dt} = w_2, \qquad \frac{dw_2}{dt} = \frac{k^2}{m_2} (-2u_2 + u_1), \qquad (10.27)$$

and can therefore be solved by exponential substitution.   However, it is simpler to work directly with the equations (10.26) as in Section 10.1. We verify that these equations can be written as

$$m_i \frac{d^2 u_i}{dt^2} + \frac{\partial V}{\partial u_i} = 0, \qquad i = 1, 2, \qquad (10.28)$$

where $V = \frac{1}{2} \Sigma a_{ij} u_i u_j$ and $A = (a_{ij})$ is symmetric, with positive eigenvalues, so that V is positive definite (Problem 4 below). To find the solutions, we set

$$u_i = A_i \sin (\lambda t + \epsilon_i), \qquad i = 1, 2$$

in (10.26). We obtain the equations

$$(2k^2 - m_1 \lambda^2) A_1 - k^2 A_2 = 0, \qquad -k^2 A_1 + (2k^2 - m_2 \lambda^2) A_2 = 0$$

$$(10.29)$$

and a characteristic equation

$$\begin{vmatrix} 2k^2 - m_1\lambda^2 & -k^2 \\ -k^2 & 2k^2 - m_2\lambda^2 \end{vmatrix} = 0. \tag{10.30}$$

The roots are the four distinct real numbers $\pm\alpha, \pm\beta$, where

$$\alpha = k\sqrt{p_1 + p_2 + \sqrt{p_1^2 - p_1 p_2 + p_2^2}},$$

$$\beta = k\sqrt{p_1 + p_2 - \sqrt{p_1^2 - p_1 p_2 + p_2^2}} \tag{10.31}$$

and $p_1 = 1/m_1$, $p_2 = 1/m_2$ (see Problem 1 below). When $\lambda = \pm\alpha$, Eqs. (10.29) are satisfied by $A_1 = k^2$, $A_2 = 2k^2 - m_1\alpha^2$; when $\lambda = \pm\beta$, they are satisfied by $A_1 = k^2$, $A_2 = 2k^2 - m_1\beta^2$. Accordingly, the general solution is a superposition of normal modes:

$$u_1 = C_1 k^2 \sin(\alpha t + \epsilon_1) + C_2 k^2 \sin(\beta t + \epsilon_2),$$

$$u_2 = C_1(2k^2 - m_1\alpha^2)\sin(\alpha t + \epsilon_1) + C_2(2k^2 - m_1\beta^2)\sin(\beta t + \epsilon_2). \tag{10.32}$$

The frequencies α, β are called *resonant frequencies*.

c) *Exponential decay.* We assume $m_1 = m_2 = 0$, $F_1 \equiv 0$, $F_2 \equiv 0$, $u_0 = u_3 = 0$, $h_1 > 0$, $h_2 > 0$. The equations (10.25) become

$$h_1 \frac{du_1}{dt} - k^2(u_2 - 2u_1) = 0,$$

$$h_2 \frac{du_2}{dt} - k^2(-2u_2 + u_1) = 0. \tag{10.33}$$

We seek solutions:

$$u_1 = A_1 e^{\lambda t}, \qquad u_2 = A_2 e^{\lambda t} \tag{10.34}$$

and proceed as in Section 10.1. The characteristic equation is of second degree and has *distinct* real roots $-a, -b$, which are both negative (Problem 5 below):

$$a = k^2\left(q_1 + q_2 + \sqrt{q_1^2 - q_1 q_2 + q_2^2}\right),$$

$$b = k^2\left(q_1 + q_2 - \sqrt{q_1^2 - q_1 q_2 + q_2^2}\right), \tag{10.35}$$

where $q_1 = 1/h_1$, $q_2 = 1/h_2$. The general solution is found (Problem 5 below) to be

$$u_1 = C_1 k^2 e^{-at} + C_2 k^2 e^{-bt},$$

$$u_2 = C_1(2k^2 - h_1 a)e^{-at} + C_2(2k^2 - h_1 b)e^{-bt}. \tag{10.36}$$

We can say: *The general motion of the system is an exponential approach to the equilibrium state:* $u_1 = u_2 = 0$. Both "normal modes" are transients for this

Figure 10.3   Model for heat conduction.

case, but we can consider the general motion as a linear combination of two such normal modes.

The physical model of Fig.10.1 is not very appropriate here, since we are considering the limiting case $m_1 = 0$, $m_2 = 0$. A better model is suggested in Fig. 10.3. Here we consider a rod made of four pieces, each of which is a perfect conductor of heat, so that the temperature u is constant throughout each piece. The two end pieces are maintained at temperatures u_0 and u_3. It is assumed that heat can flow only across the faces of adjacent pieces according to Newton's Law, so that the rate of flow is proportional to the temperature difference. If h_1 and h_2 denote total specific heats for the middle pieces, we obtain differential equations

$$h_1 \frac{du_1}{dt} = -k_1^2(u_1 - u_0) - k_2^2(u_1 - u_2),$$

$$h_2 \frac{du_2}{dt} = -k_2^2(u_2 - u_1) - k_3^2(u_2 - u_3). \tag{10.37}$$

If $k_1 = k_2 = k_3$ and $u_3 = u_0 = 0$, we obtain (10.33). It is physically clear that when the end temperatures are maintained at 0, the temperatures of the inner pieces will also gradually approach 0.

d) *Equilibrium.* We assume that u_0 and u_3 have constant values u_0^* and u_3^* and that constant forces F_1 and F_2 are applied. Equations (10.25) then have an equilibrium solution, obtained by setting all derivatives with respect to t equal to 0:

$$-k^2(u_2 - 2u_1 + u_0^*) = F_1, \qquad -k^2(u_3^* - 2u_2 + u_1) = F_2. \tag{10.38}$$

Accordingly, the equilibrium values of u_1, u_2 are

$$u_1^* = \frac{1}{3}\left(2u_0^* + u_3^* + \frac{2F_1}{k^2} + \frac{F_2}{k^2}\right), \qquad u_2^* = \frac{1}{3}\left(u_0^* + 2u_3^* + \frac{F_1}{k^2} + \frac{2F_2}{k^2}\right).$$

$$\tag{10.39}$$

e) *Approach to equilibrium.* Let $w_1 = u_1 - u_1^*$, $w_2 = u_2 - u_2^*$, where u_1^* and u_2^* are defined by (10.39). Then substitution in (10.25), with $u_0 = u_0^*$, $u_3 = u_3^*$, $F_1 = \text{const}$, $F_2 = \text{const}$, yields the equations

$$m_1 \frac{d^2 w_1}{dt^2} + h_1 \frac{dw_1}{dt} - k^2(w_2 - 2w_1) = 0,$$

$$m_2 \frac{d^2 w_2}{dt^2} + h_2 \frac{dw_2}{dt} - k^2(-2w_2 + w_1) = 0. \tag{10.40}$$

If, for example, $m_1 = m_2 = 0$, then these are the same as (10.33), so that

$$w_1 = c_1 k^2 e^{-at} + c_2 k^2 e^{-bt},$$

$$w_2 = c_1(2k^2 - h_1 a)e^{-at} + c_2(2k^2 - h_1 b)e^{-bt}. \tag{10.41}$$

Accordingly,

$$u_1 = w_1 + u_1^* = c_1 k^2 e^{-at} + c_2 k^2 e^{-bt} + u_1^*,$$

$$u_2 = w_2 + u_2^* = c_1(2k^2 - h_1 a)e^{-at} + c_2(2k^2 - h_1 b)e^{-bt} + u_2^*. \tag{10.42}$$

Just as for one particle, the solution is an exponential approach to equilibrium.

f) *Forced motion.* External forces can be applied both by varying the wall positions u_0, u_3 and through the forces F_1, F_2. We allow for both effects but neglect the masses by considering the equations

$$h_1 \frac{du_1}{dt} - k^2[u_2 - 2u_1 + u_0(t)] = F_1(t),$$

$$h_2 \frac{du_2}{dt} - k^2[u_3(t) - 2u_2 + u_1] = F_2(t). \tag{10.43}$$

We now use the method of variation of parameters; we replace the constants C_1, C_2 in the solutions (10.36) of the homogeneous equations by variables $v_1(t)$, $v_2(t)$:

$$u_1 = v_1 k^2 e^{-at} + v_2 k^2 e^{-bt},$$

$$u_2 = v_1(2k^2 - h_1 a)e^{-at} + v_2(2k^2 - h_1 b)e^{-bt}. \tag{10.44}$$

If we substitute in (10.43) and use the fact that (10.44) satisfies (10.33) when the v's are treated as constants, then we obtain equations

$$h_1(v_1' k^2 e^{-at} + v_2' k^2 e^{-bt}) = F_1 + k^2 u_0,$$

$$h_2[v_1'(2k^2 - h_1 a)e^{-at} + v_2'(2k^2 - h_1 b)e^{-bt}] = F_2 + k^2 u_3$$

for v_1', v_2'. The solutions are

$$v_1 = \int e^{at}[(2k^2 - h_1 b)G_1 - G_2] \, dt,$$

$$v_2 = -\int e^{bt}[(2k^2 - h_1 a)G_1 - G_2] \, dt;$$

$$G_1 = \frac{F_1(t) + k^2 u_0(t)}{(a - b)h_1^2 k^2}, \qquad G_2 = \frac{F_2(t) + k^2 u_3(t)}{h_1 h_2(a - b)}. \tag{10.45}$$

Substitution in (10.44) yields a solution $u_1^*(t)$, $u_2^*(t)$. The general solution is then

$$u_1 = c_1 k^2 e^{-at} + c_2 k^2 e^{-bt} + u_1^*(t),$$

$$u_2 = c_1(2k^2 - h_1 a)e^{-at} + c_2(2k^2 - h_1 b)e^{-bt} + u_2^*(t). \tag{10.46}$$

The terms in e^{-at}, e^{-bt} can be regarded as transients. The solutions can be interpreted as following an *input*, defined by (10.39) with u_0^*, u_3^*, F_1, F_2 replaced by the given functions of t; the speed of follow-up depends on the size of h_1, h_2.

PROBLEMS

1. Show that the roots of (10.30) are given by $\pm \alpha$, $\pm \beta$, where α and β are defined by (10.31). Show that α and β are real and that $\alpha > \beta > 0$.

2. Solve (10.27) as in Section 10.1 by setting $u_1 = A_1 e^{\lambda t}$, $u_2 = A_2 e^{\lambda t}$, $w_1 = B_1 e^{\lambda t}$, $w_2 = B_2 e^{\lambda t}$. Show that (10.23) is again obtained.

3. Let $m_1 = m_2 = 1$ and $k^2 = 1$, in appropriate units, in (10.26).

   a) Write out the general solution.

   b) Obtain the particular solution for which $u_1 = u_2 = 0$ for $t = 0$ and $du_1/dt = 1$, $du_2/dt = 0$ for $t = 0$. Graph u_1 and u_2 as functions of t. Also plot the curve $u_1 = u_1(t)$, $u_2 = u_2(t)$ in the $u_1 u_2$ plane; this is a "Lissajous figure."

4. a) Show that Eqs. (10.26) can be written in the form (10.28) with $V = k^2(u_1^2 - u_1 u_2 + u_2^2)$ and that V is positive definite.

   b) Prove that the expression $E = \frac{1}{2} m_1 (du_1/dt)^2 + \frac{1}{2} m_2 (du_2/dt)^2 + V(u_1, u_2)$ is constant for each solution of (10.26). [Hint: Differentiate E with respect to t and use (10.26).] The first two terms give the total kinetic energy; the third term is the potential energy. E is the total energy and remains constant (Section 5.15).

5. Obtain the general solution (10.36) of (10.33) and verify that $0 < b < a$.

6. Let $h_1 = h_2 = 1$, $k^2 = 1$, in appropriate units, in (10.33).

   a) Obtain the general solution.

   b) Obtain the particular solution for which $u_1 = 1$, $u_2 = 3$ when $t = 0$. Graph u_1 and u_2 as functions of t. Also plot the curve $u_1 = u_1(t)$, $u_2 = u_2(t)$ in the $u_1 u_2$ plane.

7. Let $h_1 = h_2 = 1$, $k^2 = 1$, $F_1(t) \equiv 0$, $F_2(t) \equiv 0$, in appropriate units in (10.43). Let further $u_0(t) = \sin t$ and $u_3(t) \equiv 0$. Obtain the particular solution for which $u_1 = u_2 = 0$ for $t = 0$. Plot u_1 and u_2 as functions of t and also graph the curve $u_1 = u_1(t)$, $u_2 = u_2(t)$ in the $u_1 u_2$-plane.

8. Let $m_1 = m_2 = 1$, $h_1 = h_2 = 0$, $k^2 = 1$, $u_0(t) = u_3(t) = 0$, $F_1(t) = 4 \sin t$, $F_2(t) = 4a \sin t$ in (10.25). Find a particular solution. Does resonance occur? [Hint: use (10.27) and variation of parameters.]

9. a) Let $m_1 = m_2 = 4$, $h_1 = h_2 = 1$, $k^2 = 1$, $u_0 = u_3 = 0$, $F_1(t) \equiv F_2(t) \equiv 0$ in (10.25). Show that the solutions represent damped vibrations.

   b) Can the values of h_1, h_2 be modified so that one normal mode is undercritically damped, while the other mode is overcritically damped?

10. In Eq. (10.6) let the m_i be positive constants and let $A = (a_{ij})$ be symmetric and positive definite. Let $N = \text{diag}(m_1^{1/2}, \dots, m_n^{1/2})$ and set $y = Nx$. Show that (10.6) becomes

    a) $d^2 y / dt^2 = -By$, where $B = N^{-1} A N^{-1}$ is also symmetric and positive definite, so that $B = C^{-1} D C$, where $D = \text{diag}(\lambda_1, \dots, \lambda_n)$ and $\lambda_1, \dots, \lambda_n$ are the eigenvalues of B. Let $z = Cy$ and show that (a) becomes (b) $d^2 z_i / dt^2 + \omega_i^2 z_i = 0$, $i = 1, \dots, n$. Deduce that the general solution of (10.6) has form (10.7).

Figure 10.4   Linear system with N particles.

10.4   CASE OF N PARTICLES

We now consider the general case of N particles $P_1,\ldots,P_N$ of masses $m_1,\ldots,m_N$ moving on the x-axis as in Fig. 10.4. The particle P_σ is attached to the particles $P_{\sigma-1}$ and $P_{\sigma+1}$ by springs; the particle P_1 is attached to a wall at P_0 and to P_2; the particle P_N is attached to P_{N-1} and to a wall at P_{N+1}. In general, x_σ is the x-coordinate of P_σ. For simplicity we assume that all springs have the same natural length l and spring constant k^2. We assume that P_σ is subject to a resistance $-h_\sigma \dfrac{dx_\sigma}{dt}$ and an outside force $F_\sigma(t)$. The differential equations corresponding to (10.21) are then the following:

$$m_\sigma \frac{d^2 x_\sigma}{dt^2} + h_\sigma \frac{dx_\sigma}{dt} - k^2(x_{\sigma+1} - 2x_\sigma + x_{\sigma-1}) = F_\sigma(t). \qquad (10.47)$$

If the walls are fixed: $x_0 = x_0^*$, $x_{N+1} = x_{N+1}^*$ and $F_\sigma(t) \equiv 0$ for $\sigma = 1,\ldots,N$, then there is an equilibrium state, determined by the N equations:

$$x_{\sigma+1} - 2x_\sigma + x_{\sigma-1} = 0, \qquad \sigma = 1,\ldots,N. \qquad (10.48)$$

Equations (10.48) can be written as follows:

$$x_\sigma = \tfrac{1}{2}(x_{\sigma+1} + x_{\sigma-1}).$$

They assert that P_σ is halfway between $P_{\sigma-1}$ and $P_{\sigma+1}$. Hence at equilibrium, all particles are equally spaced between x_0 and x_{N+1}:

$$x_1 = x_1^* = x_0^* + \frac{1}{N+1}(x_{N+1}^* - x_0^*),\ldots,$$

$$x_N = x_N^* = x_0^* + \frac{N}{N+1}(x_{N+1}^* - x_0^*). \qquad (10.49)$$

We now refer the motion of the particles to the equilibrium positions (10.49) by introducing new coordinates:

$$u_\sigma = x_\sigma - x_\sigma^* \qquad (\sigma = 0,\ldots,N+1). \qquad (10.50)$$

The differential equations (10.47) are then replaced by the following:

$$m_\sigma \frac{d^2 u_\sigma}{dt^2} + h_\sigma \frac{du_\sigma}{dt} - k^2(u_{\sigma+1} - 2u_\sigma + u_{\sigma-1}) = F_\sigma(t). \qquad (10.51)$$

A second physical model that leads to Eqs. (10.51) is suggested in Fig. 10.5. Here the particles $P_1,\ldots,P_N$ are constrained to move along lines $x = x_1,\ldots,x = x_N$ in the xu-plane. Again P_σ is attached to $P_{\sigma+1}$ and to $P_{\sigma-1}$ by springs; P_0 and P_{N+1} are points on the "walls": $x = x_0$, $x = x_{N+1}$. We assume that the lines

Figure 10.5   *N* particle model for the vibrating string.

$x = x_\sigma$ are equally spaced, at distance Δx and that all particles have the same spring constant k_0^2 and natural length l, where $l < \Delta x$. If P_σ has coordinates (x_σ, u_σ) and is subject to a resistance $-h_\sigma(du_\sigma/dt)$ and an outside force $F_\sigma(t)$, then the differential equations governing the motion are as follows:

$$m_\sigma \frac{d^2 u_\sigma}{dt^2} + h_\sigma \frac{du_\sigma}{dt} = k_0^2(u_{\sigma+1} - 2u_\sigma + u_{\sigma-1}) - k_0^2 l(\sin\alpha_\sigma - \sin\alpha_{\sigma-1}) + F_\sigma(t),$$

$$(10.52)$$

where $\sigma = 1, \ldots, N$ and α_σ is the angle from the positive x-direction to $\overrightarrow{P_\sigma P_{\sigma+1}}$. If the angles α_σ remain so small that an approximation $\sin\alpha_\sigma \sim \tan\alpha_\sigma$ is justified, the equations become

$$m_\sigma \frac{d^2 u_\sigma}{dt^2} + h_\sigma \frac{du_\sigma}{dt} - k^2(u_{\sigma+1} - 2u_\sigma + u_{\sigma-1}) = F_\sigma(t), \qquad (10.53)$$

where $k^2 = k_0^2[1 - (l/\Delta x)]$. The derivation of (10.52) and (10.53) is left to Problem 1 below. Equations (10.53) are identical with (10.51).

When $m_\sigma = 0$ for $\sigma = 1, \ldots, N$, a natural model can be devised by generalizing the heat conduction model of Fig. 10.3. Other models can be constructed, for example, by use of electric circuits.

Equations (10.51) can be written in a form that suggests further generalizations. We write

$$V(u_1, \ldots, u_N) = k^2(u_1^2 + \cdots + u_N^2 - u_1 u_2 - \cdots - u_{N-1} u_N). \quad (10.54)$$

V is the *potential energy* associated with the system. Equations (10.51) then become

$$m_\sigma \frac{d^2 u_\sigma}{dt^2} + h_\sigma \frac{du_\sigma}{dt} + \frac{\partial V}{\partial u_\sigma} = F_\sigma(t), \qquad (10.55)$$

except for $\sigma = 1$ and $\sigma = N$; the exceptional cases can be included if we modify the definitions of $F_1(t)$ and $F_N(t)$ to include $k^2 u_0(t)$ and $k^2 u_{N+1}(t)$, respectively.

If we allow V to be a general function of $u_1, \ldots, u_N$, instead of the special function (10.54), then an extremely broad class of physical problems is included in (10.55). Of great importance is the case in which V is a general quadratic expression:

$$V = \sum_{i=1}^{N} \sum_{j=1}^{N} a_{ij} u_i u_j. \qquad (10.56)$$

As observed in Section 10.1, this case arises in consideration of problems of "small vibrations," that is, problems concerning the vibrations of a system of particles that remain close to equilibrium positions. The problems of Figs. 10.4 and 10.5 are of this sort. The approximation $\sin \alpha \sim \tan \alpha$ was based on the smallness of the departure from equilibrium.

We now state briefly the results corresponding to (a),...,(f) for (10.51). Actually, many of the statements apply equally well to Eqs. (10.55), where V is defined by (10.56) and is positive definite. The positive definiteness of V in (10.54) is proved in Problem 2 below.

a) *Harmonic motion.* Here $h_\sigma = 0$, $u_0 = u_{N+1} = 0$, $F_\sigma(t) \equiv 0$. We find that there are N normal modes of vibration, with resonant frequencies $\lambda_1, \ldots, \lambda_N$, and that the general motion consists of a linear combination of these:

$$u_\sigma = \sum_{n=1}^{N} c_n A_{n,\sigma} \sin(\lambda_n t + \epsilon_n);$$ (10.57)

the $A_{n,\sigma}$ are certain constants, and $c_1, \ldots, c_N, \epsilon_1, \ldots, \epsilon_N$ are arbitrary constants.

b) *Damped vibrations.* In this case, $m_\sigma > 0$, $h_\sigma > 0$, $u_0 = u_{N+1} = 0$, $F_\sigma(t) \equiv 0$. The oscillations of (a) are replaced by damped oscillations, of form $e^{-at} \sin(\lambda t + \epsilon)$, or exponential decay terms: e^{-at}, te^{-at}.

c) *Exponential decay.* Here $m_\sigma = 0$, $F_\sigma \equiv 0$, $u_0 = u_{N+1} = 0$, $h_\sigma > 0$. We find that there are N "normal modes of decay" and that the general motion is a linear combination of the normal modes:

$$u_\sigma = \sum_{n=1}^{N} c_n A_{n,\sigma} e^{-a_n t}.$$ (10.58)

d) *Equilibrium.* If $u_0 = u_n^*$, $u_{N+1} = u_{N+1}^*$, $F_\sigma = F_\sigma^*$, where the starred values are constants, then the equilibrium values $u_\sigma = u_\sigma^*$ are equally spaced between $u_0^* + (F_1^*/k^2)$ and $u_{N+1}^* + (F_N^*/k^2)$.

e) *Approach to equilibrium.* Under the assumptions of case (d) the general motion $u_\sigma(t)$ is a solution of the homogeneous equations corresponding to (10.51) plus the equilibrium solution u_σ^*. When the homogeneous equations are of type (c), one therefore has the exponential approach to equilibrium:

$$u_\sigma = \sum_{n=1}^{N} c_n A_{n,\sigma} e^{-a_n t} + u_\sigma^* \qquad (\sigma = 1, \ldots, N).$$ (10.59)

f) *Forced motion.* Here u_0, u_{N+1}, and all F_σ are allowed to depend on t. The general motion $u_\sigma(t)$ is a solution of the homogeneous equations corresponding to (10.51) plus a particular solution $u_\sigma^*(t)$. The particular solution can always be found by variation of parameters; it can always be interpreted as a follow-up of an input. The total effect of all $F_\sigma(t)$ on the solution can be considered as a superposition of the effects of the $F_\sigma(t)$ individually. Addition of a sinusoidal term of frequency λ to one F_σ leads to the addition of a sinusoidal term of frequency λ to all $u_\sigma^*(t)$, unless all h_σ are 0 and λ is a resonant frequency, in which case resonance may occur (cf. Problem 9 following Section 10.3).

PROBLEMS

1. a) Obtain the equations (10.52) for the model of Fig. 10.5.

   b) Show that when the angles α_σ remain small, the equations (10.53) are justified as approximations to (10.52).

2. a) Show that the potential energy V of (10.54) is positive definite. [Hint: Write $V = (k^2/2)[u_1^2 + (u_1 - u_2)^2 + \cdots + u_N^2]$.

   b) Show that when $h_\sigma = 0$ and $F_\sigma(t) \equiv 0$ for $\sigma = 1,\dots,N$, every solution of (10.55) has the property that the *total energy*

$$E = \tfrac{1}{2}m_1(du_1/dt)^2 + \cdots + \tfrac{1}{2}m_N(du_N/dt)^2 + V(u_1,\dots,u_N)$$

   is constant. [Hint: See Problem 4 following Section 10.3.]

3. Let $N = 3$ in (10.51) and let $m_1 = m_2 = m_3 = 1$, $k^2 = 1$, $h_1 = h_2 = h_3 = 0$, $u_0 = u_4 = 0$, $F_1 = F_2 = F_3 = 0$, so that one has case (a). Seek solutions: $u_1 = A_1 e^{\lambda t}$, $u_2 = A_2 e^{\lambda t}$, $u_3 = A_3 e^{\lambda t}$ and obtain the general solution in the form (10.57).

4. Let $N = 3$ in (10.51) and let $m_1 = m_2 = m_3 = 0$, $k^2 = 1$, $h_1 = h_2 = h_3 = 1$, $u_0 = u_4 = 0$, $F_1 = F_2 = F_3 = 0$, so that one has case (c). Seek solutions: $u_1 = A_1 e^{\lambda t}$, $u_2 = A_2 e^{\lambda t}$, $u_3 = A_3 e^{\lambda t}$ and obtain the general solution in the form (10.58).

5. (*Difference equations*) We consider functions $f(\sigma)$ of an integer variable σ: $\sigma = 0, \pm 1, \pm 2,\dots$. Let $f(\sigma)$ be defined for $\sigma = m$, $\sigma = m + 1,\dots,\sigma = n$. Then the *first difference* $\Delta_+ f(\sigma)$ is the function $f(\sigma + 1) - f(\sigma)$ $(m \le \sigma < n)$; the *first difference* $\Delta_- f(\sigma)$ is the function $f(\sigma) - f(\sigma - 1)$ $(m < \sigma \le n)$. The *second difference* is the first difference of the first difference; this could mean $\Delta_+\Delta_- f, \Delta_-\Delta_+ f, \Delta_+\Delta_+ f, \Delta_-\Delta_- f$; we shall use only

$$\Delta^2 f = \Delta_+\Delta_- f = \Delta_-\Delta_+ f = f(\sigma + 1) - 2f(\sigma) + f(\sigma - 1);$$

   see Problem 9 following Section 6.18. Differences of higher order can be defined in analogous fashion. A *difference equation* is an identity to be satisfied by $f(\sigma)$, and its differences of various orders. In general, the "linear difference equation" has a theory quite analogous to that of the linear differential equation, and the functions $e^{r\sigma}$ play a similar role. We consider only two cases:

   a) $\Delta^2 f(\sigma) = 0$. Show that the general solution is given by $f(\sigma) = c_1\sigma + c_2$, where c_1, c_2 are constants. Obtain the solution satisfying the *boundary conditions*: $f(0) = u_0$, $f(N + 1) = u_{N+1}$, where u_0, u_{N+1} are given constants.

   b) $\Delta^2 f(\sigma) + p^2 f(\sigma) = 0$. Show that the functions $c_1 \cos q\sigma + c_2 \sin q\sigma$ are solutions, provided that $\cos q = 1 - \tfrac{1}{2}p^2$ and $0 < p^2 < 4$. Show that when $p^2 > 4$, distinct numbers a_1 and a_2 can be chosen so that the functions $c_1 a_1^\sigma + c_2 a_2^\sigma$ are solutions and that when $p^2 = 4$, the functions $c_1(-1)^\sigma + c_2\sigma(-1)^\sigma$ are solutions.

   c) Show that if $f(0)$ and $f(1)$ are given, then the difference equation of part (b) successively determines $f(2), f(3),\dots$. Hence for such "initial conditions" there is one and only one solution. Determine the constants c_1, c_2 for each of the cases of part (b) to match the given initial conditions. The fact that this is possible ensures that each expression gives the general solution for the corresponding value of p.

d) Show that the only solutions of the difference equation of part (b) which satisfy the boundary conditions: $f(0) = 0$, $f(N + 1) = 0$ are constant multiples of the N functions

$$\phi_n(\sigma) = \sin\left(\frac{n\pi}{N + 1}\sigma\right) \qquad (n = 1,\ldots,N)$$

and that $\phi_n(\sigma)$ is a solution only when

$$p^2 = 2\left(1 - \cos\frac{n\pi}{N + 1}\right).$$

6. Show that the solution of (10.48), where $x_0 = x_0^*$, $x_{N+1} = x_{N+1}^*$ is equivalent to solution of the difference equation $\Delta^2 f(\sigma) = 0$ of Problem 5(a), subject to boundary conditions, and compare with the solution of Problem 5(a).

7. Let $m_1 = m_2 = \cdots = m_N = m$, $h_\sigma = 0$ and $F_\sigma(t) \equiv 0$ for $\sigma = 1,\ldots,N$ and $u_0 = u_{N+1} = 0$ in (10.51), so that one has case (a), with *equal masses*. Show that the substitution $u_\sigma = A(\sigma)\sin(\lambda t + \epsilon)$ leads to the difference equation with boundary conditions:

$$\Delta^2 A(\sigma) + p^2 A(\sigma) = 0, \qquad p^2 = m\lambda^2/k^2,$$
$$A(0) = 0, \qquad A(N + 1) = 0.$$

Use the result of Problem 5(d) to obtain the N normal modes

$$u_\sigma(t) = \sin\left(\frac{n\pi}{N + 1}\sigma\right)\sin(\lambda_n t + \epsilon_n),$$
$$\lambda_n = \frac{2k}{\sqrt{m}}\sin\frac{n\pi}{2(N + 1)}, \qquad n = 1,\ldots,N.$$

Show that $0 < \lambda_1 < \lambda_2 < \cdots < \lambda_N$.

8. For two functions $f(\sigma)$, $g(\sigma)$ defined for $\sigma = 0,1,2,\ldots,N + 1$, we define an *inner product* (f, g) by the equation (cf. Section 7.10)

$$(f, g) = f(0)g(0) + f(1)g(1) + \cdots + f(N)g(N) + f(N + 1)g(N + 1);$$

the *norm* $\|f\|$ is then defined as $(f, f)^{\frac{1}{2}}$. In the following we consider *only functions that equal 0 for $\sigma = 0$ and $\sigma = N + 1$*. In particular, we use the functions $\phi_n(\sigma)$ of Problem 5(d):

$$\phi_n(\sigma) = \sin(n\alpha\sigma), \qquad \alpha = \pi/(N + 1).$$

a) Graph the functions $\phi_n(\sigma)$ for the case $N = 5$.

b) Show that $(\phi_m, \phi_n) = 0$ for $m \neq n$ and that $\|\phi_n\|^2 = \frac{1}{2}(N + 1)$. [Hint: Write

$$\phi_n(\sigma) = \frac{r^\sigma - s^\sigma}{2i}, \qquad r = e^{\alpha ni}, \qquad s = e^{-\alpha ni}$$

and evaluate inner product and norm with the aid of the formula for sum of a geometric progression.]

c) Show that if we associate with each function $f(\sigma)$ the vector $\mathbf{v} = [v_1, v_2, \ldots, v_N]$, where $v_1 = f(1)$, $v_2 = f(2),\ldots,v_N = f(N)$, then the operations $f + g$, cf, (f, g) correspond to the vector operations $\mathbf{u} + \mathbf{v}$, $c\mathbf{u}$, $\mathbf{u} \cdot \mathbf{v}$ of Section 1.13. Accordingly, the space of functions considered forms *an N-dimensional Euclidean vector space.* The vectors corresponding to the functions $\phi_n(\sigma)/\|\phi_n(\sigma)\|$ form *a system of base vectors.*

d) Show that if $f(\sigma)$ is defined for $\sigma = 0,\dots,N + 1$ and $f(0) = f(N + 1) = 0$, then $f(\sigma)$ can be represented in one and only one way as a linear combination of the functions $\phi_n(\sigma)$, namely, as follows:

$$f(\sigma) = \sum_{n=1}^{N} b_n\phi_n(\sigma), \qquad b_n = \frac{2}{N + 1} \sum_{\sigma=0}^{N+1} f(\sigma)\phi_n(\sigma).$$

Compare with the Fourier sine series (Section 7.5).

9. Write the general solution of Problem 7 in the form

$$u_\sigma(t) = \sum_{n=1}^{N} \phi_n(\sigma)(\alpha_n \sin \lambda_n t + \beta_n \cos \lambda_n t),$$

where α_n and β_n are arbitrary constants. Use the result of Problem 8(d) to show that the constants α_n and β_n can be chosen in one and only one way so that $u_\sigma(t)$ satisfies given initial conditions:

$$u_\sigma(0) = f(\sigma), \qquad \frac{du_\sigma}{dt}(0) = g(\sigma).$$

This shows that one has indeed obtained *all* solutions.

10. Let $h_1 = h_2 = \cdots = h_N = h$, $m_\sigma = 0$ and $F_\sigma(t) \equiv 0$ for $\sigma = 1,\dots,N$, $u_0 = u_{N+1} = 0$ in (10.51), so that one has case (c), with *equal friction coefficients.* Show that the substitution $u_\sigma = A(\sigma)e^{\lambda t}$ leads to the difference equation with boundary conditions:

$$\Delta^2 A(\sigma) + p^2 A(\sigma) = 0, \qquad p^2 = -h\lambda/k^2,$$

$$A(0) = 0, \qquad A(N + 1) = 0.$$

Use the result of Problem 5(d) to obtain the "modes of decay":

$$u_\sigma(t) = \sin\left(\frac{n\pi}{N + 1}\sigma\right)e^{-a_n t},$$

$$a_n = \frac{2k^2}{h}\left(1 - \cos \frac{n\pi}{N + 1}\right).$$

Show that $0 < a_1 < a_2 < \cdots a_N$.

11. Prove that constants c_n can be chosen in one and only one way so that

$$u_\sigma(t) = \sum_{n=1}^{N} c_n\phi_n(\sigma)e^{-a_n t}$$

is a solution of the exponential decay problem (Problem 10) and matches given initial conditions: $u_\sigma(0) = f(\sigma)$ (cf. Problem 9).

12. In (10.51), let $h_1 = h_2 = \cdots = h$, $m_\sigma = 0$, for $\sigma = 1,\dots,N$ as in Problem 10, but let $F_\sigma(t)$ be permitted to depend on t. We assume $u_0 = 0$, $u_{N+1} = 0$, as any variation in the "walls" can be absorbed in $F_0(t)$ and $F_{N+1}(t)$. Use the method of variation of parameters to obtain a particular solution. [Hint: The "complementary function" is given in Problem 11. If we replace c_n by $v_n(t)$ for $n = 1,\dots,N$ and substitute in

(10.51), we obtain equations

$$h \sum_{n=1}^{N} \frac{dv_n}{dt} \phi_n(\sigma) e^{-a_n t} = F_\sigma(t).$$

Now use the result of Problem 8(d) to conclude that

$$h \frac{dv_n}{dt} e^{-a_n t} = \frac{2}{N+1} \sum_{\sigma=1}^{N} F_\sigma(t) \phi_n(\sigma).]$$

10.5 CONTINUOUS MEDIUM · FUNDAMENTAL PARTIAL DIFFERENTIAL EQUATION

We now consider the limiting case: $N \to \infty$. Rather than attempt to carry out a precise passage to the limit, we allow ourselves to be guided by physical intuition. The natural limiting case of the system of N particles moving on a line (Fig. 10.4) is that of a *rod* that is permitted to vibrate longitudinally, as suggested in Fig. 10.6. The individual particle is replaced by a cross section of the rod, which can be thought of as a thin layer of molecules that move together parallel to the axis of the rod. When no external forces are applied, this layer has an equilibrium position x. Just as for the particles, we can measure the displacement u of the layer from its equilibrium position x; u then becomes a function of x and t.

Figure 10.6 Longitudinal vibrations of a rod.

If we pass to the limit in the model of Fig. 10.5, we obtain the *vibrating string* —for example, a violin string. To first approximation, each "molecule" of such a string executes vibrations perpendicular to the line represented by the equilibrium position of the string. The displacement of the molecule at position x from its equilibrium position is measured by u, which is a function of x and t. The vibrations are assumed to take place in an xu-plane; one could consider the more general case in which the vibrations are not confined to a plane.

In order to obtain a differential equation for $u(x, t)$ we return to the basic equations (10.51) and write them as follows:

$$\frac{m_\sigma}{\Delta x} \frac{d^2 u_\sigma}{dt^2} + \frac{h_\sigma}{\Delta x} \frac{du_\sigma}{dt} - k^2 \Delta x \frac{u_{\sigma+1} - 2u_\sigma + u_{\sigma-1}}{(\Delta x)^2} = \frac{F_\sigma(t)}{\Delta x}. \qquad (10.60)$$

We assume x_0 and x_{N+1} to be fixed and let $L = x_{N+1} - x_0$, $\Delta x = L/(N+1)$.

We then let N increase indefinitely. The ratio $m_\sigma/\Delta x$ represents an "average density" at position x; it is reasonable to postulate that this approaches as limit a function $\rho(x)$ representing density (mass per unit length) at x. The simplest law of friction would make h_σ proportional to m_σ, so that $h_\sigma/\Delta x$ would approach a function $H(x)$ of dimensions force per unit of length per unit of velocity. For the model of Fig. 10.4 the product $k^2 \Delta x$ represents the tension in one of the springs when it is stretched a distance Δx. But precisely the same tension must hold in each half of the spring, which is stretched only a distance $\frac{1}{2}\Delta x$. Hence if we always use springs of the same stiffness, $k^2 \Delta x$ will approach as limit a constant force K^2. For the model of Fig. 10.5, one indeed has $k^2 \Delta x = k_0^2(\Delta x - l)$, where k_0^2 is the actual spring constant for each spring; therefore $k^2 \Delta x$ represents precisely the tension in each spring when all displacements u_σ are 0; the limiting value K^2 is precisely the tension in the string.

One can write

$$\frac{u_{\sigma+1} - 2u_\sigma + u_{\sigma-1}}{(\Delta x)^2} = \frac{u(x_\sigma + \Delta x, t) - 2u(x_\sigma, t) + u(x_\sigma - \Delta x, t)}{(\Delta x)^2},$$

where x_σ is the equilibrium position of P_σ. In the limit, x_σ becomes a continuous variable, and, as in Problem 9 following Section 6.18, the ratio of the second difference to $(\Delta x)^2$ approaches as "limit" the derivative

$$\frac{\partial^2 u}{\partial x^2}(x, t).$$

The right-hand members we assume to approach as limit a function $F(x,t)$ representing applied force per unit length at x. We are thus led to the partial differential equation

$$\rho(x)\frac{\partial^2 u}{\partial t^2} + H(x)\frac{\partial u}{\partial t} - K^2\frac{\partial^2 u}{\partial x^2} = F(x, t), \tag{10.61}$$

with $\rho(x) \geq 0, H(x) \geq 0, K > 0$. This is the fundamental partial differential equation to be studied. Certain generalizations will be introduced in later sections, notably by replacement of $\partial^2 u/\partial x^2$ by the Laplacian $\nabla^2 u$:

$$\rho\frac{\partial^2 u}{\partial t^2} + H\frac{\partial u}{\partial t} - K^2\nabla^2 u = F(x, y, z, t). \tag{10.62}$$

This corresponds to a generalization to motion in 2- or 3-dimensional space. One can easily construct an N particle model for this. From the general equation (10.55), one obtains a broader class of equations in which the term $-K^2\nabla^2 u$ is replaced by a more complicated expression, possibly nonlinear, in u and its derivatives. Two- and three-dimensional problems can lead to *simultaneous partial differential equations*.

While the generalizations do introduce complications, the principal problems and methods reveal themselves in Eq. (10.61) and, in fact, in the N particle approximation to this of the preceding section; as was pointed out in the introductory section, even the single particle displays the properties that are crucial.

The limiting process by which we arrived at (10.61) was based on physical intuition, and the basic test of the validity of the result is its accuracy in explaining the behavior of continuous media. This is a problem of physics, by no means simple, with which we shall not concern ourselves. However, one can ask the purely mathematical question: Do the solutions of difference equations converge to the solutions of the corresponding differential equations, when the basic interval (for example, Δx) approaches 0? This question has been made precise and answered in a generally affirmative fashion in recent research. We refer to Chapter 9 of the book by Isaacson and Keller listed at the end of the chapter for a discussion of the problem and further references to the literature.

10.6 CLASSIFICATION OF PARTIAL DIFFERENTIAL EQUATIONS ▪ BASIC PROBLEMS

Equations (10.61) and (10.62) are linear in u and its derivatives and are hence *linear partial differential equations*. They involve derivatives of u up to the second order and are hence partial differential equations of *second order*. The most general linear partial differential equation of second order in two independent variables has the form

$$A\frac{\partial^2 u}{\partial x^2} + 2B\frac{\partial^2 u}{\partial x\,\partial y} + C\frac{\partial^2 u}{\partial y^2} + D\frac{\partial u}{\partial x} + E\frac{\partial u}{\partial y} + Fu + G = 0, \quad (10.63)$$

where $A,\dots,G$ are functions of x and y. Equations (10.63) are classified into three types:

elliptic: $B^2 - AC < 0$, $A\xi^2 + 2B\xi\eta + C\eta^2 = 1$ is an ellipse;

parabolic: $B^2 - AC = 0$, $A\xi^2 + 2B\xi\eta + C\eta^2 + D\xi + E\eta = 0$ is a parabola;

hyperbolic: $B^2 - AC > 0$, $A\xi^2 + 2B\xi\eta + C\eta^2 = 1$ is a hyperbola.

An equation can be of one type in part of the xy-plane and of another type in a second part. An analogous classification is made for equations in three or more independent variables. The three types are illustrated respectively by the equations

$$\frac{\partial^2 u}{\partial x^2} + \frac{\partial^2 u}{\partial y^2} = 0, \qquad \frac{\partial u}{\partial x} - \frac{\partial^2 u}{\partial y^2} = 0, \qquad \frac{\partial^2 u}{\partial x^2} - \frac{\partial^2 u}{\partial y^2} = 0.$$

The first of these is the equation $\nabla^2 u = 0$ and occurs naturally in connection with the equilibrium problem for *two* dimensions. The second corresponds to *exponential decay*; the third to *harmonic motion*.

The differential equation (10.61) was proposed as a natural one for the longitudinal oscillations of a rod, as in Fig. 10.6, or for the transverse vibrations of a string. There are other 1-dimensional problems to which the equation is applicable: Planar sound waves and electromagnetic waves, diffusion of heat, and other diffusion processes ($\rho = 0$). The equation can also be applied to *infinite* intervals on the x-axis; although the vibration of a string infinite in length may appear to be an artificial concept, such an ideal case is useful in applications. Equation (10.62) has analogous applications in two and three dimensions, including the basic hyperbolic, parabolic, and elliptic equations:

wave equation:
$$\rho \frac{\partial^2 u}{\partial t^2} - K^2 \nabla^2 u = 0,$$

heat equation:
$$H \frac{\partial u}{\partial t} - K^2 \nabla^2 u = 0,$$

Laplace equation:
$$\nabla^2 u = 0;$$

here ρ, H, and K^2 are usually considered to be positive constants. As was pointed out in Section 5.15, the Laplace equation is satisfied by the velocity potential of an irrotational, incompressible fluid motion. The complete equations of hydrodynamics are simultaneous equations that are *nonlinear* (see Lamb's *Hydrodynamics,* Cambridge University Press, 1932).

The basic problems associated with (10.61) are simply the analogues for the continuous medium of the problems studied in the previous sections. For example, problem (a) concerns the case for which $\rho(x) > 0$, $H(x) = 0$, $F(x, t) = 0$, and the "walls" are fixed: $u(0, t) = 0$, $u(L, t) = 0$; we expect to show that there is one and only one solution to the initial value problem: $u = f(x)$, $\partial u / \partial t = g(x)$ for $t = 0$. Such a solution $u(x, t)$ would be defined and continuous for $0 \leq x \leq L$ and for $t \geq 0$ and would be required to have partial derivatives through the second order for $0 < x < L$ and $t > 0$ and to satisfy the differential equation in the *domain* just described. One can also consider the possibility of discontinuities on the boundary; this requires care, but significant results can be obtained. The word *solution* will refer to functions $u(x, t)$ continuous for $0 \leq x \leq L$, $t \geq 0$, unless otherwise indicated. Problems (b) and (c) (damped vibrations and exponential decay) are formulated in a similar manner.

The equilibrium problem (d) now becomes an *ordinary* differential equation:

$$-K^2 \frac{d^2 u}{dx^2} = F(x),$$

with boundary conditions: $u = u_0$ for $x = 0$, $u = u_1$ for $x = L$. In two dimensions the analogous equation is the *Poisson equation*

$$-K^2 \nabla^2 u = F(x, y),$$

where u has given values on the boundary of a 2-dimensional region; when $F \equiv 0$, this is the Dirichlet problem. In all cases we wish to show that there is a unique equilibrium state $u^*(x)$ [in two dimensions, $u^*(x, y)$]. As for the N particle problem, the approach to equilibrium, problem (e), is described by a function $u^*(x) + u(x, t)$ where $u(x, t)$ is a solution of a *homogeneous* problem (a), (b), or (c).

The problem (f) of forced motion includes the other five problems as special cases. Boundary conditions: $u(0, t) = u_0(t)$, $u(L, t) = u_1(t)$ and initial values of u are given; we wish to show that there is one and only one corresponding solution.

The *methods* to be used are a natural extension of those used for the N particle problem. The homogeneous problems are handled by a substitution: $u(x, t) = A(x)e^{\lambda t}$, which gives the normal modes; the "general solution" is again obtained as a linear combination of normal modes. The nonhomogeneous problem of forced motion is solved by variation of parameters; the general solution is the sum of a particular solution $u^*(x, t)$ and the general solution of the homogeneous problem.

We have considered only the cases of fixed walls or walls moving in a prescribed manner. There are other natural boundary conditions; for example, one could require that $\partial u/\partial x$ be 0 for $x = 0$. For the N particle case this would correspond to the requirement that the wall P_0 move in such a fashion that $u_0 = u_1$; accordingly, the distance between P_0 and P_1 is fixed, and *no energy can be transmitted*. For the heat conduction problem this corresponds to an *insulated* boundary at $x = 0$.

10.7 THE WAVE EQUATION IN ONE DIMENSION ▪ HARMONIC MOTION

In the basic equation (10.61) we assume that $H(x) \equiv 0$ and $F(x, t) \equiv 0$ and that ρ is constant, independent of x. The differential equation becomes

$$\rho \frac{\partial^2 u}{\partial t^2} - K^2 \frac{\partial^2 u}{\partial x^2} = 0, \qquad 0 < x < L, \qquad t > 0. \tag{10.64}$$

The equation is to be applied to a rod or string occupying the portion of the x-axis between $x = 0$ and $x = L$. We assume that the ends are fixed:

$$u(0, t) = 0, \qquad u(L, t) = 0.$$

By a change of scale, $x' = \pi x/L$, we can reduce the problem to the case for which $L = \pi$. The equation becomes

$$\rho L^2 \frac{\partial^2 u}{\partial t^2} - \pi^2 K^2 \frac{\partial^2 u}{\partial x'^2} = 0.$$

For simplicity we drop the prime in the following. We introduce the abbreviation:

$$a = \frac{\pi K}{L\sqrt{\rho}}. \tag{10.65}$$

The equation and boundary conditions now read

$$\frac{\partial^2 u}{\partial t^2} - a^2 \frac{\partial^2 u}{\partial x^2} = 0, \qquad 0 < x < \pi, \qquad t > 0, \tag{10.66}$$

$$u(0, t) = 0, \qquad u(\pi, t) = 0. \tag{10.67}$$

To determine the normal modes, we can now set

$$u(x) = A(x)e^{\lambda t} \tag{10.68}$$

in (10.66) and (10.67). However, we find as in the N particle case that λ would have to be pure imaginary (Problem 5 below), and we simplify the process by replacing (10.68) by the substitution

$$u(x) = A(x)\sin(\lambda t + \epsilon). \tag{10.68'}$$

Equations (10.66) and (10.67) then become the two equations

$$-A(x)\lambda^2 - a^2 A''(x) = 0, \tag{10.69}$$
$$A(0) = A(\pi) = 0. \tag{10.70}$$

The simultaneous linear equations of the N particle problem are therefore replaced by a differential equation with boundary conditions. This is anticipated in Problems 5 to 12 following Section 10.4, in which it is shown that the simultaneous equations of the N particle problem can be treated as a *difference* equation with boundary conditions. The solutions obtained there are very closely related to those to be obtained for (10.69) and (10.70).

The general solution of (10.69) is

$$A(x) = c_1 \sin\left(\frac{\lambda x}{a}\right) + c_2 \cos\left(\frac{\lambda x}{a}\right). \tag{10.71}$$

Equations (10.70) are satisfied only if

$$c_2 = 0, \qquad \sin\left(\frac{\lambda \pi}{a}\right) = 0.$$

One obtains the *characteristic values* (resonant frequencies or *eigenvalues*)

$$\lambda_n = an \qquad (n = 1, 2, \ldots) \tag{10.72}$$

and associated *characteristic functions* (or *eigenfunctions*)

$$A_n(x) = \sin nx \qquad (n = 1, 2, \ldots). \tag{10.73}$$

We restrict to positive λ, since a change in sign can be absorbed in the phase constant ϵ. The normal modes are

$$\sin nx \sin(ant + \epsilon_n) \tag{10.74}$$

and constant multiples thereof; there are *infinitely many normal modes*. The set of frequencies λ_n occurring is called the *spectrum*.

We now attempt to construct the general solution $u(x, t)$ as a linear combination of normal modes:

$$u(x, t) = \sum_{n=1}^{\infty} c_n \sin nx \sin(ant + \epsilon_n). \tag{10.75}$$

However, we face a new difficulty: *The infinite series (10.75) may fail to converge.* Even if it does converge, it may fail to satisfy the differential equation (10.66), for this requires the existence of second derivatives. Now the series (10.75) can be regarded as a Fourier sine series in x, with coefficients dependent on t. From the theory of Fourier series (Chapter 7) we easily obtain conditions on the constants c_n such that the series converge for all x and can be differentiated twice with respect to x and t.

The choice of the constants in (10.75) depends on the initial conditions, for we can write (10.75) in the form:

$$u(x, t) = \sum_{n=1}^{\infty} \sin nx \left[\alpha_n \sin (ant) + \beta_n \cos (ant) \right],$$

$$\alpha_n = c_n \cos \epsilon_n, \qquad \beta_n = c_n \sin \epsilon_n. \tag{10.75'}$$

Then, if we assume that the series concerned are uniformly convergent,

$$u(x, 0) = \sum_{n=1}^{\infty} \beta_n \sin nx, \qquad \frac{\partial u}{\partial t}(x, 0) = \sum_{n=1}^{\infty} na\alpha_n \sin nx. \tag{10.76}$$

Thus β_n and $na\alpha_n$ are the Fourier sine coefficients of the initial displacement and velocity, respectively.

THEOREM If the constants c_n are such that $c_n n^4$ is bounded:

$$|c_n| < \frac{M}{n^4} \qquad (n = 1, 2, \ldots), \tag{10.77}$$

then the series (10.75) converges uniformly for all x and t and defines a solution of the wave equation (10.66) for all x and t. Let $f(x)$ and $g(x)$ be defined for $0 \le x \le \pi$; let $f(x)$ have continuous derivatives through the fourth order and let $f(0) = f(\pi) = f''(0) = f''(\pi) = 0$; let $g(x)$ have continuous derivatives through the third order and let $g(0) = g(\pi) = g''(0) = g''(\pi) = 0$. Then there exists a solution $u(x, t)$ of the wave equation (10.66) with boundary conditions (10.67) such that

$$u(x, 0) = f(x), \qquad \frac{\partial u}{\partial t}(x, 0) = g(x); \tag{10.78}$$

namely, the series (10.75'), where

$$\beta_n = \frac{2}{\pi} \int_0^{\pi} f(x) \sin nx \, dx, \qquad \alpha_n = \frac{2}{na\pi} \int_0^{\pi} g(x) \sin nx \, dx. \tag{10.79}$$

The solution is unique; that is, if $u(x, t)$ satisfies (10.66), (10.67), and (10.78) and the partial derivatives u_{xx}, u_{tt} are continuous for $0 \le x \le \pi$, $t \ge 0$, then $u(x, t)$ is necessarily represented by the series (10.75'), with coefficients given by (10.79).

Proof. If (10.77) holds, then the Weierstrass M-test (Section 6.13) shows that the series (10.75) converges uniformly for all x and t. Similarly, the series

$$- \sum n^2 c_n \sin nx \sin (ant + \epsilon_n), \qquad - \sum n^2 a^2 c_n \sin nx \sin (ant + \epsilon_n)$$

obtained by differentiating (10.75) twice with respect to x and t converge uniformly for all x and t; for by (10.77) $|n^2 c_n| < Mn^{-2}$. Accordingly, these series represent u_{xx} and u_{tt}, respectively. By substitution in (10.66) we verify that the wave equation is satisfied.

If $f(x)$ and $g(x)$ satisfy the conditions stated, then (Problem 4 following Section 10.8) $n^4 \alpha_n$ and $n^4 \beta_n$ are bounded so that

$$n^4 c_n = \sqrt{\left(n^4 \alpha_n\right)^2 + \left(n^4 \beta_n\right)^2}$$

is bounded. Accordingly, (10.77) holds, so that (10.75) or (10.75′) represents a solution $u(x, t)$ that is continuous for all x and t. When $t = 0$, the series for u and u_t reduce to the Fourier sine series of $f(x)$ and $g(x)$; these series converge to $f(x)$ and $g(x)$ (Problem 1 following Section 7.13).

The proof of uniqueness is left to Problem 6 following Section 10.8. ☐

10.8 PROPERTIES OF SOLUTIONS OF THE WAVE EQUATION

We consider the solutions in the form

$$u(x, t) = \sum_{n=1}^{\infty} \sin nx \left[\alpha_n \sin (nat) + \beta_n \cos (nat) \right]. \qquad (10.80)$$

For each fixed x the series is a Fourier series in t, with period $2\pi/a$ (Section 7.5). Accordingly, *each point of the rod or string considered moves in a periodic fashion*, with period $2\pi/a$.

The normal mode for which $n = 1$ is called the *fundamental mode*. Here

$$u(x, t) = \sin x \left[\alpha_1 \sin (at) + \beta_1 \cos (at) \right].$$

This is easily visualized in the case of a vibrating string, for which the displacement has the shape of a sine curve at all times (Fig. 10.7). The string is therefore vibrating in this shape with frequency $a/2\pi$ (cycles per unit time). The modes corresponding to $n = 2, 3, \ldots$ are called the *first overtone*, *second overtone*, and so on; musically, they give the octave, octave plus a fifth, and so on. These are suggested in Fig. 10.7. They are easily demonstrated on a stringed instrument, especially on low notes such as low C on a violoncello. We remark that the shapes of the normal modes are precisely the characteristic functions $A_n(x) = \sin nx$ and that these functions form a *complete orthogonal system* for the interval $0 \le x \le \pi$.

Figure 10.7 Normal modes for vibrating string.

The relation between solution and initial conditions can be shown in a striking way by the following observation. Let us first assume that $g(x) \equiv 0$, so that the rod (or vibrating string) is initially at rest but has an initial displacement

$f(x)$. By (10.79), $\alpha_n = 0$ for all n. We can now write

$$u(x, t) = \sum_{n=1}^{\infty} \beta_n \sin nx \cos nat = \tfrac{1}{2} \sum_{n=1}^{\infty} \beta_n [\sin n(x + at) + \sin n(x - at)].$$

Since

$$f(x) = \sum_{n=1}^{\infty} \beta_n \sin nx, \tag{10.81}$$

we can write:

$$u(x, t) = \tfrac{1}{2}[f(x + at) + f(x - at)]. \tag{10.82}$$

This representation is at first valid only for

$$0 \le x + at \le \pi, \qquad 0 \le x - at \le \pi.$$

However, if we extend the definition of $f(x)$ to all x by (10.81), then (10.82) has meaning for all x and t and, under the assumptions of the theorem above, represents a solution of the wave equation for all x and t.

The term $f(x + at)$ represents the initial displacement translated at units to the left; the second term represents this displacement translated at units to the right; this is suggested in Fig. 10.8. In Fig. 10.9, $f(x)$ is chosen as a displacement confined almost entirely to an interval $\tfrac{1}{2}\pi - \delta < x < \tfrac{1}{2}\pi + \delta$, where δ is small; the solution can then be plotted as a function of x and t. The disturbance is seen to split into two disturbances that travel in opposite directions until they reach the walls, where they are reflected, with a change in sign, and move back together. This can be demonstrated experimentally in various ways—by displacing and releasing a violin string or by sound echoes, for example.

If the initial displacement $f(x)$ has a jump discontinuity, for example, at x_0, but is piecewise very smooth, then (10.82) continues to define a solution of the wave equation except for $x \pm at = x_0 \pm k\pi$. These lines are the paths of "propagation of discontinuities"; they are called *characteristics*.

The constant a appears as the velocity with which the disturbance or discontinuity is propagated to the left and right; it is termed the *wave velocity*. If we choose as u a single mode of vibration:

$$u(x, t) = \tfrac{1}{2}\beta_n [\sin n(x + at) + \sin n(x - at)],$$

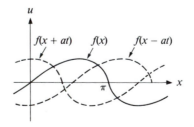

Figure 10.8 $f(x)$ versus $f(x + at), f(x - at)$.

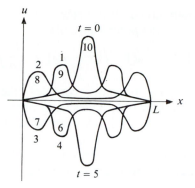

Figure 10.9   Solution $u(x, t)$ of the wave equation. The curves show the wave forms for $t = 0, 1, \ldots, 10$. The units are chosen so that the wave velocity a is $0.2\ L$ per unit of time.

then the initial disturbance $f(x)$ is simply $\beta_n \sin nx$, a wave repeating itself at intervals of $2\pi/n$, the *wavelength*. The oscillations at each x have a frequency of $na/2\pi$ (cycles per unit time). Hence

$$(wavelength) \cdot (frequency) = \frac{2\pi}{n} \cdot \frac{na}{2\pi}$$

$$= a = wave\ velocity.$$

This is one of the fundamental rules of physics.

We now consider the case in which $f(x) = 0$ but the initial velocity $g(x)$ is different from zero. Now, as before,

$$g(x) = \sum_{n=1}^{\infty} na\alpha_n \sin nx. \tag{10.83}$$

Hence we can integrate (Section 7.13):

$$\int g(x)\, dx = \sum_{n=1}^{\infty} (-a\alpha_n \cos nx) + \text{const.} \tag{10.84}$$

The solution $u(x, t)$ is therefore

$$u(x, t) = \sum_{n=1}^{\infty} \alpha_n \sin nx \sin nat$$

$$= \tfrac{1}{2} \sum_{n=1}^{\infty} \alpha_n \{\cos n(x - at) - \cos n(x + at)\}$$

$$= \frac{1}{2a} \int_{x-at}^{x+at} g(s)\, ds, \tag{10.85}$$

where s is a dummy variable of integration. The representation (10.85) is valid for all x and t if we use (10.83) to extend the definition of $g(x)$ to all x. Equation (10.85) can be interpreted as the *difference* of two disturbances moving to the left and right with velocity a.

If we allow for both initial displacement and initial velocity, the same reasoning as before leads to the general formula

$$u(x, t) = \tfrac{1}{2}[f(x + at) + f(x - at)] + \frac{1}{2a} \int_{x-at}^{x+at} g(s)\, ds. \qquad (10.86)$$

Remark. For the wave equation one may consider other natural boundary conditions, as at the end of Section 10.6. The shapes of the normal modes continue to yield complete orthogonal systems. This can be verified in particular cases as in the problems following Section 7.13 or as a consequence of a general property of Sturm-Liouville boundary value problems, discussed in Section 10.13 below. A similar statement applies to the heat equation (Section 10.9).

PROBLEMS

1. Let $f(x)$ be an odd periodic function of period 2π and let $f(x) = 0$ for $0 < x < \pi/3$ and for $2\pi/3 < x < \pi$, $f(x) = 1$ for $\pi/3 < x < 2\pi/3$. Show that (10.82) defines a solution of the wave equation for a certain portion of the xt-plane; analyze the solution graphically as in Figs. 10.8 and 10.9. Is the Fourier series solution valid with this $f(x)$ as the initial displacement?

2. Let $g(x)$ be an odd periodic function of period 2π and let $g(x) = 0$ for $0 < x < \pi/3$ and for $2\pi/3 < x < \pi$, $g(x) = 1$ for $\pi/3 < x < 2\pi/3$. Show that (10.85) defines a solution of the wave equation for a certain portion of the xt-plane; analyze the solution graphically. Is the Fourier series solution valid with this $g(x)$ as the initial velocity?

3. Show that the change of variables $r = x + at$, $s = x - at$, converts the wave equation (10.66) into the equation

$$\frac{\partial^2 u}{\partial r\, \partial s} = 0. \qquad (a)$$

Interpret the change of variables geometrically. Show that the "general solution" of (a) has the form

$$u = F(r) + G(s).$$

Discuss the relation between this representation and (10.86).

4. **a)** Prove that if $f(x)$ satisfies the conditions stated in the theorem of Section 10.7, then $\beta_n n^4$ is bounded. [Hint: Use integration by parts as in Section 7.8.]

   b) Prove that if $g(x)$ satisfies the conditions stated in the theorem of Section 10.7, then $\alpha_n n^4$ is bounded.

5. Show that the substitution (10.68) in (10.66) and (10.67) leads to equations that are satisfied only when $\lambda_n = ani$ and $A = A_n(x) = c(e^{inx} - e^{-inx})$. From these expressions, obtain the normal modes (10.74).

6. Prove that under the conditions stated in the theorem of Section 10.7 the solution $u(x, t)$ satisfying initial conditions (10.78) must have the form (10.75') and is hence uniquely determined. [Hint: Under the assumptions made, $u(x, t)$ has a representation as a Fourier sine series in x:

$$u = \sum_{n=1}^{\infty} \phi_n(t) \sin nx, \qquad \phi_n(t) = \frac{2}{\pi} \int_0^{\pi} u(x, t) \sin nx\, dx.$$

Differentiate the second equation twice with respect to t, using Leibnitz's Rule (Section 4.9) and integration by parts to show that $\phi_n''(t) + a^2 n^2 \phi_n(t) = 0$. Hence $\phi_n(t) = \alpha_n \sin(nat) + \beta_n \cos(nat)$.]

7. To raise the pitch of a violin string, one can (a) increase the tension, (b) decrease the density, (c) shorten the string. Show how these conclusions follow from (10.72) and (10.65).

8. Find the general solutions of the following partial differential equations for $0 < x < \pi$, $t > 0$, with boundary conditions $u(0, t) = u(\pi, t) = 0$:

a) $\dfrac{\partial^2 u}{\partial t^2} - \dfrac{\partial^2 u}{\partial x^2} - u = 0$ b) $\dfrac{\partial^2 u}{\partial t^2} - \dfrac{\partial^2 u}{\partial x^2} - 2\dfrac{\partial u}{\partial x} = 0$

9. Find the general solution of the wave equation

$$\frac{\partial^2 u}{\partial t^2} - a^2 \frac{\partial^2 u}{\partial x^2} = 0, \qquad 0 < x < 2\pi, \qquad t > 0,$$

such that $u(0, t) = u(2\pi, t)$, $u_x(0, t) = u_x(2\pi, t)$.

10. Find the general solution of the wave equation

$$\frac{\partial^2 u}{\partial t^2} - a^2 \frac{\partial^2 u}{\partial x^2} = 0, \qquad 0 < x < \pi, \qquad t > 0,$$

such that $u(0, t) = 0$, $u_x(\pi, t) = 0$.

10.9   THE 1-DIMENSIONAL HEAT EQUATION · EXPONENTIAL DECAY

We return to the basic equation of Section 10.5:

$$\rho \frac{\partial^2 u}{\partial t^2} + H \frac{\partial u}{\partial t} - K^2 \frac{\partial^2 u}{\partial x^2} = F(x, t).$$

We neglect masses; that is, we set $\rho = 0$. We assume that H is a constant, that there is no outside force F, and that the "walls" are fixed. The differential equation and boundary conditions become

$$H \frac{\partial u}{\partial t} - K^2 \frac{\partial^2 u}{\partial x^2} = 0, \qquad 0 < x < L, \qquad t > 0; \tag{10.87}$$

$$u(0, t) = 0, \qquad u(L, t) = 0. \tag{10.88}$$

Equation (10.87) is the 1-*dimensional heat equation*. The 3-dimensional heat equation

$$\frac{\partial T}{\partial t} - c^2 \nabla^2 T = 0 \tag{10.89}$$

was derived in Section 5.15 [Eq. (5.144)]; Eq. (10.87) can be regarded as the special case of heat conduction in an infinite slab bounded by planes $x = 0$, $x = L$ in space, the boundary conditions being such that the temperature depends only on x. One can also interpret the equation as describing conduction of heat in a thin rod, insulated except at the ends.

The transition from a system of masses attached by springs to the problem of heat conduction may at first seem artificial. However, the solutions of (10.87) do indeed display the properties of a limiting case of the system of masses as the friction increases and the total mass approaches 0. Because masses are neglected, disturbances can be propagated *instantaneously*; the wave velocity is *infinite*.

As in the preceding sections, we introduce a new variable $x' = \pi x / L$ and make the substitution:

$$c = \left(\frac{\pi^2 K^2}{L^2 H} \right)^{\frac{1}{2}}. \tag{10.90}$$

After dropping the primes the equation and boundary conditions become

$$\frac{\partial u}{\partial t} - c^2 \frac{\partial^2 u}{\partial x^2} = 0, \qquad t > 0, \qquad 0 < x < \pi, \tag{10.91}$$

$$u(0, t) = 0, \qquad u(\pi, t) = 0. \tag{10.92}$$

The substitution

$$u = A(x) e^{\lambda t} \tag{10.93}$$

leads to the characteristic value problem:

$$A''(x) - \frac{\lambda}{c^2} A = 0, \tag{10.94}$$

$$A(0) = A(\pi) = 0. \tag{10.95}$$

If λ is positive or zero, the only solution of (10.94) and (10.95) is the trivial solution: $A(x) \equiv 0$ (Problem 9 following Section 10.11). If λ is negative, the solutions are the *characteristic functions*

$$A_n(x) = \sin nx \qquad (n = 1, 2, \dots), \tag{10.96}$$

with associated *characteristic values*

$$\lambda_n = -n^2 c^2. \tag{10.97}$$

Accordingly, the "normal modes" are the functions

$$\sin nx \, e^{-c^2 n^2 t} \tag{10.98}$$

and constant multiples thereof. We expect the general solution to be given by the linear combinations

$$u(x, t) = \sum_{n=1}^{\infty} b_n \sin nx \, e^{-c^2 n^2 t}, \tag{10.99}$$

where the b_n are arbitrary constants.

Again we have to be careful about convergence. The problem is simpler than that for the wave equation, for if the coefficients b_n are bounded—$|b_n| < M$ for $n = 1, 2, \dots$—then the series (10.99) converges uniformly in x and t in each half-plane: $t \geq t_1$, $-\infty < x < \infty$, provided that $t_1 > 0$. This is a consequence of the Weierstrass M-test (Section 6.13), for

$$|b_n \sin nx \, e^{-c^2 n^2 t}| < M e^{-c^2 n^2 t_1} = M_n;$$

the series M_n is a series of constants whose convergence follows from the ratio test. In the same way we verify that the series (10.99) remains uniformly convergent in the range stated, when differentiated as often as desired with respect to x and t (Problem 5 following Section 10.11). In particular,

$$\frac{\partial u}{\partial t} = \sum_{n=1}^{\infty} b_n(-c^2 n^2) \sin nx \, e^{-c^2 n^2 t} = c^2 \frac{\partial^2 u}{\partial x^2},$$

so that (10.99) is a solution of the heat equation in the range $t > 0$, $-\infty < x < \infty$. The boundary conditions (10.92) are satisfied, since each term in the series is 0 when $x = 0$ or $x = \pi$. Every sufficiently smooth solution of the heat equation and boundary conditions in the range described must have the form (10.99) (Problem 7 following Section 10.11).

For $t = 0$ the series (10.99), if convergent, reduces to

$$u(x, 0) = \sum_{n=1}^{\infty} b_n \sin nx. \tag{10.100}$$

Thus, just as for the wave equation, the initial values of u, $u(x, 0) = f(x)$, are represented by a Fourier sine series. If, for example, $f(0) = f(\pi) = 0$ and $f(x)$ has a continuous second derivative for $0 \le x \le \pi$, then its sine series will converge uniformly to $f(x)$, and the series (10.99) will converge uniformly for $0 \le x \le \pi$ and $t \ge 0$. Accordingly, under the assumptions stated, the series (10.99) defines a function $u(x, t)$ that is continuous for $t \ge 0$, $0 \le x \le \pi$ and that satisfies the heat equation (10.91), the boundary conditions (10.92), and the initial condition $u(x, 0) = f(x)$. The solution is furthermore unique (Problem 7 following Section 10.11). These results show that with minor modifications the theorem of Section 10.7 holds for the heat equation.

If $f(x)$ is merely piecewise continuous, the constants

$$b_n = \frac{2}{\pi} \int_0^{\pi} f(x) \sin nx \, dx \tag{10.101}$$

are bounded, since the sequence b_n converges to 0 [Eq. (7.14), Section 7.4]. Accordingly, the series (10.99) defines a solution of the heat equation for $t > 0$. The series converges to $f(x)$ *in the mean* (Section 7.11) for $t = 0$; in fact, it can be shown that

$$\lim_{t \to 0+} u(x, t) = f(x), \qquad 0 < x < \pi$$

for each x at which $f(x)$ is continuous. (See pages 279–290 of the book by Doetsch listed at the end of the chapter.)

10.10 PROPERTIES OF SOLUTIONS OF THE HEAT EQUATION

For each fixed x, each term of the series (10.99) describes an exponential approach to 0. A similar statement applies to the sum of the series; that is,

$$\lim_{t \to \infty} u(x, t) = 0$$

(Problem 6 following Section 10.11). The rate of decay varies with n; the high-frequency terms in (10.99) have smaller exponential coefficients than the lower-frequency terms and are accordingly damped out more rapidly. This corresponds to the observed fact that abrupt variations in temperature in an object disappear rapidly, whereas differences in temperature at large distances are evened out slowly.

If we write

$$\sin nx = nx - \frac{(nx)^3}{3!} + \frac{(nx)^5}{5!} - \cdots$$

in (10.99) and collect terms in powers of x, we obtain a series:

$$\sum_{k=1}^{\infty} \phi_k(t)x^{2k-1}, \qquad \phi_k(t) = \sum_{n=1}^{\infty} b_n \frac{(-1)^{k-1}(n)^{2k-1}}{(2k-1)!} e^{-c^2 n^2 t}. \quad (10.102)$$

This operation can be justified by theorems on series or, more simply, by complex variables as in Problem 4 following Section 10.11. Accordingly, for each fixed $t > 0$ the solution $u(x, t)$ can be represented by a power series in x; the power series has infinite convergence radius, so that *the series converges for all x*. The solutions $u(x, t)$ are *analytic in x*.

From this last result we deduce another property of the solutions—the infinite speed of propagation of disturbances. For example, let the initial function $f(x)$ be 0 for $0 \le x \le \pi$ and greater than 0 for $x_1 < x < x_2$, where $0 < x_1 < x_2 < \pi$. In the case of the wave equation the solution $u(x, t)$ would remain identically 0 near $x = 0$ until the "wave" starting at $x = x_1$ could move across. For the heat equation the solution $u(x, t)$ takes on nonzero values in every x interval for every $t > 0$. To establish this, we consider a fixed positive t. Then $u(x, t)$ can be considered as an analytic function of a complex variable x. By Theorem 26 of Section 9.13, if $u(x, t) \equiv 0$ when x ranges over a whole interval of real values, then $u \equiv 0$ for this value of t. Hence necessarily, $b_n = 0$ for $n = 1, 2, \ldots$, so that $f(x) \equiv 0$, contrary to assumption. Accordingly, for each positive t, $u(x, t)$ takes on nonzero values in every interval of x. The disturbance is propagated *instantaneously*.

10.11　EQUILIBRIUM AND APPROACH TO EQUILIBRIUM

By analogy with case (d) of Sections 10.2 and 10.3 we allow for applied forces $F(x)$ and "wall displacements" u_0 and u_1, which do not vary with time. The differential equation and boundary conditions are as follows:

$$\rho(x)\frac{\partial^2 u}{\partial t^2} + H(x)\frac{\partial u}{\partial t} - K^2 \frac{\partial^2 u}{\partial x^2} = F(x), \qquad (10.103)$$

$$u(0) = u_0, \qquad u(L) = u_1. \qquad (10.104)$$

We seek a solution $u^*(x)$ independent of time; $u^*(x)$ then describes the equilibrium state of the system. Accordingly, we replace derivatives with respect to t by 0 in (10.103) and are led to the problem

$$-K^2\frac{d^2u}{dx^2} = F(x), \qquad 0 < x < L, \tag{10.105}$$

$$u(0) = u_0, \qquad u(L) = u_1. \tag{10.106}$$

Equation (10.105) is an *ordinary* differential equation whose general solution is obtained by integrating twice:

$$u = \int\left\{\int\frac{-F(x)}{K^2}dx\right\}dx.$$

If a particular choice $G(x)$ of the indefinite integral is made, so that $G''(x) = -F(x)/K^2$, then the general solution is

$$u = Ax + B + G(x). \tag{10.107}$$

The boundary conditions give the equations

$$B + G(0) = u_0, \qquad AL + B + G(L) = u_1.$$

These are easily solved for A and B. The equilibrium state is then

$$u^*(x) = [u_1 - u_0 + G(0) - G(L)]\frac{x}{L} + u_0 - G(0) + G(x). \tag{10.108}$$

We have assumed $F(x)$ to be continuous for $0 \le x \le L$, so that the integrals above have meaning.

To describe the *approach to equilibrium*, we let $u(x, t)$ be an arbitrary solution of (10.103) and (10.104) for $t > 0$, $0 < x < L$. Then

$$y(x, t) = u(x, t) - u^*(x)$$

satisfies the *homogeneous problem*:

$$\rho(x)\frac{\partial^2 y}{\partial t^2} + H(x)\frac{\partial y}{\partial t} - K^2\frac{\partial^2 y}{\partial x^2} = 0, \tag{10.109}$$

$$y(0) = 0, \qquad y(L) = 0. \tag{10.110}$$

This we verify by substitution in these equations and use of the fact that $u(x, t)$ satisfies (10.103) and (10.104), while $u^*(x)$ is the solution of the equilibrium problem (10.105) and (10.106). Accordingly,

$$u(x, t) = y(x, t) + u^*(x);$$

the general solution is formed of "complementary function" and particular solution in the usual manner.

If, in particular, $H(x) = 0$ and $\rho(x)$ is a positive constant ρ, then the complementary function $y(x, t)$ is a solution of the wave equation; the solutions $u(x, t)$ consist of oscillations about the equilibrium position. If H is also a positive constant, the oscillations are damped (cf. Problem 8 below). If H is so large that ρ can be neglected, the function $y(x, t)$ is a solution of the heat equation; equilibrium is approached exponentially.

PROBLEMS

1. Determine the solution for $t \geq 0$, $0 \leq x \leq \pi$ of the heat equation (10.91), with $c = 1$, such that $u(0, t) = 0$, $u(\pi, t) = 0$, $u(x,0) = \sin x + 5 \sin 3x$. Plot the solution as a function of x and t and compare the rates of decay of the terms in $\sin x$ and $\sin 3x$.

2. Determine the solution for $t > 0$, $0 < x < \pi$ of the heat equation (10.91) that is continuous for $t \geq 0$, $0 \leq x \leq \pi$ and has a continuous derivative $\partial u/\partial x$ in this region and that furthermore satisfies the conditions: $\partial u/\partial x = 0$ for $x = 0$ and $x = \pi$, $u(x,0) = f(x)$, where $f(x)$ has continuous first and second derivatives for $0 \leq x \leq \pi$. This can be interpreted as a problem in heat conduction in a slab whose faces are insulated.

3. Determine the solution, for $t > 0$, $0 < x < \pi$, of the equation

$$\frac{\partial u}{\partial t} - \frac{\partial^2 u}{\partial x^2} + 4u = 5 \sin x + 4x,$$

such that $u(0, t) = 0$, $u(\pi, t) = \pi$, $u(x,0) = x + 2 \sin x$.

4. Prove that if the constants b_n are bounded, then the series (10.99) can be written for each $t > 0$ as a power series in x, converging for all x. [Hint: Let $t > 0$ be fixed and let

$$v(x, y) = \sum_{n=1}^{\infty} b_n \sin nx \cosh ny \, e^{-n^2 c^2 t}.$$

Show that the series for v converges uniformly for $-\infty < x < \infty$, $-y_1 \leq y \leq y_1$ by applying the M-test with

$$M_n = Me^{-\frac{1}{2}n^2 c^2 t}$$

for n sufficiently large. Show that the series remains uniformly convergent after differentiation any number of times with respect to x and y. Each term of the series is *harmonic* in x and y; hence conclude that $v(x, y)$ is harmonic for all x and y. By Section 9.11, $v(x, y)$ can be expanded as a power series in x and y; put $y = 0$ to obtain the desired series for $u(x, t)$.]

5. Prove that if the constants b_n are bounded, then the series (10.99) remains uniformly convergent for $t \geq t_1 > 0$, $-\infty < x < \infty$, after differentiation any number of times with respect to x and t.

6. Prove that if the constants b_n are bounded, $|b_n| < M$, then the function $u(x, t)$ defined by (10.99) converges *uniformly* to 0 as $t \to \infty$; that is, given $\epsilon > 0$, a t_0 can be found such that $|u(x, t)| < \epsilon$ for $t > t_0$ and $-\infty < x < \infty$. [Hint: Show that $|u(x, t)|$ is less than the sum of the geometric series $M \sum_{n=1}^{\infty} (e^{-c^2 t})^n$.]

7. Let $u(x, t)$ have continuous derivatives through the second order in x and t for $t \geq 0$ and $0 \leq x \leq \pi$ and let $u(0, t) = u(\pi, t) = 0$. Prove that if $u(x, t)$ satisfies the heat equation (10.91) for $t > 0$, $0 < x < \pi$, then $u(x, t)$ has the form (10.99). [Hint: See Problem 6 following Section 10.8.]

8. Discuss the nature of the solutions for $0 < x < \pi$, $t > 0$ of the equation

$$\rho \frac{\partial^2 u}{\partial t^2} + H \frac{\partial u}{\partial t} - K^2 \frac{\partial^2 u}{\partial x^2} = 0,$$

with boundary conditions $u(0, t) = u(\pi, t) = 0$, if ρ, H, and K are positive constants.

9. Prove that if $\lambda \geq 0$, the equations (10.94) and (10.95) have no solution other than the trivial one: $A(x) \equiv 0$.

10.12   FORCED MOTION

We now consider the general problem of type (f):

$$\rho(x)\frac{\partial^2 u}{\partial t^2} + H(x)\frac{\partial u}{\partial t} - K^2\frac{\partial^2 u}{\partial x^2} = F(x, t), \qquad 0 < x < L, \qquad t > 0,$$

$$\tag{10.111}$$

$$u(0, t) = a(t), \qquad u(L, t) = b(t). \tag{10.112}$$

This is the problem of response of the 1-dimensional system to outside forces varying both in position and time. The fact that the boundary conditions (10.112) are variable shows that the motion is being forced at the ends $x = 0$, $x = L$ also.

Just as in the case of the equilibrium problem of Section 10.11, we reason that if $u(x, t)$ is a particular solution of (10.111) and (10.112) for $0 < x < L$, $t > 0$, then the general solution $u(x, t)$ in this domain is

$$u(x, t) = y(x, t) + u^*(x, t), \tag{10.113}$$

where $y(x, t)$ is a solution of the homogeneous problem (10.109), (10.110).

Accordingly, the problem is that of determining a particular solution. We can further concentrate attention on the case of fixed "walls," that is, $a(t) = b(t) = 0$. To show this, we let

$$g(x, t) = a(t)\left(1 - \frac{x}{L}\right) + \frac{x}{L}b(t);$$

$g(x, t)$ is simply the linear function of x that interpolates between the values $a(t)$ at $x = 0$, $b(t)$ at $x = L$. Now (if we assume $a(t)$ and $b(t)$ to have the requisite derivatives)

$$\rho(x)\frac{\partial^2 g}{\partial t^2} + H(x)\frac{\partial g}{\partial t} - K^2\frac{\partial^2 g}{\partial x^2} = G(x, t),$$

$$G(x, t) = \left(1 - \frac{x}{L}\right)[\rho a''(t) + Ha'(t)] + \frac{x}{L}[\rho b''(t) + Hb'(t)].$$

Accordingly, if $u^*(x, t)$ satisfies (10.111) and (10.112), then $w(x, t) = u^*(x, t) - g(x, t)$ satisfies

$$\rho(x)\frac{\partial^2 w}{\partial t^2} + H(x)\frac{\partial w}{\partial t} - K^2\frac{\partial^2 w}{\partial x^2} = F(x, t) - G(x, t) = F_1(x, t),$$

$$\tag{10.114}$$

$$w(0, t) = a(t) - a(t) = 0, \qquad w(L, t) = 0. \tag{10.115}$$

Conversely, if $w(x, t)$ satisfies (10.114) and (10.115), then we verify that $w(x, t) + g(x, t) = u^*(x, t)$ satisfies (10.111) and (10.112). Therefore we have reduced our problem to the case of fixed walls.

The determination of a function $w(x, t)$ is now accomplished by *variation of parameters*. The applicability of this method is not dependent on the fact that the coefficients ρ, H are constant. However, we here confine attention to the case of constant coefficients; the procedure in the general case differs only slightly.

Let us suppose first that $\rho \equiv 0$ and $H = \text{const} > 0$. Then the "complementary function" is the general solution of the heat equation. We change scale and introduce the abbreviations of Section 10.9, so that the solution is

$$\sum_{n=1}^{\infty} b_n \sin nx \, e^{-c^2 n^2 t}.$$

We now replace the constants b_n by functions $v_n(t)$ and seek a solution

$$w(x, t) = \sum_{n=1}^{\infty} v_n(t) \sin nx \, e^{-c^2 n^2 t}. \tag{10.116}$$

We proceed formally and then determine conditions under which the solution obtained is valid. Substitution of (10.116) in (10.114) (with $\rho = 0$) leads to the equation

$$H \sum_{n=1}^{\infty} \left\{ v_n'(t) e^{-n^2 c^2 t} \sin nx \right\} = F_1(x, t);$$

the other terms cancel out. This equation is simply a Fourier sine series in x for $F_1(x, t)$. Hence

$$H \frac{dv_n}{dt} e^{-n^2 c^2 t} = \frac{2}{\pi} \int_0^{\pi} F_1(x, t) \sin nx \, dx,$$

$$v_n(t) = \frac{2}{\pi H} \int e^{n^2 c^2 t} \left\{ \int_0^{\pi} F_1(x, t) \sin nx \, dx \right\} dt. \tag{10.117}$$

Since we seek only one particular solution, we can choose the indefinite integral here so that $v_n(0) = 0$; that is, we choose

$$v_n(t) = \frac{2}{\pi H} \int_0^t e^{n^2 c^2 s} \left\{ \int_0^{\pi} F_1(x, s) \sin nx \, dx \right\} ds,$$

where s is a dummy variable of integration. Accordingly,

$$v_n(t) = \frac{2}{\pi H} \int_0^t \int_0^{\pi} F_1(x, s) e^{n^2 c^2 s} \sin nx \, dx \, ds, \tag{10.118}$$

and the particular solution sought is

$$w(x, t) = \frac{2}{\pi H} \sum_{n=1}^{\infty} \left\{ \sin nx \, e^{-n^2 c^2 t} \int_0^t \int_0^{\pi} F_1(r, s) e^{n^2 c^2 s} \sin nx \, dx \, ds \right\}. \tag{10.119}$$

We now study the validity of the result. If $F_1(x, t)$ is defined for $t \geq 0$ and $0 \leq x \leq \pi$ and is continuous in both variables in this region, then $v_n(t)$ is well defined by (10.118). Furthermore, let $\partial^2 F_1/\partial x^2$ be continuous in x and t, so that

$\partial^2 F_1/\partial x^2$ has a maximum M (t_1) in each rectangle: $0 \le x \le \pi$, $0 \le t \le t_1$; let also $F_1(0, t) = F_1(\pi, t) = 0$. Then, as in Section 7.8, we conclude that for $0 \le t \le t_1$

$$\left| \frac{2}{\pi} \int_0^\pi F_1(x, t) \sin nx \, dx \right| \le \frac{2M(t_1)}{n^2} .$$

Hence by (10.118), in this rectangle,

$$|v_n(t)| \le \frac{2M(t_1)}{Hn^2} \int_0^t e^{n^2 c^2 s} \, ds = \frac{2M(t_1)}{Hn^4 c^2} \left(e^{n^2 c^2 t} - 1 \right),$$

and each term of the series (10.119) is bounded by $2M(t_1)/c^2 n^4 H$. The series therefore converges uniformly and continues to do so after differentiation once with respect to t or twice with respect to x. We therefore find

$$H \frac{\partial w}{\partial t} - K^2 \frac{\partial^2 w}{\partial x^2} = \frac{2}{\pi} \sum_{n=1}^\infty \sin nx \int_0^\pi F_1(r, t) \sin nr \, dr.$$

The right-hand side is the Fourier sine series of $F_1(x, t)$ and, under the assumptions made, converges to $F_1(x, t)$. Hence (10.114) is satisfied, as is (10.115).

Therefore (10.119) is indeed the particular solution sought. The function

$$u^*(x, t) = w(x, t) + g(x, t)$$

is then a particular solution of (10.111) and (10.112), and

$$u(x, t) = y(x, t) + u^*(x, t) = y(x, t) + w(x, t) + g(x, t)$$

is the general solution, where

$$y(x, t) = \sum_{n=1}^\infty b_n \sin nx \, e^{-n^2 c^2 t},$$

the b_n being arbitrary constants. To determine a solution satisfying the initial condition $u(x, 0) = f(x)$, we have to determine the constants b_n so that

$$f(x) = y(x, 0) + w(x, 0) + g(x, 0)$$

$$= \sum_{n=1}^\infty b_n \sin nx + a(0)\left(1 - \frac{x}{L}\right) + \frac{x}{L} b(0);$$

that is, the series $\Sigma b_n \sin nx$ must be the Fourier sine series of

$$f(x) - a(0)\left(1 - \frac{x}{L}\right) - \frac{x}{L} b(0).$$

We have throughout assumed $\rho = 0$ and $H = $ const. If ρ is a positive constant and $H = 0$, then the complementary function is the general solution of the wave equation:

$$\sum_{n=1}^\infty \sin nx \, (\alpha_n \sin nat + \beta_n \cos nat).$$

The trial function (10.116) is now replaced by the function:

$$w(x, t) = \sum_{n=1}^{\infty} \sin nx \left[p_n(t) \sin nat + q_n(t) \cos nat \right]. \qquad (10.116')$$

Substitution in (10.114) does not yield enough conditions to determine $p_n(t)$ and $q_n(t)$, because the equation is now of *second* order in t. We choose an extra set of conditions:

$$p_n'(t) \sin nat + q_n'(t) \cos nat = 0 \qquad (n = 1, 2, \ldots). \qquad (10.120)$$

These conditions can be obtained by replacing (10.114) by a system of equations:

$$\frac{\partial w}{\partial t} = z, \qquad \rho \frac{\partial z}{\partial t} - K^2 \frac{\partial^2 w}{\partial x^2} = F_1(x, t) \qquad (10.121)$$

(Problem 5 below). Proceeding as before, we find

$$p_n(t) = \frac{2}{na\pi\rho} \int_0^t \int_0^\pi F_1(x, s) \cos nas \sin nx \, dx \, ds,$$

$$q_n(t) = \frac{-2}{na\pi\rho} \int_0^t \int_0^\pi F_1(x, s) \sin nas \sin nx \, dx \, ds. \qquad (10.122)$$

A similar procedure applies when ρ and H are both positive constants.

PROBLEMS

1. Find the solution of the partial differential equation

$$\frac{\partial u}{\partial t} - \frac{\partial^2 u}{\partial x^2} = x^2 \cos t - 2 \sin t, \qquad 0 < x < \pi, \qquad t > 0,$$

satisfying boundary conditions $u(0, t) = 0$, $u(\pi, t) = \pi^2 \sin t$, and initial conditions $u(x, 0) = \pi x - x^2$.

2. Let $u(x, t)$ be a solution of the partial differential equation

$$\frac{\partial^2 u}{\partial t^2} - \frac{\partial^2 u}{\partial x^2} = \sin x \sin \omega t, \qquad 0 < x < \pi, \qquad t > 0,$$

and boundary conditions $u(0, t) = 0$, $u(\pi, t) = 0$, $u(x, 0) = 0$, $\partial u/\partial t(x, 0) = 0$. Show that resonance occurs only when $\omega = \pm 1$ and determine the form of the solution in the two cases: $\omega = \pm 1$, $\omega \neq \pm 1$. [The other resonant frequencies $2, 3, \ldots$ are not excited because the force $F(x, t)$ is orthogonal to the corresponding "basis vectors" $\sin 2x, \sin 3x, \ldots$; cf. Problem 9 following Section 10.3.]

3. Let the outside force $F(x, t)$ be given as a Fourier sine series:

$$F(x, t) = \sum_{n=1}^{\infty} F_n(t) \sin nx, \qquad t \geq 0, \qquad 0 \leq x \leq \pi.$$

Obtain a particular solution of the partial differential equation

$$H \frac{\partial u}{\partial t} - K^2 \frac{\partial^2 u}{\partial x^2} = F(x, t)$$

with boundary conditions $u(0, t) = 0$, $u(\pi, t) = 0$, by setting

$$u(x, t) = \sum_{n=1}^{\infty} \phi_n(t) \sin nx, \qquad \phi_n(t) = 0,$$

substituting in the differential equation, and comparing coefficients of $\sin nx$ (corollary to Theorem 1, Section 7.2). Show that the result obtained agrees with (10.119).

4. Show that the substitution of (10.116′) in (10.114), with $\rho > 0$, $H = 0$, and application of (10.120) leads to the equations (10.122) for $p_n(t)$, $q_n(t)$.

5. Show that the solution of the homogeneous problem $(F = 0)$ corresponding to (10.121), with boundary conditions $w(0, t) = 0$, $w(\pi, t) = 0$, is given by

$$w = \sum_{n=1}^{\infty} \sin nx [\alpha_n \sin nat + \beta_n \cos nat],$$

$$z = \sum_{n=1}^{\infty} \sin nx [na\alpha_n \cos nat - na\beta_n \sin nat].$$

Show that replacement of α_n by $p_n(t)$, β_n by $q_n(t)$, and substitution in (10.121) lead to the equations:

$$p_n' \sin nat + q_n' \cos nat = 0,$$

$$nap_n' \cos nat - naq_n' \sin nat = \frac{2}{\pi\rho} \int_0^{\pi} F_1 \sin nx \, dx,$$

and hence one obtains (10.122).

6. Let $u_1(x, t)$, $u_2(x, t)$, $u_3(x, t)$, respectively, be solutions of the problems (for $0 < x < \pi$, $t > 0$):

$$u_t - u_{xx} = F(x, t), \qquad u(0, t) = 0, \qquad u(\pi, t) = 0;$$

$$u_t - u_{xx} = 0, \qquad u(0, t) = a(t), \qquad u(\pi, t) = 0;$$

$$u_t - u_{xx} = 0, \qquad u(0, t) = 0, \qquad u(\pi, t) = b(t).$$

Show that $u_1(x, t) + u_2(x, t) + u_3(x, t)$ is a solution of the problem

$$u_t - u_{xx} = F(x, t), \qquad u(0, t) = a(t), \qquad u(\pi, t) = b(t).$$

This shows that the effects of the different ways of forcing the system combine by *superposition*.

10.13   EQUATIONS WITH VARIABLE COEFFICIENTS · STURM-LIOUVILLE PROBLEMS

In order to determine the normal modes in the problem

$$\rho(x)\frac{\partial^2 u}{\partial t^2} - K^2\frac{\partial^2 u}{\partial x^2} = 0,$$

$$u(0, t) = 0, \qquad u(L, t) = 0, \tag{10.123}$$

we make the substitution:

$$u = A(x) \sin(\lambda t + \epsilon)$$

and are led to the equations:

$$K^2 A''(x) + \rho(x)\lambda^2 A(x) = 0,$$
$$A(0) = A(L) = 0. \tag{10.124}$$

When $\rho(x)$ is constant, we know that the only solutions of (10.124) are the functions $A_n(x) = c\sin(n\pi x/L)$; the associated frequencies λ_n are of form an. What is the nature of the solutions when ρ is variable?

A similar question arises if we consider the problem for the heat equation with variable coefficient $H(x)$:

$$H(x)\frac{\partial u}{\partial t} - K^2 \frac{\partial^2 u}{\partial x^2} = 0,$$
$$u(0, t) = 0, \qquad u(L, t) = 0. \tag{10.125}$$

The substitution

$$u = A(x)e^{\lambda t}$$

leads to the equations

$$K^2 A''(x) - \lambda H(x)A(x) = 0,$$
$$A(0) = 0, \qquad A(L) = 0. \tag{10.126}$$

Except for a change in notation, these equations are the same as (10.124).

The problems (10.124) and (10.126) are special cases of the class of *Sturm-Liouville boundary value problems*. A more general case is as follows:

$$\frac{d}{dx}\left[r(x)\frac{dy}{dx}\right] + [\lambda p(x) + q(x)]\, y = 0,$$
$$\alpha y(a) + \beta y'(a) = 0, \qquad |\alpha| + |\beta| > 0, \tag{10.127}$$
$$\gamma y(b) + \delta y'(b) = 0, \qquad |\gamma| + |\delta| > 0.$$

Here the function $y(x)$ is to be a solution of the differential equation for $a \le x \le b$ and is to satisfy the given boundary conditions at a and b. We further assume that $r(x)$, $p(x)$, $q(x)$ have continuous derivatives over the interval and that $r(x) > 0$, $p(x) > 0$. A value of λ for which (10.127) has a solution other than $y(x) \equiv 0$ is called a *characteristic value*. One could allow for complex characteristic values, but it can be shown that under the assumptions made this does not arise; hence we restrict to real characteristic values. For each characteristic value λ there is an associated solution $y(x)$, called a *characteristic function*; the functions $cy(x)$, where c is a constant, are also characteristic functions. It could conceivably happen that there are functions besides $cy(x)$ that have the same λ; it can be shown that under the assumptions made this cannot arise:

THEOREM   The characteristic values of the Sturm-Liouville problem (10.127) can be numbered to form an increasing sequence: $\lambda_1 < \lambda_2 < \cdots < \lambda_n < \cdots$. The corresponding characteristic functions can be numbered similarly to form a sequence: $y_n(x)$; each $y_n(x)$ is determined only up to a constant multiplier. The functions $y_n(x)$ are orthogonal with respect to the weight function $p(x)$:

$$\int_a^b y_m(x)\,y_n(x)\,p(x)\,dx = \begin{cases} 0, & m \neq n, \\ B_n > 0, & m = n. \end{cases}$$

The Fourier series of a function $F(x)$ with respect to the orthogonal system $\{\sqrt{p(x)}\,y_n(x)\}$ converges uniformly to $F(x)$ for every function $F(x)$ having a continuous derivative for $a \leq x \leq b$ and such that $F(a) = 0$, $F(b) = 0$.

For a proof of this theorem and of theorems for more general Sturm-Liouville boundary value problems, we refer to the books of Hellwig, Titchmarsh and Kamke listed at the end of this chapter. (See also Problem 13 following Section 2.18.)

Because of the theorem, we can be assured that except for minor changes in form, the statements about the wave equation and heat equation in Sections 10.7 and 10.9 continue to hold when the coefficients $\rho(x)$, $H(x)$ are variable $[\rho(x) > 0,$ $H(x) > 0]$. For example, the characteristic functions $A_n(x)$ of (10.124) provide normal modes:

$$A_n(x) \sin (\lambda_n t + \epsilon)$$

and constant multiples thereof. The "general solution" of (10.123) is again a series

$$\sum_{n=1}^\infty c_n A_n(x) \sin (\lambda_n t + \epsilon_n) = \sum_{n=1}^\infty A_n(x)[\alpha_n \sin \lambda_n t + \beta_n \cos \lambda_n t].$$

To satisfy initial conditions:

$$u(x,0) = f(x), \qquad \frac{\partial u}{\partial t}(x,0) = g(x),$$

one has only to choose the constants α_n, β_n so that

$$f(x) = \sum_{n=1}^\infty \beta_n A_n(x), \qquad g(x) = \sum_{n=1}^\infty \lambda_n \alpha_n A_n(x);$$

this requires expansion of the functions $\sqrt{p(x)}\,f(x)$, $\sqrt{p(x)}\,g(x)$ in Fourier series:

$$\sqrt{p(x)}\,f(x) = \sum_{n=1}^\infty \beta_n \sqrt{p(x)}\,A_n(x), \qquad \sqrt{p(x)}\,g(x) = \sum_{n=1}^\infty \lambda_n \alpha_n \sqrt{p(x)}\,A_n(x).$$

By the preceding theorem these expansions have the same properties as the sine series used above, so that there is no change in the results. The theory of equilibrium states, approach to equilibrium, and forced motion can also be repeated.

The only difficulty is in effective determination of characteristic values and functions. For various special equations, infinite series are effective. For others, one is forced to use numerical methods. These are discussed in Sections 10.16 and 10.17.

The previous theorem on the Sturm-Liouville problem can be extended under appropriate assumptions to the "singular case" in which the function $r(x)$ is 0 at a or b or both, while remaining positive for $a < x < b$. This case includes in particular the important problem:

$$\frac{d}{dx}\left[(1-x^2)\frac{dy}{dx}\right] + \lambda y = 0, \qquad -1 \leq x \leq 1, \tag{10.128}$$

whose solutions are the Legendre polynomials $P_n(x)$, with $\lambda_n = n(n+1)$ (Section 7.14. No boundary condition is imposed at $x = \pm 1$, but it is required that the solutions remain continuous at these points. The theory can also be extended to include the problem:

$$(xy')' + \left(\lambda x - \frac{m^2}{x}\right)y = 0, \qquad 0 \leq x \leq 1,$$

$$y(1) = 0, \qquad (m \geq 0). \tag{10.129}$$

The solution is required to be continuous at $x = 0$. The solutions of (10.129) are the functions $J_m(s_{mn}x)$, where s_{mn} ranges over the positive roots of the Bessel function $J_m(x)$; the corresponding $\lambda_{mn} = s_{mn}^2$. The extensions to these cases are covered in the book of Titchmarsh listed at the end of the chapter.

10.14 EQUATIONS IN TWO AND THREE DIMENSIONS · SEPARATION OF VARIABLES

The generalization of our basic problem to two and three dimensions brings no basic change in the results, though the determination of normal modes is in general more complicated.

As an example we consider the wave equation for a rectangle:

$$\frac{\partial^2 u}{\partial t^2} - a^2\left(\frac{\partial^2 u}{\partial x^2} + \frac{\partial^2 u}{\partial y^2}\right) = 0, \qquad 0 < x < \pi, \qquad 0 < y < \pi,$$

$$u(x, y, t) = 0 \quad \text{for} \quad x = 0, \qquad x = \pi, \qquad y = 0, \qquad y = \pi. \tag{10.130}$$

The substitution

$$u(x, y, t) = A(x, y)\sin(\lambda t + \epsilon)$$

leads to the characteristic value problem

$$a^2\left(\frac{\partial^2 A}{\partial x^2} + \frac{\partial^2 A}{\partial y^2}\right) + \lambda^2 A = 0,$$

$$A(x, y) = 0 \quad \text{for} \quad x = 0, \qquad x = \pi, \qquad y = 0, \qquad y = \pi. \tag{10.131}$$

In order to determine the characteristic functions $A(x, y)$ we seek particular characteristic functions having the form of a product of a function of x by a function of y:

$$A(x, y) = X(x)Y(y).$$

Accordingly,

$$a^2[X''(x)Y + XY''(y)] + \lambda^2 X(x)Y(y) = 0,$$

$$\frac{X''}{X} + \frac{Y''}{Y} + \frac{\lambda^2}{a^2} = 0. \tag{10.132}$$

Now if we vary x, the second and third terms of the last equation cannot vary; hence the first term is a constant. Similarly, the second term is a constant:

$$X'' = -\mu X, \qquad Y'' = \left(\mu - \frac{\lambda^2}{a^2}\right)Y,$$

$$X'' + \mu X = 0, \qquad Y'' + \left(\frac{\lambda^2}{a^2} - \mu\right)Y = 0.$$

Because of the boundary conditions in (10.131), we are led to two new boundary value problems:

$$X'' + \mu X = 0, \qquad X(0) = X(\pi) = 0;$$

$$Y'' + \left(\frac{\lambda^2}{a^2} - \mu\right)Y = 0, \qquad Y(0) = Y(\pi) = 0. \tag{10.133}$$

The first is satisfied by $X_n(x) = \sin nx$, for $\mu = n^2$; for this value of μ the second is satisfied by the functions $Y_m(y) = \sin my$, provided that $(\lambda^2/a^2) - \mu = m^2$. Accordingly,

$$A_{mn}(x, y) = \sin nx \sin my$$

is a solution of (10.131), for $\lambda^2 = a^2(m^2 + n^2)$, where m and n are integers. For a given characteristic value λ there may be several characteristic functions; in fact, if $A_{mn}(x, y)$ is one function, then $A_{nm}(x, y)$ provides a second one, unless $m = n$. This causes no trouble because we put all linear combinations into the "general solution":

$$u(x, y, t) = \sum_{m, n} c_{mn} \sin nx \sin my \sin(\lambda_{mn}t + \epsilon_n),$$

$$\lambda_{mn} = a\sqrt{m^2 + n^2}. \tag{10.134}$$

The series is a "double series," to be summed over all combinations of positive integral values of m and n; the series is a Fourier series in the two variables x, y and, as was pointed out in Section 7.16, can be rearranged to form a single series or summed as an "iterated" series:

$$\sum_{m=1}^{\infty} \left\{ \sum_{n=1}^{\infty} c_{mn} \sin nx \sin my \sin(\lambda_{mn}t + \epsilon_n) \right\}.$$

Since only sine functions appear, the series is really a *Fourier sine series in two variables*. As in Section 7.16, the functions $\sin nx \sin my$ form a *complete orthogonal system* for the region: $0 \leq x \leq \pi, 0 \leq y \leq \pi$. Accordingly, the results achieved in one dimension can all be generalized to two dimensions. Because of the more complicated characteristic values, the solutions are more difficult to analyze; in particular, the solutions are in general no longer periodic in t.

The crucial step was the replacement of $A(x, y)$ by the product $X(x)Y(y)$. This led to a so-called separation of variables in (10.133) and determination of particular characteristic functions $A(x, y)$, which together form a complete orthogonal system. The fact that such a procedure can be successful was already indicated in our method for determining normal modes for the problems in one dimension; the substitution $u = A(x)e^{\lambda t}$ could have been replaced by a substitution $u = A(x)T(t)$, and a separation of variables would then have led to the same results.

The method of separation of variables thus appears as a general method for attacking homogeneous linear partial differential equations. The method may in some cases provide only certain particular solutions; in a wide variety of cases these particular solutions have been shown to provide a complete set of orthogonal functions. These cases include the Laplace equation in cylindrical and spherical coordinates (Sections 2.17 and 3.8); see Problems 4 and 7 following Section 10.15.

The *equilibrium problem*: $\nabla^2 u = -F(x, y)$, with values of u prescribed on the boundary of a region R of the xy-plane, can be attacked in several ways. It can be shown that under appropriate assumptions on $F(x, y)$ the function (logarithmic potential)

$$u_0(x, y) = -\frac{1}{4\pi} \int_R \int F(r, s) \log\left[(x - r)^2 + (y - s)^2\right] dr \, ds$$

$$(10.135)$$

satisfies the Poisson equation $\nabla^2 u_0 = -F$ inside R and is continuous in R plus boundary. The function $v = u - u_0$ will then satisfy the condition $\nabla^2 v = -F + F = 0$ inside R and have certain new boundary values on the boundary of R. Determination of v is then a *Dirichlet* problem, which can be attacked by conformal mapping (Section 9.3). While conformal mapping is not available as a tool for the corresponding problems in three dimensions, methods based on *potential theory* can be used; see especially the book of Kellogg listed at the end of the chapter. (An introduction to potential theory is given in Sections 5.16 to 5.18.)

One can in general reduce the equilibrium problem $\nabla^2 u = -F$ to the case of zero boundary values by the following procedure (cf. Section 10.12): Let $u(x, y)$ be required to have values $h(x, y)$ on the boundary C of R. If $h(x, y)$ is sufficiently smooth, one can then find a function $h_1(x, y)$ that has continuous first and second derivatives inside R, is continuous in R plus C, and equals $h(x, y)$ on C. The function $v = u - h_1$ is then zero on C, and $\nabla^2 v = -F - \nabla^2 h_1 = -F_1(x, y)$. The determination of v can be carried out with the aid of a *Green's function*, as will be indicated in Section 10.18. (See also Sections 5.17 and 5.18.)

10.15   UNBOUNDED REGIONS · CONTINUOUS SPECTRUM

In many physical problems it is natural to regard the continuous medium as being *infinite* in extent. For example, in one dimension, one can consider the wave equation

$$\frac{\partial^2 u}{\partial t^2} - a^2 \frac{\partial^2 u}{\partial x^2} = 0 \tag{10.136}$$

for the infinite interval $x > 0$. If one seeks normal modes, one is led to the characteristic value problem:

$$a^2 A''(x) + \lambda^2 A(x) = 0, \qquad x > 0; \qquad A(0) = 0. \tag{10.137}$$

This problem has solutions for every value of λ, namely, the functions $\sin \alpha x$, for $a^2 \alpha^2 = \lambda^2$. Thus the resonant frequencies λ form a "continuous" set of numbers, and one has a "continuous spectrum." [There also exist *unbounded* "normal modes": $u = \sinh \alpha x\, e^{a \alpha t}$. These are of less physical interest.]

One can construct linear combinations of the normal modes in order to obtain a "general solution" of the homogeneous problem. Since there is a continuous sequence of λ's, an *integration* is called for rather than a summation. For $\alpha \geq 0$ we must integrate expressions of form

$$\sin x \left[p(\alpha) \cos (a \alpha t) + q(\alpha) \sin (a \alpha t) \right],$$

where $p(\alpha)$ and $q(\alpha)$ are "arbitrary" functions of α. We obtain the integral

$$\int_0^\infty \sin \alpha x \left[p(\alpha) \cos (a \alpha t) + q(\alpha) \sin (a \alpha t) \right]\, d\alpha.$$

For each fixed t this can be considered as a *Fourier integral* (the Fourier sine integral, Section 7.18). In particular, for $t = 0$ we obtain a Fourier integral representation of the initial displacement $u(x, 0)$:

$$\int_0^\infty p(\alpha) \sin \alpha x\, d\alpha.$$

Since the theory of the Fourier integral has been highly developed, most of the results for the finite interval can be extended to the infinite case. Similar statements can be made for problems in two or three dimensions for unbounded regions. For more information we refer to the books of the following authors, as listed at the end of the chapter: Sneddon, Titchmarsh, Wiener, Courant and Hilbert, Frank and von Mises, and Tamarkin and Feller. The Laplace transform (Section 6.25) can also be used to represent solutions over infinite intervals. This is discussed in the texts mentioned and in the second book by Churchill.

PROBLEMS

1. a) Let a vibrating string be stretched between $x = 0$ and $x = 1$; let the tension K^2 be $(x + 1)^2$ and the density ρ be 1 in appropriate units. Show that the normal modes are given by the functions

$$A_n(x) = \sqrt{x+1} \sin\left[n\pi \frac{\log(x+1)}{\log 2} \right] \sin(\lambda_n t + \epsilon_n), \qquad \lambda_n = \left(\frac{n^2\pi^2}{\log^2 2} + \frac{1}{4} \right)^{\frac{1}{2}}.$$

[Hint: Make the substitution $x + 1 = e^u$ in the boundary value problem for $A_n(x)$.]

b) Show directly that every function $f(x)$ having continuous first and second derivatives for $0 \le x \le 1$ and such that $f(0) = f(1) = 0$ can be expanded in a uniformly convergent series in the characteristic functions $A_n(x)$ of part (a). [Hint: Let $x + 1 = e^u$ as in part (a). Then expand $F(u) = f(e^u - 1)e^{-\frac{1}{2}u}$ in a Fourier sine series for the interval $0 \le u \le \log 2$.]

2. Show that the general second-order linear equation

$$p_0(x)y'' + p_1(x)y' + [\lambda p_2(x) + p_3(x)]y = 0,$$

where $p_0(x) \neq 0$, takes on the form of a Sturm-Liouville equation (10.127) if the equation is multiplied by $r(x)/p_0(x)$, where $r(x)$ is chosen so that $r'/r = p_1/p_0$. In general, an equation of form: $(ry')' + h(x)y = 0$ is called *self-adjoint*.

3. Obtain the general solution, for $t > 0$, $0 < x < \pi$, $0 < y < \pi$, of the heat equation with boundary conditions:

$$\frac{\partial u}{\partial t} - c^2 \left(\frac{\partial^2 u}{\partial x^2} + \frac{\partial^2 u}{\partial y^2} \right) = 0,$$

$$u(x, y, t) = 0 \quad \text{for} \quad x = 0, \qquad x = \pi, \qquad y = 0, \qquad y = \pi.$$

4. Show that separation of variables $u(r, \theta) = R(r)\Theta(\theta)$ in the problem in polar coordinates for the domain $r < 1$:

$$\nabla^2 u + \lambda u \equiv \frac{1}{r^2} \left[r \frac{\partial}{\partial r} \left(r \frac{\partial u}{\partial r} \right) + \frac{\partial^2 u}{\partial \theta^2} \right] + \lambda u = 0, \qquad u(1, \theta) = 0,$$

leads to the problems:

$$(rR')' + \left(\lambda r - \frac{\mu}{r} \right) R = 0, \qquad R(1) = 0,$$

$$\Theta'' + \mu\Theta = 0.$$

If we require that $u(r, \theta)$ be continuous in the circle $r \le 1$, then $\Theta(\theta)$ must be periodic in θ, with period 2π. Show that this implies that $\mu = m^2$ ($m = 0, 1, 2, 3, \ldots$) and that $R(r) = J_m(\sqrt{\lambda_{mn}} t)$, for $\lambda = \lambda_{mn}$; cf. (10.129). Hence one obtains the characteristic functions

$$J_m(\sqrt{\lambda_{mn}} r) \cos m\theta, \qquad J_m(\sqrt{\lambda_{mn}} r) \sin m\theta,$$

and linear combinations thereof. It can be shown that these characteristic functions form a complete orthogonal system for the circle $r \le 1$.

5. Using the results of Problem 4, determine the normal modes for the vibrations of a circular membrane; that is, find normal modes for the equation

$$\frac{\partial^2 u}{\partial t^2} - a^2 \nabla^2 u = 0, \qquad x^2 + y^2 < 1,$$

$$u(x, y, t) = 0 \quad \text{for} \quad x^2 + y^2 = 1.$$

6. Using the results of Problem 4, determine the general solution of the heat conduction problem:

$$\frac{\partial u}{\partial t} - c^2 \nabla^2 u = 0, \qquad x^2 + y^2 < 1,$$

$$u(x, y, t) = 0 \quad \text{for} \quad x^2 + y^{2\cdot} = 1.$$

7. Show that the substitution $u = R(\rho)\Phi(\phi)\Theta(\theta)$ in the problem in spherical coordinates for the domain $\rho < 1$:

$$\nabla^2 u + \lambda u \equiv \frac{1}{\rho^2 \sin^2 \phi} \left[\sin^2 \phi \frac{\partial}{\partial \rho} \left(\rho^2 \frac{\partial u}{\partial \rho} \right) + \sin \phi \frac{\partial}{\partial \phi} \left(\sin \phi \frac{\partial u}{\partial \phi} \right) + \frac{\partial^2 u}{\partial \theta^2} \right] + \lambda u = 0,$$

$$u(\rho, \phi, \theta) = 0 \quad \text{for} \quad \rho = 1,$$

leads to the separate Sturm-Liouville problems:

$$(\rho^2 R')' + (\lambda \rho^2 - \alpha) R = 0, \qquad R = 0 \quad \text{for} \quad \rho = 1;$$

$$(\sin \phi \, \Phi')' + (\alpha \sin \phi - \beta \csc \phi) \Phi = 0, \qquad \Theta'' + \beta \Theta = 0.$$

Here α, β, and λ are characteristic values to be determined. The condition that u be continuous throughout the sphere requires Θ to have period 2π, so that $\beta = k^2$ ($k = 0, 1, 2, \ldots$) and $\Theta_k(\theta)$ is a linear combination of $\cos k\theta$ and $\sin k\theta$. When $\beta = k^2$, it can be shown that continuous solutions of the second problem for $0 \le \phi \le \pi$ are obtainable only when $\alpha = n(n + 1)$, $k = 0, 1, \ldots, n$, and Φ is a constant times $P_{n,k}(\cos \phi)$, where

$$P_{n,k}(x) = (1 - x^2)^{\frac{1}{2}n} \frac{d^k}{dx^k} P_n(x)$$

and $P_n(x)$ is the nth Legendre polynomial. When $\alpha = n(n + 1)$ ($n = 0, 1, 2, \ldots$), the first problem has a solution continuous for $\rho = 0$ only when λ is one of the roots $\lambda_{n+\frac{1}{2},1}, \lambda_{n+\frac{1}{2},2}, \ldots$ of the function $J_{n+\frac{1}{2}}(\sqrt{x})$, where $J_{n+\frac{1}{2}}(x)$ is the Bessel function of order $n + \frac{1}{2}$; for each such λ the solution is a constant times $\rho^{-\frac{1}{2}}J_{n+\frac{1}{2}}(\sqrt{\lambda}\rho)$. Thus one obtains the characteristic functions $\rho^{-\frac{1}{2}}J_{n+\frac{1}{2}}(\sqrt{\lambda}\rho)P_{n,k}(\cos \phi)\cos k\theta$, $\rho^{-\frac{1}{2}}J_{n+\frac{1}{2}}(\sqrt{\lambda}\rho)P_{n,k}(\cos \phi)\sin k\theta$, where $n = 0, 1, 2, \ldots, k = 0, 1, \ldots, n$, and λ is chosen as above for each n. It can be shown that these functions form a complete orthogonal system for the spherical region $\rho \le 1$.

8. Show that the wave equation (10.136) for the infinite interval $-\infty < x < \infty$ has a continuous spectrum and find the characteristic functions for the bounded normal modes.

10.16   NUMERICAL METHODS

For problems with variable coefficients or problems in two or three dimensions concerning regions of inappropriate shape, the methods described before will in general fail to produce solutions in a form suitable for numerical applications. Similar remarks apply to classes of differential equations more general than those considered here, in particular nonlinear equations. Although the theoretical aspects of the subject are highly developed and one can often establish existence of solutions, this is not always sufficient for the needs of physics. (Such an existence theorem is the famous Cauchy-Kowaleski theorem for initial value problems; see

Chapter 3 of the second book by John listed at the end of the chapter.)

Accordingly, a variety of numerical methods have been devised for explicit determination of solutions satisfying given boundary conditions and initial conditions. We consider briefly some of these methods.

The first method consists simply in *a reversal of the limit process of Section 10.5.* We replace the derivative $\partial^2 u/\partial x^2$ by the difference expression

$$\frac{u_{\sigma+1} - 2u_{\sigma} + u_{\sigma-1}}{(\Delta x)^2},$$

where $u_{\sigma} = u(x_{\sigma})$. From the differential equation

$$\rho(x)\frac{\partial^2 u}{\partial t^2} + H(x)\frac{\partial u}{\partial t} - K^2\frac{\partial^2 u}{\partial x^2} = F(x, t) \tag{10.138}$$

we are thus led to the system of equations

$$m_{\sigma}\frac{d^2 u_{\sigma}}{dt^2} + h_{\sigma}\frac{du_{\sigma}}{dt} - k^2(u_{\sigma+1} - 2u_{\sigma} + u_{\sigma-1}) = F_{\sigma}(t), \tag{10.139}$$

where $\sigma = 1,\ldots,N$ and

$$m_{\sigma} = \rho(x_{\sigma})\,\Delta x, \qquad h_{\sigma} = H(x_{\sigma})\,\Delta x, \qquad k^2 = \frac{K^2}{\Delta x}, \qquad F_{\sigma}(t) = F(x_{\sigma}, t)\,\Delta x.$$

$$\tag{10.140}$$

Equations (10.139) can be handled completely by the methods of Section 10.1. The tools required are basically algebraic, in particular the solution of simultaneous equations. In order that (10.139) be an accurate approximation to (10.138) it is necessary that N be large; this makes the algebraic problems far from trivial, at least as far as time requirements are concerned.

Initial value problems. If one seeks a particular solution of (10.138) satisfying given initial conditions and boundary conditions (values of u_0 and u_{N+1}), one can set up the approximating equations (10.139) and solve these numerically (see Chapter 8 of the book by Isaacson and Keller listed at the end of the chapter). These measures are even more appropriate if nonlinearity is present, for example, if $\partial^2 u/\partial x^2$ is replaced by its square.

Characteristic value problems. Determination of normal modes for a wave equation or heat equation obtained from (10.138) leads in general to a Sturm-Liouville problem (10.127). This can be attacked by considering the approximating problem (10.139), for which determination of normal modes is an algebraic problem; one can also use *difference* equations, as in Problems 5 to 10 following Section 10.4. A *variational* method is also helpful; this is described in Section 10.17.

One can treat the problem as an initial value problem, in the following way: To solve the equations:

$$A''(x) + \lambda\rho(x)A(x) = 0, \qquad A(0) = A(1) = 0,$$

one constructs particular solutions of the initial value problem: $A(0) = 0, A'(0) = 1$ for different values of λ. As λ is gradually increased, the solutions vary in a simple manner, and one can by trial and error determine solutions for which the condition $A(1) = 0$ is satisfied. These are precisely the characteristic functions sought.

Equilibrium problems. The equilibrium problem for (10.138) is solved in all generality in Section 10.11; the only difficult step is an integration, which might have to be carried out numerically as in Section 4.2. Another way of writing the solutions, with the aid of a Green's function, is explained in Section 10.18.

Problems in two dimensions. If $\partial^2 u/\partial x^2$ is replaced by a two-dimensional Laplacian $\nabla^2 u$ in (10.138), so that one has a problem for $u(x, y, t)$ in a region R of the xy-plane, an approximating system analogous to (10.139) can be devised. If R is a rectangle $a \leq x \leq b$, $c \leq y \leq d$, one can divide R into squares (if the sides of R are commensurable) of side h. We then consider the values of u only at the corners of the squares. At each such corner (x, y) the Laplacian is computed approximately (Problem 5 following Section 6.21) as the expression

$$\frac{u(x + h, y) + u(x, y + h) + u(x - h, y) + u(x, y - h) - 4u(x, y)}{h^2}.$$

A system of equations analogous to (10.138) is obtained. If R is not a rectangle, one can approximate R by a figure pieced together of rectangles and proceed similarly.

The statements concerning the *initial value problem* above can now be repeated without change. The *characteristic value problem* can also be replaced by an approximating algebraic problem in the same way; the variational approach of Section 10.17 below is also useful.

The *equilibrium problem* can be attacked numerically by considering the approximating system of equations in variables u_σ as above. One has then N simultaneous linear equations; if N is large, these may be far from simple to handle. One can also regard the equilibrium problem as a special case of a *heat equation*: $u_t - K^2 \nabla^2 u = F(x, y)$, with u given on the boundary of R, for all solutions of the heat equation tend exponentially to the equilibrium solution. One can assign arbitrary *initial* values and obtain a particular solution; for t large this will approximate the equilibrium solution sought. The variational methods of Section 10.17 are also of use for the equilibrium problem.

Most of the remarks made can be generalized to problems in *three dimensions*.

The accuracy of the approximate methods described has been investigated, and in general the processes described can be carried out to yield the accuracy desired for the solutions. Some details of this are given in the books of Tamarkin and Feller, Isaacson and Keller, and John listed at the end of the chapter. It is worth remarking that the crucial step of replacing (10.138) by (10.139) can be regarded as the *replacement of one physical model by another*. According to modern physics, both models are an oversimplification of what is observed in nature. If either one serves to describe the phenomena concerned with sufficient accuracy, then it can be regarded as a useful one.

10.17 VARIATIONAL METHODS

The equilibrium solution for systems (10.139) and the analogous ones in two and three dimensions can be considered as problems of minimizing a function $\phi(u_1, \ldots, u_N)$. For, as was remarked in Section 10.4, equations (10.139) can be written in the form

$$m_\sigma \frac{d^2 u_\sigma}{dt^2} + h_\sigma \frac{du_\sigma}{dt} + \frac{\partial V}{\partial u_\sigma} = F_\sigma(t) \qquad (\sigma = 1, \ldots, N). \tag{10.141}$$

The equilibrium problem is then the problem

$$\frac{\partial V}{\partial u_\sigma} = F_\sigma, \tag{10.142}$$

where the F_σ are constants. The end-values u_0 and u_{N+1} are also given as constants; we can consider them to be 0 by modifying the definition of F_1 and F_N. If we now let

$$\phi(u_1, \ldots, u_N) = V(u_1, \ldots, u_N) - (F_1 u_1 + \cdots + F_N u_N), \tag{10.143}$$

then (10.142) is simply the condition that

$$\frac{\partial \phi}{\partial u_\sigma} = 0 \qquad (\sigma = 1, \ldots, N). \tag{10.144}$$

For (10.139) we have

$$\phi = k^2 \left[u_1^2 + \cdots + u_N^2 - u_1 u_2 - u_2 u_3 - \cdots - u_{N-1} u_N \right]$$
$$- F_1 u_1 - \cdots - F_N u_N, \tag{10.145}$$

and we can verify that (10.144) has precisely one solution $u_1^*, \ldots, u_N^*$ and that this critical point is a minimum point is also easily verified (Problem 6 following Section 10.18). A similar statement applies to the analogous problems in two or three dimensions. Indeed, the physical picture that leads to equations of form (10.141) is almost always that of a system of particles capable of an equilibrium state, at which the potential energy V has its smallest value; equations (10.141) then describe the forced oscillations about this equilibrium state. When the applied forces are *constant*, the potential energy V is replaced by a modified one ϕ; the equations (10.141) then become

$$m_\sigma \frac{d^2 u_\sigma}{dt^2} + h_\sigma \frac{du_\sigma}{dt} + \frac{\partial \phi}{\partial u_\sigma} = 0; \tag{10.146}$$

the new equilibrium state is now the minimum of ϕ.

By appropriate passage to the limit it can be shown that the equilibrium problem for (10.138) is equivalent to minimizing the expression

$$\Phi_u = \int_0^L \left[\tfrac{1}{2} K^2 \{ u'(x) \}^2 - F(x) u \right] dx. \tag{10.147}$$

The expression Φ_u is a *functional*, that is, an expression whose value depends on the *function* $u(x)$ chosen. It can then be shown that the function $u^*(x)$ that satisfies the equilibrium conditions:

$$-K^2 u''(x) = F(x), \qquad u(0) = 0, \qquad u(L) = 0, \qquad (10.148)$$

assigns to Φ_u its smallest value attainable among all smooth functions $u(x)$ satisfying the boundary conditions (Problem 7 following Section 10.18).

The general problem of minimizing functionals such as Φ_u is the subject of the *calculus of variations*. Accordingly, methods based on minimizing (or maximizing) appropriate functionals or functions are called *variational methods*.

One can attack the problem of minimizing the functional Φ_u of (10.147) by the following procedure, due to Rayleigh and Ritz. One chooses a particular function $u(x)$ depending linearly on several arbitrary constants:

$$u = c_1 u_1(x) + \cdots + c_n u_n(x). \qquad (10.149)$$

The functions $u_1(x), \ldots, u_n(x)$ are chosen to satisfy the boundary conditions, that is, to be 0 at $x = 0$ and $x = L$; they are otherwise chosen as desired, though the effectiveness of the method depends greatly on the skill with which they are chosen. On substituting the expression (10.149) in (10.147), one obtains a function P whose value depends only on the constants $c_1, \ldots, c_n$. Because of the form of Φ_u, $P(c_1, \ldots, c_n)$ is also a quadratic expression, and its minimum is the unique solution of the equations

$$\frac{\partial P}{\partial c_1} = 0, \ldots, \frac{\partial P}{\partial c_n} = 0. \qquad (10.150)$$

These are simultaneous linear equations. On solving for $c_1, c_2, \ldots, c_n$, one has found a function (10.149) that gives Φ_u a smaller value than that given by certain competing functions. If the class of competing functions is large enough, one can expect the function $u(x)$ found to be close to the true minimum of Φ_u.

The functions $u_j(x)$ can be chosen as piecewise linear or piecewise polynomial functions. This is the basis of the widely used *finite element method*.

For the equilibrium problem in one dimension the procedure described is not needed, for one can solve (10.105) explicitly as in Section 10.11. However, for problems in two and three dimensions, explicit solution is usually difficult (cf. Section 10.14), and the Rayleigh-Ritz procedure can be of considerable value. It can be shown that the solution of the equilibrium problem $-K^2 \nabla^2 u = F$ in two and three dimensions is equivalent to minimizing the functionals

$$\int_R \int \left[\tfrac{1}{2} K^2 \{ u_x^2 + u_y^2 \} - F(x, y)u \right] dx\, dy,$$

$$\int \int_R \int \left[\tfrac{1}{2} K^2 \{ u_x^2 + u_y^2 + u_z^2 \} - F(x, y, z)u \right] dx\, dy\, dx, \qquad (10.151)$$

respectively.

Variational methods can also be applied to the determination of characteristic values and functions. For example, the characteristic value problem for the normal modes of (10.139), with $h_\sigma = 0$ and $F_\sigma = 0$, is the problem

$$\lambda^2 m_\sigma A_\sigma + k^2 (A_{\sigma+1} - 2A_\sigma + A_{\sigma-1}) = 0 \qquad (\sigma = 1,\dots,N), \qquad (10.152)$$

where $A_0 = A_{N+1} = 0$. These are the equations for minimizing the function

$$V(A_1,\dots,A_N) = k^2 \left(A_1^2 + \cdots + A_N^2 - A_1 A_2 - \cdots - A_{N-1} A_N \right)$$

$$(10.153)$$

subject to the *side condition*

$$g(A_1,\dots,A_N) \equiv \tfrac{1}{2}\left(m_1 A_1^2 + \cdots + m_N A_N^2 - 1 \right) = 0. \qquad (10.154)$$

Indeed, the method of Lagrange multipliers for this problem (Section 2.20) gives the equations

$$\frac{\partial V}{\partial A_\sigma} - \lambda^2 \frac{\partial g}{\partial A_\sigma} = 0,$$

which are the same as (10.152). The side condition (10.154) fixes the constant of proportionality of the A's (up to a $\pm$ sign). The characteristic values $\lambda_1,\dots,\lambda_N$ correspond to critical points of V on the "ellipsoid" defined by (10.154). In general, two of the λ's correspond to the absolute minimum and maximum of V, when (10.154) holds.

Again a limit process leads to a variational formulation of the characteristic value problem for the continuous medium. For the equation

$$\rho(x,y)\frac{\partial^2 u}{\partial t^2} - K^2 \nabla^2 A = 0,$$

the characteristic value problem concerns the "critical points" of the functional

$$Q_A = \int_R \int \tfrac{1}{2} K^2 \left(A_x^2 + A_y^2 \right) dx\, dy,$$

subject to the side condition

$$\int_R \int \rho [A(x,y)]^2 \, dx\, dy = 1.$$

The Rayleigh-Ritz method is applicable here in the same way as above.

For further information on this topic, one is referred to the article and books by Courant, Ciarlet, Gould, and Kantorovich and Krylov listed at the end of the chapter.

10.18   PARTIAL DIFFERENTIAL EQUATIONS AND INTEGRAL EQUATIONS

Given the equilibrium problem

$$-k^2 (u_{\sigma+1} - 2u_\sigma + u_{\sigma-1}) = F_\sigma \qquad (\sigma = 1,\dots,N), \qquad u_0 = u_{N+1} = 0,$$

$$(10.155)$$

one could obtain the solution by the following method. One could first solve the problem for which $F_1 = 1$ and $F_2 = F_3 = F_4 = \cdots = 0$; let the solution be $u_\sigma = g_{\sigma,1}$. One could then solve for $F_2 = 1$ and all other $F_\sigma = 0$, obtaining $g_{\sigma,2}$; in general $u_\sigma = g_{\sigma,\mu}$ is the solution when $F_\mu = 1$ and $F_\sigma = 0$ for $\sigma \neq \mu$. The linear combination

$$u_\sigma = F_1 g_{\sigma,1} + F_2 g_{\sigma,2} + \cdots + F_N g_{\sigma,N} \tag{10.156}$$

is then the desired solution of (10.155); for substitution of u_σ in the left-hand side of the first equation (10.155) gives F_1, since $g_{\sigma,1}$ gives 1 and all other g's give 0; a similar reasoning holds for the other equations. Thus the effect of all the F_σ can be built up by *superposition* of unit forces.

By a limit process a similar result is obtained for the problem:

$$u''(x) = -F(x), \qquad u(0) = u(L) = 0. \tag{10.157}$$

One finds

$$u(x) = \int_0^L g(x, s) F(s)\, ds, \tag{10.158}$$

where the $g(x, s)$ are the solutions for a force F "concentrated at a point s." The function $g(x, s)$ is called the *Green's function* for (10.157); it is 0 when $x = 0$ and $x = L$, has the value $s(L - s)/L$ when $x = s$, and is linear in x between these values. Accordingly, $g(x, s)$ has a "corner" at $x = s$, due to the concentrated force at this point, whereas $\partial^2 g/\partial x^2 = 0$ otherwise.

Similar results hold for quite general nonhomogeneous linear equations. In particular, a Green's function $g(x, y; r, s)$ can be found for the problem (Poisson equation):

$$\nabla^2 u = -F(x, y) \quad \text{inside } R,$$
$$u(x, y) = 0 \quad \text{on boundary of } R, \tag{10.159}$$

for a general region R in the plane. The solutions of (10.159) are then given by a formula:

$$u(x, y) = \int_R \!\!\int g(x, y; r, s) F(r, s)\, dr\, ds. \tag{10.160}$$

For each (r, s) the function g satisfies the equation $\nabla^2 g = 0$ except at $x = s$, $y = r$, where it has a discontinuity corresponding to a "point load." Also $g(x, y; r, s) = 0$ when (x, y) is on the boundary of R; therefore u is given as a "linear combination" of functions all 0 on the boundary of R and is therefore itself 0 on the boundary. (See Section 5.17.)

In order to solve the characteristic value problem

$$\nabla^2 u + \lambda u = 0 \quad \text{in } R,$$
$$u = 0 \quad \text{on the boundary of } R, \tag{10.161}$$

one can rewrite the problem in form (10.159): $\nabla^2 u = -\lambda u$; therefore

$$u(x, y) = \lambda \int_R \!\!\int g(x, y; r, s) u(r, s)\, dr\, ds. \tag{10.162}$$

This gives an implicit equation for u, the crucial operation being integration. The equation is called an *integral equation*.

A variety of other problems in partial differential equations can be restated as integral equations. A number of methods are available for obtaining solutions of integral equations, and they must be considered among the most powerful ways of attacking partial differential equations. The following features are of great importance: The theory of integral equations is much more *unified* than that of differential equations, problems in one, two, or three dimensions being treated in the same way; the treatment of *boundary values* is simpler, for example, in (10.162) the boundary condition on u is automatically taken care of, since $g = 0$ on the boundary; the methods of solution are much better adaptable to *nonlinear* problems.

For a full discussion of integral equations and their applications refer to the books of Tamarkin and Feller, Frank and von Mises, Courant, Kellogg, and Mikhlin listed at the end of the chapter.

PROBLEMS

1. Let the equilibrium problem $\nabla^2 u(x, y) = 0$ be given for the square $0 \le x \le 3$, $0 \le y \le 3$, with boundary values $u = x^2$ for $y = 0$, $u = x^2 - 9$ for $y = 3$, $u = -y^2$ for $x = 0$, $u = 9 - y^2$ for $x = 3$. Obtain the solution by considering the heat equation $u_t - \nabla^2 u = 0$. Use only integer values of x, y so that only four points $(1, 1)$, $(2, 1)$, $(1, 2)$, $(2, 2)$ inside the rectangle are concerned. Let u_1, u_2, u_3, u_4, respectively, be the four values of u at these points. Using the given boundary values, show that the approximating equations are

$$u_1'(t) - (u_2 + u_3 - 4u_1) = 0, \qquad u_2'(t) - (12 + u_4 + u_1 - 4u_2) = 0,$$

$$u_3'(t) - (u_4 - 12 + u_1 - 4u_3) = 0, \qquad u_4'(t) - (u_3 + u_2 - 4u_4) = 0.$$

Replace by difference equations in t: $\Delta u_1 = (u_2 + u_3 - 4u_1)\,\Delta t, \ldots$, where $\Delta u_i = u_i(t + \Delta t) - u_i(t)$. These equations can be used to obtain $u_1, \ldots, u_4$ numerically at $t_0 + \Delta t, t_0 + 2\Delta t, \ldots$ from given initial values at t_0 (Euler method). Take $t_0 = 0$, $\Delta t = 0.1$, and $u_i(0) = 1$ for $i = 1, \ldots, 4$ to find $u_i(1)$. Verify that the values found are close to the equilibrium values: $u_1 = 0$, $u_2 = 3$, $u_3 = -3$, $u_4 = 0$.

2. The method of *relaxation* or *Liebmann's method*, as applied to Problem 1, consists in choosing initial values of u_1, u_2, u_3, u_4, then correcting each one in turn by replacing it by the average of the *four neighboring values*. Thus at $(1, 1)$ the value u_1 would be replaced by the average of u_2, u_3, -1, and 1. The value u_2 at $(2, 1)$ would then be replaced by the average of 8, u_4, u_1 (new value), and 4. Apply this process repeatedly, starting with $u_1 = u_2 = u_3 = u_4 = 1$ and show that the corrected values gradually approach the equilibrium sought. This technique is discussed in the two books of Southwell listed at the end of the chapter.

3. Let the wave equation problem $u_{tt} - \nabla^2 u = 0$, $u(x, y, t) = 0$ on the boundary, be given for the square of Problem 1. Determine the resonant frequencies by using difference expressions for $\nabla^2 u$ as in Problem 1, so that one has the equations:

$$u_1''(t) - (u_2 + u_3 - 4u_1) = 0, \ldots .$$

The exact frequencies are found as in Section 10.14 to be $\frac{1}{3}\pi(m^2 + n^2)^{\frac{1}{2}}$ ($m = 1, 2, \ldots, n = 1, 2, \ldots$). Show that the four lowest frequencies are fairly well approximated.

4. Let (10.138) be the wave equation $u_{tt} - a^2 u_{xx} = 0$ for the interval $0 < x < \pi$ as in Section 10.7. The corresponding approximating system (10.139) was considered in Problem 7 following Section 10.4. In the notation used here the characteristic values and functions found were

$$\lambda_n = \frac{2a(N + 1)}{\pi} \sin\frac{n\pi}{2(N + 1)}, \qquad A_n(x_\sigma) = \sin(nx_\sigma),$$

where $\sigma = 0, 1, \ldots, N + 1$, $n = 1, \ldots, N$; compare with the exact solutions of the wave equation. Show that for each fixed n, $\lambda_n \to an$ as $N \to \infty$.

5. Study the behavior of the solutions of the *initial value* problem:

$$u''(x) + \lambda u(x) = 0, \qquad u(0) = 0, \qquad u'(0) = 1,$$

as λ increases from 0 to ∞; note in particular the appearance of values of λ for which the condition $u(1) = 0$ is satisfied. It can be shown that the same qualitative picture holds for the general Sturm-Liouville problem of Section 10.13.

6. Show that the function ϕ defined by (10.145) has precisely one critical point at which ϕ takes on its absolute minimum. [Hint: Show that by proper choice of the constants $\alpha_1, \ldots, \alpha_N$, the substitution

$$w_1 = u_1 - u_2 + \alpha_1, \qquad w_2 = u_2 - u_3 + \alpha_2, \ldots,$$
$$w_{N-1} = u_{N-1} - u_N + \alpha_{N-1}, \qquad w_N = u_N + \alpha_N$$

transforms ϕ into an expression

$$\tfrac{1}{2}k^2 \left[w_1^2 + \cdots + w_N^2 + (w_1 + \cdots + w_N - \alpha_1 - \cdots - \alpha_N)^2 \right] + \text{const}.$$

This shows that $\phi \to \infty$ as $w_1^2 + \cdots + w_N^2 \to \infty$, so that ϕ has at least one critical point that gives ϕ its absolute minimum. The equations in $u_1, \ldots, u_N$ for the critical point are simultaneous linear equations. The equations have a unique solution, if the corresponding homogeneous equations (all F_σ zero) have a unique solution; in the homogeneous case $u_1 = u_2 = \cdots = u_N = 0$ is one critical point; if $u_1^*, \ldots, u_N^*$ were a second one, then $\partial \phi / \partial u_\sigma$ would be 0 for all σ when $u_1 = u_1^* t, \ldots, u_N = u_N^* t$ and $-\infty < t < \infty$. This contradicts the fact that $\phi \to \infty$ as $w_1^2 + \cdots + w_N^2 \to \infty$. The uniqueness of the critical point can also be established by using difference equations, as in Problem 5 following Section 10.4.]

7. Prove that the function Φ_u defined by (10.147) attains its minimum value, among smooth functions $u(x)$ satisfying the boundary conditions $u(0) = u(L) = 0$, when u is the solution of the equation $-K^2 u''(x) = F(x)$. [Hint: Take $L = \pi$ for convenience. Then express the integral in terms of Fourier sine coefficients of $u(x)$, $u'(x)$, and $F(x)$, using Theorem 14 of Section 7.13. This gives a separate minimum problem for each n, which is solved precisely when $-K^2 u'' = F(x)$.]

8. **a)** Determine the function $u(x)$ that minimizes

$$\int_0^1 \left\{ [u'(x)]^2 + 6xu \right\} dx,$$

if $u(0) = u(1) = 0$.

b) Use the Rayleigh-Ritz procedure to solve the problem of part (a), using as trial functions the functions

$$u = c_1(x - x^2) + c_2 \sin 2\pi x.$$

9. Verify that the Green's function for (10.157), as described in the text, is the following function when $L = 1$:

$$g(x, s) = x(1 - s), \quad 0 \leq x \leq s \leq 1; \quad g(x, s) \equiv s(1 - x), \quad 0 \leq s \leq x \leq 1.$$

Verify that the function

$$u(x) = \int_0^1 g(x, s)s \, ds$$

solves the problem $u''(x) = -x$, $u(0) = u(1) = 0$.

Suggested References

Bateman, H., Partial Differential Equations of Mathematical Physics. New York: Dover, 1944.

Churchill, R. V., and Brown, James W., Fourier Series and Boundary Value Problem, 4th ed. New York: McGraw-Hill, 1987.

Churchill, R. V., Operational Mathematics, 3rd ed. New York: McGraw-Hill, 1972.

Ciarlet, P. G., Numerical Analysis of the Finite Element Method. Montreal: Les Presses de l'Université de Montréal, 1976.

Courant, R., Advanced Methods in Applied Mathematics. Lithoprinted notes of lectures at New York University, 1941.

Courant, R., "Variational methods for the solution of problems of equilibrium and vibrations," Bulletin of the American Mathematical Society, Vol. 49, pp. 1–23. New York: American Mathematical Society, 1943.

Courant, R., and D. Hilbert, Methods of Mathematical Physics. Vol. 1, New York: Interscience, 1953. Vol. 2, New York: Interscience, 1961.

Doetsch, G., Theory and Application of the Laplace Transformation, transl. by W. Nader. Berlin: Springer, 1974.

Frank, P., and R. von Mises, Die Differentialgleichungen und Integralgleichungen der Mechanik und Physik (2 vols.), 2nd ed. New York: Dover Publications, 1961.

Goldstein, Herbert, Classical Mechanics. Cambridge: Addison-Wesley Press, 1950.

Gould, S. H., Variational Methods for Eigenvalue Problems, 2nd ed. Toronto: University of Toronto Press, 1966.

Hellwig, Günther, Differential Operators of Mathematical Physics. Reading, Mass.: Addison-Wesley, 1964.

Hildebrand, F. B., Finite-Difference Equations and Simulations. Englewood Cliffs, N. J.: Prentice-Hall 1968.

Isaacson, E., and H. B. Keller, Analysis of Numerical Methods. New York: Wiley, 1966.

John, F., Lectures on Advanced Numerical Analysis. New York: Gordon and Breach, 1967.

John, F., Partial Differential Equations, 3rd ed. New York: Springer-Verlag, 1978.

Kamke, E., Differentialgleichungen reeller Funktionen. Leipzig: Akademische Verlagsgesellschaft, 1930.

Kantorovich, L. V., and V. I. Krylov, Approximate Methods of Higher Analysis, 2nd ed., transl. by C. D. Benster. Groningen: P. Noordhoff Ltd., 1958.

Kármán, T. V., and M. A. Biot, Mathematical Methods in Engineering. New York: McGraw-Hill, 1940.

Kellogg, O. D., Foundations of Potential Theory. New York: Springer, Berlin, 1967.

Mikhlin, S. G., Integral Equations, transl. by A. H. Armstrong, 2nd ed. New York: Pergamon Press, 1964.

Lord Rayleigh, The Theory of Sound (2 vols.), 2nd ed. New York: Dover Publications, 1945.

Sneddon, I. N., Fourier Transforms. New York: McGraw-Hill, 1951.

Sommerfeld, A., Partial Differential Equations in Physics (transl. by E. G. Straus). New York: Academic Press, 1949.

Southwell, R. V., Relaxation Methods in Engineering Science. Oxford: Oxford University Press, 1946.

Southwell, R. V., Relaxation Methods in Theoretical Physics. Oxford: Oxford University Press, 1946.

Tamarkin, J. D., and W. Feller, Partial Differential Equations. Mimeographed notes of lectures at Brown University, 1941.

Titchmarsh, E. C., Eigenfunction Expansions Associated with Second-order Differential Equations, 2nd ed. Oxford: Oxford University Press, 1962.

Titchmarsh, E. C., Theory of Fourier Integrals. Oxford: Oxford University Press, 1937.

Wiener, N., The Fourier Integral. Cambridge: Cambridge University Press, 1933.

Answers
to Problems

Chapter 1

Section 1.4, page 15
1. b) $|\overrightarrow{P_1 P_2}| = \sqrt{3}$, $|\overrightarrow{P_1 P_3}| = \sqrt{29}$, $|\overrightarrow{P_2 P_3}| = 3\sqrt{2}$;
c) $\cos^{-1} 7/\sqrt{87} = 0.72$, $\cos^{-1}(-2\sqrt{6}/9) = 2.15$, $\cos^{-1}(11\sqrt{58}/87) = 0.27$;
d) $\sqrt{38}/2$; **e)** $\sqrt{38/3}$; **f)** $(3/2, 1/2, 5/2)$; **g)** $(4/3, 2, 3)$.
2. a) 0; **b)** $-\mathbf{i} + 7\mathbf{j} + 4\mathbf{k}$; **c)** $\sqrt{6}$; **d)** $\pi/2$;
e) $-38\mathbf{i} + 14\mathbf{j} + 26\mathbf{k}$; **f)** $42, -42$; **g)** 0; **h)** $1/\sqrt{6}, -1/\sqrt{6}, 2/\sqrt{6}$; **i)** $\sqrt{30}/5$.
3. b) $3x + 2y - 5z + 7 = 0$. **4. b)** $20/3$. **5. a)** dep., **b)** dep., **c)** dep.
6. a) $x = 2 + t, y = 1 + t, z = 5t$; **b)** $x = 1 - 5t, y = 1 + 2t, z = 2 + 3t$;
c) $x = 5t, y = -t, z = t$; **d)** $x = 1 - 6t, y = 2 - 3t, z = 2 + 3t$.
7. a) $y = 2x$; **b)** $x - 5y + 4z + 1 = 0$; **c)** $8x - 11y + 5z + 9 = 0$.
9. a) $(4, 5, 7), (7, 8, 11)$; **b)** $(10 + 9k, 11 + 9k, 15 + 12k), k = \sqrt{34}/51$.
10. a) -17; **b)** -7; **c)** 2; **d)** -1; **e)** -36; **f)** $abc(b - a)(c - a)(c - b)$.
11. a) negative; **b)** neither.
13. a) $x = 2, y = 1$; **b)** $x = 1, y = 1$; **c)** $x = 3/2, y = -2, z = -5/2$;
d) $x = t, y = -t, z = -2t, -\infty < t < \infty$.

Section 1.6, page 20
1. a) Number of rows first: $A: 2$ and 1, $F: 2$ and 3, $H: 1$ and 3, $L: 3$ and 3, $P: 3$ and 2;
b) $a_{11} = 1, a_{21} = 3, c_{21} = 4, c_{22} = 1, d_{12} = -1, e_{21} = 2, f_{11} = 1, g_{23} = -1, g_{21} = -1$, $h_{12} = 0, m_{23} = 1$;
c) $C: (2, 3)$ and $(4, 1)$, $G: (3, 1, 4,)$ and $(-1, 0, -1)$, $L: (3, 1, 0), (2, 5, 6)$ and $(1, 4, 3)$, $P: (2, 2,), (-1, -1), (3, 3)$;
d) $D:$ $\text{col}(1, 2)$ and $\text{col}(-1, 0)$, $F:$ $\text{col}(1, 2), \text{col}(4, 0)$ and $\text{col}(5, 7)$, $L:$ $\text{col}(3, 2, 1)$, $\text{col}(1, 5, 4)$ and $\text{col}(0, 6, 3)$, $N:$ $\text{col}(1, 0, 7)$ and $\text{col}(4, 3, 1)$.

2. a) $\begin{bmatrix} 3 \\ 3 \end{bmatrix}$; **b)** $\begin{bmatrix} 3 & 2 \\ 6 & 1 \end{bmatrix}$; **c)** meaningless; **d)** $\begin{bmatrix} 5 & 0 & 0 \\ 3 & 7 & 7 \\ 4 & 6 & 2 \end{bmatrix}$;

e) $\begin{bmatrix} -1 & 2 \\ 1 & 4 \\ 4 & -2 \end{bmatrix}$; **f)** $\begin{bmatrix} 2 & -3 & -1 \\ -3 & 0 & -8 \end{bmatrix}$; **g)** $\begin{bmatrix} 10 & 15 \\ 20 & 5 \end{bmatrix}$; **h)** $\begin{bmatrix} 2 & 4 \\ 4 & 8 \end{bmatrix}$;

i) $\begin{bmatrix} 7 & 2 \\ 14 & 12 \end{bmatrix}$; **j)** $\begin{bmatrix} 4 & 3 \\ 8 & -2 \end{bmatrix}$; **k)** meaningless.

3. a) $D - C = \begin{bmatrix} -1 & -4 \\ -2 & -1 \end{bmatrix}$; **b)** $(\frac{1}{5})(F - G) = \begin{bmatrix} -\frac{2}{5} & \frac{3}{5} & \frac{1}{5} \\ \frac{3}{5} & 0 & \frac{8}{5} \end{bmatrix}$.

4. a) $X = \frac{1}{2}(N + P) = \begin{bmatrix} \frac{3}{2} & 3 \\ -\frac{1}{2} & 1 \\ 5 & 2 \end{bmatrix}$, $Y = \frac{1}{2}(N - P) = \begin{bmatrix} -\frac{1}{2} & 1 \\ \frac{1}{2} & 2 \\ 2 & -1 \end{bmatrix}$;

b) $X = 2L - 3M = \begin{bmatrix} 0 & 5 & 0 \\ 1 & 4 & 9 \\ -7 & 2 & 9 \end{bmatrix}$, $Y = L - 2M = \begin{bmatrix} -1 & 3 & 0 \\ 0 & 1 & 4 \\ -5 & 0 & 5 \end{bmatrix}$.

Section 1.7, page 25

1. a) meaningless; **b)** $\begin{bmatrix} 11 \\ 7 \end{bmatrix}$; **c)** meaningless;

d) $\begin{bmatrix} 8 & -2 \\ 6 & -4 \end{bmatrix}$ and $\begin{bmatrix} -2 & 2 \\ 4 & 6 \end{bmatrix}$; **e)** $\begin{bmatrix} 8 & 16 \\ 6 & 12 \end{bmatrix}$ and $\begin{bmatrix} 10 & 5 \\ 20 & 10 \end{bmatrix}$;

f) A; **g)** L; **h)** undefined; **i)** O; **j)** O; **k)** $\begin{bmatrix} 3 & 15 \\ 44 & 29 \\ 22 & 19 \end{bmatrix}$, **l)** $\begin{bmatrix} 10 & 5 & 15 \\ 6 & 3 & 9 \\ 2 & 1 & 3 \end{bmatrix}$;

m) $(7, 10, 5)$; **n)** (18); **o)** N (O is 3×2); **p)** $\begin{bmatrix} 5 & 10 \\ 10 & 20 \end{bmatrix} = 5E$;
q) $25E$; **r)** $125E$; **s)** meaningless.

2. a) $\begin{bmatrix} 3 & 1 & 9 & 1 & 9 \\ 5 & 2 & 16 & 2 & 16 \end{bmatrix}$; **b)** $\begin{bmatrix} 14 & 9 & 9 & 14 & 9 \\ 7 & 4 & 4 & 7 & 4 \\ 10 & 5 & 5 & 10 & 5 \end{bmatrix}$.

3. a) $y_1 = 17x_1 + x_2, y_2 = 31x_1 + 7x_2$; **b)** $y_1 = x_2 - 3x_3, y_2 = 7x_1 + 13x_2 - 4x_3$.
6. It is true precisely when $AB = BA$.

Section 1.8, page 29

1. a) $\frac{1}{2}\begin{bmatrix} 4 & -5 \\ -2 & 3 \end{bmatrix}$; **b)** $\frac{1}{17}\begin{bmatrix} 6 & -7 \\ -1 & 4 \end{bmatrix}$;

c) $\begin{bmatrix} 3 & -1 & 2 \\ -2 & 1 & -1 \\ -2 & 1 & -2 \end{bmatrix}$; **d)** $\frac{1}{2}\begin{bmatrix} 3 & 0 & -1 \\ -1 & 2 & -1 \\ -4 & 0 & 2 \end{bmatrix}$.

2. a) $\frac{1}{2}\begin{bmatrix} -5 & 4 \\ 3 & -2 \end{bmatrix}$; **b)** $\begin{bmatrix} -13 & 8 \\ 8 & -2 \end{bmatrix}$; **c)** $\frac{1}{34}\begin{bmatrix} 214 & -279 \\ -64 & 89 \end{bmatrix}$;

d) $\frac{1}{1156}\begin{bmatrix} -1714 & 1553 \\ 1458 & -1981 \end{bmatrix}$; **e)** $\frac{1}{2} \text{col}(1, 1)$; **f)** $(\frac{1}{17}) \text{col}(-43, 27)$;

g) $\dfrac{1}{17}\begin{bmatrix} -29 & 36 & 33 \\ 19 & -6 & -14 \end{bmatrix}$; **h)** $\begin{bmatrix} -18 & 9 & -13 \\ 13 & -4 & 8 \end{bmatrix}$; **i)** $(-7, 4, -6)$;

j) $\frac{1}{2}(5, 2, -3)$.

3. a) $A^2 B$; **b)** I; **c)** B^4.

8. All matrices whose inverses appear in the answers are assumed to be nonsingular.

a) $X = \frac{1}{2}(A + B)$, $Y = \frac{1}{2}(A - B)$;

b) $X = (B - I)^{-1}(BA - C)$, $Y = (B - I)^{-1}(C - A)$;

c) $X = B - A(A - C)^{-1}(B - D)$, $Y = (A - C)^{-1}(B - D)$;

d) $X = (B^{-1}A - E^{-1}D)^{-1}(B^{-1}C - E^{-1}F)$; $Y = (A^{-1}B - D^{-1}E)^{-1}(A^{-1}C - D^{-1}F)$;

e) $X = (CB^{-1} - FE^{-1})(AB^{-1} - DE^{-1})^{-1}$, $Y = (CA^{-1} - FD^{-1})(BA^{-1} - ED^{-1})^{-1}$.

Section 1.9, page 33

1. a) $x = -6$, $y = 12$, $z = 5$; **b)** $x = 3$, $y = 1$, $z = -1$.

c) $x = \frac{13}{7}$, $y = \frac{10}{7}$, $z = \frac{2}{7}$;

d) $x = 1$, $y = 1$, $z = 1$.

2. a) 21; **b)** 0.

3. a) $x = 2$, $y = 2$, $z = 3$; **b)** $x = 4$, $y = 1$, $z = 1$, $w = 1$.

4. a) $\begin{bmatrix} -1 & -1 & 1 \\ 4 & 5 & -\frac{7}{2} \\ -2 & -3 & 2 \end{bmatrix}$; **b)** $\begin{bmatrix} -1 & \frac{5}{2} & -1 \\ 0 & -\frac{3}{2} & 1 \\ 1 & -2 & 1 \end{bmatrix}$; **c)** $\dfrac{1}{7}\begin{bmatrix} 1 & -3 & 4 & 1 \\ 2 & 1 & 1 & -5 \\ -4 & 5 & -2 & 3 \\ 3 & -2 & -2 & 3 \end{bmatrix}$.

5. a) $x = (7 - t)/5$, $y = (3t - 1)/5$, $z = t$, $-\infty < t < \infty$; **b)** no solution;

c) $x = -3t$, $y = 6t$, $z = t$, $w = 8t$, $-\infty < t < \infty$;

d) $x = t - 2u$, $y = 5t - 7u$, $z = 3t$, $w = 3u$, $-\infty < t < \infty$, $-\infty < u < \infty$.

Section 1.10, page 38

Throughout, k is an arbitrary nonzero real scalar; c is an arbitrary nonzero complex scalar.

1. a) $\lambda = 1$, $k(1, -2)$, and $\lambda = 5$, $k(1, 2)$; **b)** $\lambda = 0$, $k(3, -1)$, and $\lambda = 7$, $k(1, 2)$;

c) $\lambda = 1$, $k(0, 2, 1)$; $\lambda = 2$, $k(1, 2, 0)$; $\lambda = 3$, $k(1, 1, -1)$;

d) $\lambda = 1$, $k(3, -2, -2)$; $\lambda = -1$, $k(-1, 1, 1)$; $\lambda = -2$, $k(2, -1, -2)$.

2. a) $C = \begin{bmatrix} 1 & 1 \\ -2 & 2 \end{bmatrix}$, $B = \begin{bmatrix} 1 & 0 \\ 0 & 5 \end{bmatrix}$;

b) $C = \begin{bmatrix} 3 & 1 \\ -1 & 2 \end{bmatrix}$, $B = \begin{bmatrix} 0 & 0 \\ 0 & 7 \end{bmatrix}$;

c) $C = \begin{bmatrix} 0 & 1 & 1 \\ 2 & 2 & 1 \\ 1 & 0 & -1 \end{bmatrix}$, $B = \begin{bmatrix} 1 & 0 & 0 \\ 0 & 2 & 0 \\ 0 & 0 & 3 \end{bmatrix}$;

d) $C = \begin{bmatrix} 3 & -1 & 2 \\ -2 & 1 & -1 \\ -2 & 1 & -2 \end{bmatrix}$, $B = \begin{bmatrix} 1 & 0 & 0 \\ 0 & -1 & 0 \\ 0 & 0 & -2 \end{bmatrix}$;

3. a) $\lambda = 1 + 2i$, $c(1, -2i)$ and $\lambda = 1 - 2i$, $c(1, 2i)$, $C = \begin{bmatrix} 1 & 1 \\ -2i & 2i \end{bmatrix}$,

$B = \begin{bmatrix} 1 + 2i & 0 \\ 0 & 1 - 2i \end{bmatrix}$;

b) $\lambda = 4 + i$, $c(2, -1 - i)$ and $\lambda = 4 - i$, $c(2, -1 + i)$, $C = \begin{bmatrix} 2 & 2 \\ -1 - i & -1 + i \end{bmatrix}$,

$B = \begin{bmatrix} 4 + i & 0 \\ 0 & 4 - i \end{bmatrix}$;

c) $\lambda = 0$, $c(1,0,0)$, $\lambda = i$, $c(0,2,1-i)$, $\lambda = -i$, $c(0,2,1+i)$,

$$C = \begin{bmatrix} 1 & 0 & 0 \\ 0 & 2 & 2 \\ 0 & 1-i & 1+i \end{bmatrix}, \qquad B = \begin{bmatrix} 0 & 0 & 0 \\ 0 & i & 0 \\ 0 & 0 & -i \end{bmatrix}.$$

4. a) $\lambda = 1$, all nonzero vectors;   **b)** $\lambda = 0$, all nonzero vectors;
c) $\lambda = -1$, $k(1,1)$;   **d)** $\lambda = 2$, $k(1,2,0)$ and $\lambda = 3$, $a(1,3,0) + b(0,2,1)$, with a and b not both 0.

Section 1.12, page 44

1. a) $\begin{bmatrix} 1 & 3 \\ 2 & 0 \\ 3 & 5 \end{bmatrix}$;   **b)** $\begin{bmatrix} 3 & 0 & 1 \\ 1 & 2 & 0 \end{bmatrix}$;   **c)** $\mathrm{col}\,(1,5,0,4)$;   **d)** $(1,0,7)$.

2. a) $a = 1$;   **b)** $a = 4$, $b = 8$.

3. a) $\begin{bmatrix} 5 & 2 \\ 2 & 3 \end{bmatrix}$;   **b)** $\begin{bmatrix} 7 & 1 \\ 1 & -1 \end{bmatrix}$;   **c)** $\begin{bmatrix} 1 & 2 & 3 \\ 2 & 3 & 1 \\ 3 & 1 & -1 \end{bmatrix}$;   **d)** $\begin{bmatrix} 2 & 0 & 1 \\ 0 & 1 & -2 \\ 1 & -2 & 1 \end{bmatrix}$.

Section 1.14, page 53

1. a) $(4,2,2,2)$, $(8,6,2,-2)$, $(6,4,2,0)$, $(-3,0,-3,-6)$, $(0,0,0,0) = \mathbf{0}$;
b) $(7,6,1,-4)$, $(17,12,5,-2)$, $(-2,-2,0,2)$, $(-6,-6,0,6)$;
c) $4, 24, \sqrt{14}, \sqrt{6}$.
2. b) No;   **c)** $(6,2,-1,3,6)$;   **d)** $(10,0,8,7,4)$, $(13,-1,9,5,2)$.
5. a) $\sqrt{38}$, $\sqrt{15}$, $\sqrt{15}$, $\cos^{-1}\left(-\frac{4}{15}\right)$, $\cos^{-1}(19/\sqrt{570})$, $\cos^{-1}(19/\sqrt{570})$.

Section 1.15, page 59
1. a) $(2,3)$, $(1,5)$, $(3,1)$, $(-1,2)$;
b) Kernel is $\mathbf{0}$ alone, T is one-to-one, $T(\mathbf{x}) = (2,3)$ only for $\mathbf{x} = (1,0)$;
c) Range is V^2, T maps V^2 onto V^2.
2. a) $(2,4)$, $(3,6)$, $(-1,-2)$, $(-5,-10)$;
b) Kernel is all $t(3,-2)$, T is not one-to-one, $T(\mathbf{x}) = (2,4)$ for $\mathbf{x} = (1,0) + t(3,-2)$,
$-\infty < t < \infty$;   **c)** Range is all $t(1,2)$; T does not map V^2 onto V^2.
3. a) $n = 3$, $m = 2$;   **b)** all $t(0,2,-1)$, not one-to-one,
c) Range is V^2; T maps V^3 onto V^2.
4. a) $n = 3$, $m = 2$;   **b)** All $t_1(-4,1,0) + t_2(-3,0,1)$, not one-to-one;
c) Range is all $t(1,2)$, T does not map V^3 onto V^2.
5. a) $n = 2$, $m = 3$;   **b)** All $t(1,-2)$, not one-to-one;
c) Range is all $t(1,2,-3)$, T does not map V^2 onto V^3.
6. a) $n = 2$, $m = 3$;   **b)** $\mathbf{0}$ alone, one-to-one;   **c)** Range is all $t_1(2,1,1) + t_2(1,2,2)$,
T does not map V^2 onto V^3.
7. a) $n = 3$, $m = 3$;   **b)** All $t(1,-1,-1)$, not one-to-one;
c) Range is all $t_1(3,1,5) + t_2(1,0,2)$, T does not map V^3 onto V^3.
8. a) $n = 3$, $m = 3$;   **b)** $\mathbf{0}$ alone, one-to-one;
c) Range is V^3, T maps V^3 onto V^3.
9. Angle is $\pi/4$, $|\mathbf{x}| = |T(\mathbf{x})|$.   **10.** Reflection in x-axis.
11. a) Reflection in origin;   **b)** All vectors stretched in ratio 2 to 1.

Section 1.16, page 66
1. a) 2;   **b)** 3;   **c)** 3;   **d)** 2
2. a) $h = 1$, basis $(-10/13, 4/13, 1)$;

b) $h = 0$;   **c)** $h = 1$, basis $(-5, -22, 6, 1)$;
d) $h = 3$, basis $(-1, 2, 1, 0, 0)$, $(-1, 1, 0, 1, 0)$, $(-3, 4, 0, 0, 1)$.
3. a) $(1, 3, 2)$, $(0, 0, -5)$, $(0, 1, 1)$, or e_1, e_2, e_3 $(W = V^3)$;
b) $(1, 3, 1)$, $(0, -14, 2)$;   **c)** $(1, 2, 2, 3)$, $(0, 1, 1, 2)$;
d) $(0, 1, -1, -3, 1)$, $(1, 0, 2, 1, 0)$, $(0, 0, 1, -2, 1)$.

Section 1.17, page 70
1. a) 3, basis: $1, x, x^2$;   **b)** infinite;   **c)** infinite;   **d)** 2, basis: e^x, e^{-x};
e) 3, basis: E_{11}, E_{22}, E_{33};
f) 10, basis: $E_{11}, E_{22}, E_{33}, E_{44}, E_{12} + E_{21}, E_{13} + E_{31}, E_{14} + E_{41}, E_{23} + E_{32}, E_{24} + E_{42}, E_{34} + E_{43}$;   **g)** 2, basis: $\cos x$, $\sin x$;
h) 3, basis: $1, e^x, e^{-x}$;   **i)** infinite;   **j)** infinite;   **k)** infinite;   **l)** infinite.

Chapter 2

Section 2.4, page 82
4. a) 0;   **b)** no limit;   **c)** 1;   **d)** ∞.
6. a) all (x, y), a domain;   **b)** $x^2 + y^2 > 1$, a domain;
c) $x^2 + y^2 \leq 1$, a closed region;
d) all (x, y, z) except the points of the xy-plane, an open set, not a domain.

Section 2.6, page 89
1. a) $\dfrac{\partial z}{\partial x} = \dfrac{-2xy}{(x^2 + y^2)^2}$, $\dfrac{\partial z}{\partial y} = \dfrac{x^2 - y^2}{(x^2 + y^2)^2}$;

b) $\dfrac{\partial z}{\partial x} = y^2 \cos xy$, $\dfrac{\partial z}{\partial y} = \sin xy + xy \cos xy$;

c) $\dfrac{\partial z}{\partial x} = \dfrac{3x^2 + 2xy - 2xz}{x^2 - 3z^2}$, $\dfrac{\partial z}{\partial y} = \dfrac{x^2}{x^2 - 3z^2}$;

d) $\dfrac{\partial z}{\partial x} = \dfrac{e^{x+2y}}{2\sqrt{e^{x+2y} - y^2}}$, $\dfrac{\partial z}{\partial y} = \dfrac{e^{x+2y} - y}{\sqrt{e^{x+2y} - y^2}}$.

e) $z_x = 3x(x^2 + y^2)^{1/2}$, $z_y = 3y(x^2 + y^2)^{1/2}$;
f) $z_x = [1 - (x + 2y)^2]^{-1/2}$, $z_y = 2[1 - (x + 2y)^2]^{-1/2}$;
g) $z_x = e^x(e^z + 1)^{-1}$, $z_y = 2e^y(e^z + 1)^{-1}$;
h) $z_x = -(y + z)(x + 2z)^{-1}$, $z_y = -(2xy + z^2 + xz)(2yz + xy)^{-1}$.

2. Approximately $f_x(1, 1) = \dfrac{f(2,1) - f(1,1)}{1} = 3$,

or $f_x(1, 1) = \dfrac{f(1,1) - f(0,1)}{1} = 1$,   or $f_x(1, 1) = \dfrac{f(2,1) - f(0,1)}{2} = 2$.

It can be shown that the last value is "in general" the best estimate. For $f_y(1, 1)$ the analogous formulas give the estimates $-2, -2, -2$. This topic is discussed in Problem 9 following Section 6.18.

3. a) $\left(\dfrac{\partial u}{\partial x}\right)_y = 2x$, $\left(\dfrac{\partial v}{\partial y}\right)_x = -2$;

b) $\left(\dfrac{\partial x}{\partial u}\right)_v = e^u \cos v$, $\left(\dfrac{\partial y}{\partial v}\right)_u = e^u \cos v$;

c) $\left(\dfrac{\partial x}{\partial u}\right)_y = 1, \left(\dfrac{\partial y}{\partial v}\right)_u = -\dfrac{1}{2}$;

d) $\left(\dfrac{\partial r}{\partial x}\right)_y = x(x^2 + y^2)^{-1/2}, \left(\dfrac{\partial r}{\partial \theta}\right)_x = x\sec\theta\tan\theta$.

4. a) $\dfrac{y\,dx - x\,dy}{y^2}$;   b) $\dfrac{x\,dx + y\,dy}{x^2 + y^2}$;   c) $\dfrac{(y - y^2)\,dx + (x - x^2)\,dy}{(1 - x - y)^2}$;

d) $(x - 2y)^4 e^{xy}\,[(xy - 2y^2 + 5)\,dx + (x^2 - 2xy - 10)\,dy]$;

e) $\dfrac{-y\,dx + x\,dy}{x^2 + y^2}$,   f) $\dfrac{-(x\,dx + y\,dy + z\,dz)}{(x^2 + y^2 + z^2)^{3/2}}$.

5. a) $\Delta z = 4\Delta x + 2\Delta y + \overline{\Delta x}^2 + 2\Delta x\,\Delta y$, $dz = 4\Delta x + 2\Delta y$;

b) $\Delta z = \dfrac{\Delta x - \Delta y}{2(2 + \Delta x + \Delta y)}$, $dz = \dfrac{\Delta x - \Delta y}{4}$, so $\Delta z = dz - \dfrac{(\Delta x - \Delta y)(\Delta x + \Delta y)}{4(2 + \Delta x + \Delta y)}$.

6. 2.2, 2.4, 2.6.

Section 2.7, page 96

1. a) $\begin{bmatrix} 5 & 2 \\ 2 & 3 \end{bmatrix}$;   b) $\begin{bmatrix} 4x_1 & 2x_2 \\ 3x_2 & 3x_1 \end{bmatrix}$;   c) $\begin{bmatrix} x_2x_3 & x_1x_3 & x_1x_2 \\ 2x_1x_3 & 0 & x_1^2 \end{bmatrix}$;

d) $\begin{bmatrix} \cos y & -x\sin y \\ \sin y & x\cos y \\ 2x & 0 \end{bmatrix}$;   e) $(2xyz, x^2z, x^2y)$;   f) $(2x, 2y, -2z)$;

g) $\operatorname{col}(2t, 3t^2, 4t^3)$.

2. a) $\begin{bmatrix} dy_1 \\ dy_2 \end{bmatrix} = \begin{bmatrix} 4 & 2 \\ 1 & 2 \end{bmatrix}\begin{bmatrix} dx_1 \\ dx_2 \end{bmatrix}$, $(5.18, 2.06)$;

b) $\begin{bmatrix} dy_1 \\ dy_2 \end{bmatrix} = \begin{bmatrix} 2 & 3 & -2 \\ 3 & 3 & 3 \end{bmatrix}\begin{bmatrix} dx_1 \\ dx_2 \\ dx_3 \end{bmatrix}$, $(4.93, 9.09)$;

c) $\begin{bmatrix} du \\ dv \\ dw \end{bmatrix} = \begin{bmatrix} 0 & -1 \\ 1 & 0 \\ 2 & 0 \end{bmatrix}\begin{bmatrix} dx \\ dy \end{bmatrix}$, $(-0.03, 1.1, 2.2)$;

d) $d\mathbf{y} = \begin{bmatrix} 0 & 0 & \cdots & 0 \\ 2 & 0 & \cdots & 0 \\ 2 & 0 & \cdots & 0 \\ \vdots & \vdots & & \vdots \\ 2 & 0 & \cdots & 0 \end{bmatrix} d\mathbf{x}$, $(0, 1, \dots, 1)$.

3. a) $9(x^2 + y^2)^2$,   b) 0;   c) $2u^3v^2w$;   d) $4z^2(x^2 - y)$.

4. a) $e^2 = 7.39$;   b) 7.44;   c) $du = e\,dx$; $dv = e\,dy$, $e^2 = 7.39$.

Section 2.8, page 100

2. $\dfrac{dy}{dx} = vu^{v-1}\dfrac{du}{dx} + u^v \log u \dfrac{dv}{dx}$.

3. $\dfrac{dy}{dx} = -\dfrac{\log v}{u\log^2 u}\dfrac{du}{dx} + \dfrac{1}{v\log u}\dfrac{dv}{dx}$.

4. 1.   **5.** 336.   **6.** 197.

10. In all cases $dz = \dfrac{\partial z}{\partial x}dx + \dfrac{\partial z}{\partial y}dy$:

a) $dz = 2\cot(x^2y^2 - 1)(xy^2\,dx + x^2y\,dy)$;

b) $dz = \dfrac{(2xy^2 - 3x^3y^2 - 2xy^4)\,dx + (2x^2y - 3x^2y^3 - 2x^4y)\,dy}{\sqrt{1 - x^2 - y^2}}$;

c) $dz = \dfrac{x\,dx + 2y\,dy}{z}$.

Section 2.9, page 105

1. a) $\begin{bmatrix} u_2 - 3 & u_1 \\ 2u_2 + 2 & 2u_2 + 2u_1 - 1 \end{bmatrix}\begin{bmatrix} \cos 3x_2 & -3x_1 \sin 3x_2 \\ \sin 3x_2 & 3x_1 \cos 3x_2 \end{bmatrix},\begin{bmatrix} -3 & 0 \\ 2 & 0 \end{bmatrix}$;

b) $\begin{bmatrix} 2u_1 - 3 & 2u_2 & 1 \\ 2u_1 + 2 & -2u_2 & -3 \end{bmatrix}\begin{bmatrix} x_2x_3^2 & x_1x_3^2 & 2x_1x_2x_3 \\ x_2^2x_3 & 2x_1x_2x_3 & x_1x_3^2 \\ 2x_1x_2x_3 & x_1^2x_3 & x_1^2x_2 \end{bmatrix},\begin{bmatrix} 3 & 4 & 1 \\ -4 & -3 & 3 \end{bmatrix}$;

c) $\begin{bmatrix} e^{u_2} & u_1e^{u_2} \\ e^{-u_2} & -u_1e^{-u_2} \\ 2u_1 & 0 \end{bmatrix}\begin{bmatrix} 2x_1 & 1 \\ 4x_1 & -1 \end{bmatrix},\begin{bmatrix} 6e^2 & 0 \\ -2e^{-2} & 2e^{-2} \\ 4 & 2 \end{bmatrix}$;

d) $\begin{bmatrix} 0 & 2u_2 & \cdots & 2u_n \\ 2u_1 & 0 & \cdots & 2u_n \\ \vdots & \vdots & & \vdots \\ 2u_1 & & \cdots & 0 \end{bmatrix}\begin{bmatrix} 2x_1 + x_2 & x_1 \\ 2x_1 + 2x_2 & 2x_1 \\ \vdots & \vdots \\ 2x_1 + nx_2 & nx_1 \end{bmatrix},\begin{bmatrix} 4(n-1) & n^2 + n - 2 \\ 4(n-1) & n^2 + n - 4 \\ \vdots & \vdots \\ 4(n-1) & n^2 + n - 2n \end{bmatrix}$.

2. a) 31; **b)** $-\dfrac{3}{2}$.

4. $\begin{bmatrix} 13 & 26 \\ -8 & 1 \end{bmatrix}$.   **5.** $\begin{bmatrix} -5 & -9 \\ 16 & 2 \end{bmatrix}$.

Section 2.11, page 118

1. a) $2x/z,\ y/z$;

b) $-(yz + 4xz + 3z^2)/(xy + 2x^2 + 6xz),\ -z/(y + 2x + 6z)$;

c) $-z/(3z^2 + x + 2y),\ -2z/(3z^2 + x + 2y)$;

d) $-ze^{xz}/(xe^{xz} + ye^{yz} + 1),\ -ze^{yz}/(xe^{xz} + ye^{yz} + 1)$.

2. $\left(\dfrac{\partial x}{\partial y}\right)_z = -\dfrac{5}{4},\ \left(\dfrac{\partial y}{\partial x}\right)_u = -\dfrac{5}{7},\ \left(\dfrac{\partial z}{\partial u}\right)_x = -\dfrac{5}{7},\ \left(\dfrac{\partial y}{\partial z}\right)_x = \dfrac{1}{5}$.

3. a) $3x/u,\ y/u$;

b) $(xu + vy - ue^v)/(e^{u+v} - xe^u + xe^v - x^2 + y^2),\ (ve^v - xv - yu)/(e^{u+v} - xe^u + xe^v - x^2 + y^2)$.

c) $-(4xy + 2yu + u^2 - 2yuv)/(2xy + 2vy + xu - 2xyv)$, $(2uv - 2yv^2)/(2xy + 2vy$ $+ xu - 2xyv)$.

4. a) $du = \frac{1}{9}(dx + 3\,dy + 2\,dz)$, $dv = -\frac{1}{3}(dx + 2\,dz)$;

b) $\left(\dfrac{\partial u}{\partial x}\right)_{y,\,z} = \dfrac{1}{9}$, $\left(\dfrac{\partial v}{\partial y}\right)_{x,\,z} = 0$; c) $u = 3.033$, $v = 2.1$.

5. $(3yu - 4xu + 4x^2 + 9xy + 8xv)^{-1}\begin{bmatrix} a & b \\ c & d \end{bmatrix}$, where $a = 8uv + 4u^2 - 8xy - 4xu$,

$b = -3u^2 - 8x^2 - 9xu$, $c = 4xy - 3y^2 - 6yu - 4xv - 9yv - 4uv - 8v^2$,
$d = 4x^2 - 3xy + 6xu + 3uv$.

6. a) $\frac{1}{2}, -1$; b) $-1, \frac{3}{2}$; c) $-1, -1$.

Section 2.12, page 122

2. a) $u = \frac{1}{5}(x + 2y)$, $v = \frac{1}{5}(y - 2x)$; b) $J = 5$, J for inverse $= \frac{1}{5}$.

3. a) $J = 4(u^2 + v^2)$, b) $\left(\dfrac{\partial u}{\partial x}\right)_y = \dfrac{u}{2(u^2 + v^2)}$, $\left(\dfrac{\partial v}{\partial x}\right)_y = -\dfrac{v}{2(u^2 + v^2)}$.

6. a) $J = \rho^2 \sin\phi$; b) $\dfrac{\partial \rho}{\partial y} = \sin\phi \sin\theta$, $\dfrac{\partial \phi}{\partial z} = -\dfrac{\sin\phi}{\rho}$, $\dfrac{\partial \theta}{\partial x} = -\dfrac{\sin\theta}{\rho \sin\phi}$.

Section 2.13, page 129

1. b) $x = \dfrac{\sqrt{3}}{2} + \dfrac{1}{2}\left(t - \dfrac{\pi}{3}\right)$, $y = \dfrac{1}{2} - \dfrac{\sqrt{3}}{2}\left(t - \dfrac{\pi}{3}\right)$, $z = \dfrac{3}{4} + \dfrac{\sqrt{3}}{2}\left(t - \dfrac{\pi}{3}\right)$;

c) $x - \sqrt{3}\,y + \sqrt{3}\,z = 3\sqrt{3}\,/4$.
2. b) $\sqrt{3}\,x + 2y + z = 13/4$.
5. b) $3\sqrt{3}\,x - y - 4z = 1$.
8. a) $2x + 2y + z = 9$, $\dfrac{x - 2}{2} = \dfrac{y - 2}{2} = \dfrac{z - 1}{1}$; b) $z = 1$, $x = 0$, $y = 0$;

c) $2x - y - z = 0$, $\dfrac{x - 1}{2} = \dfrac{y - 1}{-1} = \dfrac{z - 1}{-1}$; d) none; e) $y_1 x + x_1 y - z = x_1 y_1$;

f) $(y_1 + z_1)x + (x_1 + z_1)y + (x_1 + y_1)z = 2$.

10. a) $z - 2 = 2(x - 1) + 2(y - 1)$, $\dfrac{x - 1}{2} = \dfrac{y - 1}{2} = \dfrac{z - 2}{-1}$;

b) $z - \frac{1}{3} = -2(x - \frac{2}{3}) - 2(y - \frac{2}{3})$, $\dfrac{x - \frac{2}{3}}{2} = \dfrac{y - \frac{2}{3}}{2} = \dfrac{z - \frac{1}{3}}{1}$;

c) $z = x - 2y + 2$, $x = 2 + t$, $y = 1 - 2t$, $z = 2 - t$;
d) $5z = 6x + 8y - 10$, $x = \frac{3}{5} + \frac{6}{5}t$, $y = \frac{4}{5} + \frac{8}{5}t$, $z = -t$.
11. a) $x = 3 + 4t$, $y = 1 - 5t$, $z = 1 + 3t$; b) $x = 2 + t$, $y = 2 - t$, $z = 1$;
c) $x = 1$, $y = -1 + t$, $z = -t$.

13. $\begin{vmatrix} x - x_1 & y - y_1 & z - z_1 \\ \dfrac{\partial F}{\partial x} & \dfrac{\partial F}{\partial y} & \dfrac{\partial F}{\partial z} \\ \dfrac{\partial G}{\partial x} & \dfrac{\partial G}{\partial y} & \dfrac{\partial G}{\partial z} \end{vmatrix} = 0$, $\begin{vmatrix} x - 2 & y - 2 & z - 1 \\ 4 & 4 & 2 \\ 4 & 4 & -16 \end{vmatrix} = 0$,

$\begin{vmatrix} x - 1 & y & z + 1 \\ 2 & 0 & 0 \\ 1 & 1 & 1 \end{vmatrix} = 0$.

14. $y + 2z = 0.$ **15. a)** $2x\mathbf{i} + 2y\mathbf{j} + 2z\mathbf{k};$ **b)** $4x\mathbf{i} + 2y\mathbf{j}.$
16. c) $x + y + \sqrt{2}\, z = 2.$

17. $\dfrac{x - x_1}{\dfrac{\partial g}{\partial y} - \dfrac{\partial f}{\partial y}} = \dfrac{y - y_1}{\dfrac{\partial f}{\partial x} - \dfrac{\partial g}{\partial x}} = \dfrac{z - z_1}{\dfrac{\partial(f, g)}{\partial(x, y)}}.$

18. $\dfrac{x - x_1}{\dfrac{\partial(F, G, H)}{\partial(t, y, z)}} = \dfrac{y - y_1}{\dfrac{\partial(F, G, H)}{\partial(x, t, z)}} = \dfrac{z - z_1}{\dfrac{\partial(F, G, H)}{\partial(x, y, t)}}.$

Section 2.14, page 138

1. a) $-\sqrt{22}$;   **b)** 0;   **c)** $\frac{1}{2}$;   **d)** $-\dfrac{4}{\sqrt{5}}$;   **e)** $-\frac{2}{3}$;   **f)** $0.$

2. a) $x^2 - y^2$;   **b)** $\dfrac{7xyz}{\sqrt{x^2 + 4y^2 + 16z^2}}.$

6. $du/ds = -2xy.$ Minimum is -4 at $(0, \pm 2).$
8. Maximum is 6 in direction $\mathbf{i}$ at $(1, 0)$, in direction $-\mathbf{i}$ at $(-1, 0).$
9. $3, \dfrac{7}{\sqrt{2}}, \dfrac{15}{\sqrt{3}}.$

Section 2.18, page 146

1. a) $\dfrac{2x^2 - y^2}{\left(x^2 + y^2\right)^{\frac{5}{2}}}, \dfrac{2y^2 - x^2}{\left(x^2 + y^2\right)^{\frac{5}{2}}};$ **b)** $\dfrac{2xy}{\left(x^2 + y^2\right)^2}, \dfrac{-2xy}{\left(x^2 + y^2\right)^2};$

c) $4xe^{x^2 - y^2}(2y^2 - 1), -4ye^{x^2 - y^2}(2x^2 + 1);$ **d)** $m! n!.$
4. c) $a = -c;$ **d)** $c = -3a, b = -3d.$

6. $\dfrac{\partial z}{\partial x} \dfrac{\partial^2 x}{\partial u\, \partial v} + \dfrac{\partial z}{\partial y} \dfrac{\partial^2 y}{\partial u\, \partial v}$

$\qquad + \dfrac{\partial^2 z}{\partial x^2} \dfrac{\partial x}{\partial u} \dfrac{\partial x}{\partial v} + \dfrac{\partial^2 z}{\partial x\, \partial y}\left(\dfrac{\partial x}{\partial u} \dfrac{\partial y}{\partial v} + \dfrac{\partial x}{\partial v} \dfrac{\partial y}{\partial u} \right) + \dfrac{\partial^2 z}{\partial y^2} \dfrac{\partial y}{\partial u} \dfrac{\partial y}{\partial v}.$

10. $\dfrac{2\left(u^2 - y^2\right)}{(1 - 2ux)^3}(2u - 3u^2x - xy^2).$

12. a) $\dfrac{dy}{du} = \dfrac{1}{y + u};$ **b)** $\dfrac{dv}{du} = \dfrac{2u - v}{u + v};$ **c)** $\dfrac{d^3y}{dt^3} = 0;$ **d)** $\dfrac{d^2x}{dy^2} - 1 = 0;$

e) $\dfrac{d^2v}{dx^2} - x^2v = 0;$ **f)** $\dfrac{\partial u}{\partial w} = 0;$ **g)** $\dfrac{\partial^2 u}{\partial z\, \partial w} = 0.$

Section 2.21, page 163

1. a) max. at -1, min. at 1;
b) max. at $\pi/3 + 2n\pi$, min. at $-\pi/3 + 2n\pi$, horiz. infl. at $\pi + 2n\pi\,(n = 0, \pm1, \pm2, \dots);$
c) max. at $\log 2.$
2. min. for $n = 2, 4, 6, \dots$, horiz. infl. for $n = 3, 5, 7, \dots$.
3. a) max. $= 1$, min. $= 0$;   **b)** max. $= 0$;   **c)** no max. or min.;
d) max. $= \frac{1}{2}$, min. $= -\frac{1}{2}.$
4. a) max. at $(0, 0)$;   **b)** min. at $(0, 0)$;   **c)** saddle point at $(1, 1)$;
d) saddle point at $(0, 0)$;   **e)** critical point at every point of the line $y = x$, each point
giving a relative min.;   **f)** triple saddle point at $(0, 0)$;

g) min. at $\left(\sqrt{2}, \frac{\pi}{4} + 2n\pi\right)$, $\left(-\sqrt{2}, \frac{5\pi}{4} + 2n\pi\right)$, neither at $\left(0, n\pi - \frac{\pi}{4}\right)$;

h) min. at $(\frac{1}{3}, \frac{1}{3})$, saddle points at $(0, 0)$, $(1, 0)$, $(0, 1)$;

i) $(0, 0)$, neither max. nor min.;

j) min. at $(0, 0)$;

k) min at $(0, \pm 1)$, saddle point at $(0, 0)$.

5. a) abs. max. at $(0, 0)$;   **b)** saddle point at $(0, 0)$;

c) saddle points at $\pi/2 + 2n\pi$ $(n = 0, \pm 1, \pm 2, \ldots)$;

d) no critical points [discontinuity at $(0, 0)$];   **e)** abs. min. at $(0, 0)$;

f) abs. max. at $(\sqrt{3}/3, \sqrt{3}/3)$.

6. a) abs. min. at $(-3/5, -4/5)$, abs. max. at $(3/5, 4/5)$;

b) abs. max. at $(\pm 2^{-1/4}, \pm 2^{-1/4})$ (four points), abs. min. at $(\pm 1, 0)$ and $(0, \pm 1)$;

c) max. at $(\pm 3, \pm 4)$, min. at $(\pm 4, \mp 3)$;

d) max. at $\left(\frac{1}{\sqrt{2}}, 0, \frac{1}{\sqrt{2}}\right)$, min. at $\left(-\frac{1}{\sqrt{2}}, 0, -\frac{1}{\sqrt{2}}\right)$;

e) max. at $(\pm \sqrt{\frac{2}{3}}, \sqrt{\frac{1}{3}}, \pm \sqrt{\frac{2}{3}})$ and $(0, -1, 0)$, min. at $(\pm \sqrt{\frac{2}{3}}, -\sqrt{\frac{1}{3}}, \pm \sqrt{\frac{2}{3}})$ and $(0, 1, 0)$.

f) Abs. min. at $(1 - \sqrt{2}/2, 1 - \sqrt{2}/2, \sqrt{2} - 1)$, rel. min. at $(1 + \sqrt{2}/2, 1 + \sqrt{2}/2, -\sqrt{2} - 1)$, no abs. max.

7. $(0, \pm 1, 0)$ and $(\pm 1, 0, 0)$.

8. a) max. $= 1$;   **b)** max. $= \frac{1}{2}$, min. $= -\frac{1}{2}$;   **c)** max. $= 3$, min. $= -3$;

d) max. $= 1$.

9. (a) and (c) are positive definite.

11. $a = \frac{1}{14}(2e_1 - e_2 - 2e_3 - e_4 + 2e_5)$, $b = \frac{1}{10}(-2e_1 - e_2 + e_4 + 2e_5)$, $c = \frac{1}{35}(-3e_1 + 12e_2 + 17e_3 + 12e_4 - 3e_5)$.

12. $\sqrt{3}/3$.

Section 2.22, page 172

1. $1, 0, \sin x$.

3. $ax + by + c$, a, b, c arbitrary constants.

5. $f(x, y) = g(x) + h(y)$, $g(x)$ and $h(y)$ being "arbitrary functions."

Chapter 3

Section 3.3, page 186

9. a) $\begin{bmatrix} 6xy & 3x^2 & 0 \\ 3x^2 & -6yz & -3y^2 \\ 0 & -3y^2 & 0 \end{bmatrix}$, $\begin{bmatrix} 2 & 4 & 10 \\ 4 & 8 & 2 \\ 10 & 2 & 4 \end{bmatrix}$.

Section 3.6, page 191

5. a) $x^2yz + \text{const}$;   **b)** $(2y + z^2)e^{xy} + \text{const}$.

7. a) $yz\mathbf{i} - 2xz\mathbf{j} + \text{grad } f$, f arbitrary;

b) $\frac{1}{2}[z^2\mathbf{i} + (x^2 - 2yz)\mathbf{j}] + \text{grad } f$, f arbitrary.

12. b) $\dfrac{\partial f}{\partial x} - \dfrac{\partial f}{\partial y}$;   **c)** $2x^2\mathbf{i} + 2y^2\mathbf{j}$.   **15.** $-\frac{2}{3}$.

16. $\text{div } \mathbf{v} = 0$, $\text{curl } \mathbf{v} = 2\boldsymbol{\omega}$.   **17.** Vol. $= 1$.   **18.** Vol. $= e$.

Section 3.11, page 218

1. a) $u_1 = -36x^1x^1 + 51x^1x^2 - 18x^2x^2 - 28x^1 + 20x^2$, $u_2 = 24x^1x^1 - 34x^1x^2 + 12x^2x^2 + 21x^1 - 15x^2$;

b) $v^1 = (3x^1 - 2x^2)[3\cos(-4x^1 + 3x^2) + 2\sin(-4x^1 + 3x^2)],$
$v^2 = (3x^1 - 2x^2)[(4\cos(-4x^1 + 3x^2) + 3\sin(-4x^1 + 3x^2)];$
c) $w_{11} = w_{22} = 0, w_{12} = -w_{21} = -12x^1x^1 + 17x^1x^2 - 6x^2x^2;$
d) $z_1^1 = -71x^1 + 49x^2, z_2^1 = 53x^1 - 36x^2, z_1^2 = -103x^1 + 72x^2, z_2^2 = 77x^1 - 53x^2.$
2. In standard coordinates, all reduce to δ_{ij}. In (x^i), $g_{11} = g^{22} = 25, g_{12} = g_{21} = -g^{12} = -g^{21} = -18, g_{22} = g^{11} = 13, g_j^i = \delta_{ij}$ always.
3. a) $u_1 = 13u_1 + 18u_2, u^2 = 18u_1 + 25u_2,$ where u_1, u_2 are as in the answer to Problem 1(a);
b) $v_1 = 25v^1 - 18v^2, v^2 = -18v^1 + 13v^2,$ where v^1, v^2 are as in the answer to Problem 1(b);
c) $w^{11} = w^{22} = 0, w^{12} = -w^{21} = w_{12},$ where w_{12} is as in the answer to Problem 1(c).

Chapter 4

Section 4.1, page 229
1. a) $2x\sin x - (x^2 - 2)\cos x + c;$ **b)** $\frac{1}{2}\arctan x^2 + c;$
c) $\log\dfrac{2 - x}{x - 1} + C;$ **d)** $2(\sqrt{x - 1} - \log(1 + \sqrt{x - 1})) + C.$
2. a) $\pi/4;$ **b)** $0;$ **c)** $3e - 8;$ **d)** $(\pi/4) + \log(1/\sqrt{2}).$
3. a) $\pi/2;$ **b)** $1;$ **c)** $-1;$ **d)** $\log(1 + \sqrt{2});$ **e)** $2;$ **f)** $1.$
4. a) $0;$ **b)** div;   **c)** $\pi/2;$ **d)** div;   **e)** div;   **f)** $\log 2.$
5. a) $4/3;$ **b)** $1;$ **c)** $14 - \pi - 6\sqrt{3}.$
6. a) $2/\pi;$ **b)** $-2/\pi;$ **c)** $1/2,$ **d)** $b + \frac{1}{2}a(x_1 + x_2).$
8. a) value 0.3095, error at most 0.0082 (worst error at $x = 1$);
b) value 0.7667, error at most 0.1321 (worst error at $x = 1$).

Section 4.5, page 244
1. a) $1/3;$ **b)** $\pi/48;$ **c)** $(-15\sqrt{2})/8;$ **d)** $1.$
2. a) $e - 1;$ **b)** $(2 - 5e^{-1})(1 - e^{-2});$ **c)** $3/4;$ **d)** $\pi/6.$
3. a) $\int_1^2 \int_{1-x}^{1+x} f\, dy\, dx$ or $\int_{-1}^3 \int_{\phi(y)}^2 f\, dx\, dy,$ where $\phi(y) = |1 - y|$ for $-1 \leq y \leq 0$ and for $2 \leq y \leq 3, \phi(y) = 1$ for $0 \leq y \leq 2;$
b) $\int_0^1 \int_{-\phi(x)}^{\phi(x)} f\, dy\, dx$ or $\int_{-1/2}^{1/2} \int_{1/2-\psi(y)}^{1/2+\psi(y)} f\, dx\, dy,$ where $\phi(x) = (x - x^2)^{1/2}, \psi(y) = \frac{1}{2}(1 - 4y^2)^{1/2}.$
4. a) $(8/35)(9\sqrt{3} - 8\sqrt{2} + 1);$ **b)** $2/5.$
5. a) $\int_0^{1/2} \int_{1/2}^{1-y} f\, dx\, dy;$ **b)** $\int_0^1 \int_0^1 \sqrt{1-y^2} f\, dx\, dy;$ **c)** $\int_{-1}^0 \int_0^{x+1} f\, dy\, dx;$ **d)** $\int_0^2 \int_{|y-1|}^1 f\, dx\, dy.$
10. a) $\frac{1}{3}i - (e - 1)j + \log 2k;$ **b)** $\frac{1}{60}(i + j).$

Section 4.6, page 252
1. a) $\frac{3}{16}\pi;$ **b)** $2\sqrt{2} - 2 + 2\log(2\sqrt{2} - 2);$
c) $\sqrt{2}(\tan^{-1}\sqrt{2} - \tan^{-1}\dfrac{1}{\sqrt{2}});$ **d)** $\log c + 1 - c, c = \pi/(4\sqrt{2}).$
3. b) $\cosh 1 - 1;$ **c)** $\log 2.$
4. a) $\pi/2;$ **b)** $(14\sqrt{2} - 7)/9;$ **c)** $\pi^4/3;$ **d)** $\frac{11}{18} - \frac{1}{3}\log\frac{3}{2};$ **e)** $\dfrac{2\pi}{9}.$
5. $\int_0^e \int_{f(u)}^{g(u)} \dfrac{1}{1 + u^2 v^2}\, dv\, du,$ where $f(u) = \sqrt{1 - u^2}$ for $0 \leq u \leq 1, f(u) = 0$ for $1 \leq u \leq e,$ and $g(u) = \sqrt{e^2 - u^2}$ for $0 \leq u \leq e.$

7. a) $2\int_0^{\frac{1}{2}}\int_v^{1-v}\log(1 + 2u^2 + 2v^2)\, du\, dv;$ **b)** $\int_0^1\int_{1-2u}^1\sqrt{1 + u^2(u + v)^2}\, dv\, du.$

11. a) $\int_0^{2\pi}\int_0^1\int_0^1 r^4\cos^2\theta\sin\theta\, dz\, dr\, d\theta;$ **b)** $\int_0^{\frac{\pi}{2}}\int_0^1\int_0^{1+r(\cos\theta+\sin\theta)} r^3\cos 2\theta\, dz\, dr\, d\theta.$

12. a) $\int_0^{2\pi}\int_0^{\pi}\int_0^a \rho^5\sin^4\phi\cos^2\theta\sin\theta\, d\rho\, d\phi\, d\theta;$ **b)** $\int_0^{2\pi}\int_0^{\frac{\pi}{4}}\int_0^{\sec\phi}\rho^4\sin\phi\, d\rho\, d\phi\, d\theta.$

Section 4.7, page 259

3. a) $4\pi^2 ab;$ **b)** $\pi(5^{3/2} - 1)/24.$

Section 4.8, page 264

2. $-\pi/2.$ **3. a)** $4\pi/(3 - p);$ **b)** $4\pi/(p - 3).$
4. a) div;   **b)** conv;   **c)** div;   **d)** conv;   **e)** conv.

Section 4.9, page 268

1. a) $-\int_{\frac{\pi}{2}}^{\pi}\sin(xt)\, dx;$ **b)** $\int_1^2 \dfrac{2x^3}{(1 - tx)^3}\, dx;$ **c)** $\int_1^2 \dfrac{1}{u}\, dx;$ **d)** $n!\int_1^2 \dfrac{\sin x}{(x - y)^{n+1}}\, dx.$

2. a) $x^2;$ **b)** $2t\sin t^4;$ **c)** $-3t^2\log(1 + t^6);$ **d)** $\sec^2 xe^{-\tan^2 x} - e^{-x^2}.$

4. a) $\dfrac{-1}{(n + 1)^2};$ **b)** $\dfrac{n!}{a^{n+1}};$ **c)** $\dfrac{\pi}{2}\dfrac{1\cdot 3\cdots(2n - 3)}{2\cdot 4\cdots(2n - 2)}\dfrac{1}{x^{2n-1}},\ x > 0.$

Section 4.11, page 277

1. b) uniformly continuous.
2. All but b) are uniformly continuous.
4. a) f must be uniformly continuous;
b) f need not be uniformly continuous.

Chapter 5

Section 5.3, page 290

1. a) $\frac{8}{3};$ **b)** $-\frac{3}{2};$ **c)** $0.$
2. a) $\frac{4}{3};$ **b)** $8;$ **c)** $-\pi/2.$
3. a) $0;$ **b)** $-2\pi;$ **c)** $-\frac{1}{4}.$
4. a) $0;$ **b)** $\sqrt{2}/2;$ **c)** $\frac{1}{2}\sqrt{5} + \frac{1}{4}\log(2 + \sqrt{5}).$
6. a) $7;$ **b)** $5;$ **c)** $8;$ **d)** $5.5;$ **e)** $0.$

Section 5.5, page 298

1. a) $\frac{4}{3}$;   **b)** $\frac{4}{3}$;   **c)** $\frac{4}{3}$.
2. a) 0;   **b)** $\frac{1}{3}$;   **c)** $\frac{4}{3}$.
5. a) $(b - a)$ times area enclosed by C;   **b)** 0;   **c)** $\frac{3}{2}\pi$;   **d)** 0;   **e)** 0;   **f)** 4π;   **g)** 0;
h) 0.

Section 5.7, page 313

1. a) $F = x^2 y$, integral $= 1$;   **b)** $F = e^{xy}$, integral $= 0$;
c) $F = -1/\sqrt{x^2 + y^2}$, integral $= 1 - e^{-2\pi}$.
2. a) $-\frac{10}{3}$;   **b)** $\frac{1}{3}$;   **c)** 2;   **d)** 1.
3. a) 0;   **b)** -2π;   **c)** -2;   **d)** 12π;   **e)** 0;   **f)** 0.

4. $\dfrac{\pi}{4} + 2n\pi$ $(n = 0, \pm 1, \pm 2, \ldots)$.

5. a) 0;   **b)** 2π;   **c)** 4π;   **d)** 0.

6. a) $x^2 y - \frac{1}{3}(y^3 + 2)$;   **b)** $x \sin y$.

7. $-\pi$.   **8.** $K + n_1\alpha_1 + n_2\alpha_2 + \cdots + n_k\alpha_k$, where the numbers $n_1, \ldots, n_k$ are
positive or negative integers or 0.   **9.** 7.   **10. a)** -60;   **b)** 0.

Section 5.10, page 326

1. a) 3π;   **b)** $-\frac{5}{3}$;   **c)** $\frac{1}{2}$;   **d)** 0;   **e)** $-163/15$.
5. a) $\frac{1}{2}$;   **b)** π;   **c)** 2π;   **d)** $\dfrac{\pi}{2}$.

7. a) $\pi/48$;   **b)** $-(e - 1) - \dfrac{3\pi}{4}(e^2 - 1)$;   **c)** $\pi/8$;   **d)** $(e^2 - 1)/2$;   **e)** 0.

Section 5.11, page 333

1. a) 4π;   **b)** 3;   **c)** 0;   **d)** $6V$;   **e)** 0;   **f)** 0.

Section 5.13, page 344

1. a) 6π;   **b)** 0.   **2. a)** -2;   **b)** 0.   **4.** $\dfrac{\pi}{2} \pm 2n\pi$.
5. b) $\mathbf{v} + \operatorname{grad} f$, where $\mathbf{v} = x^2 y^2 z\mathbf{i} - xy^3 z\mathbf{j} + xy^2 z^2\mathbf{k}$.

Section 5.14, page 350

1. a) $\displaystyle\int_0^1 \int_0^u (u^2 + v^2)\, dv\, du$;   **b)** $\displaystyle\int\int_{R_{uv}} (u - v)(1 - 2u - 2v)\, du\, dv$;

c) $\displaystyle 4\int\int_{R_{uv}} uv(u^4 - v^4)\, du\, dv$.

3. a) -1;   **b)** 2;   **c)** 3.   **6. a)** -1;   **b)** 2.

Section 5.15, page 361

1. a) Potential energy is $\frac{1}{2}k^2x^2$; $\frac{1}{2}mv^2 + \frac{1}{2}k^2x^2 = $ const.

b) Potential energy is $\frac{1}{2}(a^2x^2 + b^2y^2)$; $\frac{1}{2}(mv^2 + a^2x^2 + b^2y^2) = $ const.

5. $T = T_1 + \dfrac{T_2 - T_1}{d}x$.

Section 5.18, page 377

3. (a) and (b) $2\pi ab \log 1/a$ for $R < a$, $2\pi ab \log 1/R$ for $R > a$;

c) $\mathbf{0}$ for $R < a$, $-2\pi abR^{-2}\mathbf{R}$ for $R > a$.

4. a) $-2\pi\gamma$ for $R < a$, 0 for $R > a$;

b) normal derivatives have limits 0 everywhere.

6. $U = \pi k(4a^3 - R^3)/3$ for $R \le a$, $U = \pi ka^4/R$ for $R \ge a$.

Section 5.22, page 388

1. a) $x^1(x^4)^2dx^1 - (x^1)^2dx^4$;

b) $-x^1dx^2dx^3 - x^2dx^3dx^1 - x^3dx^1dx^2$;

c) $2x^1x^4dx^1dx^2dx^3 - (2x^1)^2dx^2dx^3dx^4 + 2x^1x^2dx^1dx^3dx^4 - 2x^1x^3dx^1dx^2dx^4$;

d) 0 (4-form);

e) $x^4dx^1 + x^1dx^4$;

f) 0;

g) $3dx^1dx^2dx^3$.

5. a) $(\bar{x}^3 + 3\bar{x}^4)d\bar{x}^1d\bar{x}^2 + (6\bar{x}^4 - \bar{x}^2)d\bar{x}^1d\bar{x}^3 + (-3\bar{x}^2 - 6\bar{x}^3)d\bar{x}^1d\bar{x}^4 + (\bar{x}^1 + 6\bar{x}^4)d\bar{x}^2d\bar{x}^3 + (3\bar{x}^1 - 6\bar{x}^3)d\bar{x}^2d\bar{x}^4 + (6\bar{x}^1 + 6\bar{x}^2)d\bar{x}^3d\bar{x}^4$;   **b)** $(\bar{x}^1)^3\bar{x}^3d\bar{x}^1d\bar{x}^2 - (\bar{x}^1)^3\bar{x}^2d\bar{x}^1d\bar{x}^3 + (\bar{x}^1)^4d\bar{x}^2d\bar{x}^3$.

6. a) $\int(\bar{X}_1 + \bar{X}_2 + \bar{X}_3)d\bar{x}^1 + \bar{X}_2d\bar{x}^2 + \bar{X}_3d\bar{x}^3$ on $\overline{C}$;

b) $\int\int \bar{X}_1d\bar{x}^2d\bar{x}^3 + (\bar{X}_2 - \bar{X}_1)d\bar{x}^3d\bar{x}^1 + (\bar{X}_3 - \bar{X}_1)d\bar{x}^1d\bar{x}^2$ on $\overline{S}$.

8. $C_{ijk} = A_iB_{jk} + A_kB_{ij} + A_jB_{ki}$

9. a) $\int_C X_1dx^1 + \cdots + X_4dx^4 = \int\int_S \sum_{1 \le i < j \le 4}\left(\dfrac{\partial X_j}{\partial x^i} - \dfrac{\partial X_i}{\partial x^j}\right)dx^idx^j$;

b) $\int\int_S \sum_{1 \le i < j \le 4}A_{ij}dx^idx^j = \int\int_R\int \left(\dfrac{\partial A_{23}}{\partial x^4} - \dfrac{\partial A_{24}}{\partial x^3} + \dfrac{\partial A_{34}}{\partial x^2}\right)dx^2dx^3dx^4 + \left(\dfrac{\partial A_{34}}{\partial x^1} - \dfrac{\partial A_{14}}{\partial x^3} + \dfrac{\partial A_{13}}{\partial x^4}\right)dx^1dx^3dx^4 + (\ldots)dx^1dx^2dx^4 + (\ldots)dx^1dx^2dx^3$.

Chapter 6

Section 6.4, page 399

1. a) 0;   **b)** 0;   **c)** 0;   **d)** 1;   **e)** 1.

2. a) $1, -1$;   **b)** $0.951, -0.951$;   **c)** $\infty, -\infty$.

6. Upper limit is ∞ for $|x| > 1$, 1 for $x = \pm 1$ and 0 for $|x| < 1$; lower limit is $-\infty$ for $x < -1$, -1 for $x = -1$, 0 for $-1 < x < 1$, 1 for $x = 1$, ∞ for $x > 1$.

Section 6.7, page 414

12. a) conv;   **b)** conv;   **c)** div;   **d)** conv;   **e)** conv;   **f)** conv;   **g)** div;   **h)** conv;   **i)** div;   **j)** conv.

14. a) div;   **b)** conv;   **c)** div;   **d)** conv.

Section 6.9, page 423
1. a) 1 term, 1; **b)** 3 terms, 0.86; **c)** 5 terms, 0.51; **d)** 2 terms, 0.70;
e) 3 terms, 1.287; **f)** 4 terms, 1.708; **g)** 3 terms, 0.8417; **h)** 1 term, 0.72;
i) 1 term, 0.18; **j)** 8 terms, 1.70.
2. a) 8 terms, 66 terms, 918 terms.

Section 6.10, page 428
1. a) π^2; **b)** $\dfrac{\pi^2}{6} + \dfrac{\pi^4}{90}$; **c)** $\dfrac{\pi^2}{3} - \dfrac{\pi^4}{30}$; **d)** $\dfrac{5\pi^2}{6} + \dfrac{\pi^4}{30} + \dfrac{\pi^6}{105}$;
e) $\dfrac{\pi^2}{6} - \dfrac{\pi^6}{945} - \dfrac{15}{64}$; **f)** $\dfrac{\pi^2}{6} - \dfrac{5}{8}$.
7. b) 0.9729; **c)** 1.0766.

Section 6.13, page 438
1. a) $|x| \le 1$; **b)** $|x| < 2$; **c)** $x > 1$ and $x < -1$; **d)** $x > 0$; **e)** $x < \frac{1}{2}$; **f)** $|x - n\pi| \le \pi/6$ $(n = 0, \pm 1, \pm 2, \dots)$; **g)** $0 \le x \le 2$; **h)** $x > 1$ and $x \le -1$; **i)** all x; **j)** all x.

Section 6.16, page 451
3. a) $\sum\limits_{n=0}^{\infty} (-1)^n (x - 1)^n,\ 0 < x < 2$; **b)** $\sum\limits_{n=0}^{\infty} (-1)^n \dfrac{x^n}{2^{n+1}},\ -2 < x < 2$;

c) $\sum\limits_{n=0}^{\infty} \dfrac{(-1)^n 3^n}{5^{n+1}} x^n,\ -\dfrac{5}{3} < x < \dfrac{5}{3}$; **d)** $\sum\limits_{n=0}^{\infty} \dfrac{(-1)^n 3^n}{8^{n+1}} (x - 1)^n,\ -\dfrac{5}{3} < x < \dfrac{11}{3}$;

e) $\sum\limits_{n=0}^{\infty} \dfrac{(-1)^n a^n}{(ac + b)^{n+1}} (x - c)^n,\ c - \left|\dfrac{ac + b}{a}\right| < x < c + \left|\dfrac{ac + b}{a}\right|$;

f) $\sum\limits_{n=0}^{\infty} x^{2n},\ -1 < x < 1$;

g) $\sum\limits_{n=0}^{\infty} \left(\dfrac{1}{2^{n+1}} - \dfrac{1}{3^{n+1}}\right) x^n,\ -2 < x < 2$;

h) $\sum\limits_{n=0}^{\infty} (-1)^n (n + 1)(x - 1)^n,\ 0 < x < 2$;

i) $\sum\limits_{n=0}^{\infty} \dfrac{(-1)^n 3^n (n + 1)}{8^{n+2}} (x - 1)^n,\ -\dfrac{5}{3} < x < \dfrac{11}{3}$;

j) $\dfrac{1}{(ac + b)^k} + \sum\limits_{n=1}^{\infty} (-1)^n \dfrac{k(k + 1) \cdots (k + n - 1)}{1 \cdot 2 \cdots n} \dfrac{a^n (x - c)^n}{(ac + b)^{n+k}},\ c - \left|\dfrac{ac + b}{a}\right| < x < c + \left|\dfrac{ac + b}{a}\right|$.

4. b) $f(0) = 0,\ f(1) = 1.29,\ f'(0) = 1,\ f'(1) = 1.63,\ f''(0) = \frac{1}{2}$;

c) $f'(x) = \sum\limits_{n=0}^{\infty} \dfrac{x^n}{(n + 1)^n},\ f''(x) = \sum\limits_{n=0}^{\infty} \dfrac{n + 1}{(n + 2)^{n+1}} x^n$.

Section 6.18, page 458

2. a) $x + x^2 + \frac{1}{3}x^3$;   **b)** $x + \frac{1}{3}x^3 + \frac{2}{15}x^5$;   **c)** $x^2 - x^3 + \frac{11}{12}x^4$;

d) $-x^2 - \frac{1}{2}x^4 - \frac{1}{3}x^6$;   **e)** $15 + 15(x-2) + 6(x-2)^2$;   **f)** $1 + x + \frac{1}{2}x^2$;

g) $x - \frac{1}{6}x^3 + \frac{3}{40}x^5$;   **h)** $x - \frac{1}{3}x^3 + \frac{2}{15}x^5$;   **i)** $x + \frac{1}{3}x^3 + \frac{1}{5}x^5$;

j) $\frac{1}{2}x^2 + \frac{1}{12}x^4 + \frac{1}{45}x^6$.

5. a) 0.747;   **b)** 0.497.

Section 6.19, page 465

1. a) 0;   **b)** $1 - i$.   **2. a)** absolutely convergent;   **b)** divergent;

c) convergent, not absolutely;   **d)** absolutely convergent.

Section 6.21, page 471

1. a) $1 + (x^2 - y^2) + \frac{1}{2!}(x^4 - 2x^2y^2 + y^4) + \cdots + \frac{1}{n!}(x^2 - y^2)^n + \cdots$, all (x, y);

b) $xy - \frac{1}{3!}x^3y^3 + \cdots + (-1)^{n-1}\frac{(xy)^{2n-1}}{(2n-1)!} + \cdots$, all (x, y);

c) $1 + (x + y) + (x^2 + 2xy + y^2) + \cdots + (x + y)^n + \cdots$, $-1 < x + y < 1$;

d) $1 + (x + y + z) + (x + y + z)^2 + \cdots + (x + y + z)^n + \cdots$, $-1 < x + y + z < 1$.

4. 0.240 (three significant figures).

Section 6.25, page 484

1. a) divergent;   **b)** convergent;   **c)** convergent;

d) convergent for $s > 0$;   **e)** convergent.

12. a), c), d), e), g), h) principal value 0, no usual value;   **b)** both values 0;   **f)** both values $(3/2)(5^{2/3} - 1)$.

Chapter 7

Section 7.4, page 502

1. a) $\frac{1}{2} + \frac{2}{\pi}\left(\sin x + \frac{\sin 3x}{3} + \frac{\sin 5x}{5} + \cdots\right)$;

b) $\frac{\pi}{4} - \frac{2}{\pi}\sum_{n=1}^{\infty}\frac{\cos(2n-1)x}{(2n-1)^2} - \sum_{n=1}^{\infty}(-1)^n\frac{\sin nx}{n}$;

c) $\frac{\pi}{2} - \frac{4}{\pi}\sum_{n=1}^{\infty}\frac{\cos(2n-1)x}{(2n-1)^2}$;

d) $\frac{\pi^2}{3} + 4\sum_{n=1}^{\infty}\frac{(-1)^n\cos nx}{n^2}$;

e) $\frac{1}{\pi}\sum_{n=1}^{\infty}\frac{\sin nx}{n}$;

f) $\frac{2\pi}{3} - \frac{1}{\pi}\sum_{n=1}^{\infty}\frac{\cos nx}{n^2}$.

3. a) $\pi - 2\sum_{n=1}^{\infty}\frac{\sin nx}{n}$;   **b)** $\frac{2}{\pi} + \frac{4}{\pi}\sum_{n=1}^{\infty}\frac{(-1)^{n+1}}{4n^2 - 1}\cos 2nx$.

4. a) $\frac{2\pi}{5}$;   **b)** 6π;   **c)** 2;   **d)** not periodic;   **e)** 2π;   **f)** 30π;

g) not periodic.

Section 7.5, page 509

1. a) $1 - 4 \sum_{n=1}^{\infty} \frac{(-1)^n}{n} \sin nx;$ **b)** $1 + 2\pi - 4 \sum_{n=1}^{\infty} \frac{\sin nx}{n};$

c) $1 + \pi - \frac{8}{\pi} \sum_{n=1}^{\infty} \frac{\cos (2n-1) x}{(2n-1)^2};$ **d)** $\frac{2}{\pi} \sum_{n=1}^{\infty} \frac{1-(-1)^n(2\pi+1)}{n} \sin nx;$

e) $\pi + 1 - 2 \sum_{n=1}^{\infty} \frac{\sin 2nx}{n}.$

2. a) $\frac{13\pi^2}{3} + 4 \sum_{n=1}^{\infty} \frac{(-1)^n}{n^2} \cos nx - 8\pi \sum_{n=1}^{\infty} \frac{(-1)^n}{n} \sin nx;$

b) $\frac{7}{3} + \frac{1}{\pi^2} \sum_{n=1}^{\infty} \frac{\cos 2n\pi x}{n^2} - \frac{3}{\pi} \sum_{n=1}^{\infty} \frac{\sin 2n\pi x}{n}.$

3. a) $\sin x;$ **b) and c)** $: \frac{2}{\pi} - \frac{4}{\pi} \sum_{n=1}^{\infty} \frac{\cos 2nx}{4n^2 - 1}.$

4. a) $\frac{\pi}{2} - \frac{4}{\pi} \sum_{n=1}^{\infty} \frac{\cos (2n-1) x}{(2n-1)^2};$ **b)** $-2 \sum_{n=1}^{\infty} (-1)^n \frac{\sin nx}{n};$

c) $\frac{1}{2} - \frac{4}{\pi^2} \sum_{n=1}^{\infty} \frac{\cos (2n-1)\pi x}{(2n-1)^2};$ **d)** $-\frac{2}{\pi} \sum_{n=1}^{\infty} (-1)^n \frac{\sin n\pi x}{n}.$

5. a) $b_n = 0;$ **b)** $a_n = 0;$ **c)** $a_{2n+1} = 0, b_{2n} = 0;$
d) $a_{4n+2} = 0, b_{4n} = 0, a_{4n+1} = b_{4n+1}, a_{4n+3} = -b_{4n+3};$
e) $b_n = 0; a_n = 0$ except for $n = 0, 4, 8, \ldots, 4k, \ldots;$
f) $a_{2n} = 0, b_{2n+1} = 0;$ **g)** $a_{2n+1} = 0, b_{2n+1} = 0;$
h) $a_n = 0$ and $b_n = 0$ except for $n = 0, 4, 8, \ldots, 4k, \ldots;$
i) $a_n = 0$ and $b_n = 0$ except for $n = 0, 6, 12, \ldots, 6k, \ldots;$
j) $a_n = 0$ and $b_n = 0$ for $n = 1, 2, \ldots.$

Section 7.13, page 531

13. a) $\cos 2nt, 0 \leq t \leq \pi/2;$
b) $(-1)^{n/2} \cos nt$ for $n = 0, 2, \ldots, (-1)^{(n-1)/2} \sin nt$ for $n = 1, 3, \ldots, -\pi/2 \leq t \leq \pi/2;$
c) $(-1)^{n/2} \sin nt,$ for $n = 2, 4, \ldots, (-1)^{(n-1)/2} \cos nt$ for $n = 1, 3, \ldots, -\pi/2 \leq t \leq \pi/2.$

15. a) $(1-t^2)^{-1/4} \cos(n \cos^{-1} t) = T_n(t)(1-t^2)^{-1/4};$
b) $(1-t^2)^{-1/4} \sin(n \cos^{-1} t) = U_n(t)(1-t^2)^{1/4};$
c) $(1-t^2)^{-1/4} \sin[(2n+1) \cos^{-1} t] = U_{2n+1}(t)(1-t^2)^{1/4};$
d) $(1-t^2)^{-1/4} \cos[(2n+1) \cos^{-1} t] = T_{2n+1}(t)(1-t^2)^{-1/4}.$

Section 7.19, page 549

1. b) $\frac{\pi^4}{9} + \frac{4\pi^2}{3} \sum_{n=1}^{\infty} (-1)^n \frac{\cos nx}{n^2} + \frac{4\pi^2}{3} \sum_{m=1}^{\infty} (-1)^n \frac{\cos my}{m^2}$

$$+ 16 \sum_{n=1}^{\infty} \sum_{m=1}^{\infty} (-1)^{m+n} \frac{\cos nx \cos my}{n^2 m^2}.$$

2. a) $\frac{1}{\pi} \int_0^{\infty} \frac{\cos xt + t \sin xt}{1 + t^2} dt;$ **b)** $\frac{1}{\pi} \int_0^{\infty} \frac{\sin t \cos xt + (1 - \cos t) \sin xt}{t} dt.$

4. a) $(2 \sin t)/t;$ **b)** $(1 + it)^{-2};$ **c)** $(2t^{-1} - 4t^{-3}) \sin t + 4t^{-2} \cos t.$

6. b) for $\sigma > 0$, $(2\pi)^{-1} \int_{-\infty}^{\infty} k! e^{(\sigma+it)x} (\sigma + it)^{-k-1} dt = x^k$ for $x > 0$, $= 0$ for $x < 0$.

7. a) $h(x) = x$; **b)** $h(x) = x^2/2$; **c)** $h(x) = e^x - 1$.

Section 7.20, page 555

3. a) $p(x) + q(x) + 15\delta(x) - 2\delta(x + \pi) + 2\delta(x - \pi)$, where $p(x) = -1$ for $x > 0$, $q(x) = -\sin x$, $-\pi \le x \le \pi$, $p = q = 0$ otherwise; **b)** $3\delta'(x) - \delta(x)$;
c) $2\delta(x - 1)$; **d)** $\delta'(x)$; **e)** $-f(x)g(x) + (3 - \pi)\delta'(x - \pi) + \delta(x - \pi)$;
f) 1; **g)** -1; **h)** $3s - 1$; **i)** $2t^3 \sin \pi t/(1 - t^2)$ (value π^3 for $t = \pm 1$);
j) $p(x) + 3\delta(x)$, with $p(x)$ as in answer to (a); **k)** $9\delta(x) + r(x)$, where $r(x) = x - 6$ for $x \ge 0$, $r = 0$ otherwise.

Chapter 8

Section 8.2, page 562

5. a) $1 - i$; **b)** $4t^3 + 2it$; **c)** $3ie^{3it}$; **d)** $(-2 + 4i)e^{(-1+2i)t}$;
e) $e^{(-1+2i)t}[1 + t(-1 + 2i)]$; **f)** $-e^{-t} + e^{it}$.
6. $w' = 6ie^{2it}$, $w'' = -12e^{2it}$.
7. a) $(3 + i)/2$; **b)** $(3 + 20i)/15$; **c)** $(e^{3i} - 1)/3i$;
d) $2(e^{-1+2i} - 1)/(-1 + 2i)$; **e)** $[1 + (2i - 2)e^{-1+2i}]/(-3 - 4i)$;
f) $2 - e^{-1} - e^i$.
8. a) $(t^2 - 2it - 3)\cos t + (2i - 2t)\sin t + c$;

b) $-e^{-at}\left(\dfrac{t^n}{a} + \dfrac{nt^{n-1}}{a^2} + \cdots + \dfrac{n!}{a^{n+1}}\right) + c$;

c) $\cos bt \left[-\dfrac{t^n}{b} + \dfrac{n(n-1)t^{n-2}}{b^3} - \dfrac{n(n-1)(n-2)(n-3)t^{n-4}}{b^5} + \cdots\right]$

$\qquad + \sin bt \left[\dfrac{nt^{n-1}}{b^2} - \dfrac{n(n-1)(n-2)t^{n-3}}{b^4} + \cdots\right] + c$;

d) $\text{Re}\left\{-e^{ait}\left[\dfrac{t^n}{-ai} + \dfrac{nt^{n-1}}{(-ai)^2} + \cdots + \dfrac{n!}{(-ai)^{n+1}}\right]\right\} + c$;

e) $-\dfrac{1}{8}\sum_{k=1}^{8}\left[e^{-a_k t}\left(\dfrac{t^n}{a_k} + \dfrac{nt^{n-1}}{a_k^2} + \cdots + \dfrac{n!}{a_k^{n+1}}\right)\right] + C$,

where the a_k are the eight numbers $(\pm a \pm b \pm c)i$.

Section 8.4, page 567

1. a) $u = x^2 - y^2 - 2xy$, $v = x^2 - y^2 + 2xy$, all z;
b) $u = (x^2 + y^2 + y)[x^2 + (y + 1)^2]^{-1}$,
$\qquad v = -x[x^2 + (y + 1)^2]^{-1}$ ($z \ne -i$);
c) $u = \tan x \, \text{sech}^2 y [1 + \tan^2 x \tanh^2 y]^{-1}$,
$\qquad v = \tanh y \sec^2 x [1 + \tan^2 x \tanh^2 y]^{-1}$,
$\qquad z \ne (\pi/2) + n\pi$, $n = 0, \pm 1, \pm 2, \ldots$;
d) $u = e^{-x}[(1 + x)\cos y - y \sin y][(1 + x)^2 + y^2]^{-1}$,
$\qquad v = -e^{-x}[(1 + x)\sin y + y \cos y][(1 + x)^2 + y^2]^{-1}$ ($z \ne -1$);
e) $\ldots$ i) See (8.13), continuous for all z;

j) $u = e^x(\cos x \cos y \cosh y + \sin x \sin y \sinh y)$,
   $v = e^x(\cos x \sin y \cosh y - \sin x \cos y \sinh y)$, all z.
2. a) $i(\pi + \sinh \pi)$; **b)** $-\frac{1}{2}$; **c)** ∞; **d)** 0.
3. a) $3z^2 + 5$; **b)** $-(z - 1)^{-2}$; **c)** $42[1 + (z^2 + 1)^3]^6(z^2 + 1)^2z$;
d) $(z + 1)^{-4}(2z - z^2)$.

Section 8.5, page 570
1. a) $\frac{2}{3}$; **b)** $\pi^2/2$; **c)** $\frac{1}{2}\log 2 + i(\pi/4)$.
3. a) $2\pi i$.

Section 8.6, page 576
2. a) Analytic nowhere; **b)** analytic nowhere; **c)** analytic for all z.
6. The functions are analytic except at the following points: **a)** $\frac{1}{2}\pi + n\pi$;
b) $n\pi$; **c)** $\frac{1}{2}\pi i + n\pi i$; **d)** 0; **e)** $0, \frac{1}{2}\pi + n\pi$;
f) $-\frac{1}{4}\pi + n\pi$, where $n = 0, \pm 1, \pm 2, \ldots$.

Section 8.7, page 579
1. a) $0.693 + 2n\pi i$; **b)** $i(\frac{1}{2}\pi + 2n\pi)$; **c)** $0.347 + i(\frac{7}{4}\pi + 2n\pi)$;

d) $\exp(-\frac{1}{2}\pi - 2n\pi)$; **e)** $\sqrt[3]{2}\exp(\frac{1}{6}\pi i + \frac{4n\pi}{3}i)$;

f) $\exp(\frac{\sqrt{2}}{2}\pi i + 2\sqrt{2}n\pi i)$; **g)** $\frac{1}{2}\pi + 2n\pi$; **h)** $2n\pi \pm 1.317i$;

The range of n is $0, \pm 1, \pm 2, \ldots$, except in (e), where it is $0, 1, 2$.
3. a) and **b)** $n\pi$ and $(\pi/2) + n\pi$ $(n = 0, \pm 1, \pm 2, \ldots)$.
6. a) $\log r + i\theta$, $\frac{1}{2}\pi + 2n\pi < \theta < \frac{3}{2}\pi + 2n\pi$ $(n = 0, \pm 1, \pm 2, \ldots)$;
b) $\sqrt[3]{r}\exp(i\theta/3)$, $-(\pi/2) + 2n\pi < \theta < (\pi/2) + 2n\pi$ $(n = 0, 1, 2)$.

Section 8.9, page 586
1. a) 0; **b)** 0; **c)** $(2i - 1)e^{2i}$; **d)** $-i$.
2. a) πi; **b)** πi.
3. a) $a_1 + a_2 + a_3$; **b)** $a_1 + a_2$; **c)** $a_2 + a_3$.
5. a) $6\pi i$; **b)** $-2\pi i/3$; **c)** $2\pi i$; **d)** $2\pi i \sinh 1$.

Section 8.11, page 593
1. a) 1; **b)** 3; **c)** 0; **d)** ∞.
3. a) $4\pi ei/3$; **b)** $-\pi i/3$; **c)** $\pi i/32$.

4. a) $\displaystyle\sum_{n=0}^{\infty} \frac{(-1)^n z^{2n+1}}{(2n + 1)!}$ $(r^* = R = \infty)$;

b) $\displaystyle\sum_{n=0}^{\infty} (-1)^n(z - 2)^n$ $(r^* = R = 1)$;

c) $\frac{1}{2}\log 2 + \frac{3}{4}\pi i - \displaystyle\sum_{n=1}^{\infty} \left(\frac{1 + i}{2}\right)^n \frac{(z + 1 - i)^n}{n}$ $(r^* = \sqrt{2}, R = 1)$.

Section 8.14, page 602

1. a) $\sum\limits_{n=1}^{\infty} \dfrac{z^{n-1}}{n!}$, removable;

b) $-\dfrac{1}{3z^2} - \dfrac{1}{9z} - \sum\limits_{n=0}^{\infty} \dfrac{z^n}{3^{n+3}}$, pole of order 2;

c) $-\dfrac{1}{z} + 1 + \sum\limits_{n=0}^{\infty} \dfrac{(-1)^n z^{2n+1}}{(2n+2)!}$, pole of order 1;

d) $\dfrac{1}{z} + \dfrac{z}{6} + \dfrac{7z^3}{360} + \cdots$, pole of order 1.

2. a) $\dfrac{1}{z^4} + \dfrac{3}{z^3} + \dfrac{1}{z^2};$ **b)** $-\dfrac{2}{z};$

c) $\dfrac{e \sin 1}{(z-1)^2} + \dfrac{e(\cos 1 + \sin 1)}{z - 1};$ **d)** $\dfrac{1}{z^2} - \dfrac{1}{z}.$

3. a) $\sum\limits_{n=1}^{\infty} \dfrac{-1}{z^n}$, removable, zero of first order;

b) $z - 2 + \sum\limits_{n=1}^{\infty} \dfrac{(-2)^{n+1}}{z^n}$, pole of order 1;

c) $2 + \sum\limits_{n=1}^{\infty} \dfrac{z^n}{n!} + \sum\limits_{n=1}^{\infty} \dfrac{z^{-n}}{n!}$, essential.

5. a) zero: 0, pole: 1; **b)** zeros: 1, ∞, poles: $-1, -2$;
c) zeros: $-1, -1, -1$, poles: $0, \infty, \infty$.

6. $\dfrac{a_2 b_N^2 - a_1 b_N b_{N+1} - a_0 b_{N+2} b_N + a_0 b_{N+1}^2}{b_N^3 (z - z_0)^{N-2}}$

Section 8.18, page 615

1. a) $-\pi i;$ **b)** $\pi i \sin 6;$ **c)** $-\pi i/3;$ **d)** 0; **e)** $-\pi i/8;$ **f)** 0; **g)** $4\pi i;$ **h)** $4\pi i.$

2. a) $\dfrac{1}{4} \dfrac{1}{z-2} - \dfrac{1}{4} \dfrac{1}{z+2};$ **b)** $\dfrac{1}{z-1} - \dfrac{3}{z-2} + \dfrac{2}{z-3};$

c) $-\dfrac{1}{5}\left[\dfrac{z_2}{z - z_1} + \dfrac{z_5}{z - z_2} + \dfrac{z_3}{z - z_3} + \dfrac{z_1}{z - z_4} + \dfrac{z_4}{z - z_5} \right],$

$z_k = \exp\left[(2k - 1)\pi i/5\right], k = 1,\dots,5;$

d) $\dfrac{1}{n}\left[\dfrac{z_1}{z - z_1} + \cdots + \dfrac{z_n}{z - z_n} \right], z_k = \exp\left(\dfrac{2k\pi i}{n}\right), k = 1,\dots,n;$

e) $\dfrac{1}{4} \dfrac{1}{(z-1)^2} - \dfrac{1}{4(z+1)^2};$

f) $\dfrac{c_0}{z^n} + \dfrac{c_1 - c_0}{z^{n-1}} + \cdots + \dfrac{c_{n-1} - c_{n-2} + \cdots + (-1)^{n-1} c_0}{z}$

$+ (-1)^n \phi(-1)\dfrac{1}{z+1};$

g) $\dfrac{1}{3}\dfrac{1}{(z-1)^3} - \dfrac{5}{9}\dfrac{1}{(z-1)^2} + \dfrac{7}{27}\dfrac{1}{z-1} + \displaystyle\sum_{j=1}^{4}\dfrac{g(z_j)}{z-z_j}$,

where $z_1,\dots,z_4$ are the roots of $z^4 + z + 1 = 0$ and

$$g = (-7z^3 + 5z^2 + 3z - 13)^{-1};$$

h) $\dfrac{1}{10}\left[\dfrac{-2-i}{z-i} + \dfrac{-2+i}{z+i} + \dfrac{2-i}{z+1-i} + \dfrac{2+i}{z+1+i}\right].$

7. $2A'/B''$.

Section 8.19, page 619

1. a) $\dfrac{\pi}{2}$; **b)** $\tfrac{2}{3}\pi$; **c)** $\dfrac{4\pi}{3\sqrt{3}}$; **d)** $\dfrac{7\pi\sqrt{3}}{72}$.

2. a) $\dfrac{2\pi\sqrt{3}}{3}$; **b)** $\dfrac{\pi}{6}$; **c)** $\dfrac{2\pi}{3}$; **d)** $\dfrac{\pi}{4}$.

3. a) $\tfrac{1}{2}\pi e^{-2}$; **b)** $-2\dfrac{\sqrt{3}}{3}\pi e^{-\sqrt{3}}\sin 1$; **c)** $\tfrac{1}{2}\pi e^{-\frac{1}{2}\sqrt{2}}\cos\left(\tfrac{1}{2}\sqrt{2}\right)$;

d) $-\tfrac{1}{2}\pi e^{-3}$.

Chapter 9

Section 9.2, page 632

7. a) none, **b)** $\tfrac{1}{2}\pi + n\pi$ $(n = 0, \pm 1,\dots)$, **c)** ± 1.

9. d) (i) $w = \dfrac{z-1}{(1-2i)z-1}$, **(ii)** $w = \dfrac{i}{z}$.

Section 9.4, page 642

a) $\dfrac{1}{\pi}\arg\dfrac{z^2-1}{z^2+1}$, **b)** $\dfrac{1}{\pi}(\pi - \arg z^4)$, **c)** $1 - \dfrac{\theta}{\pi}$ $(0 < \theta < 2\pi)$,

1. d) $2 - \dfrac{2}{\pi}\arg\dfrac{e^{\pi z}-1}{e^{\pi z}+1}$, **e)** $1 - \dfrac{2}{\pi}\arg\left(\dfrac{z-1}{z+1}\right)^2$.

The argument functions are all taken between 0 and π.

2. b) $U = \dfrac{U_0}{\pi}\arg z$, where z is the inverse of the function

$$w = \dfrac{a}{2\pi}(2\mathrm{Log}\,z - z^2 + 1), \qquad \mathrm{Im}(z) > 0;$$

3. $T_1 + \dfrac{T_2 - T_1}{\pi}\arg(\sqrt{4w - 1} - i)$; the square root is chosen to have imaginary part > 1 and the arg is chosen between 0 and π.

5. b) $[\alpha h_j + (\pi - \alpha)h_{j+1}]/\pi$.

Section 9.7, page 649

2. $F(z) = z + \dfrac{1}{z} + i\log z$ (not single-valued). **3.** $T_0 + x(T_1 - T_0)$.

Section 9.8, page 655

1. a) The domain bounded by the lines $v = \pm\pi$ and the ray $v = 0$, $\log 4 \le u < \infty$.

  b) the domain bounded by the lines $v = \pi$, $v = -2\pi$ and the rays: $v = -\pi$, $1.89 \le u < \infty$; $v = 0$, $1.28 \le u < \infty$; $v = 0$, $-\infty < u \le 3.17$;

  c) the upper half of the w plane minus the portions of the imaginary axis between 0 and $0.27i$ and between $3.7i$ and ∞.

2. a) $\frac{1}{\pi}\left[\text{Log}\frac{z-1}{z} + 2\pi i - 2\text{Log}(z - 1) - \log 4\right]$;

  b) $\frac{1}{\pi}\left[\text{Log}\frac{z}{z+1} + 2\text{Log}\frac{z-5.1}{z} + \pi i - \text{Log}(z - 5.1) - 0.03\right]$;

  c) $(1 + i)z^{-1/2}(z - 1)^{1/4}$.

Chapter 10

Section 10.3, page 667

3. a) $u_1 = c_1 \sin(\sqrt{3}\,t + \epsilon_1) + c_2 \sin(t + \epsilon_2)$,
  $u_2 = -c_1 \sin(\sqrt{3}\,t + \epsilon_1) + c_2 \sin(t + \epsilon_2)$;
b) $u_1 = \frac{1}{6}\sqrt{3}\,\sin\sqrt{3}\,t + \frac{1}{2}\sin t$, $u_2 = -\frac{1}{6}\sqrt{3}\,\sin\sqrt{3}\,t + \frac{1}{2}\sin t$.
6. a) $u_1 = c_1 e^{-t} + c_2 e^{-3t}$, $u_2 = c_1 e^{-t} - c_2 e^{-3t}$;
b) $u_1 = 2e^{-t} - e^{-3t}$, $u_2 = 2e^{-t} + e^{-3t}$.
7. $u_1 = 0.25e^{-t} + 0.05e^{-3t} - 0.1(3\cos t - 4\sin t)$,
  $u_2 = 0.25e^{-t} - 0.05e^{-3t} - 0.1(2\cos t - \sin t)$.
8. $u_1 = (1 + a)(-t\cos t)$, $u_2 = (1 + a)(-t\cos t) + (2a - 2)\sin t$. Resonance occurs except when $a = -1$.
9. a) $u_1 = e^{-at}(c_1 \cos \beta t + c_2 \sin \beta t + c_3 \cos \gamma t + c_4 \sin \gamma t)$,
  $u_2 = e^{-at}(c_1 \cos \beta t + c_2 \sin \beta t - c_3 \cos \gamma t - c_4 \sin \gamma t)$,
  $a = \frac{1}{8}$, $\beta = \sqrt{15}/8$, $\gamma = \sqrt{47}/8$.

Section 10.4, page 671

3. $u_1 = c_1 \sin(\alpha t + \epsilon_1) + c_2 \sin(\beta t + \epsilon_2) + c_3 \sin(\gamma t + \epsilon_3)$,
  $u_2 = \sqrt{2}\,c_1 \sin(\alpha t + \epsilon_1) - \sqrt{2}\,c_3 \sin(\gamma t + \epsilon_3)$,
  $u_3 = c_1 \sin(\alpha t + \epsilon_1) - c_2 \sin(\beta t + \epsilon_2) + c_3 \sin(\gamma t + \epsilon_3)$,
where $\alpha = (2 - \sqrt{2})^{\frac{1}{2}}$, $\beta = \sqrt{2}$, $\gamma = (2 + \sqrt{2})^{\frac{1}{2}}$.
4. $u_1 = c_1 e^{-at} + c_2 e^{-bt} + c_3 e^{-ct}$, $u_2 = \sqrt{2}\,c_1 e^{-at} - \sqrt{2}\,c_3 e^{-ct}$,
  $u_3 = c_1 e^{-at} - c_2 e^{-bt} + c_3 e^{-ct}$, where $a = 2 - \sqrt{2}$, $b = 2$, $c = 2 + \sqrt{2}$.
5. a) $u_0 + (u_{N+1} - u_0)\sigma/(N + 1)$.

Section 10.8, page 684

8. a) $\displaystyle\sum_{n=1}^{\infty} c_n \sin nx \sin(\sqrt{n^2 - 1}\,t + \epsilon_n)$;

  b) $\displaystyle\sum_{n=1}^{\infty} c_n e^{-x} \sin nx \sin(\sqrt{n^2 + 1}\,t + \epsilon_n)$.

9. $\displaystyle\sum_{n=1}^{\infty} (a_n \cos nx + b_n \sin nx) \sin (nat + \epsilon_n).$

10. $\displaystyle\sum_{n=1}^{\infty} c_n \sin (n + \tfrac{1}{2})x \sin \left[(n + \tfrac{1}{2})at + \epsilon_n\right].$

Section 10.11, page 690

1. $e^{-t} \sin x + 5e^{-9t} \sin 3x.$

2. $u = \tfrac{1}{2}a_0 + \displaystyle\sum_{n=0}^{\infty} a_n e^{-n^2 c^2 t} \cos nx, \; a_n = \dfrac{2}{\pi} \displaystyle\int_0^\pi f(x) \cos nx \, dx.$

3. $x + \sin x + e^{-5t} \sin x.$

8. Solutions have form $e^{-at}\sum_{n=1}^{\infty} \sin nx (\alpha_n e^{\gamma_n t} + \beta_n e^{-\gamma_n t})$, where $a = \tfrac{1}{2}H/\rho$, $b = K^2/\rho$, $\gamma_n = (a^2 - bn^2)^{\frac{1}{2}}$. If $\gamma_n = 0$ for $n = m$, $e^{-\gamma_m t}$ is replaced by t.

Section 10.12, page 694

1. $x^2 \sin t + \dfrac{4}{\pi} \displaystyle\sum_{n=1}^{\infty} \dfrac{1 - (-1)^n}{n^3} \sin nx \, e^{-n^2 t}.$

2. For $\omega = \pm 1$, $u = \pm \tfrac{1}{2} \sin x (\sin t - t \cos t).$
For $\omega \neq \pm 1$, $u = \sin x (\sin \omega t - \omega \sin t)/(1 - \omega^2).$

Section 10.15, page 701

3. $\displaystyle\sum_{m=1}^{\infty} \left\{ \sum_{n=1}^{\infty} c_{mn} \sin nx \sin my \, e^{-c^2(m^2 + n^2)t} \right\}.$

5. $J_m(\sqrt{\lambda_{mn}}\, r) \sin (a\sqrt{\lambda_{mn}}\, t + \epsilon)(c_1 \cos m\theta + c_2 \sin m\theta)$, where c_1 and c_2 are constants.

6. $\displaystyle\sum_{n=1}^{\infty} \left\{ \sum_{m=0}^{\infty} J_m(\sqrt{\lambda_{mn}}\, r) e^{-c^2 \lambda_{mn} t} (\alpha_{mn} \cos m\theta + \beta_{mn} \sin m\theta) \right\}.$

8. $c_1 \cos \alpha x + c_2 \sin \alpha x$, $0 \le \alpha < \infty$, $\lambda = a\alpha.$

Index